Selected Keys of the Graphing Calculator

Magnifies or reduces a portion of the curve being viewed and can "square" the graph to reduce distortion.

Controls the values that are used when creating a table.

Determines the portion of the curve(s) shown and the scale of the graph.

Used to enter the equation(s) that is to be graphed.

Controls whether graphs are drawn sequentially or simultaneously.

Activates the secondary functions printed above many keys.

Used to delete previously entered characters.

Used to write the variable, x.

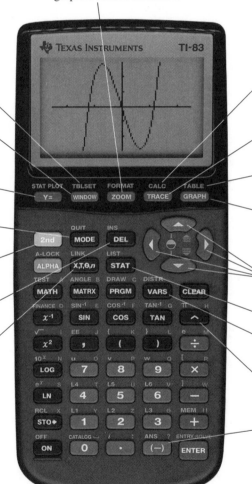

Used to determine certain important values associated with a graph.

Used to display the coordinates of points on a curve.

Used to display x- and y-values in a table.

Used to graph equations that were entered using the Y= key.

Used to move the cursor and adjust contrast.

Used to fit curves to data.

Used to access a previously named function or equation.

Used to raise a base to a power.

Used as a negative sign.

Intermediate Algebra

Algebra

GRAPHS AND MODELS

Intermediate Algebra

GRAPHS AND MODELS

Marvin L. Bittinger

Indiana University–Purdue University at Indianapolis

David J. Ellenbogen

Community College of Vermont

Barbara L. Johnson

Indiana University–Purdue University at Indianapolis

 **ADDISON-WESLEY**

An imprint of Addison Wesley Longman, Inc.

Reading, Massachusetts • Menlo Park, California • New York • Harlow, England
Don Mills, Ontario • Sydney • Mexico City • Madrid • Amsterdam

Publisher	Jason A. Jordan
Project Manager	Susan Connors Estey
Managing Editor	Ron Hampton
Production Supervisor	Kathleen A. Manley
Text Designer	Geri Davis/The Davis Group, Inc.
Art Editor	Geri Davis/The Davis Group, Inc.
Copy Editor	Martha Morong, Quadrata, Inc.
Marketing Managers	Craig Bleyer and Laura Rogers
Illustrators	Scientific Illustrators, Jim Bryant, and Maria Sas
Compositor	The Beacon Group
Cover Designer	Susan Carsten
Manufacturing Supervisor	Evelyn Beaton

Photo Credits

Cover, L'Image Magick, Inc./FPG International LLC
227, Corbis/Bettmann   **272,** Simon Wilkinson/The Image Bank
513, Reuters/Ira Strickstein Archive Photos   **534,** National Park Service
629, Comstock   **638,** AP/Wide World Photos   **642,** From *Classic Baseball Cards*, by Bert Randolph Sugar, copyright © 1977 by Dover Publishing, Inc.   **642,** AP/Wide World Photos

Library of Congress Cataloging-in-Publication Data

Bittinger, Marvin L.
    Intermediate algebra: graphs and models/Marvin L. Bittinger, David J. Ellenbogen, Barbara L. Johnson.—1st ed.
       p. cm.
    ISBN 0-201-35994-4
    1. Algebra. I. Ellenbogen, David. II. Johnson, Barbara L. III. Title.
    QA154.2.B543   1999
    512.9—dc21   99-15726
                                                                   CIP

Reprinted with corrections, July 2000

2 3 4 5 6 7 8 9 10—WCT—020100

Contents

Preface

Appropriate for a one-term course in intermediate algebra, *Intermediate Algebra: Graphs and Models* is intended for those students who have completed a first course in algebra. This text is more interactive than most other intermediate algebra texts. Our goal is to enhance the learning process through the use of technology and to provide as much support and help for students as possible in their study of algebra.

Content Features

• **Integrated Technology** The technology of the graphing calculator is completely integrated throughout the text to provide a visual means of increasing understanding. In this text, we use the term "grapher" to refer to all graphing calculator technology. The use of the grapher is woven throughout the exposition, the exercise sets, and the testing program without sacrificing algebraic skills. We use the grapher technology to enhance, not to replace, the students' mathematical skills and to alleviate the tedium associated with certain procedures. It is assumed that each student is required to have a grapher (or at least access to one) while enrolled in this course.

• **Learning to Use the Technology** To minimize the need to spend valuable class time teaching students how to use a grapher, we have included an introduction to the graphing calculator as well as explanations of the grapher features used throughout the text. The features are introduced on an "as-needed" basis. Specific keystrokes for various graphers are given in the *Graphing Calculator Manual,* which is bundled free with every new copy of this text.

• **Interactive Discoveries** The grapher provides an exciting teaching opportunity in which a student can discover and further investigate mathematical concepts. This unique Interactive Discovery feature is used to introduce new topics and provides a vehicle for students to "see" a concept quickly. This feature reinforces the idea that grapher technology is an integral part of the course as well as an important learning tool. It invites the student to develop analytic and reasoning skills while taking an active role in the learning process. (See pp. 120, 244, and 294.)

- **Function Emphasis** The use of technology with its immediate visualization of a concept encourages the early presentation of functions. Functions are introduced in the first section of Chapter 2. The study of the family of functions (linear, quadratic, higher-degree polynomial, rational, exponential, and logarithmic) has been enhanced and streamlined with the inclusion of the grapher. Applications with graphs are incorporated throughout to amplify and add relevance to the study of functions. (See pp. 108, 273, and 632.)

- **Variety in Approaches to Solutions** Skill in solving mathematical problems is expanded when a student is exposed to a variety of approaches to finding a solution. We have carefully incorporated three solution approaches throughout the text: algebraic, graphical, and numerical. Chapter openers illustrate an application with a concurrent grapher presentation of both a table and a graph (see pp. 67 and 487). The TABLE feature on a grapher provides a numerical display or check of the solution (see pp. 299 and 553).

  To highlight both the algebraic- and graphical-solution approaches in solving equations, we have used a two-column solution format in numerous examples (see pp. 170, 308, and 623). In the algebraic/graphical side-by-side features, both methods are presented together; each method provides a complete solution. This feature emphasizes that there is more than one way to obtain a result and illustrates the comparative efficiency and accuracy of the two methods.

- **Real-Data Applications** Throughout the writing process, we conducted an energetic search for real-data applications. The result of that effort is a variety of examples and exercises that connect the mathematical content with the real world. Source lines appear with most real-data applications and charts and graphs are frequently included. Many applications are drawn from the fields of health, business and economics, life and physical sciences, social science, and areas of general interest such as sports and daily life. We encourage students to "see" and interpret the mathematics that appears around them every day. (See pp. 52, 510, and 585.)

- **Regression** Using regression or curve fitting to model data is introduced in Chapter 2 with linear functions. This visual theme is continued with quadratic, cubic, quartic, exponential, and logarithmic functions. Although the theoretic aspects of curve fitting cannot be developed in this course, the power of the grapher is very apparent in this area as the technique is applied to real data. Students can quickly make the "what is this used for?" connection between real data and the extrapolated results of the curve fitting, thus giving them a better conceptual understanding of the material. (See pp. 130–131 and 336.)

- **Verifying Identities** Identities can be partially verified with a grapher using both the GRAPH and TABLE features (see pp. 276, 284, and 446). This content feature allows a visual answer to such frequent questions as "Why isn't $(x + 2)^2$ equal to $x^2 + 4$?" This approach also provides a unique lead-in to the development of the properties of exponents and logarithms.

Pedagogical Features

- **Use of Color** The text uses full color in an extremely functional way, as seen in the design elements and artwork on nearly every page. The choice of color has been carried out in a methodical and precise manner so that its use carries a consistent meaning, which enhances the readability of the text for both student and instructor. (See pp. 73 and 299.)

- **Art Package** The text contains nearly 1000 art pieces. The exceptional situational art and statistical graphs throughout the text highlight the abundance of real-world applications while helping students visualize the mathematics (see pp. 55, 303, and 562). The design and use of color with the grapher windows exemplifies the impact that technology has in today's mathematical curriculum (see pp. 75, 84, and 234).

- **Annotated Examples** Over 470 examples fully prepare the student for the exercise sets. Learning is carefully guided with numerous color-coded art pieces and step-by-step annotations, with substitutions and annotations highlighted in red (see pp. 162 and 590). The basis for problem solving is a five-step process established early in the text to aid the student in strategically approaching and solving applications (see pp. 91 and 389–390).

- **Collaborative Corners** In today's professional world, teamwork is essential. We have included optional Collaborative Corner features throughout the text to allow students to work in groups to solve problems. There is an average of three Collaborative Corner activities per chapter, each one appearing after the appropriate section's exercise set (see pp. 101, 341–342, and 588). Additional Collaborative Corner activities and suggestions for directing collaborative learning appear in the *Printed Test Bank/Instructor's Resource Guide*.

- **Variety of Exercises** There are over 6400 exercises in this text. The exercise sets are enhanced not only by the inclusion of real-data applications with source lines, detailed art pieces, and technology windows that include both tables and graphs, but also by the following features.

  Technology Exercises Since use of the grapher is totally integrated in this text, exercise sets include both grapher and nongrapher exercises. In some cases, detailed instruction lines indicate the approach the student is expected to use. In others, the student is left to choose the approach that seems best, thereby encouraging critical thinking. (See pp. 47 and 328.)

Skill Maintenance The exercises in this section have been specifically selected to review concepts previously taught in the text that are foundations for the material presented in the following section. They are often chosen to prepare the student for the new concept(s) that will be covered next. (See p. 543.)

Synthesis Exercises These exercises, which appear at the end of each exercise set, encourage critical thinking by requiring students to synthesize concepts from several sections or to take a concept a step further than in the regular exercises. (See pp. 60–61.)

Thinking and Writing Exercises for thinking and writing, at the beginning of the synthesis exercises, are denoted with a maze icon ◈. They encourage students to both consider and write about key mathematical ideas in the chapter. Many of these exercises are open-ended, making them particularly suitable for use in class discussions or as collaborative activities. (See p. 124.)

Stop and Think Exercises Throughout the text certain problems are marked with a magnifying glass icon ⌕ —these problems can usually be solved without lengthy computation or use of a grapher. They are intended to remind the student to look at the problem carefully to help develop the habit of thinking about every problem before attempting to solve it. (See p. 20 and 132.)

- **Chapter Openers** Each chapter opens with an application illustrated with both technology windows and situational art. The openers also include a table of contents listing section titles. (See pp. 151 and 579.)

- **Highlighted Information** Important definitions, properties, and rules are displayed in screened boxes. Summaries and procedures are listed in color-outlined boxes. Both of these design features present and organize the material for efficient learning and review. (See pp. 104 and 387.)

- **Summary and Review** The Summary and Review at the end of each chapter provides an extensive set of review exercises along with a list of important properties and formulas covered in that chapter. This feature provides an excellent preparation for chapter tests and the final examination. Answers to all review exercises appear in the text along with section references that direct students to material to reexamine if they have difficulty with a particular exercise. (See pp. 145 and 342.)

- **Cumulative Review** After every three chapters, and at the end of the text, we have included a Cumulative Review, which reviews skills and concepts from all preceding chapters of the text. (See pp. 216, 414, 651, and 695.)

Supplements for the Instructor

Instructor's Solutions Manual

The *Instructor's Solutions Manual* by Judith A. Penna contains worked-out solutions to all exercises in the exercise sets, including the thinking and writing exercises.

Printed Test Bank/Instructor's Resource Guide

This supplement contains the following:

- Extra practice problems.
- Black-line masters of grids and number lines for transparency masters or test preparation.
- A videotape index.
- Additional collaborative learning activities and suggestions.

The test bank portion contains the following:

- Six free-response test forms for each chapter, following the format and level of difficulty of the chapter tests in the text.
- Two multiple-choice test forms for each chapter.
- Eight alternative forms of the final examination, three with questions organized by type, three with questions organized by chapter, and two with multiple-choice questions.

Testgen-EQ CD-ROM

Testgen-EQ is a computerized test generator that allows instructors to select test questions manually or randomly from selected topics or to use a ready-made test for each chapter. The test questions are algorithm-driven so that regenerated number values maintain problem types and provide a large number of test items in both multiple-choice and open-ended formats for one or more test forms. Test items can be viewed on screen, and the built-in question editor lets instructors modify existing questions or add new ones that include pictures, graphs, accurate math symbols, and variable text and numbers.

Additional features in the new Testgen-EQ CD-ROM allow the instructor to customize both the look and content of test banks and tests. Test questions are easily transferred from the test bank to a test and can be sorted, searched, and displayed in various ways. Testgen-EQ is available on a dual platform (Macintosh/Windows) CD-ROM.

Course Management and Testing System

InterAct Math Plus for Windows and Macintosh (available from Addison Wesley Longman) combines course management and on-line testing with the features of the basic tutorial software (see "Supplements for the Student") to create an invaluable teaching resource. Consult your local Addison Wesley Longman sales consultant for details.

Supplements for the Student

Graphing Calculator Manual

The *Graphing Calculator Manual* by Judith A. Penna, with the assistance of Daphne Bell, contains keystroke level instruction for the Texas Instruments TI-83/83+, TI-86, and TI-89.

Bundled free with every copy of the text, the *Graphing Calculator Manual* uses actual examples and exercises from *Intermediate Algebra: Graphs and Models* to help teach students to use their graphing calculator. The order of topics in the *Graphing Calculator Manual* mirrors that of the text, providing a just-in-time mode of instruction. Keystroke moduals for other calculators are available upon request.

Student's Solutions Manual

The *Student's Solutions Manual* by Judith A. Penna contains completely worked-out solutions with step-by-step annotations for all the odd-numbered exercises in the exercise sets in the text, with the exception of the thinking and writing exercises. It also includes answers for the even-numbered exercises.

The *Student's Solutions Manual* can be purchased by your students from Addison Wesley Longman.

InterAct Math Tutorial Software CD-ROM

InterAct Math Tutorial Software CD-ROM has been developed and designed by professional software engineers working closely with a team of experienced math educators.

InterAct Math Tutorial Software includes exercises that are linked with every objective in the textbook and require the same computational and problem-solving skills as their companion exercises in the text. Each exercise has an example and an interactive guided solution that are designed to involve students in the solution process and to help them identify precisely where they are having trouble. In addition, the software recognizes common student errors and provides students with appropriate customized feedback.

With its sophisticated answer recognition capabilities, *InterAct Math Tutorial Software* recognizes appropriate forms of the same answer for any kind of input. It also tracks student activity and scores for each section, which can then be printed out.

InterAct Math Tutorial Software is available in CD-ROM for both Windows and Macintosh computers.

World Wide Web Supplement

http://www.GraphsModels.com
This specially developed Web site provides additional practice and learning resources. For each chapter, students can find additional practice exercises, Web links for further exploration, and expanded Summary and Review pages that reinforce the concepts and skills learned throughout the chapter. Students can also download a plug-in for Addison Wesley Longman's *InterAct Math Tutorial Software* that allows students to access tutorial problems directly through their Web browser.

Videotapes

Developed and produced especially for this text, these videotapes feature an engaging team of instructors presenting material and concepts from every section of the text in a student-interactive format. The lecturer's presentations include examples and problems from the text and support an approach that emphasizes the use of technology, visualization, and problem solving.

Math Tutor Center

The Math Tutor Center is a service provided by Addison Wesley Longman. Staffed by qualified mathematics instructors, this service offers live tutoring for students via telephone, fax, and/or e-mail, five days a week, seven hours a day. Students can receive tutoring on examples, exercises, and problems contained in their text. This service is free to any student who purchases a new Bittinger/Ellenbogen/Johnson text bundled with a valid registration number. A valid registration number can be purchased separately for those students with used texts.

Acknowledgments

No book can be produced without a team of professionals who take pride in their work and are willing to put in long hours. Laurie A. Hurley and Dr. Richard Semmler deserve special thanks for their careful accuracy checks and fine suggestions. Judy Penna's work in preparing the *Student's Solutions Manual,* the *Instructor's Solutions Manual,* and the *Graphing Calculator Manual,* as well as her helpful comments, have been invaluable. We are also grateful to Daphne Bell for her fine work on the *Graphing Calculator Manual.*

We'd like to thank Jason Jordan, our publisher, who shared our vision and provided encouragement and direction when it was needed. Kathy Manley, our production supervisor, expertly guided this book to its publication. Martha Morong, of Quadrata, Inc., provided outstanding copyediting services. We also thank Geri Davis, of the Davis Group Inc., for her excellent art direction; George and Brian Morris, of Scientific Illustrators, who skillfully generated the graphs, charts, and many of the illustrations; and Maria Sas, who designed and sketched the many hand-drawn illustrations. And finally a special thank you to Susan Estey for coordinating the reviews and managing so many of the day-to-day details professionally and courteously.

In addition, we thank the following professors for their thoughtful reviews and insightful comments.

Diane Adams, *Hazard Community College*

Jose Alonso, *Montgomery College*

Dan Anderson, *University of Iowa*

Roger Angevine, *Somerset Community College*

Dorothy Anway, *College of Saint Scholastica*

Rick Armstrong, *St. Louis Community College at Florissant Valley*

Eldon Baldwin, *Prince George's Community College*

Jane Baldwin, *Capital University*

Judy Becker, *Santa Fe Community College*

Shirley Beil, *Normandale Community College*

Joann Bossenbroek, *Columbus State Community College*

Joseph Brown, *Ferrum College*
Sally Copeland, *Johnson County Community College*
Sherry Crabtree, *Northwest Shoals Community College*
Gerald Davey, *University of Utah*
David Ebert, *Peninsula College*
Fawzi Emad, *University of Maryland*
Jeanne Fitzgerald, *Phoenix College*
Karen Froelich, *William Rainey Harper College*
Thomas Gregory, *Ohio State University*
Julie Guelich, *Normandale Community College*
Pamela Harris, *Gulf Coast Community College*
Judy Hector, *Walters State Community College*
Beth Hempleman, *Mira Costa College*
Celeste Hernandez, *Richland College*
Diane Hillyer, *Manchester Technical College*
Anita Johnston, *Jackson Community College*
Anne Landry, *Dutchess Community College*
Mitzi Logan, *Pitt Community College*
Carol Lucas, *University of Kansas*
Lew Ludwig, *Ohio University*
Amy Madden, *Moraine Valley Community College*
Janice McFatter, *Gulf Coast Community College*
Beverly Michael, *University of Pittsburgh*
Dave Olsen, *Santa Rosa Junior College*
Nancy Olson, *Johnson County Community College*
Jeanette O'Rourke, *Middlesex County College*
Charanjit Rangi, *Central State University*
Russell O. Reich, Jr., *Sierra Nevada College*
Alan Russell, *Elon College*
Wayne Schmidt, *North Hennepin Community College*
Ron Seater, *University of Alaska—Juneau*
Ellen Shatto, *Harrisburg Area Community College*

M.L.B.
D.J.E.
B.L.J.

Basics of Algebra and Graphing

1

The theme of this text is problem solving in algebra. Both symbolic algebra and graphs are used to develop models and solve applications. Models are used extensively in many fields, ranging from sociology to medicine to meteorology. In this chapter, we see what is meant by a mathematical model, and we introduce the basics of algebra, graphing, and the graphing calculator.

APPLICATION

CHILD EXPENDITURE. The following table shows the annual expenditures on a child in 1996 by families with an average income of $46,100. Use the data to draw a line graph.

AGE OF CHILD	ANNUAL EXPENDITURE
1	$7860
4	8060
7	8130
10	8100
13	8830
16	8960

Source: The Wall Street Journal Almanac, 1998

This situation can be *modeled* by *graphing the data*.

This problem appears as Exercise 49 in Section 1.6.

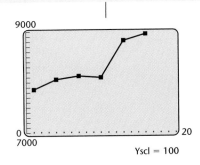

1.1
Some Basics of Algebra

- *Algebraic Expressions*
- *Evaluating Algebraic Expressions*
- *Equations and Inequalities*
- *Sets of Numbers*
- *Absolute Value and Order*
- *Introduction to the Graphing Calculator*

This section introduces some important concepts of algebra. We will examine different types of numbers and certain expressions that arise in problem solving.

Algebraic Expressions

We are all familiar with expressions like

$$73 + 21, \qquad 18 \times 34, \qquad 9 - 5, \quad \text{and} \quad \frac{21}{34}.$$

In algebra, we use these as well as expressions like

$$x + 21, \qquad l \cdot w, \qquad 9 - s, \quad \text{and} \quad \frac{d}{t}.$$

When a letter is used to stand for various numbers, it is called a **variable**. If a letter represents one particular number, it is called a **constant**. Let $d =$ the number of hours in a day. Then d is a constant. If $t =$ the number of hours that a jet has been flying, then t is a variable since t changes as the flight progresses.

An **algebraic expression** consists of variables, numbers, and operation signs. All the expressions above are examples of algebraic expressions. Algebraic expressions can indicate operations other than addition, subtraction, multiplication, and division. Another type of notation used in some algebraic expressions is *exponential notation*. Many different kinds of numbers can be used as *exponents*. Here we establish the meaning of a^n when n is a counting number, $1, 2, 3, \ldots$.

Exponential Notation

The expression a^n, in which n is a counting number greater than 1, means

$$\underbrace{a \cdot a \cdot a \cdot \cdots \cdot a \cdot a}_{n \text{ factors}}.$$

In a^n, a is called the *base* and n is the *exponent*, or *power*. The symbol a^1 means a.

The expression a^n is read "a raised to the nth power" or simply "a to the nth." We read s^2 as "s-squared" and x^3 as "x-cubed." This terminology comes from the fact that the area of a square of side s is $s \cdot s = s^2$ and the volume of a cube of side x is $x \cdot x \cdot x = x^3$.

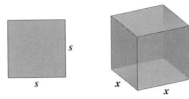

Some examples of algebraic expressions containing exponential notation are

$$3x^2 + 2x - 5, \qquad x^3 y^5, \quad \text{and} \quad \frac{x^5 - y^2}{z^5}.$$

Evaluating Algebraic Expressions

When we replace a variable with a number, we say that we are **substituting** for the variable. This process is called **evaluating the expression**.

Geometric formulas are often evaluated. In the following example, we use the formula for the area A of a triangle with a base of length b and a height of length h:

$$A = \tfrac{1}{2} \cdot b \cdot h.$$

Example 1 *Area of a Sail.* The base of a triangular sail is 8 m and the height is 6.4 m. Find the area of the sail.

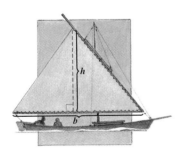

SOLUTION   We substitute 8 for b and 6.4 for h and multiply:

$$\begin{aligned}
\tfrac{1}{2} \cdot b \cdot h &= \tfrac{1}{2} \cdot 8 \cdot 6.4 \\
&= 25.6 \text{ square meters (m}^2\text{).}
\end{aligned}$$

When evaluating algebraic expressions, the order in which we perform the operations may affect the result. Consider the expression $2 + 3 \cdot 5$. If we add first and then multiply, we get $5 \cdot 5$, or 25. If we multiply first and then add, we get $2 + 15$, or 17. The following convention indicates that the second approach is correct.

Rules for Order of Operations

1. Calculate within grouping symbols before calculating outside.

2. Simplify all exponential expressions.

3. Do all multiplication and division in order from left to right.

4. Do all addition and subtraction in order from left to right.

Example 2   Evaluate $5 + 2(a - 1)^2$ for $a = 4$.

SOLUTION

$$
\begin{aligned}
5 + 2(a - 1)^2 &= 5 + 2(4 - 1)^2 && \text{Substituting} \\
&= 5 + 2(3)^2 && \text{Working within parentheses} \\
& && \text{first} \\
&= 5 + 2(9) && \text{Simplifying } 3^2 \\
&= 5 + 18 && \text{Multiplying} \\
&= 23 && \text{Adding} \quad\blacksquare
\end{aligned}
$$

Step (3) in the rules for order of operations tells us to divide before we multiply when division appears first, reading left to right. Similarly, if subtraction appears before addition, we subtract before we add.

Example 3   Evaluate $9 - x^3 + 6 \div 2y^2$ for $x = 2$ and $y = 5$.

SOLUTION

$$
\begin{aligned}
9 - x^3 + 6 \div 2y^2 &= 9 - 2^3 + 6 \div 2(5)^2 && \text{Substituting} \\
&= 9 - 8 + 6 \div 2 \cdot 25 && \text{Simplifying } 2^3 \\
& && \text{and } 5^2 \\
&= 9 - 8 + 3 \cdot 25 && \text{Dividing} \\
&= 9 - 8 + 75 && \text{Multiplying} \\
&= 1 + 75 && \text{Subtracting} \\
&= 76 && \text{Adding} \quad\blacksquare
\end{aligned}
$$

Equations and Inequalities

When an equals sign is placed between two algebraic expressions, an **equation** is formed. An equation can be true or false. For example, the equation

$$13 + 3 = 16$$

is true, and the equation

$$4 + 25 = 30$$

is false. The equation

$$x + 12 = 35$$

is neither true nor false. However, when x is replaced with a number, the equation becomes true or false, depending on the replacement for x. A replacement that makes the equation true is called a *solution* of the equation.

Example 4   Tell whether each number is a solution of the equation $x + 12 = 35$: **(a)** 15; **(b)** 23.

SOLUTION

a) Replacing x with 15, we have

$$15 + 12 = 35.$$

Since this is false, 15 is not a solution of the equation.

b) Replacing x with 23, we have

$$23 + 12 = 35.$$

Since this is true, 23 is a solution of the equation.

The symbols $<$ (is less than), $>$ (is greater than), $\leq$ (is less than or equal to), and $\geq$ (is greater than or equal to) are inequality symbols. An **inequality** is formed when an inequality symbol is placed between two algebraic expressions. Like equations, inequalities can be true, false, or neither true nor false. A *solution* of an inequality is a replacement that makes the inequality true.

Example 5   Tell whether each number is a solution of the inequality $3y + 2 \leq 11$:   **(a)** 0;   **(b)** 5;   **(c)** 3.

SOLUTION

a) Replacing y with 0, we have

$$3 \cdot 0 + 2 \leq 11$$
$$0 + 2 \leq 11$$
$$2 \leq 11.$$

This is read "2 is less than or equal to 11." Since 2 is less than 11, this inequality is true, and 2 is a solution of the inequality.

b) Replacing y with 5, we have

$$3 \cdot 5 + 2 \leq 11$$
$$15 + 2 \leq 11$$
$$17 \leq 11.$$

Since "17 is less than or equal to 11" is false, 5 is not a solution of the inequality.

c) Replacing y with 3, we have

$$3 \cdot 3 + 2 \leq 11$$
$$9 + 2 \leq 11$$
$$11 \leq 11.$$

Since "11 is less than or equal to 11" is true (11 is equal to 11), 3 is a solution of the inequality.

Sets of Numbers

When evaluating algebraic expressions, and in problem solving in general, we often must examine the *type* of numbers used. For example, if a formula is used to determine an optimal class size, any fractional results must be rounded off, since it is impossible to have a fractional part of a student. Three frequently used sets of numbers are listed below.

Natural Numbers, Whole Numbers, and Integers

Natural Numbers (or Counting Numbers)
Those numbers used for counting: $\{1, 2, 3, \ldots\}$

Whole Numbers
The set of natural numbers with 0 included: $\{0, 1, 2, 3, \ldots\}$

Integers The set of all whole numbers and their opposites:

$$\{\ldots, -4, -3, -2, -1, 0, 1, 2, 3, 4, \ldots\}$$

The dots mean that the pattern continues without end.

The integers correspond to the points on a number line as follows:

To fill in the rest of the points on our number line, we must describe two more sets of numbers. To do so, we must first discuss set notation.

The set containing the numbers $-2, 1,$ and 3 can be written $\{-2, 1, 3\}$. This method of writing a set is known as **roster notation**. Roster notation was used for the sets listed above. A second type of set notation, **set-builder notation**, specifies conditions under which a number is in the set. The following example of set-builder notation is read as shown:

$$\{x \mid x \text{ is an odd number less than } 5\}$$

"The set of all x"   "x is an odd number less than 5"

such that

Example 6 Using both roster notation and set-builder notation, name the set consisting of the first four even natural numbers.

SOLUTION

Using roster notation: $\{2, 4, 6, 8\}$

Using set-builder notation: $\{n \mid n \text{ is an even number between } 1 \text{ and } 9\}$

The symbol $\in$ is used to indicate that an element belongs to a set. Thus if $A = \{2, 4, 6, 8\}$, we can write $4 \in A$ to indicate that 4 *is an element of A*. We can also write $5 \notin A$ to indicate that 5 *is not an element of A*.

Example 7 Classify the statement $8 \in \{x \mid x \text{ is an integer}\}$ as true or false.

SOLUTION Since 8 *is* an integer, the statement is true. In other words, since 8 is an integer, it belongs to the set of all integers. ▬

With set-builder notation, we can describe the set of all *rational numbers*.

Rational Numbers

Numbers that can be expressed as an integer divided by a nonzero integer are called *rational numbers*:

$$\left\{\frac{p}{q} \mid p \text{ is an integer, } q \text{ is an integer, and } q \neq 0\right\}.$$

Rational numbers can be written using fractional or decimal notation. *Fractional notation* uses symbolism like the following:

$$\frac{5}{8}, \quad \frac{12}{-7}, \quad \frac{-17}{15}, \quad -\frac{9}{7}, \quad \frac{39}{1}, \quad \frac{0}{6}.$$

In *decimal notation,* rational numbers either *terminate* or *repeat.*

Example 8 When written in decimal form, does each of the following numbers terminate or repeat? **(a)** $\frac{5}{8}$; **(b)** $\frac{6}{11}$.

SOLUTION

a) Since $\frac{5}{8}$ means $5 \div 8$, we perform long division to find that $\frac{5}{8} = 0.625$, a decimal that ends, or terminates. Thus, $\frac{5}{8}$ can be written as a terminating decimal.

b) Using long division, we find that $6 \div 11 = 0.5454\ldots$, so we can write $\frac{6}{11}$ as a repeating decimal. Repeating decimal notation can be abbreviated by writing a bar over the repeating part—in this case, $0.\overline{54}$. ▬

Many numbers, like π, $\sqrt{2}$, and $-\sqrt{15}$, can be only approximated by rational numbers. For example, $\sqrt{2}$ is the number for which $\sqrt{2} \cdot \sqrt{2} = 2$. A calculator's representation of $\sqrt{2}$ as 1.414213562 is an approximation since $(1.414213562)^2$ is not exactly 2.

To see that $\sqrt{2}$ is a "real" point on the number line, recall that when a right triangle has two legs of length 1 unit, the remaining side has length $\sqrt{2}$ units. Thus we can "measure" $\sqrt{2}$ and locate it precisely on a number line.

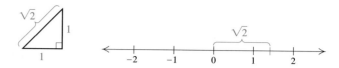

Numbers like π, $\sqrt{2}$, and $-\sqrt{15}$ are said to be **irrational**. Decimal notation for irrational numbers neither terminates nor repeats.

The set of all rational numbers, combined with the set of all irrational numbers, gives us the set of all **real numbers**.

Real Numbers

Numbers that are either rational or irrational are called *real numbers*:

$$\{x \mid x \text{ is rational or } x \text{ is irrational}\}.$$

Every point on the number line represents some real number and every real number is represented by a point on the number line.

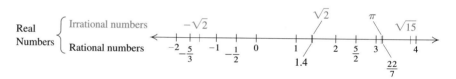

The figure at the top of the next page shows the relationships among various kinds of numbers.

When all members of one set are found in a second set, the first set is a **subset** of the second set. Thus if $A = \{2, 4, 6\}$ and $B = \{1, 2, 4, 5, 6\}$, we write $A \subseteq B$ to indicate that *A is a subset of B*. Similarly, if $\mathbb{N}$ represents the set of all natural numbers and $\mathbb{Z}$ the set of all integers, we can write $\mathbb{N} \subseteq \mathbb{Z}$. Additional statements can be made using other sets in the diagram on the next page.

Absolute Value and Order

It is convenient to have a notation that represents a number's distance from zero on the number line.

Absolute Value

We write $|a|$, read "the absolute value of a," to represent the number of units that a is from zero.

```
                    ┌─────────────────────────────┐
                    │       Real numbers:          │
                    │  −19, −√10, 0, 2/3, π, 17.8  │
                    └─────────────────────────────┘
                       │                        │
        ┌──────────────────────────┐   ┌──────────────────────────┐
        │   Rational numbers:      │   │  Irrational numbers:     │
        │  −5, −4/7, 0, 7/8, 9.45  │   │       −√10,              │
        └──────────────────────────┘   │        √3,               │
           │                    │      │         π,               │
┌──────────────────┐ ┌──────────────────────┐  │        √15               │
│    Integers:     │ │  Rational numbers     │  └──────────────────────────┘
│                  │ │  that are not integers:│
│ −19, −1, 0, 5, 23│ │  −5/8, 2/3, 19/7, 3.4 │
└──────────────────┘ └──────────────────────┘
   │         │         │
┌──────────┐ ┌──────┐ ┌──────────────────┐
│ Negative │ │Zero: │ │ Positive integers│
│ integers:│ │      │ │ or natural numbers:│
│−19, −3,−1│ │  0   │ │  1, 2, 3, 29     │
└──────────┘ └──────┘ └──────────────────┘
```

Example 9 Find the absolute value: **(a)** $|-3|$; **(b)** $|2.5|$; **(c)** $|0|$.

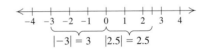

$$|-3| = 3 \qquad |2.5| = 2.5$$

SOLUTION

a) $|-3| = 3$ −3 is 3 units from 0.

b) $|2.5| = 2.5$ 2.5 is 2.5 units from 0.

c) $|0| = 0$ 0 is 0 units from itself.

Note that whereas the absolute value of a nonnegative number is the number itself, the absolute value of a negative number is its opposite.

We use inequality symbols to indicate how two real numbers compare with each other. For any two numbers on the number line, the one to the left is said to be less than, or smaller than, the one to the right. In the figure below, note that although $|-6| > |-1|$, we have $-6 < -1$ because -6 is to the left of -1.

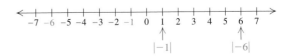

Example 10   Determine whether each inequality is a true statement: **(a)** $-7 < -2$; **(b)** $1 > -4$; **(c)** $-3 \geq -2$.

SOLUTION

a) $-7 < -2$ is *true* because -7 is to the left of -2.

b) $1 > -4$ is *true* because 1 is to the right of -4.

c) $-3 \geq -2$ is *false* because -3 is to the left of -2.   ▬

Introduction to the Graphing Calculator

Graphing calculators and computers equipped with graphing software can be valuable aids in understanding and applying algebra. Features and keystrokes vary among the many brands and models of graphing calculators available. In this text, we use features that are common to most graphing calculators. Specific keystrokes and instructions for certain calculators are included in the Graphing Calculator Manual that accompanies this book. For other procedures, you should consult your instructor or the user's manual for your particular calculator.

There are two important things to keep in mind as you proceed through this text:

1. You will not learn to use a graphing calculator by simply reading about it; you must in fact *use* the calculator. Press the keys on your own calculator as you read the text, do the calculator exercises in the exercise set, and experiment with new options.

2. Your user's manual contains more information about your calculator than appears in this text. If you need additional explanation and examples, be sure to consult the manual.

KEYPAD   A diagram of the keypad of a graphing calculator appears at the front of this text. The organization and labeling of the keys may differ for different calculators. Note that there are options written above keys as well as on the keys. To access those options, press $\boxed{\text{2nd}}$ and then the key below the desired option.

SCREEN   After you have turned the calculator on, you should see a blinking rectangle, or **cursor**, at the top left corner of the screen. If you do not see anything, try adjusting the **contrast**. To perform computations, you should be in the **home screen**.

EXPONENTS   On most graphing calculators, you can enter exponents using the $\boxed{\wedge}$ key before the exponent. For an exponent of 2, you can often use an $\boxed{x^2}$ key.

ORDER OF OPERATIONS   Graphing calculators generally follow the rules for order of operations, so expressions can be entered as they are written. Some calculators will not always multiply in order from left to right. For example $1/2x$ is interpreted as $(1/2)x$ by some calculators and $1/(2x)$ by others. To make the meaning clear, we use parentheses.

Example 11 Evaluate $2(y - 3)^2 + 7$ for $y = 5$.

SOLUTION We replace y with 5 and enter the expression.

```
2(5−3)²+7
                                    15
▮
```

The result is 15.

1.1 Exercise Set

Evaluate each expression using the values provided.

1. $4x - y$, for $x = 3$ and $y = 2$
2. $3a + b$, for $a = 5$ and $b = 4$
3. $2c \div 3b$, for $b = 4$ and $c = 6$
4. $3z \div 2y$, for $y = 1$ and $z = 6$
5. $25 - r^2 + s$, for $r = 3$ and $s = 7$
6. $n^3 - 2 + p$, for $n = 2$ and $p = 5$
7. $3n^2p + 2p^4$, for $n = 5$ and $p = 3$
8. $2a^3b - 2b^2$, for $a = 3$ and $b = 7$
9. $5x \div (2 + x - y)$, for $x = 6$ and $y = 2$
10. $29 - (a - b)^2$, for $a = 7$ and $b = 2$
11. $m + n(5 + n^2)$, for $m = 15$ and $n = 3$
12. $a^2 - 3(a - b)$, for $a = 7$ and $b = 4$

Find the area of a triangular window with the given base and height.

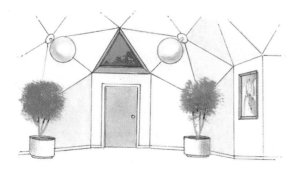

13. Base = 5 ft, height = 6 ft
14. Base = 2.9 m, height = 2.1 m
15. Base = 4 m, height = 3.2 m
16. Base = 4.6 ft, height = 4 ft

Tell whether each number is a solution of the given equation or inequality.

17. $12 - y = 5$; **(a)** 7; **(b)** 5; **(c)** 12
18. $2x + 4 = 14$; **(a)** 3; **(b)** 7; **(c)** 5
19. $5 - x \le 2$; **(a)** 0; **(b)** 4; **(c)** 3
20. $3y - 5 > 10$; **(a)** 3; **(b)** 5; **(c)** 7
21. $3m - 8 < 13$; **(a)** 6; **(b)** 7; **(c)** 9
22. $15 - 3n = 6$; **(a)** 3; **(b)** 2; **(c)** 5

Use roster notation to name each set.

23. The set of all vowels in the English alphabet
24. The set of all days of the week
25. The set of all odd natural numbers
26. The set of all even natural numbers
27. The set of all natural numbers that are multiples of 7
28. The set of all natural numbers that are multiples of 10

Use set-builder notation to name each set.

29. The set of all odd numbers between 10 and 30
30. The set of all multiples of 4 between 22 and 45
31. {0, 1, 2, 3, 4}

32. $\{-3, -2, -1, 0, 1, 2\}$

33. The set of all multiples of 5 between 7 and 79

34. The set of all even numbers between 9 and 99

Classify each statement as true or false. The following sets are used:

$\mathbb{N}$ = the set of natural numbers;

$\mathbb{W}$ = the set of whole numbers;

$\mathbb{Z}$ = the set of integers;

$\mathbb{Q}$ = the set of rational numbers;

$\mathbb{H}$ = the set of irrational numbers;

$\mathbb{R}$ = the set of real numbers.

35. $7.3 \in \mathbb{N}$ **36.** $8 \in \mathbb{N}$

37. $\mathbb{N} \subseteq \mathbb{W}$ **38.** $\mathbb{W} \subseteq \mathbb{Z}$

39. $\sqrt{8} \in \mathbb{Q}$ **40.** $4.1 \in \mathbb{H}$

41. $\mathbb{H} \subseteq \mathbb{R}$ **42.** $\sqrt{10} \in \mathbb{R}$

43. $4.3 \notin \mathbb{Z}$ **44.** $\mathbb{Z} \not\subseteq \mathbb{N}$

45. $\mathbb{Q} \subseteq \mathbb{R}$ **46.** $\mathbb{Q} \subseteq \mathbb{Z}$

Find each absolute value.

47. $|-8|$ **48.** $|-7|$ **49.** $|9|$

50. $|12|$ **51.** $|-6.2|$ **52.** $|-7.9|$

53. $|0|$ **54.** $\left|3\frac{3}{4}\right|$ **55.** $\left|1\frac{7}{8}\right|$

56. $|0.91|$ **57.** $|-4.21|$ **58.** $|-5.309|$

Determine whether each inequality is true or false.

59. $-8 \leq -2$ **60.** $-1 \leq -5$

61. $-7 > 1$ **62.** $7 \geq -2$

63. $3 \geq -5$ **64.** $-9 \leq -9$

65. $-9 < -4$ **66.** $7 \geq -8$

67. $-4 \geq -4$ **68.** $-2 > -2$

69. $-5 < -5$ **70.** $-2 > -12$

Evaluate each of the following using a graphing calculator.

71. $13 - (y - 4)^3 + 10$, for $y = 6$

72. $(t + 4)^2 - 12 \div (19 - 17) + 68$, for $t = 5$

73. $3.86 + 2.7(2.1x + 1.7)$, for $x = 5.82$

74. $1.5(3.982 - a) + 2.3^2$, for $a = 1.99$

75. $3(m + 2n) \div m$, for $m = 1.6$ and $n = 5.9$

76. $1.5 + (2x - y)^2$, for $x = 9.25$ and $y = 1.7$

77. $\frac{1}{2}(x + 5/z)^2$, for $x = 141$ and $z = 0.2$

78. $a - \frac{3}{4}(2a - b)$, for $a = 213$ and $b = 165$

Synthesis

To the student and the instructor: *The synthesis exercises in each section are designed to challenge students to extend the concepts or skills studied in that section. Some synthesis exercises will require the assimilation of skills and concepts from several sections.*

Writing exercises, denoted by ◈ *, should be answered using one or more complete English sentences. In nearly every section, several writing exercises appear as the first synthesis exercises. These exercises are not as challenging as those exercises appearing later in the exercise set and can be assigned to students who might not otherwise attempt synthesis exercises. Because many writing exercises have a variety of correct answers, solutions are not listed in the answer section.*

79. ◈ Explain the difference between an algebraic expression and an equation.

80. ◈ Explain the difference between rational numbers and irrational numbers.

81. ◈ What advantage does set-builder notation have over roster notation?

82. ◈ Werner insists that $15 - 4 + 1 \div 2 \cdot 3$ is 2. What error is he making?

83. Write an equation that has 14 as a solution.

84. Write an inequality that has 4 and 0 as solutions.

Tell whether each number is a solution of the given equation or inequality.

85. $2 + |x| = 5$; **(a)** 3; **(b)** -3; **(c)** 5

86. $|x| - 9 \geq 7$; **(a)** -5; **(b)** 20; **(c)** -16

Tell whether each equation or inequality is true or false.

87. $|-6| \leq 6$ **88.** $|-2| = |2|$

89. $|-3| + |2| = |-5|$ **90.** $|-3 + 2| = |-5|$

Use roster notation to write each set.

91. The set of all whole numbers that are not natural numbers

92. The set of all integers that are not whole numbers

93. $\{x \mid x = 5n,\ n \text{ is a natural number}\}$

94. $\{x \mid x = 3n,\ n \text{ is a natural number}\}$

95. $\{x \mid x = 2n,\ n \text{ is an integer}\}$

96. $\{x \mid x = 2n + 1,\ n \text{ is a whole number}\}$

97. Draw a right triangle that could be used to measure $\sqrt{13}$ units.

1.2
Operations with Real Numbers

- *Addition, Subtraction, and Opposites*
- *Multiplication, Division, and Reciprocals*

In this section, we review how real numbers are added, subtracted, multiplied, and divided.

Addition, Subtraction, and Opposites

We first review the addition of real numbers.

Addition of Two Real Numbers

1. *Positive numbers:* Add the numbers. The result is positive.

2. *Negative numbers:* Add absolute values. Make the answer negative.

3. *A negative and a positive number:* If the numbers have the same absolute value, the answer is 0. Otherwise, subtract the smaller absolute value from the larger one:

   a) If the positive number is further from 0, make the answer positive.

   b) If the negative number is further from 0, make the answer negative.

4. *One number is zero:* The sum is the other number.

Example 1 Add:

a) $-9 + (-5)$;

b) $-3.2 + 9.7$;

c) $-\frac{3}{4} + \frac{1}{3}$.

SOLUTION

a) $-9 + (-5)$ — We add the absolute values, getting 14. The answer is *negative,* -14.

b) $-3.2 + 9.7$ — The absolute values are 3.2 and 9.7. Subtract 3.2 from 9.7 to get 6.5. The positive number is further from 0, so the answer is *positive,* 6.5.

c) $-\frac{3}{4} + \frac{1}{3} = -\frac{9}{12} + \frac{4}{12}$ — The absolute values are $\frac{9}{12}$ and $\frac{4}{12}$. Subtract to get $\frac{5}{12}$. The negative number is further from 0, so the answer is *negative,* $-\frac{5}{12}$.

When numbers like 7 and -7 are added, the result is 0. Such numbers are called **opposites**, or **additive inverses**, of one another.

The Law of Opposites

For any two numbers a and $-a$,

$$a + (-a) = 0.$$

(When opposites are added, their sum is 0.)

Example 2   Find the opposite: **(a)** -17.5; **(b)** $\frac{4}{5}$; **(c)** 0.

SOLUTION

a) The opposite of -17.5 is 17.5 because $-17.5 + 17.5 = 0$.

b) The opposite of $\frac{4}{5}$ is $-\frac{4}{5}$ because $\frac{4}{5} + \left(-\frac{4}{5}\right) = 0$.

c) The opposite of 0 is 0 because $0 + 0 = 0$.   ▬

To name the opposite, we use the symbol "$-$" and read the symbolism $-a$ as "the opposite of a."

Note that $-a$ does not necessarily denote a negative number. In fact, when a is *negative*, $-a$ is *positive*.

Example 3   Find $-x$ for the following: **(a)** $x = -2$; **(b)** $x = \frac{3}{4}$.

SOLUTION

a) If $x = -2$, then $-x = -(-2) = 2$.   The opposite of -2 is 2.

b) If $x = \frac{3}{4}$, then $-x = -\frac{3}{4}$.   The opposite of $\frac{3}{4}$ is $-\frac{3}{4}$.   ▬

Using the notation of opposites, we can formally define absolute value.

Absolute Value

$$|x| = \begin{cases} x & \text{if } x \geq 0, \\ -x & \text{if } x < 0 \end{cases}$$

(When x is nonnegative, the absolute value of x is x. When x is negative, the absolute value of x is the opposite of x. Thus, $|x|$ is never negative.)

Using this definition of absolute value, we have

$$|3| = 3$$

and

$$|-3| = -(-3) = 3.$$

A negative number is said to have a negative "sign" and a positive number a positive "sign." To subtract, we can add an opposite. Thus we sometimes say that we "change the sign of the number being subtracted and then add."

Example 4 Subtract:

a) $5 - 9$;

b) $-1.2 - (-3.7)$;

c) $-\frac{4}{5} - \frac{2}{3}$.

SOLUTION

a) $5 - 9 = 5 + (-9)$ **Change the sign and add.**

$\qquad\quad = -4$

b) $-1.2 - (-3.7) = -1.2 + 3.7$ **Instead of *subtracting* -3.7, we add 3.7.**

$\qquad\qquad\qquad\quad = 2.5$

c) $-\frac{4}{5} - \frac{2}{3} = -\frac{4}{5} + \left(-\frac{2}{3}\right)$

$\qquad\quad = -\frac{12}{15} + \left(-\frac{10}{15}\right)$ **Finding a common denominator**

$\qquad\quad = -\frac{22}{15}$

Multiplication, Division, and Reciprocals

In this text, we direct exploration of mathematical concepts using a graphing calculator in Interactive Discovery features like the one that follows. Such explorations are a part of the development of the material presented and should be performed as you read the text.

When real numbers are multiplied, the sign of the product depends on the signs of the factors.

Interactive Discovery

Consider the following table. Calculate each product and fill in all the columns.

	PRODUCT	SIGN OF FIRST FACTOR	SIGN OF SECOND FACTOR	SIGN OF PRODUCT
$3 \cdot 4$	12	+	+	+
$(2.1)(-3.2)$	-6.72	$+$	$-$	$-$
$(-17)(19)$	-323	$-$	$+$	$-$
$(-12.1)(-22)$	266.2	$-$	$-$	$+$

How can we determine the sign of a product from the signs of the factors?

Division is defined in terms of multiplication. For example, $10 \div (-2) = -5$ because $(-5)(-2) = 10$. Thus the rules for division are the same as those for multiplication.

Multiplication or Division of Two Real Numbers

   1. To multiply or divide two numbers with *unlike signs*, multiply or divide their absolute values. The answer is *negative*.
   2. To multiply or divide two numbers that have the *same sign*, multiply or divide their absolute values. The answer is *positive*.

Example 5   Multiply or divide: **(a)** $\left(-\frac{2}{3}\right)\left(-\frac{3}{8}\right)$; **(b)** $20 \div (-4)$; **(c)** $\dfrac{-45}{-15}$.

SOLUTION

a) $\left(-\frac{2}{3}\right)\left(-\frac{3}{8}\right) = \frac{6}{24} = \frac{1}{4}$ Multiply absolute values. The answer is positive.

b) $20 \div (-4) = -5$ Divide absolute values. The answer is negative.

c) $\dfrac{-45}{-15} = 3$ Divide absolute values. The answer is positive.

Note that since

$$\frac{-8}{2} = \frac{8}{-2} = -\frac{8}{2} = -4,$$

we have the following generalization.

The Sign of a Fraction

For any number a and any nonzero number b,

$$\frac{-a}{b} = \frac{a}{-b} = -\frac{a}{b}.$$

Recall that

$$\frac{a}{b} = \frac{a}{1} \cdot \frac{1}{b} = a \cdot \frac{1}{b}.$$

That is, if we prefer, we can multiply by $1/b$ rather than divide by b. Provided that b is not 0, the numbers b and $1/b$ are called **reciprocals**, or **multiplicative inverses**, of each other.

The Law of Reciprocals

For any two numbers a and $1/a$ $(a \neq 0)$,

$$a \cdot \frac{1}{a} = 1.$$

(When reciprocals are multiplied, their product is 1.)

Example 6   Find the reciprocal: **(a)** $\frac{7}{8}$; **(b)** $-\frac{3}{4}$; **(c)** -8.

SOLUTION

a) The reciprocal of $\frac{7}{8}$ is $\frac{8}{7}$ because $\frac{7}{8} \cdot \frac{8}{7} = 1$.

b) The reciprocal of $-\frac{3}{4}$ is $-\frac{4}{3}$.

c) The reciprocal of -8 is $\frac{1}{-8}$, or $-\frac{1}{8}$.   ▬

To divide, we can multiply by a reciprocal. We sometimes say that we "invert and multiply."

Example 7   Divide: **(a)** $-\frac{1}{4} \div \frac{3}{5}$; **(b)** $-\frac{6}{7} \div (-10)$.

SOLUTION

a) $-\frac{1}{4} \div \frac{3}{5} = -\frac{1}{4} \cdot \frac{5}{3}$ **"Inverting" $\frac{3}{5}$ and changing division to multiplication**

$= -\frac{5}{12}$

b) $-\frac{6}{7} \div (-10) = -\frac{6}{7} \cdot \left(-\frac{1}{10}\right) = \frac{6}{70}$, or $\frac{3}{35}$ ▬

Thus far, we have never divided by 0 or, equivalently, had a denominator of 0. There is a reason for this. Suppose 5 were divided by 0. The answer would have to be a number that, when multiplied by 0, gave 5. But any number times 0 is 0. Thus we cannot divide 5 or any other nonzero number by 0.

What if we divide 0 by 0? In this case, our solution would need to be some number that, when multiplied by 0, gave 0. But then *any* number would work as a solution to $0 \div 0$. This could lead to contradictions so we agree to exclude division of 0 by 0 also.

Division by Zero

We never divide by 0. If asked to divide a nonzero number by 0, we say that the answer is *undefined*. If asked to divide 0 by 0, we say that the answer is *indeterminate*.

The rules for order of operations discussed in Section 1.1 apply to *all* real numbers, regardless of their signs.

Example 8   Simplify: $7 - 5^2 + 6 \div 2(-5)^2$.

SOLUTION

$$7 - 5^2 + 6 \div 2(-5)^2 = 7 - 25 + 6 \div 2 \cdot 25 \quad \text{**Simplifying 5^2 and $(-5)^2$**}$$

$$= 7 - 25 + 3 \cdot 25 \quad \text{**Dividing**}$$

$$= 7 - 25 + 75 \quad \text{**Multiplying**}$$

$$= -18 + 75 \quad \text{**Subtracting**}$$

$$= 57 \quad \text{**Adding**}$$ ▬

In addition to the usual grouping symbols—parentheses, brackets, and braces—a fraction bar, absolute-value symbol, or radical sign ($\sqrt{\ }$) may indicate groupings.

Example 9 Calculate: $\dfrac{12|7-9|+4\cdot5}{(-3)^4+2^3}$.

SOLUTION   We simplify the numerator and the denominator and divide the results:

$$\frac{12|7-9|+4\cdot5}{(-3)^4+2^3} = \frac{12|-2|+4\cdot5}{81+8}$$

$$= \frac{12(2)+20}{89}$$

$$= \frac{44}{89}. \qquad \textbf{Multiplying and then adding}$$

When entering a negative number into a calculator, we use the $\boxed{(-)}$ key, not the subtraction key. Also, some grouping symbols, such as a fraction bar and an absolute-value symbol, may require parentheses to indicate the grouping. The absolute-value option is accessed using a **menu**.

MENUS   A menu is a list of options that appears when a key is pressed. Thus multiple operations can be accessed by pressing one key. For example, pressing $\boxed{\text{MATH}}$ may result in a screen like the following. Four menu titles, or **submenus**, are listed across the top of the screen. We refer to the submenu NUM, shown in the figure, as MATH NUM, which means that we must first press $\boxed{\text{MATH}}$ in order to access the NUM menu.

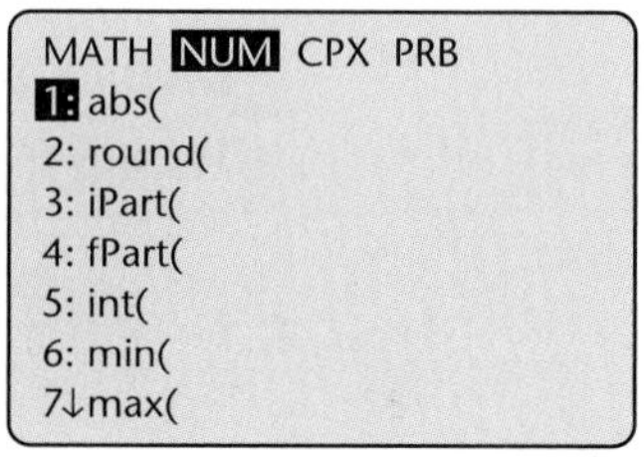

```
MATH NUM CPX PRB
1: abs(
2: round(
3: iPart(
4: fPart(
5: int(
6: min(
7↓max(
```

We use the left and right arrow keys to highlight the desired menu. In the screen shown above, the NUM menu is highlighted. The options in the highlighted menu appear on the screen. Note that the NUM menu contains more options than can fit on a screen, as indicated by the arrow in entry 7. The remaining options will appear as the down arrow is pressed.

To copy the item on the home screen, we highlight its number using the up or down arrow keys and press $\boxed{\text{ENTER}}$, or simply press the number of the item.

Example 10 Calculate: $\dfrac{14-3|-16+38|}{4|-2^4-3^2|}$.

$\dfrac{16+9}{25}$

SOLUTION There is no long fraction bar on the calculator, so the expression must be rewritten using parentheses:

$$(14 - 3|-16 + 38|) \div (4|-2^4 - 3^2|).$$

For most graphers, $|x|$ is written abs(x) and is accessed through a menu. The expressions within the absolute-value symbols must be enclosed in parentheses. Some calculators will supply the left parenthesis; for these, you must close the expression with a right parenthesis.

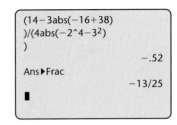

Many graphers can convert decimal notation to fractional notation, as shown in the screen on the right above. The answer is -0.52, or $-\frac{13}{25}$.

Example 10 demonstrates that being able to use a calculator does not make it any less important to understand the mathematics being studied. A solid understanding of the rules for order of operations is necessary in order to locate parentheses properly in an expression.

1.2 Exercise Set

Add.

1. $5 + 12$

2. $9 + 7$

3. $-4 + (-7)$

4. $-8 + (-3)$

5. $-5.9 + 2.7$

6. $-1.9 + 7.3$

7. $\frac{2}{7} + \left(-\frac{3}{5}\right)$

8. $\frac{3}{8} + \left(-\frac{2}{5}\right)$

9. $-4.9 + (-3.6)$

10. $-2.1 + (-7.5)$

11. $-\frac{1}{9} + \frac{2}{3}$

12. $-\frac{1}{2} + \frac{4}{5}$

13. $0 + (-4.5)$

14. $-3.19 + 0$

15. $-7.24 + 7.24$

16. $-9.46 + 9.46$

17. $15.9 + (-22.3)$

18. $21.7 + (-28.3)$

Find the opposite, or additive inverse.

19. 7.29

20. 5.43

21. $-4\frac{1}{3}$

22. $2\frac{3}{5}$

23. 0

24. $-2\frac{3}{4}$

Find $-x$ for each of the following.

25. $x = 7$

26. $x = 3$

27. $x = -2.7$

28. $x = -1.9$

29. $x = 1.79$

30. $x = 3.14$

31. $x = 0$

32. $x = -1$

Subtract.

33. $9 - 7$

34. $8 - 3$

35. $4 - 9$

36. $3 - 10$

37. $-6 - (-10)$

38. $-3 - (-9)$

39. $-4 - 13$

40. $-7 - 8$

41. $2.7 - 5.8$

42. $3.7 - 4.2$

43. $-\frac{3}{5} - \frac{1}{2}$

44. $-\frac{2}{3} - \frac{1}{5}$

45. $-3.9 - (-6.8)$

46. $-5.4 - (-4.3)$

47. $0 - (-7.9)$

48. $0 - 5.3$

Multiply.

49. $(-4)7$

50. $(-5)9$

51. $(-3)(-8)$

52. $(-7)(-8)$

53. $(4.2)(-5)$

54. $(3.5)(-8)$

55. $\frac{3}{7}(-1)$

56. $-1 \cdot \frac{2}{5}$

57. $(-17.45) \cdot 0$

58. 15.2×0

59. $(-3.2) \times (-1.7)$

60. $(1.9) \cdot (4.3)$

Divide.

61. $\frac{-10}{-2}$

62. $\frac{-15}{-3}$

63. $\frac{-100}{20}$

64. $\frac{-50}{5}$

65. $\frac{73}{-1}$

66. $\frac{-62}{1}$

67. $\frac{0}{-7}$

68. $\frac{0}{-11}$

Find the reciprocal, or multiplicative inverse.

69. 5

70. 3

71. -9

72. -7

73. $\frac{2}{3}$

74. $\frac{4}{7}$

75. $-\frac{3}{11}$

76. $-\frac{7}{3}$

Divide.

77. $\frac{2}{3} \div \frac{4}{5}$

78. $\frac{2}{7} \div \frac{6}{5}$

79. $-\frac{3}{5} \div \frac{1}{2}$

80. $\left(-\frac{4}{7}\right) \div \frac{1}{3}$

81. $\left(-\frac{2}{9}\right) \div (-8)$

82. $\left(-\frac{2}{11}\right) \div (-6)$

83. $\frac{12}{7} \div (-1)$

84. $\left(-\frac{2}{7}\right) \div (-1)$

Match the algebraic expressions with the correct series of keystrokes. Check your answer by calculating the expression both by hand and by using the calculator.

85. $\dfrac{5(3-7) + 4^3}{(-2-3)^2}$

86. $(5(3-7) + 4)^3 \div (-2) - 3^2$

87. $5(3-7) + 4^3 \div (-2-3)^2$

88. $\dfrac{5(3-7) + 4^3}{(-2)-3^2}$

a) (5 (3 − 7) + 4 ∧ 3) ÷
(((−) 2 − 3) x^2) ENTER

b) (5 (3 − 7) + 4 ∧ 3) ÷
(((−) 2) − 3 x^2) ENTER

c) (5 (3 − 7) + 4) ∧ 3 ÷
((−) 2) − 3 x^2 ENTER

d) 5 (3 − 7) + 4 ∧ 3 ÷
((−) 2 − 3) x^2 ENTER

Calculate using the rules for order of operations.

89. $12 - (9 - 3 \cdot 2^3)$

90. $19 - (4 + 2 \cdot 3^2)$

91. $\dfrac{5 \cdot 2 - 4^2}{27 - 2^4}$

92. $\dfrac{7 \cdot 3 - 5^2}{9 + 4 \cdot 2}$

93. $\dfrac{3^4 - (5 - 3)^4}{1 - 2^3}$

94. $\dfrac{4^3 - (7 - 4)^2}{3^2 - 7}$

95. $5^3 - [2(4^2 - 3^2 - 6)]^3$

96. $7^2 - [3(5^2 - 4^2 - 7)]^2$

97. $|2^2 - 7|^3 + 1$

98. $|-2 - 3| \cdot 4^2 - 1$

99. $\dfrac{13.4 - 5|1.2 + 4.6|}{(9.3 - 5.4)^2}$

100. $|13.5 + 8(-4.7)|^3$

101. $134 - \sqrt{16 \cdot 35 - 48 \div 160}$

102. $-35 \div 700 \cdot 3 - \sqrt{-50 - (12 - 185)}$

To the student and the instructor: *It is always a good idea to examine a problem before attempting to solve it. Throughout the text, certain problems such as Exercises 103 and 104 are marked with a magnifying glass to remind you to look at them carefully. These problems in particular can often be solved without either lengthy computations or the use of a grapher. They are intended to help you develop the habit of thinking about every problem before attempting to solve it.*

103. 🔍 $-12.86 - 5.2(-1.7 - 3.8)^2 \cdot 0$

104. 🔍 $\dfrac{0 \cdot [(-1.2)^2 + (9.2 - 1.78)^3]}{-2.5 - (1.7 + 0.9)}$

Skill Maintenance

To the student and the instructor: *Exercises included for Skill Maintenance review skills previously studied in the text. You can expect such exercises in every exercise set. Often these serve as preparation for an upcoming section. The numbers in brackets immediately following the directions or exercise indicate the section in which the skill was introduced.*

Evaluate each expression using the values provided. [1.1]

105. $13 - (x - y)^3$, for $x = 10$ and $y = 8$

106. $a^2 + 3(a + b)^4$, for $a = 2$ and $b = 1$

Synthesis

107. ◆ Describe in your own words a method for determining the sign of the sum of a positive number and a negative number.

108. ◆ Explain in your own words the difference between the opposite of a number and the reciprocal of a number.

109. ◆ Explain in your own words why $\frac{7}{0}$ is undefined.

110. ◈ Explain in your own words why 0 has an opposite but not a reciprocal.

Insert one pair of parentheses to change each false statement into a true statement.

111. $3 - 8^2 + 9 = 34$

112. $2 \cdot 7 + 3^2 \cdot 5 = 104$

113. $5 \cdot 2^3 \div 3 - 4^4 = 40$

114. $2 - 7 \cdot 2^2 + 9 = -11$

115. Find the greatest value of a for which $|a| \geq 6.2$ and $a < 0$.

1.3
Equivalent Algebraic Expressions

- *The Commutative, Associative, and Distributive Laws*
- *Combining Like Terms*

In algebra, we are often interested in manipulating expressions without changing their value. In this section, we look at some properties of real numbers that allow us to write *equivalent expressions*.

Equivalent Expressions

Two expressions that have the same value for all possible replacements are called *equivalent expressions*.

The Commutative, Associative, and Distributive Laws

As was the case with the numbers used in arithmetic, when a pair of real numbers are added or multiplied, the order in which the numbers are written does not affect the result.

The Commutative Laws

For any real numbers a and b,

$$a + b = b + a; \qquad a \cdot b = b \cdot a.$$
(for Addition)   (for Multiplication)

The commutative laws provide one way of writing equivalent expressions.

Example 1 Use a commutative law to write an expression equivalent to $7x + 9$.

SOLUTION Using the commutative law of addition, we have

$$7x + 9 = 9 + 7x.$$

We can also use the commutative law of multiplication to write

$$7x + 9 = x7 + 9.$$

The expressions $7x + 9$, $9 + 7x$, and $x7 + 9$ are all equivalent. They name the same number for any replacement of x.   ▬

The *associative laws* also enable us to form equivalent expressions.

The Associative Laws

For any real numbers a, b, and c,

$$a + (b + c) = (a + b) + c;$$
(for Addition)

$$a \cdot (b \cdot c) = (a \cdot b) \cdot c.$$
(for Multiplication)

Example 2 Write an expression equivalent to $(3x + 7y) + 9z$, using the associative law of addition.

SOLUTION   We have

$$(3x + 7y) + 9z = 3x + (7y + 9z).$$

The expressions $(3x + 7y) + 9z$ and $3x + (7y + 9z)$ are equivalent. They name the same number for any replacements of x, y, and z.   ▬

Example 3 Use the commutative and associative laws to write an expression equivalent to

$$\frac{5}{x} \cdot (yz).$$

SOLUTION   Answers may vary. We use the associative law first and then the commutative law.

$$\frac{5}{x} \cdot (yz) = \left(\frac{5}{x} \cdot y \right) \cdot z \qquad \text{Using the associative law of multiplication}$$

$$= \left(y \cdot \frac{5}{x} \right) \cdot z \qquad \text{Using the commutative law in the parentheses}$$ ▬

The *distributive law* that follows provides another way of forming equivalent expressions. In essence, the distributive law allows us to re-write the *product* of a and $b + c$ as the *sum* of ab and ac.

The Distributive Law

For any numbers a, b, and c,

$$a(b + c) = ab + ac.$$

Since $b - c = b + (-c)$, it is also true by the distributive law that $a(b - c) = ab - ac$.

Example 4   Obtain an expression equivalent to $5x(y + 4)$ by multiplying.

SOLUTION   We use the distributive law to get

$$5x(y + 4) = 5xy + 5x \cdot 4 \qquad \text{Using the distributive law}$$
$$= 5xy + 5 \cdot 4 \cdot x \qquad \text{Using the commutative law of multiplication}$$
$$= 5xy + 20x. \qquad \text{Simplifying}$$

The expressions $5x(y + 4)$ and $5xy + 20x$ are equivalent. They name the same number for any replacements of x and y.

When we do the opposite of what we did in Example 4, we say that we are **factoring** an expression. This allows us to rewrite a sum as a product.

Example 5   Obtain an expression equivalent to $3x - 6$ by factoring.

SOLUTION   We use the distributive law to get

$$3x - 6 = 3(x - 2).$$

In Example 5, since the product of 3 and $x - 2$ is $3x - 6$, we say that 3 and $x - 2$ are **factors** of $3x - 6$. Thus "factor" can act as a noun or as a verb.

Combining Like Terms

A **term** is a number, a variable, a product of numbers and/or variables, or a quotient of numbers and/or variables. Thus, $2x$, $-y$, and $3z$ are terms in the expression $2x - y + 3z$. When terms have variable factors that are exactly the same, we refer to those terms as **like**, or similar, **terms**. Thus, $3x^2$ and $-6x^2$ are like terms. We can often simplify expressions by **combining**, or collecting, **like terms.**

Example 6   Combine like terms: $3a + 5b + 4a + b$.

SOLUTION

$$3a + 5b + 4a + b = 3a + 4a + 5b + b \qquad \text{Using a commutative law}$$
$$= (3 + 4)a + (5 + 1)b \qquad \text{Using the distributive law. Note that } b = 1b.$$
$$= 7a + 6b$$

Sometimes we must use the distributive law to remove grouping symbols before combining like terms.

Example 7   Simplify: $3x + 2[4 + 5(x + 2y)]$.

SOLUTION

$$3x + 2[4 + 5(x + 2y)] = 3x + 2[4 + 5x + 10y] \qquad \text{Using the distributive law}$$

$$= 3x + 8 + 10x + 20y \qquad \text{Using the distributive law}$$

$$= 13x + 8 + 20y \qquad \text{Combining like terms}$$

The product of a number and -1 is its opposite, or additive inverse. For example,

$$-1 \cdot 8 = -8 \qquad \text{(the opposite of 8)}.$$

Thus we have $-8 = -1 \cdot 8$, and in general, $-x = -1 \cdot x$. We can use this fact along with the distributive law when parentheses are preceded by a negative sign or a subtraction symbol.

Example 8   Simplify $-(a - b)$, using multiplication by -1.

SOLUTION   We have

$$-(a - b) = -1 \cdot (a - b) \qquad \text{Replacing } - \text{ with multiplication by } -1$$

$$= -1 \cdot a - (-1) \cdot b \qquad \text{Using the distributive law}$$

$$= -a - (-b) \qquad \text{Replacing } -1 \cdot a \text{ with } -a \text{ and } (-1) \cdot b \text{ with } -b$$

$$= -a + b, \text{ or } b - a. \qquad \text{Try to go directly to this step.}$$

The expressions $-(a - b)$ and $b - a$ are equivalent. They name the same number for all replacements of a and b.

Example 8 illustrates a useful shortcut worth remembering:

The opposite of $a - b$ is $-a + b$, or $b - a$.

Example 9   Simplify: $9x - 5y - (5x + y - 7)$.

SOLUTION

$$9x - 5y - (5x + y - 7) = 9x - 5y - 5x - y + 7 \qquad \text{Taking the opposite of } 5x + y - 7$$

$$= 4x - 6y + 7 \qquad \text{Combining like terms}$$

If an expression contains nested grouping symbols, we work from the inside out when simplifying, using the rules for order of operations.

Example 10   Simplify by combining like terms:

$$3x^2 - 2\{3[x - 5x^2 - 4(x + 3) + 7] - x\}.$$

SOLUTION

$$3x^2 - 2\{3[x - 5x^2 - 4(x + 3) + 7] - x\}$$
$$= 3x^2 - 2\{3[x - 5x^2 - 4x - 12 + 7] - x\}$$

Multiplying to re-move the innermost parentheses using the distributive law

$$= 3x^2 - 2\{3[-5x^2 - 3x - 5] - x\}$$

Combining like terms in the brackets

$$= 3x^2 - 2\{-15x^2 - 9x - 15 - x\}$$

Multiplying to remove the brackets using the distributive law

$$= 3x^2 - 2\{-15x^2 - 10x - 15\}$$

Combining like terms

$$= 3x^2 + 30x^2 + 20x + 30$$

Removing the braces

$$= 33x^2 + 20x + 30$$

Combining like terms

1.3 Exercise Set

Write an equivalent expression using a commutative law. Answers may vary.

1. $3x + 8y$ **2.** $ab + 9$
3. $(7x)y$ **4.** $-9(ab)$

Write an equivalent expression using an associative law.

5. $(3x)y$ **6.** $-7(ab)$
7. $x + (2y + 5)$ **8.** $(3y + 4) + 10$

Write an equivalent expression using the distributive law.

9. $3(a + 1)$ **10.** $8(x + 1)$
11. $4(x - y)$ **12.** $9(a - b)$
13. $-5(2a + 3b)$ **14.** $-2(3c + 5d)$
15. $2a(b - c + d)$ **16.** $5x(y - z + w)$
17. $2\pi r(h + 1)$ **18.** $P(1 + rt)$

Find an equivalent expression by factoring.

19. $5x + 5y$ **20.** $7a + 7b$
21. $3p - 9$ **22.** $12x - 3$
23. $7x - 21y$ **24.** $6y - 9$
25. $2x - 2y + 2z$ **26.** $3x + 3y - 3z$
27. $xy + x$ **28.** $ab + a$
29. $ab + ac - ad$ **30.** $xy - xz + xw$

List the terms of each expression.

31. $4a - 5b + 6$ **32.** $5x - 9y + 12$
33. $2x^2 - 6x + 7$ **34.** $-5y^2 - 7y - 8$

Simplify by combining like terms. Use the distributive law as needed.

35. $4a + 5a$ **36.** $9x + 3x$
37. $7rt - 9rt$ **38.** $3ab + 7ab$
39. $8x^2 + x^2$ **40.** $7a^2 + a^2$
41. $12a - a$ **42.** $15x - x$
43. $t - 9t$ **44.** $x - 6x$
45. $5x - 3x + 8x$ **46.** $3x - 11x + 2x$
47. $5x - 2x^2 + 3x$ **48.** $9a - 5a^2 + 4a$
49. $3a + 5a^2 - a + 4a^2$
50. $9x + 2x^3 + 5x - 6x^2$
51. $4x - 7 + 18x + 25$
52. $13p + 5 - 4p + 7$
53. $-7t^2 + 3t + 5t^3 - t^3 + 2t^2 - t$
54. $-9n + 8n^2 + n^3 - 2n^2 - 3n + 4n^3$
55. $a - (2a + 5)$ **56.** $x - (5x + 9)$
57. $4m - (3m - 1)$ **58.** $5a - (4a - 3)$
59. $3d - 7 - (5 - 2d)$ **60.** $8x - 9 - (7 - 5x)$
61. $-2(x + 3) - 5(x - 4)$
62. $-9(y + 7) - 6(y - 3)$
63. $5x - 7(2x - 3)$
64. $8y - 4(5y - 6)$
65. $9a - [7 - 5(7a - 3)]$

66. $12b - [9 - 7(5b - 6)]$

67. $5\{-2a + 3[4 - 2(3a + 5)]\}$

68. $7\{-7x + 8[5 - 3(4x + 6)]\}$

69. $2y + \{7[3(2y - 5) - (8y + 7)] + 9\}$

70. $7b - \{6[4(3b - 7) - (9b + 10)] + 11\}$

Skill Maintenance

Find each absolute value. [1.1]

71. $|-3|$

72. $|3.59|$

73. Find the opposite of -35. [1.2]

74. Find the reciprocal of -35. [1.2]

Synthesis

75. ◆ What is the difference between the associative law of multiplication and the distributive law?

76. ◆ Write a sentence in which the word "factor" appears once as a verb and once as a noun.

77. ◆ Explain how the distributive and commutative laws can be used to rewrite $3x + 6y + 4x + 2y$ as $7x + 8y$.

78. ◆ Lee simplifies the expression $a(b + c)$ by omitting the parentheses and writing $ab + c$. Are these expressions equivalent? Why or why not?

79. Use the commutative, associative, and distributive laws to show that $5(a + bc)$ is equivalent to $c(b5) + a5$. Use only one law in each step of your work.

Simplify.

80. $11(a - 3) + 12a - \{6[4(3b - 7) - (9b + 10)] + 11\}$

81. $-3[9(x - 4) + 5x] - 8\{3[5(3y + 4)] - 12\}$

82. $z - \{2z + [3z - (4z + 5x) - 6z] + 7z\} - 8z$

83. $x + [f - (f + x)] + [x - f] + 3x$

84. $x - \{x + 1 - [x + 2 - (x - 3 - \{x + 4 - [x - 5 + (x - 6)]\})]\}$

$\dfrac{1.4}{}$

Exponential and Scientific Notation

- *The Product and Quotient Rules*
- *The Zero Exponent*
- *Negative Integers as Exponents*
- *Raising Powers to Powers*
- *Raising a Product or a Quotient to a Power*
- *Scientific Notation*

In Section 1.1, we discussed how whole-number exponents are used. We now develop rules for manipulating exponents and define what zero and negative integers will mean as exponents.

The Product and Quotient Rules

Interactive Discovery

Calculate the value of each of the following expressions. Then match each numbered expression with the lettered expression having the same value.

b **1.** $3^4 \cdot 3^5$ C **2.** $3^2 \cdot 3^5$ **a.** 3^{11} **b.** 3^9

a **3.** $3^5 \cdot 3^6$ d **4.** $3^4 \cdot 3$ **c.** 3^7 **d.** 3^5

How are the exponents of the equivalent expressions related?

The result in the Interactive Discovery above is generalized in the *product rule*.

秊言文

Multiplying with Like Bases: The Product Rule

For any number a and any positive integers m and n,

$$a^m \cdot a^n = a^{m+n}.$$

(When multiplying with exponential notation, if the bases are the same, keep the base and add the exponents.)

Example 1   Multiply and simplify: (a) $m^5 \cdot m^7$; (b) $(5a^2b^3)(3a^4b^5)$.

SOLUTION

a) $m^5 \cdot m^7 = m^{5+7} = m^{12}$

b) $(5a^2b^3)(3a^4b^5) = 5 \cdot 3 \cdot a^2 \cdot a^4 \cdot b^3 \cdot b^5$ 組を式。   Using the associative and commutative laws

$\qquad\qquad = 15a^{2+4}b^{3+5}$ Multiplying; using the product rule

$\qquad\qquad = 15a^6b^8$

Interactive Discovery

Calculate the value of each of the following expressions. Then match each numbered expression with the lettered expression having the same value.

1. $\dfrac{3^{11}}{3^{10}}$ **2.** $\dfrac{3^7}{3^2}$ **a.** 3^5 **b.** 3^4

3. $\dfrac{3^5}{3}$ **c.** 3^1

How are the exponents of the equivalent expressions related?

The generalization of the result in the above Interactive Discovery is the *quotient rule*.

分式文

Dividing with Like Bases: The Quotient Rule

For any nonzero number a and any positive integers m and n, $m > n$,

$$\frac{a^m}{a^n} = a^{m-n}.$$

(When dividing with exponential notation, if the bases are the same, keep the base and subtract the exponent of the denominator from the exponent of the numerator.)

分用

Example 2 Divide and simplify: **(a)** $\dfrac{r^9}{r^3}$; **(b)** $\dfrac{10x^{11}y^5}{2x^4y^3}$.

SOLUTION

a) $\dfrac{r^9}{r^3} = r^{9-3} = r^6$ Using the quotient rule

b) $\dfrac{10x^{11}y^5}{2x^4y^3} = 5 \cdot x^{11-4} \cdot y^{5-3}$ Dividing; using the quotient rule

  $= 5x^7y^2$

The Zero Exponent

Suppose now that the bases in the numerator and the denominator are both raised to the same power:

$$\frac{x^5}{x^5} = 1 \quad \text{or} \quad \frac{8^3}{8^3} = 1.$$

These results follow from the fact that any (nonzero) expression, divided by itself, is equal to 1. On the other hand, were we to subtract exponents, we would obtain

$$\frac{x^5}{x^5} = x^{5-5} = x^0 \quad \text{or} \quad \frac{8^3}{8^3} = 8^{3-3} = 8^0.$$

Thus, in order to continue subtracting exponents when dividing like bases raised to the same power, we must have $x^0 = 1$ and $8^0 = 1$. This leads to the following definition.

The Zero Exponent

For any real number a, $a \neq 0$,

  $a^0 = 1.$

(Any nonzero number raised to the zero power is 1.)

Example 3 Evaluate each of the following for $x = 2.9$:
(a) x^0; **(b)** $-x^0$; **(c)** $(-x)^0$.

SOLUTION

a) $x^0 = 2.9^0 = 1$ Using the definition of 0 as an exponent

b) $-x^0 = -2.9^0 = -1$ The exponent is used before the negative sign.

c) $(-x)^0 = (-2.9)^0 = 1$ The negative sign is used before the exponent.

Parts (b) and (c) of Example 3 illustrate an important result:

  Since $-a^n$ means $-1 \cdot a^n$, $-a^n$ and $(-a)^n$ are not equivalent expressions.

Negative Integers as Exponents

Later in this text, we will explain what numbers like $\frac{2}{9}$ or $\sqrt{2}$ mean as exponents. Until then, integer exponents will suffice.

To develop a definition for negative integer exponents, we simplify $5^3/5^7$ two ways. First we proceed as in arithmetic:

$$\frac{5^3}{5^7} = \frac{5 \cdot 5 \cdot 5}{5 \cdot 5 \cdot 5 \cdot 5 \cdot 5 \cdot 5 \cdot 5} = \frac{5 \cdot 5 \cdot 5 \cdot 1}{5 \cdot 5 \cdot 5 \cdot 5 \cdot 5 \cdot 5 \cdot 5}$$

$$= \frac{5 \cdot 5 \cdot 5}{5 \cdot 5 \cdot 5} \cdot \frac{1}{5 \cdot 5 \cdot 5 \cdot 5}$$

$$= \frac{1}{5^4}.$$

Were we to apply the quotient rule, we would have

$$\frac{5^3}{5^7} = 5^{3-7}$$

$$= 5^{-4}.$$

These two expressions for $5^3/5^7$ suggest that

$$5^{-4} = \frac{1}{5^4}.$$

This leads to the definition of negative exponents.

Negative Exponents

For any nonzero real number a and any integer n,

$$a^{-n} = \frac{1}{a^n}.$$

(The numbers a^{-n} and a^n are reciprocals of each other.)

The definitions above preserve the following pattern:

$$4^3 = 4 \cdot 4 \cdot 4,$$
$$4^2 = 4 \cdot 4, \qquad \text{Dividing by 4 on both sides}$$
$$4^1 = 4, \qquad \text{Dividing by 4 on both sides}$$
$$4^0 = 1, \qquad \text{Dividing by 4 on both sides}$$
$$4^{-1} = \frac{1}{4}, \qquad \text{Dividing by 4 on both sides}$$
$$4^{-2} = \frac{1}{4 \cdot 4} = \frac{1}{4^2}. \qquad \text{Dividing by 4 on both sides}$$

Example 4   Express using positive exponents and then simplify:
(a) 3^{-2}; **(b)** $5x^{-4}y^3$; **(c)** $\dfrac{1}{5^{-2}}$.

SOLUTION

a) $3^{-2} = \dfrac{1}{3^2} = \dfrac{1}{9}$

b) $5x^{-4}y^3 = 5\left(\dfrac{1}{x^4}\right)y^3 = \dfrac{5y^3}{x^4}$

c) $\dfrac{1}{5^{-2}} = \dfrac{1}{\frac{1}{5^2}} = 1 \cdot \dfrac{5^2}{1} = 25$

Example 4(c) reveals that when a factor of the numerator or the denominator is raised to any power, the factor can be moved to the other side of the fraction bar provided the sign of the exponent is changed. Thus, for example,

$$\frac{a^{-2}b^3}{c^{-4}} = \frac{c^4}{a^2b^{-3}}.$$

The product and quotient rules apply for all integer exponents.

Example 5   Simplify: **(a)** $7^{-3} \cdot 7^8$; **(b)** $\dfrac{b^{-5}}{b^{-4}}$.

SOLUTION

a) $7^{-3} \cdot 7^8 = 7^{-3+8}$ ⎫ ——— Adding exponents
        $= 7^5$ ⎬

Check: $7^{-3} \cdot 7^8 = \dfrac{1}{7^3} \cdot 7^8$
$= \dfrac{7^8}{7^3} = 7^5$

b) $\dfrac{b^{-5}}{b^{-4}} = b^{-5-(-4)} = b^{-1}$ ——— Subtracting exponents

$= \dfrac{1}{b}$ Writing the answer without a negative exponent

Example 5(b) can also be simplified as follows:

$$\frac{b^{-5}}{b^{-4}} = \frac{b^4}{b^5} = b^{4-5} = b^{-1} = \frac{1}{b}.$$

Raising Powers to Powers

Next, consider an expression like $(3^4)^2$:

$(3^4)^2 = (3^4)(3^4)$ **We are raising 3^4 to the second power**

$= (3 \cdot 3 \cdot 3 \cdot 3)(3 \cdot 3 \cdot 3 \cdot 3)$

$= 3 \cdot 3 \cdot 3 \cdot 3 \cdot 3 \cdot 3 \cdot 3 \cdot 3$ **Using the associative law**

$= 3^8.$

Note that in this case, we could have multiplied the exponents:

$$(3^4)^2 = 3^{4 \cdot 2} = 3^8.$$

Likewise, $(y^8)^3 = (y^8)(y^8)(y^8) = y^{24}$. Once again, we get the same result if we multiply the exponents:

$$(y^8)^3 = y^{8 \cdot 3} = y^{24}.$$

Raising a Power to a Power: The Power Rule

For any real number a and any integers m and n,

$$(a^m)^n = a^{mn}.$$

(To raise a power to a power, multiply the exponents.)

Example 6   Simplify: **(a)** $(3^5)^4$; **(b)** $(y^{-5})^7$; **(c)** $(a^{-3})^{-7}$.

SOLUTION

a) $(3^5)^4 = 3^{5 \cdot 4} = 3^{20}$

b) $(y^{-5})^7 = y^{-5 \cdot 7} = y^{-35}$

c) $(a^{-3})^{-7} = a^{(-3)(-7)} = a^{21}$

Raising a Product or a Quotient to a Power

When an expression inside parentheses is raised to a power, the inside expression is the base. Let's compare $2a^3$ and $(2a)^3$.

$$2a^3 = 2 \cdot a \cdot a \cdot a \qquad\qquad \begin{aligned} (2a)^3 &= (2a)(2a)(2a) \\ &= (2 \cdot 2 \cdot 2)(a \cdot a \cdot a) \\ &= 2^3 a^3 \\ &= 8a^3 \end{aligned}$$

We see that $2a^3$ and $(2a)^3$ are *not* equivalent. Note also that to simplify $(2a)^3$ we can raise each factor to the power 3. This leads to the following rule.

Raising a Product to a Power

For any real numbers a and b and any integer n (provided $ab \neq 0$ when $n \leq 0$),

$$(ab)^n = a^n b^n.$$

(To raise a product to the nth power, raise each factor to the nth power.)

Example 7 Simplify: **(a)** $(-2x)^3$; **(b)** $(-2x^3y^{-1})^{-4}$.

SOLUTION

a) $(-2x)^3 = (-2)^3 \cdot x^3$ **Raising each factor to the third power**

$\qquad\qquad = -8x^3$

b) $(-2x^3y^{-1})^{-4} = (-2)^{-4}(x^3)^{-4}(y^{-1})^{-4}$ **Raising each factor to the negative fourth power**

$$= \frac{1}{(-2)^4} \cdot x^{-12}y^4 \qquad \text{\textbf{Multiplying powers; writing} } (-2)^{-4}$$
$$\text{\textbf{as} } \frac{1}{(-2)^4}$$

$$= \frac{y^4}{16x^{12}} \qquad\qquad \text{\textbf{Note that } } x^{-12} = \frac{1}{x^{12}}.$$

There is a similar rule for raising a quotient to a power.

Raising a Quotient to a Power

For any integer n, and any real numbers a and b for which a/b, a^n, and b^n exist,

$$\left(\frac{a}{b}\right)^n = \frac{a^n}{b^n}.$$

(To raise a quotient to a power, raise the numerator to the power and divide by the denominator to the power.)

Example 8 Simplify: **(a)** $\left(\dfrac{x^2}{3}\right)^4$; **(b)** $\left(\dfrac{y^2z^3}{5}\right)^{-3}$.

SOLUTION

a) $\left(\dfrac{x^2}{3}\right)^4 = \dfrac{(x^2)^4}{3^4} = \dfrac{x^8}{81}$

b) $\left(\dfrac{y^2z^3}{5}\right)^{-3} = \dfrac{(y^2z^3)^{-3}}{5^{-3}}$

$$= \frac{5^3}{(y^2z^3)^3} \qquad \text{\textbf{Moving factors to the other side of the fraction}}$$
$$\text{\textbf{bar and changing each } } -3 \text{ \textbf{to} } 3$$

$$= \frac{125}{y^6z^9}$$

The rule for raising a quotient to a power allows us to derive a useful result for manipulating negative exponents:

$$\left(\frac{a}{b}\right)^{-n} = \frac{a^{-n}}{b^{-n}} = \frac{b^n}{a^n} = \left(\frac{b}{a}\right)^n.$$

Using this result, we can simplify Example 8(b) as follows:

$$\left(\frac{y^2z^3}{5}\right)^{-3} = \left(\frac{5}{y^2z^3}\right)^3 \quad \text{\textbf{Taking the reciprocal of the base and changing the exponent's sign}}$$

$$= \frac{5^3}{(y^2z^3)^3} = \frac{125}{y^6z^9}.$$

Scientific Notation

Very large and very small numbers that occur in science and other fields are often written using exponents in **scientific notation**.

Scientific Notation

Scientific notation for a number is an expression of the type $M \times 10^n$, where M is in decimal notation, $1 \le M < 10$, and n is an integer.

To convert a number to scientific notation, we can multiply by 1, writing 1 in the form $10^b/10^b$, or $10^b \cdot 10^{-b}$.

Example 9 *Population Projections.* It has been estimated that in the year 2025, the world population will be 8,504,000,000 (*Source: The Universal Almanac*). Write scientific notation for this number.

SOLUTION To write 8,504,000,000 as 8.504×10^n for some integer n, we must move the decimal point 9 places to the left. This is accomplished by dividing—and then multiplying—by 10^9:

$$8,504,000,000 = \frac{8,504,000,000}{10^9} \cdot 10^9 \quad \text{\textbf{Multiplying by 1:}} \ \frac{10^9}{10^9} = 1$$

$$= 8.504 \times 10^9. \quad \text{\textbf{This is scientific notation.}}$$

Example 10 Write scientific notation for the mass of a grain of sand:

0.0648 gram (g).

SOLUTION To write 0.0648 as 6.48×10^n for some integer n, we must move the decimal 2 places to the right. To do this, we multiply—and then divide—by 10^2:

$$0.0648 = \frac{0.0648 \cdot 10^2}{10^2} \quad \text{\textbf{Multiplying by 1:}} \ \frac{10^2}{10^2} = 1$$

$$= \frac{6.48}{10^2}$$

$$= 6.48 \times 10^{-2} \text{ g.} \quad \text{\textbf{Writing scientific notation}}$$

Try to make conversions to scientific notation mentally as often as possible. In doing so, remember that negative powers of 10 are used for small numbers and positive powers of 10 are used for large numbers.

Example 11 Convert mentally to scientific notation: **(a)** 82,500,000; **(b)** 0.0000091.

SOLUTION

a) $82,500,000 = 8.25 \times 10^7$ *Check*: **Multiplying 8.25 by 10^7 moves the decimal point 7 places to the right.**

b) $0.0000091 = 9.1 \times 10^{-6}$ *Check*: **Multiplying 9.1 by 10^{-6} moves the decimal point 6 places to the left.** ▬

Example 12 Convert mentally to decimal notation: **(a)** 4.371×10^7; **(b)** 1.73×10^{-5}.

SOLUTION

a) $4.371 \times 10^7 = 43,710,000$ **Moving the decimal point 7 places to the right**

b) $1.73 \times 10^{-5} = 0.0000173$ **Moving the decimal point 5 places to the left** ▬

We can use the associative and commutative laws to multiply and divide using scientific notation.

Example 13 Multiply and write scientific notation for the answer: $(7.2 \times 10^5)(4.3 \times 10^9)$.

SOLUTION We have

$$(7.2 \times 10^5)(4.3 \times 10^9) = (7.2 \times 4.3)(10^5 \times 10^9)$$ **Using the commutative and associative laws**

$$= 30.96 \times 10^{14}.$$ **Adding exponents**

The result is not written in scientific notation (because $30.96 > 10$), so we convert 30.96 to scientific notation and simplify:

$$30.96 \times 10^{14} = (3.096 \times 10^1) \times 10^{14} = 3.096 \times 10^{15}.$$ ▬

Example 14 Divide and write scientific notation for the answer:

$$\frac{3.48 \times 10^{-7}}{4.64 \times 10^6}.$$

SOLUTION

$$\frac{3.48 \times 10^{-7}}{4.64 \times 10^6} = \frac{3.48}{4.64} \times \frac{10^{-7}}{10^6}$$ **Separating factors**

$$= 0.75 \times 10^{-13}$$ **Subtracting exponents; simplifying**

$$= (7.5 \times 10^{-1}) \times 10^{-13}$$ **Converting 0.75 to scientific notation**

$$= 7.5 \times 10^{-14}$$ **Adding exponents** ▬

A graphing calculator will write very large or very small numbers in scientific notation. For the number shown on the screen below, the notation E22 stands for $\times 10^{22}$.

```
1.05E22

              1.05E22
```

Example 15 Check Example 13 using a graphing calculator.

SOLUTION   We enter the numbers using scientific notation.

```
7.2E5*4.3E9

              3.096E15
```

As we see in the screen above, we have 3.096×10^{15}, which checks.

Definitions and Rules for Exponents

For any integers m and n (assuming 0 is not raised to a nonpositive power):

Zero as an exponent:	$a^0 = 1$
Negative integers as exponents:	$a^{-n} = \dfrac{1}{a^n}$
Multiplying with like bases:	$a^m \cdot a^n = a^{m+n}$
Dividing with like bases:	$\dfrac{a^m}{a^n} = a^{m-n}, a \neq 0$
Raising a product to a power:	$(ab)^n = a^n b^n$
Raising a power to a power:	$(a^m)^n = a^{mn}$
Raising a quotient to a power:	$\left(\dfrac{a}{b}\right)^n = \dfrac{a^n}{b^n}, b \neq 0$

1.4 Exercise Set

Multiply and simplify. Leave the answer in exponential notation.

1. $7^5 \cdot 7^2$

2. $2^3 \cdot 2^8$

3. $a^3 \cdot a^0$

4. $x^0 \cdot x^5$

5. $6x^5 \cdot 3x^2$

6. $4a^3 \cdot 2a^7$

7. $(-3m^4)(-7m^9)$

8. $(-2a^5)(7a^4)$

9. $(x^3y^4)(x^7y^6z^0)$

10. $(m^6n^5)(m^4n^7p^0)$

Divide and simplify.

11. $\dfrac{a^9}{a^3}$

12. $\dfrac{x^{12}}{x^3}$

13. $\dfrac{8x^7}{4x^4}$

14. $\dfrac{20a^{20}}{5a^4}$

15. $\dfrac{m^7 n^9}{m^2 n^5}$

16. $\dfrac{m^{12} n^9}{m^4 n^6}$

17. $\dfrac{18 x^6 y^7 z^9}{-6 x^2 y z^3}$

18. $\dfrac{28 x^8 y^{10} z^{12}}{-7 x^4 y^2 z}$

Simplify.

19. $(-3)^4$ **20.** $(-2)^6$ **21.** -3^4

22. -2^6 **23.** $(-5)^{-2}$ **24.** $(-4)^{-2}$

25. -5^{-2} **26.** -4^{-2} **27.** -1^{-8}

Write an equivalent expression without negative exponents.

28. a^{-3} **29.** n^{-6} **30.** $(5x)^{-3}$

31. $(4xy)^{-5}$ **32.** $x^2 y^{-3}$ **33.** $2a^2 b^{-5}$

34. $\dfrac{y^{-5}}{x^2}$ **35.** $\dfrac{z^{-4}}{3x^5}$ **36.** $\dfrac{x^{-2} y^7}{z^{-4}}$

Write an equivalent expression with negative exponents.

37. $\dfrac{1}{3^4}$ **38.** $\dfrac{1}{(-8)^6}$ **39.** x^5

40. n^3 **41.** $6x^2$ **42.** $-4y^5$

43. $\dfrac{1}{(5y)^3}$ **44.** $\dfrac{1}{(5x)^5}$ **45.** $\dfrac{1}{3y^4}$

Simplify. Should negative exponents appear in the answer, write a second answer using only positive exponents.

46. $8^{-2} \cdot 8^{-4}$ **47.** $9^{-1} \cdot 9^{-6}$

48. $b^2 \cdot b^{-5}$ **49.** $a^4 \cdot a^{-3}$

50. $a^{-3} \cdot a^4 \cdot a^2$ **51.** $x^{-8} \cdot x^5 \cdot x^3$

52. $(5a^{-2} b^{-3})(2a^{-4} b)$ **53.** $(3a^{-5} b^{-7})(2ab^{-2})$

54. $\dfrac{10^{-3}}{10^6}$ **55.** $\dfrac{12^{-4}}{12^8}$

56. $\dfrac{2^{-7}}{2^{-5}}$ **57.** $\dfrac{9^{-4}}{9^{-6}}$

58. $\dfrac{y^4}{y^{-5}}$ **59.** $\dfrac{a^3}{a^{-2}}$

60. $\dfrac{-5 x^{-2} y^4 z^7}{30 x^{-5} y^6 z^{-3}}$ **61.** $\dfrac{9 a^6 b^{-4} c^7}{27 a^{-4} b^5 c^9}$

62. $(x^4)^3$ **63.** $(a^3)^2$

64. $(9^3)^{-4}$ **65.** $(8^4)^{-3}$

66. $(7^{-8})^{-5}$ **67.** $(6^{-4})^{-3}$

68. $(a^3 b)^4$ **69.** $(x^3 y)^5$

70. $5(x^2 y^2)^3$ **71.** $7(a^3 b^4)^2$

72. $(7 x^3 y^{-4})^{-2}$ **73.** $(3 x^2 y^{-5})^{-2}$

74. $\dfrac{(3 x^3 y^4)^3}{6 x y^3}$ **75.** $\dfrac{(5 a^3 b)^2}{10 a^2 b}$

76. $\left(\dfrac{-4 x^4 y^{-2}}{5 x^{-1} y^4}\right)^{-4}$ **77.** $\left(\dfrac{2 x^3 y^{-2}}{3 y^{-3}}\right)^3$

78. $\left(\dfrac{4 a^3 b^{-9}}{2 a^{-2} b^5}\right)^0$ **79.** $\left(\dfrac{5 x^0 y^{-7}}{2 x^{-2} y^4}\right)^0$

Convert to scientific notation.

80. $47{,}000{,}000{,}000$ **81.** $2{,}600{,}000{,}000{,}000$

82. 0.000000016 **83.** 0.000000263

84. $407{,}000{,}000{,}000$ **85.** $3{,}090{,}000{,}000{,}000$

86. 0.000000603 **87.** 0.00000000802

Convert to decimal notation.

88. 4×10^{-4} **89.** 5×10^{-5}

90. 6.73×10^8 **91.** 9.24×10^7

92. 8.923×10^{-10} **93.** 7.034×10^{-2}

94. 9.03×10^{10} **95.** 9.001×10^{10}

Simplify and write scientific notation for the answer.

96. $(2.3 \times 10^6)(4.2 \times 10^{-11})$

97. $(6.5 \times 10^3)(5.2 \times 10^{-8})$

98. $(2.34 \times 10^{-8})(5.7 \times 10^{-4})$

99. $(3.26 \times 10^{-6})(8.2 \times 10^{-6})$

100. $(1.507 \times 10^3)(4.369 \times 10^{-15})$

101. $(4.316 \times 10^{-20})(1.0765 \times 10^{-10})$

102. $(2.506 \times 10^{-7})(1.408 \times 10^{10})$

103. $(1.6158 \times 10^6)(9.075 \times 10^{-8})$

104. $\dfrac{5.1 \times 10^6}{3.4 \times 10^3}$ **105.** $\dfrac{8.5 \times 10^8}{3.4 \times 10^5}$

106. $\dfrac{7.5 \times 10^{-9}}{2.5 \times 10^{-4}}$ **107.** $\dfrac{12.6 \times 10^8}{4.2 \times 10^{-3}}$

108. $\dfrac{1.23 \times 10^8}{6.87 \times 10^{-13}}$ **109.** $\dfrac{4.95 \times 10^{-3}}{1.64 \times 10^{10}}$

110. $\dfrac{4.905 \times 10^{-1}}{2.058 \times 10^{-3}}$ **111.** $\dfrac{1.695 \times 10^{12}}{9.137 \times 10^{11}}$

112. $\dfrac{780{,}000{,}000 \times 0.00071}{0.000005}$

113. $\dfrac{830{,}000{,}000 \times 0.12}{3{,}100{,}000}$

114. $5.9 \times 10^{23} + 2.4 \times 10^{23}$

115. $1.8 \times 10^{-34} + 5.4 \times 10^{-34}$

116. $4.6 \times 10^{20} + 3.8 \times 10^{15}$

117. $9.8 \times 10^{-10} + 4.3 \times 10^{-6}$

Skill Maintenance

118. Subtract: $-\frac{5}{6} - \left(-\frac{3}{4}\right)$. [1.2]

119. Multiply: $(-7.2)(-4.3)$. [1.2]

120. Multiply: $-2(4x - 6y)$. [1.2]

121. Factor: $8x - 10$. [1.3]

Synthesis

122. ◈ Is 5^{-9} greater or less than 4^{-9}? Why?

123. ◈ Explain why $(-1)^n = 1$ for any even number n.

124. ◈ Explain why $(-17)^{-8}$ is positive.

125. ◈ Some numbers exceed the limits of the calculator. Enter 1.3×10^{-1000} and 1.3×10^{1000} and explain the results.

Simplify. Assume that all variables represent nonzero integers.

126. $(3^{a+2})^a$

127. $(12^{3-a})^{2b}$

128. $\dfrac{4x^{2a+3}y^{2b-1}}{2x^{a+1}y^{b+1}}$

129. $\dfrac{25x^{a+b}y^{b-a}}{-5x^{a-b}y^{b+a}}$

130. $\dfrac{(2^{-2})^a \cdot (2^b)^{-a}}{(2^{-2})^{-b}(2^b)^{-2a}}$

131. $\dfrac{3^{q+3} - 3^2(3^q)}{3(3^{q+4})}$

132. Compare $8 \cdot 10^{-90}$ and $9 \cdot 10^{-91}$. Which is the larger value? How much larger? Write scientific notation for the difference.

133. Write the reciprocal of 8.00×10^{-23} in scientific notation.

134. Evaluate: $(4096)^{0.05}(4096)^{0.2}$.

135. What is the ones digit in 513^{128}?

COLLABORATIVE CORNER

Focus: Estimation and properties of exponents

Time: 20–30 minutes

Group size: 2

According to legend, the inventor of chess was offered compensation for his game by an enthusiastic, chess-playing king. The inventor slyly asked that the king place *one* grain of gold on a corner square of his chessboard, *two* grains on the next square, *four* grains on the next, *eight* grains on the next, and so on, doubling the number of grains in each square until each of the 64 squares of the chessboard had been accounted for.

Activity

1. Express, as a power of 2, the number of grains of gold on each of the first 7 squares and also on the last (64th) square. Each group member should then estimate the volume of gold that would be used for the 64th square. Would it fill a shoebox? a refrigerator? a garage? something smaller? bigger?

2. When 32 grains of gold are placed side by side, they form a line that is about 1 cm long. Determine the number of grains of gold in one cubic centimeter. How can this be represented as a power of 2?

3. Determine how many cubic centimeters are in one cubic meter. Approximate this number with a power of 2, using the fact that $2^{10} \approx 10^3$.

4. Use the results of parts (2) and (3) to estimate the number of grains of gold in one cubic meter.

5. Use the results of parts (1) and (4) to determine the volume—in cubic meters—of gold that the king would have needed for the 64th square of the chessboard. (Use powers of 2.)

6. Las Vegas, Nevada, a city known for gold-seekers, has about 128, or 2^7, square kilometers of land that has not yet been built upon. Using the calculations from part (5), how deep a layer of gold could you spread on this land? Which estimate from part (1) comes closest to this volume?

7. Gold recently sold for $360 per ounce, which is the same as 75.6¢ per grain. Use this to estimate the value of the gold needed for the 64th square. (Can you see why, according to legend, the king had the inventor beheaded?)

1.5
Graphs

- *Points and Ordered Pairs*
- *Quadrants and Windows*
- *Graphs of Equations*
- *Nonlinear Equations*

It has often been said that a picture is worth a thousand words. As we turn our attention to the study of graphs, we discover that in mathematics this is quite literally the case. Graphs are a compact means of displaying information and provide a visual approach to problem solving.

Points and Ordered Pairs

On a number line, each point corresponds to a number. On a plane, each point corresponds to a pair of numbers. The idea of using two perpendicular number lines, called **axes**, to identify points in a plane is commonly attributed to the great French mathematician and philosopher René Descartes (1596–1650). Because the variable x is normally represented on the horizontal axis and the variable y is normally represented on the vertical axis, we often refer to the **x, y-coordinate system**. In honor of Descartes, this representation is also called the **Cartesian coordinate system**.

Note in the figure below that (2, 3) and (3, 2) are different points. These pairs of numbers are called **ordered pairs** because the order in which the numbers are listed is important. The ordered pair (0, 0) is called the **origin**.

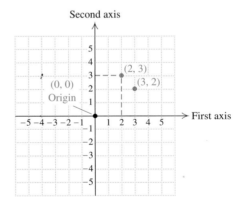

Example 1   Plot the points $(-4, 3)$, $(-5, -3)$, $(0, 4)$, and $(2.5, 0)$.

SOLUTION   To plot $(-4, 3)$, we note that the first number, -4, tells us the distance in the first, or horizontal, direction. We go 4 units *left* of the origin. The second number tells us the distance in the second, or vertical, direction. We go 3 units *up*. The point $(-4, 3)$ is then marked, or "plotted."

The points $(-5, -3)$, $(0, 4)$, and $(2.5, 0)$ are also plotted below.

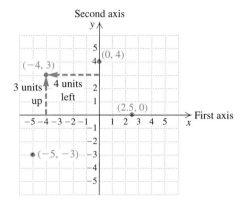

The numbers in an ordered pair are called **coordinates**. In $(-4, 3)$, the *first coordinate* is -4 and the *second coordinate** is 3.

Quadrants and Windows

The axes divide the plane into four regions called **quadrants**, as shown here.

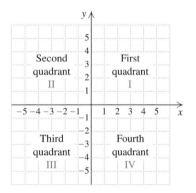

In region I (the *first* quadrant), both coordinates of a point are positive. In region II (the *second* quadrant), the first coordinate is negative and the second coordinate is positive. In the third quadrant, both coordinates are negative, and in the fourth quadrant, the first coordinate is positive and the second coordinate is negative.

Points with one or more 0's as coordinates, such as $(0, -6)$, $(4, 0)$, and $(0, 0)$, are on axes and *not* in quadrants.

The coordinate plane extends without end in all directions. We draw only part of it when we plot points. Although it is standard to show portions of all four quadrants, as in the graphs above, it may be more practical to show a different portion of the plane.

---

*The first coordinate is sometimes called the **abscissa** and the second coordinate the **ordinate**.

Example 2   Plot the points (10, 44), (95, 120), (55, 130), and (70, 15).

SOLUTION   The points are all in the first quadrant. The first coordinates range from 10 to 95 and the second coordinates range from 15 to 130. Thus we need show only the first quadrant. We must show at least 95 units of the first axis and at least 130 units of the second axis. Because it would be impractical to label the axes with all the natural numbers, we will label only every tenth unit. We say that we use a *scale* of 10.

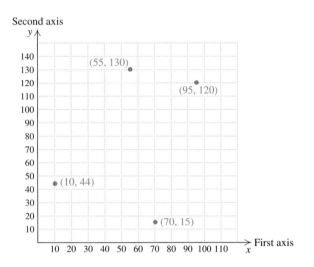

In any graph, we draw the axes in order to display the information clearly. The portion of the axes drawn is often a matter of choice. When appropriate, a different scale can be used for each axis.

On a graphing calculator, the portion of the coordinate plane shown is called the **viewing window**. The window settings are often abbreviated in the form [L, R, B, T], with the letters standing for **L**eft, **R**ight, **B**ottom, and **T**op endpoints. The **standard viewing window** is the window determined by the settings [−10, 10, −10, 10].

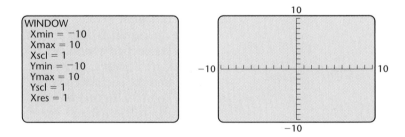

The left and right endpoints of the horizontal axis shown are determined by setting Xmin and Xmax. Similarly, the values of Ymin and Ymax determine the bottom and top endpoints of the vertical axis shown. The scales for the axes are set using Xscl and Yscl. In this text, the window dimensions are written outside the graphs.

Example 3 Plot the points (10, 44), (95, 120), (55, 130), and (70, 15) with a graphing calculator.

SOLUTION As we noted in Example 2, the points are all in the first quadrant. We choose a viewing window of $[-10, 110, -10, 140]$ with Xscl = 10 and Yscl = 10.

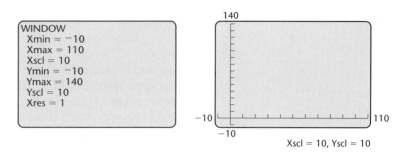

Coordinates of ordered pairs are entered as lists, usually from a STAT menu. (Any old lists stored in the calculator should be cleared before entering new numbers.) We enter the first coordinates of the ordered pairs as one list and the second coordinates as another list. The coordinates of each point should be at the same position on both lists. In order to plot the points, you must turn on the STAT PLOT feature. Then graph the points.

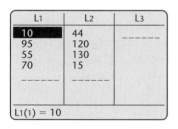
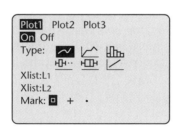
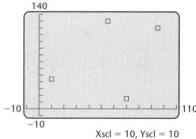

Graphs of Equations

If an equation has two variables, its solutions are pairs of numbers. When such a solution is written as an ordered pair, the first number listed in the pair generally replaces the variable that occurs first alphabetically.

Example 4 Determine whether the pairs (4, 2), $(-1, -4)$, and (2, 5) are solutions of the equation $y = 3x - 1$.

SOLUTION To determine whether each pair is a solution, we replace x with the first coordinate and y with the second coordinate. When the replacements make the equation true, we say that the ordered pair is a solution.

$$y = 3x - 1$$

2 ? 3(4) − 1	
	12 − 1
2	11

$$y = 3x - 1$$

−4 ? 3(−1) − 1	
	−3 − 1
−4	−4

$$y = 3x - 1$$

5 ? 3(2) − 1	
	6 − 1
5	5

Since 2 = 11 is *false*, the pair (4, 2) *is not* a solution.

Since −4 = −4 is *true*, the pair (−1, −4) *is* a solution.

Since 5 = 5 is *true*, the pair (2, 5) *is* a solution.

In fact, there is an infinite number of solutions of $y = 3x - 1$. Rather than attempt to list all these solutions, we will use a graph as a convenient representation. Thus to *graph* an equation means to make a drawing that represents its solutions. Note: Be sure that the STAT PLOT feature is turned off before graphing equations.

Example 5 Graph the equation $y = x$.

SOLUTION   We label the horizontal axis as the *x*-axis and the vertical axis as the *y*-axis.

Next, we find some ordered pairs that are solutions of the equation. In this case, it is easy. Here are a few pairs that satisfy the equation $y = x$:

$$(0, 0), \quad (1, 1), \quad (5, 5), \quad (-1, -1), \quad (-6, -6).$$

Now we plot these points. We can see that if we were to plot a million solutions, the dots that we drew would merge into a solid line. Observing the pattern, we can draw the line with a ruler. The line is the graph of the equation $y = x$. We label the line $y = x$.

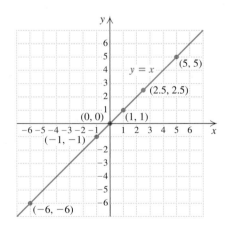

Note that the coordinates of *any* point on the line—for example, (2.5, 2.5)—satisfy the equation $y = x$. Note too that the line continues indefinitely in both directions—only part of it is shown.

Example 6 Graph $y = 2x$ using a graphing calculator.

SOLUTION Equations are entered using the equation editor screen. The first part of each equation, "$y =$," is already written. We first clear any other equations present and then enter $y = 2x$. The standard $[-10, 10, -10, 10]$ window is a good choice for this graph. We check the window dimensions and then graph the equation.

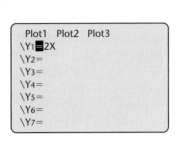

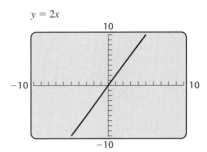

The equation in the next example is graphed both by hand and by using a graphing calculator. Let's compare the processes and the results.

Example 7 Graph the equation $y = -\frac{1}{2}x$.

SOLUTION

┌─ *BY HAND*

We find some ordered pairs that are solutions. This time we list the pairs in a table. To find an ordered pair, we can choose *any* number for x and then determine y. By choosing even integers for x, we can avoid fractional values when calculating y. For example, if we choose 4 for x, we get $y = \left(-\frac{1}{2}\right)(4)$, or -2. If x is -6, we get $y = \left(-\frac{1}{2}\right)(-6)$, or 3. We find several ordered pairs, plot them, and draw the line.

x	y	(x, y)
4	-2	$(4, -2)$
-6	3	$(-6, 3)$
0	0	$(0, 0)$
2	-1	$(2, -1)$

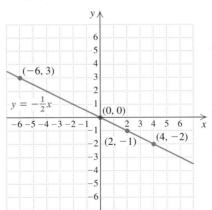

┌─ *WITH A GRAPHING CALCULATOR*

To enter the equation, we access the equation editor and clear any equations present. We then enter the equation as $y = -(1/2)x$. (Remember to use the $\boxed{(-)}$ key for the negative sign.) A standard viewing window is a good choice for this graph.

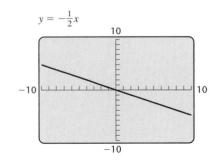

We need not create a table of values in order to graph an equation using a graphing calculator. However, such tables are useful in many situations. To list ordered pairs that are solutions, we can use the TABLE feature of a graphing calculator.

Example 8 Create a table of ordered pairs that are solutions of the equation $y = -\frac{1}{2}x$.

SOLUTION We first enter the equation and then set up the table. You can choose to have the calculator supply the values for x, or you can supply them yourself. Since the value of y depends on the choice of the value for x, we say that y is the **dependent** variable and x is the **independent** variable.

To choose the values for the independent variable, we set Indpnt to Ask.

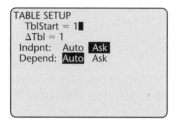

Next, we view the table and choose values for x. In the table shown above on the right, the values entered for x were 4, -6, 0, and 2. The corresponding y-values appear in the second column.

If Indpnt is set to Auto, the calculator will provide values for x, beginning with the value specified as TblStart and continuing by adding the value of ΔTbl to the preceding value for x. Referring to the equation in Example 8, if TblStart $= -3$ and ΔTbl $= 1$, the following table results.

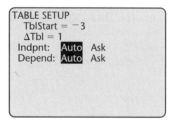

Nonlinear Equations

As you can see, the graphs in Examples 5–7 are straight lines. We refer to any equation whose graph is a straight line as a **linear equation**. Linear equations are discussed in more detail in Chapter 2. Many equations, however, are not linear. When ordered pairs that are solutions of such an equation are plotted, the pattern formed is not a straight line. Let's look at some of these **nonlinear equations**.

Example 9 For each of the following equations, create a table of solutions for integer values of x beginning at -3. Then graph the equation.

a) $y = x^2 - 5$

b) $y = \dfrac{1}{x}$

c) $y = |x|$

SOLUTION

a) We first enter the equation and then create a table of values by setting Indpnt to Auto, letting TblStart $= -3$ and ΔTbl $= 1$. We use a standard viewing window for the graph.

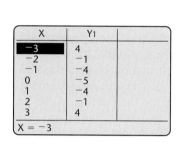

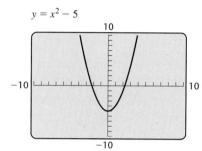

The graph rises steeply on either side of the y-axis since the value of $x^2 - 5$ grows rapidly as x moves away from the origin.

b) The graph of $y = 1/x$ is shown in the middle below. Because the graph lies very close to the axes, it is difficult to determine its shape. Making the viewing window smaller has the effect of magnifying a portion of the graph. The graph is shown again on the right below using a viewing window of $[-3, 3, -3, 3]$. We can see that there are two "branches" to this graph—one in the first quadrant and one in the third quadrant. For x-values far to the right or far to the left of 0, the graph approaches, but does not touch, the x-axis. The table shows that $1/x$ is undefined for $x = 0$. Thus there is no point on the graph for which $x = 0$; in other words, the graph does not touch the y-axis, even though it may appear to do so.

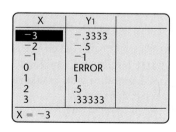

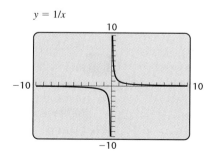

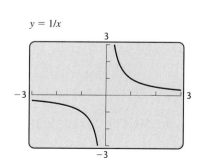

c) For the graph of $y = |x|$, note that the absolute value of a positive number is the same as the absolute value of its opposite. Thus, for example, the x-values 3 and -3 both are paired with the y-value 3. Note that the graph is V-shaped and centered at the origin.

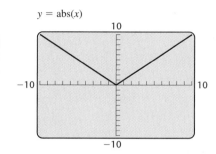

$y = \text{abs}(x)$

As we saw in Example 9(b), a standard viewing window is not always the best window to use. Choosing an appropriate viewing window for a graph can be challenging. There is generally no one "correct" window; the choice can vary according to personal preference and can also be dictated by the portion of the graph that you need to see. It often involves trial and error, but as you learn more about graphs of equations, you will be able to determine an appropriate viewing window more directly. For now, start with a standard window, and change the dimensions if needed. The ZOOM menu can help in setting window dimensions; for example, choosing ZStandard from the menu will graph the selected equations using the standard viewing window.

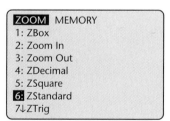

1.5   *Exercise Set*

Plot the points. Label each point with the indicated letter.

1. $A(5, 3)$, $B(2, 4)$, $C(0, 2)$, $D(0, -6)$, $E(3, 0)$, $F(-2, 0)$, $G(1, -3)$, $H(-5, 3)$, $J(-4, 4)$

2. $A(3, 5)$, $B(1, 5)$, $C(0, 4)$, $D(0, -4)$, $E(5, 0)$, $F(-5, 0)$, $G(1, -5)$, $H(-7, 4)$, $J(-5, 5)$

3. $A(3, 0)$, $B(4, 2)$, $C(5, 4)$, $D(6, 6)$, $E(3, -4)$, $F(3, -3)$, $G(3, -2)$, $H(3, -1)$

4. $A(1, 1)$, $B(2, 3)$, $C(3, 5)$, $D(4, 7)$, $E(-2, 1)$, $F(-2, 2)$, $G(-2, 3)$, $H(-2, 4)$, $J(-2, 5)$, $K(-2, 6)$

5. Plot the points $M(2, 3)$, $N(5, -3)$, and $P(-2, -3)$. Draw $\overline{MN}$, $\overline{NP}$, and $\overline{MP}$. ($\overline{MN}$ means the line segment from M to N.) What kind of geometric figure is formed? What is its area?

6. Plot the points $Q(-4, 3)$, $R(5, 3)$, $S(2, -1)$, and $T(-7, -1)$. Draw $\overline{QR}$, $\overline{RS}$, $\overline{ST}$, and $\overline{TQ}$. What kind of figure is formed? What is its area?

Name the quadrant in which each point is located.

7. $(-3, -5)$ **8.** $(2, 17)$ **9.** $(-6, 1)$

10. $(4, -8)$ **11.** $\left(3, \frac{1}{2}\right)$ **12.** $(-1, -8)$

13. $(7, -0.2)$ **14.** $(-4, 31)$

Determine whether each ordered pair is a solution of the given equation.

15. $(1, -1)$; $y = 2x - 3$ **16.** $(2, 5)$; $y = 4x - 3$

17. $(3, 4)$; $3s + t = 4$ **18.** $(2, 3)$; $2p + q = 5$

19. $(3, 5)$; $4x - y = 7$ **20.** $(2, 7)$; $5x - y = 3$

21. $\left(0, \frac{3}{5}\right)$; $2a + 5b = 3$ **22.** $\left(0, \frac{3}{2}\right)$; $3f + 4g = 6$

23. $(2, -1)$; $4r + 3s = 5$ **24.** $(2, -4)$; $5w + 2z = 2$

25. $(3, 2)$; $3x - 2y = -4$ **26.** $(1, 2)$; $2x - 5y = -6$

27. $(-1, 3)$; $y = 3x^2$ **28.** $(2, 4)$; $2r^2 - s = 5$

29. $(2, 3)$; $5s^2 - t = 7$ **30.** $(2, 3)$; $y = x^3 - 5$

Graph by hand.

31. $y = -2x$ **32.** $y = \frac{1}{2}x$

33. $y = x + 3$ **34.** $y = x - 2$

35. $y = 3x - 2$ **36.** $y = -4x + 1$

37. $y = -2x + 3$ **38.** $y = -3x + 1$

39. $y = \frac{2}{3}x + 1$ **40.** $y = \frac{1}{3}x + 2$

Using a graphing calculator, create a table of solutions for integer values of x beginning at -3. Then graph.

41. $y = -\frac{3}{2}x + 1$ **42.** $y = -\frac{2}{3}x - 2$

43. $y = \frac{3}{4}x + 1$ **44.** $y = x^2$

45. $y = -x^2$ **46.** $y = x^2 + 2$

47. $y = x^2 - 2$ **48.** $y = 4x^2$

49. $y = |x| + 2$ **50.** $y = -|x|$

51. $y = 3 - x^2$ **52.** $y = x^3 - 2$

53. $y = -\dfrac{1}{x}$ **54.** $y = \dfrac{3}{x}$

Graph each equation using both viewing windows indicated. Determine which window best shows the shape of the graph and where it crosses the x- and y-axes.

55. $y = x - 15$

a) $[-10, 10, -10, 10]$, Xscl = 1, Yscl = 1
b) $[-20, 20, -20, 20]$, Xscl = 5, Yscl = 5

56. $y = -3x + 30$

a) $[-10, 10, -10, 10]$, Xscl = 1, Yscl = 1
b) $[-20, 20, -20, 40]$, Xscl = 5, Yscl = 5

57. $y = 5x^2 - 8$

a) $[-10, 10, -10, 10]$, Xscl = 1, Yscl = 1
b) $[-3, 3, -3, 3]$, Xscl = 1, Yscl = 1

58. $y = \frac{1}{10}x^2 + \frac{1}{3}$

a) $[-10, 10, -10, 10]$, Xscl = 1, Yscl = 1
b) $[-0.5, 0.5, -0.5, 0.5]$, Xscl = 0.1, Yscl = 0.1

59. $y = 4x^3 - 12$

a) $[-10, 10, -10, 10]$, Xscl = 1, Yscl = 1
b) $[-5, 5, -20, 10]$, Xscl = 1, Yscl = 5

60. $y = |4x^3 - 12|$

a) $[-10, 10, -10, 10]$, Xscl = 1, Yscl = 1
b) $[-3, 3, 0, 20]$, Xscl = 1, Yscl = 5

61. Determine which of the equations in the odd-numbered exercises 31–59 are linear.

62. Determine which of the equations in the even-numbered exercises 32–60 are linear.

Skill Maintenance

Tell whether the number is a solution of the given equation or inequality. [1.1]

63. $3x - 5 = 10$; 5   **64.** $4y \geq 18$; 6

65. $3n - 7 < 5$; 4   **66.** $2t + 3 = 5$; 2

Synthesis

67. ◆ Using the equation $y = |x|$, explain why it is "dangerous" to draw a graph after plotting just two points.

68. ◆ Without making a drawing, how can you tell that the graph of $y = x - 30$ passes through three quadrants?

69. ◆ At what point will the line passing through $(a, -1)$ and $(a, 5)$ intersect the line that passes through $(-3, b)$ and $(2, b)$? Why?

70. ◆ Graph $y = 6x$, $y = 3x$, $y = \frac{1}{2}x$, $y = -6x$, $y = -3x$, and $y = -\frac{1}{2}x$ using the same set of axes or viewing window, and compare the slants of the lines. Describe the pattern that relates the slant of the line to the multiplier of x.

71. ◆ Using the same set of axes or viewing window, graph $y = 2x$, $y = 2x - 3$, and $y = 2x + 3$. Describe the pattern relating each line to the number that is added to $2x$.

72. Which of the following equations have $\left(-\frac{1}{3}, \frac{1}{4}\right)$ as a solution?

a) $-\frac{3}{2}x - 3y = -\frac{1}{4}$
b) $8y - 15x = \frac{7}{2}$
c) $0.16y = -0.09x + 0.1$
d) $2(-y + 2) - \frac{1}{4}(3x - 1) = 4$

73. If $(2, -3)$ and $(-5, 4)$ are the endpoints of a diagonal of a square, what are the coordinates of the other two vertices? What is the area of the square?

74. If $(-10, -2)$, $(-3, 4)$, and $(6, 4)$ are the coordinates of three consecutive vertices of a parallelogram, what are the coordinates of the fourth vertex?

75. One value of y for the equation $y = 3.2x - 5$ is -11.4. Use the TABLE feature of a graphing calculator to determine the x-value that is paired with -11.4.

76. Graph each of the following equations and determine which appear to be linear. Try to find a way to tell if an equation is linear without graphing it.

a) $y = 3x + 2$
b) $y = \frac{1}{2}x^2 - 5$
c) $y = 8$
d) $y = 4 - \frac{1}{5}x$
e) $y = |3 - x|$
f) $y = 4x^3$

77. The graph of $y = 0.5x^2 - 15x + 64$ crosses the x-axis twice. Determine a viewing window that shows both intersections of $y = 0.5x^2 - 15x + 64$ and the x-axis.

1.6
Mathematical Models

- *The Five-Step Strategy*
- *Translating to Algebraic Expressions*
- *Translating to Equations*
- *Models*
- *Graphs as Models*

We now begin to study and practice the "art" of problem solving. Although we are interested mainly in problems that can be solved using mathematics, much of what we say here applies to solving all kinds of problems.

What do we mean by a *problem*? Perhaps you have already used algebra to solve some "real-world" problems. What procedure did you use? Was there anything in your approach that could be used to solve problems of a more general nature? These are some questions that we will answer in this section.

In this text, we do not restrict the use of the word "problem" to computational situations involving arithmetic or algebra, such as $589 + 437 = a$ or $3x + 5x = 9$. We mean instead some question to which we wish to find an answer. Perhaps this can best be illustrated with some sample problems:

1. Can I afford to buy a new car?

2. If I exercise twice a week and consume 3000 calories a day, will I lose weight?

3. Do I have enough time to take 4 courses while working 20 hr a week?

4. I have a piece of wood trim 100 in. long, all of which I want to use to make two square frames. The length of a side of one square must be $1\frac{1}{2}$ times the length of a side of the other. How should I cut the wood?

Although these problems are all different, there are some similarities. While there are no rules for problem solving, a general *strategy* can be used.

The Five-Step Strategy

The following steps constitute a good strategy for problem solving in general.

Five Steps for Problem Solving with Algebra

1. *Familiarize* yourself with the problem situation.
2. *Translate* to mathematical language.
3. *Carry out* some mathematical manipulation.
4. *Check* your possible answer in the original problem.
5. *State* the answer clearly.

Of the five steps, probably the most important is the first: becoming familiar with the problem situation. Here are some hints for familiarization.

The First Step in Problem Solving with Algebra

Familiarize yourself with a problem situation.

1. If a problem is given in words, read it carefully.
2. Reread the problem, perhaps aloud. Verbalize the problem to yourself.
3. List the information given and restate the question being asked. Select a variable(s) to represent any unknown(s) and clearly state what each variable represents. Be descriptive! For example, let t = time, in seconds; p = Paul's weight, in kilograms; and so on.
4. Find further information. Look up formulas or definitions with which you are not familiar. (Geometric formulas appear at the very end of this text.) Consult an expert in the field or a reference librarian.
5. Create a table using both variables and known information. Look for possible patterns.
6. Make and label a drawing.
7. Estimate or guess an answer and check to see if it's correct.

Example 1 How might you familiarize yourself with the situation of Problem 1 on p. 48: "Can I afford to buy a new car?"

SOLUTION Clearly more information is needed to solve this problem. You might:

a) Estimate the cost of various cars in which you are interested.

b) Examine what your savings are and how your income is budgeted.

c) Find out what payment plans are available.

When enough information is known, it might be wise to make a chart or table to help you reach an answer.

Example 2 How might you familiarize yourself with the situation of Problem 4 on p. 48: "How should I cut the wood?"

SOLUTION First, read the question *very* carefully. This may even involve speaking aloud. You may need to reread the problem several times to understand fully what information is given and what information is required. A sketch is often helpful.

We see from the sketch that the lengths of the sides of the squares are unknown. If s is used to represent the length of a side of the smaller square, then $\left(1\frac{1}{2}\right)s$ will represent the length of a side of the larger square.

The length of wood needed for each frame is the distance around, or *perimeter*, of the square. Recall or look up the formula for the perimeter of a square:

$$\text{Perimeter of a square} = 4 \cdot \text{length of a side}.$$

We can make several guesses and list the results in a table. If the length of a side of the smaller square were 6 in., then its perimeter would be $4 \cdot 6$, or 24 in. The length of a side of the larger square would then be $1\frac{1}{2} \cdot 6$, or 9 in., so its perimeter would be $4 \cdot 9$, or 36 in. The total amount of wood needed would then be $24 + 36 = 60$ in., which is less than the length of the wood trim available.

If we try using 12 in. for the length of a side of the smaller square, we see that the total amount of wood needed is 120 in., which is more than the length of the wood trim available.

Although neither guess was correct, we now know that the sum of the perimeters of the squares must be 100. Since our guess of 6 in. was too small and our guess of 12 in. was too large, we also know that the length of a side of the smaller square will be between 6 and 12 in. We can now fill in a row of the table using the variable s to represent the length, in inches, of a side of the smaller square.

s

$1\frac{1}{2}s$

100 in.

SIDE OF SMALLER SQUARE	PERIMETER OF SMALLER SQUARE	SIDE OF LARGER SQUARE	PERIMETER OF LARGER SQUARE	TOTAL AMOUNT OF WOOD NEEDED
6 in.	24 in.	9 in.	36 in.	60 in.
12 in.	48 in.	18 in.	72 in.	120 in.
s	$4s$	$\left(1\frac{1}{2}\right)s$	$4\left(1\frac{1}{2}\right)s$	100 in.

In the second step of the problem-solving process, we translate the problem to mathematical language.

Translating to Algebraic Expressions

One way to model a problem situation is to use an equation. In order to translate problems to equations, we must first be able to translate phrases to algebraic expressions. We know that certain words correspond to certain symbols, as shown in the following tables.

Key Words

ADDITION	SUBTRACTION	MULTIPLICATION	DIVISION
add	subtract	multiply	divide
sum	difference	product	divided by
plus	minus	times	quotient
increased by	decreased by	twice	ratio
more than	less than	of	per

PHRASE	ALGEBRAIC EXPRESSION
Five *more than* some number	$n + 5$
Half *of* a number	$\frac{1}{2}t$ or $\frac{t}{2}$
Five *more than* three *times* some number	$3p + 5$
The *difference* of two numbers	$x - y$
Six *less than* the *product* of two numbers	$rs - 6$
Seventy-six percent *of* some number	$0.76z$ or $\frac{76}{100}z$

Note that expressions like rs represent products and can also be written as $r \cdot s$, $r \times s$, or $(r)(s)$. The multipliers r and s are also called *factors*.

Example 3 Translate to an algebraic expression:

Five less than forty-three percent of the quotient of two numbers.

SOLUTION We let r and s represent the two numbers.

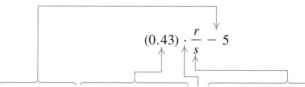

Five less than forty-three percent of the quotient of two numbers

Translating to Equations

Sometimes a problem can be translated to an equation. It is often helpful to reword the problem using phrases that can be translated directly to algebraic expressions. The key word "is" translates to an equals sign, =. Remember to familiarize yourself with each problem situation before attempting to translate it.

Example 4   Translate to an equation:

> The following graph shows the price paid by recyclers for a ton of waste paper in 1994, 1995, 1996, and 1997. By how much did the price increase from 1994 to 1995?

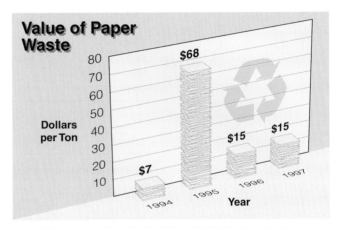

Source: Chittenden Solid Waste District, Williston VT, 1996 telephone interview

SOLUTION

1. Familiarize.   Read the problem carefully. The problem asks by how much the price increased from 1994 to 1995. We let

$x =$ the increase in price from 1994 to 1995.

From the graph, we see that the price increased from \$7 to \$68.

2. Translate.   We reword the problem and translate.

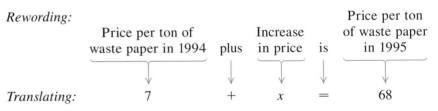

The translation of the problem is the equation

$7 + x = 68,$

where x is the increase in price from 1994 to 1995.

Example 5   Translate to an equation:

> I have a piece of wood trim 100 in. long, all of which I want to use to make two square frames. The length of a side of one square must be $1\frac{1}{2}$ times the length of a side of the other. How should I cut the wood?

SOLUTION

1. Familiarize.   This was done in Example 2.

2. Translate.   We reword the problem and translate.

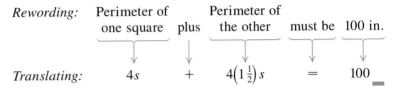

Rewording:	Perimeter of one square	plus	Perimeter of the other	must be	100 in.
Translating:	$4s$	$+$	$4\left(1\frac{1}{2}\right)s$	$=$	100

Models

When we translate a problem into mathematical language, we *model* the problem. A **mathematical model** is a representation, using mathematics, of a real-world situation. Following are some examples of models.

1. REAL-WORLD SITUATION

Landscaping. Grass seed is being spread on a triangular traffic island. If the grass seed can cover an area of 200 ft^2 and the base of the island is 16 ft long, how tall a triangle can the seed cover?

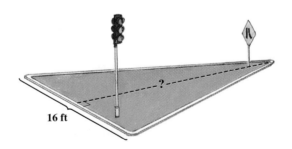

16 ft

MATHEMATICAL MODEL

A *formula* can be a mathematical model. We use the formula for the area of a triangle,

$$A = \tfrac{1}{2}bh,$$

to model this situation. Since $A = 200$ and $b = 16$, we have the model $200 = \frac{1}{2} \cdot 16 \cdot h$.

2. REAL-WORLD SITUATION

Price of paper. The price paid by recyclers for a ton of waste paper was $7 in 1994, $68 in 1995, $15 in 1996, and $15 in 1997. By how much did the price increase from 1994 to 1995?

MATHEMATICAL MODEL

An *equation* or *inequality* can be used to model a situation. In Example 4, we translated this problem to the equation

$$7 + x = 68.$$

3. REAL-WORLD SITUATION

Part-time workers. Approximately 5 million Americans worked part time in 1990. That number rose to 5.7 million in 1991, 6.1 million in 1992, and 6.3 million in 1993 (*Source*: *The Macmillan Visual Almanac, 1996*). How many Americans worked part time in 1998?

MATHEMATICAL MODEL

We can visualize this situation by plotting the points (1990, 5), (1991, 5.7), (1992, 6.1), and (1993, 6.3) as shown in the graph on the left below. The graph shows that the number of part-time workers is increasing steadily.

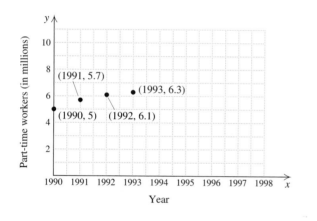

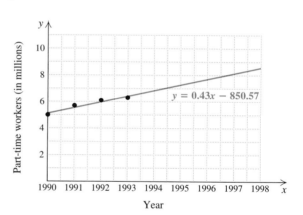

As we will see later in this course, the number of part-time workers for a given year can also be modeled by the equation

$$y = 0.43x - 850.57,$$

where x is the year and y is the number of Americans, in millions, employed part time. The graph of this equation, along with the points, is shown on the right above.

Throughout the remainder of this text, we will be learning how to develop and use such models.

Graphs as Models

Graphs can represent a large amount of information in a concise way. They also help to visualize problem situations.

Example 6 *Ground-Beef Consumption.* The following graph shows the annual per capita consumption of ground beef in the United States for the years 1980–1997.

a) What was the annual ground-beef consumption per person in 1980?

b) In what year was the ground-beef consumption the greatest?

c) Between the years 1982 and 1984, did ground-beef consumption increase or decrease?

d) What was the amount of decrease in ground-beef consumption between 1996 and 1997?

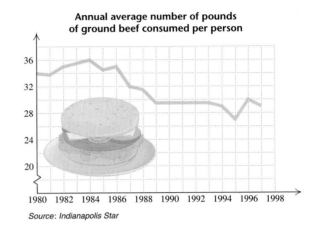

Annual average number of pounds
of ground beef consumed per person

Source: Indianapolis Star

SOLUTION

a) We begin at 1980 on the horizontal axis. We then move up to the graph and observe the corresponding value on the vertical axis. We see that in 1980, the annual per capita ground-beef consumption was about 34 lb.

b) We find the highest point on the graph and then move down to the horizontal axis. We see that the per capita ground-beef consumption was greatest in 1984.

c) We note that the point on the graph corresponding to 1984 is higher than the point corresponding to 1982, so we know that the ground-beef consumption increased between 1982 and 1984.

d) We note that the per capita ground-beef consumption was about 30 lb in 1996 and 29 lb in 1997, so the amount of decrease was 1 lb.   ▬

A graphing calculator can plot data points and form line graphs from those points.

Example 7 *Motor-Vehicle Safety.* The table at left shows the number of motor-vehicle deaths in the United States for a number of years from 1975 to 1993. Use the data to draw a graph.

SOLUTION   We first enter the years in list L1. When all 8 years have been entered, we press the right arrow key to move to list L2. We then enter the number of motor-vehicle deaths in the second list. Use the arrow keys to move up and down through the list.

YEAR	NUMBER OF MOTOR-VEHICLE DEATHS
1975	45,900
1980	53,500
1985	45,500
1987	48,500
1988	49,000
1989	47,000
1990	46,500
1993	42,000

Source: National Safety Council; *The Macmillan Visual Almanac, 1996*

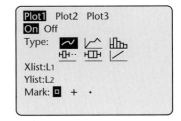

L₁	L₂	L₃
1975	45900	------
1980	53500	
1985	45500	
1987	48500	
1988	49000	
1989	47000	
1990	46500	

$L_2(1) = 45900$

Next, we turn on the PLOT feature. The type of plot we want is a scatter diagram, or scattergram, the first of the six types shown. You may choose the mark used to plot each point from the options shown.

We set the viewing window so that all the points will be shown. We use a viewing window of [1970, 2000, 40000, 55000], with Yscl = 1000. (Many graphing calculators have a ZOOMSTAT option that will select a window that shows all the data points.) Also, we clear any equations from the equation editor screen to ensure that they will not be graphed along with the points. The points are plotted as shown on the left below. Note that there are no axes shown on the screen. Only a portion of the first quadrant is shown.

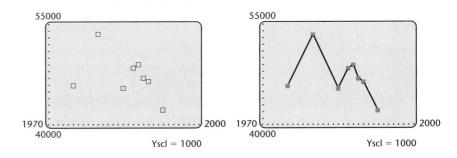

A line graph can be formed by connecting the points. We simply change the type of graph for Plot1 to a line by choosing the second of the six types shown. The line graph is shown on the right above.

A graphing calculator can display the coordinates of points on the graph of an equation. The TRACE feature displays the coordinates of the point indicated by a cursor. The VALUE feature displays the y-value associated with a particular x-value.

Example 8   *Model Rockets.* Suppose that a model rocket is launched upward with an initial velocity of 96 ft/sec. Its height, h, in feet, after t seconds is given by

$$h = -16t^2 + 96t.$$

a) For how long will the rocket climb?

b) How high will the rocket go?

c) After how long will the rocket reach the ground?

SOLUTION We graph the equation, using the viewing window [0, 10, 0, 200], with Yscl = 10.

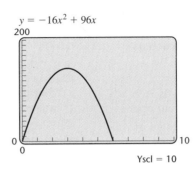

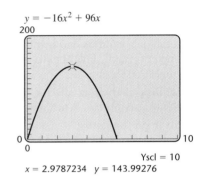

a) The graph shows the height of the rocket. We press TRACE and use the left and right arrow keys to move the cursor along the graph to the greatest y-value. The greatest y-value occurs when x is about 3, as shown in the graph on the right above. Later in this text, we will consider methods to determine more precisely where such a *maximum* value occurs. The rocket will climb for about 3 sec.

b) To approximate how high the rocket will go, we use the VALUE option in the CALC menu. At the bottom of the graph, we are asked for an x-value. Since we want the height when $x = 3$, we enter 3, as shown in the graph at left. The rocket will climb about 144 ft.

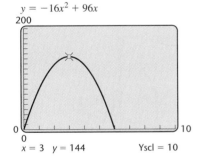

c) The rocket is at the ground when the height y is 0. Using TRACE, we see that y is 0 when x is about 6. The rocket will reach the ground about 6 sec after it has been launched.

1.6 Exercise Set

Use mathematical symbols to translate each phrase.

1. Seven more than some number
2. Two less than some number
3. Twelve times a number
4. Twice a number
5. Sixty-five percent of some number
6. Thirty-nine percent of some number
7. Nine less than twice a number
8. Four more than half of a number
9. Eight more than ten percent of some number
10. Five less than six percent of some number
11. One less than the difference of two numbers
12. Two more than the product of two numbers
13. Ninety miles per every four gallons of gas
14. One hundred words per every sixty seconds

For each problem, familiarize yourself with the situation. Then translate to mathematical language. You need not actually solve the problem; just carry out the first two steps of the five-step strategy.

15. The sum of two numbers is 65. One of the numbers is 7 more than the other. What are the numbers?
16. The sum of two numbers is 83. One of the numbers is 11 more than the other. What are the numbers?
17. The number 128 is 0.4 of what number?
18. The number 456 is $\frac{1}{3}$ of what number?
19. The quotient of two numbers is 12.3. If the divisor is 4, find the other number.

20. One number is less than another by 65. The sum of the numbers is 92. What is the smaller number?

21. A rectangle's length is twice its width and its perimeter is 21 m. Find the dimensions of the rectangle.

22. A rectangle's width is one-third its length and its perimeter is 32 m. Find the dimensions of the rectangle.

23. A reclining chair is on sale for $377. This is 35% off the original price. What was the original price?

24. A 150-lb person burns 5.8 calories per minute by walking at 4 mph. This is 45% more than the number of calories per minute burned by walking at 3 mph. How many calories will a 150-lb person burn each minute by walking at 3 mph? (*Source*: Home and Garden Bulletin No. 62, U.S. Government Printing Office; *The Macmillan Visual Almanac, 1996*)

25. Rhonda bicycled 25 mi at a rate of 15 mph. For how long did she ride?

26. Brock ran 6 mi in 45 min. Logan ran 9 mi at the same rate. For how long did Logan run?

27. *Angles in a Triangle.* The degree measures of the angles in a triangle are three consecutive integers. Find the measures of the angles.

28. *Pricing.* The Sound Connection prices its blank audiotapes by raising the wholesale price 50% and adding 25 cents. What must a tape's wholesale price be if the tape is to sell for $1.99?

29. *Cruising Altitude.* A commercial jet has been instructed to climb from its present altitude of 8000 ft to a cruising altitude of 29,000 ft. If the plane ascends at a rate of 3500 ft/min, how long will it take to reach the cruising altitude?

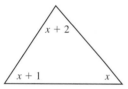

30. A piece of wire 10 m long is to be cut into two pieces, one of them $\frac{2}{3}$ as long as the other. How should the wire be cut?

31. *Angles in a Triangle.* One angle of a triangle is three times as great as a second angle. The third angle measures 12° less than twice the second angle. Find the measures of the angles.

32. *Angles in a Triangle.* One angle of a triangle is four times as great as a second angle. The third angle measures 5° more than twice the second angle. Find the measures of the angles.

33. Find three consecutive odd integers such that the sum of the first, two times the second, and three times the third is 70.

34. Find two consecutive even integers such that two times the first plus three times the second is 76.

35. A piece of wire 100 cm long is to be cut into two pieces, each to be bent to make a square. The length of a side of one square is to be twice the length of a side of the other. How should the wire be cut?

36. A piece of wire 100 cm long is to be cut into two pieces, and those pieces are each to be bent to make a square. The area of one square is to be 144 cm² greater than that of the other. How should the wire be cut? (*Remember:* Do not solve.)

37. Three numbers are such that the second is 6 less than 3 times the first, and the third is 2 more than $\frac{2}{3}$ of the second. The sum of the three numbers is 172. Find the largest number.

38. *Pricing.* Whitney's Appliances is having a sale on 13 TV sets. They are displayed in order of increasing price from left to right. The price of each set differs by $20 from either set next to it. For the price of the set at the extreme right, a customer can buy both the second and seventh sets. What is the price of the least expensive set?

39. *Test Scores.* Deirdre's scores on five tests are 93, 89, 72, 80, and 96. What must the score be on her next test so that the average will be 88?

40. *Population Growth.* The population of Newcastle grew 12% for each of three consecutive years. At the end of that time, the population was 50,577. Find Newcastle's population at the start of the three-year period.

Heart Attacks and Cholesterol. *For Exercises 41 and 42, use the following graph, which shows the annual heart attack rate per 10,000 men and their blood cholesterol level**

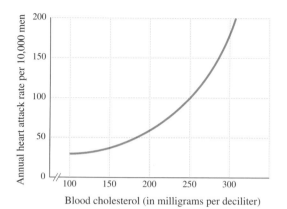

41. Approximate the annual heart attack rate per 10,000 men for those whose blood cholesterol level is 225 mg/dl.

42. Approximate the annual heart attack rate per 10,000 men for those whose blood cholesterol level is 275 mg/dl.

Minivan Sales. *For Exercises 43–46, use the following graph, which shows the number of minivans sold for various years (Source: Indianapolis Star, May 8, 1994).*

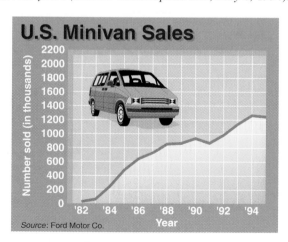

---

*Copyright 1989, CSPI. Adapted from *Nutrition Action Healthletter* (1875 Connecticut Avenue, N.W., Suite 300, Washington, DC 20009-5728. $24 for 10 issues).

43. Approximate the number of minivans sold in 1992.

44. Approximate the number of minivans sold in 1989.

45. In what years was the number of minivans sold less than the number sold the year before?

46. The number of minivans sold increased from 1982 to 1983 and from 1983 to 1984. In which year was the increase greater?

Blood Alcohol Level. *The following table can be used to predict the number of drinks required for a person of a specified weight to be legally intoxicated (blood alcohol level of 0.08 or above) in many states. One 12-oz glass of beer, a 5-oz glass of wine, or a cocktail containing 1 oz of a distilled liquor all count as one drink. Assume that all drinks are consumed within one hour.*

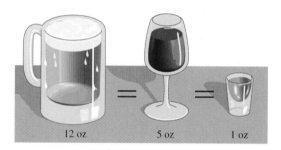

BODY WEIGHT (IN POUNDS)	NUMBER OF DRINKS
100	2.5
160	4
180	4.5
200	5

47. Use the data in the table above to draw a graph and to estimate the number of drinks that a 140-lb person must have had in order to be considered intoxicated.

48. Use the graph from Exercise 47 to estimate the number of drinks a 120-lb person must have had in order to be considered intoxicated.

49. *Child Expenditure.* The table on p. 60 shows the annual expenditures on a child in 1996 by families with an average income of $46,100 (*Source: The*

Wall Street Journal Almanac, 1998). Use the data to draw a line graph with a graphing calculator.

AGE OF CHILD	ANNUAL EXPENDITURE
1	$7860
4	8060
7	8130
10	8100
13	8830
16	8960

50. *Residential Remodeling.* The following table shows the amount spent in the United States on residential remodeling (*Source: The Wall Street Journal Almanac, 1998*). Use the data to draw a line graph with a graphing calculator.

YEAR	REMODELING EXPENSES (IN BILLIONS)
1980	$ 46.3
1985	80.3
1990	106.8
1991	97.5
1992	103.7
1993	108.3
1994	115.0
1995	112.6
1996 (projected)	118.5
1997 (projected)	125.3
2000 (projected)	143.0

U.S. Farms. The number of U.S. farms f, in millions, can be approximated by the equation

$$f = -\frac{1}{50}t + 2.5,$$

where t is the number of years after 1975 (Source: The Wall Street Journal Almanac, 1998).

51. Use the graph of the equation to estimate how many farms there were in the United States in 1985.

52. Use the graph of the equation to approximate in what year there will be 2.0 million farms in the United States.

Tax Refunds. The number of federal income-tax refunds n, in millions, can be approximated by the equation

$$n = 0.2x^2 - 1.3x + 81.3,$$

where x is the number of years since 1990 (Source: Internal Revenue Service).

53. Use the graph of the equation to estimate in what year or years there were 80 million refunds.

54. Use the graph of the equation to estimate the number of refunds in 1996.

Skill Maintenance

Simplify. [1.4]

55. $(-2)^4$ **56.** -2^4 **57.** $z^{-2} \cdot z^6$ **58.** $\dfrac{z^{-2}}{z^6}$

Synthesis

59. ◈ How can a guess or estimate help prepare you for the *Translate* step in problem solving?

60. ◈ Write a problem for a classmate to translate to mathematical language. Devise the problem so that it translates to the equation $x - 5 = 21$.

61. ◈ Write a problem for a classmate to translate to mathematical language. Devise the problem so that it can be modeled using the formula for the circumference C of a circle: $C = 2\pi r$.

62. ◈ Describe at least two benefits of using a graph to model a situation.

63. Match each sentence with the most appropriate graph.

a) Carpooling to work, Terry spent 10 min on local streets, then 20 min cruising on the freeway, and then 5 min on local streets to his office.

b) For her commute to work, Sharon drove 10 min to the train station, rode the express for 20 min, and then walked for 5 min to her office.

c) For his commute to school, Roger walked 10 min to the bus stop, rode the express for 20 min, and then walked for 5 min to his class.

d) Coming home from school, Kristy waited 10 min for the school bus, rode the bus for 20 min, and then walked 5 min to her house.

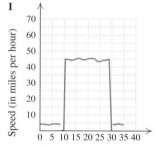

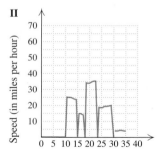

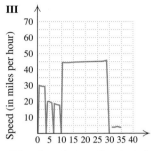

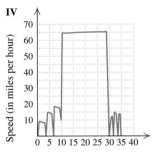

64. Match each sentence with the most appropriate graph.

a) Roberta worked part time until September, full time until December, and overtime until Christmas.

b) Clyde worked full time until September, half time until December, and full time until Christmas.

c) Clarissa worked overtime until September, full time until December, and overtime until Christmas.

d) Doug worked part time until September, half time until December, and full time until Christmas.

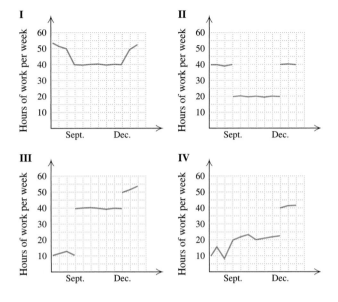

Focus: Models

Time: 15–20 minutes

Group size: 3

Many manufacturers issue cents-off coupons to encourage customers to buy their products. The following table shows the number of coupons issued and the number redeemed for various years.

YEAR	NUMBER OF COUPONS DISTRIBUTED (IN BILLIONS)	NUMBER OF COUPONS REDEEMED (IN BILLIONS)	PERCENTAGE OF COUPONS REDEEMED
1985	199.9	6.5	
1988	247.4	7.1	
1990	279.4	7.1	
1992	310.0	7.7	
1994	309.7	6.2	
1996	268.5	5.3	

Source: The Wall Street Journal Almanac, 1998

Activity

1. Calculate, as a group, the percentage of coupons distributed each year that were redeemed.

2. Each group member should choose a different one of the last three columns and create a line graph, using the figures in that column on the vertical axis and the corresponding years on the horizontal axis.

3. Compare the graphs for the number of coupons distributed and the number redeemed. Determine the year for which the greatest number of coupons were distributed and for which the greatest number were redeemed. How are they related?

4. Compare both graphs showing the numbers of coupons with the graph showing the percentage of coupons redeemed. Determine the year in which the highest percentage of coupons was redeemed. Compare this result with the results from step (3).

5. Decide, as a group, which, if any, of the graphs shows a *linear* relationship, and compare your conclusion with other groups in the class.

CHAPTER 1

Summary and Review

Key Terms

Variable, p. 2
Constant, p. 2
Algebraic expression, p. 2
Exponential notation, p. 2
Exponent, or power, p. 2
Base, p. 2
Substituting, p. 3
Evaluating the expression, p. 3
Equation, p. 4
Solution, p. 4
Inequality, p. 5
Natural numbers, p. 6
Whole numbers, p. 6
Integers, p. 6
Roster notation, p. 6
Set-builder notation, p. 6
Element, p. 7
Rational numbers, p. 7

Fractional notation, p. 7
Decimal notation, p. 7
Irrational numbers, p. 8
Real numbers, p. 8
Subset, p. 8
Absolute value, p. 8
Cursor, p. 10
Contrast, p. 10
Home screen, p. 10
Opposites, p. 13
Additive inverses, p. 13
Reciprocals, p. 15
Multiplicative inverses, p. 16
Indeterminate, p. 17
Menu, p. 18
Submenus, p. 18
Equivalent expressions, p. 21
Factoring, p. 23
Term, p. 23

Like terms, p. 23
Combining like terms, p. 23
Scientific notation, p. 33
Axes, p. 38
x, y-coordinate system, p. 38
Cartesian coordinate system, p. 38
Ordered pairs, p. 38
Origin, p. 38
Coordinates, p. 39
Quadrants, p. 39
Viewing window, p. 40
Standard viewing window, p. 40
Graph, p. 42
Linear equation, p. 44
Nonlinear equations, p. 44
Mathematical model, p. 53

Important Properties and Formulas

Area of a rectangle: $A = lw$

Area of a square: $A = s^2$

Area of a parallelogram: $A = bh$

Area of a trapezoid: $A = \dfrac{h}{2}(b_1 + b_2)$

Area of a triangle: $A = \frac{1}{2}bh$

Area of a circle: $A = \pi r^2$

Circumference of a circle: $C = \pi d$

Volume of a cube: $V = s^3$

Volume of a right circular cylinder: $V = \pi r^2 h$

Perimeter of a square: $P = 4s$

Distance traveled: $d = rt$

Simple interest: $I = Prt$

Addition of Two Real Numbers

1. *Positive numbers:* Add the numbers. The result is positive.

2. *Negative numbers:* Add absolute values. Make the answer negative.

3. *A negative and a positive number:* If the numbers have the same absolute value, the answer is 0. Otherwise, subtract the smaller absolute value from the larger one:

   a) If the positive number is further from 0, make the answer positive.

   b) If the negative number is further from 0, make the answer negative.

4. *One number is zero:* The sum is the other number.

Multiplication of Two Real Numbers

1. To multiply two numbers with *unlike signs,* multiply their absolute values. The answer is *negative.*

2. To multiply two numbers with the *same sign,* multiply their absolute values. The answer is *positive.*

Division of Two Real Numbers

1. To divide two numbers with *unlike signs,* divide their absolute values. The answer is *negative.*

2. To divide two numbers with the *same sign,* divide their absolute values. The answer is *positive.*

The law of opposites: $a + (-a) = 0$

The law of reciprocals: $a \cdot \dfrac{1}{a} = 1,\ a \neq 0$

Absolute value: $|x| = \begin{cases} x, & \text{if } x \geq 0, \\ -x, & \text{if } x < 0 \end{cases}$

For any number a and any nonzero number b,

$$\frac{-a}{b} = \frac{a}{-b} = -\frac{a}{b}.$$

Rules for Order of Operations

1. Calculate within grouping symbols before calculating outside.

2. Simplify all exponential expressions.

3. Do all multiplication and division, in order, from left to right.

4. Do all addition and subtraction, in order, from left to right.

Commutative laws: $a + b = b + a,$ $ab = ba$

Associative laws: $a + (b + c) = (a + b) + c,$ $a(bc) = (ab)c$

Distributive law: $a(b + c) = ab + ac$

Definitions and Rules for Exponents

For any integers m and n (assuming 0 is not raised to a nonpositive power):

Zero as an exponent: $a^0 = 1$

Negative integers as exponents: $a^{-n} = \dfrac{1}{a^n}$

Multiplying with like bases:
$a^m \cdot a^n = a^{m+n}$ (Product Rule)

Dividing with like bases:
$\dfrac{a^m}{a^n} = a^{m-n}; \ a \neq 0$ (Quotient Rule)

Raising a product to a power: $(ab)^n = a^n b^n$

Raising a power to a power:
$(a^m)^n = a^{mn}$ (Power Rule)

Raising a quotient to a power:
$\left(\dfrac{a}{b}\right)^n = \dfrac{a^n}{b^n}; \ b \neq 0$

Scientific notation for a number is an expression of the type $M \times 10^n$, where $1 \leq M < 10$, M is in decimal notation, and n is an integer.

REVIEW EXERCISES

The following review exercises are for practice. Answers are at the back of the book. If you need to, restudy the section indicated alongside the answer.

Evaluate each expression using the values provided.

1. $3x - (4 - y)$, for $x = 10$ and $y = 2$

2. $7x^2 - 5y \div zx$, for $x = -2.78$, $y = 1.5$, and $z = 3.2$

3. Name the set consisting of the first six even natural numbers using both roster notation and set-builder notation.

4. Find the area of a triangular sign that has a base of 90 cm and a height of 70 cm.

Tell whether each number is a solution of the given equation or inequality.

5. $10 - 3x = 1$; **(a)** 7; **(b)** 3

6. $5a + 2 \leq 7$; **(a)** 0; **(b)** 1

Find the absolute value.

7. $|-7.3|$ **8.** $|4.09|$ **9.** $|0|$

Perform the indicated operation.

10. $-9.4 + (-3.7)$

11. $\left(-\frac{4}{5}\right) + \left(\frac{1}{7}\right)$

12. $\left(-\frac{1}{3}\right) + \frac{4}{5}$

13. $-7.9 - 3.6$

14. $-\frac{2}{3} - \left(-\frac{1}{2}\right)$

15. $12.5 - 17.9$

16. $(-2.1)(-3)$

17. $\left(-\frac{2}{3}\right)\left(\frac{5}{8}\right)$

18. $\dfrac{72.8}{-8}$

19. $-7 \div \dfrac{4}{3}$

20. Find $-a$ if $a = -4.01$.

Use a commutative law to write an equivalent expression.

21. $5 + a$ **22.** $7y$ **23.** $5x + y$

Use an associative law to write an equivalent expression.

24. $(4 + a) + b$ **25.** $(xy)7$

26. Obtain an expression that is equivalent to $7mn + 14m$ by factoring.

27. Combine like terms: $5x^3 - 8x^2 + x^3 + 2$.

28. Simplify: $7x - 4[2x + 3(5 - 4x)]$.

29. Multiply and simplify: $(5a^2 b^7)(-2a^3 b)$.

30. Divide and simplify: $\dfrac{12x^3 y^8}{3x^2 y^2}$.

31. Evaluate a^0, a^2, and $-a^2$ for $a = -5.3$.

Simplify. Do not use negative exponents in the answer.

32. $3^{-4} \cdot 3^7$ **33.** $(5a^2)^3$

34. $(-2a^{-3} b^2)^{-3}$ **35.** $\left(\dfrac{x^2 y^3}{z^4}\right)^{-2}$

36. $\left(\dfrac{2a^{-2} b}{4a^3 b^{-3}}\right)^4$

Simplify.

37. $\dfrac{7(5 - 2 \cdot 3) - 3^2}{4^2 - 3^2}$

38. $2.1 - |-3.8 + 4.65|^2 \div 2.5 \times 4.8$

39. Convert 0.000000103 to scientific notation.

40. One *parsec* (a unit that is used in astronomy) is 30,860,000,000,000 km. Write scientific notation for this number.

Simplify and write scientific notation for each answer.

41. $(8.7 \times 10^{-9}) \times (4.3 \times 10^{15})$ **42.** $\dfrac{1.2 \times 10^{-12}}{6.1 \times 10^{-7}}$

Determine whether the ordered pair is a solution.

43. $(3, 7)$; $4p - q = 5$

44. $(-2, 4)$; $x - 2y = 12$

45. $\left(0, -\tfrac{1}{2}\right)$; $3a - 4b = 2$

46. $(8, -2)$; $3c + 2d = 28$

Graph.

47. $y = -3x + 2$ **48.** $y = -x^2 + 1$

49. $y = 3 - |x|$ **50.** $y = 6$

51. Using a graphing calculator, complete the following table for the equation

$$y = 2|x| - 3.$$

Then graph.

X	Y₁	
-3		
-2		
-1		
0		
1		
2		
3		
X = -3		

Translate to an equation.

52. 13 less than twice a number is 21.

53. A number is 17 less than another number. The sum of the numbers is 115. Find the smaller number.

54. One angle of a triangle measures three times the second angle. The third angle measures twice the second angle. Find the measures of the angles.

55. *Agriculture.* The following graph shows the cow population on United States farms from 1890 to

1990. (*Source:* U.S. Department of Agriculture; *The Macmillan Visual Almanac, 1996*).

a) What was the cow population in 1930?

b) In what year was the cow population the greatest?

c) Between the years 1890 and 1970, did the cow population increase or decrease?

d) What was the amount of decrease in the cow population between 1970 and 1990?

56. *Unemployment Rate.* The following table shows the percent unemployment rate in the United States for various years (*Source: Information Please Almanac, 1998*). Use the data to draw a line graph with a graphing calculator.

YEAR	PERCENT UNEMPLOYMENT
1928	4.2
1934	21.7
1938	19.0
1944	1.2
1948	3.8
1954	5.5
1958	6.8
1964	5.2
1968	3.6
1974	5.6
1978	6.0
1984	7.5
1988	5.4
1994	6.1
1997	5.3

Synthesis

57. ◈ Explain the difference between a solution of an equation like $y = 2x + 1$ and a solution of an equation like $3x + 5 = 2$.

58. ◈ Explain why it is necessary to have a standard set of rules for order of operations.

59. Evaluate $a + b(c - a^2)^0 + (abc)^{-1}$ for $a = 2$, $b = -3$, and $c = -4$.

60. Simplify:

$$\frac{(3^{-2})^a \cdot (3^b)^{-2a}}{(3^{-2})^b \cdot (9^{-b})^{-3a}}.$$

61. Use the commutative law for addition once and the distributive law twice to show that

$$a2 + cb + cd + ad = a(d + 2) + c(b + d).$$

62. Find an irrational number between $\tfrac{1}{2}$ and $\tfrac{3}{4}$.

Source: U.S. Department of Agriculture; *The Macmillan Visual Almanac, 1996*

CHAPTER 1 TEST

1. Evaluate $a^3 - 5b + b \div ac$ for $a = -2$, $b = 6$, and $c = 3$.

2. The base of a triangular stamp measures 3 cm and its height 2.5 cm. Find the area of the stamp.

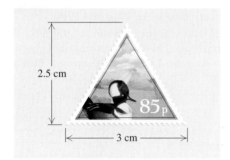

2.5 cm

3 cm

3. Tell whether each number is a solution of the equation $14 - 5x = 4$.

a) 1

b) 2

c) 0

Perform the indicated operation.

4. $-25 + (-16)$

5. $-10.5 + 6.8$

6. $6.21 + (-8.32)$

7. $29.5 - 43.7$

8. $-17.8 - 25.4$

9. $-6.4(5.3)$

10. $-\frac{7}{3} - \left(-\frac{3}{4}\right)$

11. $-\frac{2}{7}\left(-\frac{5}{14}\right)$

12. $\frac{-42.6}{-7.1}$

13. $\frac{2}{5} \div \left(-\frac{3}{10}\right)$

14. Simplify: $5 + (1 - 3)^2 - 7 \div 2^2 \cdot 6$.

15. Use a commutative law to write an expression equivalent to $7x + y$.

16. Combine like terms: $4y - 10 - 7y - 19$.

17. Simplify: $9x - 3(2x - 5) - 7$.

Simplify. Do not use negative exponents in the answer.

18. $(12x^{-4}y^{-7})(-6x^{-6}y)$

19. -3^{-2}

20. $(-6x^2y^{-4})^{-2}$

21. $\left(\dfrac{2x^3y^{-6}}{-4y^{-2}}\right)^2$

22. $(5x^3y)^0$

Simplify and write scientific notation for the answer.

23. $(9.05 \times 10^{-3})(2.22 \times 10^{-5})$

24. $\dfrac{5.6 \times 10^7}{2.8 \times 10^{-3}}$

25. $\dfrac{1.067 \times 10^{-5}}{3.49 \times 10^{-10}}$

Determine whether the ordered pair is a solution.

26. $(0, -5); \ x + 4y = -20$

27. $(1, -4); \ -2p + 5q = 18$

Graph.

28. $y = -5x + 4$

29. $y = -2x^2 + 3$

30. Create a table of solutions of the equation
$$y = 10 - x^2$$
for integer values of x from -3 to 3. Then graph.

31. Translate to an algebraic expression:

   Three more than the product of two numbers.

32. Translate to an equation:

   Greg's scores on five tests are 94, 80, 76, 91, and 75. What must Greg score on the sixth test so that his average will be 85?

Gas Mileage.   The following graph shows the gas mileage of a truck traveling at different speeds.

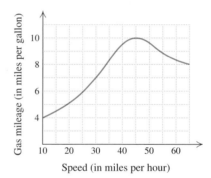

33. At what speed is the gas mileage highest?

34. What is the gas mileage when the truck is traveling at 30 mph?

Synthesis

Simplify.

35. $(4x^{3a}y^{b+1})^{2c}$

36. $\dfrac{-27a^{x+1}}{3a^{x-2}}$

37. $\dfrac{(-16x^{x-1}y^{y-2})(2x^{x+1}y^{y+1})}{(-7x^{x+2}y^{y+2})(8x^{x-2}y^{y-1})}$

Functions, Linear Equations, and Models 2

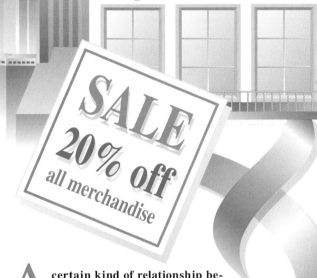

A certain kind of relationship between sets is known as a *function*. Functions are very important in mathematics in general, and in problem solving in particular. In this chapter, you will learn what a function is and how to use functions to solve problems.

A function that can be described by a linear equation is called a *linear function*. We will study graphs of linear equations in detail, and will use linear equations and functions to model and solve applications.

APPLICATION

SHOPPING CENTERS. The number of shopping centers in the United States has grown in recent years, as shown in the following table. Use linear regression to fit a linear function to the data, and graph the line and the data.

YEAR	NUMBER OF CENTERS
1964	7,600
1972	13,174
1976	17,523
1980	22,050
1984	22,508
1988	32,563
1992	38,966
1996	42,130

Source: International Council of Shopping Centers; *The Wall Street Journal Almanac*, 1998, p. 285.

A linear function that fits the data is

$$f(x) = 1.128261905x + 4.819666667,$$

where x is the number of years since 1964 and f is the number of shopping centers, in thousands.

This problem appears as Example 4 in Section 2.6.

$y = 1.128261905x + 4.819666667$

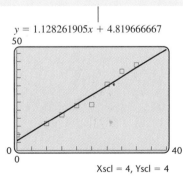

Xscl = 4, Yscl = 4

2.1
Functions

- *Functions and Graphs*
- *Function Notation and Equations*
- *Applications*

The idea of a *function* is one of the most important concepts in mathematics. A function is a special kind of correspondence from one set to another. For example:

To each person in a class	there corresponds	his or her mother.
To each item in a store	there corresponds	its price.
To each real number	there corresponds	the cube of that number.

In each example, the first set is called the **domain**. The second set is called the **range**. For any member of the domain, there is *just one* member of the range to which it corresponds. This kind of correspondence is called a **function**.

Example 1   Determine whether each correspondence is a function.

a) −3 ⟶ 5
        1 ⟹ 2
        4 ⟶

b) San Francisco ⟶ Giants
      New York ⟶ Mets
      Atlanta ⟶ Falcons

SOLUTION

a) The correspondence *is* a function because each member of the domain corresponds to *just one* member of the range.

b) The correspondence *is not* a function because a member of the domain (New York) corresponds to more than one member of the range.   ▬

Function

A *function* is a correspondence between a first set, called the *domain*, and a second set, called the *range*, such that each member of the domain corresponds to *exactly one* member of the range.

Example 2   Determine whether each correspondence is a function.

DOMAIN	CORRESPONDENCE	RANGE
a) A family	Each person's weight	A set of positive numbers
b) {−2, 0, 1, 2}	Each number's square	{0, 1, 4}
c) The set of all states	Each state's members of the U.S. Senate	A set of U.S. senators

SOLUTION

a) The correspondence *is* a function, because each person has *only one* weight.

b) The correspondence *is* a function, because every number has *only one* square.

c) The correspondence *is not* a function, because each state has *two* U.S. senators.

Functions and Graphs

The functions in Examples 1(a) and 2(b) can be expressed as sets of ordered pairs. Example 1(a) can be written {(−3, 5), (1, 2), (4, 2)} and Example 2(b) can be written {(−2, 4), (0, 0), (1, 1), (2, 4)}. We can graph these functions as follows.

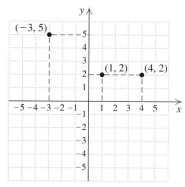

The function {(−3, 5), (1, 2), (4, 2)}
Domain is {−3, 1, 4}
Range is {5, 2}

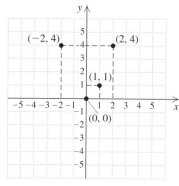

The function {(−2, 4), (0, 0), (1, 1), (2, 4)}
Domain is {−2, 0, 1, 2}
Range is {4, 0, 1}

When a function is given as a set of ordered pairs, the domain is simply the set of all first coordinates and the range is the set of all second coordinates.

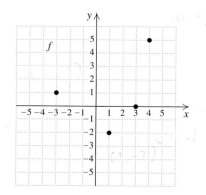

Example 3 Find the domain and the range of the function *f* shown at left.

SOLUTION Here *f* can be written {(−3, 1), (1, −2), (3, 0), (4, 5)}. The domain is the set of all first coordinates, {−3, 1, 3, 4}, and the range is the set of all second coordinates, {1, −2, 0, 5}. We can also find the domain and the range directly by observing the *x*- and *y*-values used in the graph.

Example 4   For the function f shown here, determine each of the following.

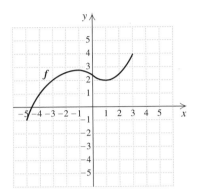

a) The member of the range that is paired with 1

b) The domain of f

c) The member of the domain that is paired with 1

d) The range of f

SOLUTION

a) To determine what member of the range is paired with 1, we locate 1 on the horizontal axis. Next, we find the point on the graph of f for which 1 is the first coordinate. (See the graph on the left below.) From that point, we can look to the vertical axis to find the corresponding y-coordinate, 2. The "input" 1 has the "output" 2.

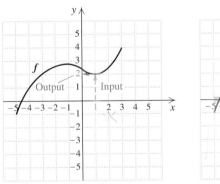   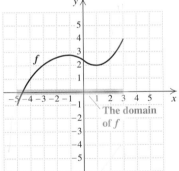

b) The domain of the function is the set of all x-values that are in the graph. (See the graph on the right above.) These extend from -5 to 3 and can be viewed as the curve's shadow, or *projection*, on the x-axis. Thus the domain is $\{x\,|\,-5 \le x \le 3\}$.

c) To determine what member of the domain is paired with 1, we locate 1 on the vertical axis. (See the graph on the left at the top of the next page.) From there we look left and right to the graph of f to find any points for which 1 is the second coordinate. One such point exists, $(-4, 1)$. We note that -4 is the only element of the domain paired with 1.

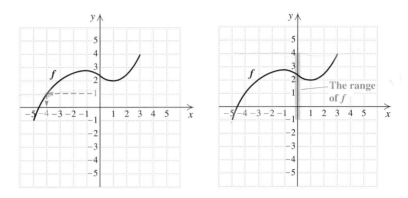

d) The range of the function is the set of all y-values that are in the graph. (See the graph on the right above.) These extend from -1 to 4 and can be viewed as the curve's projection on the y-axis. Thus the range is $\{y \mid -1 \leq y \leq 4\}$. ▬

Note that if a graph contains two or more points with the same first coordinate, that graph cannot represent a function (otherwise one member of the domain would correspond to more than one member of the range). This observation is the basis of the *vertical-line test*.

The Vertical-Line Test

A graph represents a function if it is not possible to draw a vertical line that intersects the graph more than once.

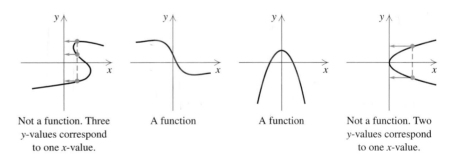

Not a function. Three A function A function Not a function. Two
y-values correspond y-values correspond
to one x-value. to one x-value.

Graphs that do not represent functions still do represent *relations*.

Relation

A *relation* is a correspondence between a first set, called the *domain*, and a second set, called the *range*, such that each member of the domain corresponds to *at least one* member of the range.

Thus, although the correspondences and graphs above are not all functions, they *are* all relations.

Function Notation and Equations

To understand function notation, it helps to imagine a "function machine." Think of putting a member of the domain (an *input*) into the machine. The machine knows the correspondence and produces the appropriate member of the range (the *output*).

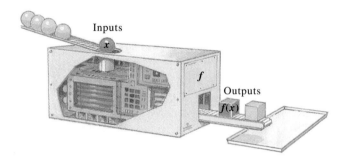

The function pictured has been labeled f. Here x represents an arbitrary input, and $f(x)$—read "f of x," "f at x," or "the value of f at x"—represents the corresponding output. You should check that in Example 3, $f(4)$ is 5, $f(-3)$ is 1, and so on. Similarly, in Example 4(a), $f(1) = 2$. Note that $f(x)$ *does not mean f times x.*

Most functions are described by equations. For example, $f(x) = 2x + 3$ describes the function that takes an input x, multiplies it by 2, and then adds 3.

$$\underset{\text{Double}}{\overset{\text{Input}}{f(x) \;\; = \;\; 2x \;\; + 3}}$$
$$\text{Double} \qquad \text{Add 3}$$

To calculate the output $f(4)$, we take the input 4, double it, and add 3 to get 11. That is, we substitute 4 into the formula for $f(x)$:

$$f(4) = 2 \cdot 4 + 3$$
$$= 11.$$

Sometimes, in place of $f(x) = 2x + 3$, we write $y = 2x + 3$, where it is understood that the value of y, the *dependent variable*, is calculated after first choosing a value for x, the *independent variable*. To understand why $f(x)$ notation is so useful, consider two equivalent statements:

a) If $f(x) = 2x + 3$, then $f(4) = 11$.

b) If $y = 2x + 3$, then the value of y is 11 when x is 4.

The notation used in part (a) is far more concise.

Example 5   Find the indicated function value.

a) $f(5)$, for $f(x) = 3x + 2$

b) $g(-2)$, for $g(r) = 5r^2 + 3r$

c) $h(4)$, for $h(x) = 7$

d) $F(a + 1)$, for $F(x) = 3x + 2$

SOLUTION

a) $f(5) = 3 \cdot 5 + 2 = 17$

b) $g(-2) = 5(-2)^2 + 3(-2)$
$= 5 \cdot 4 - 6 = 14$

c) For the function given by $h(x) = 7$, all inputs share the same output, 7. Therefore, $h(4) = 7$. The function h is an example of a *constant function*.

d) $F(a + 1) = 3(a + 1) + 2$
$= 3a + 3 + 2 = 3a + 5$ ▬

Note that whether we write $f(x) = 3x + 2$, or $f(t) = 3t + 2$, or $f(\square) = 3\square + 2$, we still have $f(5) = 17$. Thus the independent variable can be thought of as a *dummy variable*. The letter chosen for the dummy variable is not as important as the algebraic manipulations to which it is subjected.

Values of functions defined by an equation can be found using a grapher in several ways. Many graphers use function notation directly. Alternatively, we can store a value for x in the grapher's memory and calculate the corresponding function value, or use a table of values. We can also calculate the value by carrying out the indicated operations.

Example 6   For $f(a) = 2a^2 - 3a + 1$, find $f(3)$ and $f(-5.1)$.

SOLUTION   We first enter the function into the grapher, replacing a with x and the notation $f(a)$ with Y1. The equation $Y_1 = 2x^2 - 3x + 1$ represents the same function, with the understanding that the Y1-values are the outputs and the x-values are the inputs; $Y_1(x)$ is equivalent to $f(a)$.

To find $f(3)$ directly, from the home screen, we enter Y1(3). We see that $f(3) = 10$.

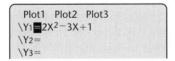

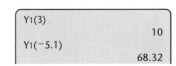

We can find $f(-5.1)$ by **editing** the previous entry. First, we place a copy of the previous entry on the screen. Then, using the arrow keys to move the cursor, we replace 3 with -5.1. When the cursor is a solid rectangle, pressing a key will replace the character under the cursor. If we change the cursor to an underscore, pressing a key will insert a character to the left of the cursor. We press ENTER after the entry has been edited. We see that $f(-5.1) = 68.32$.

Another way to calculate a function value once Y1 has been entered is to store the input value in the grapher's memory. We store 3 to the variable X and press Y1. The grapher will calculate the value of Y1 when $x = 3$.

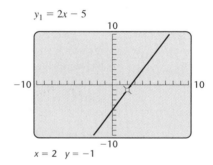

We can also find function values from the graph of the function.

Example 7   Find $g(2)$ for $g(x) = 2x - 5$.

SOLUTION   We enter and graph the function, using the standard viewing window. Some graphers have a VALUE option in the CALC menu that will calculate a function's value given a value of x. In this case, when $x = 2$, $y = -1$, so $g(2) = -1$.

$$y_1 = 2x - 5$$

A table is useful for finding several values of the same function.

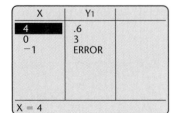

Example 8   If $f(x) = \dfrac{3}{x + 1}$, determine whether $4, 0$, and -1 are in the domain of the function.

SOLUTION   The fraction bar in the expression for the function acts as a grouping symbol. When entering this type of function, be sure to use parentheses when necessary. We enter the function as $y_1 = 3/(x + 1)$ and set up a table, with Indpnt set to Ask. To find the function values, we enter the values $4, 0$, and -1 for x.

From the table, we see that $f(4) = 0.6$, $f(0) = 3$, and $f(-1)$ is undefined. This tells us 4 and 0 are in the domain of f and -1 is not.

We can use a grapher to estimate the domain and the range of a function from its graph.

Example 9 Graph each of the following functions using a $[-5, 5, -5, 5]$ viewing window and estimate the domain and the range from the graph.

a) $f(x) = |x + 2|$

b) $f(x) = \sqrt{x + 2}$

SOLUTION

a) Since the absolute-value symbols act as grouping symbols, parentheses are necessary when entering this function. We enter and graph $y = \text{abs}(x + 2)$.

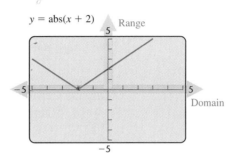

It appears from the graph that the function is defined for all values of x, but that the value of the function is never negative. We estimate the domain to be all real numbers and the range to be all nonnegative real numbers. We write this as Domain $= \{x \,|\, x$ is a real number$\}$, Range $= \{y \,|\, y \geq 0\}$. The domain and range are indicated on the graph.

b) The function adds 2 to a number and takes the square root of the result. We indicate this order by using parentheses and entering $y = \sqrt{\ } (x + 2)$.

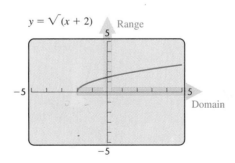

The graph indicates that the domain is $\{x \,|\, x \geq -2\}$ and the range is $\{y \,|\, y \geq 0\}$. ▬

It can be difficult to estimate a function's domain and range from a graph. As you learn more about different kinds of functions throughout this and other courses, you will be able to use other methods to better determine domains and ranges.

Applications

Function notation is often used in formulas. For example, to emphasize that the area A of a circle is a function of its radius r, instead of

$$A = \pi r^2,$$

we can write

$$A(r) = \pi r^2.$$

When a function is given as a graph in a problem-solving situation, we are often asked to determine certain quantities on the basis of the graph. Often models can be developed and used for calculations. Such a model for Example 10 is developed in Chapter 9. For now, we simply use the graph itself.

Example 10   *Spread of AIDS.* According to the Federal Centers for Disease Control, there were 30,657 newly reported cases of AIDS in the United States in 1988, 41,639 cases in 1990, 45,839 cases in 1992, 102,605 cases in 1993, 77,561 cases in 1994, and 56,730 cases in 1996.* Estimate the number of newly reported cases for the years 1989 and 1995.

SOLUTION

1., 2. Familiarize and **Translate.**   The given information enables us to plot and connect six points on a graph. We let the horizontal axis represent the year and the vertical axis the number of reported cases. We label the function itself c.

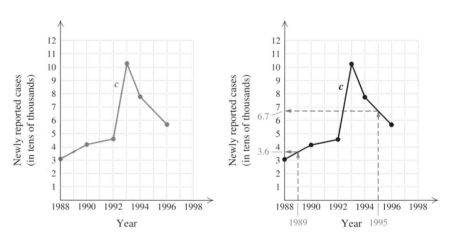

3. Carry out.   To estimate the reported number of cases in 1989, we locate the point on the graph that is directly above the year 1989. After doing so, we estimate its second coordinate by moving horizontally from that point to the vertical axis. We see from the graph that

---

*The great increase in the number of cases for 1993 resulted in part from implementation of an expanded definition of AIDS.

$c(1989) \approx 36,000$. Following a similar procedure, we find that $c(1995) \approx 67,000$.

4. Check. Since 36,000 is between 30,657 and 41,639 and 67,000 is between 77,561 and 56,730, our estimates seem plausible.

5. State. There were about 36,000 newly reported cases of AIDS in 1989 and about 67,000 newly reported cases in 1995. ▬

2.1 Exercise Set 1-85 2 9

Determine whether each correspondence is a function.

1. 3 → a
5 → b
7 → c
9 → d
e

2. 1 → a
2 → b
3 → c
4 → d
5

3.

Girl's Age (in months)	Average Daily Weight Gain (in grams)
2	→ 21.8
9	→ 11.7
16	→ 8.5
23	→ 7.0

Source: American Family Physician, December 1993, p. 1435.

4.

Boy's Age (in months)	Average Daily Weight Gain (in grams)
2	→ 24.3
9	→ 11.7
16	→ 8.2
23	→ 7.0

Source: American Family Physician, December 1993, p. 1435.

5. cat → dog
fish → worm
dog → cat
tiger → fish
teacher → student

6.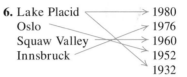
Lake Placid — 1980
Oslo — 1976
Squaw Valley — 1960
Innsbruck — 1952
1932

Determine whether each of the following is a function. Identify any relations that are not functions.

	Domain	Correspondence	Range
7.	A family	Each person's eye color	A set of colors
8.	A textbook	An even-numbered page in the book	A set of pages
9.	A set of avenues	An intersecting road	A set of cross streets
10.	A math class	Each person's seat number	A set of numbers
11.	A set of numbers	Square each number and then add 4.	A set of numbers
12.	A set of shapes	The area of each shape	A set of numbers

For each graph of a function, determine **(a)** $f(1)$; **(b)** *the domain;* **(c)** *any x-values for which* $f(x) = 2$; *and* **(d)** *the range.*

13.

14.

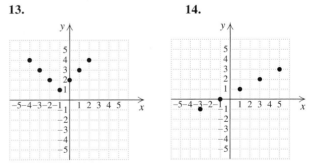

15.

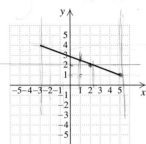

16.

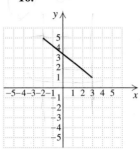

Determine whether each of the following is the graph of a function.

23.

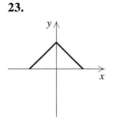

24.

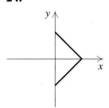

17.

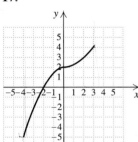

18.

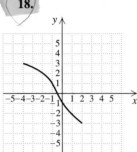

25.

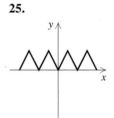

26.

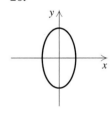

19.

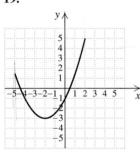

20.

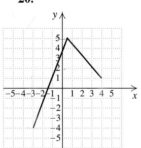

27.

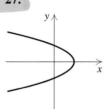

28.

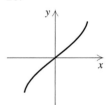

29.

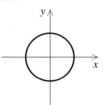

30.

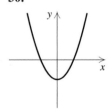

Hint for Exercises 21 and 22: An open circle indicates that the point is not included in the graph of the function.

21.

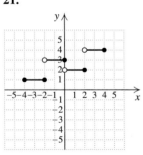

22.

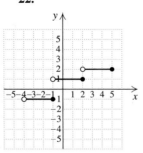

Find the function values.

31. $g(x) = x + 1$

    a) $g(0)$ **b)** $g(-4)$ **c)** $g(-7)$
    d) $g(8)$ **e)** $g(-0.815)$ **f)** $g(a + 2)$

32. $h(x) = x - 4$

    a) $h(4)$ **b)** $h(8)$ **c)** $h(-3)$
    d) $h(-4)$ **e)** $h(2.73)$ **f)** $h(a - 1)$

33. $f(n) = 5n^2 + 4n$

    a) $f(0)$ **b)** $f(-1)$ **c)** $f(3)$
    d) $f(1.06)$ **e)** $f(t)$ **f)** $f(2a)$

34. $g(n) = 3n^2 - 2n$

   a) $g(0)$ **b)** $g(-1)$ **c)** $g(3)$
   d) $g(-3.72)$ **e)** $g(t)$ **f)** $g(2a)$

35. $f(x) = \dfrac{x - 3}{2x - 5}$

   a) $f(0)$ **b)** $f(4)$ **c)** $f(-1)$
   d) $f(3)$ **e)** $f(42.7)$ **f)** $f(x + 2)$

36. $s(x) = \dfrac{3x - 4}{2x + 5}$

   a) $s(10)$ **b)** $s(2)$ **c)** $s\left(\frac{1}{2}\right)$
   d) $s(-1)$ **e)** $s(-12.9)$ **f)** $s(x + 3)$

Determine whether each number is in the domain of the function described by the given equation.

37. $f(x) = \dfrac{x + 1}{2x - 3}$

   a) 2.6   **b)** -1 **c)** 1.48   **d)** 1.5

38. $g(x) = \dfrac{3x^2}{x + 2}$

   a) 0   **b)** -2 **c)** -1.79 **d)** -5

39. $f(x) = \dfrac{1}{|4x - 9|}$

   a) 2.25   **b)** -2.25 **c)** -57 **d)** 0

40. $f(t) = \dfrac{t - 2}{t^2 - 5t + 4}$

   a) 0   **b)** 1   **c)** 4   **d)** 2

Graph each function using a standard viewing window and use the graph to estimate the domain and the range.

41. $f(x) = 2x + 1$ **42.** $f(x) = 3 - x$
43. $f(x) = \sqrt{x - 4}$ **44.** $g(x) = \sqrt{5 - x}$
45. $r(x) = |x| - 6$ **46.** $g(x) = x^2 - 3$
47. $h(x) = x^2 + 1$ **48.** $f(x) = |x^2 - 3|$

The function A described by $A(s) = s^2\dfrac{\sqrt{3}}{4}$ gives the area of an equilateral triangle with side s.

49. Find the area when a side measures 4 cm.

50. Find the area when a side measures 6 in.

The function V described by $V(r) = 4\pi r^2$ gives the surface area of a sphere with radius r.

51. Find the area when the radius is 3 in.

52. Find the area when the radius is 5 cm.

Chemistry. *The function F described by*

$$F(C) = \tfrac{9}{5}C + 32$$

gives the Fahrenheit temperature corresponding to the Celsius temperature C.

53. Find the Fahrenheit temperature equivalent to $-10°C$.

54. Find the Fahrenheit temperature equivalent to $5°C$.

Archaeology. *The function H described by*

$$H(x) = 2.75x + 71.48$$

can be used to estimate the height, in centimeters, of a woman whose humerus *(the bone from the elbow to the shoulder) is x cm long. Estimate the height of a woman whose humerus is the length given.*

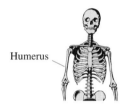

Humerus

55. 32.15 cm   **56.** 35.4 cm

Stopping Distance. *The function*

$$d(s) = 0.084s^2 - 0.851s + 24.142$$

can be used to estimate the average minimum stopping distance, in feet, of a car traveling at speed s, in miles per hour (Source: The Handy Science Answer Book. Detroit: Visible Ink Press, 1994, p. 400).

57. Graph the function and use the graph to estimate the average minimum stopping distance of a car traveling at 25 mph.

58. Use the graph from Exercise 57 to estimate the average minimum stopping distance of a car traveling at 55 mph.

Electronics. *An older video cassette recorder has a revolution counter and a booklet with a table relating the counter reading and the time for which a tape has run.*

COUNTER READING	TIME FOR TAPE (IN HOURS)
000	0
300	1
500	2
675	3
800	4

59. Use the data in the table above to draw a graph of the time that a tape has run as a function of the counter reading and then estimate the time elapsed when the counter has reached 600.

60. Use the graph from Exercise 59 to estimate the time elapsed when the counter has reached 200.

Population Growth. The town of Falconburg recorded the following dates and populations.

INPUT, YEAR	OUTPUT, POPULATION (IN TENS OF THOUSANDS)
1991	5.8
1993	6
1995	7
1997	10

61. Use the data in the table above to draw a graph of the population as a function of time. Then estimate what the population was in 1994.

62. Use the graph in Exercise 61 to estimate Falconburg's population in 1996.

63. *Retailing.* Shoreside Gifts is experiencing constant growth. They recorded a total of $250,000 in sales in 1992 and $285,000 in 1998. Use a graph that displays the store's total sales as a function of time to estimate total sales for the year 2001.

64. Use the graph constructed in Exercise 63 to estimate what the total sales were in 1995.

Skill Maintenance

Simplify by combining like terms. [1.3]

65. $3x - 5 - 9x + 15$

66. $x - (9x + 3)$

67. $2x + 4 - 6(5 - 7x)$

68. $9y - 3(2y + 7)$

69. $3 - 2[5(x - 7) + 1]$

70. $2x - \{3[4 - 2(1 - x)] + 7x\}$

Synthesis

Researchers at Yale University have suggested that the following graphs* may represent three different aspects of love.

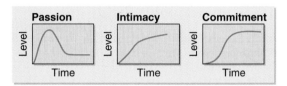

71. ◈ In what unit would you measure time if the horizontal length of each graph were ten units? Why?

72. ◈ Do you agree with the researchers that these graphs should be shaped as they are? Why or why not?

73. ◈ For the function given by $n(z) = ab + wz$, what is the independent variable? How can you tell?

74. ◈ Explain in your own words why every function is a relation, but not every relation is a function.

For Exercises 75 and 76, let $f(x) = 3x^2 - 1$ *and* $g(x) = 2x + 5$.

75. Find $f(g(-4))$ and $g(f(-4))$.

76. Find $f(g(-1))$ and $g(f(-1))$.

The function $V(r) = \frac{4}{3}\pi r^3$ *gives the volume of a sphere with radius r. Use a table of values or a graph to find the radius of a sphere with the given volume.*

77. 50 cm^3

78. 1.2 in^3

Pregnancy. For Exercises 79–82, use the graph of a woman's stress test on the following page. This graph shows the size of a pregnant woman's contractions as a function of time.

79. How large is the largest contraction that occurred during the test?

---

*From "A Triangular Theory of Love," by R. J. Sternberg, 1986, *Psychological Review*, **93**(2), 119–135. Copyright 1986 by the American Psychological Association, Inc. Reprinted by permission.

80. At what time during the test did the largest contraction occur?

81. ◈ On the basis of the information provided, how large a contraction would you expect 60 seconds after the end of the test? Why?

82. What is the frequency of the largest contraction?

83. The *greatest integer function* $f(x) = [\![x]\!]$ is defined as follows: $[\![x]\!]$ is the greatest integer that is less than or equal to x. For example, if $x = 3.74$, then $[\![x]\!] = 3$; and if $x = -0.98$, then $[\![x]\!] = -1$. Graph the greatest integer function for $-5 \le x \le 5$. (The notation $f(x) = \text{INT}[x]$ is used for many graphers.)

84. Suppose that a function g is such that $g(-1) = -7$ and $g(3) = 8$. Find a formula for g if $g(x)$ is of the form $g(x) = mx + b$, where m and b are constants.

85. *Energy Expenditure.* On the basis of the information given in the following table, what burns more energy: walking $4\frac{1}{2}$ mph for two hours or bicycling 14 mph for one hour?

Approximate Energy Expenditure by a 150-Pound Person in Various Activities

ACTIVITY	CALORIES PER HOUR
Walking, $2\frac{1}{2}$ mph	210
Bicycling, $5\frac{1}{2}$ mph	210
Walking, $3\frac{3}{4}$ mph	300
Bicycling, 13 mph	660

Source: Based on material prepared by Robert E. Johnson, M.D., Ph.D., and colleagues, University of Illinois

- - - - - - - - - - -

COLLABORATIVE CORNER

Focus: Functions

Time: 15–20 minutes

Group size: 3–4

The California Department of Motor Vehicles calculates automobile registration fees (VLF) according to the schedule shown on the next page.

Activity

1. Determine the original sale price of the oldest vehicle owned by a member of your group. Approximate and/or refer to the age of a family member's vehicle, if necessary. Be sure to note the year in which the car was purchased.

2. Use the schedule on the next page to calculate part (1) above for each year from the year of purchase to the present. To speed your work, each group member can find the fee for a few different years.

3. Graph the results from part (2). On the x-axis, plot years beginning with the year of purchase, and on the y-axis, plot $V(x)$, the VLF as a function of year.

4. What is the lowest VLF that the owner of this car will ever have to pay, according to this schedule? Compare your group's answer with other groups' answers.

5. Does your group feel that California's method for calculating registration fees is fair? Why or why not? How could it be improved?

DMV VEHICLE LICENSE FEE INFORMATION

A Public Service Agency

The 2% **Vehicle License Fee (VLF)** is in lieu of a personal property tax on vehicles. Most VLF revenue is returned to City and County Local Governments (see reverse side). The license fee charged is based upon the sale price or vehicle value when initially registered in California. The vehicle value is adjusted for any subsequent sale or transfer, that occurred 8/19/91 or later, excluding sales or transfers between specified relatives.

The VLF is calculated by rounding the sale price to the nearest **odd** hundred dollar. That amount is reduced by a percentage utilizing an eleven year schedule (shown to the right), and 2% of that amount is the fee charged. See the accompanying example for a vehicle purchased last year for $9,199. This would be the second registration year following that purchase.

WHERE DO YOUR DMV FEES GO? SEE REVERSE SIDE.

DMV77 8(REV.8/95)   95 30123

PERCENTAGE SCHEDULE
Rev. & Tax. Code Sec. 10753.2
(Trailer coaches have a different schedule)

1st Year	100%	7th Year	40%
2nd Year	90%	8th Year	30%
3rd Year	80%	9th Year	25%
4th Year	70%	10th year	20%
5th Year	60%	11th Year	
6th Year	50%	onward	15%

VLF CALCULATION EXAMPLE

Purchase Price:	$9,199
Rounded to:	$9,100
Times the Percentage:	90%
Equals Fee Basis of:	$8,190
Times 2% Equals:	$163.80
Rounded to:	$164

6. Try, as a group, to find an algebraic form for the function $y = V(x)$.

7. *Optional out-of-class extension*: Create a program for a grapher that accepts two inputs (initial value of the vehicle and year of purchase) and produces $V(x)$ as the output.

2.2

Solving Linear Equations

- *Solving Equations Graphically*
- *Equivalent Equations*
- *The Addition and Multiplication Principles*
- *Types of Equations*

To *solve* an equation means to find all the replacements for the variable that make the equation true. In this section, we solve *linear equations* using graphical and algebraic methods.

Solving Equations Graphically

To see how solutions of equations are related to graphs, consider the graphs of the functions given by $f(x) = 2x + 5$ and $g(x) = -3$.

At the point where the graphs intersect, $f(x) = g(x)$. Thus, for that particular x-value, we have $2x + 5 = -3$. In other words, the solution of $2x + 5 = -3$ is the x-coordinate of the point of intersection of the graphs of $f(x) = 2x + 5$ and $g(x) = -3$. Careful inspection suggests that -4 is that x-value. To check, note that $f(-4) = 2(-4) + 5 = -3$.

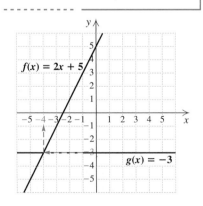

Example 1 Solve graphically: $\frac{1}{2}x + 3 = 2$.

SOLUTION To find the *x*-value for which $\frac{1}{2}x + 3$ will equal 2, we graph $f(x) = \frac{1}{2}x + 3$ and $g(x) = 2$ on the same set of axes. Since the intersection appears to be $(-2, 2)$, the solution is apparently -2.

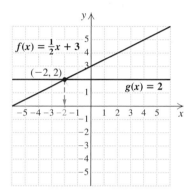

CHECK:

$$\frac{1}{2}x + 3 = 2$$
$$\overline{\frac{1}{2}(-2) + 3 \ ? \ 2}$$
$$-1 + 3 \ \Big| $$
$$2 \ \Big| \ 2 \quad \text{TRUE}$$

The solution is -2. ▬

Example 2 Solve graphically: $-\frac{3}{4}x + 6 = 2x - 1$.

SOLUTION We graph $f(x) = -\frac{3}{4}x + 6$ and $g(x) = 2x - 1$ on the same set of axes. It appears that the lines intersect at $(2.5, 4)$.

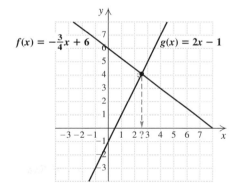

CHECK:

$$-\frac{3}{4}x + 6 = 2x - 1$$
$$\overline{-\frac{3}{4}(2.5) + 6 \ ? \ 2(2.5) - 1}$$
$$-1.875 + 6 \ \Big| \ 5 - 1$$
$$4.125 \ \Big| \ 4 \qquad \text{FALSE}$$

Our check shows that 2.5 is *not* the solution, although it may not be off by much. To find the exact solution, we need either a more precise way of determining the coordinates of the point of intersection or a different approach altogether. ▬

A grapher can be used to determine the point of intersection of graphs. Often it provides a more accurate solution than graphs drawn by hand. We can trace along the graph of one of the functions to find the co-ordinates of the point of intersection, enlarging the graph using a ZOOM feature if necessary. Many graphers can find the point of intersection directly using an INTERSECT feature, often found in the CALC menu.

Example 3   Solve using a grapher: $-\frac{3}{4}x + 6 = 2x - 1$.

SOLUTION   In Example 2, we saw that the graphs of $f(x) = -\frac{3}{4}x + 6$ and $g(x) = 2x - 1$ intersect near the point (2.5, 4). To determine more precisely the point of intersection, we graph $y_1 = -\frac{3}{4}x + 6$ and $y_2 = 2x - 1$ using the same viewing window. To use the INTERSECT feature, we indicate which two graphs, or *curves*, we are considering. Then we make a guess. The coordinates of the point of intersection appear at the bottom of the screen.

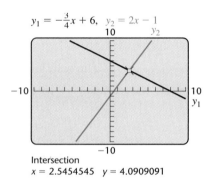

It appears from the screen above that the solution is 2.5454545. To check, we evaluate both sides of the equation $-\frac{3}{4}x + 6 = 2x - 1$ for this value of x. The first coordinate of the point of intersection is stored as X, so we evaluate Y1 and Y2, as shown at left.

Although the check shows that 2.5454545 is the solution, it is actually an approximation of the solution. To find the exact solution, we need an algebraic approach.   ▬

Y1	
	4.090909091
Y2	
	4.090909091

Equivalent Equations

Algebraic methods of solving equations use principles of algebra to write *equivalent equations* from which solutions can be more readily found.

Equivalent Equations

Two equations are said to be *equivalent* if they have the same solution(s).

Example 4   Determine whether $2x = 6$ and $10x = 30$ are equivalent equations.

SOLUTION   The equation $2x = 6$ is true only when x is 3. Similarly,

$10x = 30$ is true only when x is 3. Since both equations have the same solution, they are equivalent.

Example 5 Determine whether $x + 4 = 9$ and $x = 5$ are equivalent equations.

SOLUTION Each equation has only one solution, the number 5. Thus the equations are equivalent.

Example 6 Determine whether $3x = 4x$ and $3/x = 4/x$ are equivalent equations.

SOLUTION When x is replaced with 0, neither $3/x$ nor $4/x$ is defined, so 0 is *not* a solution of $3/x = 4/x$. Since 0 *is* a solution of $3x = 4x$, the equations are not equivalent.

The Addition and Multiplication Principles

Suppose that a and b represent the same number and that some number c is added to a. If c is also added to b, we will get two equal sums, since a and b are the same number. The same is true if we multiply both a and b by c. In this manner, we can produce equivalent equations.

The Addition and Multiplication Principles for Equations

For any real numbers a, b, and c:

a) $a = b$ is equivalent to $a + c = b + c$;

b) $a = b$ is equivalent to $a \cdot c = b \cdot c$, provided $c \neq 0$.

Example 7 Solve: $y - 4.7 = 13.9$.

SOLUTION

$$y - 4.7 = 13.9$$
$$y - 4.7 + 4.7 = 13.9 + 4.7 \qquad \text{Using the addition principle; adding 4.7}$$
$$y + 0 = 13.9 + 4.7 \qquad \text{The law of opposites}$$
$$y = 18.6$$

CHECK:
$$\begin{array}{c|c} y - 4.7 = 13.9 \\ \hline 18.6 - 4.7 \ ? \ 13.9 & \text{Substituting 18.6 for } y \\ 13.9 \ | \ 13.9 & \text{TRUE} \end{array}$$

The solution is 18.6.

In Example 7, why did we add 4.7 on both sides? Because we wanted the variable y alone on one side of the equation. Adding 4.7 gave us $y + 0$, or just y, on the left side. This led to the equivalent equation, $y = 18.6$, from which the solution, 18.6, is immediately apparent.

Example 8   Solve: $\frac{2}{5}x = \frac{9}{10}$.

SOLUTION   We have

$$\frac{2}{5}x = \frac{9}{10}$$

$$\frac{5}{2} \cdot \frac{2}{5}x = \frac{5}{2} \cdot \frac{9}{10} \qquad \text{Using the multiplication principle, we multiply by } \frac{5}{2}, \text{ the reciprocal of } \frac{2}{5}.$$

$$1x = \frac{45}{20} \qquad \text{The law of reciprocals}$$

$$x = \frac{9}{4} \qquad \text{Simplifying}$$

The check is left to the student. The solution is $\frac{9}{4}$.

In Example 8, why did we multiply by $\frac{5}{2}$? Because we wanted x alone on one side of the equation. When we multiplied by $\frac{5}{2}$, we got $1x$, or just x, on the left side. This led to the equivalent equation $x = \frac{9}{4}$, from which the solution, $\frac{9}{4}$, was obvious.

There is no need for a subtraction or division principle since subtraction can be regarded as adding opposites and division can be regarded as multiplying by reciprocals.

The addition and multiplication principles can be used together to solve equations. We combine all the terms containing the variable on one side of the equation and then use the multiplication principle.

Example 9   Solve: $-\frac{3}{4}x + 6 = 2x - 1$.

SOLUTION

$$-\frac{3}{4}x + 6 = 2x - 1$$

$$-\frac{3}{4}x + 6 - 6 = 2x - 1 - 6 \qquad \text{Using the addition principle; subtracting 6 on both sides}$$

$$-\frac{3}{4}x = 2x - 7 \qquad \text{Simplifying}$$

$$-\frac{3}{4}x - 2x = 2x - 7 - 2x \qquad \text{Using the addition principle; subtracting } 2x \text{ on both sides}$$

$$-\frac{3}{4}x - \frac{8}{4}x = -7 \qquad \text{Simplifying. All the terms containing } x \text{ are on the left side of the equation.}$$

$$-\frac{11}{4}x = -7 \qquad \text{Combining like terms}$$

$$-\frac{4}{11}\left(-\frac{11}{4}x\right) = -\frac{4}{11}(-7) \qquad \text{Using the multiplication principle; multiplying by } -\frac{4}{11} \text{ on both sides}$$

$$x = \frac{28}{11} \qquad \text{The law of reciprocals; simplifying}$$

CHECK:

$$\frac{-\frac{3}{4}x + 6 = 2x - 1}{-\frac{3}{4}\left(\frac{28}{11}\right) + 6 \;?\; 2\left(\frac{28}{11}\right) - 1}$$

$$-\frac{21}{11} + \frac{66}{11} \;\Big|\; \frac{56}{11} - \frac{11}{11}$$

$$\frac{45}{11} \;\Big|\; \frac{45}{11} \qquad \text{TRUE}$$

The solution is $\frac{28}{11}$.

Compare the solutions of Examples 2, 3, and 9. Only the algebraic approach gave the exact solution; the graphical solutions found were approximations. This does not mean that one method is "better" than another. Often in applications, an approximate solution is sufficient. Also, graphical methods provide a visualization of the problem that algebraic methods do not.

Example 10 Solve both algebraically and graphically:
$5x - 2(x - 5) = 7x - 2$.

ALGEBRAIC SOLUTION

We have

$5x - 2(x - 5) = 7x - 2$

$5x - 2x + 10 = 7x - 2$	Using the distributive law
$3x + 10 = 7x - 2$	Combining like terms
$3x + 10 - 3x = 7x - 2 - 3x$	Using the addition principle; adding $-3x$ or subtracting $3x$ on both sides
$10 = 4x - 2$	Combining like terms
$10 + 2 = 4x - 2 + 2$	Using the addition principle
$12 = 4x$	Simplifying
$\frac{1}{4} \cdot 12 = \frac{1}{4} \cdot 4x$	Using the multiplication principle; multiplying by $\frac{1}{4}$ or dividing by 4 on both sides
$3 = x.$	The law of reciprocals; simplifying

The solution is 3.

GRAPHICAL SOLUTION

We first graph the functions given by

$f(x) = 5x - 2(x - 5)$ and $g(x) = 7x - 2.$

We begin by using a standard viewing window.

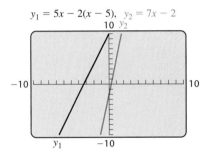

This window does not show the intersection of the graphs. It appears that the intersection is in the first quadrant and that the y-coordinate is greater than 10. Changing the viewing window to [0, 10, 0, 30] shows the intersection, and we can then determine the point of intersection of the graphs.

The x-coordinate of the point of intersection is 3, so the solution of the equation is 3.

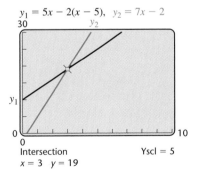

Note in Example 10 that the same solution was found using both methods, thus providing a check.

Parallel

Types of Equations

Any equation falls into one of three categories. An **identity** is an equation that is true for all replacements that can be used on both sides of the equation (for example, $x + 3 = 2 + x + 1$ or $3a - 5 = 3a - 5$). A **contradiction** is an equation, like $n + 5 = n + 7$, that is *never* true. A **conditional equation**, like $2x + 5 = 17$, is sometimes true and sometimes false, depending on what the replacement of x is. Most of the equations examined in this text are conditional.

Example 11 Solve each of the following equations and classify the equation as an identity, a contradiction, or a conditional equation.

a) $2x + 7 = 7(x + 1) - 5x$

b) $3x - 5 = 3(x - 2) + 4$

c) $3 - 8x = 5 - 7x$

a)

┌── ALGEBRAIC SOLUTION

We have

$2x + 7 = 7(x + 1) - 5x$

$2x + 7 = 7x + 7 - 5x$ Using the distributive law

$2x + 7 = 2x + 7.$ Combining like terms

The equation $2x + 7 = 2x + 7$ is true regardless of the replacement for x, so all real numbers are solutions.

The equation is an identity.

┌── GRAPHICAL SOLUTION

We graph $y_1 = 2x + 7$ and $y_2 = 7(x + 1) - 5x$. The graphs are the same line.

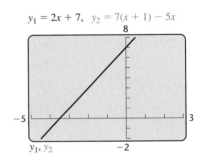

$y_1 = 2x + 7, \quad y_2 = 7(x + 1) - 5x$

We can also see this by looking at the table of values for Y1 and Y2.

X	Y1	Y2
0	7	7
1	9	9
2	11	11
3	13	13
4	15	15
5	17	17
6	19	19
X = 0		

Since $y_1 = y_2$ for any choice of x, all real numbers are solutions.

The equation is an identity.

b)

ALGEBRAIC SOLUTION

We have

$$3x - 5 = 3(x - 2) + 4$$
$$3x - 5 = 3x - 6 + 4 \qquad \text{Using the distributive law}$$
$$3x - 5 = 3x - 2$$
$$-3x + 3x - 5 = -3x + 3x - 2 \qquad \text{Using the addition principle}$$
$$-5 = -2.$$

Since our original equation is equivalent to $-5 = -2$, which is false for any choice of x, there is no solution to this problem. There is no choice of x that is a solution of the original equation.

The equation is a contradiction.

GRAPHICAL SOLUTION

We graph $y_1 = 3x - 5$ and $y_2 = 3(x - 2) + 4$.

It does not appear as though the lines intersect. If we attempt to find a point of intersection using INTERSECT, we get an error. There is no solution.

The equation is a contradiction.

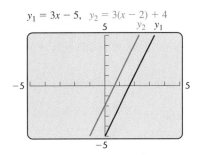

$y_1 = 3x - 5, \quad y_2 = 3(x - 2) + 4$

c)

ALGEBRAIC SOLUTION

We have

$$3 - 8x = 5 - 7x$$
$$3 - 8x + 7x = 5 - 7x + 7x \qquad \text{Using the addition principle}$$
$$3 - x = 5 \qquad \text{Simplifying}$$
$$-3 + 3 - x = -3 + 5 \qquad \text{Using the addition principle}$$
$$-x = 2 \qquad \text{Simplifying}$$
$$x = \frac{2}{-1}, \text{ or } -2. \qquad \begin{array}{l}\text{Dividing by } -1 \text{ or} \\ \text{multiplying by } \frac{1}{-1} \\ \text{on both sides}\end{array}$$

There is one solution, -2. For other choices of x, the equation is false.

This equation is conditional since it can be true or false, depending on the replacement for x.

GRAPHICAL SOLUTION

We graph $y_1 = 3 - 8x$ and $y_2 = 5 - 7x$ and determine the coordinates of any point of intersection.

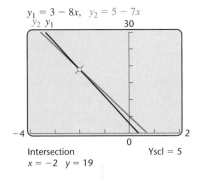

$y_1 = 3 - 8x, \quad y_2 = 5 - 7x$

Intersection
$x = -2 \quad y = 19$

The only solution is -2.

This equation is conditional.

We will sometimes refer to the set of solutions, or **solution set**, of a particular equation. Thus the solution set for Example 11(c) is $\{-2\}$. The solution set for Example 11(a) is simply $\mathbb{R}$, the set of all real numbers, and the solution set for Example 11(b) is the **empty set**, denoted $\varnothing$ or $\{\ \}$. As its name suggests, the empty set is the set containing no elements.

2.2 Exercise Set

Solve graphically.

1. $x - 3 = 4$ **2.** $x + 4 = 6$

3. $2x + 1 = 7$ **4.** $3x - 5 = 1$

5. $\frac{1}{3}x - 2 = 1$ **6.** $\frac{1}{2}x + 3 = -1$

7. $x + 3 = 5 - x$ **8.** $x - 7 = 3x - 3$

9. $5 - \frac{1}{2}x = x - 4$ **10.** $3 - x = \frac{1}{2}x - 3$

11. $2x - 1 = -x + 3$ **12.** $-3x + 4 = 3x - 4$

Determine whether the two equations in each pair are equivalent.

13. $3x = 15$ and $2x = 10$ Yes

14. $5x = 20$ and $15x = 60$

15. $x + 5 = 11$ and $3x = 18$

16. $x - 3 = 7$ and $3x = 24$

17. $13 - x = 4$ and $2x = 20$ No

18. $3x - 4 = 8$ and $3x = 12$

19. $5x = 2x$ and $\frac{4}{x} = 3$ No

20. $6 = 2x$ and $5 = \frac{2}{3 - x}$

Solve. Be sure to check.

21. $x - 5.2 = 9.4$ **22.** $y + 4.3 = 11.2$

23. $9y = 72$ **24.** $7x = 63$

25. $4x - 12 = 60$ **26.** $4x - 6 = 70$

27. $5y + 3 = 28$ **28.** $7t + 11 = 74$

29. $5x + 2x = 56$ **30.** $3x + 7x = 120$

31. $9y - 7y = 42$ **32.** $8t - 3t = 65$

33. $-6y - 10y = -32$ **34.** $-9y - 5y = 28$

35. $2(x + 6) = 8x$ **36.** $3(y + 5) = 8y$

37. $80 = 10(3t + 2)$ **38.** $27 = 9(5y - 2)$

39. $180(n - 2) = 900$ **40.** $210(x - 3) = 840$

41. $5y - (2y - 10) = 25$

42. $8x - (3x - 5) = 40$

43. $7y - 1 = 23 - 5y$

44. $15x + 20 = 8x - 22$

45. $\frac{1}{5} + \frac{3}{10}x = \frac{4}{5}$

46. $-\frac{5}{2}x + \frac{1}{2} = -18$

47. $0.9y - 0.7 = 4.2$

48. $0.8t - 0.3t = 6.5$

49. $4.23x - 17.898 = -1.65x - 42.454$

50. $-0.00458y + 1.7787 = 13.002y - 1.005$

51. $5r - 2 + 3r = 2r + 6 - 4r$

52. $5m - 17 - 2m = 6m - 1 - m$

53. $\frac{1}{8}(16y + 8) - 17 = -\frac{1}{4}(8y - 16)$

54. $\frac{1}{6}(12t + 48) - 20 = -\frac{1}{8}(24t - 144)$

55. $5 + 2(x - 3) = 2[5 - 4(x + 2)]$

56. $3[2 - 4(x - 1)] = 3 - 4(x + 2)$

Find each solution set. Then classify each equation as a conditional equation, an identity, or a contradiction.

57. $4x - 2x - 2 = 2x$

58. $2x + 4 + x = 4 + 3x$

59. $2 + 9x = 3(3x + 1) - 1$

60. $4 + 7x = 7(x + 1)$

61. $-8x + 5 = 5 - 10x$

62. $-8x + 5 = 5 - 8x$

63. $2\{9 - 3[-2x - 4]\} = 12x + 42$

64. $3\{7 - 2[7x - 4]\} = -40x + 45$

Skill Maintenance

65. Write the set consisting of the positive integers less than 10, using both roster notation and set-builder notation. [1.1]

66. Write the set consisting of the negative integers greater than -9, using both roster notation and set-builder notation. [1.1]

Translate to an algebraic expression. [1.6]

67. Three less than the product of two numbers

68. Ten more than the difference of two numbers

Synthesis

69. ◆ Explain the difference between equivalent expressions and equivalent equations.

70. ◆ Karl solves the equation $x + 5 = 2x + 8$ graphically and states that the solution is $(-3, 2)$. Is this correct? Why or why not?

71. ◆ When an equation is solved graphically, why is it important that the solution be checked algebraically?

72. ◈ As the first step in solving

$$2x + 5 = -3,$$

a student multiplies by $\frac{1}{2}$ on both sides. Is this incorrect? Why or why not?

Solve and check.

73. $x - \{3x - [2x - (5x - (7x - 1))]\} = x + 7$

74. $3x - \{5x - [7x - (4x - (3x + 1))]\} = 3x + 5$

75. $17 - 3\{5 + 2[x - 2]\} + 4\{x - 3(x + 7)\}$
$$= 9\{x + 3[2 + 3(4 - x)]\}$$

76. $23 - 2\{4 + 3[x - 1]\} + 5\{x - 2(x + 3)\}$
$$= 7\{x - 2[5 - (2x + 3)]\}$$

Solve graphically. Be sure to check.

77. $2x = |x + 1|$ **78.** $x - 1 = |4 - 2x|$

79. $\frac{1}{2}x = 3 - |x|$ **80.** $2 - |x| = 1 - 3x$

81. $x^2 = x + 2$ **82.** $x^2 = x$

2.3
Applications and Formulas

- *Problem Solving*
- *Formulas*

In Chapter 1, we discussed the first two steps of the five-step problem-solving process: *familiarizing* ourselves with the problem situation and *translating* the problem to mathematical language. We now solve some applied problems using all five steps of the process.

Five Steps for Problem Solving with Algebra

1. *Familiarize* yourself with the problem situation.
2. *Translate* to mathematical language.
3. *Carry out* some mathematical manipulation.
4. *Check* your possible answer in the original problem.
5. *State* the answer clearly.

Problem Solving

At this point, our study of algebra is just beginning. Thus we have few algebraic tools with which to work problems. As the number of tools in our algebraic "toolbox" increases, so will the difficulty of the problems being solved. For now our problems may seem simple; however, to gain practice with the problem-solving process, you should try to use all five steps. Later some steps may be shortened or combined.

Example 1 *Purchasing.* Elka pays $1187.20 for a computer. If the price paid includes a 6% sales tax, what is the price of the computer itself?

SOLUTION

1. **Familiarize.** Familiarize yourself with the problem. Note that tax is calculated from, and then added to, the computer's price. We let

$$C = \text{the computer's price.}$$

Let's guess that the computer's price is $1000. To check the guess, we calculate the amount of tax, (0.06)($1000) = $60, and add it to $1000:

$$(0.06)(\$1000) + \$1000 = \$60 + \$1000$$
$$= \$1060. \qquad \mathbf{\$1060 \neq \$1187.20}$$

Our guess was wrong, but it was useful. The manner in which we checked the guess will guide us in the next step.

2. **Translate.** Translate the problem to mathematical language. Our guess leads us to the following translation:

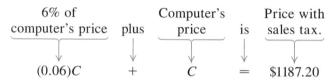

6% of computer's price	plus	Computer's price	is	Price with sales tax.
↓	↓	↓	↓	↓
$(0.06)C$	+	C	=	$1187.20

3. **Carry out.** We now carry out some mathematical manipulation:

$$0.06C + 1C = 1187.20$$
$$1.06C = 1187.20 \qquad \text{Combining like terms}$$
$$\frac{1}{1.06} \cdot 1.06C = \frac{1}{1.06} \cdot 1187.20 \qquad \text{Using the multiplication principle}$$
$$C = 1120.$$

4. **Check.** We check the answer in the original problem. To do this, we note that the tax on a computer costing $1120 would be (0.06)($1120) = $67.20. When this is added to $1120, we have

$$\$67.20 + \$1120, \text{ or } \$1187.20.$$

We see that $1120 checks in the original problem.
    We can also check by solving the equation graphically, as shown at left.

5. **State.** Finally, we state the answer clearly. The computer itself costs $1120. ▬

Example 2 *Marine Travel.* Tyla's fishing boat travels 12 km/h in still water. How long will it take her to cruise 25 km upstream if the current of the river is 3 km/h?

SOLUTION

1. **Familiarize.** First, we read the problem carefully. We are asked to find how long it will take Tyla to cruise 25 km upstream. We let

$$t = \text{the number of hours required for the boat to cruise 25 km upstream.}$$

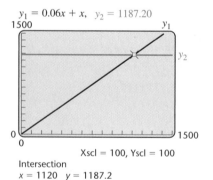

$y_1 = 0.06x + x, \quad y_2 = 1187.20$

Xscl = 100, Yscl = 100

Intersection
$x = 1120 \quad y = 1187.2$

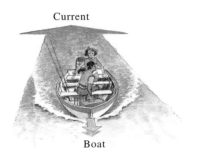

Current

Boat

In a problem such as this, involving motion, it is helpful to make a drawing. Note that the boat is traveling against the current of the river. At this point, we might consult outside references to learn that the speed of the current should be subtracted from the speed of the boat in still water to determine the speed of the boat going upstream. A physics book would tell us another important relationship involving motion:

Distance = Speed × Time.

We can now make a table listing all relevant information.

DISTANCE TO BE TRAVELED	25 km
SPEED OF BOAT IN STILL WATER	12 km/h
SPEED OF CURRENT	3 km/h
SPEED OF BOAT UPSTREAM	$12 - 3 = 9$ km/h
TIME REQUIRED	t

At this point we might try a guess. Suppose the boat traveled upstream for 2 hr. The boat would have then traveled

$$9 \quad \times \quad 2 \quad = \quad 18 \text{ km.}$$
$$\text{Speed} \times \text{Time} = \text{Distance}$$

Since $18 \neq 25$, our guess is wrong. Still, examining how we checked our guess gives added insight into how the problem might translate to an equation. We might note that a better guess, when multiplied by 9, would yield a number closer to 25.

2. **Translate.** If d is the number of kilometers traveled, r is the rate, or speed, in kilometers per hour, and t is the number of hours traveled, we know from the *Familiarize* step that

$$d = r \cdot t.$$

Since $d = 25$ km and $r = 9$ km/h, we have the equation

$$25 = 9 \cdot t.$$

3. **Carry out.** We solve the equation for t:

$$25 = 9 \cdot t$$
$$\tfrac{1}{9}(25) = \tfrac{1}{9}(9t) \qquad \textbf{Multiplying by } \tfrac{1}{9} \textbf{ on both sides}$$
$$\tfrac{25}{9} = t. \qquad \textbf{Simplifying}$$

4. **Check.** We check the answer in the original problem. If Tyla travels for $\frac{25}{9}$ hr at 9 km/h, she will have traveled $\frac{25}{9}(9) = 25$ km. The answer checks.

5. **State.** It will take Tyla $\frac{25}{9}$, or $2\frac{7}{9}$ hr to cruise 25 km upstream.

Sometimes problem solving may require the use of scientific notation.

Example 3 *Astronomy.* Alpha Centauri is the star—apart from the sun—closest to Earth. Its distance from Earth is about 2.4×10^{13} mi. How many light years is it from Earth to Alpha Centauri?

SOLUTION

1. **Familiarize.** From an astronomy text, we learn that light travels about 5.88×10^{12} mi in one year. Thus 1 light year $= 5.88 \times 10^{12}$ mi. We will let $y =$ the number of light years from Earth to Alpha Centauri. Let's guess that the answer is 3 light years. Then the distance in miles would be

$$(5.88 \times 10^{12}) \cdot 3 = 17.64 \times 10^{12} = 1.764 \times 10^{13}.$$

1 light year $= 5.88 \times 10^{12}$ mi

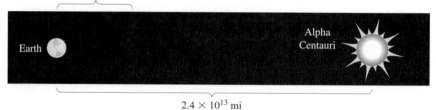

2.4×10^{13} mi

Although our guess is not correct, it does tell us that the distance is more than 3 light years. We are also better able to translate to an equation.

2. **Translate.** Note that the distance to Alpha Centauri is y light years, or $(5.88 \times 10^{12})y$ mi. We also are told that the distance is 2.4×10^{13} mi. Since the quantities $(5.88 \times 10^{12})y$ and 2.4×10^{13} both represent the number of miles to Alpha Centauri, we form the equation

$$(5.88 \times 10^{12})y = 2.4 \times 10^{13}.$$

3. **Carry out.** We solve the equation:

$$(5.88 \times 10^{12})y = 2.4 \times 10^{13}$$

$$\frac{1}{5.88 \times 10^{12}}(5.88 \times 10^{12})y = \frac{1}{5.88 \times 10^{12}} \times 2.4 \times 10^{13} \quad \begin{array}{l}\text{Multiplying by}\\ 1/(5.88 \times 10^{12})\\ \text{on both sides}\end{array}$$

$$y = \frac{2.4 \times 10^{13}}{5.88 \times 10^{12}} \quad \text{Simplifying}$$

$$= \frac{2.4}{5.88} \times \frac{10^{13}}{10^{12}} \quad \text{Factoring}$$

$$\approx 0.41 \times 10 = 4.1.$$

4. **Check.** Since light travels 5.88×10^{12} mi in one year, in 4.1 yr it will travel $4.1 \times 5.88 \times 10^{12} = 2.4108 \times 10^{13}$ mi, which is approximately the distance from Earth to Alpha Centauri.

5. **State.** The distance from Earth to Alpha Centauri is about 4.1 light years.

Formulas

A **formula** is an equation that uses letters to represent a relationship between two or more quantities. Some formulas you may recall from geometry are $A = \pi r^2$ (for the area A of a circle of radius r), $C = \pi d$ (for the circumference C of a circle of diameter d), and $A = l \cdot w$ (for the area A of a rectangle of length l and width w).* A more complete list of geometric formulas appears at the very end of this text.

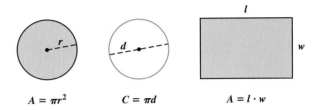

$A = \pi r^2$ $C = \pi d$ $A = l \cdot w$

Suppose we know the area and the width of a rectangular room and want to find the length. To do so, we could "solve" the formula $A = l \cdot w$ for l, using the same principles that we use for solving equations.

Example 4 *Area of a Rectangle.* Solve the formula $A = l \cdot w$ for l.

SOLUTION

$$A = l \cdot w \qquad \text{We want this letter alone.}$$

$$A \cdot \frac{1}{w} = l \cdot w \cdot \frac{1}{w} \qquad \text{Multiplying by } \frac{1}{w} \text{ on both sides}$$

$$\frac{A}{w} = l \qquad \text{Simplifying}$$

Thus to find the length of a rectangular room, we can divide the area of the room by its width. Were we to do this calculation for a variety of rectangular rooms, the formula $l = A/w$ would be more useful than repeatedly substituting into $A = l \cdot w$.

Example 5 *Simple Interest.* The formula $I = Prt$ is used to determine the simple interest, I, earned when P dollars is invested for t years at an interest rate r. Solve this formula for t.

SOLUTION

$$I = Prt \qquad \text{We want this letter alone.}$$

$$\frac{1}{Pr} \cdot I = \frac{1}{Pr} \cdot Prt \qquad \text{Multiplying by } \frac{1}{Pr} \text{ on both sides}$$

$$\frac{I}{Pr} = t \qquad \text{Simplifying}$$

---

*The Greek letter π, read "pi," is *approximately* 3.14159265358979323846264. Often 3.14 or 22/7 is used to approximate π when a calculator with a π key is unavailable.

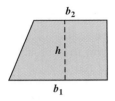

Example 6   *Area of a Trapezoid.* A trapezoid is a geometric shape with four sides, exactly two of which, the bases, are parallel to each other. The formula for calculating the area A of a trapezoid with bases b_1 and b_2 (read "b sub one" and "b sub two") and height h is given by

$$A = \frac{h}{2}(b_1 + b_2),$$

where the *subscripts* 1 and 2 distinguish one base from the other. Solve for b_1.

SOLUTION

$$A = \frac{h}{2}(b_1 + b_2)$$

$$\frac{2}{h} \cdot A = \frac{2}{h} \cdot \frac{h}{2}(b_1 + b_2) \qquad \textbf{Multiplying by } \frac{2}{h} \textbf{ on both sides}$$

$$\frac{2A}{h} = b_1 + b_2 \qquad \textbf{Simplifying. The right side is "cleared" of fractions.}$$

$$\frac{2A}{h} - b_2 = b_1 \qquad \textbf{Adding } -b_2 \textbf{ on both sides}$$

The similarities between solving formulas and solving equations can be seen below. In (a), we solve as we did in Section 2.2; in (b), we choose not to carry out all calculations; and in (c), we cannot carry out all calculations because the numbers are unknown.

a)
$$9 = \frac{3}{2}(x + 5)$$
$$\frac{2}{3} \cdot 9 = \frac{2}{3} \cdot \frac{3}{2}(x + 5)$$
$$6 = x + 5$$
$$1 = x$$

b)
$$9 = \frac{3}{2}(x + 5)$$
$$\frac{2}{3} \cdot 9 = \frac{2}{3} \cdot \frac{3}{2}(x + 5)$$
$$\frac{2 \cdot 9}{3} = x + 5$$
$$\frac{2 \cdot 9}{3} - 5 = x$$

c)
$$A = \frac{h}{2}(b_1 + b_2)$$
$$\frac{2}{h} \cdot A = \frac{2}{h} \cdot \frac{h}{2}(b_1 + b_2)$$
$$\frac{2A}{h} = b_1 + b_2$$
$$\frac{2A}{h} - b_2 = b_1$$

Example 7   *Simple Interest.* The formula $A = P + Prt$ gives the amount A that a principal of P dollars will be worth in t years when invested at simple interest rate r. Solve the formula for P.

SOLUTION

$$A = P + Prt \qquad \textbf{We want this letter alone.}$$

$$A = P(1 + rt) \qquad \textbf{Factoring (using the distributive law)}$$

$$A \cdot \frac{1}{1 + rt} = P(1 + rt) \cdot \frac{1}{1 + rt} \qquad \textbf{Multiplying by } \frac{1}{1 + rt} \textbf{ on both sides}$$

$$\frac{A}{1 + rt} = P \qquad \textbf{Simplifying}$$

This last equation can be used to determine how much should be invested at interest rate r in order to have A dollars t years later. ▬

Note in Example 7 that the factoring enabled us to write P once rather than twice. This is comparable to combining like terms when solving an equation like $16 = x + 7x$.

Most graphers graph only functions and thus require us to solve for the dependent variable. An equation written in another form must be solved for y using our formula-solving skills before it can be entered into a grapher.

Example 8 Graph: $3x - 4y = 2y + 7$.

SOLUTION   In order to graph $3x - 4y = 2y + 7$ using a grapher, we must first solve for y:

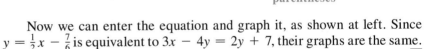

$$3x - 4y = 2y + 7 \qquad \text{We want this letter alone.}$$
$$3x - 4y - 2y = 7 \qquad \text{Adding } -2y \text{ on both sides}$$
$$3x - 6y = 7 \qquad \text{Combining like terms}$$
$$-6y = -3x + 7 \qquad \text{Adding } -3x \text{ on both sides}$$
$$\left(-\tfrac{1}{6}\right)(-6y) = \left(-\tfrac{1}{6}\right)(-3x + 7) \qquad \text{Multiplying by } -\tfrac{1}{6} \text{ on both sides}$$
$$y = \tfrac{1}{2}x - \tfrac{7}{6}. \qquad \text{Simplifying; using the distributive law to remove parentheses}$$

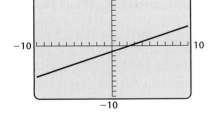

$y = \frac{1}{2}x - \frac{7}{6}$

Now we can enter the equation and graph it, as shown at left. Since $y = \frac{1}{2}x - \frac{7}{6}$ is equivalent to $3x - 4y = 2y + 7$, their graphs are the same. ▬

You may find the following summary useful.

To Solve a Formula for a Given Letter:

1. Multiply on both sides to clear fractions or decimals, if that is needed.

2. Combine like terms on each side where convenient.

3. Get all terms with the letter being solved for on one side of the equation and all other terms on the other side, using the addition principle.

4. Combine like terms again, if necessary. This may require factoring.

5. Solve for the letter in question, using the multiplication principle.

2.3

Exercise Set

Solve using all five problem-solving steps.

1. The sum of two numbers is 32. One of the numbers is 6 more than the other. What are the numbers?

2. The sum of two numbers is 91. One of the numbers is 5 more than the other. What are the numbers?

3. The product of two numbers is 72. If one number is 48, find the other number.

4. The number 596 is $\frac{2}{3}$ of what number?

5. A rectangle's length is twice its width and its perimeter is 15 ft. Find the dimensions of the rectangle.

6. A rectangle's width is $\frac{1}{4}$ its length and its perimeter is 50 m. Find the dimensions of the rectangle.

7. *Moving Sidewalks.* The moving sidewalk in O'Hare Airport is 300 ft long and moves at a rate of 5 ft/sec. If Alida walks at a rate of 4 ft/sec, how long will it take her to walk the length of the moving sidewalk?

8. *Swimming.* Fran swims at a rate of 5 km/h in still water. The Lazy River flows at a rate of 2.3 km/h. How long will it take Fran to swim 1.8 km upstream?

9. *Flight into a Headwind.* An airplane traveling 390 km/h in still air encounters a 65-km/h headwind. How long will it take the plane to travel 725 km into the wind?

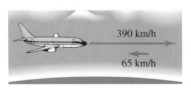

10. *Boating.* A paddleboat moves at a rate of 14 km/h in still water. How long will it take the boat to travel 56 km downstream if the river's current moves at a rate of 7 km/h?

11. *Angle Measures.* Two angles in a triangle have the same measure. The third angle is twice as great as the others. Find the measures of the angles.

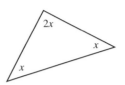

12. *Angle Measures.* One angle of a triangle is one-half as great as the second angle. The third angle measures 5° more than twice the second angle. Find the measures of the angles.

13. *Power Usage.* On a hot day, Twin City Electric set a record for power usage of 4998 megawatts of electricity. This was 2% higher than the previous record. What was the previous power usage record?

14. *Record Precipitation.* One rainy April, Greenfield received 8.16 in. of rain. This was 5% less than the record. What was the record amount of rain for April?

15. Kim and Kelly bought a $\frac{1}{2}$-lb slab of fudge. They split it so that Kim got $\frac{2}{3}$ as much fudge as Kelly. How much fudge did Kim get?

16. *Reunions.* There are about 350,000 annual family, class, and military reunions held in the United States. The number of class and military reunions is $\frac{3}{4}$ the number of family reunions. (*Source: Indianapolis Star,* June 11, 1995) How many annual family reunions are held in the United States?

17. *Strawberries.* A pint of strawberries weighs $\frac{3}{4}$ lb. Rema picked 16 lb of strawberries. How many pints did she pick?

18. *Tank Capacity.* On Monday, Karen filled the gas tank when it was $\frac{3}{4}$ empty. On Friday, she filled it when it was $\frac{1}{2}$ empty. Her total purchase of gasoline on both days was $18\frac{3}{4}$ gal. How many gallons of gasoline does the tank hold?

19. *Photography.* A photo shop charges $6.30 to develop a roll of film and $0.22 per exposure to print it. LaKenya is charged $14.66 to develop and print a roll of film. How many prints did she receive?

20. *Water Usage.* The Rosado family has budgeted $15.00 per month for home water usage. The

22. Price − 0.08p = 1007.40. #1095.⁰⁰
↓ ↓ ↓ ↓
original ducon final
price

monthly service charge is $7.20 and the volume charge is $1.04 for each 100 cubic feet of water. What is the most water the Rosados can use and stay within their budget?

21. *Interest.* Appliances Now allows customers to delay payment on a purchase for one year. It charges 12% simple interest on the purchase price. Tom bought a compact disc player and agreed to pay $504 a year later. What was the price of the compact disc player?

22. *Discount.* Elridge Furniture discounts furniture 8% to customers paying cash. Jinney paid $1007.40 cash for a roll-top desk. What was the original price of the desk?

Write the answers for Exercises 23–32 using scientific notation.

23. *Astronomy.* Venus has a nearly circular orbit of the sun. If the average distance from the sun to Venus is 1.08×10^8 km, how far does Venus travel in one orbit?

24. *Office Supplies.* A ream of copier paper weighs 2.25 kg. How much does a sheet of copier paper weigh?

25. *Printing and Engraving.* A ton of $5 bills is worth $4,540,000. How many pounds does a $5 bill weigh?

26. *Astronomy.* The average distance of Earth from the sun is about 9.3×10^7 mi. About how far does Earth travel in a yearly orbit about the sun? (Assume a circular orbit.)

27. *Astronomy.* The brightest star in the night sky, Sirius, is about 4.704×10^{13} mi from Earth. How many light years is it from Earth to Sirius?

28. *Astronomy.* The diameter of the Milky Way galaxy is approximately 5.88×10^{17} mi. How many light years is it from one end of the galaxy to the other?

29. *Biology.* An average of 4.55×10^{11} bacteria live in each pound of U.S. mud. (*Source: Harper's Magazine,* April 1996, p. 13) There are 60 drops in one teaspoon and 6 teaspoons in an ounce. How many bacteria live in a drop of U.S. mud?

30. *Astronomy.* If a star 5.9×10^{14} mi from Earth were to explode today, its light would not reach us for 100 yr. How far does light travel in 13 weeks?

31. *Astronomy.* The diameter of Jupiter is about 1.43×10^5 km. A day on Jupiter lasts about 10 hr. At what speed is Jupiter's equator spinning?

32. *Finance.* A *mil* is one thousandth of a dollar. The taxation rate in a certain school district is 5.0 mils for every dollar of assessed valuation. The assessed valuation for the district is 13.4 million dollars. How much tax revenue will be raised?

Projected Birth Weight. Ultrasonic images of 29-week-old fetuses can be used to predict weight. One model, developed by Thurnau,[*] *is $P = 9.337da - 299$; a second model, developed by Weiner,*[†] *is $P = 94.593c + 34.227a - 2134.616$. For both formulas, P is the estimated fetal weight, in grams; d the diameter of the fetal head, in centimeters; c the circumference of the fetal head, in centimeters; and a the circumference of the fetal abdomen, in centimeters.*

33. Use Thurnau's model to estimate the diameter of a fetus' head at 29 weeks when the estimated weight is 1614 g and the circumference of the fetal abdomen is 24.1 cm.

34. Use Weiner's model to estimate the circumference of a fetus' head at 29 weeks when the estimated weight is 1277 g and the circumference of the fetal abdomen is 23.4 cm.

35. *Gardening.* A garden is being constructed in the shape of a trapezoid. The dimensions are as shown

---

*Thurnau, G. R., R. K. Tamura, R. E. Sabbagha, et al. *Am. J. Obstet Gynecol* 1983; **145**:557.

†Weiner, C. P., R. E. Sabbagha, N. Vaisrub, et al. *Obstet Gynecol* 1985; **65**:812.

in the figure. The unknown dimension is to be such that the area of the garden is 90 ft². Find that unknown dimension.

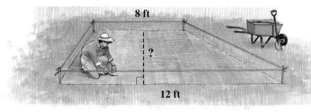

36. *Fencing.*   A rectangular garden is being constructed, and 76 ft of fencing is available. The width of the garden is to be 13 ft. What should the length be, in order to use just 76 ft of fence?

Waiting Time.   *In an effort to minimize waiting time for patients at a doctor's office without increasing a physician's idle time, Michael Goiten of Massachusetts General Hospital has developed a model. Goiten suggests that the interval time I, in minutes, between scheduled appointments be related to the total number of minutes T that a physican spends with patients in a day and the number of scheduled appointments N according to the formula I = 1.08(T/N).* (*Source: New England Journal of Medicine,* 30 August 1990: 604–608)

37. A doctor insists on an interval time of 20 min and must be able to schedule 25 appointments a day. According to Goiten's model, how many hours a day should the doctor be prepared to spend with patients?

38. A doctor determines that she has a total of 8 hr a day to see patients. If she insists on an interval time of 15 min, according to Goiten's model, how many appointments should she make in one day?

Solve.

39. $d = rt$, for t (a distance formula)

40. $d = rt$, for r

41. $F = ma$, for a (a physics formula)

42. $A = bh$, for b
(a formula for the area of a parallelogram)

43. $V = lwh$, for h (a volume formula)

44. $I = Prt$, for r (a formula for interest)

45. $L = \dfrac{k}{d^2}$, for k (a formula for intensity)

46. $L = \dfrac{k}{d^2}$, for d^2
(*Hint:* Multiply by d^2 to clear the fraction.)

47. $G = w + 150n$, for n
(a formula for the gross weight of a bus)

48. $P = b + 0.5t$, for t (a formula for parking prices)

49. $2w + 2h + l = p$, for l
(a formula used when shipping boxes)

50. $g = \dfrac{km_1m_2}{d^2}$, for d^2 (Newton's law of gravitation)

51. $Ax + By = C$, for y (a formula for graphing lines)

52. $P = 2l + 2w$, for l (a perimeter formula)

53. $C = \frac{5}{9}(F - 32)$, for F (a temperature formula)

54. $T = \frac{3}{10}(I - 12{,}000)$, for I (a tax formula)

55. $A = \dfrac{h}{2}(b_1 + b_2)$, for b_2 (an area formula)

56. $A = \dfrac{h}{2}(b_1 + b_2)$, for h (an area formula)

57. $F = \dfrac{mv^2}{r}$, for m (a physics formula)

58. $F = \dfrac{mv^2}{r}$, for v^2

59. $A = \dfrac{q_1 + q_2 + q_3}{n}$, for n (a formula for averaging)

60. $v = \dfrac{s_2 - s_1}{m}$, for m **61.** $v = \dfrac{d_2 - d_1}{t}$, for d_1

62. $v = \dfrac{s_2 - s_1}{m}$, for s_1 **63.** $r = m + mnp$, for m

64. $p = x - xyz$, for x **65.** $y = ab - ac^2$, for a

66. $d = mn - mp^3$, for m

Solve for y.

67. $3x + 6y = 9$ **68.** $4x - 7y = 6$

69. $x = y - 7$ **70.** $2 = 4x - y$

71. $y - 3(x + 2) = 4 + 2y$

72. $3x - 2(x + y) = y - x$

73. $4y + x^2 = x + 1$

74. $2x^2 - y + 3x = 0$

Skill Maintenance

Graph by hand. [1.5]

75. $y = -x$ **76.** $y = -x + 4$

Graph using a grapher. [1.5]

77. $y = |x + 1|$ **78.** $y = -2x^2$

Synthesis

79. ◆ Why is it important to check the solution from step (3) (*Carry out*) in the original wording of the problem being solved?

80. ◆ Write a problem for a classmate to solve. Devise the problem so that the solution is "The material should be cut into two pieces, one 30 cm long and the other 45 cm long."

81. ◆ A criminal claims to be carrying $5 million in $20 bills in a briefcase. Is this possible? Why or why not? (*Hint:* See Exercise 25.)

82. ◆ Predictions made using the models of Exercises 37 and 38 are often off by as much as 10%. Does this mean that the models should be discarded? Why or why not?

83. *Test Scores.* Tico's scores on four tests are 83, 91, 78, and 81. How many points above the average must Tico score on the next test in order to raise his average 2 points?

84. *Geometry.* The height and sides of a triangle are four consecutive integers. The height is the first integer, and the base is the fourth integer. The perimeter of the triangle is 42 in. Find the area of the triangle.

85. *Home Prices.* Panduski's real estate prices increased 6% from 1996 to 1997 and 2% from 1997 to 1998. From 1998 to 1999, prices dropped 1%. If a house sold for $117,743 in 1999, what was its worth in 1996? (Round to the nearest dollar.)

86. *Adjusted Wages.* Blanche's salary is reduced $n\%$ during a period of financial difficulty. By what number should her salary be multiplied in order to bring it back to where it was before the reduction?

Solve.

87. $s = v_i t + \frac{1}{2}at^2$, for a

88. $A = 4lw + w^2$, for l

89. $\frac{P_1 V_1}{T_1} = \frac{P_2 V_2}{T_2}$, for T_2

90. $\frac{P_1 V_1}{T_1} = \frac{P_2 V_2}{T_2}$, for T_1

91. $\frac{b}{a-b} = c$, for b

92. $m = \frac{(d/e)}{(e/f)}$, for d

Graph.

93. $2x^3 - y + 7 = 4y$

94. $y - 2(x^2 + y) = 0$

95. $2x^3 - 3y = 7(x - y)$

96. $|x + 2| - 3y = 4y$

COLLABORATIVE CORNER

Focus: Problem solving

Time: 15 minutes

Group size: 5

Suppose that two members in each group are celebrating their birthdays and the entire group goes out to lunch. Suppose further that each member whose birthday it is gets treated to his or her lunch by the other *four* members. Finally, suppose that all meals cost the same amount and that the bill is $40.00.*

Activity

1. Determine, as a group, how much each group member should pay for the lunch described above. Then explain how this determination was made.

2. Compare the results and methods used for part (1) with those of the other groups in the class.

*This activity was inspired by "The Birthday-Lunch Problem," *Mathematics Teaching in the Middle School* 2, no. 1, September–October 1996: 40–42.

2.4

Linear Functions: Graphs and Models

- *Slope–Intercept Form of an Equation*
- *Functions and Graphs as Models*

Different functions have different graphs. In this section, we examine functions with graphs that are straight lines. Such functions and their graphs are called *linear* and can be described by equations of the form $f(x) = mx + b$.

Slope–Intercept Form of an Equation

We can learn much about functions by examining their graphs. Also, we can tell whether a function is linear by examining the equation that describes the function.

Interactive Discovery

Graph each of the following functions using a standard viewing window:

$$f(x) = 2x,$$
$$g(x) = -2x,$$
$$h(x) = 0.3x.$$

Do the graphs appear to be linear? At what point do they cross the x-axis? At what point do they cross the y-axis?

The pattern you may have observed is true in general.

> **A function given by an equation of the form $f(x) = mx$ is a *linear function*. Its graph is a straight line passing through the origin.**

What happens to the graph of $f(x) = mx$ if we add a number b to get an equation of the form $f(x) = mx + b$?

Interactive Discovery

Graph the equation $y_1 = \frac{1}{2}x$ using a standard viewing window. On the same set of axes, graph $y_2 = \frac{1}{2}x + 5$ and $y_3 = \frac{1}{2}x - 7$. How do the latter two lines differ from the line $y_1 = \frac{1}{2}x$? By creating a table of values, explain how the values of y_2 and y_3 differ from y_1.

What do you think the graph of $y_4 = \frac{1}{2}x + 2$ will look like? Try drawing graphs for equations like $y_5 = \frac{1}{2}x + \frac{1}{4}$ and $y_6 = \frac{1}{2}x + (-3.2)$ and describe what happens to the graph of $y_1 = \frac{1}{2}x$ when a number b is added to $\frac{1}{2}x$.

The pattern you may have observed is also true for other values of m.

Example 1 Graph $y = 2x$ and $y = 2x + 3$, using the same set of axes.

SOLUTION   We first make a table of solutions of both equations.

x	y $y = 2x$	y $y = 2x + 3$
0	0	3
1	2	5
−1	−2	1
2	4	7
−2	−4	−1

We then plot these points. Drawing a blue line for $y = 2x + 3$ and a red line for $y = 2x$, we note that the graph of $y = 2x + 3$ is simply the graph of $y = 2x$ shifted, or *translated*, 3 units up. The lines are parallel.

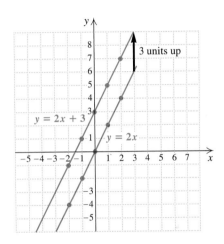

Note that the graph of $y = 2x + 3$ passes through the point $(0, 3)$. In general, we have the following.

> **The graph of $y = mx + b$, $b \neq 0$, is a line parallel to $y = mx$, passing through the point $(0, b)$.**

The point $(0, b)$ is called the **y-intercept**. Sometimes we refer to the number b as the y-intercept.

Example 2 For each equation, find the y-intercept.

a) $y = -5x + 4$ **b)** $f(x) = 5.3x - 12$

SOLUTION

a) The y-intercept is $(0, 4)$, or simply 4.

b) The y-intercept is $(0, -12)$, or simply -12.

In examining the graphs in Example 1, note that the slant of the red line seems to match the slant of the blue line. This leads us to suspect that it is the number m, in the equation $y = mx + b$, that is responsible for the slant of the line. The following definition enables us to visualize this slant, or *slope*, as a geometric ratio.

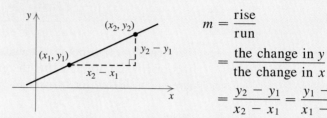

Slope

The *slope* of the line passing through (x_1, y_1) and (x_2, y_2) is given by

$$m = \frac{\text{rise}}{\text{run}}$$

$$= \frac{\text{the change in } y}{\text{the change in } x}$$

$$= \frac{y_2 - y_1}{x_2 - x_1} = \frac{y_1 - y_2}{x_1 - x_2}.$$

In the definition above, (x_1, y_1) and (x_2, y_2)—read "*x* sub-one, *y* sub-one and *x* sub-two, *y* sub-two"—represent two different points on a line. It does not matter which point is considered (x_1, y_1) and which is considered (x_2, y_2) so long as coordinates are subtracted in the same order in both the numerator and the denominator.

The letter *m* is traditionally used for slope. This usage has its roots in the French verb *monter*, to climb.

Example 3 Find the slope of the lines drawn in Example 1.

SOLUTION   To find the slope of a line, we can use the coordinates of any two points on that line. We use (1, 5) and (2, 7) to find the slope of the blue line in Example 1:

$$\text{Slope} = \frac{\text{rise}}{\text{run}} = \frac{\text{change in } y}{\text{change in } x} = \frac{y_2 - y_1}{x_2 - x_1} = \frac{7 - 5}{2 - 1} = 2.$$

To find the slope of the red line in Example 1, we use $(-2, -4)$ and $(1, 2)$:

$$\text{Slope} = \frac{\text{rise}}{\text{run}} = \frac{\text{change in } y}{\text{change in } x} = \frac{2 - (-4)}{1 - (-2)} = \frac{6}{3} = 2.$$

In Example 3, we found that the lines given by $y = 2x + 3$ and $y = 2x$ both have a slope of 2. This supports (but does not prove) the following:

The slope of any line written in the form $y = mx + b$ is m.

A proof of this result is outlined in Exercise 81 on p. 114.

Example 4 Determine the slope of the line given by $y = \frac{2}{3}x + 4$, and graph the line.

SOLUTION   Here $m = \frac{2}{3}$, so the slope is $\frac{2}{3}$. This means that from *any* point on the graph, we can locate a second point by simply going *up* 2 units and *to the right* 3 units. Where do we start? Since the y-intercept, $(0, 4)$, is known to be on the graph, we calculate that $(0 + 3, 4 + 2)$, or $(3, 6)$, is also on the graph. Knowing two points, we can draw the graph.

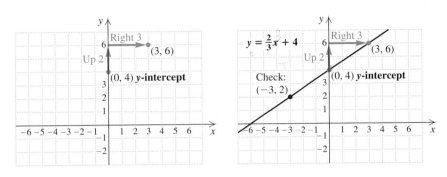

Important: To check the graph, we use some other value for x, say -3, and determine y (in this case, 2). We plot that point and see that it *is* on the line. Were it not, we would know that some error had been made.

Interactive Discovery

To see the effect of the sign of m, graph the equations $y_1 = x + 1$ and $y_2 = -x + 1$ on the same set of axes. Then graph $y_3 = 2x + 1$ and $y_4 = -2x + 1$ on those same axes. How does the sign of m affect the graph?

To see the effect of different positive values of m, graph $y_1 = x + 1$ using a standard viewing window. Then, on the same set of axes, graph the lines $y_2 = 2x + 1$ and $y_3 = 3x + 1$. What do you think the graph of $y = 4x + 1$ would look like? Try drawing graphs for equations like $y = 2.5x + 1$ and $y = \frac{3}{4}x + 1$ and describe how different positive values of m affect the graph.

When the slope of a line is positive, the line slants up from left to right. When the slope of a line is negative, the line slants down from left to right. The larger the absolute value of the slope, the steeper the slant.

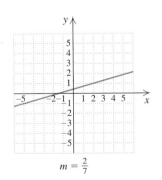

$m = \frac{2}{7}$

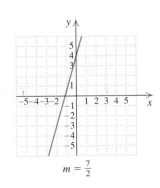

$m = \frac{7}{2}$

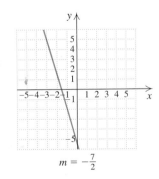

$m = -\frac{7}{2}$

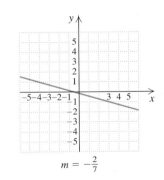

$m = -\frac{2}{7}$

y = mx + b. (0.5)

Example 5   Graph: $f(x) = -\frac{1}{2}x + 5$.

SOLUTION   The y-intercept is $(0, 5)$. The slope is $-\frac{1}{2}$, or $\frac{-1}{2}$. From the y-intercept, we go *down* 1 unit and *to the right* 2 units. That gives us the point $(2, 4)$. We can now draw the graph.

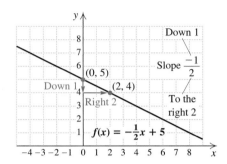

As a new type of check, we rename the slope and find another point:

$$-\frac{1}{2} = \frac{1}{-2}.$$

Thus we can go *up* 1 unit and then *to the left* 2 units. This gives the point $(-2, 6)$. Since $(-2, 6)$ is on the line, we have a check.

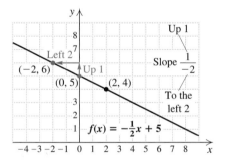

Because an equation of the form $y = mx + b$ makes use of the slope and the y-intercept, it is called the *slope–intercept* form of a linear equation.

The Slope–Intercept Equation

Any equation $y = mx + b$ has a graph that is a straight line. It goes through the *y-intercept* $(0, b)$ and has slope m. Any equation of the form $y = mx + b$ is said to be a *slope–intercept equation*.

Note that any graph of $y = mx + b$ will pass the vertical-line test and thus represents a function.

Example 6   Find an equation for a linear function whose graph has slope $-\frac{2}{3}$ and y-intercept $(0, 4)$.

SOLUTION We use the slope–intercept form, $f(x) = mx + b$:

$$f(x) = -\tfrac{2}{3}x + 4. \qquad \textbf{Substituting } -\tfrac{2}{3} \textbf{ for } m \textbf{ and 4 for } b$$ ▬

Example 7 Determine the slope and the y-intercept for the graph of $5x - 4y = 8$.

SOLUTION We convert to a slope–intercept equation:

$$5x - 4y = 8$$
$$-4y = -5x + 8 \qquad \textbf{Adding } -5x$$
$$y = -\tfrac{1}{4}(-5x + 8) \qquad \textbf{Multiplying by } -\tfrac{1}{4}$$
$$y = \tfrac{5}{4}x - 2. \qquad \textbf{Using the distributive law}$$

Because we have an equation of the form $y = mx + b$, we know that the slope is $\tfrac{5}{4}$ and the y-intercept is $(0, -2)$. ▬

Functions and Graphs as Models

We have seen that slope is the ratio of vertical change to horizontal change. Thus a slope of 3/2 can be taken to mean that the y-values increase 3 units for every 2-unit increase in x-values. Equivalently, we can say that y-values increase $\tfrac{3}{2}$ units per unit increase in x. In this way, slope is regarded as the *rate of change* of y with respect to x. This viewpoint is useful when formulating models.

Example 8 *Cost Projections.* Cleartone Communications charges $50 for a cellular phone and $40 per month for calls made under its economy plan. Formulate and graph a mathematical model for cost. Then use the model to determine the total cost for $3\tfrac{1}{2}$ months of service.

SOLUTION

1. **Familiarize.** The problem describes a situation in which a monthly fee is charged after an initial purchase has been made. After 1 month of service, the total cost will be $50 + $40 = $90. After 2 months (mos), the total cost will be $50 + $40 · 2 = $130. This can be generalized in a model if we let $C(t)$ represent the total cost, in dollars, for t months of service.

2. Translate. The total cost consists of the $50 phone purchase plus an additional $40 for each month of service. Thus,

$$C(t) = 50 + 40t,$$

where $t \geq 0$ (since there cannot be a negative number of months).

3. Carry out. Before graphing, we rewrite the model in slope–intercept form: $C(t) = 40t + 50$. We see that the vertical intercept is $(0, 50)$ and the slope—or rate—is $40 per month.

We plot $(0, 50)$ and, from there, count *up* $40 and *to the right* 1 month. This takes us to $(1, 90)$. We then draw a line passing through both points.

To find the total cost for $3\frac{1}{2}$ mos, we determine $C(3.5)$:

$$C(3.5) = 40 \cdot 3.5 + 50 \quad \text{Replacing } t \text{ with 3.5}$$
$$= 140 + 50$$
$$= 190.$$

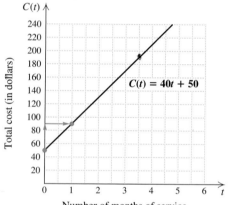

4. Check. We can check our model by seeing if it yields the values predicted in the *Familiarize* step. The cost after 1 mo of service should be $90, and the cost after 2 mos should be $130. We have

$$C(1) = 40(1) + 50 = 90 \quad \text{and}$$
$$C(2) = 40(2) + 50 = 130.$$

These function values check.

A good check of the graph is to compare it with a graph drawn using a grapher. We graph $y = 40x + 50$ using the viewing window $[0, 6, 0, 240]$, with Yscl = 20.

The graphs are the same, as you can see by comparing the y-intercepts and several other points on the lines.

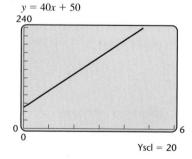

5. State. The model $C(t) = 40t + 50$ can be used to determine the total cost, in dollars, of t months of cellular phone service under Cleartone's economy plan. The total cost for $3\frac{1}{2}$ mos of service is $190.

Example 9 *Salvage Value.* Tyline Electric uses the function $S(t) = -700t + 3500$ to determine the *salvage value S(t)*, in dollars, of a photocopier t years after its purchase.

a) What do the numbers -700 and 3500 signify?

b) How long will it take the copier to *depreciate* completely?

c) What is the domain of S?

SOLUTION   Drawing, or at least visualizing, a graph can be useful here.

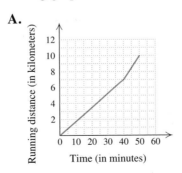

$y = -700x + 3500$

a) This function is written in slope–intercept form. Since the output is measured in dollars and the input in years, the vertical intercept $(0, 3500)$ tells us that at the time of purchase $(t = 0)$, the value of the machine is \$3500. The number -700 signifies that the value of the copier is declining at a rate of \$700 per year.

b) The copier will have depreciated completely when its value drops to 0. To learn when this occurs, we determine the value of t for which $S(t) = 0$:

$$S(t) = 0$$
$$-700t + 3500 = 0$$
$$-700t = -3500$$
$$t = 5.$$

The copier will have depreciated completely in 5 yr.

c) The number of years of service cannot be negative, nor can the salvage value be negative. In part (b), we found that after 5 yr the salvage value will have dropped to 0. Thus the domain of S is $\{t \mid 0 \leq t \leq 5\}$. The graph serves as a visual check of this result.

An equation need not be provided in order for us to determine a rate of change from a graph. In the next example, we examine the rate at which a runner's distance changes with respect to time—the *speed* at which the runner moves.

Example 10   Stephanie runs 10 km during each workout. For the first 7 km, her pace is twice as fast as it is for the last 3 km. Which of the following graphs best describes Stephanie's workout?

A.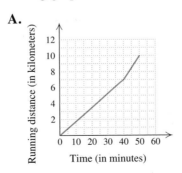

B.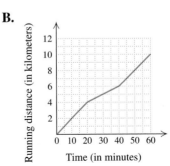

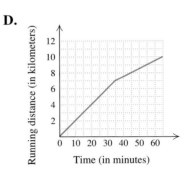

C.

D.

SOLUTION   The slope in graph A increases as we move to the right. This would indicate that Stephanie ran faster for the *last* part of her workout. Thus graph A is not the correct one.

The slopes in graph B indicate that Stephanie slowed down in the middle of her run and then resumed her original speed. Thus graph B does not correctly model the situation either.

According to graph C, Stephanie slowed down not at the 7-km mark, but at the 6-km mark. Thus graph C is also incorrect.

Graph D indicates that Stephanie ran the first 7 km in 35 min, a rate of 0.2 km/min. It also indicates that she ran the final 3 km in 30 min, a rate of 0.1 km/min. This means that Stephanie's rate was twice as fast for the first 7 km, so graph D provides a correct description of her workout.

2.4 Exercise Set

Determine the slope and the y-intercept.

1. $y = 4x + 5$

2. $y = 5x + 3$

3. $f(x) = -2x - 6$

4. $g(x) = -5x + 7$

5. $y = -\frac{3}{8}x - 0.2$

6. $y = \frac{15}{7}x + 2.2$

7. $g(x) = 0.5x - 9$

8. $f(x) = -3.1x + 5$

9. $y = 43x + 197$

10. $y = -52x + 700$

Find a linear function whose graph has the given slope and y-intercept.

11. Slope $\frac{2}{3}$, y-intercept $(0, -7)$

12. Slope $-\frac{3}{4}$, y-intercept $(0, 5)$

13. Slope -4, y-intercept $(0, 2)$

14. Slope 2, y-intercept $(0, -1)$

15. Slope $-\frac{7}{9}$, y-intercept $(0, 3)$

16. Slope $-\frac{4}{11}$, y-intercept $(0, 9)$

17. Slope 5, y-intercept $\left(0, \frac{1}{2}\right)$

18. Slope 6, y-intercept $\left(0, \frac{2}{3}\right)$

For each pair of points, find the slope of the line containing them.

19. $(6, 9)$ and $(4, 5)$

20. $(8, 7)$ and $(2, -1)$

21. $(3, 8)$ and $(9, -4)$

22. $(17, -12)$ and $(-9, -15)$

23. $(-16.3, 12.4)$ and $(-5.2, 8.7)$

24. $(14.4, -7.8)$ and $(-12.5, -17.6)$

For each graph, find the rate of change. Remember to use appropriate units. See Examples 3 and 8.

25.

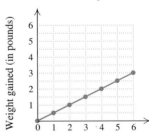

26.

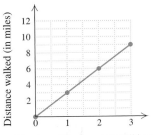

27.

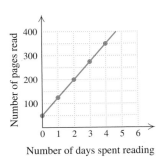

28.

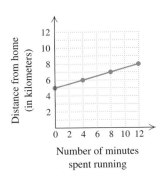

29.

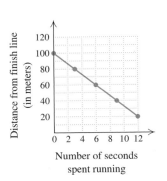

30.

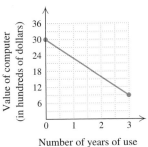

31. *Skiing Rate.* A cross-country skier reaches the 3-km mark of a race in 15 min and the 12-km mark 45 min later. Find the speed of the skier.

32. *Running Rate.* An Olympic marathoner passes the 5-km point of a race after 30 min and reaches the 25-km point 2 hr later. Find the speed of the marathoner.

33. *Rate of Sugar Production.* At the beginning of a production run, 4.5 tons of sugar had already been refined. Six hours later, the total amount of refined sugar reached 8.1 tons. Calculate the rate of production.

34. *Work Rate.* As a painter begins work, one-fourth of a house has already been painted. Eight hours later, the house is two-thirds done. Calculate the painter's work rate.

35. *Rate of Descent.* A plane descends to sea level from 12,000 ft after being airborne for $1\frac{1}{2}$ hr. The entire flight time is 2 hr and 10 min. Determine the plane's average rate of descent.

36. *Growth in Overseas Travel.*   In 1988, the number of U.S. visitors overseas was about 11.6 million. In 1995, the number grew to 18.7 million (*Source: Statistical Abstract of the United States*). Determine the rate at which the number of U.S. visitors overseas was growing.

37. Use the slope and the *y*-intercept of each line to match each equation with the correct graph.

a) $y = 3x - 5$
b) $y = 0.7x + 1$
c) $y = -0.25x - 3$
d) $y = -4x + 2$

I 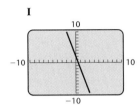   **II**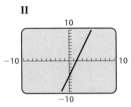

III 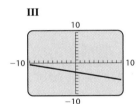   **IV**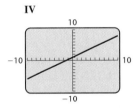

38. Use the slope and the *y*-intercept of each line to match each equation with the correct graph.

a) $y = \frac{1}{2}x - 5$
b) $y = 2x + 3$
c) $y = -3x + 1$
d) $y = -\frac{3}{4}x - 2$

I 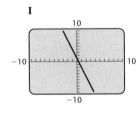   **II**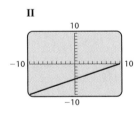

III 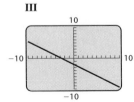   **IV**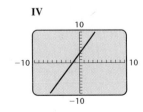

Determine the slope and the y-intercept. Then draw a graph by hand. Be sure to check as in Example 4 or 5.

39. $y = \frac{5}{2}x + 1$ **40.** $y = \frac{2}{5}x + 4$
41. $f(x) = -\frac{5}{2}x + 4$ **42.** $f(x) = -\frac{2}{5}x + 3$
43. $2x - y = 5$ **44.** $2x + y = 4$
45. $7y + 2x = 7$ **46.** $4y + 20 = x$
47. $f(x) = -0.25x + 2$ **48.** $f(x) = 1.5x - 3$
49. $4x - 5y = 10$ **50.** $5x + 4y = 4$
51. $f(x) = \frac{5}{4}x - 2$ **52.** $f(x) = \frac{4}{3}x + 2$
53. $12 - 4f(x) = 3x$ **54.** $15 + 5f(x) = -2x$

55. *Cricket Chirps per Minute.*   The number of cricket chirps per minute is given by $N(t) = 7.2t - 32$, where *t* is the temperature in degrees Celsius.

a) Graph the equation $N(t) = 7.2t - 32$.
b) Use the graph to predict the number of cricket chirps per minute when it is 5°C.

56. *Cost of a Telephone Call.*   The cost, in dollars, of a long-distance telephone call is given by $C(m) = 0.80m + 1$, where *m* is the length of the call in minutes.

a) Graph $C(m) = 0.80m + 1$.
b) Use the graph to approximate the cost of a 4-min call.
c) Use the graph to determine how long a phone call can be made for $5.00.

57. *Cable Television Charges.*   Clear County Cable Television charges a $25 installation fee and $20 per month for basic service. Formulate and graph a model that can be used to determine the total cost, $C(t)$, of *t* months of basic cable TV service. Find the total cost of 6 mos of service.

58. *Deluxe Cable Service.*   Twin Cities Cable Television charges a $35 installation fee and $30 per month for deluxe service. Formulate and graph a model that can be used to determine the total cost, $C(t)$, for *t* months of deluxe cable TV service. Find the total cost for 8 mos of service.

59. *Cellular Phone Charges.*   The Cellular Connection charges $60 for a cellular phone and $40 per month under its economy plan. Formulate and graph a model that can be used to determine the total cost, $C(t)$, of operating a Cellular Connection phone for *t* months. Find the total cost for $5\frac{1}{2}$ mos of service and the domain of *C*.

60. *Telephone Charges.*   Tubular Calling charges $50 for a telephone and $25 per month under its economy plan. Formulate and graph a model that can be used to determine the total cost, $C(t)$, of operating a Tubular Calling phone for t months. Find the total cost for $4\frac{1}{2}$ mos of service and the domain of C.

61. *Value of a Fax Machine.* FaxMax bought a multifunction fax machine for $750. The machine depreciates at a rate of $25 per month. Find and graph a function F that can be used to determine the value of the fax machine t months after purchase. Then determine the domain of F.

62. *Value of a Computer.* SendUp Graphics bought a computer for $3800. The computer depreciates at a rate of $50 per month. Find and graph a function C that can be used to determine the value of the computer t months after purchase. Then find the domain of C.

In Exercises 63–68, each model is of the form $f(x) = mx + b$. In each case, determine what m and b signify.

63. *Cost of a Taxi Ride.* The cost, in dollars, of a taxi ride in Pelham is given by $C(d) = 0.75d + 2$, where d is the number of miles traveled.

64. *Cost of a Movie Ticket.* The average price $P(t)$, in dollars, of a movie ticket can be estimated by the function

$$P(t) = 0.1522t + 4.29,$$

where t is the number of years since 1990.

65. *Sales of Cotton Goods.* The function given by $f(t) = 2.6t + 17.8$ can be used to estimate the yearly sales of cotton goods, in billions of dollars, t years after 1975.

66. *Cost of Renting a Truck.* The cost, in dollars, of a one-day truck rental is given by $C(d) = 0.3d + 20$, where d is the number of miles driven.

67. *Life Expectancy of American Women.* The life expectancy of American women t years after 1950 is given by $A(t) = \frac{3}{20}t + 72$.

68. *Natural Gas Demand.* The demand, in quadrillions of joules, for natural gas is approximated by $D(t) = \frac{1}{5}t + 20$, where t is the number of years after 1960.

69. Match each sentence with the most appropriate graph.

a) The rate at which fluids were given intravenously was doubled after 3 hr.

b) The rate at which fluids were given intravenously was gradually reduced to 0.

c) The rate at which fluids were given intravenously remained constant for 5 hr.

d) The rate at which fluids were given intravenously was gradually increased.

I
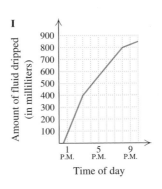

II
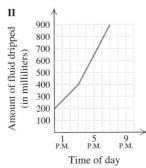

III
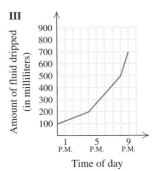

IV
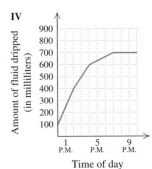

70. Match each sentence with the most appropriate graph.

a) After January 1, daily sales continued to rise, but at a slower rate.

b) After January 1, sales decreased faster than they ever grew.

c) The rate of growth in daily sales doubled after January 1.

d) After January 1, daily sales decreased at half the rate that they grew in December.

I

II
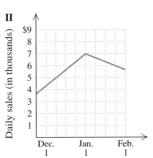

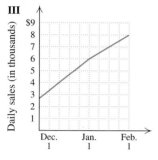

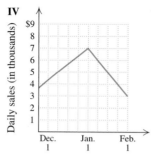

Skill Maintenance

Simplify.

71. $9\{2x - 3[5x + 2(-3x + y^0 - 2)]\}$ [1.3], [1.4]

72. $(-5a^2b^3)^3$ [1.4]

73. $(13m^2n^3)(-2m^5n)$ [1.4]

74. $2x - 3[y - 2(x + 3) + 10]$ [1.3]

Synthesis

75. ◈ A student makes a mistake when using a grapher to draw $4x + 5y = 12$ and the following screen appears.

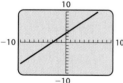

Use algebra to show that a mistake has been made. What do you think the mistake was?

76. ◈ A student makes a mistake when using a grapher to draw $5x - 2y = 3$ and the following screen appears.

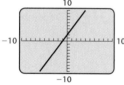

Use algebra to show that a mistake has been made. What do you think the mistake was?

77. ◈ Rosie Picshure claims that her firm's profits continue to go up, but the rate of increase is going down.

a) Sketch a graph that might represent her firm's profits as a function of time.

b) Explain why the graph can go up while the rate of increase goes down.

78. ◈ Is it true that the rate of change for a linear function is constant? Why or why not?

In Exercises 79 and 80, assume that r, p, and s are constants and that x and y are variables. Determine the slope and the y-intercept.

79. $rx + py = s$ **80.** $rx + py = s - ry$

81. Let (x_1, y_1) and (x_2, y_2) be two distinct points on the graph of $y = mx + b$. Use the fact that both pairs are solutions of the equation to prove that m is the slope of the line given by $y = mx + b$. (*Hint:* Use the slope formula.)

Given that $f(x) = mx + b$, classify each of the following as true or false.

82. $f(c + d) = f(c) + f(d)$

83. $f(cd) = f(c)f(d)$

84. $f(kx) = kf(x)$

85. $f(c - d) = f(c) - f(d)$

86. Find k such that the line containing $(-3, k)$ and $(4, 8)$ is parallel to the line containing $(5, 3)$ and $(1, -6)$.

87. *Cost of a Speeding Ticket.*   The following penalty schedule is used to determine the cost of a speeding ticket in certain states.

Use this schedule to graph the cost of a speeding ticket as a function of the number of miles per hour over the limit that a driver is going.

88. Find the slope of the line that contains the pair of points.

a) $(5b, -6c)$, $(b, -c)$

b) (b, d), $(b, d + e)$

c) $(c + f, a + d)$, $(c - f, -a - d)$

89. Match each sentence with the most appropriate graph.

a) Annie drove 2 mi to a lake, swam 1 mi, and then drove 3 mi to a store.

b) During a preseason workout, Rico biked 2 mi, ran for 1 mi, and then walked 3 mi.

c) James bicycled 2 mi to a park, hiked 1 mi over the notch, and then took a 3-mi bus ride back to the park.

d) After hiking 2 mi, Marcy ran for 1 mi before catching a bus for the 3-mi ride into town.

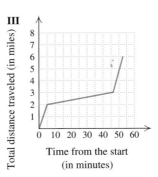

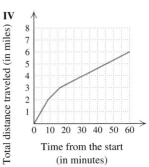

90. Graph the equations

$$y_1 = 1.4x + 2, \qquad y_2 = 0.6x + 2,$$
$$y_3 = 1.4x + 5, \quad \text{and} \quad y_4 = 0.6x + 5$$

using a grapher. If possible, use the SIMULTANEOUS mode so that you cannot tell which equation is being graphed first. Then decide which line corresponds to each equation.

State whether each of the following is a linear function.

91. $f(x) = -\frac{1}{2}x + (1.6)^2$ **92.** $f(x) = 3 - 4x$

93. $f(x) = 2x^2 + 4$ **94.** $f(x) = \frac{2}{x} + 1$

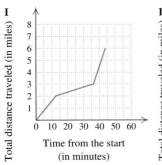

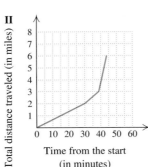

2.5
Another Look at Linear Graphs

- *Zero Slope and Lines with Undefined Slope*
- *Graphing Using Intercepts*
- *Parallel and Perpendicular Lines*
- *Recognizing Linear Equations*

In Section 2.4, we graphed linear equations using slopes and y-intercepts. We now graph lines that have slope 0 or that have an undefined slope. We also graph lines by using both the x- and y-intercepts.

Zero Slope and Lines with Undefined Slope

If two different points have the same second coordinate, what is the slope of the line joining them? In this case, we have $y_2 = y_1$, so

$$m = \frac{y_2 - y_1}{x_2 - x_1} = \frac{0}{x_2 - x_1} = 0.$$

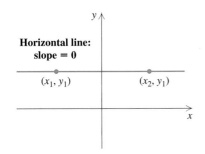

Slope of a Horizontal Line
Every horizontal line has a slope of 0.

Example 1 Graph: $f(x) = 3$.

SOLUTION Recall from Section 2.1 that a function of this type is called a *constant function.* Writing slope–intercept form,

$$f(x) = 0 \cdot x + 3,$$

we see that the y-intercept is $(0, 3)$ and the slope is 0. Thus we can graph f by plotting the point $(0, 3)$ and, from there, determining a slope of 0. Because $0 = 0/2$ (any nonzero number could be used in place of 2), we can draw the graph by going up 0 units and to the right 2 units. As a check, we also find some ordered pairs. Note that for any choice of x-value, $f(x)$ must be 3.

x	$f(x)$
-1	3
0	3
2	3

$y = 0x + 3$

We see from Example 1 that the graph of any constant function of the form $f(x) = b$ or $y = b$ is a horizontal line that crosses the y-axis at $(0, b)$.

Suppose that two different points are on a vertical line. They then have the same first coordinate. In this case, we have $x_2 = x_1$, so

$$m = \frac{y_2 - y_1}{x_2 - x_1} = \frac{y_2 - y_1}{0}.$$

Since we cannot divide by 0, this is undefined.

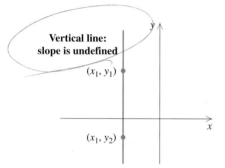

Vertical line: slope is undefined

(x_1, y_1)

(x_1, y_2)

Slope of a Vertical Line

The slope of a vertical line is undefined.

Example 2 Graph: $x = -2$.

SOLUTION With y missing, no matter which value of y is chosen, x must be -2. Thus the pairs $(-2, 3)$, $(-2, 0)$, and $(-2, -4)$ all satisfy the

$f(x) = 0x - 2$

$y = 0x$

equation. The graph is a line parallel to the y-axis. Note that since y is missing, this equation cannot be written in slope–intercept form.

x	y
-2	3
-2	0
-2	-4

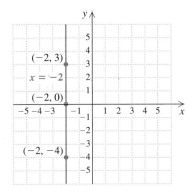

Example 2 shows us that the graph of any equation of the form $x = a$ is a vertical line that crosses the x-axis at $(a, 0)$.

Example 3   Find the slope of the graph of each equation. If the slope is undefined, state this.

a) $3y + 2 = 8$

b) $2x = 10$

SOLUTION

a) We solve for y:

$$3y + 2 = 8$$
$$3y = 6 \qquad \text{Subtracting 2 on both sides}$$
$$y = 2. \qquad \text{The graph of } y = 2 \text{ is a horizontal line.}$$

Since $3y + 2 = 8$ is equivalent to $y = 2$, the slope of the line $3y + 2 = 8$ is 0.

b) When y does not appear, we solve for x:

$$2x = 10$$
$$x = 5. \qquad \text{The graph of } x = 5 \text{ is a vertical line.}$$

Since $2x = 10$ is equivalent to $x = 5$, the slope of the line $2x = 10$ is undefined.

Graphing Using Intercepts

Any line that is not horizontal or vertical will cross both the x- and y-axes. The points at which the axes are crossed are called the *x-intercept* and the *y-intercept*. Any time these intercepts are not $(0, 0)$, they can be used to graph an equation. Note that to find the y-intercept, we replace x in the equation of the line with 0 and solve for y. To find the x-intercept, we replace y with 0 and solve for x.

> **To Determine Intercepts:**
>
> The x-intercept is $(a, 0)$. To find a, let $y = 0$ and solve the original equation for x.
>
> The y-intercept is $(0, b)$. To find b, let $x = 0$ and solve the original equation for y.

Example 4   Graph the equation $3x + 2y = 12$ by using intercepts.

SOLUTION   *To find the y-intercept, we let $x = 0$ and solve for y:*

$$3 \cdot 0 + 2y = 12$$
$$2y = 12$$
$$y = 6.$$

The y-intercept is $(0, 6)$.

To find the x-intercept, we let $y = 0$ and solve for x:

$$3x + 2 \cdot 0 = 12$$
$$3x = 12$$
$$x = 4.$$

The x-intercept is $(4, 0)$.

   We plot the two intercepts and draw the line. A third point could be calculated and used as a check.

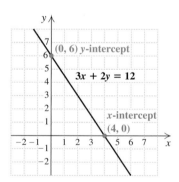

Example 5   Graph $f(x) = 2x + 5$ by using intercepts.

SOLUTION   Because the function is in slope–intercept form, we know that the y-intercept is $(0, 5)$. To find the x-intercept, we replace $f(x)$ with 0 and solve for x:

$$0 = 2x + 5$$
$$-5 = 2x$$
$$-\tfrac{5}{2} = x.$$

The x-intercept is $\left(-\tfrac{5}{2}, 0\right)$.

   We plot both intercepts and draw the line. As a check, note that the

slope of the line is 2, as expected. We might also compute and plot a third point as another check.

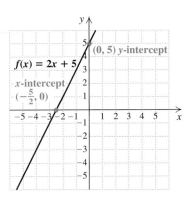

The intercepts of the graph of a function can help us determine an appropriate viewing window on a grapher.

Example 6   Determine a viewing window that shows the intercepts of the graph of the function $f(x) = 3x + 15$.

SOLUTION   In this case the function is in slope–intercept form, so we know that the y-intercept is $(0, 15)$. To find the x-intercept, we replace $f(x)$ with 0 and solve for x:

$$0 = 3x + 15$$
$$-15 = 3x \qquad \textbf{Subtracting 15 on both sides}$$
$$-5 = x.$$

The x-intercept is $(-5, 0)$.

A standard viewing window will not show the y-intercept, $(0, 15)$. Thus we adjust the Ymax value and choose a viewing window of $[-10, 10, -10, 20]$, with Yscl = 5. Other choices of window dimensions are also possible.

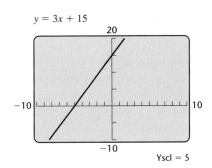

Parallel and Perpendicular Lines

If two lines are vertical, they are parallel. How can we tell whether non-vertical lines are parallel?

Interactive Discovery

Graph $y = 3x - 7$ and $y - 3x = 8$ on a grapher using a standard viewing window and determine whether the lines appear to be parallel.

Again using a standard viewing window, graph $7x + 10y = 50$ and $4x + 5y = -25$. Do these lines appear to be parallel?

Now graph each pair of equations given above using the viewing window $[-100, 100, -100, 100]$, with Xscl = 10 and Yscl = 10. Do the lines still appear to be parallel?

Graphs of nonparallel lines may appear parallel in certain viewing windows. In order to determine whether two lines are parallel, we can look at their slopes.

Slope and Parallel Lines

Two nonvertical lines are parallel if they have the same slope.

Example 7 Determine whether each of the given pairs of lines is parallel.

a) $y = 3x - 7$ and $y - 3x = 8$

b) $7x + 10y = 50$ and $4x + 5y = -25$

SOLUTION

a) The slope of the line $y = 3x - 7$ is 3. To find the slope of the line $y - 3x = 8$, we first solve for y:

$$y - 3x = 8$$
$$y = 3x + 8. \qquad \text{Adding } 3x \text{ on both sides}$$

The slope of the line $y - 3x = 8$ is also 3, so the lines are parallel.

b) We find the slope of each line:

$$7x + 10y = 50$$
$$10y = -7x + 50 \qquad \text{Adding } -7x \text{ on both sides}$$
$$y = -\tfrac{7}{10}x + 5; \qquad \text{Multiplying on both sides by } \tfrac{1}{10}$$
$$\text{The slope is } -\tfrac{7}{10}.$$

$$4x + 5y = -25$$
$$5y = -4x - 25 \qquad \text{Adding } -4x \text{ on both sides}$$
$$y = -\tfrac{4}{5}x - 5. \qquad \text{Multiplying on both sides by } \tfrac{1}{5}$$
$$\text{The slope is } -\tfrac{4}{5}.$$

Since the slopes are not the same, the lines are not parallel.   ▬

If one line is vertical and another is horizontal, they are perpendicular. There are other instances in which two lines are perpendicular.

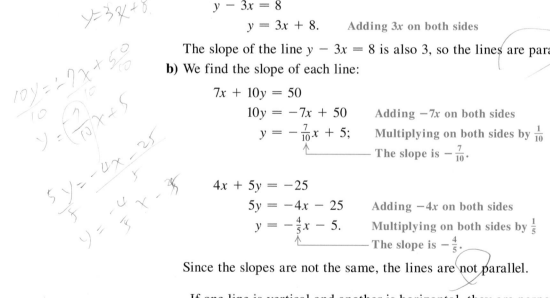

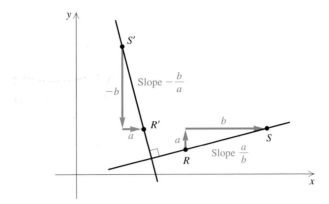

Consider a line $\overleftrightarrow{RS}$, as shown in the graph, with slope a/b. Then think of rotating the figure 90° to get a line $\overleftrightarrow{R'S'}$ perpendicular to $\overleftrightarrow{RS}$. For the new line, the rise and the run are interchanged, but the rise is now negative. Thus the slope of the new line is $-b/a$. Let's multiply the slopes:

$$\frac{a}{b}\left(-\frac{b}{a}\right) = -1.$$

This can help us determine which lines are perpendicular.

Slope and Perpendicular Lines

Two lines are perpendicular if the product of their slopes is -1. (If one line has slope m, the slope of a line perpendicular to it is $-1/m$. That is, we take the reciprocal and change the sign.) Lines are also perpendicular if one is vertical and the other is horizontal.

Example 8 Determine whether the lines given by the equations $3x - y = 7$ and $x + 3y = 1$ are perpendicular.

SOLUTION   To determine the slope of each line, we solve for y to find slope–intercept form:

$$3x - y = 7$$
$$-y = -3x + 7 \qquad \text{Adding } -3x \text{ on both sides}$$
$$y = 3x - 7; \qquad \text{Multiplying by } -1 \text{ on both sides}$$

$$x + 3y = 1$$
$$3y = -x + 1 \qquad \text{Adding } -x \text{ on both sides}$$
$$y = -\tfrac{1}{3}x + \tfrac{1}{3}. \qquad \text{Multiplying by } \tfrac{1}{3} \text{ on both sides}$$

The slopes of the lines are 3 and $-\tfrac{1}{3}$. Since $3 \cdot \left(-\tfrac{1}{3}\right) = -1$, the lines are perpendicular.

Example 9   Find a linear function g with a graph parallel to the graph of $f(x) = \frac{2}{3}x - 8$ and a y-intercept of $\left(0, \frac{3}{4}\right)$.

SOLUTION   Since g is linear, it can be written in the form $g(x) = mx + b$. To find g, we must determine its slope and y-intercept. The slope of the line given by $f(x) = \frac{2}{3}x - 8$ is $\frac{2}{3}$. Therefore, the slope of a parallel line is $\frac{2}{3}$. Since we are given that the y-intercept of the graph of g is $\left(0, \frac{3}{4}\right)$, we have $g(x) = \frac{2}{3}x + \frac{3}{4}$. ▬

Recognizing Linear Equations

Consider an equation of the form $Ax + By = C$, where A, B, and C are real numbers. Suppose that A and B are nonzero and solve for y:

$$Ax + By = C$$
$$By = -Ax + C \qquad \text{Adding } -Ax \text{ on both sides}$$
$$y = -\frac{A}{B}x + \frac{C}{B}. \qquad \text{Dividing by } B \text{ on both sides}$$

Since the last equation is a slope–intercept equation, we see that $Ax + By = C$ is a linear equation when $A \neq 0$ and $B \neq 0$.

Suppose next that A or B (but not both) is 0. If A is 0, then $By = C$ and $y = C/B$. If B is 0, then $Ax = C$ and $x = C/A$. In the first case, the graph is horizontal; in the second case, the line is vertical. Thus, $Ax + By = C$ is a linear equation when A or B (but not both) is 0. We have now justified the following result.

The Standard Form of a Linear Equation

Let A, B, and C be real numbers, with A and B not both 0.

1. Any equation of the form $Ax + By = C$ is linear.

2. Any equation of the form $Ax + By = C$ is said to be a linear equation in *standard form*.

$y = x^2 - 5$

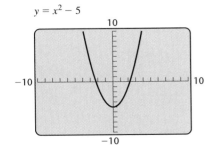

Example 10   Determine whether the equation $y = x^2 - 5$ is linear.

SOLUTION   We attempt to put the equation in standard form:

$$y = x^2 - 5$$
$$-x^2 + y = -5. \qquad \text{Adding } -x^2 \text{ on both sides}$$

This last equation is not linear because it has an x^2-term.
We can see this as well from the graph of the equation. ▬

Linear equations have graphs that are straight lines, and linear graphs have a constant slope. Were you to try to calculate the slope between several pairs of points in Example 10, you would find that the slopes vary.

2.5 | Exercise Set

For each equation, find the slope. If the slope is undefined, state this.

1. $5x - 6 = 15$ **2.** $5y - 12 = 3x$

3. $3x = 12 + y$ **4.** $-12 = 4x - 7$

5. $5y = 6$ **6.** $19 = -6y$

7. $5x - 7y = 30$ **8.** $2x - 3y = 18$

9. $12 - 4x = 9 + x$ **10.** $15 + 7x = 3x - 5$

11. $2y - 4 = 35 + x$ **12.** $2x - 17 + y = 0$

13. $3y + x = 3y + 2$ **14.** $x - 4y = 12 - 4y$

15. $4y + 8x = 6$ **16.** $5y + 6x = -3$

17. $y - 6 = 14$ **18.** $3y - 5 = 8$

19. $3y - 2x = 5 + 9y - 2x$

20. $17y + 4x + 3 = 7 + 4x$

21. $7x - 3y = -2x + 1$ **22.** $9x - 4y = 3x + 5$

Graph by hand.

23. $y = 4$ **24.** $x = -1$

25. $x = 2$ **26.** $y = 5$

27. $4 \cdot f(x) = 20$ **28.** $6 \cdot g(x) = 12$

29. $3x = -15$ **30.** $2x = 10$

31. $4 \cdot g(x) + 3x = 12 + 3x$

32. $6x - 4y + 12 = -4y$

33. $7 - 3x = 4 + 2x$ **34.** $3 - f(x) = 2$

Find the intercepts. Then graph by using the intercepts and a third point as a check.

35. $x - 2 = y$ **36.** $x - 4 = y$

37. $3x - 1 = y$ **38.** $3x - 4 = y$

39. $5x - 4y = 20$ **40.** $3x + 5y = 15$

41. $f(x) = -2 - 2x$ **42.** $g(x) = -5 - 5x$

43. $5y = -15 + 3x$ **44.** $7x = 3y - 21$

45. $g(x) = 2x - 9$ **46.** $f(x) = 3x - 8$

47. $1.4y - 3.5x = -9.8$ **48.** $3.6x - 2.1y = 22.68$

49. $5x + 2y = 7$ **50.** $3x - 4y = 10$

For each function, determine which of the given viewing windows will show both intercepts.

51. $f(x) = 20 - 4x$

    a) $[-10, 10, -10, 10]$ **b)** $[-5, 10, -5, 10]$

    c) $[-10, 10, -10, 30]$ **d)** $[-10, 10, -30, 10]$

52. $g(x) = 3x + 7$

    a) $[-10, 10, -10, 10]$ **b)** $[-1, 15, -1, 15]$

    c) $[-15, 5, -15, 5]$ **d)** $[-10, 10, -30, 10]$

53. $p(x) = -35x + 7{,}000$

    a) $[-10, 10, -10, 10]$

    b) $[-35, 0, 0, 7000]$

    c) $[-1000, 1000, -1000, 1000]$

    d) $[0, 500, 0, 10{,}000]$

54. $r(x) = 0.2 - 0.01x$

    a) $[-10, 10, -10, 10]$ **b)** $[-5, 30, -1, 1]$

    c) $[-1, 1, -5, 30]$ **d)** $[0, 0.01, 0, 0.2]$

Without graphing, tell whether the graphs of each pair of equations are parallel.

55. $x + 6 = y$,   **56.** $2x - 7 = y$,

    $y - x = -2$ $y - 2x = 8$

57. $y + 3 = 5x$,   **58.** $y + 8 = -6x$,

    $3x - y = -2$ $-2x + y = 5$

59. $f(x) = 3x + 9$,   **60.** $f(x) = -7x - 9$,

    $2y = 6x - 2$ $-3y = 21x + 7$

Without graphing, tell whether the graphs of each pair of equations are perpendicular.

61. $f(x) = 4x - 5$,   **62.** $2x - 5y = -3$,

    $4y = 8 - x$ $2x + 5y = 4$

63. $x + 2y = 5$,   **64.** $y = -x + 7$,

    $2x + 4y = 8$ $f(x) = x + 3$

Find an equation for a linear function parallel to the given line with the given y-intercept.

65. $y = 3x - 7$; $(0, 9)$

66. $y = -5x + 2$; $(0, -2)$

67. $2x + y = 3$; $(0, -1)$

68. $3x = y + 10$; $(0, 4)$

69. $2x + 5y = 8$; $\left(0, -\frac{1}{3}\right)$

70. $3y = 12$; $(0, -5)$

71. $5 = 10y$; $(0, 12)$

72. $3x - 6y = 4$; $\left(0, \frac{4}{5}\right)$

Find an equation for a linear function perpendicular to the given line with the given y-intercept.

73. $y = x - 7$; $(0, 4)$

74. $y = 2x - 5$; $(0, -3)$

75. $2x + 3y = 6$; $(0, -9)$

76. $4x + 2y = 8$; $(0, 1)$

77. $5x - y = 13$; $\left(0, \frac{1}{5}\right)$

78. $2x - 5y = 7$; $\left(0, -\frac{1}{8}\right)$

🔍 *Determine whether each equation is linear. Find the slope of any nonvertical lines.*

79. $5x - 3y = 15$

80. $3x + 5y + 15 = 0$

81. $16 + 4y = 0$

82. $3x - 12 = 0$

83. $3g(x) = 6x^2$

84. $2x + 4f(x) = 8$

85. $3y = 7xy - 5$

86. $5x - 4xy = 12$

87. $6y - \dfrac{4}{x} = 0$

88. $\dfrac{3y}{4x} = 2x$

89. $\dfrac{f(x)}{x} = x^2$

90. $\dfrac{g(x)}{2} = 3 + x$

Skill Maintenance

Multiply. [1.3]

91. $3(2x - y + 7)$

92. $-2(x + 5y - 1)$

Factor. [1.3]

93. $9x - 15y$

94. $12a + 21ab$

95. $\frac{1}{3}x + \frac{2}{3}y - \frac{4}{3}$

96. $-3x - 6y - 9$

Synthesis

97. ◈ Wind friction, or *air resistance*, increases with speed. Here are some measurements made in a wind tunnel. Plot the data and explain why a linear function does or does not give an approximate fit.

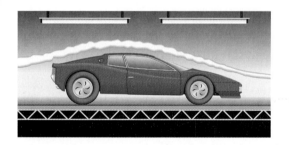

VELOCITY (IN KILOMETERS PER HOUR)	FORCE OF RESISTANCE (IN NEWTONS)
10	3
21	4.2
34	6.2
40	7.1
45	15.1
52	29.0

98. ◈ Explain why examining the graphs of two lines is not the best way to tell whether the lines are parallel.

99. ◈ Judi claims that she cannot find any linear function whose graph is parallel to the line $x = 4$. Is she correct? Why or why not?

100. ◈ Under what condition(s) will the x- and y-intercepts of a line coincide? What would the equation for such a line look like?

101. Give an equation, in standard form, for the line whose x-intercept is 5 and whose y-intercept is -4.

102. Find the x-intercept of $y = mx + b$, assuming that $m \neq 0$.

In Exercises 103–106, assume that r, p, and s are nonzero constants and that x and y are variables. Determine whether each equation is linear.

103. $rx + 3y = p - s$

104. $py = sx - ry + 2$

105. $r^2x = py + 5$

106. $\dfrac{x}{r} - py = 17$

107. Suppose that two linear equations have the same y-intercept but that equation A has an x-intercept that is half the x-intercept of equation B. How do the slopes compare?

Consider the linear equation
$$ax + 3y = 5x - by + 8.$$

108. Find a and b if the graph is a horizontal line passing through $(0, 2)$.

109. Find a and b if the graph is a vertical line passing through $(4, 0)$.

110. Since a vertical line is not the graph of a function, many graphers cannot graph equations of the form $x = a$. Some graphers can draw vertical lines using the DRAW menu. Use the VERTICAL option of the DRAW menu to graph each of the following equations.

a) $x = 3.6$ **b)** $x = -1.52$
c) $3x - 5 = 7x + 2$ **d)** $2(x - 5) = x + 10$

111. A table of values can be used to determine whether two lines are parallel. If the difference between the y-values for two lines is the same for all x-values, the lines are parallel. Use the TABLE feature of a grapher to determine whether each of the following pairs of lines is parallel.

a) $2.3x - 3.4y = 9.8$,
    $1.84x = 2.72y - 17.4$
b) $2.56y + 3.2x - 7.2 = 0$,
    $5.12y + 6.3x = 14.3$

2.6

Introduction to Curve Fitting: Point–Slope Form and Linear Regression

- *Point–Slope Equations*
- *Curve Fitting*
- *Linear Regression*
- *Which Form to Use?*

In Section 2.4, we learned how to write an equation for a line if we know the slope and the y-intercept of the line. In this section, we discuss how to find an equation for a line if we know its slope and *any* point on the line.

Point–Slope Equations

Suppose that a line of slope m passes through the point (x_1, y_1). For any other point (x, y) to lie on this line, we must have

$$\frac{y - y_1}{x - x_1} = m.$$

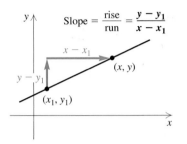

$$\text{Slope} = \frac{\text{rise}}{\text{run}} = \frac{y - y_1}{x - x_1}$$

Note that when x and y are replaced with x_1 and y_1, we have $\frac{0}{0} = m$, a false equation. To avoid this difficulty, we multiply by $x - x_1$ on both sides and simplify:

$$(x - x_1)\frac{y - y_1}{x - x_1} = m(x - x_1)$$

$$y - y_1 = m(x - x_1). \qquad \begin{array}{l}\textbf{Removing a factor equal to 1:}\\ \dfrac{x - x_1}{x - x_1} = 1\end{array}$$

This is the *point–slope* form of a linear equation.

Point–Slope Equation

The *point–slope equation* of a line with slope m, passing through (x_1, y_1), is

$$y - y_1 = m(x - x_1).$$

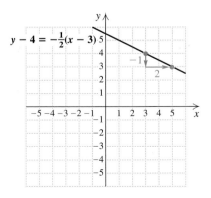

$$y - 4 = -\tfrac{1}{2}(x - 3)$$

Example 1   Find and graph an equation of the line passing through $(3, 4)$ with slope $-\frac{1}{2}$.

SOLUTION   We substitute in the equation $y - y_1 = m(x - x_1)$:

$$y - y_1 = m(x - x_1)$$

$$y - 4 = -\tfrac{1}{2}(x - 3). \qquad \textbf{Substituting}$$

To graph this point–slope equation, we plot points by counting off a slope of $-\frac{1}{2}$, starting at $(3, 4)$. Then we draw the line.   ▬

Example 2   Find a linear function that has a graph passing through the points $(-1, -5)$ and $(3, -2)$.

SOLUTION   We first determine the slope of the line and then use the point–slope equation. Note that

$$m = \frac{-5 - (-2)}{-1 - 3} = \frac{-3}{-4} = \frac{3}{4}.$$

Since the line passes through $(3, -2)$, we have

$$y - (-2) = \tfrac{3}{4}(x - 3) \qquad \text{Substituting into the point–slope equation}$$

$$y + 2 = \tfrac{3}{4}x - \tfrac{9}{4}. \qquad \text{Using the distributive law}$$

Before using function notation, we isolate y:

$$y = \tfrac{3}{4}x - \tfrac{17}{4} \qquad \text{Subtracting 2 on both sides: } -\tfrac{9}{4} - \tfrac{8}{4} = -\tfrac{17}{4}$$

$$f(x) = \tfrac{3}{4}x - \tfrac{17}{4}. \qquad \text{Using function notation}$$

You can check that using $(-1, -5)$ instead of $(3, -2)$ in $y - y_1 = m(x - x_1)$ will yield the same expression for $f(x)$.   ■

Example 3   *Medical Insurance.* According to the U.S. Bureau of the Census, the number of people living in the United States with no health insurance grew from 33.4 million in 1989 to 39.7 million in 1994 (*Source: Statistical Abstract of the United States, 1997*). Assuming constant growth since 1987, how many people living in the United States in 2001 will lack health insurance?

SOLUTION

1. **Familiarize.**   Constant growth indicates a constant rate of change, so a linear relationship can be assumed. When data contain large numbers, as in this situation, we often redefine the units. If we regard 1987 as year 0, we can form the pairs $(2, 33.4)$ and $(7, 39.7)$ as the data points, choosing suitable scales on the two axes. The jagged "break" on the vertical axis is used to avoid including a large portion of unused grid. We let $n =$ the number of people who are uninsured, in millions, and t the number of years since 1987.

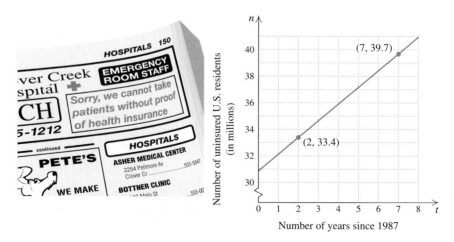

Number of years since 1987

2. **Translate.** To find an equation relating n and t, we first find the slope of the line. This corresponds to the *growth rate*:

$$m = \frac{39.7 \text{ million people} - 33.4 \text{ million people}}{7 \text{ years} - 2 \text{ years}}$$

$$= \frac{6.3 \text{ million people}}{5 \text{ years}}$$

$$= 1.26 \text{ million people per year.}$$

Next, we use the point–slope equation and solve for n:

$n - 33.4 = 1.26(t - 2)$	**Writing point–slope form**
$n - 33.4 = 1.26t - 2.52$	**Using the distributive law**
$n = 1.26t + 30.88.$	**Adding 33.4 on both sides**

3. **Carry out.** Using function notation, we have

$$n(t) = 1.26t + 30.88.$$

To predict the number of people who will be uninsured in 2001, we find

$n(14) = 1.26 \cdot 14 + 30.88$	**2001 is 14 years from 1987.**
$= 48.52.$	

4. **Check.** To check, we can repeat our calculations. We could also extend the graph to see that $(14, 48.52)$ is on the line.

5. **State.** Assuming constant growth, there will be 48.52 million uninsured residents of the United States in 2001. ▬

Curve Fitting

It is not uncommon to see tables of information, or *data*. The process of understanding and interpreting data is called *data analysis*. One helpful tool in data analysis is *curve fitting*, or finding an algebraic equation that describes the data. If the data can be plotted and appear to have a linear pattern, we can fit a linear equation to the data.

Example 4 *Shopping Centers.* The number of shopping centers in the United States has grown in recent years, as shown in the table at left (*Source:* International Council of Shopping Centers; *The Wall Street Journal Almanac, 1998*, p. 285).

a) Plot the data and determine whether they appear to be linear.

b) If the data appear to be linear, fit a linear equation to the data and graph the line.

SOLUTION

a) We plot the data, with the years on the horizontal axis and the number of shopping centers on the vertical axis. We let $t = $ the number of years since the first year that appears in the table, or 1964, and

YEAR	NUMBER OF SHOPPING CENTERS
1964	7,600
1972	13,174
1976	17,523
1980	22,050
1984	22,508
1988	32,563
1992	38,966
1996	42,130

N = the number of shopping centers, in thousands. The table is shown again along with the graph of the data.

NUMBER OF YEARS SINCE 1964	NUMBER OF SHOPPING CENTERS (IN THOUSANDS)
0	7.600
8	13.174
12	17.523
16	22.050
20	22.508
24	32.563
28	38.966
32	42.130

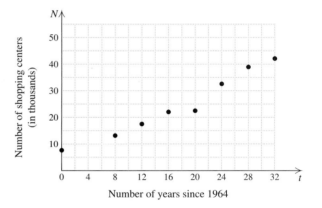

The points seem close to a straight line, so the data appear to be linear.

b) To find an equation of a line, we need to know the slope and a point on the line. To find the slope, we choose two different points on the line. We will use the points (12, 17.523) and (32, 42.13). The slope of the line is then

$$m = \frac{42.13 - 17.523}{32 - 12} = \frac{24.607}{20} = 1.23035.$$

Using the point–slope equation, we have

$$N - 42.13 = 1.23035(t - 32)$$ Substituting 1.23035 for m and (32, 42.13) for (x_1, y_1)

$$N - 42.13 = 1.23035t - 39.3712$$

$$N = 1.23035t + 2.7588.$$ Writing in slope–intercept form

We now graph the line using the two points chosen: (12, 17.523) and (32, 42.13).

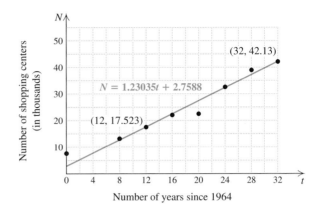

Note that the line does not go through all the data points. ▬

What if, in Example 4, we had chosen two different points? Let's find the equation for the line passing through the points (8, 13.174) and (20, 22.508). The slope of this line is

$$m = \frac{22.508 - 13.174}{20 - 8} = \frac{9.334}{12} \approx 0.778.$$

Using the point (8, 13.174) gives the equation of the line as

$$N - 13.174 = 0.778(t - 8)$$
$$N - 13.174 = 0.778t - 6.224$$
$$N = 0.778t + 6.95.$$

We graph this line, along with the line found in Example 4. Which one appears to be a better "fit" for the data?

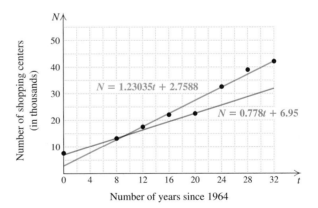

The line that *best* describes the data may not actually go through any of the given points. There are different methods for finding an equation of a line that fits a set of data. These methods generally consider all the points, not just two, when fitting an equation to data. The most commonly used method is *linear regression*.

Linear Regression

The development of the method of linear regression belongs to a later mathematics course, but most graphers offer regression as a way of fitting a line or curve to a set of data.

Plotting points and fitting curves to data are done using the STAT menu. Since graphers use the same viewing windows to plot data points as well as to graph equations, we need to take extra steps to prepare the window before plotting the points.

Example 5 *Shopping Centers.* Referring to the table given in Example 4:

a) Plot the data.

b) Use linear regression to fit a line to the data, and graph the line on the same set of axes as the data.

c) Use the equation found in part (b) to estimate the number of shopping centers in the United States in 2000.

SOLUTION

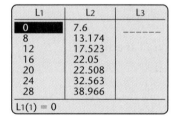

a) We first enter the data as lists of numbers using the EDIT menu, accessed by pressing STAT, after having cleared any data already in the lists. Next, we enter the years (the first column in the table) as L1 and the number of shopping centers as L2, as shown at left.

To determine the dimensions of the viewing window, look at the numbers in the table. The years range from 0 to 32, so we set $X\min = 0$ and $X\max = 40$, with $X\text{scl} = 4$. Since the number of shopping centers, in thousands, ranges from 7.6 to 42.13, we set $Y\min = 0$ and $Y\max = 50$, with $Y\text{scl} = 4$.

Finally, we clear or deselect any equations stored in the grapher, and plot the points.

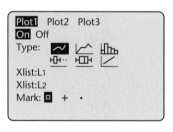
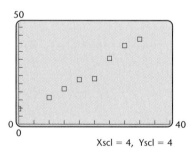

Xscl = 4, Yscl = 4

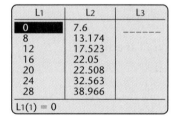

b) The linear-regression equation can be calculated using the LinReg option of the STAT CALC menu.

The equation is given in slope–intercept form $y = ax + b$. (The grapher uses a instead of m.) Substituting the given values of a and b gives the equation $y = 1.128261905x + 4.819666667$.

The values for r^2 and r shown on the screen give an indication of how well the regression line fits the data. (We call r the *coefficient of*

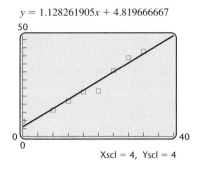

$y = 1.128261905x + 4.819666667$

Xscl = 4, Yscl = 4

correlation.) These values will appear if the *diagnostics* are turned on. For the rest of the text, regression will be done with the diagnostics turned off.

The equation can now be copied directly into the Y = screen and then graphed, as shown at left.

c) To estimate the number of shopping centers in the United States in 2000, we evaluate the linear-regression equation for $x = 36$. One way to do this is to use the Value option in the CALC menu.

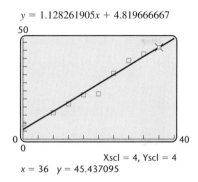

$y = 1.128261905x + 4.819666667$

Xscl = 4, Yscl = 4

$x = 36$ $y = 45.437095$

When $x = 36$, $y = 45.437095$, so we estimate the number of shopping centers in 2000 to be about 45,437.   ▬

Which Form to Use?

We have now studied the slope–intercept, point–slope, and standard forms of a linear equation. Depending on what information we are given and what information we are seeking, one form may be more useful than the others. A referenced summary is given below.

Slope–intercept form, $y = mx + b$	■ Useful when an equation is needed and the slope and y-intercept are given. See Example 6 on pp. 106–107. ■ Useful when a line's slope and y-intercept are needed. See Example 7 on p. 107. ■ Useful when solving equations graphically. See Example 1 on p. 83. ■ Commonly used for linear functions and for graphers.
Standard form, $Ax + By = C$	■ Allows for easy calculation of intercepts. See Example 4 on p. 118. ■ Will prove useful in future work. See Sections 3.1–3.3.
Point–slope form, $y - y_1 = m(x - x_1)$	■ Useful when an equation is needed and the slope and a point on the line are given. See Example 1 on p. 125. ■ Useful when a linear function is needed and two points on its graph are given. See Example 2 on p. 126. ■ Will prove useful in future work with curves and tangents in calculus.

2.6 | Exercise Set

Find an equation in point–slope form of the line having the specified slope and containing the point indicated. Then graph the line.

1. $m = 4$, $(2, 3)$ **2.** $m = 5$, $(7, 4)$

3. $m = -2$, $(4, 7)$ **4.** $m = -3$, $(7, 3)$

5. $m = 3$, $(-2, -4)$ **6.** $m = 1$, $(-5, -7)$

7. $m = \frac{2}{5}$, $(-3, 8)$ **8.** $m = \frac{3}{4}$, $(1, -5)$

For each point–slope equation listed, state the slope and a point on the graph.

9. $y - 3 = \frac{2}{7}(x - 1)$ **10.** $y - 4 = 9(x - 2)$

11. $y + 2 = -5(x - 7)$ **12.** $y - 1 = -\frac{2}{9}(x + 5)$

13. $y - 1 = -\frac{5}{3}(x + 2)$ **14.** $y + 7 = -4(x - 9)$

Find an equation of the line having the specified slope and containing the indicated point. Write your final answer as a linear function in slope–intercept form.

15. $m = 5$, $(2, -3)$ **16.** $m = -4$, $(-1, 5)$

17. $m = -\frac{2}{3}$, $(4, -7)$ **18.** $m = -\frac{1}{5}$, $(-2, 1)$

19. $m = -0.6$, $(-3, 0)$ **20.** $m = 2.3$, $(4, 0)$

21. 🔍 $m = -6$, $(0, 2.4)$ **22.** 🔍 $m = 4$, $\left(0, \frac{1}{2}\right)$

Find a function for the line containing each pair of points.

23. $(1, 4)$ and $(5, 6)$ **24.** $(2, 6)$ and $(4, 1)$

25. $(2.5, -3)$ and $(6.5, 3)$ **26.** $(2, -1.3)$ and $(7, 1.7)$

27. $(6, 1)$ and $(0, -2)$ **28.** $(-3, 0)$ and $(-1, 5)$

29. $(-2, -3)$ and $(-4, -6)$

30. $(-4, -7)$ and $(-2, -1)$

In Exercises 31–40, assume that a constant rate of change exists for each model formed.

31. *Records in the 400-Meter Run.* In 1930, the record for the 400-m run was 46.8 sec. In 1970, it was 43.8 sec. Let R = the record in the 400-m run and t the number of years since 1930.

a) Find a linear function $R(t)$ that fits the data.

b) Use the function of part (a) to predict the record in 1999; in 2002.

c) When will the record be 40 sec?

32. *Records in the 1500-Meter Run.* In 1930, the record for the 1500-m run was 3.85 min. In 1950, it was 3.70 min. Let R = the record in the 1500-m run and t the number of years since 1930.

a) Find a linear function $R(t)$ that fits the data.

b) Use the function of part (a) to predict the record in 1998; in 2002.

c) When will the record be 3.3 min?

33. *PAC Contributions.* In 1986, Political Action Committees (PACs) contributed $132.7 million to congressional candidates. In 1994, the figure rose to $179.6 million. (*Source:* Congressional Research Service and Federal Election Commission) Let A = the amount of PAC contributions and t the number of years since 1986.

a) Find a linear function $A(t)$ that fits the data.

b) Use the function of part (a) to predict the amount of PAC contributions in 2002.

34. *Consumer Demand.* Suppose that 6.5 million lb of coffee are sold when the price is $8 per pound, and 4.0 million lb are sold when the price is $9 per pound.

a) Find a linear function that expresses the amount of coffee sold as a function of the price per pound.

b) Use the function of part (a) to predict how much consumers would be willing to buy at a price of $6 per pound.

35. *Recycling.* In 1990, Americans recycled 33.9 million tons of garbage. In 1995, the figure grew to 56.2 million tons. (*Source: Statistical Abstract of the United States, 1997*) Let N = the number of tons recycled and t the number of years since 1990.

a) Find a linear function $N(t)$ that fits the data.

b) Use the function of part (a) to predict the amount recycled in 2001.

36. *Seller's Supply.* Suppose that suppliers are willing to sell 5.0 million lb of coffee at a price of $8 per pound and 7.0 million lb at $9 per pound.

a) Find a linear function that expresses the amount

suppliers are willing to sell as a function of the price per pound.

b) Use the function of part (a) to predict how much suppliers would be willing to sell at a price of $6 per pound.

37. *National Park Land.* In 1990, the National Park system consisted of about 76.4 million acres. By 1994, the figure was down to 74.9 million acres. (*Source: Statistical Abstract of the United States, 1995*) Let $A =$ the amount of land in the National Park system, in millions of acres, t years after 1990.

a) Find a linear function $A(t)$ that fits the data.

b) Use the function of part (a) to predict the amount of land in the National Park system in 2002.

38. *Pressure at Sea Depth.* The pressure 100 ft beneath the ocean's surface is approximately 4 atm (atmospheres), whereas at a depth of 200 ft, the pressure is about 7 atm.

a) Find a linear function that expresses pressure as a function of depth.

b) Use the function of part (a) to determine the pressure at a depth of 690 ft.

39. *Life Expectancy of Females in the United States.* In 1990, the life expectancy of females was 78.8 yr. In 1995, it was 78.9 yr. (*Source: Statistical Abstract of the United States, 1997*) Let $E =$ life expectancy and t the number of years since 1990.

a) Find a linear function $E(t)$ that fits the data.

b) Use the function of part (a) to predict the life expectancy of females in 2004.

40. *Life Expectancy of Males in the United States.* In 1990, the life expectancy of males was 71.8 yr. In 1995, it was 72.6 yr. (*Source: Statistical Abstract of*

the United States, 1997) Let $E =$ life expectancy and t the number of years since 1990.

a) Find a linear function $E(t)$ that fits the data.

b) Use the function of part (a) to predict the life expectancy of males in 2004.

41. *Life Expectancy of Females in the United States.* The following table shows the life expectancy of women who were born in the United States in selected years.

Life Expectancy of Women

YEAR	LIFE EXPECTANCY (IN YEARS)
1920	54.6
1930	61.6
1940	65.2
1950	71.1
1960	73.1
1970	74.7
1980	77.5
1990	78.8

Source: Statistical Abstract of the United States and The World Almanac, 1996.

a) Use linear regression to find a linear function that can be used to predict the life expectancy of a woman as a function of the year in which she was born. (Let $x =$ the number of years since 1990.) Compare this with the answer to Exercise 39.

b) Predict the life expectancy of a woman in 2004 and compare your answer with the answer to Exercise 39.

42. *Life Expectancy of Males in the United States.* The following table shows the life expectancy of males born in the United States in selected years.

Life Expectancy of Men

YEAR	LIFE EXPECTANCY (IN YEARS)
1940	60.8
1950	65.6
1960	66.6
1970	67.1
1980	70.0
1990	71.8

Source: Information Please Almanac, 1998.

a) Use linear regression to find a linear function that can be used to predict the life expectancy of a man as a function of the year in which he was born. (Let $x =$ the number of years since 1990.) Compare this with the answer to Exercise 40.

b) Predict the life expectancy of a man in 2004 and compare your answer with the answer to Exercise 40.

43. *Cost of a Cell Phone.* As the number of people using cell phones has increased, the average local monthly bill has declined. The following table shows the average monthly bill for cell phone subscribers in the United States for various years.

Cell Phone Monthly Bills

YEAR	AVERAGE LOCAL MONTHLY BILL
1987	$96.83
1988	98.02
1989	89.30
1990	80.90
1991	72.74
1992	68.68
1993	61.48
1994	56.21
1995	51.00
1996	47.70

Source: The Wall Street Almanac, 1998.

a) Use linear regression to find a linear function that can be used to predict the average local monthly cell phone bill as a function of the year. (Let $x =$ the number of years since 1987.)

b) Predict the average local monthly bill for a cell phone in 2001.

44. *Wind Chill.* Wind chill is a measure of how cold the wind makes you feel. Below are some measurements of wind chill for a 15-mph breeze.

Wind Chill

TEMPERATURE	WIND CHILL WITH A 15-MPH BREEZE
30°	9°
25°	2°
20°	−5°
15°	−11°
10°	−18°
5°	−25°
0°	−31°

Source: National Oceanic & Atmospheric Administration, as reported in the *Burlington Free Press,* 17 January 1992.

a) Use linear regression to find a linear function that can be used to determine the wind chill as a function of the temperature x.

b) Predict the wind chill for a 15-mph breeze when the temperature is 40°.

45. *Child-Rearing Costs.* The following table shows the estimated annual expenditures on a child in 1996 by a family with an annual income of between $34,700 and $58,300.

Annual Family Expenditure per Child

AGE OF CHILD	ANNUAL EXPENDITURE
1	$7860
4	8060
7	8130
13	8830
16	8960

Source: The Wall Street Journal Almanac, 1998.

a) Use linear regression to find a linear function that can be used to predict the annual expenditure on a child as a function of the age of the child.

b) Estimate the annual expenditure on a 10-year-old child in 1996.

46. *Garbage.* The following table shows the amounts of nonfood product waste, in pounds per person per day, for various years.

U.S. Garbage Generation

YEAR	NONFOOD PRODUCT WASTE (IN POUNDS PER PERSON PER DAY)
1960	1.65
1970	2.26
1980	2.57
1988	2.94

Source: The Handy Science Answer Book, Detroit: Visible Ink Press, 1994.

a) Use linear regression to find a linear function that can be used to predict the amount of nonfood product waste as a function of the year. (Let $x =$ the number of years since 1960.)

b) Estimate the amount of nonfood product waste per person per day in the United States in 2005.

Skill Maintenance

Perform the indicated operation. [1.2]

47. $-\frac{1}{3} + \frac{5}{12}$

48. $-\frac{1}{3} - \frac{5}{12}$

49. $-\frac{1}{3} \cdot \frac{5}{12}$

50. $-\frac{1}{3} \div \frac{5}{12}$

Synthesis

51. ◆ The total number of reported cases of AIDS in the United States grew from 372 in 1981 to 100,000 in 1989 and 200,000 in 1992. Can a linear function be used to predict the number of cases in 2004? Why or why not?

52. ◆ The information in Exercises 34 and 36 indicates that when suppliers charge $8 per pound for coffee, the supply will not meet the demand. How could suppliers determine a price for which their supply will exactly meet the demand?

53. ◆ Explain why the negative slope in Exercise 31 is "good," whereas the negative slope in Exercise 37 is "bad."

54. ◆ A firm offers its entering employees a starting salary with a guaranteed 7% increase each year. Can a linear function be used to express the yearly salary as a function of the number of years an employee has worked? Why or why not?

For Exercises 55–58, assume that a linear equation models each situation.

55. *Depreciation of a Computer.* Gina's computer cost $2500 new. Its value has dropped to $2150 after 5 mos. What will the computer be worth after 8 mos?

56. *Cell Phone Charges.* It cost Rick $70 to purchase and activate a cell phone. After 4 mos, his total cost for the phone is $190. Estimate Rick's total cost for the phone after 9 mos.

57. *Operating Expenses.* The total cost for operating Ming's Wings was $7500 after 4 mos and $9250 after 7 mos. Estimate the total cost after 10 mos.

58. *Depreciation of a Printer.* After 6 mos of use, the value of Pearl's printer had dropped to $900. After 8 mos, the value had gone down to $750. How much did the printer cost when new?

59. *Temperature Conversion.* Water freezes at 32° Fahrenheit and at 0° Celsius. Water boils at 212°F and at 100°C. What Celsius temperature corresponds to a room temperature of 70°F?

Write an equation of the line containing the specified point and parallel to the indicated line.

60. $(3, 7)$, $x + 2y = 6$

61. $(-1, 4)$, $3x - y = 7$

Write an equation of the line containing the specified point and perpendicular to the indicated line.

62. $(2, 5)$, $2x + y = -3$

63. $(4, 0)$, $x - 3y = 0$

64. For a linear function f, $f(-1) = 3$ and $f(2) = 4$.
  a) Find an equation for f.
  b) Find $f(3)$.
  c) Find a such that $f(a) = 100$.

65. For a linear function g, $g(3) = -5$ and $g(7) = -1$.
  a) Find an equation for g.
  b) Find $g(-2)$.
  c) Find a such that $g(a) = 75$.

66. Find the value of k such that the graph of $5y - kx = 7$ and the line containing the points $(7, -3)$ and $(-2, 5)$ are parallel.

67. Find the value of k such that the graph of $7y - kx = 9$ and the line containing the points $(2, -1)$ and $(-4, 5)$ are perpendicular.

COLLABORATIVE CORNER

Focus: Linear functions and models

Time: 15 minutes to 1 week

Group size: 3

The table on the next page shows how the political views of first-year college students have changed.

How would you describe your political views?

Political Categories	First-year College Students 1970	First-year College Students 1990
Liberal or Far Left	37%	26%
Middle of the Road	45%	54%
Conservative or Far Right	18%	20%

Source: "The American Freshman: Twenty-Five Year Trends"
(UCLA, 1992–1993)

Activity

1. Each group member should select one of the three categories and find a linear function that can be used to predict the percentage of college students who fall into that category t years after 1970. Then use the function to predict the percentage of students in that category at present.

2. After the entire group has confirmed all three calculations, survey a random sample of first-year students at your school. Survey results may not match your predictions very closely. Why?

2.7

The Algebra of Functions

- *The Sum, Difference, Product, or Quotient of Two Functions*
- *Domains and Graphs*

We now examine four ways in which functions can be combined.

The Sum, Difference, Product, or Quotient of Two Functions

Suppose that a is in the domain of two functions, f and g. The input a is paired with $f(a)$ by f and with $g(a)$ by g. The outputs can then be added to get $f(a) + g(a)$.

Interactive Discovery

Let $y_1 = x + 4$ and $y_2 = x^2 + 1$. Enter y_1, y_2, and $y_3 = y_1 + y_2$ into a grapher, and set up a table that shows values of the three functions. If $y_1 = f(x)$ and $y_2 = g(x)$, what does y_3 represent? Graph y_1, y_2, and y_3 using the same viewing window. How could you draw the graph of y_3 given the graphs of y_1 and y_2?

Now add the expressions for y_1 and y_2 algebraically:

$$(x + 4) + (x^2 + 1) = x^2 + x + 5.$$

Let $y_4 = x^2 + x + 5$. Compare the values of y_3 and y_4.

We see that if $f(x) = x + 4$ and $g(x) = x^2 + 1$, then $f(x) + g(x) = x^2 + x + 5$, which can be regarded as a "new" function, written $(f + g)(x)$.

The Algebra of Functions

If f and g are functions and x is in the domain of both functions, then:

1. $(f + g)(x) = f(x) + g(x)$;
2. $(f - g)(x) = f(x) - g(x)$;
3. $(f \cdot g)(x) = f(x) \cdot g(x)$;
4. $(f / g)(x) = f(x)/g(x)$, provided $g(x) \neq 0$.

Example 1 For $f(x) = x^2 - x$ and $g(x) = x + 2$, find the following.

a) $(f + g)(3)$ **b)** $(f - g)(x)$ and $(f - g)(-1)$
c) $(f / g)(x)$ and $(f / g)(-4)$ **d)** $(f \cdot g)(3)$

SOLUTION

a) Since $f(3) = 3^2 - 3 = 6$ and $g(3) = 3 + 2 = 5$, we have

$$(f + g)(3) = f(3) + g(3)$$
$$= 6 + 5 \qquad \textbf{Substituting}$$
$$= 11.$$

Alternatively, we could first find $(f + g)(x)$:

$$(f + g)(x) = f(x) + g(x)$$
$$= x^2 - x + x + 2$$
$$= x^2 + 2. \qquad \textbf{Combining like terms}$$

Thus,

$$(f + g)(3) = 3^2 + 2 = 11.$$

b) We have

$$(f - g)(x) = f(x) - g(x)$$
$$= x^2 - x - (x + 2) \qquad \textbf{Substituting}$$
$$= x^2 - 2x - 2. \qquad \textbf{Removing parentheses and}$$
$$\textbf{combining like terms}$$

Thus,

$$(f - g)(-1) = (-1)^2 - 2(-1) - 2 \qquad \textbf{Using } (f - g)(x) \textbf{ is faster}$$
$$\textbf{than using } f(x) - g(x).$$
$$= 1. \qquad \textbf{Simplifying}$$

c) We have

$$(f / g)(x) = f(x)/g(x)$$
$$= \frac{x^2 - x}{x + 2}.$$

Thus,

$$(f / g)(-4) = \frac{(-4)^2 - (-4)}{-4 + 2} \qquad \text{Substituting}$$

$$= \frac{20}{-2}$$

$$= -10.$$

d) Using our work in part (a), we have

$$(f \cdot g)(3) = f(3) \cdot g(3)$$

$$= 6 \cdot 5$$

$$= 30.$$

It is also possible to compute $(f \cdot g)(3)$ by first multiplying $x^2 - x$ and $x + 2$ using methods that we will discuss in Chapter 5. ▬

Domains and Graphs

Although applications involving products and quotients of functions rarely appear in newspapers, situations involving sums or differences of functions often do appear in print. Consider the following graph, reprinted from the *New York Times*.*

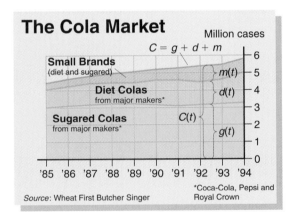

In this graph, the total number of cases of cola sold, $C(t)$, is regarded as a function of the year. The number of cases of sugared cola sold is denoted by $g(t)$, the number of cases of diet cola sold by $d(t)$, and the number of cases of "small brands" cola sold by $m(t)$. Although separate graphs for g, d, and m have not been drawn, we can see that

$$C(t) = g(t) + d(t) + m(t).$$

In the next graph, the functions *are* graphed separately before being added. Here all braces extend to the horizontal axis.

New York Times, 3/23/95, page D1, article by Glenn Collins.

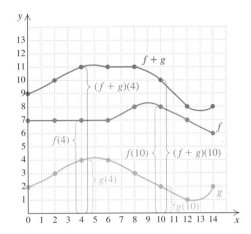

To find $(f + g)(a)$, $(f - g)(a)$, $(f \cdot g)(a)$, or $(f / g)(a)$, we must first be able to find $f(a)$ and $g(a)$. Thus we need to ensure that a is in the domain of both f and g.

When a function is described by an equation, the domain is often unspecified. In such cases, the domain is the set of all numbers for which function values can be calculated.

Example 2 For each equation, determine the domain of f.

a) $f(x) = |x|$ **b)** $f(x) = \dfrac{3}{x + 1}$

Solution

a) We ask ourselves, "Is there any number x for which we cannot compute $|x|$?" Since we can find the absolute value of *any* number, the answer is no. Thus the domain of f is $\mathbb{R}$, the set of all real numbers.

b) Is there any number x for which $\dfrac{3}{x + 1}$ cannot be computed? Since $\dfrac{3}{x + 1}$ cannot be computed when $x + 1$ is 0, the answer is yes. To determine what x-value causes the denominator to be 0, we set up and solve an equation:

$$x + 1 = 0 \qquad \text{Setting the denominator equal to 0}$$
$$x = -1. \qquad \text{Subtracting 1 on both sides}$$

Thus -1 is *not* in the domain of f, whereas all other real numbers are. The domain of f is $\{x \mid x \text{ is a real number } and \ x \neq -1\}$.

Interactive Discovery

Let

$$f(x) = \frac{5}{x} \quad \text{and} \quad g(x) = \frac{2x - 6}{x + 1}.$$

(continued)

Create a table of values for $y_1 = f(x)$, $y_2 = g(x)$, and $y_3 = y_1 + y_2$. Set up the table with TblStart $= -3$ and ΔTbl $= 1$. What number is not in the domain of f? What number is not in the domain of g? What numbers are not in the domain of $f + g$?

Now let $y_4 = y_1 - y_2$, $y_5 = y_1 \cdot y_2$, and $y_6 = y_1 / y_2$. Is the domain of $f - g$ the same as the domain of $f + g$? What about the domains of $f \cdot g$ and f / g?

In the Interactive Discovery above, the domains of $f + g$, $f - g$, and $f \cdot g$ are the same: $\{x \mid x$ is a real number *and* $x \ne 0$ *and* $x \ne -1\}$. This is the intersection of the domains of f and g. The domain of f / g also excludes the number 3, because $g(3) = 0$.

Determining the Domain

The domain of $f + g$, $f - g$, or $f \cdot g$ is the set of all values common to the domains of f and g.

The domain of f / g is the set of all values common to the domains of f and g, excluding any values for which $g(x)$ is 0.

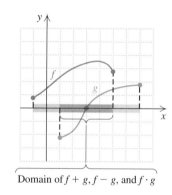

Domain of $f + g$, $f - g$, and $f \cdot g$

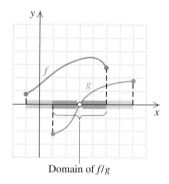

Domain of f/g

Example 3   Given $f(x) = 1/x$ and $g(x) = 2x - 7$, find the domains of $f + g$, $f - g$, $f \cdot g$, and f / g.

SOLUTION   The domain of f is $\{x \mid x \ne 0\}$, or $\{x \mid x$ is a real number *and* $x \ne 0\}$. The domain of g is $\mathbb{R}$. The domains of $f + g$, $f - g$, and $f \cdot g$ are the set of all elements common to both the domain of f and the domain of g. We have

the domain of $f + g =$ the domain of $f - g =$ the domain of $f \cdot g$
$$= \{x \mid x \text{ is a real number } and \ x \ne 0\}.$$

The domain of f / g is $\{x \mid x$ is a real number *and* $x \ne 0\}$, with the additional restriction that $g(x) \ne 0$. To determine what x-values would make $g(x) = 0$, we solve:

$$2x - 7 = 0 \qquad \textbf{Replacing } g(x) \textbf{ with } 2x - 7$$
$$2x = 7$$
$$x = \tfrac{7}{2}.$$

Since $g(x) = 0$ for $x = \tfrac{7}{2}$,

the domain of $f / g = \left\{x \mid x \text{ is a real number } and \ x \neq 0 \ and \ x \neq \tfrac{7}{2}\right\}$.

2.7 Exercise Set

Let $f(x) = -3x + 1$ and $g(x) = x^2 + 2$. Find the following.

1. $f(2) + g(2)$
2. $f(-1) + g(-1)$
3. $f(5) - g(5)$
4. $f(4) - g(4)$
5. $f(-1) \cdot g(-1)$
6. $f(-2) \cdot g(-2)$
7. $f(-4)/g(-4)$
8. $f(3)/g(3)$
9. $g(1) - f(1)$
10. $g(2)/f(2)$
11. $g(0)/f(0)$
12. $g(6) - f(6)$

Let $F(x) = x^2 - 3$ and $G(x) = 4 - x$. Find the following.

13. $(F + G)(x)$
14. $(F + G)(a)$
15. $(F + G)(-4)$
16. $(F + G)(-5)$
17. $(F - G)(3)$
18. $(F - G)(2)$
19. $(F \cdot G)(-3)$
20. $(F \cdot G)(-4)$
21. $(F / G)(0)$
22. $(F / G)(1)$
23. $(F / G)(-2)$
24. $(F / G)(-1)$

The following graph indicates the number of new cases of AIDS diagnosed each month in the United States. For Exercises 25–28, let $T(t) =$ the total number of new cases per month, $F(t)$ the number of new cases per month among people born in 1960 or later, $G(t)$ the number of new cases per month among people born before 1960, and t the number of years since January 1986.*

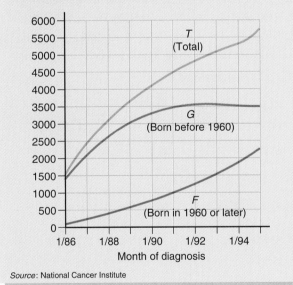

AIDS cases in the U.S. population are still on the rise among young people.

AIDS cases are increasing much more rapidly among Americans born in 1960 or later.

Source: National Cancer Institute

25. Estimate $F(7)$ and interpret its meaning.

26. Estimate $G(5)$ and interpret its meaning.

27. Estimate $T(6)$, $G(6)$, and $F(6)$. Show that $(G + F)(6) = T(6)$.

28. Estimate $T(4)$, $G(4)$, and $F(4)$. Show that $(T - F)(4) = G(4)$.

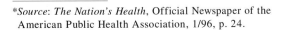

*Source: *The Nation's Health*, Official Newspaper of the American Public Health Association, 1/96, p. 24.

The following graph indicates how the three major airports servicing New York City have been utilized.

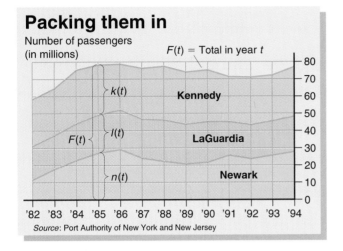

Packing them in

Number of passengers (in millions)

$F(t)$ = Total in year t

Kennedy

LaGuardia

Newark

'82 '83 '84 '85 '86 '87 '88 '89 '90 '91 '92 '93 '94

Source: Port Authority of New York and New Jersey

29. Estimate $(n + l)('92)$. What does it represent?

30. Estimate $(k + l)('94)$. What does it represent?

31. Estimate $(k - l)('92)$. What does it represent?

32. Estimate $(k - n)('89)$. What does it represent?

33. Estimate $(n + l + k)('93)$. What does it represent?

34. Estimate $(n + l + k)('83)$. What does it represent?

35. Find the domain of f.

a) $f(x) = \dfrac{2}{x - 3}$ **b)** $f(x) = \dfrac{7}{5 - x}$

c) $f(x) = 2x + 1$ **d)** $f(x) = x^2 + 3$

e) $f(x) = \dfrac{3}{2x - 5}$ **f)** $f(x) = |3x - 4|$

36. Find the domain of g.

a) $g(x) = \dfrac{3}{x - 1}$ **b)** $g(x) = |5 - x|$

c) $g(x) = \dfrac{9}{x + 3}$ **d)** $g(x) = \dfrac{4}{3x + 4}$

e) $g(x) = x^3 - 1$ **f)** $g(x) = 7x - 8$

For each pair of functions f and g, determine the domain of the sum, the difference, and the product of the two functions.

37. $f(x) = x^2$,
$g(x) = 3x - 4$

38. $f(x) = 5x - 1$,
$g(x) = 2x^2$

39. $f(x) = \dfrac{1}{x - 2}$,
$g(x) = 4x^3$

40. $f(x) = x^3 + 1$,
$g(x) = \dfrac{5}{x}$

41. $f(x) = 4x + \dfrac{2}{x - 1}$,
$g(x) = 3x^3$

42. $f(x) = 9 - x^2$,
$g(x) = \dfrac{3}{x - 5} + 2x$

43. $f(x) = \dfrac{3}{x - 2}$,
$g(x) = \dfrac{5}{4 + x}$

44. $f(x) = \dfrac{5}{x + 3}$,
$g(x) = \dfrac{1}{x - 2}$

For each pair of functions f and g, determine the domain of f / g.

45. $f(x) = x^4$,
$g(x) = x - 3$

46. $f(x) = 2x^3$,
$g(x) = 5 - x$

47. $f(x) = 3x - 2$,
$g(x) = 2x - 8$

48. $f(x) = 5 + x$,
$g(x) = 6 - 2x$

49. $f(x) = \dfrac{3}{x - 4}$,
$g(x) = 5 - x$

50. $f(x) = \dfrac{1}{2 - x}$,
$g(x) = 7 - x$

51. $f(x) = \dfrac{2x}{x + 1}$,
$g(x) = 2x + 5$

52. $f(x) = \dfrac{7x}{x - 2}$,
$g(x) = 3x + 7$

For Exercises 53–58, consider the functions F and G as shown.

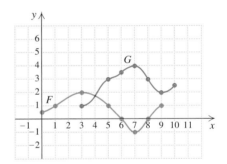

53. Determine $(F + G)(5)$ and $(F + G)(7)$.

54. Determine $(F \cdot G)(6)$ and $(F \cdot G)(9)$.

55. Find the domains of F, G, $F + G$, and F / G.

56. Find the domains of $F - G$, $F \cdot G$, and G / F.

57. Graph $F + G$.

58. Graph $G - F$.

Skill Maintenance

Evaluate each expression using the values provided. [1.1]

59. $3x^2 - 5y$, for $x = 10$ and $y = 6$

60. $3x + 10 \div 5y$, for $x = 4$ and $y = 2$

Write scientific notation. [1.4]

61. 0.00000703

62. 4,506,000,000,000

Synthesis

In the following graph, W(t) = the number of gallons of whole milk, L(t) the number of gallons of lowfat milk, and S(t) the number of gallons of skim milk consumed by the average American in a year.*

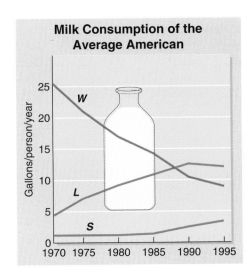

Milk Consumption of the Average American

63. ◈ Explain in words what $(W - S)(t)$ represents and what it would mean to have $(W - S)(t) < 0$.

64. ◈ Consider $(W + L + S)(t)$ and explain why you feel that total milk consumption per person has or has not changed over the years 1970–1995.

65. Find the domain of m / n, if

$$m(x) = 3x \text{ for } -1 < x < 5$$

and

$$n(x) = 2x - 3.$$

66. Find the domains of $f + g, f - g, f \cdot g,$ and f / g, if

$$f = \{(-2, 1), (-1, 2), (0, 3), (1, 4), (2, 5)\}$$

and

$$g = \{(-4, 4), (-3, 3), (-2, 4), (-1, 0), (0, 5), (1, 6)\}.$$

67. For f and g as defined in Exercise 66, find $(f + g)(-2), (f \cdot g)(0),$ and $(f / g)(1).$

---

*Copyright 1990, CSPI. Adapted from *Nutrition Action Health-letter* (1875 Connecticut Avenue, N.W., Suite 300, Washington, DC 20009-5728. $24.00 for 10 issues).

68. Find the domain of F / G, if

$$F(x) = \frac{1}{x - 4} \quad \text{and} \quad G(x) = \frac{x^2 - 4}{x - 3}.$$

69. Find the domain of f / g, if

$$f(x) = \frac{3x}{2x + 5} \quad \text{and} \quad g(x) = \frac{x^4 - 1}{3x + 9}.$$

70. Write equations for two functions f and g such that the domain of $f + g$ is

$\{x|\ x \text{ is a real number } and\ x \neq -2\ and\ x \neq 5\}.$

71. Sketch the graph of two functions f and g such that the domain of f / g is

$\{x|\ -2 \leq x \leq 3\ and\ x \neq 1\}.$

72. ◈ If $f(x) = c$, where c is some nonzero constant, describe how the graphs of $y = g(x)$ and $y = (f + g)(x)$ will differ.

73. ◈ Refer to the graph accompanying Exercises 25–28. Compute

$$\frac{F(7) - F(4)}{7 - 4} \quad \text{and} \quad \frac{G(7) - G(4)}{7 - 4}$$

and explain how these figures support the claim that the number of AIDS cases is increasing most rapidly among Americans born in 1960 or later.

74. Using the window $[-5, 5, -1, 9]$, graph $y_1 = 5$, $y_2 = x + 2$, and $y_3 = \sqrt{x}$. Then predict what shape the graphs of $y_1 + y_2, y_1 + y_3,$ and $y_2 + y_3$ will take. Use a grapher to check each prediction.

75. Let $y_1 = 2.5x + 1.5, y_2 = x - 3,$ and $y_3 = y_1 / y_2$. Depending on whether the CONNECTED or DOT mode is used, the graph of y_3 appears as follows.

CONNECTED MODE

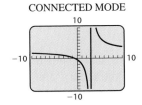

DOT MODE

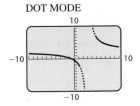

Use algebra to determine which graph more accurately represents y_3.

76. Use the TABLE feature on a grapher to check your answers to Exercises 37, 43, 45, and 51.

77. Use the graphs of f and g, shown below, to match each of $(f + g)(x)$, $(f - g)(x)$, $(f \cdot g)(x)$, and $(f / g)(x)$ with its graph.

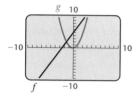

a) $(f + g)(x)$ **b)** $(f - g)(x)$

c) $(f \cdot g)(x)$ **d)** $(f / g)(x)$

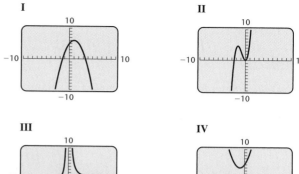

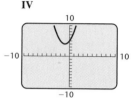

COLLABORATIVE CORNER

Focus: The algebra of functions

Time: 10–15 minutes

Group size: 2–3

The graph and data at right chart the average retirement age $R(x)$ and life expectancy $E(x)$ of U.S. citizens in year x.

Activity

1. Working as a team, perform the appropriate calculations and then graph $E - R$.

2. What does $(E - R)(x)$ represent? In what fields of study or business might the function $E - R$ prove useful?

3. Using only the data from 1955 through 1985, use linear regression to find and graph:

   a) a linear function $r(x)$ that could be used to predict the average retirement age in year x, and

   b) a linear function $e(x)$ that could be used to predict the average life expectancy in year x.

4. Use the functions $r(x)$ and $e(x)$ to predict the average retirement age and life expectancy in 1995 and in 2005. Compare these predictions with the estimates given in the table.

5. Calculate and graph $e - r$ using the functions found in part (3). Compare this function with $E - R$ found in part (1). Which function might be more useful for making predictions?

6. Should E and R really be calculated separately for men and women? Why or why not?

7. What advice would you give to someone considering early retirement?

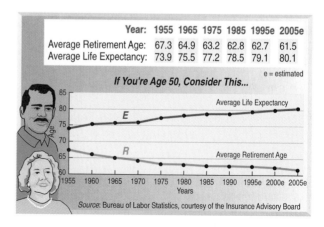

Year:	1955	1965	1975	1985	1995e	2005e
Average Retirement Age:	67.3	64.9	63.2	62.8	62.7	61.5
Average Life Expectancy:	73.9	75.5	77.2	78.5	79.1	80.1

e = estimated

Source: Bureau of Labor Statistics, courtesy of the Insurance Advisory Board

CHAPTER

2 Summary and Review

Key Terms

Domain, p. 68
Range, p. 68
Function, p. 68
Projection, p. 70
Relation, p. 71
Input, p. 72
Output, p. 72
Dependent variable, p. 72
Independent variable, p. 72
Constant function, p. 73

Dummy variable, p. 74
Equivalent equations, p. 84
Identity, p. 88
Contradiction, p. 88
Conditional equation, p. 88
Solution set, p. 89
Empty set, p. 89
Formula, p. 95
Linear function, p. 102
y-intercept, p. 103

Slope, p. 104
Rise, p. 104
Run, p. 104
Rate of change, p. 107
Salvage value, p. 109
Depreciate, p. 109
Zero slope, p. 115
Undefined slope, p. 116
x-intercept, p. 117
Linear regression, p. 130

Important Properties and Formulas

The Vertical-Line Test

A graph represents a function if it is not possible to draw a vertical line that intersects the graph more than once.

The addition principle for equations:

$a = b$ is equivalent to $a + c = b + c$.

The multiplication principle for equations:

For $c \neq 0$, $a = b$ is equivalent to
$a \cdot c = b \cdot c$.

Five Steps for Problem Solving with Algebra

1. *Familiarize* yourself with the problem situation.
2. *Translate* to mathematical language.
3. *Carry out* some mathematical manipulation.
4. *Check* your possible answer in the original problem.
5. *State* the answer clearly.

To solve a formula for a given letter, identify the letter, and:

1. Multiply on both sides to clear fractions or decimals, if that is needed.
2. Combine like terms on each side where convenient.
3. Get all terms with the letter being solved for on one side of the equation and all other terms on the other side, using the addition principle.
4. Combine like terms again, if necessary. This may require factoring.
5. Solve for the letter in question, using the multiplication principle.

$$\text{Slope} = m = \frac{\text{rise}}{\text{run}} = \frac{\text{change in } y}{\text{change in } x} = \frac{y_2 - y_1}{x_2 - x_1}$$

Every horizontal line has a slope of 0.
The slope of a vertical line is undefined.

The *x*-intercept is $(a, 0)$. To find *a*, let $y = 0$ and solve the original equation for *x*.

The *y*-intercept is $(0, b)$. To find *b*, let $x = 0$ and solve the original equation for *y*.

The slope–intercept equation of a line is

$$y = mx + b.$$

The point–slope equation of a line is

$$y - y_1 = m(x - x_1).$$

The standard form of a linear equation is

$$Ax + By = C.$$

Parallel lines: The slopes are equal.
Perpendicular lines: The product of the slopes is -1.

The Algebra of Functions

1. $(f + g)(x) = f(x) + g(x)$
2. $(f - g)(x) = f(x) - g(x)$
3. $(f \cdot g)(x) = f(x) \cdot g(x)$
4. $(f / g)(x) = f(x)/g(x)$, provided $g(x) \neq 0$

REVIEW EXERCISES

1. For the following graph of *f*, determine **(a)** $f(2)$; **(b)** the domain of *f*; **(c)** any *x*-values for which $f(x) = 2$; and **(d)** the range of *f*.

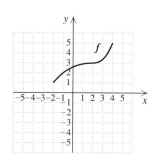

2. Estimate the domain and the range of the function whose graph is shown below.

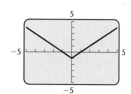

3. The function $C(t) = 645t + 9800$ can be used to estimate the average cost of tuition at a state university *t* years after 1997. (Figures reflect prices paid by both in-state and out-of-state students.) Predict the average cost of tuition at a state university in 2010.

Solve each equation graphically and check.

4. $\frac{2}{3}x - 5 = 3$ **5.** $3x - 7 = 2(x + 1)$

Solve. If the solution set is $\varnothing$ or $\mathbb{R}$, classify the equation as a contradiction or as an identity.

6. $x - 4.9 = 1.7$
7. $\frac{2}{3}a = 9$
8. $-9x + 4(2x - 3) = 5(2x - 3) + 7$
9. $3(x - 4) + 2 = x + 2(x - 5)$
10. $5t - (9 - t) = 4t + 2(3 + t)$
11. $1.7x - 4.03 = -12.4(x - 0.2)$

Solve.

12. Three numbers are such that the second is 6 less than three times the first and the third is 2 more than two-thirds the first. The sum of the three numbers is 150. Find the largest of the three numbers.

13. *Landscaping.* A mature tree can be moved using a tree spade. Standard tree spades range from 20 in. to 92 in. in diameter. The diameter of the largest tree that can be moved successfully is one-tenth the diameter of the spade. (*Source: Popular Mechanics,*

April 1998: 102) How large a tree spade would be needed to move a tree that is $5\frac{1}{2}$ in. in diameter?

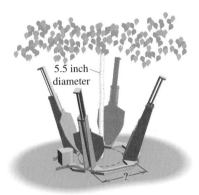

5.5 inch
diameter

?

14. *Angle Measures.*   One angle of a triangle measures 10° more than twice the measure of the second angle. The third angle measures 39° more than the sum of the other angles. What are the measures of the angles of the triangle?

15. A sheet of plastic has a thickness of 0.00015 mm. The sheet is 1.2 m by 79 m. Use scientific notation to find the volume of the sheet.

Solve each equation for the given letter.

16. $P = m/S$, for m

17. $c = mx - rx$, for x

18. $5x - 6y = 7$, for y

Find the slope and the y-intercept.

19. $g(x) = -4x - 9$

20. $-6y + 2x = 7$

Find the slope of each line. If the slope is undefined, state this.

21. Containing the points $(4, 5)$ and $(-3, 1)$

22. Containing $(-16.4, 2.8)$ and $(-16.4, 3.5)$

23. Find the rate of change for the graph below. Use appropriate units.

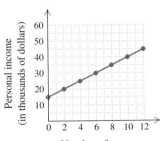

Number of years
since high school

24. Find a linear function whose graph has slope $\frac{2}{7}$ and y-intercept $(0, -6)$.

Graph by hand.

25. $f(x) = 5$

26. $3 - x = 9$

27. $-2x + 4y = 7$

28. $x + 7y = 14$

29. Determine an appropriate viewing window for the graph of $f(x) = -\frac{1}{7}x + 14$. Answers may vary.

Determine whether each pair of lines is parallel, perpendicular, or neither.

30. $y + 5 = -x,$
$x - y = 2$

31. $3x - 5 = 7y,$
$7y - 3x = 7$

Determine whether each of these is a linear equation.

32. $2x - 7 = 0$

33. $3x - 8f(x) = 7$

34. $2a + 7b^2 = 3$

35. $2p - \dfrac{7}{q} = 1$

36. Find an equation in point–slope form of the line with slope -2 and containing $(-3, 4)$.

37. Use function notation to write an equation for the line containing $(2, 5)$ and $(-4, -3)$.

38. In 1955, the U.S. minimum wage was $0.75, and in 1995, it was $4.25. Let $W =$ the minimum wage, in dollars, t years after 1955.

a) Find a linear function $W(t)$ that fits the data.

b) Use the function of part (a) to predict the minimum wage in 2005.

39. *Second-Hand Shopping.*   The following table shows the annual retail revenue for Goodwill Industries for various years.

Goodwill Revenue

YEAR	RETAIL REVENUE (IN MILLIONS)
1992	$382.9
1993	421.6
1994	470.3
1995	513.5
1996	598.7

Source: The Wall Street Journal Almanac, 1998

a) Use linear regression to find a linear function that predicts the annual retail revenue of Goodwill Industries as a function of x years since 1992.

b) Estimate the retail revenue of Goodwill in 2008.

40. Find the domain of f.

a) $f(x) = \dfrac{2}{x - 7}$ **b)** $f(x) = \dfrac{x - 7}{2}$

Let $g(x) = 3x - 6$ and $h(x) = x^2 + 1$. Find the following.

41. $g(0)$ **42.** $h(-5)$

43. $(g \cdot h)(4)$ **44.** $(g - h)(-2)$

45. $(g / h)(-1)$

46. The domains of $g + h$ and $g \cdot h$

47. The domain of h / g

Skill Maintenance

48. Simplify: $-\frac{2}{3} - \left(-\frac{4}{5}\right)$.

49. Simplify: $(5a^3b)^2$.

50. Graph: $y = -\frac{1}{2}x^2$.

51. Factor: $2xy - 8x + 4$.

Synthesis

52. ◈ Explain why every function is a relation, but not every relation is a function.

53. ◈ Explain why the slope of a vertical line is undefined whereas the slope of a horizontal line is 0.

54. Find the y-intercept of the function given by
$$f(x) + 3 = 0.17x^2 + (5 - 2x)^x - 7.$$

55. Determine the value of a such that the lines
$$3x - 4y = 12 \quad \text{and} \quad ax + 6y = -9$$
are parallel.

56. Homespun Jellies charges $2.49 for each jar of preserves. Shipping charges are $3.75 for handling, plus $0.60 per jar. Find a linear function for determining the cost of shipping x jars of preserves.

57. Match each sentence with the most appropriate graph.

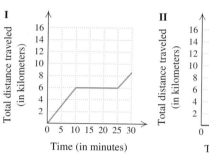

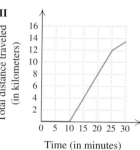

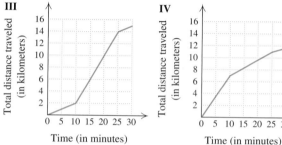

a) Joni walks for 10 min to the train station, rides the train for 15 min, and then walks 5 min to the office.

b) During a workout, Phil bikes for 10 min, runs for 15 min, and then walks for 5 min.

c) Sam pilots his motorboat for 10 min to the middle of the lake, fishes for 15 min, and then motors for another 5 min to another spot.

d) Patti waits 10 min for her train, rides the train for 15 min, and then runs for 5 min to her job.

CHAPTER 2 TEST

1. For the following graph of f, determine

a) $f(-2)$;

b) the domain of f;

c) any x-value for which $f(x) = \frac{1}{2}$; and

d) the range of f.

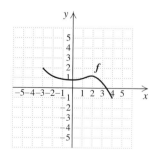

2. Estimate the domain and the range of the function whose graph is shown below.

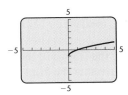

3. The function $S(t) = 1.2t + 21.4$ can be used to estimate the total U.S. sales of books, in billions of dollars, t years after 1992. Predict the total U.S. sales of books in 2002.

4. Solve graphically and check:

$$1.2x - 5 = 3.6 + x$$

Solve. If the solution set is $\mathbb{R}$ or $\varnothing$, classify the equation as an identity or a contradiction.

5. $13x - 7 = 41x + 49$

6. $8t - (5 - 2t) = 5(2t - 1)$

Solve.

7. Find three consecutive even integers such that the sum of the first, two times the second, and three times the third is 124.

8. The changes in the salary of a vice president of a corporation for three consecutive years are, respectively, a 10% increase, a 15% increase, and a 5% decrease. What is the percent of total change for those three years?

9. *Astronomy.* The average distance from the planet Venus to the sun is 6.7×10^7 mi. About how far does Venus travel in one orbit around the sun? (Assume a circular orbit.)

10. The surface area of a rectangular solid of length l, width w, and height h is given by $S = 2lh + 2lw + 2wh$. Solve for l.

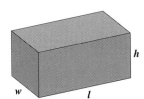

11. Solve for y: $3x^2 + 2y = x - 7$.

Find the slope and the y-intercept.

12. $f(x) = -\frac{3}{5}x + 12$

13. $-5y - 2x = 7$

Find the slope of the line containing the following points. If the slope is undefined, state this.

14. $(-2, -2)$ and $(6, 3)$

15. $(-3.1, 5.2)$ and $(-4.4, 5.2)$

16. Find a linear function whose graph has slope -5 and y-intercept $(0, -1)$.

Graph by hand.

17. $f(x) = 5x - 2$

18. $3 - x = 5 + x$

19. $-2x + 5y = 12$

20. Determine whether the standard viewing window shows the x- and y-intercepts of the graph of $f(x) = 2x + 9$.

Determine without graphing whether each pair of lines is parallel, perpendicular, or neither.

21. $4y + 2 = 3x$,
$-3x + 4y = -12$

22. $y = -2x + 5$,
$2y - x = 6$

23. Which of these are linear equations?

a) $8x - 7 = 0$

b) $4b - 9a^2 = 2$

c) $2x - 5y = 3$

24. Find an equation in point–slope form of the line with slope 4 and containing $(-2, -4)$.

25. Use function notation to write an equation for the line containing $(3, -1)$ and $(4, -2)$.

26. If you rent a van for one day and drive it 250 mi, the cost is $100. If you drive it 300 mi, the cost is $115. Let $C(m)$ = the cost, in dollars, of driving m miles.

a) Find a linear function that fits the data.

b) Use the function to find how much it will cost to rent the van for one day and drive it 500 mi.

27. *Answering Machines.* The following table shows the sales of telephone answering machines for the years 1989–1993.

Sales of Answering Machines

YEAR	SALES (IN MILLIONS)
1989	12.5
1990	13.6
1991	15.4
1992	14.6
1993	16.3

Source: U.S. Consumer Electronics Industry, in *The Macmillan Visual Almanac, 1996*

a) Use linear regression to find a linear function that can be used to predict the number of answering machines sold as a function of x years since 1989.

b) Predict the number of answering machines sold in 2005.

28. Find the domain of $g(x) = \dfrac{x + 3}{2x - 5}$.

29. Find the following, given that $g(x) = -3x - 4$ and $h(x) = x^2 + 1$.

a) $h(-2)$
b) $(g \cdot h)(3)$
c) The domain of h / g

Skill Maintenance

30. Simplify: $-\frac{4}{5} - \frac{2}{15}$.
31. Simplify: $(a^4 b^5)(a^2 b^7)$.
32. Graph: $y = -|2x|$.
33. Factor: $3x - 6xy$.

Synthesis

34. The function $f(t) = 5 + 15t$ can be used to determine a bicycle racer's location, in miles from the starting line, measured t hours after passing the 5-mi mark.

a) How far from the start will the racer be 1 hr and 40 min after passing the 5-mi mark?
b) Assuming a constant rate, how fast is the racer traveling?

Find an equation of the line.

35. Containing $(-3, 2)$ and parallel to the line $2x - 5y = 8$

36. Containing $(-3, 2)$ and perpendicular to the line $2x - 5y = 8$

37. The graph of the function $f(x) = mx + b$ contains the points $(r, 3)$ and $(7, s)$. Express s in terms of r if the graph is parallel to the line $3x - 2y = 7$.

38. Given that $f(x) = 5x^2 + 1$ and $g(x) = 4x - 3$, find an expression for $h(x)$ such that the domain of $f / g / h$ is $\left\{x \mid x \text{ is a real number } and \ x \neq \frac{3}{4} \ and \ x \neq \frac{2}{7}\right\}$. Answers may vary.

Systems of Equations and Problem Solving 3

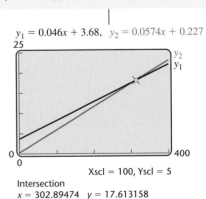

We have seen how to solve linear equations using both algebraic and graphical approaches. In this chapter, we study *systems of equations* and how to solve them using graphing, substitution, and elimination.

Systems of equations have extensive application to many fields, such as psychology, sociology, business, education, engineering, and science. This chapter also includes a brief introduction to *matrices*, which can provide another method for solving systems of equations.

APPLICATION

RECYCLING. In the United States, the amount of solid waste being recycled is slowly catching up to the amount being generated, as shown by the data in the following table.

YEAR	AMOUNT OF WASTE GENERATED (IN POUNDS PER PERSON PER DAY)	AMOUNT OF WASTE RECYCLED (IN POUNDS PER PERSON PER DAY)
1980	3.7	0.35
1985	3.8	0.38
1990	4.3	0.7
1995	4.3	1.2

Source: The Statistical Abstract of the United States, 1997

Predict the year in which the amount of waste recycled will equal the amount generated.

We let $x =$ the number of years since 1980, and use linear regression to fit a line to each set of data, forming a *system of equations*. The solution can be read from the point of intersection of the graphs.

This problem appears as Example 7 in Section 3.3.

$y_1 = 0.046x + 3.68, \quad y_2 = 0.0574x + 0.227$

Xscl = 100, Yscl = 5

Intersection
$x = 302.89474 \quad y = 17.613158$

3.1

Systems of Equations in Two Variables

- *Translating*
- *Identifying Solutions*
- *Solving Systems Graphically*

Translating

Problems involving more than one unknown quantity are often solved most easily by translating to two or more equations.

Example 1   *Real Estate.* Translate the following problem situation to mathematical language, using two equations.

> In 1996, the Simon Property Group and the DeBartolo Realty Corporation merged to form the largest real estate company in the United States, owning 183 shopping centers in 32 states. Prior to merging, Simon owned twice as many properties as DeBartolo. How many properties did each company own before the merger?

DeBartolo shareholders would receive 0.68 share of Simon common stock for each share of DeBartolo common stock. Simon also would agree to repay $1.5 billion in DeBartolo debt. At Tuesday's closing price of $23.625 a share for common stock, the transaction is valued at roughly $3 billion.

Executives say the proposed company, Simon DeBartolo Group, would be the largest real estate company in the United States, worth $7.5 billion. **Not included in the deal:** DeBartolo's ownership stake in the San Francisco 49ers, or the Indiana Pacers, owned separately by the Simon

SOLUTION

1. Familiarize.   We have already seen problems in which we need to look up certain formulas or the meaning of certain words. Here we need only observe that the words shopping centers and properties are being used interchangeably.

Often problems contain information that has no bearing on the problem being solved. In this case, the number 32 is irrelevant to the question being asked. Instead we focus on the number of properties owned and the phrase "twice as many." Rather than guess and check, let's proceed to the next step, using x for the number of properties originally owned by Simon and y for the number of properties originally owned by DeBartolo.

2. Translate.   There are two statements to translate. First we look at the total number of properties involved:

Rewording: The number of Simon properties   plus   the number of DeBartolo properties   total   183.

Translating: x $+$ y $=$ 183

The second statement compares the number of properties that each company held before merging:

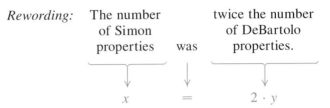

Rewording: The number of Simon properties   was   twice the number of DeBartolo properties.

$$x = 2 \cdot y$$

We have now translated the problem to a pair, or **system, of equations**:

$$x + y = 183,$$
$$x = 2y.$$

Problems like Example 1 *can* be solved using one variable; however, as problems become complicated, you will find that using more than one variable (and more than one equation) is often the preferable approach.

Example 2 *Retail Sales.* In one day, Glovers, Inc., sold 20 pairs of gloves. Fleece gloves sold for $24.95 a pair and Gore-Tex® gloves for $37.50. Receipts totaled $687.25. Write a system of equations that could be used to find how many of each kind were sold.

SOLUTION

1. **Familiarize.** To familiarize ourselves with this problem, let's guess that Glovers sold 12 pairs of fleece gloves and 11 pairs of Gore-Tex gloves. Does this guess check? Since a total of 20 pairs of gloves was sold, and since $12 + 11 \neq 20$, our guess cannot be right.

   As a second guess, suppose that 12 pairs of fleece gloves and 8 pairs of Gore-Tex gloves were sold. Now the total sold is 20, so our guess is right in that respect. How much money would have been taken in? Since fleece gloves sold for $24.95 and Gore-Tex gloves for $37.50, the total received would have been

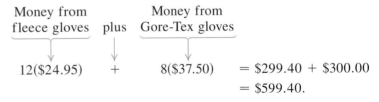

   Money from fleece gloves   plus   Money from Gore-Tex gloves

$$12(\$24.95) \quad + \quad 8(\$37.50) \quad = \$299.40 + \$300.00$$
$$= \$599.40.$$

   Although the total number of pairs is correct, our guess is incorrect because the problem states that the total amount received was $687.25. Since $599.40 is less than $687.25, more of the expensive gloves were sold than we had guessed. We could now adjust our guess accordingly. Instead, let's work toward an algebraic approach that avoids guessing.

2. **Translate.** We let f = the number of pairs of fleece gloves sold and g = the number of pairs of Gore-Tex gloves sold. The information can be organized in a table, which will help with the translating.

KIND OF GLOVE	FLEECE	GORE-TEX	TOTAL	
NUMBER SOLD	f	g	20	$\longrightarrow f + g = 20$
PRICE	$24.95	$37.50		
AMOUNT TAKEN IN	$24.95f$	$37.50g$	687.25	$\longrightarrow 24.95f + 37.50g = 687.25$

The first row of the table and the first sentence of the problem indicate that a total of 20 pairs of gloves was sold:

$$f + g = 20.$$

Since each pair of fleece gloves cost $24.95 and f pairs were sold, $24.95f$ represents the amount taken in from the sale of fleece gloves. Similarly, $37.50g$ represents the amount taken in from the sale of g pairs of Gore-Tex gloves. From the third row of the table and the third sentence of the problem, we get the second equation:

$$24.95f + 37.50g = 687.25.$$

Multiplying by 100 on both sides, we can clear the decimals. This gives the following system of equations as the translation:

$$f + g = 20,$$
$$2495f + 3750g = 68{,}725.$$

$\blacksquare$

Identifying Solutions

A *solution* of a system of equations in two variables is an ordered pair of numbers that makes *both* equations true.

Example 3   Determine whether $(-4, 7)$ is a solution of the system

$$x + y = 3,$$
$$5x - y = -27.$$

SOLUTION   We use alphabetical order of the variables. Thus we replace x with -4 and y with 7:

$$\frac{x + y = 3}{-4 + 7 \;?\; 3}$$
$$3 \mid 3 \quad \text{TRUE}$$

$$\frac{5x - y = -27}{5(-4) - 7 \;?\; -27}$$
$$-20 - 7$$
$$-27 \mid -27 \quad \text{TRUE}$$

The pair $(-4, 7)$ makes both equations true, so it is a solution of the system. We sometimes describe such a solution by saying, in this case, that $x = -4$ and $y = 7$. Set notation can also be used to list the solution set $\{(-4, 7)\}$.

$\blacksquare$

graph
addition
substitution

Solving Systems Graphically

Recall that the graph of an equation is a drawing that represents its solution set. If we graph the equations in Example 3, we find that $(-4, 7)$ is the only point common to both lines. Thus one way to solve a system of two equations is to graph both equations and identify any points of intersection. The coordinates of each point of intersection represent a solution of that system.

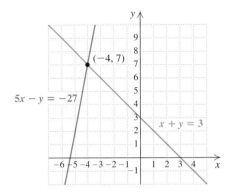

In Section 2.2, we solved linear equations by graphing two equations and looking for points of intersection. There, the solution was the x-coordinate; here, we need both coordinates for the solution.

Example 4 Solve each system graphically.

a) $y - x = 1,$
   $y + x = 3$

b) $y = -3x + 5,$
   $y = -3x - 2$

c) $3y - 2x = 6,$
   $-12y + 8x = -24$

SOLUTION

a) We graph each equation. All ordered pairs from line L_1 are solutions of the first equation. All ordered pairs from line L_2 are solutions of the second equation. The point of intersection has coordinates that make *both* equations true. Apparently, $(1, 2)$ is the solution. Graphing is not perfectly accurate, so solving by graphing may yield approximate answers. Our check below shows that $(1, 2)$ is indeed the solution.

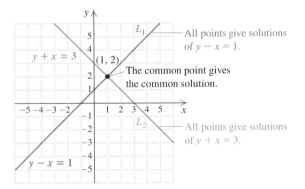

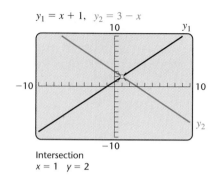

$y_1 = x + 1, \quad y_2 = 3 - x$

Intersection
$x = 1 \quad y = 2$

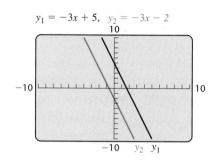

$y_1 = -3x + 5, \quad y_2 = -3x - 2$

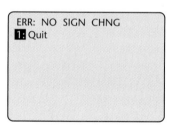

ERR: NO SIGN CHNG
1: Quit

CHECK:

$$
\begin{array}{c}
y - x = 1 \\
\hline
2 - 1 \ ? \ 1 \\
1 \ | \ 1 \quad \text{TRUE}
\end{array}
\qquad
\begin{array}{c}
y + x = 3 \\
\hline
2 + 1 \ ? \ 3 \\
3 \ | \ 3 \quad \text{TRUE}
\end{array}
$$

We can also solve the system using the INTERSECT feature on a grapher, as shown in the figure at left.

b) We graph the equations. The lines have the same slope, -3, and different y-intercepts, so they are parallel. There is no point at which they cross, so the system has no solution.

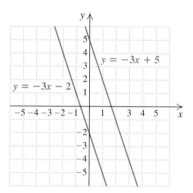

What happens when we try to solve this system using a grapher? We enter and graph both equations at left, noting that the graphs appear to be parallel. When we attempt to find the intersection using the INTERSECT feature, we get an error message.

c) We graph the equations and find that the same line is drawn twice. Thus any solution of one of the equations is a solution of the other. Each equation has an infinite number of solutions, some of which are listed on the graph. Each of these is also a solution of the other equation. We check one solution, $(0, 2)$, which is the y-intercept of each equation.

CHECK:

$$
\begin{array}{c}
3y - 2x = 6 \\
\hline
3(2) - 2(0) \ ? \ 6 \\
6 - 0 \\
6 \ | \ 6 \quad \text{TRUE}
\end{array}
$$

$$
\begin{array}{c}
-12y + 8x = -24 \\
\hline
-12(2) + 8(0) \ ? \ -24 \\
-24 + 0 \\
-24 \ | \ -24 \quad \text{TRUE}
\end{array}
$$

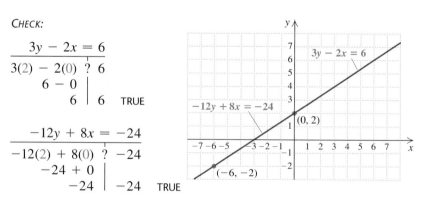

You can check that $(-6, -2)$ is another solution of both equations. In fact, any pair that is a solution of one equation is a solution of the other equation as well. Thus the solution set is $\{(x, y) \mid 3y - 2x = 6\}$ or, in words, "the set of all pairs (x, y) for which $3y - 2x = 6$." Since the two equations are equivalent, we could have used $-12y + 8x = -24$ in place of $3y - 2x = 6$.

If we attempt to find the intersection of the graphs of the equations in this system using INTERSECT, the grapher will return as the intersection whatever point we choose as the guess. This is a point of intersection, as is any other point on the graph of the lines. ▬

Example 4 illustrates that when we graph a system of two linear equations in two variables, one of the following three outcomes will occur.

1. The lines have one point in common, and that point is the only solution of the system (see part (a)).
2. The lines are parallel, with no point in common, and the system has no solution (see part (b)). This system is called **inconsistent**.
3. The lines coincide, sharing the same graph. Since every solution of one equation is a solution of the other, the system has an infinite number of solutions (see part (c)). The equations are said to be **dependent**.

Any system of equations that has at least one solution is said to be **consistent**. The systems of Examples 4(a) and 4(c) are both consistent.

Graphers are especially useful when equations contain fractions or decimals or when the coordinates of the intersection are not integers.

Example 5 Solve graphically:

$$3.45x + 4.21y = 8.39,$$
$$7.12x - 5.43y = 6.18.$$

SOLUTION First, we solve for y in each equation:

$$3.45x + 4.21y = 8.39$$

$$4.21y = 8.39 - 3.45x \qquad \text{Subtracting } 3.45x \text{ on both sides}$$

$$y = (8.39 - 3.45x)/4.21; \qquad \text{Dividing by } 4.21 \text{ on both sides}$$

$$7.12x - 5.43y = 6.18$$

$$-5.43y = 6.18 - 7.12x \qquad \text{Subtracting } 7.12x \text{ on both sides}$$

$$y = (6.18 - 7.12x)/(-5.43). \qquad \text{Dividing by } -5.43 \text{ on both sides}$$

It is not necessary to simplify further. We have the system

$$y = (8.39 - 3.45x)/4.21,$$
$$y = (6.18 - 7.12x)/(-5.43).$$

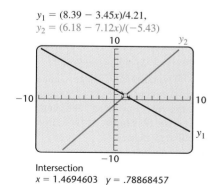

$y_1 = (8.39 - 3.45x)/4.21,$
$y_2 = (6.18 - 7.12x)/(-5.43)$

Intersection
$x = 1.4694603 \quad y = .78868457$

Next, we enter both equations and graph using the same viewing window. By using the INTERSECT feature in the CALC menu, we see that, to the nearest hundredth, the solution is $(1.47, 0.79)$. ▬

Graphing is helpful when solving systems because it allows us to "see" the solution. It can also be used on systems of nonlinear equations, and in many applications, it provides a satisfactory answer. However, graphing has the disadvantage of often yielding inexact answers when fractional or decimal solutions are involved. In Section 3.2, we will develop two algebraic methods of solving systems. Both methods produce exact answers.

3.1 Exercise Set

Translate each problem situation to a system of equations. Do not attempt to solve, but save for later use.

1. The difference between two numbers is 11. Twice the smaller plus three times the larger is 123. What are the numbers?

2. The sum of two numbers is −42. The first number minus the second number is 52. What are the numbers?

3. *Retail Sales.*   Paint Town sold 45 paintbrushes, one kind at $8.50 each and another at $9.75 each. In all, $398.75 was taken in for the brushes. How many of each kind were sold?

4. *Retail Sales.*   Mountainside Fleece sold 40 neckwarmers. Solid color neckwarmers sold for $9.90 each and print ones sold for $12.75 each. In all, $421.65 was taken in for the neckwarmers. How many of each type were sold?

5. *Geometry.*   Two angles are supplementary.* One angle is 3° less than twice the other. Find the measures of the angles.

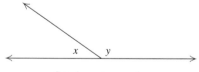

Supplementary angles

6. *Geometry.*   Two angles are complementary.† The sum of the measures of the first angle and half the second angle is 64°. Find the measures of the angles.

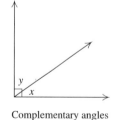

Complementary angles

---

*The sum of the measures of two supplementary angles is 180°.

†The sum of the measures of two complementary angles is 90°.

7. *Basketball Scoring.*   Amma scored 18 times during one basketball game. She scored a total of 30 points, two for each field goal and one for each free throw. How many field goals did she make? How many free throws?

8. *Fundraising.*   The St. Mark's Community Barbecue served 250 dinners. A child's plate cost $3.50 and an adult's plate cost $7.00. A total of $1347.50 was collected. How many of each type of plate was served?

9. *Sales of Pharmaceuticals.*   The Diabetic Express recently charged $15.75 for a vial of Humulin insulin and $12.95 for a vial of Novolin insulin. If a total of $959.35 was collected for 65 vials of insulin, how many vials of each type were sold?

10. *Court Dimensions.*   The perimeter of a standard basketball court is 288 ft. The length is 44 ft longer than the width. Find the dimensions.

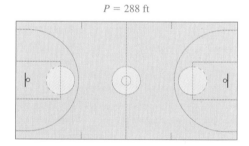

$P = 288$ ft

11. *Court Dimensions.*   The perimeter of a standard tennis court used for doubles is 228 ft. The width is 42 ft less than the length. Find the dimensions.

12. *Basketball Scoring.*   The Central College Cougars made 40 field goals in a recent basketball game, some 2-pointers and the rest 3-pointers. Altogether the 40 baskets counted for 89 points. How many of each type of field goal was made?

13. *Lumber Production.*   Snookers Lumber can convert logs into either lumber or plywood. In a given day, the mill turns out twice as many units of plywood as lumber. It makes a profit of $25 on a unit of lumber and $40 on a unit of plywood. How many units of each type must be produced and sold in order to make a profit of $10,920?

14. *Video Rentals.* J. P.'s Video rents general interest films for \$3.00 each and children's films for \$1.50 each. In one day, a total of \$213 was taken in from the rental of 77 videos. How many of each type of video was rented?

15. *Hockey Rankings.* Hockey teams receive 2 points for a win and 1 point for a tie. The Wildcats once won a championship with 60 points. They won 9 more games than they tied. How many wins and how many ties did the Wildcats have?

16. *Radio Airplay.* Omar must play 12 commercials during his 1-hr radio show. Each commercial is either 30 sec or 60 sec long. If the total commercial time during that hour is 10 min, how many commercials of each type does Omar play?

17. *Nontoxic Floor Wax.* A nontoxic floor wax can be made from lemon juice and food-grade linseed oil. The amount of oil should be twice the amount of lemon juice. How much of each ingredient is needed to make 32 oz of floor wax? (The mix should be spread with a rag and buffed when dry.)

18. *Airplane Seating.* An airplane has a total of 152 seats. The number of coach-class seats is 5 more than six times the number of first-class seats. How many of each type of seat are there on the plane?

Determine whether the ordered pair is a solution of the given system of equations. Remember to use alphabetical order of variables.

19. $(1, 2)$; $4x - y = 2$,
$10x - 3y = 4$

20. $(-1, -2)$; $2x + y = -4$,
$x - y = 1$

21. $(2, 5)$; $y = 3x - 1$,
$2x + y = 4$

22. $(-1, -2)$; $x + 3y = -7$,
$3x - 2y = 12$

23. $(1, 5)$; $x + y = 6$,
$y = 2x + 3$

24. $(5, 2)$; $a + b = 7$,
$2a - 8 = b$

25. $(3, 1)$; $3x + 4y = 13$,
$5x - 4y = 11$

26. $(4, -2)$; $-3x - 2y = -8$,
$y = 2x - 5$

Solve each system graphically. Be sure to check your solution. If a system has an infinite number of solutions, use set-builder notation to write the solution set. If a system has no solution, state this.

27. $x - y = 3$,
$x + y = 5$

28. $x + y = 4$,
$x - y = 2$

29. $3x + y = 5$,
$x - 2y = 4$

30. $2x - y = 4$,
$5x - y = 13$

31. $4y = x + 8$, $(4, 3)$
$3x - 2y = 6$

32. $4x - y = 9$,
$x - 3y = 16$

33. $x = y - 1$, $(-3, -2)$
$2x = 3y$

34. $a = 1 + b$,
$b = 5 - 2a$

35. $2u + v = 3$,
$2u = v + 7$

36. $x = 4$,
$y = -5$

37. $x = -3$,
$y = 2$

38. $2a + b = 4$,
$b = 4a + 1$

39. $y = -5.43x + 10.89$,
$y = 6.29x - 7.04$

40. $y = 123.52x + 89.32$,
$y = -89.22x + 33.76$

41. $2b + a = 11$,
$a - b = 5$

42. $3x - 10y = -40$,
$6x - 20y = 20$

43. $y = -\frac{1}{4}x + 1$,
$2y = x - 4$

44. $y = -\frac{1}{3}x - 1$,
$4x - 3y = 18$

45. $2.18x + 7.81y = 13.78$,
$5.79x - 3.45y = 8.94$

46. $2.6x - 1.1y = 4$,
$1.32y = 3.12x - 5.04$

47. $y - x = 5$,
$2x - 2y = 10$

48. $6x - 2y = 2$,
$9x - 3y = 1$

49. $57y - 45x = 33$,
$95y - 22 = 30x$

50. $-9.25x - 12.94y = -3.88$,
$21.83x + 16.33y = 13.69$

51. $y = 3 - x$,
$2x + 2y = 6$

52. $y = -x - 1$,
$4x - 3y = 24$

53. $1.9x = 4.8y + 1.7$,
$12.92x + 23.8 = 32.64y$

54. $2x - 3y = 6$,
$3y - 2x = -6$

55. For the systems in the odd-numbered exercises 27–53, which are consistent?

56. For the systems in the even-numbered exercises 28–54, which are consistent?

57. For the systems in the odd-numbered exercises 27–53, which contain dependent equations?

58. For the systems in the even-numbered exercises 28–54, which contain dependent equations?

Skill Maintenance

Solve. [2.2]

59. $3x + 4 = x - 2$

60. $\frac{3}{5}x + 2 = \frac{2}{5}x - 5$

61. $4x - 5x = 8x - 9 + 11x$

62. Solve $Q = \frac{1}{4}(a - b)$ for b. [2.3]

Factor. [1.3]

63. $3x - 21$

64. $2a + ab - ac$

Synthesis

65. ◈ Write a problem for a classmate to solve that can be translated into a system of two equations. Devise the problem so that the solution is "Shelly gave 9 haircuts and 5 shampoos."

66. ◈ Write a problem for a classmate to solve that requires writing a system of two equations. Devise the problem so that the solution is "The Lakers made 6 three-point baskets and 31 two-point baskets."

67. ◈ Explain the difference between solving a linear equation graphically and solving a system of equations graphically.

Arms Sales. *For Exercises 68–70, consider the following graph.**

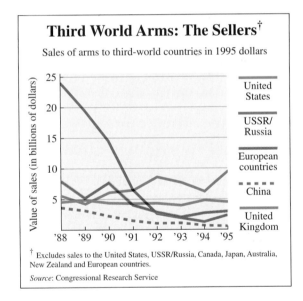

Third World Arms: The Sellers†

Sales of arms to third-world countries in 1995 dollars

† Excludes sales to the United States, USSR/Russia, Canada, Japan, Australia, New Zealand and European countries.

Source: Congressional Research Service

68. ◈ Does the graph above support the statement, "The United States has always sold twice as many arms to third-world countries as China has"? Why or why not?

69. In what years did U.S. arms sales to third-world countries exceed USSR/Russia arms sales to third-world countries?

*Based on information provided by the Congressional Research Service, 1996.

70. Determine the most recent year in which USSR/Russia arms sales to third-world countries exceeded the combined sales of the United States, Europe, and China.

71. Write a system of equations for which:
   a) (5, 1) is a solution,
   b) there is no solution, and
   c) there is an infinite number of solutions.

72. A system of linear equations has $(1, -1)$ and $(-2, 3)$ as solutions. Determine:
   a) a third point that is a solution, and
   b) how many solutions there are.

73. The solution of the following system is $(4, -5)$. Find A and B.
$$Ax - 6y = 13,$$
$$x - By = -8.$$

Translate to a system of equations. Do not solve.

74. *Ages.*   Burl is twice as old as his son. Ten years ago, Burl was three times as old as his son. How old are they now?

75. *Work Experience.*   Lou and Juanita are mathematics professors at a state university. Together, they have 46 years of service. Two years ago, Lou had taught 2.5 times as many years as Juanita. How long has each taught at the university?

76. *Design.*   A piece of posterboard has a perimeter of 156 in. If you cut 6 in. off the width, the length becomes four times the width. What are the dimensions of the original piece of posterboard?

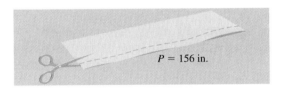

$P = 156$ in.

77. *Nontoxic Scouring Powder.*   A nontoxic scouring powder is made up of 4 parts baking soda and 1 part vinegar. How much of each ingredient is needed for a 16-oz mixture?

Solve graphically.

78. $y = |x|,$
   $x + 4y = 15$

79. $x - y = 0,$
   $y = x^2$

In Exercises 80–83, match each system with one of the graphs shown below.

a)

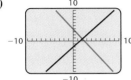

b)

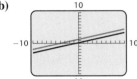

c)

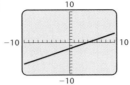

d)

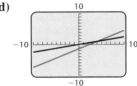

80. $x = 4y,$
$3x - 5y = 7$

81. $2x - 8 = 4y,$
$x - 2y = 4$

82. $8x + 5y = 20,$
$4x - 3y = 6$

83. $x = 3y - 4,$
$2x + 1 = 6y$

COLLABORATIVE CORNER

Focus: Systems of linear equations

Time: 20 minutes

Group size: 3

The box score at right, from a basketball game between the Boston Celtics and the Detroit Pistons, contains information on how many field goals and free throws each player attempted and made. For example, the line "Hill 9–17 7–9 25" means that Detroit's Grant Hill made 9 field goals out of 17 attempts and 7 free throws out of 9 attempts, for a total of 25 points. (Each free throw is worth 1 point and each field goal is worth either 2 or 3 points, depending on how far from the basket it was shot.)

Activity

1. Work as a group to develop a system of two equations in two unknowns that can be used to determine how many 2-pointers and how many 3-pointers were made by Boston's David Wesley.

2. Each group member should solve the system from part (1) in a different way: one person by graphing, one person by making a table and methodically checking all combinations of 2- and 3-pointers, and one person by guesswork. Compare answers when this has been completed.

■ **Pistons 99, Celtics 89:** In Auburn Hills, Grant Hill had 25 points and 11 rebounds, leading the Detroit Pistons over the Boston Celtics.

Hill also had eight assists. Otis Thorpe added 16 points and Joe Dumars had 15.

David Wesley led the Celtics with 25.

BOSTON (89)
Fox 6-9 3-4 17, Williams 5-9 4-4 14, Radja 4-10 2-2 10, Minor 0-3 0-0 0, Wesley 9-13 3-5 25, Walker 5-14 6-8 16, Barros 1-5 0-0 3, Day 1-6 1-2 4, Conlon 0-0 0-0 0, Szabo 0-0 0-0 0. Totals 31-69 19-25 89.

Detroit (99)
Hill 9-17 7-9 25, Long 3-5 2-2 8, Thorpe 7-8 2-3 16, Hunter 2-14 2-2 7, Dumars 5-10 3-3 15, Mills 5-11 0-0 13, Curry 3-3 0-0 7, Ratliff 2-3 0-0 4, Green 1-1 2-2 4, Mahorn 0-0 0-2 0. Totals 37-72 18-23 99.
Boston 26 16 21 26—89
Detroit 27 29 17 26—99

3. Determine, as a group, how many 2- and 3-pointers the Detroit Pistons made as a team.

3.2
Solving by Substitution or Elimination

- *The Substitution Method*
- *The Elimination Method*
- *Comparing Methods*

The Substitution Method

One algebraic and nongraphical method for solving systems of equations, the *substitution method,* relies on having one variable isolated.

Example 1   Solve the system

$$x + y = 4, \qquad (1)$$
$$x = y + 1. \qquad (2)$$

SOLUTION   Equation (2) says that x and $y + 1$ name the same number. Thus we can substitute $y + 1$ for x in equation (1):

$$x + y = 4 \qquad \text{Equation (1)}$$
$$(y + 1) + y = 4. \qquad \text{Substituting } y + 1 \text{ for } x$$

We solve this last equation, using methods learned earlier:

$$(y + 1) + y = 4$$
$$2y + 1 = 4 \qquad \text{Removing parentheses and combining like terms}$$
$$2y = 3 \qquad \text{Subtracting 1 on both sides}$$
$$y = \tfrac{3}{2}. \qquad \text{Dividing by 2}$$

We now return to the original pair of equations and substitute $\tfrac{3}{2}$ for y in either equation so that we can solve for x. For this problem, calculations are slightly easier if we use equation (2):

$$x = y + 1 \qquad \text{Equation (2)}$$
$$= \tfrac{3}{2} + 1 \qquad \text{Substituting } \tfrac{3}{2} \text{ for } y$$
$$= \tfrac{3}{2} + \tfrac{2}{2} = \tfrac{5}{2}.$$

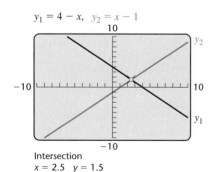

$y_1 = 4 - x, \; y_2 = x - 1$

Intersection
$x = 2.5 \quad y = 1.5$

We obtain the ordered pair $\left(\tfrac{5}{2}, \tfrac{3}{2}\right)$. A check ensures that it is a solution:

CHECK:

$$\begin{array}{c|c} x + y = 4 \\ \hline \tfrac{5}{2} + \tfrac{3}{2} \; ? \; 4 \\ \tfrac{8}{2} \\ 4 \; | \; 4 \quad \text{TRUE} \end{array} \qquad \begin{array}{c|c} x = y + 1 \\ \hline \tfrac{5}{2} \; ? \; \tfrac{3}{2} + 1 \\ \tfrac{3}{2} + \tfrac{2}{2} \\ \tfrac{5}{2} \; | \; \tfrac{5}{2} \quad \text{TRUE} \end{array}$$

Since $\left(\tfrac{5}{2}, \tfrac{3}{2}\right)$ checks, it is the solution. ▬

The system in Example 1 can also be solved by using a grapher, with the coordinates given in decimal notation. The graph shown in the margin above serves as a check and provides a visualization of the problem.

If neither equation in a system has a variable alone on one side, we first isolate a variable in one equation and then substitute.

Example 2 Solve the system

$$2x + y = 6, \qquad (1)$$
$$3x + 4y = 4. \qquad (2)$$

SOLUTION   First, we select an equation and solve for one variable. To isolate y, we can add $-2x$ on both sides of equation (1) to get

$$y = 6 - 2x. \qquad (3)$$

Then we substitute $6 - 2x$ for y in equation (2) and solve for x:

$$3x + 4(6 - 2x) = 4 \qquad \text{Substituting } 6 - 2x \text{ for } y \text{ in equation (2).}$$
$$\text{Use parentheses!}$$
$$3x + 24 - 8x = 4 \qquad \text{Distributing to remove parentheses}$$
$$3x - 8x = 4 - 24$$
$$-5x = -20$$
$$x = 4.$$

Next, we substitute 4 for x in either equation (1), (2), or (3). It is easiest to use equation (3) since it has already been solved for y:

$$y = 6 - 2x$$
$$= 6 - 2(4)$$
$$= 6 - 8 = -2.$$

The pair $(4, -2)$ appears to be the solution.

CHECK:

$$\frac{2x + y = 6}{2(4) + (-2) \; ? \; 6}$$
$$8 - 2 \mid$$
$$6 \mid 6 \quad \text{TRUE}$$

$$\frac{3x + 4y = 4}{3(4) + 4(-2) \; ? \; 4}$$
$$12 - 8 \mid$$
$$4 \mid 4 \quad \text{TRUE}$$

We can also check by solving graphically, as shown at left. Since $(4, -2)$ checks, it is the solution.

$y_1 = 6 - 2x, \quad y_2 = (4 - 3x)/4$

Intersection
$x = 4 \quad y = -2$

Some systems have no solution, as we saw graphically in Section 3.1. How do we recognize such systems if we are solving by an algebraic method?

Example 3 Solve the system

$$y = -3x + 5,$$
$$y = -3x - 2.$$

SOLUTION   We solved this system graphically in Example 4(b) of Section 3.1, and found that the lines are parallel and the system has no solution. Let's now try to solve the system by substitution. We substitute $-3x - 2$ for y in the first equation:

$$-3x - 2 = -3x + 5 \qquad \text{Substituting } -3x - 2 \text{ for } y$$
$$-2 = 5. \qquad \text{Adding } 3x \text{ on both sides; } -2 = 5 \text{ is a contradiction.}$$

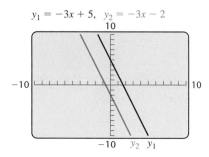

$y_1 = -3x + 5, \quad y_2 = -3x - 2$

When we add $3x$ to get the x-terms on one side, the x-terms drop out and we end up with a contradiction. When solving algebraically yields a false equation, the system has no solution. Note from the graph at left that the graphs of the equations do not intersect.

The Elimination Method

The *elimination method* for solving systems of equations makes use of the *addition principle*: If $a = b$, then $a + c = b + c$. Consider the following system:

$$2x - 3y = 0, \qquad (1)$$
$$-4x + 3y = -1. \qquad (2)$$

The key to the advantage of the elimination method for solving this system involves the $-3y$ in one equation and the $3y$ in the other. These terms are opposites. If we add all terms on the left side of the equations, $-3y$ and $3y$ add to 0, and in effect, the variable y is "eliminated."

To use the addition principle for equations, note that according to equation (2), $-4x + 3y$ and -1 are the same number. Thus we can use a vertical form and add $-4x + 3y$ on the left side of equation (1) and -1 on the right side:

$$
\begin{array}{ll}
2x - 3y = 0 & (1) \\
\underline{-4x + 3y = -1} & (2) \\
-2x + 0y = -1. & \textbf{Adding}
\end{array}
$$

This eliminates the variable y, which is why this process is called the elimination method. We now have an equation with just one variable, x, for which we solve:

$$-2x = -1$$
$$x = \tfrac{1}{2}.$$

Next, we substitute $\tfrac{1}{2}$ for x in equation (1) and solve for y:

$$2 \cdot \tfrac{1}{2} - 3y = 0 \qquad \textbf{Substituting. We also could have used equation (2).}$$
$$1 - 3y = 0$$
$$-3y = -1, \text{ so } y = \tfrac{1}{3}.$$

CHECK:

$$
\begin{array}{c|c}
2x - 3y = 0 \\
\hline
2\left(\tfrac{1}{2}\right) - 3\left(\tfrac{1}{3}\right) \ ? \ 0 \\
1 - 1 \ \Big| \\
0 \ \Big| \ 0 \quad \text{TRUE}
\end{array}
\qquad
\begin{array}{c|c}
-4x + 3y = -1 \\
\hline
-4\left(\tfrac{1}{2}\right) + 3\left(\tfrac{1}{3}\right) \ ? \ -1 \\
-2 + 1 \ \Big| \\
-1 \ \Big| \ -1 \quad \text{TRUE}
\end{array}
$$

Since $\left(\tfrac{1}{2}, \tfrac{1}{3}\right)$ checks, it is the solution.

To eliminate a variable, we must sometimes multiply before adding.

Example 4   Solve the system

$$5x + 4y = 22, \qquad (1)$$
$$-3x + 8y = 18. \qquad (2)$$

SOLUTION   If we add, we will not eliminate a variable. However, if the $4y$ in equation (1) were $-8y$, we would. Thus we multiply by -2 on both sides of the first equation:

$$
\begin{array}{ll}
-10x - 8y = -44 & \text{Multiplying by } -2 \text{ on both sides of} \\
\underline{-3x + 8y = 18} & \text{equation (1)} \\
-13x + 0 = -26 & \text{Adding} \\
 x = 2. & \text{Solving for } x
\end{array}
$$

Then

$$
\begin{array}{ll}
-3 \cdot 2 + 8y = 18 & \text{Substituting 2 for } x \text{ in equation (2)} \\
-6 + 8y = 18 & \\
8y = 24 & \\
y = 3. & \text{Solving for } y
\end{array}
$$

We obtain $(2, 3)$, or $x = 2$, $y = 3$. We can check either by substituting in the original equations or by solving graphically, as shown in the figure at left. The solution is $(2, 3)$.

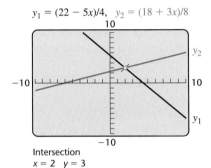

$y_1 = (22 - 5x)/4, \quad y_2 = (18 + 3x)/8$

Intersection
$x = 2 \quad y = 3$

Sometimes we must multiply twice in order to make two terms become opposites.

Example 5   Solve the system

$$2x + 3y = 17, \qquad (1)$$
$$5x + 7y = 29. \qquad (2)$$

SOLUTION   We multiply so that the x-terms are eliminated.

$$
\begin{array}{lll}
2x + 3y = 17, & \longrightarrow \text{Multiplying by 5} \longrightarrow & 10x + 15y = 85 \\
5x + 7y = 29 & \longrightarrow \text{Multiplying by } -2 \longrightarrow & \underline{-10x - 14y = -58} \\
& & 0 + y = 27 \quad \text{Adding} \\
& & y = 27.
\end{array}
$$

Next, we substitute to find x:

$$
\begin{array}{ll}
2x + 3 \cdot 27 = 17 & \text{Substituting 27 for } y \text{ in equation (1)} \\
2x + 81 = 17 & \\
2x = -64 & \\
x = -32. & \text{Solving for } x
\end{array}
$$

To check, we use a grapher, as shown in the figure at left. We obtain $(-32, 27)$, or $x = -32$, $y = 27$, as the solution.

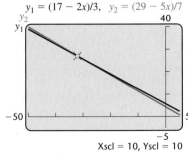

$y_1 = (17 - 2x)/3, \quad y_2 = (29 - 5x)/7$

Xscl = 10, Yscl = 10

Intersection
$x = -32 \quad y = 27$

Example 6 Solve the system

$$3y - 2x = 6, \qquad (1)$$
$$-12y + 8x = -24. \qquad (2)$$

$y_1 = (6 + 2x)/3,$
$y_2 = (-24 - 8x)/(-12)$

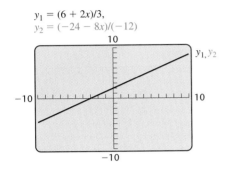

SOLUTION We graphed this system in Example 4(c) of Section 3.1, and found that the lines coincide and the system has an infinite number of solutions. (See the figure at left.) Suppose we were to solve this system using the elimination method:

$$12y - 8x = 24 \qquad \text{Multiplying by 4 on both sides of}$$
$$\underline{-12y + 8x = -24} \qquad \text{equation (1)}$$
$$0 = 0. \qquad \text{We obtain an identity; } 0 = 0 \text{ is always true.}$$

Note that both variables have been eliminated and what remains is an identity. Any pair that is a solution of equation (1) is also a solution of equation (2). The equations are dependent and the solution set is infinite: $\{(x, y) \mid 3y - 2x = 6\}$.

Special Cases

When solving a system of two linear equations in two variables:

1. If an identity is obtained, such as $0 = 0$, then the system has an infinite number of solutions. The equations are dependent and, since a solution exists, the system is consistent.*

2. If a contradiction is obtained, such as $0 = 7$, then the system has no solution. The system is inconsistent.

Should decimals or fractions appear, it often helps to *clear* before solving.

Example 7 Solve the system

$$0.2x + 0.3y = 1.7,$$
$$\tfrac{1}{7}x + \tfrac{1}{5}y = \tfrac{29}{35}.$$

SOLUTION We have

$$0.2x + 0.3y = 1.7, \longrightarrow \text{Multiplying by 10} \longrightarrow 2x + 3y = 17$$
$$\tfrac{1}{7}x + \tfrac{1}{5}y = \tfrac{29}{35} \longrightarrow \text{Multiplying by 35} \longrightarrow 5x + 7y = 29.$$

We multiplied by 10 to clear the decimals. Multiplication by 35, the least common denominator, clears the fractions. The problem is now identical to Example 5. The solution is $(-32, 27)$, or $x = -32$, $y = 27$.

---

*Consistent systems and dependent equations are discussed in greater detail in Section 3.4.

Comparing Methods

The following table is a summary that compares the graphical, substitution, and elimination methods for solving systems of equations.

METHOD	STRENGTHS	WEAKNESSES
Graphical	Can "see" solutions. Works with any system that can be graphed.	Inexact when solutions involve numbers that are not integers. Solution may not appear on the part of the graph drawn.
Substitution	Yields exact solutions. Easy to use when a variable is alone on one side.	Introduces extensive computations with fractions when solving more complicated systems. Cannot "see" solutions quickly.
Elimination	Yields exact solutions. Easy to use when fractions or decimals appear in the system. The preferred method for systems of three or more equations in three or more variables (see Section 3.4).	Cannot "see" solutions quickly.

Often it is helpful to use both an algebraic and a graphical method to solve a system of equations. In Examples 1–6, we solved each system algebraically and checked by solving graphically. A graph helps us visualize a solution that can be found exactly using algebra. On the other hand, algebra can help us determine an appropriate viewing window for a graph. Note that the graph of the system in Example 5 is shown in the viewing window $[-50, 5, -5, 40]$. If that system were graphed using a standard viewing window, the lines would appear to be parallel. When graphing a system of equations, we generally choose a viewing window that shows any points of intersection.

3.2 | *Exercise Set*

For Exercises 1–44, if a system has an infinite number of solutions, use set-builder notation to write the solution set. If a system has no solution, state this.

Solve using the substitution method.

1. $3x + 5y = 3,$
$\quad x = 8 - 4y$

2. $2x - 3y = 13,$
$\quad y = 5 - 4x$

3. $3x - 6 = y,$
$9x - 2y = 3$

4. $x = 3y - 3,$
$x + 2y = 9$

5. $4x + y = 1,$
$x - 2y = 16$

6. $5m + n = 8,$
$3m - 4n = 14$

7. $-3b + a = 7,$
$5a + 6b = 14$

8. $4x + 12y = 4,$
$-5x + y = 11$

9. $5p + 7q = 1,$
$4p - 2q = 16$

10. $3x - y = 1,$
$2x + 2y = 2$

11. $5x + 3y = 4,$
$x - 4y = 3$

12. $3x - y = 7,$
$2x + 2y = 5$

13. $y - 2x = 1,$
$2x - 3 = y$

14. $x + 2y = 6,$
$x = 4 - 2y$

Solve using the elimination method.

15. $x + 3y = 7,$
$-x + 4y = 7$

16. $x + y = 9,$
$2x - y = -3$

17. $2x + y = 6,$
$x - y = 3$

18. $x - 2y = 6,$
$-x + 3y = -4$

19. $6x - 3y = 18,$
$6x + 3y = -12$

20. $9x + 3y = -3,$
$2x - 3y = -8$

21. $3x + 2y = 3,$
$9x - 8y = -2$

22. $5x + 3y = 19,$
$2x - 5y = 11$

23. $5x - 7y = -16,$
$2x + 8y = 26$

24. $5r - 3s = 24,$
$3r + 5s = 28$

25. $0.7x - 0.3y = 0.5,$
$-0.4x + 0.7y = 1.3$

26. $0.3x - 0.2y = 4,$
$0.2x + 0.3y = 1$

27. $6x + 7y = 9,$
$8x + 9y = 11$

28. $8x + 9y = 15,$
$9x + 6y = 21$

29. $\frac{1}{3}x + \frac{1}{5}y = 7,$
$\frac{1}{6}x - \frac{2}{5}y = -4$

30. $\frac{2}{5}x + \frac{1}{2}y = 2,$
$\frac{1}{2}x - \frac{1}{6}y = 3$

31. $6x + 10y = 14,$
$3x + 5y = 7$

32. $12x - 6y = -15,$
$-4x + 2y = 5$

Solve using any appropriate method.

33. $a - 2b = 16,$
$b + 3 = 3a$

34. $5x - 9y = 7,$
$7y - 3x = -5$

35. $10x + y = 306,$
$10y + x = 90$

36. $3(a - b) = 15,$
$4a = b + 1$

37. $3y = x - 2,$
$x = 2 + 3y$

38. $x + 2y = 6,$
$x = 4 - 2y$

39. $2x - 7y = 9,$
$2x - 7y = -5$

40. $4x - 7y = 6,$
$4x - 7y = 2$

41. $0.05x + 0.25y = 22,$
$0.15x + 0.05y = 24$

42. $1.3x - 0.2y = 12,$
$0.4x + 17y = 89$

43. $2x + 3y = 5,$
$-2x - 3y = -5$

44. $x - y = 3,$
$y - x = -3$

In Exercises 45–48, determine which of the given viewing windows below shows the point of intersection of the graphs of the equations in the given system. Check by graphing.

a) $[-5, 5, -5, 5]$
b) $[25, 50, 0, 10]$
c) $[0, 20, 0, 10]$
d) $[100, 200, 0, 100]$

45. The system of Exercise 41

46. The system of Exercise 34

47. The system of Exercise 35

48. The system of Exercise 42

Skill Maintenance

49. *Athletic Directors.* The number of female collegiate athletic directors is increasing, as shown by the following graph. [2.6]

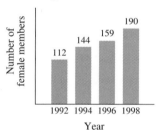

National Association of Collegiate Athletic Directors

Source: Sears Collegiate Champions, NADCA

a) Use regression to find a linear function $a(t)$ that can be used to predict the number of female members of the National Association of Collegiate Athletic Directors t years after 1992.
b) Use the function from part (a) to predict the number of female athletic directors in 2002.

50. What simple interest rate is given if a principal of $320 earns $17.60 in $\frac{1}{2}$ year? [2.3]

Find each solution set. [2.2]

51. $3x - 14 = x + 2(x - 7)$

52. $x + 2(3x + 5) = 7x - 3$

Synthesis

53. ◈ A student solving the system

$$17x + 19y = 102,$$
$$136x + 152y = 826$$

graphs both equations on a grapher and gets the following screen.

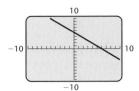

The student then (incorrectly) concludes that the equations are dependent and the solution set is infinite. How can algebra be used to convince the student that a mistake has been made?

54. ◆ Write a system of linear equations that would be most easily solved using the substitution method. Explain why substitution would be easier to use than the elimination method.

55. ◆ Write a system of linear equations that would be most easily solved using the elimination method. Explain why elimination would be easier to use than the substitution method.

56. ◆ Can Exercises 32 and 38 be solved mentally (with no writing)? If so, how? If not, why not?

57. If (1, 2) and (−3, 4) are two solutions of $f(x) = mx + b$, find m and b.

58. If (0, −3) and $\left(-\frac{3}{2}, 6\right)$ are two solutions of $px − qy = −1$, find p and q.

59. Determine a and b for which (−4, −3) will be a solution of the system
$$ax + by = −26,$$
$$bx − ay = 7.$$

60. Solve for x and y in terms of a and b:
$$5x + 2y = a,$$
$$x − y = b.$$

Solve.

61. $\dfrac{x + y}{2} − \dfrac{x − y}{5} = 1,$

 $\dfrac{x − y}{2} + \dfrac{x + y}{6} = −2$

62. $3.5x − 2.1y = 106.2,$
$4.1x + 16.7y = −106.28$

Each of the following is a system of nonlinear equations. However, each is reducible to linear, since an appropriate substitution (say, u for 1/x and v for 1/y) yields a linear system. Make such a substitution, solve for the new variables, and then solve for the original variables.

63. $\dfrac{2}{x} + \dfrac{1}{y} = 0,$

 $\dfrac{5}{x} + \dfrac{2}{y} = −5$

64. $\dfrac{1}{x} − \dfrac{3}{y} = 2,$

 $\dfrac{6}{x} + \dfrac{5}{y} = −34$

- - - - - - - - - - -
COLLABORATIVE CORNER

Focus: Systems of linear equations

Time: 20–30 minutes

Group size: 2–4

For aerobic exercise, a person's "target heart rate" T, in beats per minute, is given by $T = \frac{3}{4}(220 − a)$, where a is the person's age, in years.

Activity

1. Working independently, each group member should calculate the target heart rate for someone whose age is known only by that group member. That target heart rate should then be shared with the other group members.

2. Each group member should then calculate the age associated with each heart rate in two ways:

(1) by using the formula above to check a series of guesses; and (2) by substituting into the formula above and solving for a.

3. Express the formula above in the form $y = mx + b$, where x represents age. What does a negative value for m signify?

4. Formulate a system of equations that could be used to determine the age of a person whose target heart rate is twice his or her age. Then have one group member solve the system using the substitution method while the other group members solve the system graphically. Check by comparing your answers.

3.3

Applications and Models Using Systems of Two Equations

• *Total-Value and Mixture Problems*
• *Motion Problems*
• *Models*

You are in a much better position to solve problems now that systems of equations can be used. Using systems often makes the *Translate* step easier.

Example 1   *Real Estate.*   In 1996, the Simon Property Group and the DeBartolo Realty Corporation merged to form the largest real estate company in the United States, owning 183 shopping centers in 32 states. Prior to merging, Simon owned twice as many properties as DeBartolo. How many properties did each company own before the merger?

SOLUTION   The *Familiarize* and *Translate* steps have been done in Example 1 of Section 3.1. The resulting system of equations is

$$x + y = 183, \qquad (1)$$
$$x = 2y, \qquad (2)$$

where $x =$ the number of properties originally owned by Simon and $y =$ the number of properties originally owned by DeBartolo.

3. Carry out.   We solve the system of equations both algebraically and graphically.

ALGEBRAIC SOLUTION

Since one equation already has a variable isolated, let's use the substitution method:

$$x + y = 183$$
$$2y + y = 183 \qquad \text{Substituting } 2y \text{ for } x$$
$$3y = 183 \qquad \text{Combining like terms}$$
$$y = 61. \qquad \text{Dividing by 3 on both sides}$$

Returning to the second equation, we substitute 61 for y and compute x:

$$x = 2y = 2 \cdot 61 = 122.$$

We have $x = 122$, $y = 61$.

GRAPHICAL SOLUTION

We graph $y_1 = 183 - x$ and $y_2 = \frac{1}{2}x$ and find the point of intersection of the graphs. Writing $y_1 = 183 - x$ in slope–intercept form—that is, $y_1 = -x + 183$—we see that the y-intercept is $(0, 183)$. Only positive values of x and y make sense in this problem, so we choose a viewing window of $[0, 200, 0, 200]$, with Xscl = 10 and Yscl = 10.

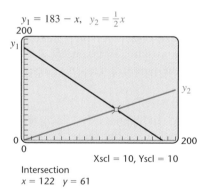

$y_1 = 183 - x, \ \ y_2 = \frac{1}{2}x$

Xscl = 10, Yscl = 10

Intersection
$x = 122 \quad y = 61$

We have a solution of $(122, 61)$.

4. Check.   Our solution says that Simon owned 122 properties and DeBartolo 61. Checking in the original problem, we see that the sum of 122 and 61 is 183, so the total number of properties is correct. Since 122 is twice 61, the numbers check.

5. State.   Prior to merging, Simon owned 122 properties and DeBartolo owned 61.

Total-Value and Mixture Problems

Example 2   *Retail Sales.* In one day, Glovers, Inc., sold 20 pairs of gloves. Fleece gloves sold for $24.95 a pair and Gore-Tex gloves for $37.50. The company took in $687.25. How many of each kind were sold?

SOLUTION   The *Familiarize* and *Translate* steps were done in Example 2 of Section 3.1.

3. Carry out.   We are to solve the system of equations

$$f + \quad g = 20, \qquad (1)$$
$$2495f + 3750g = 68{,}725 \qquad (2)$$

where f = the number of pairs of fleece gloves sold and g = the number of pairs of Gore-Tex gloves sold.

╭── ALGEBRAIC SOLUTION

Since no variable appears alone and the equations are in the form $Ax + By = C$, let's use the elimination method. We eliminate f by multiplying equation (1) by -2495 and adding it to equation (2):

$$
\begin{array}{rl}
-2495f - 2495g = -49{,}900 & \textbf{Multiplying equation (1)} \\
\underline{2495f + 3750g = \quad 68{,}725} & \textbf{by } -2495 \\
1255g = \quad 18{,}825 & \textbf{Adding} \\
g = 15. & \textbf{Solving for } g
\end{array}
$$

To find f, we substitute 15 for g in equation (1) and solve for f:

$$
\begin{array}{ll}
f + g = 20 & \textbf{Equation (1)} \\
f + 15 = 20 & \textbf{Substituting 15 for } g \\
f = 5. & \textbf{Solving for } f
\end{array}
$$

We obtain (5, 15), or $f = 5$, $g = 15$.

╭── GRAPHICAL SOLUTION

We replace f with x and g with y and graph $y_1 = 20 - x$ and $y_2 = (68725 - 2495x)/3750$ to find the point of intersection of the graphs. What would be an appropriate viewing window? The number of each kind of glove is between 0 and 20, so we select the viewing window [0, 20, 0, 20] (other choices are possible).

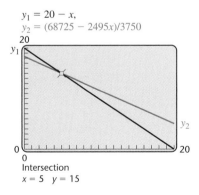

$y_1 = 20 - x$,
$y_2 = (68725 - 2495x)/3750$

Intersection
$x = 5$ $y = 15$

The point of intersection is (5, 15). Since f was replaced with x and g with y, we have $f = 5$ and $g = 15$.

4. Check.   Since the algebraic and graphical solutions match, we have a partial check. For a final check, we check in the original problem. Recall that f is the number of pairs of fleece gloves and g the number of pairs of Gore-Tex gloves:

Number of gloves: $f + g = 5 + 15 = 20$

Money from fleece gloves: $\$24.95f = 24.95 \times 5 = \$124.75$

Money from Gore-Tex gloves: $\$37.50g = 37.50 \times 15 = \underline{\$562.50}$

Total $= \$687.25$

The numbers check.

5. State.   Glovers sold 5 pairs of fleece gloves and 15 pairs of Gore-Tex gloves.

Example 2 involved two types of items (gloves), the quantity of each type sold, and the total value of the items. We refer to this type of problem as a *total-value problem.*

Example 3   *Blending Teas.* Tara's Tea Terrace sells loose Black tea for 95¢ an ounce and Lapsang Souchong for $1.43 an ounce. Tara wants to make a 1-lb mixture of the two types, called Imperial Blend, that sells for $1.10 an ounce. How much tea of each type should Tara use?

SOLUTION

1. Familiarize.   This problem is similar to Example 2. Rather than pairs of fleece or Gore-Tex gloves, we have ounces of Black tea and ounces of Lapsang Souchong. Rather than two different prices per pair, we have two different prices per ounce. Finally, rather than knowing the total value of the gloves, we know the weight and the price per ounce of the Imperial Blend. It is important to note that we can find the total value of the blend by multiplying 16 ounces (1 lb) times $1.10 per ounce. Although we could make and check a guess, we proceed to let b = the number of ounces of Black tea and l = the number of ounces of Lapsang Souchong.

2. Translate.   Since a 16-oz batch is being made, we must have

$$b + l = 16.$$

To find a second equation, note that the total value of the 16-oz blend must match the combined value of the separate ingredients:

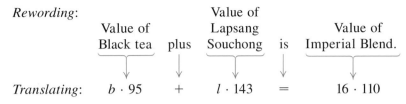

Rewording:

| Value of Black tea | plus | Value of Lapsang Souchong | is | Value of Imperial Blend. |

Translating: $b \cdot 95$ $+$ $l \cdot 143$ $=$ $16 \cdot 110$

These equations can also be obtained from a table.

	BLACK TEA	LAPSANG SOUCHONG	IMPERIAL BLEND
NUMBER OF OUNCES	b	l	16
PRICE PER OUNCE	95¢	143¢	110¢
VALUE OF TEA	95b	143l	$16 \cdot 110$, or 1760

$\longrightarrow b + l = 16$

$\longrightarrow 95b + 143l = 1760$

We have translated to a system of equations:

$$b + l = 16, \qquad (1)$$
$$95b + 143l = 1760. \qquad (2)$$

3. Carry out. We solve the system both algebraically and graphically.

ALGEBRAIC SOLUTION

We can solve using the substitution method. When equation (1) is solved for b, we have $b = 16 - l$. Substituting $16 - l$ for b in equation (2), we find l:

$95(16 - l) + 143l = 1760$	Substituting
$1520 - 95l + 143l = 1760$	Using the distributive law
$48l = 240$	Combining like terms; subtracting 1520 on both sides
$l = 5.$	Dividing by 48 on both sides

We have $l = 5$ and, from equation (1) above, $b + l = 16$. Thus, $b = 11$.

GRAPHICAL SOLUTION

We replace b with x and l with y and solve for y, giving us the system of equations

$$y = 16 - x,$$
$$y = (1760 - 95x)/143.$$

We know from the problem that the number of ounces of each kind of tea is between 0 and 16, so we choose the viewing window $[0, 16, 0, 16]$, graph the equations, and find the point of intersection.

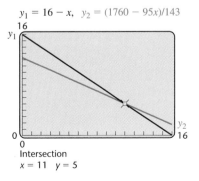

$y_1 = 16 - x, \quad y_2 = (1760 - 95x)/143$

Intersection
$x = 11 \quad y = 5$

The point of intersection is $(11, 5)$. Since b was replaced with x and l with y, we have the solution $b = 11$, $l = 5$.

4. **Check.** The fact that both approaches yield the same solution serves as a partial check. For a final check, note that if 11 oz of Black tea and 5 oz of Lapsang Souchong are combined, a 16-oz, or 1-lb, blend will result. The value of 11 oz of Black tea is 11($0.95), or $10.45. The value of 5 oz of Lapsang Souchong is 5($1.43), or $7.15, so the combined value of the blend is $10.45 + $7.15 = $17.60. A 16-oz batch, priced at $1.10 an ounce, would also be worth $17.60, so our answer checks.

5. **State.** The Imperial Blend should be made by combining 11 oz of Black tea and 5 oz of Lapsang Souchong. ▬

Before proceeding to Example 4, briefly scan Examples 2 and 3 for similarities. Note that in each case, one of the equations in the system is a simple sum while the other equation represents a sum of products. Example 4 continues this pattern with what is commonly called a *mixture problem.*

Problem-Solving Tip

When solving a problem, see if it is patterned or modeled after a problem that you have already solved.

Example 4 *Mixing Fertilizers.* Yardbird Gardening, Inc., carries two brands of fertilizer containing nitrogen and water. "Gently Green" is 5% nitrogen and "Sun Saver" is 15% nitrogen. Yardbird Gardening needs to combine the two types of solutions in order to make 100 L of a solution that is 12% nitrogen. How much of each brand should be used?

SOLUTION

1. **Familiarize.** Let's look for similarities between this example and Examples 2 and 3. Instead of two types of gloves or tea, we have two types of fertilizer. Instead of being told the total value, we are told the nitrogen concentration of the mixture. Let's try a guess to gain familiarity with the problem.

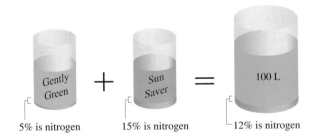

Suppose that 40 L of Gently Green and 60 L of Sun Saver are mixed. The resulting mixture will be the right size, 100 L, but will it be the right strength? To find out, note that 40 L of Gently Green would contribute $0.05(40) = 2$ L of nitrogen to the mixture whereas 60 L of Sun Saver would contribute $0.15(60) = 9$ L of nitrogen to the mixture. Altogether, 40 L of Gently Green and 60 L of Sun Saver would make 100 L of a mixture that has $2 + 9 = 11$ L of nitrogen. Since this would mean that the final mixture is only 11% nitrogen, our guess of 40 L and 60 L is incorrect. Still, the process of checking our guess has familiarized us with the problem.

2. **Translate.**   Let g = the number of liters of Gently Green and s = the number of liters of Sun Saver. The information can be organized in a table.

	GENTLY GREEN	SUN SAVER	MIXTURE	
NUMBER OF LITERS	g	s	100	$\longrightarrow g + s = 100$
PERCENT OF NITROGEN	5%	15%	12%	
AMOUNT OF NITROGEN	$0.05g$	$0.15s$	0.12×100, or 12 liters	$\longrightarrow 0.05g + 0.15s = 12$

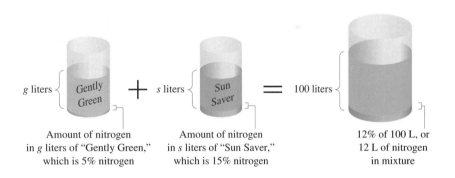

g liters { Gently Green $+$ s liters { Sun Saver $=$ 100 liters {

Amount of nitrogen in g liters of "Gently Green," which is 5% nitrogen

Amount of nitrogen in s liters of "Sun Saver," which is 15% nitrogen

12% of 100 L, or 12 L of nitrogen in mixture

If we add g and s in the first row of the table, we get one equation. It represents the total amount of mixture: $g + s = 100$.

If we add the amounts of nitrogen listed in the third row, we get a second equation. This equation represents the amount of nitrogen in the mixture: $0.05g + 0.15s = 12$.

After clearing decimals, we have translated the problem to the system

$$g + s = 100, \quad (1)$$
$$5g + 15s = 1200. \quad (2)$$

3. Carry out. We solve the system both algebraically and graphically.

ALGEBRAIC SOLUTION

We use the elimination method to solve the system:

$$-5g - 5s = -500 \qquad \text{Multiplying equation (1)}$$
$$\text{by } -5 \text{ on both sides}$$

$$\underline{5g + 15s = 1200}$$

$$10s = 700 \qquad \text{Adding}$$

$$s = 70; \qquad \text{Solving for } s$$

$$g + 70 = 100 \qquad \text{Substituting into equation (1)}$$

$$g = 30. \qquad \text{Solving for } g$$

We obtain $g = 30$, $s = 70$.

GRAPHICAL SOLUTION

We replace g with x and s with y and graph the equations. The number of liters of each solution is between 0 and 100, so we use the viewing window [0, 100, 0, 100].

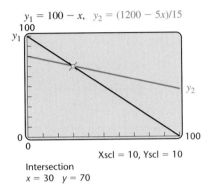

$y_1 = 100 - x$, $y_2 = (1200 - 5x)/15$

Xscl = 10, Yscl = 10

Intersection
$x = 30$ $y = 70$

Since g was replaced with x and s with y, we have the possible solution $g = 30$, $s = 70$.

4. Check. Since both approaches yield the same solution, we are probably correct. Let's check in the original problem. Remember, g is the number of liters of Gently Green and s is the number of liters of Sun Saver.

Total amount of mixture: $\qquad g + s = 30 + 70 = 100$

Total amount of nitrogen: $\quad$ 5% of 30 + 15% of 70 = 1.5 + 10.5 = 12

Percentage of nitrogen in mixture: $\qquad \dfrac{\text{Total amount of nitrogen}}{\text{Total amount of mixture}} = \dfrac{12}{100} = 12\%$

The numbers do check in the original problem.

5. State. Yardbird Gardening should mix 30 L of Gently Green with 70 L of Sun Saver.

Motion Problems

When a problem deals with distance, speed (rate), and time, recall the following.

Distance, Rate, and Time Equations

If r represents rate, t represents time, and d represents distance, then:

$$d = rt, \qquad r = \frac{d}{t}, \quad \text{and} \quad t = \frac{d}{r}.$$

Be sure to remember at least one of these equations. The others can be obtained by using algebraic manipulations as needed.

Example 5 *Train Travel.* A freight train leaves Ames traveling east at a speed of 60 km/h. Two hours later, a passenger train leaves Ames traveling in the same direction on a parallel track at 90 km/h. At what point will the passenger train catch up to the freight train?

SOLUTION

1. Familiarize. Let's make a guess—say, 180 km—and check to see if it is correct. The freight train, traveling 60 km/h, would reach a point 180 km from Ames in $\frac{180}{60} = 3$ hr. The passenger train, traveling 90 km/h, would cover 180 km in $\frac{180}{90} = 2$ hr. Since 3 hr is *not* two hours more than 2 hr, our guess of 180 km is incorrect. Although our guess is wrong, we see that the time that the trains are running and the point at which they meet are both unknown. Let $t =$ the number of hours that the freight train is running before they meet and $d =$ the distance at which the trains meet. Since the freight train has a 2-hr head start, the passenger train runs for $t - 2$ hours before catching up to the freight train.

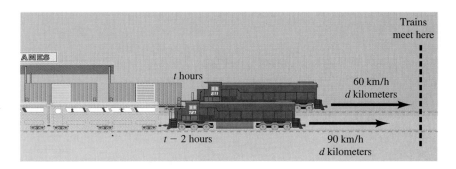

2. Translate.   We can organize the information in a chart. Each row is determined by the formula *Distance = Rate · Time*.

	DISTANCE	RATE	TIME	
FREIGHT TRAIN	d	60	t	$\longrightarrow d = 60t$
PASSENGER TRAIN	d	90	$t - 2$	$\longrightarrow d = 90(t - 2)$

Using *Distance = Rate · Time* twice, we get two equations:

$$d = 60t, \qquad (1)$$
$$d = 90(t - 2). \qquad (2)$$

3. Carry out.   We solve the system both algebraically and graphically.

ALGEBRAIC SOLUTION

We solve the system using the substitution method:

$60t = 90(t - 2)$ **Substituting 60*t* for *d* in equation (2)**

$60t = 90t - 180$ **Using the distributive law**

$-30t = -180$

$t = 6.$

The time for the freight train is 6 hr, which means that the time for the passenger train is $6 - 2$, or 4 hr. Remember that it is distance, not time, that the problem asked for. Thus for $t = 6$, we have $d = 60 \cdot 6 = 360$ km.

GRAPHICAL SOLUTION

The variables used in the equations are d and t. Note that in both equations, d is given in terms of t. If we replace d with y and t with x, we can enter the equations directly. Since $y =$ distance and $x =$ time, we use a viewing window of $[0, 10, 0, 500]$. We graph $y_1 = 60x$ and $y_2 = 90(x - 2)$ and find the point of intersection.

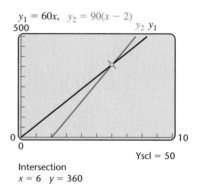

The point of intersection is (6, 360). The problem asks for the distance that the trains travel. Recalling that $y =$ distance, we have the distance that both trains travel is 360 km.

4. Check.   At 60 km/h, the freight train will travel $60 \cdot 6$, or 360 km, in 6 hr. At 90 km/h, the passenger train will travel $90 \cdot (6 - 2)$, or 360 km in 4 hr. The numbers check.

5. State.   The trains will meet at a point 360 km east of Ames.   ▬

Example 6 *Marine Travel.* A coast guard patrol boat travels 4 hr downstream with a 6-mph current. Returning, against the current, takes 5 hr. Find the speed of the boat in still water.

SOLUTION

1. Familiarize. We imagine the situation and make a drawing. Note that the current *speeds up* the boat traveling downstream and *slows down* the boat traveling upstream. Since the distances traveled each way must be the same, we can easily check a guess of the boat's speed in still water. Suppose the speed of the boat in still water is 40 mph. The boat would then go $40 + 6 = 46$ mph downstream and $40 - 6 = 34$ mph upstream. The boat would travel $46 \cdot 4 = 184$ mi downstream and $34 \cdot 5 = 170$ mi upstream. Since $184 \neq 170$, our guess of 40 mph is incorrect. Rather than guess again, let's have $r =$ the speed, in miles per hour, of the boat in still water. Then $r + 6 =$ the boat's speed downstream, and $r - 6 =$ the boat's speed upstream. We also let $d =$ the distance traveled, in miles.

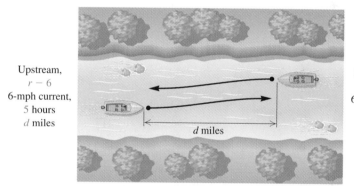

Upstream, $r - 6$ 6-mph current, 5 hours d miles

Downstream, $r + 6$ 6-mph current, 4 hours d miles

d miles

2. Translate. The information can be organized in a chart. The distances traveled are the same, so we use

$$Distance = Rate \text{ (or } Speed) \cdot Time.$$

Each row of the chart gives an equation.

	DISTANCE	RATE	TIME	
DOWNSTREAM	d	$r + 6$	4	$\longrightarrow d = (r + 6)4$
UPSTREAM	d	$r - 6$	5	$\longrightarrow d = (r - 6)5$

The two equations constitute a system:

$$d = (r + 6)4, \quad (1)$$
$$d = (r - 6)5. \quad (2)$$

3. Carry out.   We solve the system both algebraically and graphically.

We solve the system using the substitution method:

$(r - 6)5 = (r + 6)4$ **Substituting $(r - 6)5$ for d in equation (1)**

$5r - 30 = 4r + 24$ **Using the distributive law**

$5r - 4r = 24 + 30$ **Adding $-4r$ and 30 on both sides**

$\quad\quad r = 54.$ **Solving for r**

Thus we know that the speed of the boat in still water is 54 mph.

We replace d with y and r with x, graph the equations $y_1 = (x + 6)4$ and $y_2 = (x - 6)5$, and look for a point of intersection.

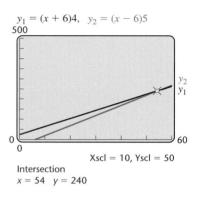

Intersection
$x = 54$ $y = 240$

Since $x =$ the speed of the boat in still water, we have a possible solution of 54 mph.

4. Check.   Both solutions match. As a final check, note that when $r = 54$, the speed downstream is $54 + 6 = 60$ mph, and the speed upstream is $54 - 6 = 48$ mph. The distance downstream, $60 \cdot 4 = 240$ mi, matches the distance upstream, $48 \cdot 5 = 240$ mi, so we have a check.

5. State.   The speed of the boat in still water is 54 mph. ▬

Tips for Solving Motion Problems

1. Make a drawing using an arrow or arrows to represent distance and the direction of each object in motion.
2. Organize the information in a chart.
3. Look for times, distances, or rates that are the same. These often can lead to an equation.
4. Translating to a system of equations allows for the use of two variables.
5. Always make sure that you have answered the question asked.

Models

Often, after we have found two lines using regression, we want to know at what point they intersect.

Example 7 *Recycling.* In the United States, the amount of solid waste being recycled is slowly catching up to the amount being generated, as shown by the data in the following table.

YEAR	AMOUNT OF WASTE GENERATED (IN POUNDS PER PERSON PER DAY)	AMOUNT OF WASTE RECYCLED (IN POUNDS PER PERSON PER DAY)
1980	3.7	0.35
1985	3.8	0.38
1990	4.3	0.7
1995	4.3	1.2

Source: The Statistical Abstract of the United States, 1997

a) Find linear functions w and r, where $w(t)$ is the number of pounds of waste generated per person per day t years after 1980 and $r(t)$ is the number of pounds of waste recycled per person per day t years after 1980.

b) Predict the year in which the amount recycled will equal the amount generated.

SOLUTION

a) First, we enter and plot the data using a grapher. To avoid confusion, we choose a different type of mark for each list. The amount generated is plotted using □ and the amount recycled using +. Both sets of data appear to be linear.

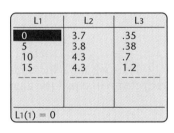 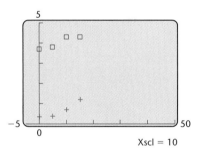

```
LinReg(ax+b) L1,
L3▮
```

Next, we use linear regression to fit a line to each set of data. To use lists other than L1 and L2, we enter the name of the list after the linear regression command. The screen at left shows the command used to find $r(t)$.

For the amount of waste generated, we have

$$w(t) = 0.046t + 3.68.$$

We copy this to the Y= screen as Y1.

For the amount of waste recycled, we have

$$r(t) = 0.0574t + 0.227,$$

which we copy as Y2.

b) To predict the year in which the amount of waste recycled will equal the amount generated, we solve the system

$$y = 0.046x + 3.68,$$
$$y = 0.0574x + 0.227.$$

First, we graph both equations using the same viewing window. The window that we used for the data, $[-5, 50, 0, 5]$, does not show a point of intersection. We need to show much more of the x-axis as well as more of the y-axis. We find, perhaps through a trial-and-error process, that we can see a point of intersection in the window $[-5, 400, 0, 25]$.

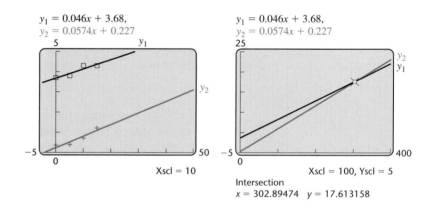

Thus we have the approximate solution (303, 17.61). The amount recycled will equal the amount generated 303 yr after 1980, or in 2283.

3.3 *Exercise Set*

1.–18. *For Exercises 1–18, solve Exercises 1–18 from Exercise Set 3.1.*

19. *Inventory.*   The Everton College store paid $1728 for an order of 45 calculators. The store paid $9 for each scientific calculator. The others, all graphing calculators, cost the store $58 each. How many of each type of calculator was ordered?

20. *Sales of Food.*   High Flyin' Wings charges $12 for a bucket of chicken wings and $7 for a chicken dinner. After filling 28 orders for buckets and dinners, High Flyin' Wings had collected $281. How many buckets and how many dinners did they sell?

21. *Blending Coffees.*   The Coffee Counter charges $9.00 per pound for Kenyan French Roast coffee and $8.00 per pound for Sumatran coffee. How much of each type should be used to make a 20-lb blend that sells for $8.40 per pound?

22. *Mixed Nuts.* The Nutty Professor sells cashews for $6.75 per pound and Brazil nuts for $5.00 per pound. How much of each type should be used to make a 50-lb mixture that sells for $5.70 per pound?

23. *Blending Granola.* Deep Thought Granola is 25% nuts and dried fruit. Oat Dream Granola is 10% nuts and dried fruit. How much of Deep Thought and how much of Oat Dream should be mixed to form a 20-lb batch of granola that is 19% nuts and dried fruit?

24. *Catering.* Casella's Catering is planning a wedding reception. The bride and groom would like to serve a nut mixture containing 25% peanuts. Casella has available mixtures that are either 40% or 10% peanuts. How much of each type should be mixed to get a 10-lb mixture that is 25% peanuts?

25. *Ink Remover.* Etch Clean Graphics uses one cleanser that is 25% acid and a second that is 50% acid. How many liters of each should be mixed to get 10 L of a solution that is 40% acid?

26. *Livestock Feed.* Soybean meal is 16% protein and corn meal is 9% protein. How many pounds of each should be mixed to get a 350-lb mixture that is 12% protein?

27. *Student Loans.* Lomasi's two student loans totaled $12,000. One of her loans was at 6% simple interest and the other at 9%. After one year, Lomasi owed $855 in interest. What was the amount of each loan?

28. *Investments.* An executive nearing retirement made two investments totaling $15,000. In one year, these investments yielded $1432 in simple interest. Part of the money was invested at 9% and the rest at 10%. How much was invested at each rate?

29. *Food Science.* The following bar graph shows the milk fat percentages in three dairy products. How many pounds each of whole milk and cream should be mixed to form 200 lb of milk for cream cheese?

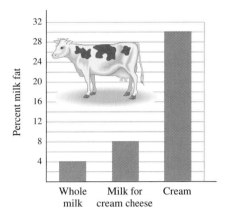

30. *Automotive Maintenance.* "Arctic Antifreeze" is 18% alcohol and "Frost No-More" is 10% alcohol. How many liters of each should be mixed to get 20 L of a mixture that is 15% alcohol?

31. *Architecture.* The rectangular ground floor of the John Hancock building has a perimeter of 860 ft. The length is 100 ft more than the width. Find the length and the width.

32. *Real Estate.*   The perimeter of a lot is 190 m. The width is one-fourth of the length. Find the dimensions.

33. *Teller Work.*   Ashford goes to a bank and gets change for a $50 bill consisting of all $5 bills and $1 bills. There are 22 bills in all. How many of each kind are there?

34. *Making Change.*   Cecilia makes a $9.25 purchase at the bookstore with a $20 bill. The store has no bills and gives her the change in quarters and fifty-cent pieces. There are 30 coins in all. How many of each kind are there?

35. *Train Travel.*   A train leaves Danville Junction and travels north at a speed of 75 km/h. Two hours later, a second train leaves on a parallel track and travels north at 125 km/h. How far from the station will they meet?

36. *Car Travel.*   Two cars leave Denver traveling in opposite directions. One car travels at a speed of 80 km/h and the other at 96 km/h. In how many hours will they be 528 km apart?

37. *Canoeing.*   Alvin paddled for 4 hr with a 6-km/h current to reach a campsite. The return trip against the same current took 10 hr. Find the speed of Alvin's canoe in still water.

38. *Boating.*   Mia's motorboat took 3 hr to make a trip downstream with a 6-mph current. The return trip against the same current took 5 hr. Find the speed of the boat in still water.

39. *Point of No Return.*   A plane flying the 3458-mi trip from New York City to London has a 50-mph tailwind. The flight's *point of no return* is the point at which the flight time required to return to New York is the same as the time required to continue to London. If the speed of the plane in still air is 360 mph, how far is New York from the point of no return?

40. *Point of No Return.*   A plane is flying the 2553-mi trip from Los Angeles to Honolulu into a 60-mph headwind. If the speed of the plane in still air is 310 mph, how far from Los Angeles is the plane's point of no return? (See Exercise 39.)

41. *Outpatient Surgery.*   The percentage of surgeries that are done as outpatient procedures has increased in recent years. The following table shows the percentage of outpatient and inpatient hospital surgeries performed in the United States for various years (*Source: The Macmillan Visual Almanac, 1996,* p. 77).

YEAR	INPATIENT SURGERY (IN PERCENT)	OUTPATIENT SURGERY (IN PERCENT)
1980	83	17
1982	80	20
1984	72	28
1986	60	40
1988	55	45
1990	49	51

a) Find linear functions n and u, where $n(x)$ can be used to predict the percentage of inpatient surgeries x years after 1980 and $u(x)$ can be used to predict the percentage of outpatient surgeries x years after 1980.

b) Estimate the year in which the number of inpatient surgeries was the same as the number of outpatient surgeries.

42. *Recycling.*   Some experts estimate that only 80% of solid waste generated in the United States is recyclable (*Source:* Saign, Geoffrey C., *Green Essentials.* San Francisco: Mercury House, 1994, p. 356). Use the data in Example 7 to predict the year in which the amount of waste recycled will equal 80% of the amount generated.

43. *Work Force.*   The following table shows the percentage of the population in the civilian labor force for various years. Predict the year in which the percentage of women in the work force will be the same as the percentage of men in the work force.

YEAR	PERCENTAGE OF WOMEN IN WORK FORCE	PERCENTAGE OF MEN IN WORK FORCE
1955	35.7	85.4
1965	39.3	80.7
1975	46.3	77.9
1985	54.5	76.3
1996	59.3	74.9

Source: U.S. Bureau of Labor Statistics

44. *Eating Out.*   The number of meals per person annually purchased in a restaurant and the number annually purchased for takeout are given in the following table for various years. Predict the year in which the number of meals eaten on-premise will equal the number eaten off-premise.

YEAR	NUMBER OF MEALS (PER PERSON ANNUALLY) EATEN ON-PREMISE	NUMBER OF MEALS (PER PERSON ANNUALLY) EATEN OFF-PREMISE
1990	64	55
1991	64	56
1992	63	57
1993	62	59

Source: The Wall Street Journal Almanac, 1998

Skill Maintenance

45. Simplify: $-3(x-7) - 2[x - (4 + 3x)]$. [1.3]

46. Solve: $3(x - 5) = 7(x - 6)$. [2.2]

47. Write an equation of the line containing $(2, -5)$ with slope $-\frac{3}{4}$. [2.4]

48. Graph: $y = -\frac{1}{2}x + 5$. [2.4]

49. Solve: $0.45x + 6.82 = 1.5 - 4.38x$. [2.2]

50. If $f(x) = 3x^2 - 7x$, find $f(-1)$. [2.1]

Synthesis

51. ◈ List three or four study tips of your own for someone beginning this exercise set.

52. ◈ Suppose that in Example 3 you are asked only for the amount of Black tea needed for the Imperial Blend. Would the method of solving the problem change? Why or why not?

53. ◈ In what ways are Examples 3 and 4 alike? In what sense are the systems of equations similar?

54. ◈ Write a problem similar to Example 2 for a classmate to solve. Design the problem so that the solution is "The florist sold 14 hanging plants and 9 flats of petunias."

55.–58. *For Exercises 55–58, solve Exercises 74–77 from Exercise Set 3.1.*

59. *Automotive Maintenance.* The radiator in Michelle's car contains 16 L of antifreeze and water. This mixture is 30% antifreeze. How much of this mixture should she drain and replace with pure antifreeze so that there will be a mixture of 50% antifreeze?

60. *Exercise.* Natalie jogs and walks to school each day. She averages 4 km/h walking and 8 km/h jogging. From home to school is 6 km and Natalie makes the trip in 1 hr. How far does she jog in a trip?

61. *Book Sales.* A limited edition of a book published by a historical society was offered for sale to members. The cost was one book for $12 or two books for $20. The society sold 880 books, for a total of $9840. How many members ordered two books?

62. The tens digit of a two-digit positive integer is 2 more than three times the units digit. If the digits are interchanged, the new number is 13 less than half the given number. Find the given integer. (*Hint:* Let $x =$ the tens-place digit and $y =$ the units-place digit; then $10x + y$ is the number.)

63. *Train Travel.* A train leaves Union Station for Central Station, 216 km away, at 9 A.M. One hour later, a train leaves Central Station for Union Station. They meet at noon. If the second train had started at 9 A.M. and the first train at 10:30 A.M., they would still have met at noon. Find the speed of each train.

64. *Fuel Economy.* Ellen Jordan's station wagon gets 18 miles per gallon (mpg) in city driving and 24 mpg in highway driving. The car is driven 465 mi on 23 gal of gasoline. How many miles were driven in the city and how many were driven on the highway?

65. *Wood Stains.* Williams' Custom Flooring has 0.5 gal of stain that is 20% brown and 80% neutral. A customer orders 1.5 gal of a stain that is 60% brown and 40% neutral. How much pure brown stain and how much neutral stain should be added to the original 0.5 gal in order to make up the order?*

66. *Gender.* Phil and Phyllis are siblings. Phyllis has twice as many brothers as she has sisters. Phil has the same number of brothers as sisters. How many girls and how many boys are in the family?

67. See Exercise 65 above. Let $x =$ the amount of pure brown stain added to the original 0.5 gal. Find a function $P(x)$ that can be used to determine the percentage of brown stain in the 1.5-gal mixture. Using a grapher, draw the graph of P and confirm the answer to Exercise 65.

*This problem was suggested by Chris Burditt of Yountville, California.

3.4
Systems of Equations in Three Variables

- *Identifying Solutions*
- *Solving Systems in Three Variables*
- *Dependency, Inconsistency, and Geometric Considerations*

Some problems translate easily to two equations. Others more naturally call for a translation to three or more equations. In this section, we learn how to solve systems of three linear equations. Later, we will use such systems in problem-solving situations.

Identifying Solutions

A **linear equation in three variables** is an equation equivalent to one in the form $Ax + By + Cz = D$, where A, B, C, and D are real numbers. We refer to the form $Ax + By + Cz = D$ as *standard form* for a linear equation in three variables.

A solution of a system of three equations in three variables is an ordered triple (p, q, r) that makes *all three* equations true.

Example 1   Determine whether $\left(\frac{3}{2}, -4, 3\right)$ is a solution of the system

$$4x - 2y - 3z = 5,$$
$$-8x - y + z = -5,$$
$$2x + y + 2z = 5.$$

SOLUTION   We substitute $\left(\frac{3}{2}, -4, 3\right)$ into the three equations, using alphabetical order:

$$
\begin{array}{c|c}
\underline{4x - 2y - 3z = 5} \\
4 \cdot \frac{3}{2} - 2(-4) - 3 \cdot 3 \ ?\ 5 \\
6 + 8 - 9 \\
5 \ \big|\ 5 \quad \text{TRUE}
\end{array}
\qquad
\begin{array}{c|c}
\underline{-8x - y + z = -5} \\
-8 \cdot \frac{3}{2} - (-4) + 3 \ ?\ -5 \\
-12 + 4 + 3 \\
-5 \ \big|\ -5 \quad \text{TRUE}
\end{array}
$$

$$
\begin{array}{c|c}
\underline{2x + y + 2z = 5} \\
2 \cdot \frac{3}{2} + (-4) + 2 \cdot 3 \ ?\ 5 \\
3 - 4 + 6 \\
5 \ \big|\ 5 \quad \text{TRUE}
\end{array}
$$

The triple makes all three equations true, so it is a solution.   ▬

Solving Systems in Three Variables

Graphical methods for solving linear equations in three variables are problematic, because a three-dimensional coordinate system is required and the graph of a linear equation in three variables is a plane. The substitution method *can* be used but becomes very cumbersome unless one or more of the equations has only two variables. Fortunately, the elimination method allows us to manipulate a system of three equations in three variables so that a simpler system of two equations in two variables is formed.

Once that simpler system has been solved, we can substitute into one of the three original equations and solve for the third variable.

Example 2   Solve the following system of equations:

$$x + y + z = 4, \qquad (1)$$
$$x - 2y - z = 1, \qquad (2)$$
$$2x - y - 2z = -1. \qquad (3)$$

SOLUTION   We select *any* two of the three equations and work to get one equation in two variables. Let's add equations (1) and (2):

$$
\begin{array}{ll}
x + y + z = 4 & (1) \\
\underline{x - 2y - z = 1} & (2) \\
2x - y \quad\;\; = 5. & (4) \qquad \text{Adding to eliminate } z
\end{array}
$$

Next, we select a different pair of equations and eliminate the *same variable* that we did above. Let's use equations (1) and (3) to again eliminate z. Be careful here! A common error is to eliminate a different variable in this step.

$$
\begin{array}{l}
x + y + z = 4, \\
2x - y - 2z = -1
\end{array}
\xrightarrow[\text{by 2}]{\text{Multiplying equation (1)}}
\begin{array}{l}
2x + 2y + 2z = 8 \\
\underline{2x - y - 2z = -1} \\
4x + y \quad\;\; = 7 \\
\qquad\qquad\qquad (5)
\end{array}
$$

Now we solve the resulting system of equations (4) and (5). That solution will give us two of the numbers in the solution of the original system.

$$
\begin{array}{ll}
2x - y = 5 & (4) \\
\underline{4x + y = 7} & (5) \\
6x \quad\;\; = 12 & \text{Adding} \\
\quad x = 2 &
\end{array}
$$

Note that we now have two equations in two variables. Had we eliminated different variables above, this would not be the case.

We can use either equation (4) or (5) to find y. We choose equation (5):

$$
\begin{array}{ll}
4x + y = 7 & (5) \\
4 \cdot 2 + y = 7 & \text{Substituting 2 for } x \text{ in equation (5)} \\
8 + y = 7 & \\
y = -1. &
\end{array}
$$

We now have $x = 2$ and $y = -1$. To find the value for z, we use any of the original three equations and substitute to find the third number, z. Let's use equation (1) and substitute our two numbers in it:

$$
\begin{array}{ll}
x + y + z = 4 & (1) \\
2 + (-1) + z = 4 & \text{Substituting 2 for } x \text{ and } -1 \text{ for } y \\
1 + z = 4 & \\
z = 3. &
\end{array}
$$

We have obtained the triple $(2, -1, 3)$. It should check in *all three* equations:

$$\begin{array}{c} x + y + z = 4 \\ \hline 2 + (-1) + 3 \; ? \; 4 \\ 4 \; | \; 4 \quad \text{TRUE} \end{array} \qquad \begin{array}{c} x - 2y - z = 1 \\ \hline 2 - 2(-1) - 3 \; ? \; 1 \\ 1 \; | \; 1 \quad \text{TRUE} \end{array}$$

$$\begin{array}{c} 2x - y - 2z = -1 \\ \hline 2 \cdot 2 - (-1) - 2 \cdot 3 \; ? \; -1 \\ -1 \; | \; -1 \quad \text{TRUE} \end{array}$$

The solution is $(2, -1, 3)$.

Solving Systems of Three Linear Equations

To use the elimination method to solve systems of three linear equations:

1. Write all equations in the standard form $Ax + By + Cz = D$.
2. Clear any decimals or fractions.
3. Choose a variable to eliminate. Then select two of the three equations and work to get one equation in two variables.
4. Next, use a different pair of equations and eliminate the same variable that you did in step (3).
5. Solve the system of equations that resulted from steps (3) and (4).
6. Substitute the solution from step (5) into one of the original three equations and solve for the third variable. Then check.

Example 3   Solve the system

$$\begin{aligned} 4x - 2y - 3z &= 5, & (1) \\ -8x - y + z &= -5, & (2) \\ 2x + y + 2z &= 5. & (3) \end{aligned}$$

SOLUTION

1., 2. The equations are already in standard form with no fractions or decimals.

3. First, we select a variable to eliminate. We choose y because the y-terms are opposites of each other in equations (2) and (3). We add:

$$\begin{array}{lr} -8x - y + z = -5 & (2) \\ \underline{2x + y + 2z = 5} & (3) \\ -6x + 3z = 0. & (4) \qquad \textbf{Adding} \end{array}$$

4. We use another pair of equations to create a second equation in x and z. That is, we eliminate the same variable, y, as in step (3). We use equations (1) and (3):

$$4x - 2y - 3z = 5,$$
$$2x + y + 2z = 5 \xrightarrow[\text{by } 2]{\text{Multiplying equation (3)}}$$

$$4x - 2y - 3z = 5$$
$$\underline{4x + 2y + 4z = 10}$$
$$8x \qquad + z = 15. \qquad (5)$$

5. Now we solve the resulting system of equations (4) and (5). That allows us to find two parts of the ordered triple.

$$-6x + 3z = 0,$$
$$8x + z = 15 \xrightarrow[\text{by } -3]{\text{Multiplying equation (5)}}$$

$$-6x + 3z = 0$$
$$\underline{-24x - 3z = -45}$$
$$-30x \qquad = -45$$
$$x = \frac{-45}{-30}$$
$$= \frac{3}{2}$$

We use equation (5) to find z:

$$8x + z = 15$$
$$8 \cdot \tfrac{3}{2} + z = 15 \qquad \text{Substituting } \tfrac{3}{2} \text{ for } x$$
$$12 + z = 15$$
$$z = 3.$$

6. Finally, we use any of the original equations and substitute to find the third number, y. We choose equation (3):

$$2x + y + 2z = 5 \qquad (3)$$
$$2 \cdot \tfrac{3}{2} + y + 2 \cdot 3 = 5 \qquad \text{Substituting } \tfrac{3}{2} \text{ for } x \text{ and } 3 \text{ for } z$$
$$3 + y + 6 = 5$$
$$y + 9 = 5$$
$$y = -4.$$

The solution is $\left(\tfrac{3}{2}, -4, 3\right)$. The check was performed as Example 1.

Sometimes, certain variables are missing at the outset.

Example 4 Solve the system

$$x + y + z = 180, \qquad (1)$$
$$x \qquad - z = -70, \qquad (2)$$
$$2y - z = 0. \qquad (3)$$

SOLUTION

1., 2. The equations appear in standard form with no fractions or decimals.

3., 4. Note that there is no y in equation (2). Thus, at the outset, we already have y eliminated from one equation. We need another equation with y eliminated. We use equations (1) and (3):

$$x + y + z = 180, \xrightarrow{\substack{\textbf{Multiplying equation} \\ \textbf{(1) by } -2}} \begin{array}{r} -2x - 2y - 2z = -360 \\ 2y - z = 0 \\ \hline -2x - 3z = -360. \ (4) \end{array}$$

5., 6. Now we solve the resulting system of equations (2) and (4):

$$\begin{array}{r} x - z = -70, \\ -2x - 3z = -360 \end{array} \xrightarrow{\substack{\textbf{Multiplying equation} \\ \textbf{(2) by } 2}} \begin{array}{r} 2x - 2z = -140 \\ -2x - 3z = -360 \\ \hline -5z = -500 \\ z = 100. \end{array}$$

Continuing as in Examples 2 and 3, we get the solution $(30, 50, 100)$. The check is left to the student.

Dependency, Inconsistency, and Geometric Considerations

Each equation in Examples 2, 3, and 4 has a graph that is a plane in three dimensions. The solutions are points common to the planes of each system. Since three planes can have an infinite number of points in common or no points at all in common, we need to generalize the concept of *consistency.*

One solution: planes intersecting in exactly one point. System is consistent.

The planes intersect along a common line. An infinite number of points are common to the three planes. System is consistent.

Three parallel planes. There is no common point of intersection. System is inconsistent.

Planes intersect two at a time, but there is no point common to all three. System is inconsistent.

Consistency

A system of equations that has at least one solution is said to be **consistent**.

A system of equations that has no solution is said to be **inconsistent**.

Example 5   Solve:

$$y + 3z = 4, \qquad (1)$$
$$-x - y + 2z = 0, \qquad (2)$$
$$x + 2y + z = 1. \qquad (3)$$

SOLUTION   The variable x is missing in equation (1). By adding equations (2) and (3), we can find a second equation in which x is missing:

$$-x - y + 2z = 0 \qquad (2)$$
$$\underline{x + 2y + z = 1} \qquad (3)$$
$$y + 3z = 1. \qquad (4) \qquad \text{Adding}$$

Equations (1) and (4) form a system in y and z. We solve as before:

$$\begin{array}{l} y + 3z = 4, \\ y + 3z = 1 \end{array} \xrightarrow[\text{by } -1]{\text{Multiplying equation (1)}} \begin{array}{l} -y - 3z = -4 \\ \underline{y + 3z = 1} \end{array}$$

$$\text{This is a contradiction.} \longrightarrow 0 = -3. \quad \text{Adding}$$

Since we end up with a *false* equation, or contradiction, we know that the system has no solution. It is *inconsistent*.   ▬

The notion of *dependency* can also be extended.

Dependency

If a system of n linear equations is equivalent to a system of fewer than n of them, we say that the equations are *dependent*. If such is not the case, we call the equations *independent*.

Example 6   Solve:

$$2x + y + z = 3, \qquad (1)$$
$$x - 2y - z = 1, \qquad (2)$$
$$3x + 4y + 3z = 5. \qquad (3)$$

SOLUTION   Using equations (1) and (2), we add to eliminate z:

$$2x + y + z = 3$$
$$\underline{x - 2y - z = 1}$$
$$3x - y = 4. \qquad (4)$$

Next, we use equations (2) and (3) to eliminate z again:

$$\begin{array}{l} x - 2y - z = 1, \\ 3x + 4y + 3z = 5 \end{array} \xrightarrow[\text{by 3}]{\text{Multiplying equation (2)}} \begin{array}{l} 3x - 6y - 3z = 3 \\ \underline{3x + 4y + 3z = 5} \\ 6x - 2y = 8. \qquad (5) \end{array}$$

We now try to solve the resulting system of equations (4) and (5):

$$3x - y = 4, \quad \xrightarrow{\substack{\text{Multiplying equation (4)} \\ \text{by } -2}} \quad \begin{array}{r} -6x + 2y = -8 \\ 6x - 2y = 8 \\ \hline 0 = 0. \quad (6) \end{array}$$

$$6x - 2y = 8$$

Equation (6), which is an identity, indicates that equations (1), (2), and (3) are *dependent.* This means that the original system of three equations is equivalent to a system of two equations. One way to see this is to observe that two times equation (1), minus equation (2), is equation (3). Thus removing equation (3) from the system does not affect the solution of the system. In writing an answer to this problem, we simply state that "the equations are dependent."

Recall that when dependent equations appeared in Section 3.1, the solution sets were always infinite in size and were written in set-builder notation. There, all systems of dependent equations were *consistent.* This is not always the case for systems of three or more equations. The following figures illustrate some possibilities geometrically.

The planes intersect along a common line. The equations are dependent and the system is consistent. There is an infinite number of solutions.

The planes coincide. The equations are dependent and the system is consistent. There is an infinite number of solutions.

Two planes coincide. The third plane is parallel. The equations are dependent and the system is inconsistent. There is no solution.

3.4 Exercise Set

1. Determine whether $(2, -1, -2)$ is a solution of the system
$$\begin{aligned} x + y - 2z &= 5, \\ 2x - y - z &= 7, \\ -x - 2y + 3z &= 6. \end{aligned}$$

2. Determine whether $(1, -2, 3)$ is a solution of the system
$$\begin{aligned} x + y + z &= 2, \\ x - 2y - z &= 2, \\ 3x + 2y + z &= 2. \end{aligned}$$

Solve each system. If a system's equations are dependent or if there is no solution, state this.

3. $\begin{aligned} 2x - y + z &= 10, \\ 4x + 2y - 3z &= 10, \\ x - 3y + 2z &= 8 \end{aligned}$

4. $\begin{aligned} x + y + z &= 6, \\ 2x - y + 3z &= 9, \\ -x + 2y + 2z &= 9 \end{aligned}$

5. $\begin{aligned} x - y + z &= 6, \\ 2x + 3y + 2z &= 2, \\ 3x + 5y + 4z &= 4 \end{aligned}$

6. $\begin{aligned} 2x - y - 3z &= -1, \\ 2x - y + z &= -9, \\ x + 2y - 4z &= 17 \end{aligned}$

7. $\begin{aligned} 6x - 4y + 5z &= 31, \\ 5x + 2y + 2z &= 13, \\ x + y + z &= 2 \end{aligned}$

8. $\begin{aligned} 2x - 3y + z &= 5, \\ x + 3y + 8z &= 22, \\ 3x - y + 2z &= 12 \end{aligned}$

9. $\begin{aligned} x + y + z &= 0, \\ 2x + 3y + 2z &= -3, \\ -x + 2y - 3z &= -1 \end{aligned}$

10. $\begin{aligned} 3a - 2b + 7c &= 13, \\ a + 8b - 6c &= -47, \\ 7a - 9b - 9c &= -3 \end{aligned}$

11. $2x + y - 3z = -4,$
$\quad 4x - 2y + z = 9,$
$\quad 3x + 5y - 2z = 5$

12. $2x + 3y + z = 17,$
$\quad x - 3y + 2z = -8,$
$\quad 5x - 2y + 3z = 5$

13. $2x + y + 2z = 11,$
$\quad 3x + 2y + 2z = 8,$
$\quad x + 4y + 3z = 0$

14. $2x + y + z = -2,$
$\quad 2x - y + 3z = 6,$
$\quad 3x - 5y + 4z = 7$

15. $-2x + 8y + 2z = 4,$
$\quad x + 6y + 3z = 4,$
$\quad 3x - 2y + z = 0$

16. $x - y + z = 4,$
$\quad 5x + 2y - 3z = 2,$
$\quad 4x + 3y - 4z = -2$

17. $a + 2b + c = 1,$
$\quad 7a + 3b - c = -2,$
$\quad a + 5b + 3c = 2$

18. $4x - y - z = 4,$
$\quad 2x + y + z = -1,$
$\quad 6x - 3y - 2z = 3$

19. $5x + 3y + \frac{1}{2}z = \frac{7}{2},$
$\quad 0.5x - 0.9y - 0.2z = 0.3,$
$\quad 3x - 2.4y + 0.4z = -1$

20. $r + \frac{3}{2}s + 6t = 2,$
$\quad 2r - 3s + 3t = 0.5,$
$\quad r + s + t = 1$

21. $3p + 2r = 11,$
$\quad q - 7r = 4,$
$\quad p - 6q = 1$

22. $4a + 9b = 8,$
$\quad 8a + 6c = -1,$
$\quad 6b + 6c = -1$

23. $x + y + z = 105,$
$\quad 10y - z = 11,$
$\quad 2x - 3y = 7$

24. $x + y + z = 57,$
$\quad -2x + y = 3,$
$\quad x - z = 6$

25. $2a - 3b = 2,$
$\quad 7a + 4c = \frac{3}{4},$
$\quad 2c - 3b = 1$

26. $a - 3c = 6,$
$\quad b + 2c = 2,$
$\quad 7a - 3b - 5c = 14$

27. $x + y + z = 180,$
$\quad y = 2 + 3x,$
$\quad z = 80 + x$

28. $l + m = 7,$
$\quad 3m + 2n = 9,$
$\quad 4l + n = 5$

29. $x + y = 0,$
$\quad x + z = 1,$
$\quad 2x + y + z = 2$

30. $x + z = 0,$
$\quad x + y + 2z = 3,$
$\quad y + z = 2$

31. $y + z = 1,$
$\quad x + y + z = 1,$
$\quad x + 2y + 2z = 2$

32. $x + y + z = 1,$
$\quad -x + 2y + z = 2,$
$\quad 2x - y = -1$

Skill Maintenance

33. If $f(x) = 2x + 7$, find $f(a + 1)$. [2.1]

34. If $g(x) = 1/(x - 7)$, find the domain of g. [2.1]

35. Solve $K = \frac{1}{2}t(a - b)$ for b. [2.3]

36. Solve $K = \frac{1}{2}t(a - b)$ for a. [2.3]

Synthesis

37. ◆ Explain in your own words what it means for the equations of a system to be dependent.

38. ◆ Describe a method for writing an inconsistent system of three equations in three variables.

39. ◆ Is it possible for a system of three equations to have exactly two ordered triples in its solution set? Why or why not?

40. ◆ Suggest a procedure that could be used to solve a system of four equations in four variables.

Solve.

41. $\dfrac{x + 2}{3} - \dfrac{y + 4}{2} + \dfrac{z + 1}{6} = 0,$

$\quad \dfrac{x - 4}{3} + \dfrac{y + 1}{4} - \dfrac{z - 2}{2} = -1,$

$\quad \dfrac{x + 1}{2} + \dfrac{y}{2} + \dfrac{z - 1}{4} = \dfrac{3}{4}$

42. $w + x + y + z = 2,$
$\quad w + 2x + 2y + 4z = 1,$
$\quad w - x + y + z = 6,$
$\quad w - 3x - y + z = 2$

43. $w + x - y + z = 0,$
$\quad w - 2x - 2y - z = -5,$
$\quad w - 3x - y + z = 4,$
$\quad 2w - x - y + 3z = 7$

For Exercises 44 and 45, let u represent $1/x$, v represent $1/y$, and w represent $1/z$. Solve for u, v, and w, and then solve for x, y, and z.

44. $\dfrac{2}{x} - \dfrac{1}{y} - \dfrac{3}{z} = -1,$

$\quad \dfrac{2}{x} - \dfrac{1}{y} + \dfrac{1}{z} = -9,$

$\quad \dfrac{1}{x} + \dfrac{2}{y} - \dfrac{4}{z} = 17$

45. $\dfrac{2}{x} + \dfrac{2}{y} - \dfrac{3}{z} = 3,$

$\quad \dfrac{1}{x} - \dfrac{2}{y} - \dfrac{3}{z} = 9,$

$\quad \dfrac{7}{x} - \dfrac{2}{y} + \dfrac{9}{z} = -39$

Determine k such that the equations in each system are dependent.

46. $x - 3y + 2z = 1,$
$\quad 2x + y - z = 3,$
$\quad 9x - 6y + 3z = k$

47. $5x - 6y + kz = -5,$
$\quad x + 3y - 2z = 2,$
$\quad 2x - y + 4z = -1$

In each case, three solutions of an equation in x, y, and z are given. Find the equation.

48. $Ax + By + Cz = 12;$
$\quad \left(1, \frac{3}{4}, 3\right), \left(\frac{4}{3}, 1, 2\right),$ and $(2, 1, 1)$

49. $z = b - mx - ny;$
$\quad (1, 1, 2), (3, 2, -6),$ and $\left(\frac{3}{2}, 1, 1\right)$

3.5
Applications Using Systems of Three Equations

• *Applications of Three Equations in Three Unknowns*

Applications of Three Equations in Three Unknowns

Solving systems of three or more equations is important in many applications. Such systems arise in the natural and social sciences, business, and engineering. In mathematics, purely numerical applications also arise.

Example 1   The sum of three numbers is 4. The first number minus twice the second, minus the third is 1. Twice the first number minus the second, minus twice the third is −1. Find the numbers.

SOLUTION

1. **Familiarize.**   There are three statements involving the same three numbers. Let's label these numbers x, y, and z.

2. **Translate.**   We can translate directly as follows.

The sum of the three numbers is 4.
$$x + y + z = 4$$

The first number minus twice the second minus the third is 1.
$$x - 2y - z = 1$$

Twice the first number minus the second minus twice the third is −1.
$$2x - y - 2z = -1$$

We now have a system of three equations:

$$x + y + z = 4,$$
$$x - 2y - z = 1,$$
$$2x - y - 2z = -1.$$

3. **Carry out.**   We need to solve the system of equations. Note that we found the solution, $(2, -1, 3)$, in Example 2 of Section 3.4.

4. **Check.**   The first statement of the problem says that the sum of the three numbers is 4. That checks. The second statement says that the first number minus twice the second, minus the third is 1: $2 - 2(-1) - 3 = 1$. That checks. The check of the third statement is left to the student.

5. **State.**   The three numbers are 2, −1, and 3.   ▬

Example 2   *Architecture.* In a triangular cross section of a roof, the largest angle is 70° greater than the smallest angle. The largest angle is twice as large as the remaining angle. Find the measure of each angle.

SOLUTION

1. Familiarize. The first thing we do is make a drawing.

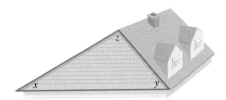

Since we don't know the size of any angle, we use x, y, and z for the measures of the angles. Recall that the measures of the angles in any triangle add up to $180°$.

2. Translate. This geometric fact about triangles gives us one equation:

$$x + y + z = 180.$$

Two of the statements can be translated almost directly.

The largest angle is 70° greater than the smallest angle.
$$z = x + 70$$

The largest angle is twice as large as the remaining angle.
$$z = 2y$$

We now have a system of three equations:

$$x + y + z = 180, \qquad x + y + z = 180,$$
$$x + 70 = z, \quad \text{or} \quad x \qquad - z = -70, \qquad \text{**Rewriting in standard form**}$$
$$2y = z; \qquad \qquad 2y - z = 0.$$

3. Carry out. The system was solved in Example 4 of Section 3.4. The solution is (30, 50, 100).

4. Check. The sum of the numbers is 180, so that checks. The measure of the largest angle, $100°$, is $70°$ greater than the measure of the smallest angle, $30°$, so that checks. The measure of the largest angle is also twice the measure of the remaining angle, $50°$. Thus we have a check.

5. State. The angles in the triangle measure $30°$, $50°$, and $100°$. ▬

Example 3 *Cholesterol Levels.* Americans have become very conscious of their cholesterol levels. Recent studies indicate that a child's intake of cholesterol should be no more than 300 mg per day. By eating 1 egg, 1 cupcake, and 1 slice of pizza, a child consumes 302 mg of cholesterol. If the child eats 2 cupcakes and 3 slices of pizza, he or she takes in 65 mg of cholesterol. If the child eats 2 eggs and 1 cupcake, he or she consumes 567 mg of cholesterol. How much cholesterol is in each item?

SOLUTION

1. **Familiarize.** After we have read the problem a few times, it becomes clear that an egg contains considerably more cholesterol than the other foods. Let's guess that one egg contains 200 mg of cholesterol and one cupcake contains 50 mg. Because of the third sentence in the problem, it would follow that a slice of pizza contains 52 mg of cholesterol since $200 + 50 + 52 = 302$.

   To see if our guess satisfies the other statements in the problem, we find the amount of cholesterol that 2 cupcakes and 3 slices of pizza would contain: $2 \cdot 50 + 3 \cdot 52 = 256$. Since this does not match the 65 mg listed in the fourth sentence of the problem, our guess was incorrect. Rather than guess again, we examine how we checked our guess and let e, c, and $s =$ the number of milligrams of cholesterol in an egg, a cupcake, and a slice of pizza, respectively.

2. **Translate.** By rewording some of the sentences in the problem, we can translate it into three equations.

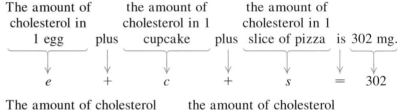

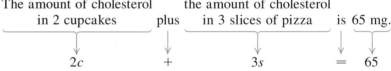

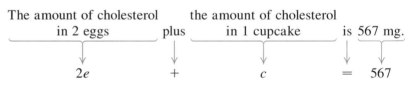

We now have a system of three equations:

$$e + c + s = 302,$$
$$2c + 3s = 65,$$
$$2e + c = 567.$$

3. **Carry out.** We solve and get $e = 274$, $c = 19$, $s = 9$, or $(274, 19, 9)$.

4. **Check.** The sum of 274, 19, and 9 is 302 so the total cholesterol in 1 egg, 1 cupcake, and 1 slice of pizza checks. Two cupcakes and three slices of pizza would contain $2 \cdot 19 + 3 \cdot 9 = 65$ mg, whereas two eggs and one cupcake would contain $2 \cdot 274 + 19 = 567$ mg of cholesterol. The answer checks.

5. **State.** An egg contains 274 mg of cholesterol, a cupcake contains 19 mg of cholesterol, and a slice of pizza contains 9 mg of cholesterol.

3.5 | Exercise Set

Solve.

1. The sum of three numbers is 57. The second is 3 more than the first. The third is 6 more than the first. Find the numbers.

2. The sum of three numbers is 5. The first number minus the second, plus the third is 1. The first minus the third is 3 more than the second. Find the numbers.

3. The sum of three numbers is 26. Twice the first minus the second is 2 less than the third. The third is the second minus three times the first. Find the numbers.

4. The sum of three numbers is 105. The third is 11 less than ten times the second. Twice the first is 7 more than three times the second. Find the numbers.

5. *Geometry.* In triangle *ABC*, the measure of angle *B* is three times that of angle *A*. The measure of angle *C* is 20° more than that of angle *A*. Find the angle measures.

6. *Geometry.* In triangle *ABC*, the measure of angle *B* is twice the measure of angle *A*. The measure of angle *C* is 80° more than that of angle *A*. Find the angle measures.

7. *Automobile Pricing.* A recent basic model of a particular automobile had a price of $12,685. The basic model with the added features of automatic transmission and power door locks was $14,070. The basic model with air conditioning (AC) and power door locks was $13,580. The basic model with AC and automatic transmission was $13,925. What was the individual cost of each of the three options?

8. *Telemarketing.* Sven, Tillie, and Isaiah can process 740 telephone orders per day. Sven and Tillie together can process 470 orders, while Tillie and Isaiah together can process 520 orders per day. How many orders can each person process alone?

9. *Lens Production.* When Sight-Rite's three polishing machines, A, B, and C, are all working, 5700 lenses can be polished in one week. When only A and B are working, 3400 lenses can be polished in one week. When only B and C are working, 4200 lenses can be polished in one week. How many lenses can be polished in a week by each machine?

10. *Welding Rates.* Elrod, Dot, and Wendy can weld 74 linear feet per hour when working together. Elrod and Dot together can weld 44 linear feet per hour, while Elrod and Wendy can weld 50 linear feet per hour. How many linear feet per hour can each weld alone?

11. *Restaurant Management.* Kyle works at Dunkin' Donuts,® where a 10-oz cup of coffee costs 95¢, a 14-oz cup costs $1.15, and a 20-oz cup costs $1.50. During one busy period, Kyle served 34 cups of coffee, emptying five 96-oz pots while collecting a total of $39.60. How many cups of each size did Kyle fill?

10 oz	14 oz	20 oz
$0.95	$1.15	$1.50

12. *Restaurant Management.* McDonald's® recently sold small soft drinks for 89¢, medium soft drinks for 99¢, and large soft drinks for $1.19. During a lunch-time rush, Chris sold 55 soft drinks for a total of $54.95. The number of small and large drinks, combined, was 5 fewer than the number of medium drinks. How many drinks of each size were sold?

small	medium	large
$0.89	$0.99	$1.19

13. *Investments.* A business class divided an imaginary investment of $80,000 among three mutual funds. The first fund grew by 10%, the second by 6%, and the third by 15%. Total earnings were $8850. The earnings from the first fund were

$750 more than the earnings from the third. How much was invested in each fund?

14. *Advertising.*   In a recent year, companies spent a total of $84.8 billion on newspaper, television, and radio ads. The total amount spent on television and radio ads was only $2.6 billion more than the amount spent on newspaper ads alone. The amount spent on newspaper ads was $5.1 billion more than what was spent on television ads. How much was spent on each form of advertising? (*Hint:* Let the variables represent numbers of billions of dollars.)

15. *Nutrition.*   A dietician in a hospital prepares meals under the guidance of a physician. Suppose that for a particular patient a physician prescribes a meal to have 800 calories, 55 g of protein, and 220 mg of vitamin C. The dietician prepares a meal of roast beef, baked potatoes, and broccoli according to the data in the following table.

	CALORIES	PROTEIN (IN GRAMS)	VITAMIN C (IN MILLIGRAMS)
ROAST BEEF, 3 OZ	300	20	0
BAKED POTATO	100	5	20
BROCCOLI, 156 G	50	5	100

How many servings of each food are needed in order to satisfy the doctor's orders?

16. *Nutrition.*   Repeat Exercise 15 but replace the broccoli with asparagus, for which one 180-g serving contains 50 calories, 5 g of protein, and 44 mg of vitamin C. Which meal would you prefer eating?

17. *Obstetrics.*   In the United States, the highest incidence of fraternal twin births occurs among Asian-Americans, then African-Americans, and then Caucasians. Of every 15,400 births, the total number of fraternal twin births for all three is 739, where there are 185 more for Asian-Americans than African-Americans and 231 more for Asian-Americans than Caucasians. How many births of fraternal twins are there for each group out of every 15,400 births?

18. *Crying Rate.*   The sum of the average number of times a man, a woman, and a one-year-old child cry each month is 71.7. A one-year-old cries 46.4 more

times than a man. The average number of times a one-year-old cries per month is 28.3 more than the average number of times combined that a man and a woman cry. What is the average number of times per month that each cries?

19. *Basketball Scoring.*   The New York Knicks recently scored a total of 92 points on a combination of 2-point field goals, 3-point field goals, and 1-point foul shots. Altogether, the Knicks made 50 baskets and 19 more 2-pointers than foul shots. How many shots of each kind were made?

20. *History.*   Find the year in which the first U.S. transcontinental railroad was completed. The following are some facts about the number. The sum of the digits in the year is 24. The ones digit is 1 more than the hundreds digit. Both the tens and the ones digits are multiples of 3.

Skill Maintenance

21. Solve for x: $3(5 - x) + 7 = 5(x + 3) - 9$. [2.2]

22. Compute: $(5 - 3^2 \div 2) \cdot (-4)^2$. [1.2]

23. Simplify: $\dfrac{(a^2 b^3)^5}{a^7 b^{16}}$. [1.4]

24. Give a slope–intercept equation for a line with slope $-\frac{3}{5}$ and y-intercept $(0, -7)$. [2.4]

25. If $g(x) = (x - 5)/(x + 7)$, find the domain of g. [2.1]

26. If $f(x) = 3x^2 - 7x$, find $f(-4)$. [2.1]

Synthesis

27. ◆ Write a problem for a classmate to solve. Design the problem so that it translates to a system of three equations in three variables.

28. ◆ Exercise 10 can be solved mentally after a careful reading of the problem. How is this possible?

29. Find a three-digit positive integer such that the sum of all three digits is 14, the tens digit is 2 more than the ones digit, and if the digits are reversed, the number is unchanged.

30. *Ages.* Tammy's age is the sum of the ages of Carmen and Dennis. Carmen's age is 2 more than the sum of the ages of Dennis and Mark. Dennis's age is four times Mark's age. The sum of all four ages is 42. How old is Tammy?

31. *Ticket Revenue.* A concert audience of 100 people consists of adults, students, and children. The ticket prices are $10 for adults, $3 for students, and 50¢ for children. The total amount of money taken in is $100. How many adults, students, and children are in attendance? Does there seem to be some information missing? Do some more careful reasoning.

32. *Sharing Raffle Tickets.* Hal gives Tom as many raffle tickets as Tom has and Gary as many as Gary has. In like manner, Tom then gives Hal and Gary as many tickets as each then has. Similarly, Gary gives Hal and Tom as many tickets as each then has. If each finally has 40 tickets, with how many tickets does Tom begin?

33. Find the sum of the angle measures at the tips of the star in this figure.

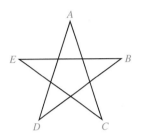

3.6
Elimination Using Matrices

- *Matrices and Systems*
- *Row-Equivalent Operations*

In solving systems of equations, we perform computations with the constants. The variables play no important role until the end. Thus we can simplify writing a system by omitting the variables. For example, the system

$$3x + 4y = 5,$$
$$x - 2y = 1$$

simplifies to

$$\begin{array}{ccc} 3 & 4 & 5 \\ 1 & -2 & 1 \end{array}$$

if we ignore the variables, the operation of addition, and the equals signs.

Matrices and Systems

In the example above, we have written a rectangular array of numbers. Such an array is called a **matrix** (plural, **matrices**). We ordinarily write brackets around matrices. The following are matrices:

$$\begin{bmatrix} -3 & 1 \\ 0 & 5 \end{bmatrix}, \quad \begin{bmatrix} 2 & 0 & -1 & 3 \\ -5 & 2 & 7 & -1 \\ 4 & 5 & 3 & 0 \end{bmatrix}, \quad \begin{bmatrix} 2 & 3 \\ 7 & 15 \\ -2 & 23 \\ 4 & 1 \end{bmatrix}$$

The individual numbers are called *elements* or *entries.*

The **rows** of a matrix are horizontal, and the **columns** are vertical. The matrix

$$A = \begin{bmatrix} 5 & -2 & -2 \\ 1 & 0 & 1 \\ 4 & -3 & 2 \end{bmatrix} \begin{matrix} \leftarrow \text{row 1} \\ \leftarrow \text{row 2} \\ \leftarrow \text{row 3} \end{matrix}$$

column 1   column 2   column 3

is called a **square matrix** because it has the same number of rows and columns. The elements of a matrix are referred to by their row number and column number. The notation a_{ij} refers to the element in row i and column j. For example, in the matrix A above,

$a_{11} = 5$ 5 is the element in the first row and first column.

and $a_{32} = -3$.   −3 is the element in the third row and second column.

The elements a_{11}, a_{22}, a_{33}, and so on, are said to be on the **main diagonal** of a square matrix. The entries 5, 0, and 2 are on the main diagonal of the matrix A above.

Let's now use matrices to solve systems of linear equations.

Example 1   Solve the system

$$5x - 4y = -1,$$
$$-2x + 3y = 2.$$

SOLUTION   We write a matrix using only coefficients and constants, listing x-coefficients in the first column and y-coefficients in the second. A dashed line separates the coefficients from the constants:

$$\begin{bmatrix} 5 & -4 & -1 \\ -2 & 3 & 2 \end{bmatrix}.$$

Our goal is to transform

$$\begin{bmatrix} 5 & -4 & -1 \\ -2 & 3 & 2 \end{bmatrix} \quad \text{into the form} \quad \begin{bmatrix} 1 & b & c \\ 0 & 1 & e \end{bmatrix}.$$

Note that the matrix contains 1's on the main diagonal of the coefficient matrix and a 0 below the main diagonal. The matrix is said to be in **row-echelon form**. The variables can then be reinserted to form equations from which we can complete the solution.

We do calculations that are similar to those that we would do if we wrote the entire equations. The first step, if possible, is to multiply and/or interchange the rows so that each number in the first column below the first number is a multiple of that number. In this case, we do so by multiplying Row 2 by 5. This corresponds to multiplying the second equation by 5 on both sides.

$$\begin{bmatrix} 5 & -4 & -1 \\ -10 & 15 & 10 \end{bmatrix} \quad \textbf{New Row 2 = 5(Row 2)}$$

Next, we multiply the first row by 2 and add the result to the second row. This corresponds to multiplying the first equation by 2 and adding the result to the second equation in order to eliminate a variable. Write out these computations as necessary—we perform them mentally.

$$\begin{bmatrix} 5 & -4 & | & -1 \\ 0 & 7 & | & 8 \end{bmatrix}$$ **New Row 2 = 2(Row 1) + (Row 2)**

Note that the part of the matrix that contains the coefficients, to the left of the dashed line, is a square matrix. To write the matrix in row-echelon form, we need 1's on the main diagonal of that square matrix. Thus we divide the first row by 5 and the second row by 7.

$$\begin{bmatrix} 1 & -\frac{4}{5} & | & -\frac{1}{5} \\ 0 & 1 & | & \frac{8}{7} \end{bmatrix}$$ **New Row 1 = $\frac{1}{5}$(Row 1)**
New Row 2 = $\frac{1}{7}$(Row 2)

If we now reinsert the variables, we have

$$x - \frac{4}{5}y = -\frac{1}{5}, \qquad (1)$$
$$y = \frac{8}{7}. \qquad (2)$$

From equation (2), we see that $y = \frac{8}{7}$. We substitute $\frac{8}{7}$ for y back in equation (1). This is called *back-substitution*.

$$x - \frac{4}{5}y = -\frac{1}{5} \qquad (1)$$
$$x - \frac{4}{5}\left(\frac{8}{7}\right) = -\frac{1}{5} \qquad \text{Substituting } \frac{8}{7} \text{ for } y \text{ in equation (1)}$$
$$x = \frac{5}{7} \qquad \text{Solving for } x$$

The solution is $\left(\frac{5}{7}, \frac{8}{7}\right)$. The check is left to the student.   ▬

Example 2 Solve the system

$$2x - y + 4z = -3,$$
$$x \qquad - 4z = 5,$$
$$6x - y + 2z = 10.$$

SOLUTION   We first write a matrix, using only the constants. Where there are missing terms, we must write 0's:

$$\begin{bmatrix} 2 & -1 & 4 & | & -3 \\ 1 & 0 & -4 & | & 5 \\ 6 & -1 & 2 & | & 10 \end{bmatrix} \begin{matrix} \text{(P1)} \\ \text{(P2)} \\ \text{(P3)} \end{matrix}$$ **(P1), (P2), and (P3) designate the equations that are in the first, second, and third position, respectively.**

Our goal is to transform the matrix to one of the form

$$\begin{bmatrix} 1 & a & b & | & c \\ 0 & 1 & d & | & e \\ 0 & 0 & 1 & | & f \end{bmatrix}.$$ **The matrix is in row-echelon form.**

The first step, if possible, is to interchange the rows so that each number in the first column below the first number is a multiple of that number. In this case, we do so by interchanging Rows 1 and 2:

$$\begin{bmatrix} 1 & 0 & -4 & \vdots & 5 \\ 2 & -1 & 4 & \vdots & -3 \\ 6 & -1 & 2 & \vdots & 10 \end{bmatrix}.$$

This corresponds to interchanging the first two equations.

Next, we multiply the first row by -2 and add it to the second row:

$$\begin{bmatrix} 1 & 0 & -4 & \vdots & 5 \\ 0 & -1 & 12 & \vdots & -13 \\ 6 & -1 & 2 & \vdots & 10 \end{bmatrix}.$$

This corresponds to multiplying new equation (P1) by -2 and adding it to new equation (P2). We perform the calculations mentally.

Now we multiply the first row by -6 and add it to the third row:

$$\begin{bmatrix} 1 & 0 & -4 & \vdots & 5 \\ 0 & -1 & 12 & \vdots & -13 \\ 0 & -1 & 26 & \vdots & -20 \end{bmatrix}.$$

This corresponds to multiplying equation (P1) by -6 and adding it to equation (P3).

Next, we multiply Row 2 by -1 and add it to the third row:

$$\begin{bmatrix} 1 & 0 & -4 & \vdots & 5 \\ 0 & -1 & 12 & \vdots & -13 \\ 0 & 0 & 14 & \vdots & -7 \end{bmatrix}.$$

This corresponds to multiplying equation (P2) by -1 and adding it to equation (P3).

Finally, we multiply Row 2 by -1 and Row 3 by $\frac{1}{14}$:

$$\begin{bmatrix} 1 & 0 & -4 & \vdots & 5 \\ 0 & 1 & -12 & \vdots & 13 \\ 0 & 0 & 1 & \vdots & -\frac{1}{2} \end{bmatrix}.$$

This corresponds to multiplying equation (P2) by -1 and equation (P3) by $\frac{1}{14}$.

Reinserting the variables gives us

$$\begin{aligned} x \qquad\quad - 4z &= 5, & \text{(P1)} \\ y - 12z &= 13, & \text{(P2)} \\ z &= -\tfrac{1}{2}. & \text{(P3)} \end{aligned}$$

From equation (P3), we see that $z = -\frac{1}{2}$. We back-substitute $-\frac{1}{2}$ for z in (P2) and solve for y:

$$y - 12\left(-\tfrac{1}{2}\right) = 13$$
$$y = 7.$$

Since there is no y-term in (P1), we need only substitute $-\frac{1}{2}$ for z to solve for x:

$$x - 4\left(-\tfrac{1}{2}\right) = 5$$
$$x = 3.$$

The solution is $\left(3, 7, -\frac{1}{2}\right)$. The check is left to the student.   ▬

The operations used in the preceding example correspond to those used to produce equivalent systems of equations. The corresponding systems of equations are all equivalent, that is, they have the same solution set. We call such matrices **row-equivalent** and the operations that produce them **row-equivalent operations**.

Row-Equivalent Operations

> **Row-Equivalent Operations**
>
> Each of the following row-equivalent operations produces an equivalent matrix:
>
> **a)** Interchanging any two rows.
>
> **b)** Multiplying all elements of a row by the same nonzero number.
>
> **c)** Multiplying all elements of a row by a nonzero number and adding the result to another row.

Many graphers can store and manipulate matrices.

Example 3   Solve the following system using a grapher:

$$2x + 5y - 8z = 7,$$
$$3x + 4y - 3z = 8,$$
$$5y - 2x = 9.$$

SOLUTION   Before writing a matrix that represents this system of equations, we rewrite the third equation in the form $ax + by + cz = d$:

$$2x + 5y - 8z = 7,$$
$$3x + 4y - 3z = 8,$$
$$-2x + 5y \quad\quad = 9.$$

The matrix that represents this system is thus

$$\begin{bmatrix} 2 & 5 & -8 & | & 7 \\ 3 & 4 & -3 & | & 8 \\ -2 & 5 & 0 & | & 9 \end{bmatrix}.$$

On many graphers, matrix operations are accessed by pressing MATRX. We enter the matrix above as matrix A using the EDIT menu. Since the matrix has 3 rows and 4 columns, its **dimensions** are 3×4 (read "3 by 4").

When entering elements of the matrix, we fill up the first row before going on to the second. Remember to write a 0 when there is a missing term. We press ENTER after each entry.

When we have entered each element of the matrix, we return to the home screen to perform matrix operations. The contents of matrix A can be displayed on the screen.

```
[A]
          [[2 5 -8 7]
           [3 4 -3 8]
           [-2 5 0 9]]
```

Matrix operations are done using the MATRX MATH menu. If you wish, you can perform each of the row-equivalent operations necessary to write the matrix in row-echelon form. The row-equivalent matrix found in each step can then be stored as the "new" matrix A. Many graphers can go to the row-echelon form with one command. The following screens show the row-echelon form of matrix A, with the elements written in fractional form. Note that the arrow keys must be used to see the entire matrix.

```
ref([A])▶Frac
[[1 4/3  −1        8...
 [0 1   −6/23      4...
 [0 0    1         1...
```

```
ref([A])▶Frac
...3    −1     8/3  ]
...    −6/23  43/23]
...     1     1/2 ]]
```

From this matrix, we can reinsert the variables and solve. We have

$$x + \tfrac{4}{3}y - 1z = \tfrac{8}{3}, \quad (1)$$
$$y - \tfrac{6}{23}z = \tfrac{43}{23}, \quad (2)$$
$$z = \tfrac{1}{2}. \quad (3)$$

We see from equation (3) that $z = \tfrac{1}{2}$. Back-substituting and solving gives us $y = 2$ and $x = \tfrac{1}{2}$. The solution is $\left(\tfrac{1}{2}, 2, \tfrac{1}{2}\right)$. ▬

The best overall method for solving systems of equations is by row-equivalent matrices; most computers and calculators are programmed to use them. Matrices are part of a branch of mathematics known as linear algebra. They are also studied in many courses in finite mathematics.

3.6 Exercise Set

Solve using matrices.

1. $5x - 3y = 13,$
$4x + y = 7$

2. $3x - 3y = 11,$
$9x - 2y = 5$

3. $x + 4y = 8,$
$3x + 5y = 3$

4. $x + 4y = 5,$
$-3x + 2y = 13$

5. $6x - 2y = 4,$
$7x + y = 13$

6. $3x + 4y = 7,$
$-5x + 2y = 10$

7. $4x - y - 3z = 1,$
$8x + y - z = 5,$
$2x + y + 2z = 5$

8. $3x + 2y + 2z = 3,$
$x + 2y - z = 5,$
$2x - 4y + z = 0$

9. $p - 2q - 3r = 3,$
$2p - q - 2r = 4,$
$4p + 5q + 6r = 4$

10. $x + 2y - 3z = 9,$
$2x - y + 2z = -8,$
$3x - y - 4z = 3$

11. $3p + 2r = 11,$
$q - 7r = 4,$
$p - 6q = 1$

12. $4a + 9b = 8,$
$8a + 6c = -1,$
$6b + 6c = -1$

13. $3x + y = 8,$
$4x + 5y - 3z = 4,$
$7x + 2y - 9z = 1$

14. $5x - 8y = 3,$
$2y + 8z = -11,$
$5x + 7z = 9$

15. $-0.01x + 0.7y = -0.9,$
$0.5x - 0.3y + 0.18z = 0.01,$
$50x + 6y - 75z = 12$

16. $4x + 2y + 5z = 1,$
$-3x + 7y + 2z = -9,$
$7x - 5y + z = 3$

17. $-w - 3y + z + 2x = -8,$
$\quad x + y - z - w = -4,$
$\quad w + y + z + x = 22,$
$\quad x - y - z - w = -14$

18. $2x + 2y - 2z - 2w = -10,$
$\quad w + y + z + x = -5,$
$\quad x - y + 4z + 3w = -2,$
$\quad w - 2y + 2z + 3x = -6$

Solve using matrices.

19. *Coin Value.* A collection of 34 coins consists of dimes and nickels. The total value is $1.90. How many dimes and how many nickels are there?

20. *Coin Value.* A collection of 43 coins consists of dimes and quarters. The total value is $7.60. How many dimes and how many quarters are there?

21. *Mixed Granola.* Grace sells two kinds of granola. One is worth $4.05 per pound and the other is worth $2.70 per pound. She wants to blend the two granolas to get a 15-lb mixture worth $3.15 per pound. How much of each kind of granola should be used?

22. *Trail Mix.* Phil mixes nuts worth $1.60 per pound with oats worth $1.40 per pound to get 20 lb of trail mix worth $1.54 per pound. How many pounds of nuts and how many pounds of oats should be used?

23. *Investments.* Elena receives $212 per year in simple interest from three investments totaling $2500. Part is invested at 7%, part at 8%, and part at 9%. There is $1100 more invested at 9% than at 8%. Find the amount invested at each rate.

24. *Investments.* Miguel receives $306 per year in simple interest from three investments totaling $3200. Part is invested at 8%, part at 9%, and part at 10%. There is $1900 more invested at 10% than at 9%. Find the amount invested at each rate.

Skill Maintenance

Solve. [2.2]

25. $0.1x - 12 = 3.6x - 2.34 - 4.9x$

26. $180 = 2x - 11x$

27. $4(9 - x) - 6(8 - 3x) = 5(3x + 4)$

28. Solve $5b = c - ab$ for b. [2.3]

Synthesis

29. ◈ Explain how you can recognize a dependent system when solving with matrices.

30. ◈ Explain how you can recognize an inconsistent system when solving with matrices.

31. The sum of the digits in a four-digit number is 10. Twice the sum of the thousands digit and the tens digit is 1 less than the sum of the other two digits. The tens digit is twice the thousands digit. The ones digit equals the sum of the thousands digit and the hundreds digit. Find the four-digit number.

32. Solve for x and y:
$\quad ax + by = c,$
$\quad dx + ey = f.$

3.7

Business and Economics Applications

- *Break-Even Analysis*
- *Supply and Demand*

Break-Even Analysis

When a company manufactures x units of a product, it invests money. This is **total cost** and can be thought of as a function C, where $C(x)$ is the total cost of producing x units. When the company sells x units of the product, it takes in money. This is **total revenue** and can be thought of as a function R, where $R(x)$ is the total revenue from the sale of x units. **Total profit** is the money taken in less the money spent, or total revenue minus total cost. Total profit from the production and sale of x units is a

function P given by

$$\text{Profit} = \text{Revenue} - \text{Cost}, \qquad \text{or} \qquad P(x) = R(x) - C(x).$$

If $R(x)$ is greater than $C(x)$, the company makes money. If $C(x)$ is greater than $R(x)$, the company has a loss. When $R(x) = C(x)$, the company breaks even.

There are two kinds of costs. First, there are costs like rent, insurance, machinery, and so on. These costs, which must be paid whether a product is produced or not, are called *fixed costs*. When a product is being produced, there are costs for labor, materials, marketing, and so on. These are called *variable costs*, because they vary according to the amount being produced. The sum of the fixed cost and the variable cost gives the *total cost* of producing a product.

Example 1 *Manufacturing Radios.* Ergs, Inc., is planning to make a new kind of radio. Fixed costs will be $90,000, and it will cost $15 to produce each radio (variable costs). Each radio sells for $26.

a) Find the total cost $C(x)$ of producing x radios.

b) Find the total revenue $R(x)$ from the sale of x radios.

c) Find the total profit $P(x)$ from the production and sale of x radios.

d) What profit or loss will the company realize from the production and sale of 3000 radios? of 14,000 radios?

e) Graph the total-cost, total-revenue, and total-profit functions using the same set of axes. Determine the break-even point.

SOLUTION

a) Total cost is given by

$$C(x) = (\text{Fixed costs}) \text{ plus } (\text{Variable costs}),$$

or $C(x) = \quad 90{,}000 \quad + \quad 15x,$

where x is the number of radios produced.

b) Total revenue is given by

$$R(x) = 26x. \qquad$$ **$26 times the number of radios sold. We assume that every radio produced is sold.**

c) Total profit is given by

$$P(x) = R(x) - C(x)$$
$$= 26x - (90{,}000 + 15x)$$
$$= 11x - 90{,}000.$$

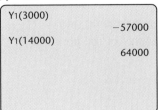

d) Profits will be

$$P(3000) = 11 \cdot 3000 - 90{,}000 = -\$57{,}000$$

when 3000 radios are produced and sold, and

$$P(14{,}000) = 11 \cdot 14{,}000 - 90{,}000 = \$64{,}000$$

when 14,000 radios are produced and sold. These values can also be found using a grapher, as shown in the figure at left. Thus the company loses money if only 3000 radios are sold, but makes money if 14,000 are sold.

e) The graphs of each of the three functions are shown below:

$$R(x) = 26x, \qquad\qquad (1)$$
$$C(x) = 90{,}000 + 15x, \qquad (2)$$
$$P(x) = 11x - 90{,}000. \qquad (3)$$

$R(x)$, $C(x)$, and $P(x)$ are all in dollars.

Equation (1) has a graph that goes through the origin and has a slope of 26. Equation (2) has an intercept on the $-axis of 90,000 and has a slope of 15. Equation (3) has an intercept on the $-axis of $-90{,}000$ and has a slope of 11. It is shown by the dashed line. The red dashed line shows a "negative" profit, which is a loss. (That is what is known as "being in the red.") The black dashed line shows a "positive" profit, or gain. (That is what is known as "being in the black.")

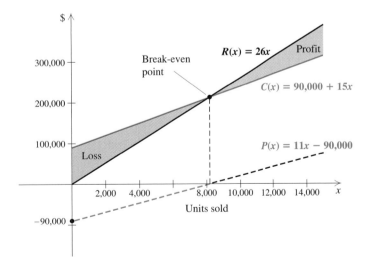

Profits occur where the revenue is greater than the cost. Losses occur where the revenue is less than the cost. The **break-even point** occurs where the graphs of R and C cross. Thus to find the break-even point, we solve a system:

$$R(x) = 26x,$$
$$C(x) = 90{,}000 + 15x.$$

ALGEBRAIC SOLUTION

Since both revenue and cost are in *dollars* and they are equal at the break-even point, the system can be rewritten as

$d = 26x,$ (1)

$d = 90{,}000 + 15x$ (2)

and solved using the substitution method:

$26x = 90{,}000 + 15x$ **Substituting $26x$ for d in equation (2)**

$11x = 90{,}000$

$x \approx 8181.8.$

Substituting in equation 1, we have
$d = 26x \approx 26 \cdot 8182 \approx 212{,}732.$

GRAPHICAL SOLUTION

We graph $y_1 = 26x$ and $y_2 = 90{,}000 + 15x$, and determine the coordinates of the point of intersection.

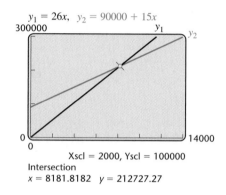

$y_1 = 26x, \quad y_2 = 90000 + 15x$

Xscl = 2000, Yscl = 100000
Intersection
$x = 8181.8182 \quad y = 212727.27$

The solution is approximately (8182, 212,727).

The firm will break even if it produces and sells about 8182 radios (8181 will yield a tiny loss and 8182 a tiny gain), and takes in a total of $R(8182) = 26 \cdot 8182 = \$212{,}732$ in revenue. Note that the x-coordinate of the break-even point can also be found by solving $P(x) = 0$. ▬

Supply and Demand

As the price of coffee varies, the amount sold varies. The table and graph below both show that consumer *demand* goes down as the price goes up. As the price goes down, demand goes up.

Demand Function, D

PRICE, p, PER KILOGRAM	QUANTITY, $D(p)$ (IN MILLIONS OF KILOGRAMS)
$ 8.00	25
9.00	20
10.00	15
11.00	10
12.00	5

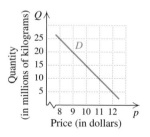

As the price of coffee varies, the amount available varies. The table and graph below both show that sellers will supply less as the price goes down, but will supply more as the price goes up.

Supply Function, S

PRICE, p, PER KILOGRAM	QUANTITY, $S(p)$ (IN MILLIONS OF KILOGRAMS)
$ 9.00	5
9.50	10
10.00	15
10.50	20
11.00	25

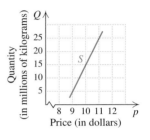

Let's look at these graphs together. We see that as price increases, demand decreases. As price increases, supply increases. The point of intersection is called the **equilibrium point**. At that price, the amount that the seller will supply is the same amount that the consumer will buy. The situation is analogous to a buyer and a seller negotiating the price of an item. The equilibrium point is the price and quantity that they finally agree on.

Any ordered pair of coordinates from the graph is (price, quantity), because the horizontal axis is the price axis and the vertical axis is the quantity axis. If D is a demand function and S is a supply function, then the equilibrium point is where demand equals supply:

$$D(p) = S(p).$$

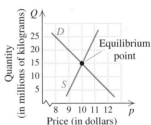

Example 2 Find the equilibrium point for the following demand and supply functions:

$$D(p) = 1000 - 60p, \quad (1)$$
$$S(p) = 200 + 4p. \quad (2)$$

SOLUTION Since both demand and supply are *quantities* and they are equal at the equilibrium point, we rewrite the system as

$$q = 1000 - 60p, \quad (1)$$
$$q = 200 + 4p. \quad (2)$$

We substitute $200 + 4p$ for q in equation (1) and solve:

$$200 + 4p = 1000 - 60p$$
$$200 + 64p = 1000 \qquad \text{Adding } 60p \text{ on both sides}$$
$$64p = 800 \qquad \text{Adding } -200 \text{ on both sides}$$
$$p = \tfrac{800}{64} = 12.5.$$

Thus the equilibrium price is $12.50 per unit.
   To find the equilibrium quantity, we substitute $12.50 into either $D(p)$ or $S(p)$. We use $S(p)$:

$$S(12.5) = 200 + 4(12.5)$$
$$= 200 + 50 = 250.$$

We graph $y_1 = 1000 - 60x$ and $y_2 = 200 + 4x$.

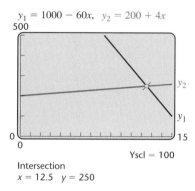

Intersection
$x = 12.5$ $y = 250$

The solution is $p = 12.5$ and $q = 250$.

Thus the equilibrium price is $12.50, the equilibrium quantity is 250 units, and the equilibrium point is ($12.50, 250).

3.7 Exercise Set

*For each of the following pairs of total cost and total-revenue functions, find **(a)** the total-profit function and **(b)** the break-even point.*

1. $C(x) = 25x + 270{,}000$,
   $R(x) = 70x$

2. $C(x) = 45x + 300{,}000$,
   $R(x) = 65x$

3. $C(x) = 10x + 120{,}000$,
   $R(x) = 60x$

4. $C(x) = 30x + 49{,}500$,
   $R(x) = 85x$

5. $C(x) = 20x + 10{,}000$,
   $R(x) = 100x$

6. $C(x) = 40x + 22{,}500$,
   $R(x) = 85x$

7. $C(x) = 22x + 16{,}000$,
   $R(x) = 40x$

8. $C(x) = 15x + 75{,}000$,
   $R(x) = 55x$

9. $C(x) = 50x + 195{,}000$,
   $R(x) = 125x$

10. $C(x) = 34x + 928{,}000$,
   $R(x) = 128x$

Find the equilibrium point for each of the following pairs of demand and supply functions.

11. $D(p) = 1000 - 10p$,
   $S(p) = 230 + p$

12. $D(p) = 2000 - 60p$,
   $S(p) = 460 + 94p$

13. $D(p) = 760 - 13p$,
   $S(p) = 430 + 2p$

14. $D(p) = 800 - 43p$,
   $S(p) = 210 + 16p$

15. $D(p) = 7500 - 25p$,
   $S(p) = 6000 + 5p$

16. $D(p) = 8800 - 30p$,
   $S(p) = 7000 + 15p$

17. $D(p) = 1600 - 53p$,
   $S(p) = 320 + 75p$

18. $D(p) = 5500 - 40p$,
   $S(p) = 1000 + 85p$

Solve.

19. *Manufacturing Lamps.* City Lights, Inc., is planning to manufacture a new type of lamp. For the first year, the fixed costs for setting up production are $22,500. The variable costs for producing each lamp are estimated to be $40. The revenue from each lamp is to be $85. Find the following.

   a) The total cost $C(x)$ of producing x lamps
   b) The total revenue $R(x)$ from the sale of x lamps
   c) The total profit $P(x)$ from the production and sale of x lamps
   d) The profit or loss from the production and sale of 3000 lamps; of 400 lamps
   e) The break-even point

20. *Computer Manufacturing.* Sky View Electronics is planning to introduce a new line of computers. For the first year, the fixed costs for setting up production are $125,100. The variable costs for producing each computer are $750. The revenue from each computer is $1050. Find the following.

a) The total cost $C(x)$ of producing x computers
b) The total revenue $R(x)$ from the sale of x computers
c) The total profit $P(x)$ from the production and sale of x computers
d) The profit or loss from the production and sale of 400 computers; of 700 computers
e) The break-even point

21. *Manufacturing Caps.* Martina's Custom Printing is planning on adding painter's caps to its product line. For the first year, the fixed costs for setting up production are $16,404. The variable costs for producing a dozen caps are $6.00. The revenue on each dozen caps will be $18.00. Find the following.

a) The total cost $C(x)$ of producing x dozen caps
b) The total revenue $R(x)$ from the sale of x dozen caps
c) The total profit $P(x)$ from the production and sale of x dozen caps
d) The profit or loss from the production and sale of 3000 dozen caps; of 1000 dozen caps
e) The break-even point

22. *Sport Coat Production.* Sarducci's is planning a new line of sport coats. For the first year, the fixed costs for setting up production are $10,000. The variable costs for producing each coat are $20. The revenue from each coat is to be $100. Find the following.

a) The total cost $C(x)$ of producing x coats
b) The total revenue $R(x)$ from the sale of x coats
c) The total profit $P(x)$ from the production and sale of x coats
d) The profit or loss from the production and sale of 2000 coats; of 50 coats
e) The break-even point

23. *Dog Food Production.* Puppy Love, Inc., will soon begin producing a new line of puppy food. The marketing department predicts that the demand function will be $D(p) = -14.97p + 987.35$ and the supply function will be $S(p) = 98.55p - 5.13$.

a) To the nearest cent, what price per unit should be charged in order to have equilibrium between supply and demand?
b) The production of the puppy food involves $87,985 in fixed costs and $5.15 per unit in variable costs. If the price per unit is the value you found in part (a), how many units must be sold in order to break even?

24. *Computer Production.* The Number Cruncher Computer Corporation is planning a new line of computers, each of which will sell for $970. The

fixed costs in setting up production are $1,235,580 and the variable costs for each computer are $697.

a) What is the break-even point? (Round to the nearest whole number.)
b) The marketing department at Number Cruncher is not sure that $970 is the best price. Their demand function for the new computers is given by $D(p) = -304.5p + 374,580$ and their supply function is given by $S(p) = 788.7p - 576,504$. What price p would result in equilibrium between supply and demand?

25. *Peanut Butter.* The following table lists the data for supply and demand of an 18-oz jar of peanut butter at various prices.

Price	Supply (in millions)	Demand (in millions)
$1.59	23.4	22.5
1.29	19.2	24.8
1.69	26.8	22.2
1.19	18.4	29.7
1.99	30.7	19.3

a) Use linear regression to find the supply function $S(p)$ for suppliers of peanut butter at price p.
b) Use linear regression to find the demand function $D(p)$ for consumers of peanut butter at price p.
c) Find the equilibrium point.

26. *Funnel Cakes.* Each year, the Harvey County Fair sets prices for concession vendors. The following table lists the data for supply and demand of a funnel cake at different prices.

Price	Supply (in thousands)	Demand (in thousands)
$1.50	3.6	5.4
1.75	4.8	5.2
2.00	6.2	5.0
2.50	7.2	4.2
2.25	7.1	4.0

a) Use linear regression to find the supply function $S(p)$ for suppliers of funnel cakes at price p.
b) Use linear regression to find the demand function $D(p)$ for consumers of funnel cakes at price p.
c) Find the equilibrium point.

Skill Maintenance

27. Graph: $y - 3 = \frac{2}{5}(x - 1)$. [2.6]

28. When two consecutive integers are added and then doubled, the result is 3 less than five times the smaller number. Find both numbers. [2.3]

29. Solve: $9x = 5x - \{3(2x - 7) - 4\}$. [2.2]

30. Solve $v - st = rw$ for t. [2.3]

Synthesis

31. ◆ Variable costs and fixed costs are often compared to the slope and the y-intercept, respectively, of an equation for a line. Explain why you feel this analogy is or is not valid.

32. ◆ In this section, we examined supply and demand functions for coffee. Does it seem realistic to you for the graph of D to have a constant slope? Why or why not?

33. *Loudspeaker Production.*   Fidelity Speakers, Inc., has fixed costs of $15,400 and variable costs of $100 for each pair of speakers produced. If the speakers sell for $250 a pair, how many pairs of speakers must be produced (and sold) in order to have enough profit to cover the fixed costs of two new facilities? Assume that all fixed costs are identical.

34. *Yo-Yo Production.*   Bing Boing Hobbies is willing to produce 100 yo-yos at $2.00 each and 500 yo-yos at $8.00 each. Research indicates that the public will buy 500 yo-yos at $1.00 each and 100 yo-yos at $9.00 each. Find the equilibrium point.

COLLABORATIVE CORNER

Focus: Cost, revenue, and profit models

Time: 20 minutes

Group size: 2–4

Dr. Bill Marks has been charting the operating cost and the revenue of his dental practice. The following table shows his data for the first six months of a year.

MONTH	NUMBER OF PATIENTS	COSTS	REVENUE
January	252	$13,948	$12,493
February	174	12,742	8,750
March	310	14,678	15,125
April	298	14,620	14,137
May	369	15,683	17,930
June	342	15,365	17,138

Activity

1. Divide the group into two smaller groups, with one examining the cost data and the other the revenue data. Each group should do the following.

   a) Enter and graph the data, treating either cost or revenue as a function of the number of patients. Confirm that the relationship appears to be linear.

   b) Use linear regression to find either the monthly cost $C(x)$ of or the monthly revenue $R(x)$ from treating x patients.

2. Working together, use the results of step (1) to find the monthly profit $P(x)$ from treating x patients.

3. Using the profit function, estimate the profit or loss from treating 310 patients and compare it with the profit for March using the data in the table.

4. Find the break-even point.

5. Use the cost function to estimate the fixed costs and the variable costs.

6. In order to accept more patients, Dr. Marks must hire another part-time hygienist at a monthly salary of $2000.

   a) Find the new monthly cost $C_1(x)$ and profit $P_1(x)$.

   b) Determine the new break-even point.

   c) At least how many patients must the hygienist see monthly in order to cover the extra costs?

CHAPTER

3 Summary and Review

Key Terms

System of equations, p. 153
Solution of a system, p. 154
Inconsistent, p. 157
Dependent, p. 157
Consistent, p. 157
Substitution method, p. 162
Elimination method, p. 164
Total-value problem, p. 172
Mixture problem, p. 174
Motion problem, p. 177
Independent, p. 191

Matrix (matrices), p. 199
Elements, p. 199
Entries, p. 199
Rows, p. 200
Columns, p. 200
Square matrix, p. 200
Main diagonal, p. 200
Row-echelon form, p. 200
Back-substitution, p. 201
Row-equivalent, p. 202

Dimensions, p. 203
Total cost, p. 205
Total revenue, p. 205
Total profit, p. 205
Fixed costs, p. 206
Variable costs, p. 206
Break-even point, p. 207
Demand function, p. 208
Supply function, p. 209
Equilibrium point, p. 209

Important Properties and Formulas

When solving a system of two linear equations in two variables:

1. If an identity is obtained, such as $0 = 0$, then the system has an infinite number of solutions. The equations are dependent and, since a solution exists, the system is consistent.

2. If a contradiction is obtained, such as $0 = 7$, then the system has no solution. The system is inconsistent.

To use the elimination method to solve systems of three linear equations:

1. Write all equations in the standard form $Ax + By + Cz = D.$

2. Clear any decimals or fractions.

3. Choose a variable to eliminate. Then select two of the three equations and work to get one equation in two variables.

4. Next, use a different pair of equations and eliminate the same variable that you did in step (3).

5. Solve the system of equations that resulted from steps (3) and (4).

6. Substitute the solution from step (5) into one of the original three equations and solve for the third variable. Then check.

Row-Equivalent Operations

Each of the following row-equivalent operations produces an equivalent matrix:

a) Interchanging any two rows.

b) Multiplying each element of a row by the same nonzero number.

c) Multiplying each element of a row by a nonzero number and adding the result to another row.

REVIEW EXERCISES

For Exercises 1–9, if a system has an infinite number of solutions, use set-builder notation to write the solution set. If a system has no solution, state this.

Solve graphically.

1. $3x + 2y = -4,$
$y = 3x + 7$

2. $5y - 2x = 7,$
$8x - 11y = 15$

Solve using the substitution method.

3. $x - 3y = -2,$
$7y - 4x = 6$

4. $y = x + 2,$
$y - x = 8$

5. $9x - 6y = 2,$
$x = 4y + 5$

Solve using the elimination method.

6. $8x + 4y = 10,$
$-4y - x = -4$

7. $4x - 7y = 18,$
$9x + 14y = 40$

8. $3x - 5y = -4,$
$5x - 3y = 4$

9. $1.5x - 3 = -2y,$
$3x + 4y = 6$

Solve.

10. Sean has $37 to spend. He can spend all the money on two CDs and a cassette, or he can buy one CD and two cassettes and have $5.00 left over. What is the price of a CD? What is the price of a cassette?

11. A train leaves Watsonville at noon traveling north at a speed of 44 mph. One hour later, another train, going 55 mph, travels north on a parallel track. How many hours will the second train travel before it overtakes the first train?

12. "Orange-Thirst" is 15% orange juice and "Quencho" is 5% orange juice. How many liters of each should be mixed together in order to get 10 L of a mixture that is 10% orange juice?

Solve. If a system's equations are dependent or if there is no solution, state this.

13. $x + 4y + 3z = 2,$
$2x + y + z = 10,$
$-x + y + 2z = 8$

14. $4x + 2y - 6z = 34,$
$2x + y + 3z = 3,$
$6x + 3y - 3z = 37$

15. $2x - 5y - 2z = -4,$
$7x + 2y - 5z = -6,$
$-2x + 3y + 2z = 4$

16. $-5x + 5y = -6,$
$2x - 2y = 4$

17. $3x + y = 2,$
$x + 3y + z = 0,$
$x + z = 2$

Solve.

18. In triangle ABC, the measure of angle A is four times the measure of angle C, and the measure of

angle B is 45° more than the measure of angle C. What are the measures of the angles of the triangle?

19. Find the three-digit number in which the sum of the digits is 11, the tens digit is 3 less than the sum of the hundreds and ones digits, and the ones digit is 5 less than the hundreds digit.

20. Lynn has $159 in her purse, consisting of $20, $5, and $1 bills. The number of $20 bills is the same as the total number of $1 and $5 bills. If she has 14 bills in her purse, how many of each denomination does she have?

Solve using matrices.

21. $3x + 4y = -13,$
$5x + 6y = 8$

22. $3x - y + z = -1,$
$2x + 3y + z = 4,$
$5x + 4y + 2z = 5$

23. $11x + 4y = 7z + 3,$
$5y + 2 = 5z,$
$8x - 3z = 11$

24. Find the equilibrium point for the demand and supply functions

$$S(p) = 60 + 7p \quad \text{and} \quad D(p) = 120 - 13p.$$

25. Kregel Furniture is planning to produce a new type of bed. For the first year, the fixed costs for setting up production are $35,000. The variable costs for producing each bed are $175. The revenue from each bed is $300. Find the following.

a) The total cost $C(x)$ of producing x beds
b) The total revenue $R(x)$ from the sale of x beds
c) The total profit $P(x)$ from the production and sale of x beds
d) The profit or loss from the production and sale of 1200 beds; of 200 beds
e) The break-even point

Skill Maintenance

26. Solve: $4x - 5x + 8 = -9x + 2x.$

27. Solve $Q = at - 4t$ for t.

28. Find a linear function $f(x)$ whose graph has slope $-\frac{1}{3}$ and y-intercept $\left(0, -\frac{9}{10}\right)$.

29. The Hammondton Hoopsters increased its score by 7 points in each of three consecutive games. In those three games, the team scored a total of 228 points. What was their score in each game?

Synthesis

30. ◆ How would you go about solving a problem that involved four variables?

31. ◈ Explain how a system of equations can be both dependent and inconsistent.

32. Solve graphically:

$$y = x + 2,$$
$$y = x^2 + 2.$$

33. The graph of $f(x) = ax^2 + bx + c$ contains the points $(-2, 3)$, $(1, 1)$, and $(0, 3)$. Find a, b, and c and give a formula for the function.

CHAPTER 3 TEST

Solve graphically.

1. $x - y = 4,$
   $y = 2x + 3$

2. $16x - 7y = 25,$
   $8x + 3y = 19$

Solve, if possible, using the substitution method.

3. $x + 3y = -8,$
   $4x - 3y = 23$

4. $2x + 4y = -6,$
   $y = 3x - 9$

Solve, if possible, using the elimination method.

5. $4x - 6y = 3,$
   $6x - 4y = -3$

6. $4y + 2x = 18,$
   $3x + 6y = 26$

7. The perimeter of a rectangle is 96. The length of the rectangle is 6 less than twice the width. Find the dimensions of the rectangle.

8. Between her home mortgage (loan), car loan, and credit card bill (loan), Rema is $75,300 in debt. Rema's credit card bill accumulates 1.5% interest, her car loan 1% interest, and her mortgage 0.6% interest each month. After one month, her total accumulated interest is $460.50. The interest on Rema's credit card bill was $4.50 more than the interest on her car loan. Find the amount of each loan.

Solve. If a system's equations are dependent or if there is no solution, state this.

9. $-3x + y - 2z = 8,$
   $-x + 2y - z = 5,$
   $2x + y + z = -3$

10. $6x + 2y - 4z = 15,$
   $-3x - 4y + 2z = -6,$
   $4x - 6y + 3z = 8$

11. $2x + 2y = 0,$
   $4x + 4z = 4,$
   $2x + y + z = 2$

12. $3x + 3z = 0,$
   $2x + 2y = 2,$
   $3y + 3z = 3$

Solve using matrices.

13. $7x - 8y = 10,$
   $9x + 5y = -2$

14. $x + 3y - 3z = 12,$
   $3x - y + 4z = 0,$
   $-x + 2y - z = 1$

15. $4y - z = 8,$
   $15x - 10 = 6z,$
   $x + 20y - 8 = 6z$

16. An electrician, a carpenter, and a plumber are hired to work on a house. The electrician earns $21 per hour, the carpenter $19.50 per hour, and the plumber $24 per hour. The first day on the job, they worked a total of 21.5 hr and earned a total of $469.50. If the plumber worked 2 more hours than the carpenter did, how many hours did the electrician work?

17. Find the equilibrium point for the demand and supply functions

$$D(p) = 79 - 8p \quad \text{and} \quad S(p) = 37 + 6p.$$

18. Sweet Spot Manufacturing is planning a new type of tennis racket. For the first year, the fixed costs for setting up production are $40,000. The variable costs for producing each racket are $30. The sales department predicts that 1500 rackets can be sold during the first year. The revenue from each racket is $80. Find the following.

a) The total cost $C(x)$ of producing x tennis rackets

b) The total revenue $R(x)$ from the sale of x tennis rackets

c) The total profit $P(x)$

d) The profit or loss if the expected sales of 1500 rackets occurs

e) The break-even point

Skill Maintenance

19. The price of a radio, including 5% sales tax, is $36.75. Find the price of the radio before the tax was added.

20. Solve $P = 4a - 3b$ for a.

21. Solve: $-3x - 5 + 6x = 8x - 14$.

22. Find an equation of the line passing through the points $(5, -1)$ and $(-3, 4)$.

Synthesis

23. The graph of the function $f(x) = mx + b$ contains the points $(-1, 3)$ and $(-2, -4)$. Find m and b.

24. At a county fair, an adult's ticket sold for $5.50, a senior citizen's ticket for $4.00, and a child's ticket for $1.50. On opening day, the number of adults' and senior citizens' tickets sold was 30 more than the number of children's tickets sold. The number of adults' tickets sold was 6 more than four times the number of senior citizens' tickets sold. Total receipts from the ticket sales were $11,219.50. How many of each type of ticket were sold?

CUMULATIVE REVIEW 1–3

Solve.

1. $-14.3 + 29.17 = x$

2. $x + 9.4 = -12.6$

3. $3.9(-11) = x$

4. $-2.4x = -48$

5. $4x + 7 = -14$

6. $-3 + 5x = 2x + 15$

7. $3n - (4n - 2) = 7$

8. $6y - 5(3y - 4) = 10$

9. $14 + 2c = -3(c + 4) - 6$

10. $5x - [4 - 2(6x - 1)] = 12$

Simplify. Do not leave negative exponents in your answers.

11. $x^4 \cdot x^{-6} \cdot x^{13}$

12. $(4x^{-3}y^2)(-10x^4y^{-7})$

13. $(6x^2y^3)^2(-2x^0y^4)^3$

14. $\dfrac{y^4}{y^{-6}}$

15. $\dfrac{-10a^7b^{-11}}{25a^{-4}b^{22}}$

16. $\left(\dfrac{3x^4y^{-2}}{4x^{-5}}\right)^4$

17. $(1.95 \times 10^{-3})(5.73 \times 10^8)$

18. $\dfrac{2.42 \times 10^5}{6.05 \times 10^{-2}}$

19. Solve $A = \frac{1}{2}h(b + t)$ for b.

20. Determine whether $(-3, 4)$ is a solution of $5a - 2b = -23$.

Graph.

21. $y = -2x + 3$

22. $y = x^2 - 1$

23. $4x + 16 = 0$

24. $-3x + 2y = 6$

25. Find the slope and the y-intercept of the line with equation $-4y + 9x = 12$.

26. Find the slope, if it exists, of the line containing the points $(2, 7)$ and $(-1, 3)$.

27. Find an equation of the line with slope -3 and containing the point $(2, -11)$.

28. Find an equation of the line containing the points $(-6, 3)$ and $(4, 2)$.

29. Determine whether the lines are parallel or perpendicular:
$$2x = 4y + 7,$$
$$x - 2y = 5.$$

30. Find a linear function perpendicular to $y = 3x + 5$ with y-intercept $(0, -11)$.

31. For the graph of f shown, determine the domain, the range, $f(-3)$, and any value of x for which $f(x) = 5$.

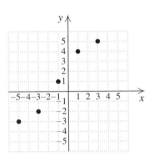

32. Determine the domain of the function given by
$$f(x) = \frac{7}{2x - 1}.$$

Given $g(x) = 4x - 3$ and $h(x) = -2x^2 + 1$, find the following function values.

33. $h(4)$

34. $-g(0)$

35. $(g \cdot h)(-1)$

36. $g(a) - h(2a)$

Solve.

37. $3x + y = 4,$
$6x - y = 5$

38. $4x + 4y = 4,$
$5x - 3y = -19$

39. $6x - 10y = -22,$
$-11x - 15y = 27$

40. $x + y + z = -5,$
$2x + 3y - 2z = 8,$
$x - y + 4z = -21$

41. $2x + 5y - 3z = -11,$
$-5x + 3y - 2z = -7,$
$3x - 2y + 5z = 12$

42. $0.5x + 1.2y = -1.2,$
$4.8x - 6.4y = 2$

43. $16x + 25y = 14 + z,$
$\quad\ x + 2y = 1,$
$\quad 6y + 12z = 9x + 7$

44. The sum of two numbers is 26. Three times the smaller plus twice the larger is 60. Find the numbers.

45. "Soakem" is 34% salt and the rest water. "Rinsem" is 61% salt and the rest water. How many ounces of each would be needed to obtain 120 oz of a mixture that is 50% salt?

46. Find three consecutive odd numbers such that the sum of four times the first number and five times the third number is 47.

47. Belinda's scores on four tests are 83, 92, 100, and 85. What must the score be on the fifth test so that the average will be 90?

48. The perimeter of a rectangle is 32 cm. If five times the width equals three times the length, what are the dimensions of the rectangle?

49. There are 4 more nickels than dimes in a piggy bank. The total amount of money in the bank is $2.45. How many of each type of coin are in the bank?

50. One month Ladi and Bo spent $680 for electricity, rent, and telephone. The electric bill was $\frac{1}{4}$ of the rent and the rent was $400 more than the phone bill. How much was the electric bill?

51. A hockey team played 64 games one season. It won 15 more games than it tied and lost 10 more games than it won. How many games did it win? lose? tie?

52. Reggie, Jenna, and Achmed are counting calories. For lunch one day, Reggie ate two cookies and a banana, for a total of 260 calories. Jenna had a cup of yogurt and a banana, for a total of 245 calories. Achmed ate a cookie, a cup of yogurt, and two bananas, for a total of 415 calories. How many calories are in each item?

53. *Construction.* The following table shows the values of new residential construction in the United States for several years.

YEAR	RESIDENTIAL CONSTRUCTION (IN MILLIONS)
1990	$182,900
1991	157,800
1992	187,800
1993	210,455
1994	238,874
1995	230,688
1996	247,177

Source: U.S. Bureau of the Census

a) Use the data to draw a line graph.
b) Use linear regression to find a linear function that can be used to estimate the value of new residential construction x years after 1990.
c) Estimate the value of new residential construction in 2000.

Synthesis

54. Simplify: $(6x^{a+2}y^{b+2})(-2x^{a-2}y^{y+1})$.

55. An automotive dealer discovers that when $1000 is spent on radio advertising, weekly sales increase by $101,000. When $1250 is spent on radio advertising, weekly sales increase by $126,000. Assuming that sales increase according to a linear equation, by what amount would sales increase when $1500 is spent on radio advertising?

56. Given that $f(x) = mx + b$ and that $f(5) = -3$ when $f(-4) = 2$, find m and b.

Inequalities and Problem Solving 4

I n many applications, we want to knowwhen one quantity is more than or less than another. These and other types of questions can be answered using *inequalities*. Principles similar to those used for solving equations enable us to solve inequalities and the problems that translate to inequalities. In this chapter, we develop procedures for solving a variety of inequalities and systems of inequalities.

$$y_1 = -0.02208405165403x + 63.445835800019,$$
$$y_2 = 19.0$$

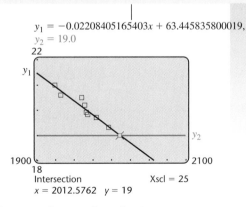

Intersection   Xscl = 25
$x = 2012.5762$ $y = 19$

4.1
Inequalities and Applications

- *Solving Inequalities*
- *Interval Notation*
- *Graphical Solutions*
- *The Addition Principle*
- *The Multiplication Principle*
- *Using the Principles Together*
- *Problem Solving*

Solving Inequalities

We can extend our equation-solving skills to the solving of inequalities. An **inequality** is any sentence containing $<$, $>$, $\leq$, $\geq$, or $\neq$ (see Section 1.1)—for example,

$$-2 < a, \qquad x > 4, \qquad x + 3 \leq 6, \qquad 6 - 7y \geq 10y - 4,$$
$$\text{and} \quad 5x \neq 10.$$

Any replacement for the variable that makes an inequality true is called a **solution**. The set of all solutions is called the **solution set**. When all solutions of an inequality are found, we say that we have **solved** the inequality.

Example 1   Determine whether the given number is a solution of the inequality.

a) $x + 3 < 6$;   5

b) $2x - 3 > -5$;   1

SOLUTION

a) We substitute to get $5 + 3 < 6$, or $8 < 6$, a false sentence. Thus, 5 is not a solution.

b) We substitute to get $2 \cdot 1 - 3 > -5$, or $-1 > -5$, a true sentence. Thus, 1 is a solution. ▬

The *graph* of an inequality is a drawing that represents its solutions. An inequality in one variable can be graphed on a number line. Inequalities in two variables can be graphed on a coordinate plane, and are considered later in this chapter.

Example 2   Graph $x < 4$ on a number line.

SOLUTION   The solutions are all real numbers less than 4, so we shade all numbers less than 4. Since 4 is not a solution, we use an open dot at 4.

We can write the solution set using *set-builder notation* (see Section 1.1):

$$\{x \mid x < 4\}.$$

This is read

"The set of all x such that x is less than 4." ▬

Interval Notation

Another way to write solutions of an inequality in one variable is to use **interval notation**. Interval notation uses parentheses, ( ), and brackets, [ ].

If a and b are real numbers such that $a < b$, we define the **open interval (a, b)** as the set of all numbers x for which $a < x < b$. Thus,

$$(a, b) = \{x \mid a < x < b\}.$$

Its graph excludes the endpoints:

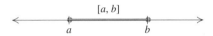

Caution! Do not confuse the *interval* (a, b) with the *ordered pair* (a, b). The context in which the notation appears usually makes the meaning clear.

The **closed interval $[a, b]$** is defined as the set of all numbers x for which $a \leq x \leq b$. Thus,

$$[a, b] = \{x \mid a \leq x \leq b\}.$$

Its graph includes the endpoints*:

There are two kinds of **half-open intervals**, defined as follows:

1. $(a, b] = \{x \mid a < x \leq b\}$. This is open on the left. Its graph is as follows:

2. $[a, b) = \{x \mid a \leq x < b\}$. This is open on the right. Its graph is as follows:

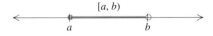

We use the symbols ∞ and $-\infty$ to represent positive and negative infinity, respectively. Thus the notation (a, ∞) represents the set of all

---

*Some books use the representations ⟶ and ⟶ instead of,

respectively, ⟶ and ⟶ .

real numbers greater than a, and $(-\infty, a)$ represents the set of all real numbers less than a.

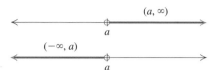

The notations $[a, \infty)$ and $(-\infty, a]$ are used when we want to include the endpoint a. Note that we use parentheses with ∞ and $-\infty$ since there is no endpoint to include.

Example 3 Graph $t \geq -2$ on a number line and write the solution set using both set-builder and interval notations.

SOLUTION Using set-builder notation, we write the solution set as $\{t \mid t \geq -2\}$.

Using interval notation, we write the solution set as $[-2, \infty)$.

To graph the solution, we shade all numbers to the right of -2 and use a solid dot to indicate that -2 is also a solution.

Graphical Solutions

Solving inequalities graphically involves first finding a point of intersection.

Example 4 Solve graphically: $16 - 7x \geq 10x - 4$.

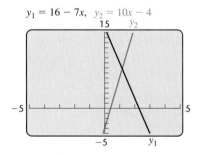

$y_1 = 16 - 7x, \quad y_2 = 10x - 4$

SOLUTION We let $y_1 = 16 - 7x$ and $y_2 = 10x - 4$, and graph y_1 and y_2 in the window $[-5, 5, -5, 15]$, as shown at left.

From the graph, we see that $y_1 = y_2$ at the point of intersection. To the left of the point of intersection, $y_1 > y_2$. You can check this by calculating the values of both y_1 and y_2 for an x-value to the left of the point of intersection—say, for $x = 0.5$. Since $12.5 > 1$, we know that $y_1 > y_2$ when $x = 0.5$. Similarly, we see that $y_1 < y_2$ to the right of the point of intersection.

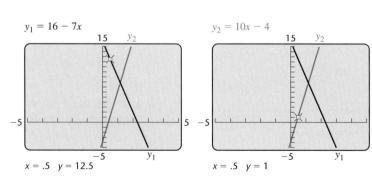

The solution set will be all x-values to the left of the point of intersection, as well as the x-coordinate of the point of intersection. Using INTERSECT, we find that the x-coordinate of the point of intersection is approximately 1.1764706. Thus the solution set is approximately $(-\infty, 1.1764706]$.

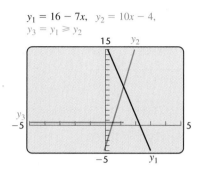

$y_1 = 16 - 7x,\ \ y_2 = 10x - 4,$
$y_3 = y_1 \geq y_2$

On many graphers, the interval on the number line that is the solution set can be found directly by using the VARS and TEST keys. To find the solution interval for Example 4 directly, we enter and graph Y3 = Y1 ≥ Y2. Where this is true, the value will be 1, and where it is false, the value will be 0. The solution set is thus displayed as an interval, shown by a horizontal line 1 unit above the x-axis. The endpoint of the interval corresponds to the intersection of the graphs of the equations.

Algebraic methods for solving inequalities make use of addition and multiplication principles that are similar to the principles used to solve equations.

The Addition Principle

Two inequalities are equivalent if they have the same solution set. For example, the inequalities $x > 4$ and $4 < x$ are equivalent. Just as the addition principle for equations produces equivalent equations, the addition principle for inequalities produces equivalent inequalities.

The Addition Principle for Inequalities

For any real numbers a, b, and c:

$a < b$ is equivalent to $a + c < b + c$;

$a > b$ is equivalent to $a + c > b + c$.

Similar statements hold for $\leq$ and $\geq$.

As with equations, we try to get the variable alone on one side in order to determine solutions easily.

Example 5 Solve and graph: **(a)** $x + 5 > 3$; **(b)** $4x - 1 \geq 5x - 2$.

SOLUTION

a)
$$x + 5 > 3$$
$$x + 5 + (-5) > 3 + (-5) \qquad \text{Using the addition principle to add } -5 \text{ on both sides}$$
$$x > -2$$

When an inequality—like this last one—has an infinite number of solutions, we cannot possibly check them all. Instead, we can perform a partial check by substituting one member of the solution set (here we use −1) into the original inequality:

$$\frac{x + 5 > 3}{\begin{array}{c} -1 + 5 \ ?\ 3 \\ 4 \ | \ 3 \quad \text{TRUE} \end{array}}$$

$y_1 = x + 5, \quad y_2 = 3, \quad y_3 = y_1 \geq y_2$

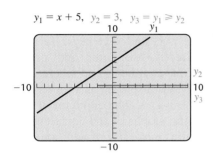

Since $4 > 3$ is true, we have our check. The graph shown at left serves as another check. The solution set is $\{x \mid x > -2\}$, or $(-2, \infty)$. The graph is as follows:

b)
$$4x - 1 \geq 5x - 2$$
$$4x - 1 + 2 \geq 5x - 2 + 2 \qquad \text{Adding 2 on both sides}$$
$$4x + 1 \geq 5x \qquad\qquad\quad \text{Simplifying}$$
$$4x + 1 - 4x \geq 5x - 4x \qquad \text{Adding } -4x \text{ on both sides}$$
$$1 \geq x \qquad\qquad\qquad\quad \text{Simplifying}$$

$y_1 = 4x - 1, \quad y_2 = 5x - 2, \quad y_3 = y_1 \geq y_2$

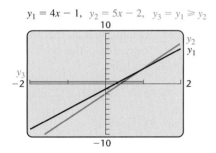

We know that $1 \geq x$ has the same meaning as $x \leq 1$. You can check that any number less than or equal to 1 is a solution. As another check, we can solve graphically, as shown in the screen at left. The solution set is $\{x \mid 1 \geq x\}$ or, more commonly, $\{x \mid x \leq 1\}$. Using interval notation, we write that the solution set is $(-\infty, 1]$. The graph is as follows:

The Multiplication Principle

The multiplication principle for inequalities differs from the multiplication principle for equations.

Consider this true inequality:

$$4 < 9.$$

If we multiply both sides of $4 < 9$ by 2, we get another true inequality:

$$4 \cdot 2 < 9 \cdot 2, \quad \text{or} \quad 8 < 18.$$

If we multiply both sides of $4 < 9$ by −2, we get a false inequality:

$$\text{FALSE} \longrightarrow 4(-2) < 9(-2), \quad \text{or} \quad -8 < -18. \longleftarrow \text{FALSE}$$

This is because negation reverses relative position on the number line. However, if the inequality symbol is reversed, we get a true inequality:

$$-8 > -18. \longleftarrow \text{TRUE}$$

The < symbol has been reversed!

The Multiplication Principle for Inequalities

For any real numbers a and b, and for any *positive* number c,

$a < b$ is equivalent to $ac < bc$;

$a > b$ is equivalent to $ac > bc$.

For any real numbers a and b, and for any *negative* number c,

$a < b$ is equivalent to $ac > bc$;

$a > b$ is equivalent to $ac < bc$.

Similar statements hold for $\leq$ and $\geq$.

Since division by c is the same as multiplication by $1/c$, there is no need for a separate division principle.

Caution! Remember that to multiply or divide both sides of an inequality by a negative number, we must reverse the inequality symbol.

Example 6 Solve and graph: **(a)** $3a < \frac{27}{2}$; **(b)** $-5x \geq -80$.

SOLUTION

a) $3a < \frac{27}{2}$

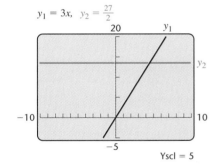

——— The symbol stays the same.

$\frac{1}{3} \cdot 3a < \frac{1}{3} \cdot \frac{27}{2}$ **Multiplying by $\frac{1}{3}$ on both sides**

$a < \frac{9}{2}$

Any number less than $\frac{9}{2}$ is a solution. The graph at left serves as a check. The solution set is $\left\{a \mid a < \frac{9}{2}\right\}$, or $\left(-\infty, \frac{9}{2}\right)$. The graph is as follows:

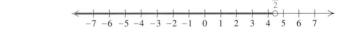

b) $-5x \geq -80$

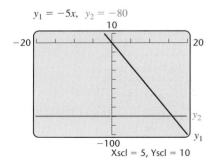

——— The symbol must be reversed.

$-\frac{1}{5} \cdot (-5x) \leq -\frac{1}{5} \cdot (-80)$ **Multiplying by $-\frac{1}{5}$, or dividing by -5, on both sides**

$x \leq 16$

We get the same solution by solving graphically, as shown at left. The solution set is $\{x \mid x \leq 16\}$, or $(-\infty, 16]$. The graph is as follows:

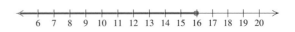

Using the Principles Together

We use the addition and multiplication principles together in solving inequalities in much the same way that we do in solving equations.

Example 7   Solve: **(a)** $16 - 7x \geq 10x - 4$; **(b)** $-3(t + 8) - 5t > 4t - 9$.

SOLUTION

a)

$$16 - 7x \geq 10x - 4$$

$$-16 + 16 - 7x \geq -16 + 10x - 4 \qquad \text{Adding } -16 \text{ on both sides}$$

$$-7x \geq 10x - 20$$

$$-10x + (-7x) \geq -10x + 10x - 20 \qquad \text{Adding } -10x \text{ on both sides}$$

$$-17x \geq -20$$

⟶ The symbol must be reversed.

$$-\tfrac{1}{17} \cdot (-17x) \leq -\tfrac{1}{17} \cdot (-20) \qquad \textbf{Multiplying by } -\tfrac{1}{17}, \textbf{ or}$$
$$\textbf{dividing by } -17, \textbf{ on both sides}$$

$$x \leq \tfrac{20}{17}$$

As a check, note that we solved this inequality in Example 4, and that $\tfrac{20}{17} \approx 1.1764706$. The solution set is $\left\{x \mid x \leq \tfrac{20}{17}\right\}$, or $\left(-\infty, \tfrac{20}{17}\right]$.

b)

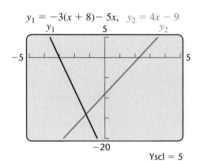

$y_1 = -3(x + 8) - 5x, \quad y_2 = 4x - 9$

y_1 5 y_2

-5 ⟶ 5

-20

Yscl = 5

$$-3(t + 8) - 5t > 4t - 9$$

$$-3t - 24 - 5t > 4t - 9 \qquad \text{Using the distributive law}$$

$$-24 - 8t > 4t - 9$$

$$-24 - 8t + 8t > 4t - 9 + 8t \qquad \text{Adding } 8t \text{ on both sides}$$

$$-24 > 12t - 9$$

$$-24 + 9 > 12t - 9 + 9 \qquad \text{Adding } 9 \text{ on both sides}$$

$$-15 > 12t$$

⟶ The symbol stays the same.

$$-\tfrac{5}{4} > t \qquad \textbf{Multiplying by } \tfrac{1}{12} \textbf{ and simplifying}$$

As a check, we solve graphically, as shown at left. The solution set is $\left\{t \mid -\tfrac{5}{4} > t\right\}$, or $\left\{t \mid t < -\tfrac{5}{4}\right\}$, or $\left(-\infty, -\tfrac{5}{4}\right)$.   ▬

Problem Solving

Many problem-solving situations translate to inequalities. In addition to "is less than" and "is more than," other phrases are commonly used.

PHRASE	TRANSLATION
a "is at most" 17	$a \leq 17$
a "is at least" 5	$a \geq 5$
a "can't exceed" 12	$a \leq 12$
a "is a better buy than" b	$a < b$

Example 8 *Records in the Men's 200-m Dash.* In the 1996 Olympics, Michael Johnson set a world record of 19.32 sec in the men's 200-m dash. Various earlier world records for the 200-m dash and the year in which they were set, along with the 1996 record, are listed in the following table. Determine (in terms of an inequality) those years for which the world record will be less than 19.0 sec.

RUNNER	YEAR	RECORD (IN SECONDS)
Michael Johnson	1996	19.32
Pietro Mennea	1979	19.72
Tommy Smith	1968	19.83
John Carlos	1966	19.92
Henry Carr	1964	20.2
Peter Radford	1960	20.5
Roland Locke	1932	20.6
Charles Paddock	1924	21.0

SOLUTION

1. **Familiarize.** We can become more familiar with the problem by first graphing the data.

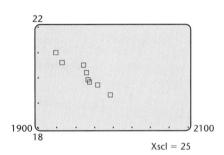

Xscl = 25

Since the data are approximately linear, we can fit a linear function to them and use that function to estimate when the world record will be less than 19.0 sec.

2. **Translate.** Using linear regression, we find a linear function $R(t)$ that fits the data:

$$R(t) = -0.02208405165403t + 63.445835800019.$$

The record $R(t)$ is to be *less than* 19.0 sec. Thus we have

$$R(t) < 19.0.$$

We replace $R(t)$ with $-0.02208405165403t + 63.445835800019$ to find the times t that solve the inequality:

$$-0.02208405165403t + 63.445835800019 < 19.0.$$

3. **Carry out.** Since we found the linear function using a grapher, it makes sense to solve the inequality graphically. Thus we graph

$y_1 = -0.02208405165403x + 63.445835800019$ and $y_2 = 19.0$ and determine those x-values for which $y_1 < y_2$.

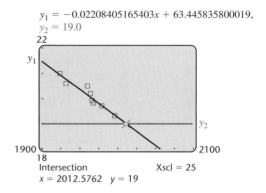

From the graph, we see that $y_1 < y_2$ to the right of the point of intersection. Rounding to the nearest tenth, we have the solution set $(2012.6, \infty)$.

4. **Check.**   As a partial check, we can trace along the line $y_1 = -0.02208405165403x + 63.445835800019$. The y-values drop below 19.0 for x-values after appoximately 2012.6. Since we are looking for the years in which the record will be less than 19.0 sec, our answer includes the years 2013 and later.

5. **State.**   The record will be less than 19.0 sec in 2013 and later, or approximately $\{t \mid t \geq 2013\}$.   ▬

Example 9   On a new job, Rose can be paid in one of two ways:

Plan A: A salary of $600 per month, plus a commission of 4% of sales;

Plan B: A salary of $800 per month, plus a commission of 6% of sales in excess of $10,000.

For what amount of monthly sales is plan A better than plan B, if we assume that sales are always more than $10,000?

SOLUTION

1. **Familiarize.**   Listing the given information in a table will be helpful.

PLAN A: MONTHLY INCOME	PLAN B: MONTHLY INCOME
$600 salary 4% of sales *Total:* $600 + 4% of sales	$800 salary 6% of sales over $10,000 *Total:* $800 + 6% of sales over $10,000

Next, suppose that Rose sold a certain amount—say, $12,000—in one month. Which plan would be better? Under plan A, she would earn $600 plus 4% of $12,000, or

$$600 + 0.04(12,000) = \$1080.$$

Since with plan B commissions are paid only on sales in excess of $10,000, Rose would earn $800 plus 6% of ($12,000 − $10,000), or

$$800 + 0.06(12,000 - 10,000) = \$920.$$

This shows that for monthly sales of $12,000, plan A is better. Similar calculations will show that for monthly sales of $30,000, plan B is better. To determine *all* values for which plan A pays more money, we must solve an inequality that is based on the calculations above.

2. **Translate.** We let S = the amount of monthly sales. Examining the calculations in the *Familiarize* step, we see that monthly income from plan A is $600 + 0.04S$ and from plan B is $800 + 0.06(S − 10,000)$. We want to find all values of S for which

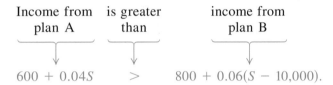

Income from plan A	is greater than	income from plan B

$$600 + 0.04S \quad > \quad 800 + 0.06(S - 10,000).$$

3. **Carry out.** We solve the problem both algebraically and graphically.

ALGEBRAIC SOLUTION

We solve the inequality:

$600 + 0.04S > 800 + 0.06(S - 10,000)$

$600 + 0.04S > 800 + 0.06S - 600$ **Using the distributive law**

$600 + 0.04S > 200 + 0.06S$ **Combining like terms**

$600 > 200 + 0.02S$ **Subtracting 0.04S on both sides**

$400 > 0.02S$ **Subtracting 200 on both sides**

$20,000 > S$, or $S < 20,000$.   **Dividing by 0.02 on both sides**

Thus plan A pays more than plan B for $S < 20,000$.

GRAPHICAL SOLUTION

We graph the equations
$y_1 = 600 + 0.04x$ and
$y_2 = 800 + 0.06(x - 10000)$.

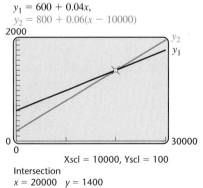

$y_1 = 600 + 0.04x$,
$y_2 = 800 + 0.06(x - 10000)$

Xscl = 10000, Yscl = 100

Intersection
$x = 20000$ $y = 1400$

We see that $y_1 > y_2$ to the left of the point of intersection, (20,000, 1400). Thus $y_1 > y_2$ for $x < 20,000$.

4. Check.   For $S = 20{,}000$, the income from plan A is

$600 + 4\% \cdot 20{,}000$, or $\$1400$.

The income from plan B is

$800 + 6\% \cdot (20{,}000 - 10{,}000)$, or $\$1400$.

} This confirms that for sales totaling $\$20{,}000$, Rose's pay is the same under either plan.

In the *Familiarize* step, we saw that for sales of $\$12{,}000$, plan A pays more. Since $12{,}000 < 20{,}000$, this is a partial check. We cannot check all possible values of S, so we will stop here.

5. State.   For monthly sales of less than $\$20{,}000$, plan A is better.

4.1 | *Exercise Set*

Determine whether the given numbers are solutions of the inequality.

1. $x - 2 \geq 6$; $-4, 0, 4, 8$

2. $3x + 5 \leq -10$; $-5, -10, 0, 27.5$

3. $t - 8 > 2t - 3$; $0, 3.3, -9, -3$

4. $5y - 7 < 5 - y$; $2, -3, 0, 3.9$

Graph each inequality, and write the solution set using both set-builder and interval notation.

5. $y < 5$ **6.** $x > 4$

7. $x \geq -4$ **8.** $t \leq 6$

9. $t > -2$ **10.** $y < -3$

11. $x \leq -5$ **12.** $x \geq -6$

Solve. Then graph.

13. $x + 8 > 3$ **14.** $x + 5 > 2$

15. $a + 7 \leq -13$ **16.** $a + 9 \leq -12$

17. $y - 9 > -18$ **18.** $y - 8 > -14$

19. $y - 18 \leq -4$ **20.** $x - 11 \leq -2$

21. $9t < -81$ **22.** $8x \geq 24$

23. $0.5x < 25$ **24.** $0.3x < -18$

25. $-8y \leq 3.2$ **26.** $-9x \geq -8.1$

27. $-\frac{5}{6}y \leq -\frac{3}{4}$ **28.** $-\frac{3}{4}x \geq -\frac{5}{8}$

29. $5y + 13 > 28$ **30.** $2x + 7 < 19$

31. $-9x + 3x \geq -24$ **32.** $5y + 2y \leq -21$

33. Let $f(x) = 8x - 9$ and $g(x) = 3x - 11$. Find all values of x for which $f(x) < g(x)$.

34. Let $f(x) = 2x - 7$ and $g(x) = 5x - 9$. Find all values of x for which $f(x) < g(x)$.

35. Let $f(x) = 0.4x + 5$ and $g(x) = 1.2x - 4$. Find all values of x for which $g(x) \geq f(x)$.

36. Let $f(x) = \frac{3}{8} + 2x$ and $g(x) = 3x - \frac{1}{8}$. Find all values of x for which $g(x) \geq f(x)$.

Assume that the graphs of $y_1 = -\frac{1}{2}x + 5$, $y_2 = x - 1$, and $y_3 = 2x - 3$ are as shown below. Solve each inequality, referring only to the figure.

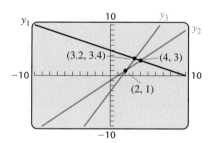

37. $-\frac{1}{2}x + 5 > x - 1$

38. $x - 1 \leq 2x - 3$

39. $2x - 3 \geq -\frac{1}{2}x + 5$

Solve.

40. $4(3y - 2) \geq 9(2y + 5)$

41. $4m + 5 \geq 14(m - 2)$

42. $3(2 - 5x) + 2x < 2(4 + 2x)$

43. $2(0.5 - 3y) + y > (4y - 0.2)8$

44. $5[3m - (m + 4)] > -2(m - 4)$

45. $[8x - 3(3x + 2)] - 5 \geq 3(x + 4) - 2x$

46. $13 - (2c + 2) \geq 2(c + 2) + 3c$

47. $\frac{1}{4}(8y + 4) - 17 < -\frac{1}{2}(4y - 8)$

48. $\frac{1}{3}(6x + 24) - 20 > -\frac{1}{4}(12x - 72)$

49. $2[4 - 2(3 - x)] - 1 \geq 4[2(4x - 3) + 7] - 25$

50. $5[3(7 - t) - 4(8 + 2t)] - 20 \leq -6[2(6 + 3t) - 4]$

Solve.

51. *Trampoline Injuries.* The following table shows the number of trampoline injuries in the United States for children 18 and younger for the years 1990–1995. Use linear regression to predict those years in which the number of trampoline injuries will exceed 100,000.

YEAR	NUMBER OF INJURIES
1990	29,600
1991	35,500
1992	39,000
1993	40,500
1994	58,500
1995	58,400

Source: The New York Times, March 3, 1998

52. *National Parks.* The number of visits to United States national parks has been increasing, as shown in the following table. Use linear regression to predict those years in which the number of visits will exceed 300 million.

YEAR	NUMBER OF VISITS (IN MILLIONS)
1976	216.6
1986	237.1
1996	265.8

Source: U.S. National Park Service

Sports Participation. The following table shows the number of people who participated more than once in the listed sports in various years. Use the information for Exercises 53 and 54.

53. Use linear regression to predict those years in which the number of people who exercise with equipment will exceed the number who swim.

54. Use linear regression to predict those years in which the number of people who play billiards or pool will exceed the number who bowl.

55. *Insurance Claims.* After a serious automobile accident, most insurance companies will replace the damaged car with a new one if repair costs exceed 80% of the N.A.D.A., or "blue-book," value of the car. Miguel's car recently sustained $9200 worth of damage but was not replaced. What was the blue-book value of his car?

56. *Truck Rentals.* Campus Entertainment rents a truck for $45 plus 20¢ per mile. A budget of $75 has been set for the rental. For what mileages will they not exceed the budget?

57. *Phone Rates.* A long-distance telephone call using Down East Calling costs 20 cents for the first minute and 16 cents for each additional minute. The same call, placed on Long Call Systems, costs 19 cents for the first minute and 18 cents for each additional minute. For what length phone calls is Down East Calling less expensive?

58. *Moving Costs.* Musclebound Movers charges $85 plus $40 an hour to move households across town. Champion Moving charges $60 an hour for cross-town moves. For what lengths of time is Champion more expensive?

59. *Wages.* Toni can be paid in one of two ways:

Plan A: A salary of $400 per month, plus a commission of 8% of gross sales;

Plan B: A salary of $610 per month, plus a commission of 5% of gross sales.

For what amount of gross sales should Toni select plan A?

60. *Checking-Account Rates.* The Hudson Bank offers two checking-account plans. Their Anywhere plan charges 20¢ per check whereas their Acu-checking plan costs $2 per month plus 12¢ per check. For

YEAR	SWIMMING (IN MILLIONS)	EXERCISING WITH EQUIPMENT (IN MILLIONS)	BOWLING (IN MILLIONS)	BILLIARDS/POOL (IN MILLIONS)
1985	73.3	32.1	35.7	23.0
1990	67.5	35.3	40.1	28.1
1995	61.5	44.4	41.9	31.1
1996	60.2	47.8	42.9	34.5

Source: The Wall Street Journal Almanac, 1998

what numbers of checks per month will the Acu-checking plan cost less?

61. *Wages.* Branford can be paid for his masonry work in one of two ways:

Plan A: $300 plus $9.00 per hour;

Plan B: Straight $12.50 per hour.

Suppose that the job takes n hours. For what values of n is plan B better for Branford?

62. *Insurance Benefits.* Bayside Insurance offers two plans. Under plan A, Giselle would pay the first $50 of her medical bills and 20% of all bills after that. Under plan B, Giselle would pay the first $250 of bills, but only 10% of the rest. For what amount of medical bills will plan B save Giselle money? (Assume that her bills will exceed $250.)

63. *Manufacturing.* Ergs, Inc., is planning to make a new kind of radio. Fixed costs will be $90,000, and variable costs will be $15 for the production of each radio. The total-cost function for x radios is

$$C(x) = 90,000 + 15x.$$

The company makes $26 in revenue for each radio sold. The total-revenue function for x radios is

$$R(x) = 26x.$$

(See Section 3.7.)

a) When $R(x) < C(x)$, the company loses money. Find the values of x for which the company loses money.

b) When $R(x) > C(x)$, the company makes a profit. Find the values of x for which the company makes a profit.

64. *Publishing.* The demand and supply functions for a locally produced poetry book are approximated by

$$D(p) = 2000 - 60p \quad \text{and}$$
$$S(p) = 460 + 94p,$$

where p is the price in dollars (see Section 3.7).

a) Find those values of p for which demand exceeds supply.

b) Find those values of p for which demand is less than supply.

Skill Maintenance

65. Graph: $y = 2x^2 - 1$. [1.5]

66. Graph: $5y - 10 = 2x$. [2.5]

67. Solve: $-3x + 5 = 11$. [2.2]

68. Solve: $-2(4 - x) = 7x$. [2.2]

69. Simplify: $|-16|$. [1.1]

70. Simplify: $-|-4|$. [1.1]

Synthesis

71. ◈ Explain in your own words why the inequality symbol must be reversed when both sides of an inequality are multiplied by a negative number.

72. ◈ A Presto photocopier costs $510 and an Exact Image photocopier costs $590. Write a problem that involves the cost of the copiers, the cost per page of photocopies, and the number of copies for which the Presto machine is the more expensive machine to own.

73. ◈ Explain how the addition principle can be used to avoid ever needing to multiply or divide on both sides of an inequality by a negative number.

74. ◈ According to the data in Exercise 54, the number of people who bowl and the number who play billiards or pool are both increasing. Using the concept of slope or rate of change, defend the prediction that the number who play billiards will someday exceed the number who bowl.

Solve for x and y. Assume that a, b, c, d, and m are positive constants.

75. $6by - 4y \le 7by + 10$

76. $a(by - 2) \ge b(2y + 5)$; assume $a > 2$

77. $c(6x - 4) < d(3 + 2x)$; assume $3c > d$

78. $c(2 - 5x) + dx > m(4 + 2x)$; assume $5c + 2m < d$

79. $a(3 - 4x) + cx < d(5x + 2)$; assume $c > 4a + 5d$

Determine whether the statement is true or false. If false, give an example that shows this.

80. For any real numbers a, b, c, and d, if $a < b$ and $c < d$, then $a - c < b - d$.

81. For all real numbers x and y, if $x < y$, then $x^2 < y^2$.

82. ◈ Are the inequalities

$$x < 3 \quad \text{and} \quad x + \frac{1}{x} < 3 + \frac{1}{x}$$

equivalent? Why or why not?

83. ◈ Are the inequalities

$$x < 3 \quad \text{and} \quad 0 \cdot x < 0 \cdot 3$$

equivalent? Why or why not?

Solve.

84. $x + 5 \le 5 + x$

85. $x + 8 < 3 + x$

86. $x^2 > 0$

4.2
Intersections, Unions, and Compound Inequalities

- *Intersections of Sets and Conjunctions of Sentences*
- *Unions of Sets and Disjunctions of Sentences*
- *More on Domains of Functions*

We now consider **compound inequalities**—that is, sentences formed by two or more inequalities, joined by the word *and* or the word *or*.

Intersections of Sets and Conjunctions of Sentences

The **intersection** of two sets A and B is the set of all members that are common to both A and B. We denote the intersection of sets A and B as

$$A \cap B.$$

The intersection of two sets is often pictured as shown here.

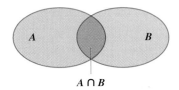

$A \cap B$

Example 1 Find the intersection:
$\{1, 2, 3, 4, 5\} \cap \{-2, -1, 0, 1, 2, 3\}$.

SOLUTION   The numbers 1, 2, and 3 are common to both sets, so the intersection is $\{1, 2, 3\}$. ▬

When two or more sentences are joined by the word *and* to make a compound sentence, the new sentence is called a **conjunction** of the sentences. The following is a conjunction of inequalities:

$$-2 < x \quad and \quad x < 1.$$

For a conjunction to be true, each individual sentence must be true. *The solution set of a conjunction is the intersection of the solution sets of the individual sentences.* Consider the conjunction

$$-2 < x \quad and \quad x < 1.$$

The graphs of each separate sentence are shown below, and the intersection is the last graph. We use both set-builder and interval notations.

$\{x \mid -2 < x\}$ ← | | | | | | ⊕ | | | | | | | → $(-2, \infty)$
           $-7 \ -6 \ -5 \ -4 \ -3 \ -2 \ -1 \ \ 0 \ \ 1 \ \ 2 \ \ 3 \ \ 4 \ \ 5 \ \ 6 \ \ 7$

$\{x \mid x < 1\}$ ← | | | | | | | | | ⊕ | | | | | → $(-\infty, 1)$
           $-7 \ -6 \ -5 \ -4 \ -3 \ -2 \ -1 \ \ 0 \ \ 1 \ \ 2 \ \ 3 \ \ 4 \ \ 5 \ \ 6 \ \ 7$

$\{x \mid -2 < x\} \cap \{x \mid x < 1\}$ ← | | | | | | ⊕━━━⊕ | | | | | → $(-2, 1)$
$= \{x \mid -2 < x \ and \ x < 1\}$ $-7 \ -6 \ -5 \ -4 \ -3 \ -2 \ -1 \ \ 0 \ \ 1 \ \ 2 \ \ 3 \ \ 4 \ \ 5 \ \ 6 \ \ 7$

Because there are numbers that are both greater than -2 and less than 1, the conjunction $-2 < x$ *and* $x < 1$ can be abbreviated by $-2 < x < 1$. Thus the interval $(-2, 1)$ can be represented as $\{x \mid -2 < x < 1\}$, the set of all numbers that are *simultaneously* greater than -2 *and* less than 1. Note that, in general, for $a < b$,

$$a < x \quad \textit{and} \quad x < b \quad \textbf{can be abbreviated} \quad a < x < b;$$

and $\quad b > x \quad \textit{and} \quad x > a \quad \textbf{can be abbreviated} \quad b > x > a.$

Example 2   Solve and graph: $-1 \le 2x + 5 < 13$.

SOLUTION   This inequality is an abbreviation for the conjunction

$$-1 \le 2x + 5 \quad \textit{and} \quad 2x + 5 < 13.$$

⌐ -- ALGEBRAIC SOLUTION

The word *and* corresponds to set *intersection*. To solve the conjunction, we solve each of the two inequalities separately and then find the intersection of the solution sets:

$-1 \le 2x + 5$	*and*	$2x + 5 < 13$
$-6 \le 2x$	*and*	$2x < 8$	Subtracting 5 on both sides of each inequality
$-3 \le x$	*and*	$x < 4.$	Dividing by 2 on both sides of each inequality

We now abbreviate the answer:

$$-3 \le x < 4.$$

The solution set is $\{x \mid -3 \le x < 4\}$, or, in interval notation, $[-3, 4)$.

The graph is the intersection of the two separate solution sets.

$\{x \mid -3 \le x\}$

$$\xleftarrow\!\!\!-\!\!\!\!\!\xrightarrow[\ -7\ -6\ -5\ -4\ -3\ -2\ -1\ \ 0\ \ 1\ \ 2\ \ 3\ \ 4\ \ 5\ \ 6\ \ 7\ ]{}\qquad [-3, \infty)$$

$\{x \mid x < 4\}$

$$\xleftarrow\!\!\!-\!\!\!\!\!\xrightarrow[\ -7\ -6\ -5\ -4\ -3\ -2\ -1\ \ 0\ \ 1\ \ 2\ \ 3\ \ 4\ \ 5\ \ 6\ \ 7\ ]{}\qquad (-\infty, 4)$$

$\{x \mid -3 \le x\} \cap \{x \mid x < 4\}$
$= \{x \mid -3 \le x < 4\}$

$$\xleftarrow\!\!\!-\!\!\!\!\!\xrightarrow[\ -7\ -6\ -5\ -4\ -3\ -2\ -1\ \ 0\ \ 1\ \ 2\ \ 3\ \ 4\ \ 5\ \ 6\ \ 7\ ]{}\qquad [-3, 4)$$

⌐ -- GRAPHICAL SOLUTION

We graph the equations $y_1 = -1$, $y_2 = 2x + 5$, and $y_3 = 13$, and determine those x-values for which $y_1 \le y_2$ *and* $y_2 < y_3$.

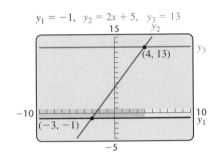

$y_1 = -1, \ y_2 = 2x + 5, \ y_3 = 13$

Using INTERSECT, we find that the graphs of y_1 and y_2 intersect at the point $(-3, -1)$ and the graphs of y_2 and y_3 intersect at the point $(4, 13)$. From the graph, we see that $y_1 < y_2$ for x-values greater than -3, as indicated by the blue shading on the x-axis. We also see that $y_2 < y_3$ for x-values less than 4, as shown by the red shading on the x-axis. The solution set, indicated by the purple shading, is the intersection of these sets, as well as the number -3. It includes all x-values for which the line $y_2 = 2x + 5$ is both on or above the line $y_1 = -1$ *and* below the line $y_3 = 13$. This can be written $\{x \mid -3 \le x < 4\}$, or $[-3, 4)$.   ▬

The steps in the algebraic solution of Example 2 are s
bined as follows:

$$-1 \leq 2x + 5 < 13$$
$$-1 - 5 \leq 2x + 5 - 5 < 13 - 5$$
$$-6 \leq 2x < 8$$
$$-3 \leq x < 4.$$

Such an approach saves some writing and will prove useful in Section 4.3.

Example 3 Solve and graph: $2x - 5 \geq -3$ *and* $5x + 2 \geq 17$.

ALGEBRAIC SOLUTION

We first solve each inequality separately:

$$2x - 5 \geq -3 \quad and \quad 5x + 2 \geq 17$$
$$2x \geq 2 \quad and \quad 5x \geq 15$$
$$x \geq 1 \quad and \quad x \geq 3.$$

Next, we find the intersection of two separate solution sets.

$\{x | x \geq 1\}$

<---+---+---+---+---+---+---+---+---+---+---+---+---+---+---> $[1, \infty)$
$\quad -7 \ -6 \ -5 \ -4 \ -3 \ -2 \ -1 \quad 0 \quad 1 \quad 2 \quad 3 \quad 4 \quad 5 \quad 6 \quad 7$

$\{x | x \geq 3\}$

<---+---+---+---+---+---+---+---+---+---+---+---+---+---+---> $[3, \infty)$
$\quad -7 \ -6 \ -5 \ -4 \ -3 \ -2 \ -1 \quad 0 \quad 1 \quad 2 \quad 3 \quad 4 \quad 5 \quad 6 \quad 7$

$\{x | x \geq 1\} \cap \{x | x \geq 3\}$
$= \{x | x \geq 3\}$

<---+---+---+---+---+---+---+---+---+---+---+---+---+---+---> $[3, \infty)$
$\quad -7 \ -6 \ -5 \ -4 \ -3 \ -2 \ -1 \quad 0 \quad 1 \quad 2 \quad 3 \quad 4 \quad 5 \quad 6 \quad 7$

The numbers common to both sets are those that are
greater than or equal to 3. Thus the solution set is
$\{x | x \geq 3\}$, or, in interval notation, $[3, \infty)$.

GRAPHICAL SOLUTION

We graph the equations $y_1 = 2x - 5$,
$y_2 = -3$, $y_3 = 5x + 2$, and $y_4 = 17$.
Since we want to determine the
x-values for which $y_1 \geq y_2$ *and*
$y_3 \geq y_4$, we find the point of
intersection of y_1 and y_2 and the point
of intersection of y_3 and y_4.

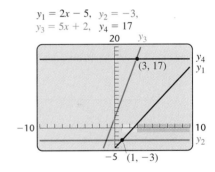

We see that $y_1 \geq y_2$ when $x \geq 1$
and that $y_3 \geq y_4$ when $x \geq 3$. *Both*
statements are true when $x \geq 3$. The
solution set is $\{x | x \geq 3\}$, or, in
interval notation, $[3, \infty)$.

You should check that any number in $[3, \infty)$ satisfies the conjunction
whereas numbers outside $[3, \infty)$ do not.

Intersection

The word "and" corresponds to "intersection" and to the symbol
"∩". Any solution of a conjunction must make each part of the
conjunction true.

Sometimes there is no way to solve both parts of a conjunction at once.

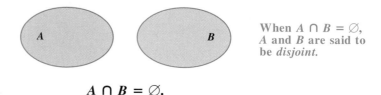

When $A \cap B = \emptyset$, A and B are said to be *disjoint*.

$$A \cap B = \emptyset.$$

Example 4 Solve and graph: $2x - 3 > 1$ *and* $3x - 1 < 2$.

┌── ALGEBRAIC SOLUTION

We solve each inequality separately:

$2x - 3 > 1$ *and* $3x - 1 < 2$
$\quad 2x > 4$ *and* $\quad\quad 3x < 3$
$\quad\quad x > 2$ *and* $\quad\quad\quad x < 1$.

The solution set is the intersection of the individual inequalities.

$\{x \mid x > 2\}$

$(2, \infty)$

$\{x \mid x < 1\}$

$(-\infty, 1)$

$\{x \mid x > 2\} \cap \{x \mid x < 1\}$
$= \{x \mid x > 2 \text{ and } x < 1\} = \emptyset$

$\emptyset$

Since no number is both greater than 2 and less than 1, the solution set is the empty set, $\emptyset$.

┌── GRAPHICAL SOLUTION

We graph $y_1 = 2x - 3$, $y_2 = 1$, $y_3 = 3x - 1$, and $y_4 = 2$, and determine the point at which y_1 and y_2 intersect and the point at which y_3 and y_4 intersect.

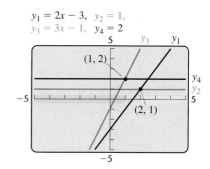

We see that $y_1 > y_2$ when $x > 2$ and that $y_3 < y_4$ when $x < 1$. Since no number is both greater than 2 and less than 1, the solution set is the empty set $\emptyset$.

Unions of Sets and Disjunctions of Sentences

The **union** of two sets A and B is the collection of elements belonging to A and/or B. We denote the union of A and B by

$$A \cup B.$$

The union of two sets is often pictured as shown at left.

Example 5 Find the union: $\{2, 3, 4\} \cup \{3, 5, 7\}$.

SOLUTION The numbers in either or both sets are 2, 3, 4, 5, and 7, so the union is $\{2, 3, 4, 5, 7\}$.

When two or more sentences are joined by the word *or* to make a compound sentence, the new sentence is called a **disjunction** of the sentences. Here are three examples:

$$x < -3 \quad or \quad x > 3;$$

$$y \text{ is an odd number} \quad or \quad y \text{ is a (prime) number;}$$

$$x < 0 \quad or \quad x = 0 \quad or \quad x > 0.$$

For a disjunction to be true, at least one of the individual sentences must be true. *The solution set of a disjunction is the union of the individual solution sets.* Consider the disjunction

$$x < -3 \quad or \quad x > 3.$$

The graphs of each separate sentence are shown below, and the union is the last graph. Again, we use both set-builder and interval notations.

$\{x \mid x < -3\}$ $(-\infty, -3)$

$\{x \mid x > 3\}$ $(3, \infty)$

$\{x \mid x < -3\} \cup \{x \mid x > 3\}$
$= \{x \mid x < -3 \text{ or } x > 3\}$ $(-\infty, -3) \cup (3, \infty)$

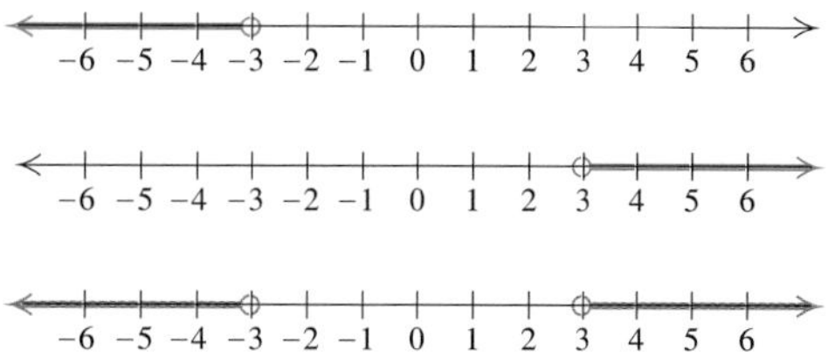

Answers to disjunctions can rarely be written concisely. The solution set of $x < -3 \; or \; x > 3$ is simply written $\{x \mid x < -3 \text{ or } x > 3\}$, or $(-\infty, -3) \cup (3, \infty)$.

Union

The word "or" corresponds to "union" and to the symbol "$\cup$". For a number to be a solution of the disjunction, it must be in *at least one* of the solution sets.

Example 6   Solve and graph: $7 + 2x < -1 \; or \; 13 - 5x \leq 3$.

ALGEBRAIC SOLUTION

We solve each inequality separately, retaining the word *or*:

$$7 + 2x < -1 \quad or \quad 13 - 5x \leq 3$$
$$2x < -8 \quad or \quad -5x \leq -10$$

reverse the symbol.

$$x < -4 \quad or \quad x \geq 2.$$

To find the solution set of the disjunction, we consider the individual graphs. We graph $x < -4$ and $x \geq 2$. Then we take the union of the graphs.

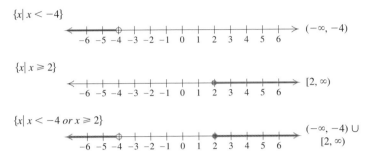

$\{x \mid x < -4\}$ $(-\infty, -4)$

$\{x \mid x \geq 2\}$ $[2, \infty)$

$\{x \mid x < -4 \text{ or } x \geq 2\}$ $(-\infty, -4) \cup [2, \infty)$

The solution set is $\{x \mid x < -4 \text{ or } x \geq 2\}$, or, in interval notation, $(-\infty, -4) \cup [2, \infty)$.

GRAPHICAL SOLUTION

We graph $y_1 = 7 + 2x$, $y_2 = -1$, $y_3 = 13 - 5x$, and $y_4 = 3$, and determine those values of x for which $y_1 < y_2$ or $y_3 \leq y_4$.

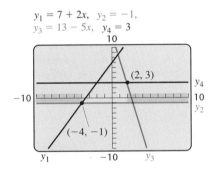

$y_1 = 7 + 2x, \quad y_2 = -1,$
$y_3 = 13 - 5x, \quad y_4 = 3$

The graphs of y_1 and y_2 intersect at $(-4, -1)$, and $y_1 < y_2$ when $x < -4$. The graphs of y_3 and y_4 intersect at $(2, 3)$, and $y_3 \leq y_4$ when $x \geq 2$. Thus, $y_1 < y_2$ or $y_3 \leq y_4$ when $x < -4$ or $x \geq 2$. The solution set is $\{x \mid x < -4 \text{ or } x \geq 2\}$, or, in interval notation, $(-\infty, -4) \cup [2, \infty)$.

Caution! A compound inequality like

$$x < -4 \quad \text{or} \quad x \geq 2,$$

as in Example 6, *cannot* be expressed as $2 \leq x < -4$ because to do so would be to say that x is *simultaneously* less than -4 and greater than or equal to 2. No number is both less than -4 *and* greater than 2, but many are less than -4 *or* greater than 2.

Example 7   Solve: $3x - 11 < 4 \ or \ 4x + 9 \geq 1$.

ALGEBRAIC SOLUTION

We solve the individual inequalities separately, retaining the word *or*:

$$3x - 11 < 4 \quad or \quad 4x + 9 \geq 1$$
$$3x < 15 \quad or \quad 4x \geq -8$$
$$x < 5 \quad or \quad x \geq -2.$$

Keep the word "or."

To find the solution set, we first look at the individual graphs.

$\{x \mid x < 5\}$

$(-\infty, 5)$

$\{x \mid x \geq -2\}$

$[-2, \infty)$

$\{x \mid x < 5\} \cup \{x \mid x \geq -2\}$
$= \{x \mid x < 5 \ or \ x \geq -2\}$

$(-\infty, \infty) = \mathbb{R}$

Since *all* numbers are less than 5 or greater than or equal to −2, the two sets fill the entire number line. Thus the solution set is $\mathbb{R}$, the set of all real numbers.

GRAPHICAL SOLUTION

We graph $y_1 = 3x - 11$, $y_2 = 4$, $y_3 = 4x + 9$, and $y_4 = 1$. Since we want to find the x-values for which $y_1 < y_2$ or $y_3 \geq y_4$, we determine the points of intersection of the graphs of y_1 and y_2 as well as those of y_3 and y_4.

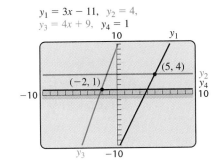

$y_1 = 3x - 11, \ y_2 = 4,$
$y_3 = 4x + 9, \ y_4 = 1$

We see that $y_1 < y_2$ when $x < 5$ and that $y_3 \geq y_4$ when $x \geq -2$. The disjunction is true when $x < 5$ *or* $x \geq -2$. The solution set is thus $\mathbb{R}$, the set of all real numbers.

More on Domains of Functions

In Section 2.7, we saw that if $g(x) = (5x - 2)/(3x - 7)$, then the domain of $g = \{x \mid x \text{ is a real number } and \ x \neq \frac{7}{3}\}$. The graph of the domain is shown at left. We can now represent such a set using interval notation:

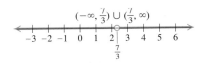

$(-\infty, \frac{7}{3}) \cup (\frac{7}{3}, \infty)$

$$\{x \mid x \text{ is a real number } and \ x \neq \tfrac{7}{3}\} = (-\infty, \tfrac{7}{3}) \cup (\tfrac{7}{3}, \infty).$$

Example 8   Find the domain of f if $f(x) = \sqrt{x + 2}$.

SOLUTION   The expression $\sqrt{x + 2}$ is not a real number when $x + 2$ is negative. Thus, if $f(x) = \sqrt{x + 2}$, the domain of f is the set of all x-values for which $x + 2 \geq 0$. Since $x + 2 \geq 0$ is equivalent to $x \geq -2$, we have

$$\text{Domain of } f = \{x \mid x \geq -2\} = [-2, \infty).$$

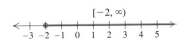

$[-2, \infty)$

The graph of the domain is shown at left.

We can check that $[-2, \infty)$ is the domain of f by looking at the graph of $f(x)$. You can confirm by tracing the curve that no y-value is given for x-values less than -2. The domain is $[-2, \infty)$.

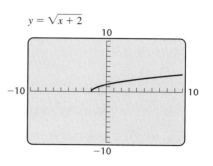

$y = \sqrt{x + 2}$

Interactive Discovery

Graph the functions $f(x) = \sqrt{3 - x}$ and $g(x) = \sqrt{x + 1}$. Determine algebraically the domains of f and g and trace each curve to verify those intervals.

Graph $y_1 + y_2$, $y_1 - y_2$, and $y_1 \cdot y_2$, and use the graphs to determine the domains of $f + g$, $f - g$, and $f \cdot g$. How can the domains of the sum, the difference, and the product of functions be found algebraically?

The domain of the sum, the difference, or the product of the functions f and g is the intersection of the domains of f and g.

Example 9   Find the domain of $f + g$ if $f(x) = \sqrt{2x - 5}$ and $g(x) = \sqrt{x + 1}$.

SOLUTION   We first find the domain of f and the domain of g. The domain of f is the set of all x-values for which $2x - 5 \geq 0$, or $\left\{x \mid x \geq \frac{5}{2}\right\}$, or $\left[\frac{5}{2}, \infty\right)$. Similarly, the domain of g is $\{x \mid x \geq -1\}$, or $[-1, \infty)$. The intersection of the domains is $\left\{x \mid x \geq \frac{5}{2}\right\}$, or $\left[\frac{5}{2}, \infty\right)$. We can confirm this at least approximately by tracing the graph of $f + g$.

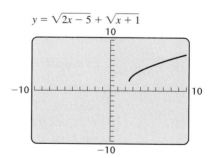

$y = \sqrt{2x - 5} + \sqrt{x + 1}$

Thus the

$$\text{Domain of } f + g = \left\{x \mid x \geq \tfrac{5}{2}\right\}, \text{ or } \left[\tfrac{5}{2}, \infty\right).$$

4.2 Exercise Set

Find each indicated intersection or union.

1. $\{9, 10, 11\} \cap \{9, 11, 13\}$

2. $\{2, 4, 8\} \cup \{8, 9, 10\}$

3. $\{1, 5, 10, 15\} \cup \{5, 15, 20\}$

4. $\{2, 5, 9, 11\} \cap \{5, 8, 12\}$

5. $\{a, b, c, d\} \cap \{b, f, g\}$

6. $\{a, b, c\} \cup \{a, c\}$

7. $\{r, s, t\} \cup \{r, u, t, s, v\}$

8. $\{m, n, o, p\} \cap \{m, o, p\}$

9. $\{3, 5, 7\} \cup \varnothing$

10. $\{3, 5, 7\} \cap \varnothing$

Graph and write interval notation.

11. $2 < x < 7$

12. $0 \le y \le 4$

13. $-6 \le y \le -2$

14. $-9 \le x < -5$

15. $x < -2$ *or* $x > 1$

16. $x < -2$ *or* $x > 3$

17. $x \le -1$ *or* $x > 4$

18. $x \le -5$ *or* $x > 2$

19. $3 \le -x < 5$

20. $x > -7$ *and* $x < -2$

21. $x > -2$ *and* $x < 4$

22. $3 > -x \ge -1$

23. $5 > a$ *or* $a > 7$

24. $t \ge 2$ *or* $-3 > t$

25. $x \ge 5$ *or* $-x \ge 4$

26. $-x < 3$ *or* $x < -6$

27. $5 > x$ *and* $x \ge -6$

28. $6 > -x \ge 0$

29. $x < 7$ *and* $x \ge 3$

30. $x \ge -3$ *and* $x < 3$

31. $t < 2$ *or* $t < 5$

32. $t > 4$ *or* $t > -1$

33. $x > -1$ *or* $x \le 3$

34. $4 > x$ *or* $x \ge -3$

35. $x \ge 5$ *and* $x > 7$

36. $x \le -4$ *and* $x < 1$

Solve and graph each solution set.

37. $-3 < t + 2 < 7$

38. $-2 < t + 1 \le 5$

39. $-5 \le 2a - 1$ *and* $3a + 1 < 7$

40. $-4 \le 3n + 2$ *and* $2n - 3 \le 5$

41. $x + 7 \le -2$ *or* $x + 7 \ge 5$

42. $x + 5 < -3$ *or* $x + 5 \ge 4$

43. $2 \le f(x) \le 8$, where $f(x) = 3x - 1$

44. $10 \ge g(x) \ge -2$, where $g(x) = 3x - 5$

45. $-18 \le f(x) < 0$, where $f(x) = -2x - 7$

46. $4 > g(t) \ge 2$, where $g(t) = -3t - 8$

47. $f(x) \le 2$ *or* $f(x) \ge 8$, where $f(x) = 3x - 1$

48. $g(x) \le -2$ *or* $g(x) \ge 10$, where $g(x) = 3x - 5$

49. $f(x) < -1$ *or* $f(x) > 1$, where $f(x) = 2x - 7$

50. $g(x) < -7$ *or* $g(x) > 7$, where $g(x) = 3x + 5$

51. $a + 4 < -1$ *and* $3a - 5 < 7$

52. $1 - a < -2$ *and* $2a + 1 > 9$

53. $3x + 2 < 2$ *or* $4 - 2x < 14$

54. $3a - 7 > -10$ *or* $5a + 2 \le 22$

55. $2t - 7 \le 5$ *or* $5 - 2t > 3$

56. $5 - 3a \le 8$ *or* $2a + 1 > 7$

Solve.

57. $3x + 1 < 5$ *or* $3x + 1 > 4$

58. $\frac{3}{4}x - 7 < 1$ *and* $\frac{3}{4}x - 7 > 10$

59. $f(x) < 3$ *and* $f(x) > 4$, where $f(x) = \frac{1}{2}x - 7$

60. $f(x) \le 2$ *or* $f(x) \ge -2$, where $f(x) = 2x - 15$

61. Use the accompanying graph of $f(x) = 2x - 5$ to solve $-7 < 2x - 5 < 7$.

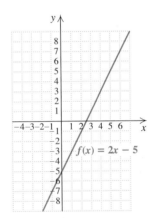

62. Use the accompanying graph of $g(x) = 4 - x$ to solve $4 - x < -2$ *or* $4 - x > 7$.

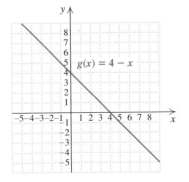

For f(x) as given, use interval notation to write the domain of f.

63. $f(x) = \dfrac{7}{x - 5}$

64. $f(x) = \dfrac{2}{x + 3}$

65. $f(x) = \sqrt{x + 4}$

66. $f(x) = \sqrt{x + 7}$

67. $f(x) = \dfrac{x + 3}{2x - 5}$

68. $f(x) = \dfrac{x - 1}{3x + 4}$

69. $f(x) = \sqrt{12 - 3x}$

70. $f(x) = \sqrt{8 - 4x}$

For f(x) and g(x) as given, use interval notation to write the domains of f + g, f − g, and f · g.

71. $f(x) = \sqrt{x + 3}, \; g(x) = \sqrt{4 - x}$

72. $f(x) = \sqrt{2x + 1}, \; g(x) = \sqrt{3 - 5x}$

73. $f(x) = \sqrt{4x - 3}, \; g(x) = \sqrt{2x + 9}$

74. $f(x) = \sqrt{5 - x}, \; g(x) = \sqrt{1 - x}$

75. $f(x) = \dfrac{x}{x - 5}, \; g(x) = \sqrt{3x + 2}$

76. $f(x) = \sqrt{4 - 5x}, \; g(x) = \dfrac{3x}{2x + 1}$

Skill Maintenance

Solve.

77. $2x - 3y = 7,$
$3x + 2y = -10$ [3.2]

78. $3x - 9(x + 4) = 20(3x + 7)$ [2.2]

79. $5(2x + 3) = 3(x - 4)$ [2.2]

Graph. [2.5]

80. $3x - 4y = -12$

81. $f(x) = 5$

82. $x = -2$

Synthesis

83. ◆ Describe the circumstances under which $A \cap B = A$.

84. ◆ Explain why the conjunction $3 < x \text{ and } x < 5$ can be rewritten as $3 < x < 5$, but the disjunction $3 < x \text{ or } x < 5$ cannot be rewritten as $3 < x < 5$.

85. ◆ When asked to graph the solution set of the compound inequality $-3 < 2x - 5 \le 4$, Joe submits the following graph.

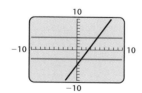

Is his solution correct? Why or why not?

86. ◆ Explain how the use of the word *or* in a compound inequality differs from the use of the word *or* in everyday English. (*Hint:* Consider the expression and/or.)

87. *Minimizing Tolls.* A $3.00 toll is charged to cross the bridge from Sanibel Island to mainland Florida. A six-month pass, costing $15.00, reduces the toll to $0.50. A one-year pass, costing $60, allows for free crossings. How many crossings per month does it take, on average, for the six-month pass to be the more economical choice?

88. *Converting Dress Sizes.* The function

$$f(x) = 2(x + 10)$$

can be used to convert dress sizes x in the United States to dress sizes $f(x)$ in Italy. For what dress sizes in the United States will dress sizes in Italy be between 32 and 46?

89. *Pressure at Sea Depth.* The function

$$P(d) = 1 + \dfrac{d}{33}$$

gives the pressure, in atmospheres (atm), at a depth of d feet in the sea. For what depths d is the pressure at least 1 atm and at most 7 atm?

90. *Temperatures of Liquids.* The formula

$$C = \tfrac{5}{9}(F - 32)$$

can be used to convert Fahrenheit temperatures F to Celsius temperatures C.

a) Gold is liquid for Celsius temperatures C such that $1063° \le C < 2660°$. Find a comparable inequality for Fahrenheit temperatures.

b) Silver is liquid for Celsius temperatures C such that $960.8° \le C < 2180°$. Find a comparable inequality for Fahrenheit temperatures.

91. *Solid Waste Generation.* The function

$$w(t) = 0.05t + 4.3$$

can be used to estimate the number of pounds of solid waste, $w(t)$, produced daily, on average, by each person in the United States, t years after 1991. For what years will waste production range from 5.0 to 5.25 lb per person per day?

92. *Records in the Women's 100-m Dash.* In 1988, Florence Griffith Joyner set a world record of 10.49 sec in the women's 100-m dash. The function

$$R(t) = -0.0433t + 10.49$$

can be used to predict the world record in the women's 100-m dash t years after 1988. Predict (in terms of an inequality) those years for which the world record was between 11.5 and 10.8 sec. (Measure from the middle of 1988.)

Solve and graph.

93. $4a - 2 \leq a + 1 \leq 3a + 4$

94. $4m - 8 > 6m + 5 \ or \ 5m - 8 < -2$

95. $x - 10 < 5x + 6 \leq x + 10$

96. $3x < 4 - 5x < 5 + 3x$

Determine whether each sentence is true or false for all real numbers a, b, and c.

97. If $-b < -a$, then $a < b$.

98. If $a \leq c$ and $c \leq b$, then $b > a$.

99. If $a < c$ and $b < c$, then $a < b$.

100. If $-a < c$ and $-c > b$, then $a > b$.

For f(x) as given, use interval notation to write the domain of f.

101. $f(x) = \dfrac{\sqrt{5 + 2x}}{x - 1}$ **102.** $f(x) = \dfrac{\sqrt{3 - 4x}}{x + 7}$

103. On many graphers, the TEST key provides access to inequality symbols, while the LOGIC option of that same key accesses the conjunction *and* and the disjunction *or*. Thus compound inequalities like Exercise 21 can be checked by forming expressions like $y_3 = y_1 \ and \ y_2$ (where $y_1 = x > -2$ and $y_2 = x < 4$). The interval(s) in the solution set are shown as a horizontal line 1 unit above the x-axis. (Be careful to "deselect" y_1 and y_2 so that only y_3 is drawn.) Use this approach to check your answers to Exercises 27 and 50.

COLLABORATIVE CORNER

Focus: Compound inequalities and solution sets

Time: 20–30 minutes

Group size: 2–3

In 1999, the U.S. Postal Service charged 22 cents per ounce plus an additional 11-cent delivery fee (1 oz or less costs 33 cents; more than 1 oz, but not more than 2 oz, costs 55 cents; and so on). Rapid Delivery charged $1.05 per pound plus an additional $2.50 delivery fee (up to 16 oz costs $3.55, more than 16 oz, but less than 32 oz, costs $4.60; and so on). Let $x =$ the weight, in ounces, of an item being mailed.*

Activity

Working together, group members should graph the function p, where $p(x)$ is the cost, in dollars, of mailing x ounces at a post office. Using the same set of axes, the group should also graph the function r, where $r(x)$ is the cost, in dollars, of mailing x ounces with Rapid Delivery. Finally, determine those weights for which the Postal Service is less expensive. Express your answer using both set-builder and interval notation.

---

*Based on an article by Michael Contino in *Mathematics Teacher*, May 1995. E-mail: mike_contino@cams.edu

4.3
Absolute-Value Equations and Inequalities

- *Equations with Absolute Value*
- *Inequalities with Absolute Value*

Equations with Absolute Value

Recall from Section 1.2 the definition of absolute value.

Absolute Value

The absolute value of x, denoted $|x|$, is defined as

$$|x| = \begin{cases} x, & \text{if } x \geq 0, \\ -x, & \text{if } x < 0. \end{cases}$$

(When x is nonnegative, the absolute value of x is x. When x is negative, the absolute value of x is the opposite of x.)

Interactive Discovery

Graph $f(x) = |x|$ and $g(x) = 4$. Use the graph to solve $|x| = 4$. What are the solutions of an equation $|x| = a$, if a is positive?

    Graph $f(x) = |x|$ and $h(x) = 0$. Use the graph to solve $|x| = 0$.

    Graph $f(x) = |x|$ and $r(x) = -7$. Use the graph to solve $|x| = -7$. What are the solutions of an equation $|x| = a$, if a is negative?

In general, the following is true:

    If a is positive, the equation $|x| = a$ has two solutions, a and $-a$.

    If a is 0, the equation $|x| = a$ has one solution, 0.

    If a is negative, the equation $|x| = a$ has no solution.

Other equations involving absolute value can be solved graphically.

Example 1 Solve: $|x - 3| = 5$.

SOLUTION   We let $y_1 = \text{abs}(x - 3)$ and $y_2 = 5$. The solution set of the equation consists of the x-coordinates of any points of intersection of the graphs of y_1 and y_2. We see that the graphs intersect at two points. To find the coordinates of one point of intersection, we move the cursor close to that point when prompted for a GUESS.

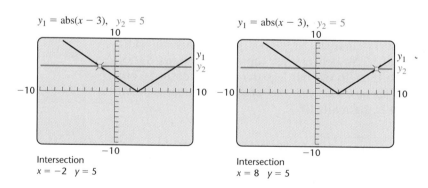

The graphs intersect at $(-2, 5)$ and $(8, 5)$. Thus the solutions are -2 and 8. To check, we substitute each in the original equation.

CHECK: *For* -2:

$$\frac{|x - 3| = 5}{|-2 - 3| \ ? \ 5}$$
$$|-5|$$
$$5 \ \Big| \ 5 \quad \text{TRUE}$$

For 8:

$$\frac{|x - 3| = 5}{|8 - 3| \ ? \ 5}$$
$$|5|$$
$$5 \ \Big| \ 5 \quad \text{TRUE}$$

The solution set is $\{-2, 8\}$.

We can solve equations involving absolute value using the following principle.

The Absolute-Value Principle for Equations

For any positive number p and any algebraic expression X:

a) The solutions of $|X| = p$ are those numbers that satisfy $X = -p$ *or* $X = p$.

b) The equation $|X| = 0$ is equivalent to the equation $X = 0$.

c) The equation $|X| = -p$ has no solution.

Example 2   Find the solution set: **(a)** $|2x + 5| = 13$; **(b)** $|4 - 7x| = -8$.

a)

ALGEBRAIC SOLUTION

We use the absolute-value principle, replacing X with $2x + 5$ and p with 13:

$$|X| = p$$
$$|2x + 5| = 13$$
$$2x + 5 = -13 \quad or \quad 2x + 5 = 13$$
$$2x = -18 \qquad or \qquad 2x = 8$$
$$x = -9 \qquad or \qquad x = 4.$$

The solutions are -9 and 4.

GRAPHICAL SOLUTION

We graph $y_1 = \text{abs}(2x + 5)$ and $y_2 = 13$ and use INTERSECT to find any points of intersection. The graphs intersect at $(-9, 13)$ and $(4, 13)$. The x-coordinates, -9 and 4, are the solutions.

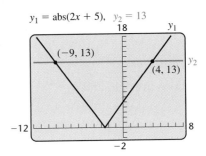

$y_1 = \text{abs}(2x + 5), \ y_2 = 13$

As one check, we note that we found the same solution using both methods. We can also check by substituting the possible solutions in the original equation.

The solution set is $\{-9, 4\}$.

b)

┌── *ALGEBRAIC SOLUTION*

The absolute-value principle reminds us that absolute value is always nonnegative. Thus the equation $|4 - 7x| = -8$ has no solution. The solution set is $\varnothing$.

┌── *GRAPHICAL SOLUTION*

We graph $y_1 = \text{abs}(4 - 7x)$ and $y_2 = -8$.

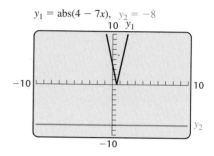

$y_1 = \text{abs}(4 - 7x), \quad y_2 = -8$

There are no points of intersection, so the solution set is $\varnothing$.

The absolute-value principle can be used together with the addition and multiplication principles to solve many types of equations with absolute value.

Example 3 Given that $f(x) = 2|x + 3| + 1$, find all x for which $f(x) = 15$.

┌── *ALGEBRAIC SOLUTION*

Since we are looking for $f(x) = 15$, we substitute:

$$f(x) = 15$$
$$2|x + 3| + 1 = 15 \qquad \text{Replacing } f(x) \text{ with } 2|x + 3| + 1$$
$$\left.\begin{array}{l} 2|x + 3| = 14 \\ |x + 3| = 7 \end{array}\right\} \quad \begin{array}{l} \text{Adding } -1 \text{ and multi-} \\ \text{plying by } \frac{1}{2} \text{ on both sides} \\ \text{to isolate } |x + 3| \end{array}$$
$$x + 3 = -7 \quad or \quad x + 3 = 7 \qquad \begin{array}{l} \text{Replacing } X \text{ with} \\ x + 3 \text{ and } p \text{ with} \\ 7 \text{ in the} \\ \text{absolute-value} \\ \text{principle} \end{array}$$
$$x = -10 \quad or \qquad x = 4.$$

The solutions are -10 and 4.

┌── *GRAPHICAL SOLUTION*

We graph $y_1 = 2\,\text{abs}(x + 3) + 1$ and $y_2 = 15$ and determine the coordinates of any points of intersection. We make certain that the viewing window is large enough to show all such points. Choosing an appropriate viewing window may involve some trial and error.

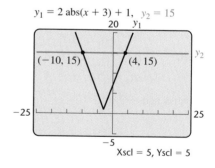

$y_1 = 2\,\text{abs}(x + 3) + 1, \quad y_2 = 15$

Xscl = 5, Yscl = 5

The solutions are the x-coordinates of the points of intersection, -10 and 4.

To check the answer, we note that we found the same solution using both methods. We can also check that $f(-10) = f(4) = 15$. The solution set is $\{-10, 4\}$.

Since distance is always nonnegative, we can think of a number's absolute value as its distance from 0 on a number line. For example, we can interpret the equation

$$|x| = 4$$

to mean that the number x is 4 units from 0 on a number line.

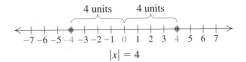

$$|x| = 4$$

There are two such numbers, 4 and -4. Thus the solution set is $\{-4, 4\}$.

Example 4   Solve: $|x - 2| = 3$.

SOLUTION   The expressions $|a - b|$ and $|b - a|$ can be used to represent the *distance between a and b* on the number line. For example, the distance between 7 and 8 is given by $|8 - 7|$ or $|7 - 8|$. From this viewpoint, the equation $|x - 2| = 3$ states that the distance between x and 2 is 3 units. We draw a number line and locate all numbers that are 3 units from 2.

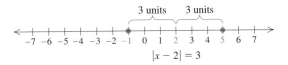

$$|x - 2| = 3$$

We can check by solving $|x - 2| = 3$ graphically. We graph $y_1 = \text{abs}(x - 2)$ and $y_2 = 3$ and determine the x-coordinates of any points of intersection. The graphs intersect at the points $(-1, 3)$ and $(5, 3)$, so the solutions are -1 and 5. Since both methods give the same solutions, the answers are probably correct.

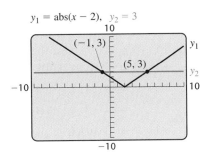

The solution set is $\{-1, 5\}$.

Sometimes an equation has two absolute-value expressions. Consider

$|a| = |b|$. This means that a and b are the same distance from 0 on a number line.

If a and b are the same distance from 0, then either they are the same number or they are opposites.

Example 5   Solve: $|2x - 3| = |x + 5|$.

ALGEBRAIC SOLUTION

Either $2x - 3 = x + 5$ (they are the same number) or $2x - 3 = -(x + 5)$ (they are opposites). We solve each equation separately:

$$2x - 3 = x + 5 \quad or \quad 2x - 3 = -(x + 5)$$
$$x - 3 = 5 \quad\quad or \quad 2x - 3 = -x - 5$$
$$x = 8 \quad\quad\quad or \quad 3x - 3 = -5$$
$$3x = -2$$
$$x = -\tfrac{2}{3}.$$

The solutions are 8 and $-\tfrac{2}{3}$. The solution set is $\left\{-\tfrac{2}{3}, 8\right\}$.

GRAPHICAL SOLUTION

We graph $y_1 = \text{abs}(2x - 3)$ and $y_2 = \text{abs}(x + 5)$ and look for any points of intersection.

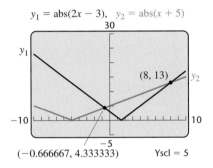

$y_1 = \text{abs}(2x - 3), \quad y_2 = \text{abs}(x + 5)$

The x-coordinates of the points of intersection are -0.6666667 and 8. The solution set is $\{-0.6666667, 8\}$.

Inequalities with Absolute Value

Our methods for solving equations with absolute value can be adapted for solving inequalities as well.

Interactive Discovery

Graph $y_1 = \text{abs}(x)$ and $y_2 = 4$. The solution set of the inequality $|x| < 4$ consists of all x-values for which the graph of $y_1 = \text{abs}(x)$ is below the horizontal line $y_2 = 4$. Similarly, the solution set of the inequality $|x| > 4$ is the set of all x-values for which (x, y_1) is above the line $y_2 = 4$.

Use the graph to write the solution set of each of the following equations using interval notation.

1. $|x| < 4$ **2.** $|x| > 4$
3. $|x| \le 4$ **4.** $|x| \ge 4$

What is the solution set of $|x| > -3$? of $|x| \le -3$?

We can also visualize solutions of inequalities involving absolute value using the number line. The solutions of $|x| < 4$ are all numbers for which the *distance from* 0 *is less than* 4. Looking at the number line, we can see that numbers like -3, -2, $-\frac{1}{2}$, 0, $\frac{1}{2}$, 2, and 3 are all solutions. In fact, the solution set is $\{x|\ -4 < x < 4\}$, or, in interval notation, $(-4, 4)$. The graph is as follows:

$$|x| < 4$$

The solutions of $|x| \geq 4$ are all numbers for which the *distance from* 0 *is greater than or equal to* 4—in other words, those numbers x such that $x \leq -4$ or $4 \leq x$. The solution set is $\{x|\ x \leq -4 \ or \ x \geq 4\}$. In interval notation, the solution is $(-\infty, -4] \cup [4, \infty)$. The graph is as follows:

$$|x| \geq 4$$

The following is a general principle for solving problems in which absolute-value symbols appear.

Principles for Solving Absolute-Value Problems

For any positive number p and any expression X:

a) The solutions of $|X| = p$ are those numbers that satisfy $X = -p \ or \ X = p$.

b) The solutions of $|X| < p$ are those numbers that satisfy $-p < X < p$.

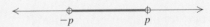

c) The solutions of $|X| > p$ are those numbers that satisfy $X < -p \ or \ p < X$.

Of course, if p is negative, any value of X will satisfy the inequality $|X| > p$ since absolute value is never negative. By the same reasoning, $|X| < p$ has no solution when p is not positive. Thus the inequality $|2x - 7| > -3$ is true for any real number x, and the inequality $|2x - 7| < -3$ has no solution.

Note that an inequality of the form $|X| < p$ corresponds to a *con*junction, whereas an inequality of the form $|X| > p$ corresponds to a *dis*junction.

Example 6   Solve: $|3x - 2| < 4$.

┌── *ALGEBRAIC SOLUTION*

We use part (b) of the principles listed on the previous page. In this case, X is $3x - 2$ and p is 4:

$$|X| < p$$

$|3x - 2| < 4$ Replacing X with $3x - 2$ and p with 4

$-4 < 3x - 2 < 4$ The number $3x - 2$ must be within 4 units of 0.

$-2 < \quad 3x \quad < 6$ Adding 2

$-\frac{2}{3} < \quad x \quad < 2.$ Multiplying by $\frac{1}{3}$

The solution set is $\{x \mid -\frac{2}{3} < x < 2\}$, or, in interval notation, $\left(-\frac{2}{3}, 2\right)$. The graph is as follows:

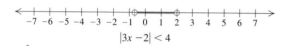

$|3x - 2| < 4$

┌── *GRAPHICAL SOLUTION*

We graph $y_1 = \text{abs}(3x - 2)$ and $y_2 = 4$.

The solution set consists of the x-values for which $y_1 < y_2$. We can see that $y_1 < y_2$ between the points of intersection of the graphs, or between $(-0.6666667, 4)$ and $(2, 4)$. The solution set of the inequality is thus the interval $(-0.6666667, 2)$, as indicated by the shading on the x-axis.

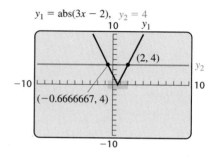

$y_1 = \text{abs}(3x - 2)$, $y_2 = 4$

Example 7   Given that $f(x) = |4x + 2|$, find all x for which $f(x) \geq 6$.

┌── *ALGEBRAIC SOLUTION*

We have

$$f(x) \geq 6,$$

or $|4x + 2| \geq 6.$ **Substituting**

To solve, we use part (c) of the principles listed on p. 249. In this case, X is $4x + 2$ and p is 6:

$$|X| \geq p$$

$|4x + 2| \geq 6$ Replacing X with $4x + 2$ and p with 6

$4x + 2 \leq -6 \quad or \quad 6 \leq 4x + 2$
  The number $4x + 2$ must be at least 6 units from 0.

$4x \leq -8 \quad or \quad 4 \leq 4x$ Adding -2

$x \leq -2 \quad or \quad 1 \leq x.$ Multiplying by $\frac{1}{4}$

The solution set is $\{x \mid x \leq -2 \ or \ x \geq 1\}$, or, in interval notation, $(-\infty, -2] \cup [1, \infty)$. The graph is as follows:

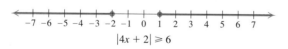

$|4x + 2| \geq 6$

┌── *GRAPHICAL SOLUTION*

To find all values of x for which $|4x + 2| \geq 6$, we graph $y_1 = \text{abs}(4x + 2)$ and $y_2 = 6$ and determine the points of intersection of the graphs.

The solution set consists of all x-values for which the graph of $y_1 = |4x + 2|$ is *on or above* the graph of $y_2 = 6$. The graph of y_1 lies *on* the graph of y_2 when $x = -2$ or $x = 1$. The graph of y_1 is *above* the graph of y_2 when $x < -2$ or $x > 1$. Thus the solution set is $(-\infty, -2] \cup [1, \infty)$, as indicated by the shading on the x-axis.

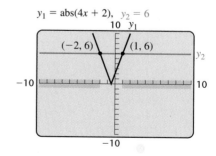

$y_1 = \text{abs}(4x + 2)$, $y_2 = 6$

4.3 Exercise Set

Use the following graph to solve Exercises 1–6.

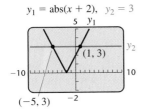

$y_1 = \text{abs}(x + 2), \quad y_2 = 3$

1. $|x + 2| = 3$

2. $|x + 2| \le 3$

3. $|x + 2| < 3$

4. $|x + 2| > 3$

5. $|x + 2| \ge 3$

6. $|x + 2| = -1$

Solve.

7. $|x| = 7$

8. $|x| = 9$

9. $|x| = -5$

10. $|x| = -3$

11. $|y| = 8.6$

12. $|p| = 0$

13. $|m| = 0$

14. $|t| = 5.5$

15. $|5x + 2| = 3$

16. $|2x - 3| = 4$

17. $|7x - 2| = -9$

18. $|3x - 10| = -8$

19. $|2y| - 5 = 13$

20. $|5x| - 3 = 37$

21. $7|z| + 2 = 16$

22. $5|q| - 2 = 9$

23. $|t - 7| + 3 = 4$

24. $|m + 5| + 9 = 16$

25. $3|2x - 5| - 7 = -1$

26. $5 - 2|3x - 4| = -5$

27. Let $f(x) = |3x - 4|$. Find all x for which $f(x) = 8$.

28. Let $f(x) = |2x - 7|$. Find all x for which $f(x) = 10$.

29. Let $f(x) = |x| - 2$. Find all x for which $f(x) = 6.3$.

30. Let $f(x) = |x| + 7$. Find all x for which $f(x) = 18$.

31. Let $f(x) = \left|\dfrac{3x - 2}{5}\right|$. Find all x for which $f(x) = 2$.

32. Let $f(x) = \left|\dfrac{1 - 2x}{3}\right|$. Find all x for which $f(x) = 1$.

Each pair of numbers represents two points on a number line. Find the distance between the points.

33. 25, 14

34. 32, 17

35. −9, 24

36. −18, −37

37. −8, −42

38. −9, −36

Solve.

39. $|x + 4| = |2x - 7|$

40. $|3x + 5| = |x - 6|$

41. $|x - 9| = |x + 6|$

42. $|x + 4| = |x - 3|$

43. $|5t + 7| = |4t + 3|$

44. $|3a - 1| = |2a + 4|$

45. $|n - 3| = |3 - n|$

46. $|y - 2| = |2 - y|$

47. $|7 - a| = |a + 5|$

48. $|6 - t| = |t + 7|$

Solve and graph.

49. $|a| \le 6$

50. $|x| < 2$

51. $|x| > 7$

52. $|a| \ge 3$

53. $|t| > 0$

54. $|t| \ge 1.7$

55. $|x - 3| < 5$

56. $|x - 1| < 3$

57. $|x - 3| + 2 \ge 7$

58. $|x - 4| + 5 \ge 10$

59. $|2y - 7| > -1$

60. $|3y - 4| > -8$

61. $|3a - 4| + 2 \ge 7$

62. $|2a - 5| + 1 \ge 8$

63. $|y - 3| < 12$

64. $|p - 2| < 3$

65. $9 - |x + 4| \le 5$

66. $12 - |x - 5| \le 9$

67. $|4 - 3y| > 8$

68. $|7 - 2y| < -6$

69. $|3 - 4x| < -5$

70. $7 + |4a - 5| \le 26$

71. $\left|\dfrac{2 - 5x}{4}\right| \ge \dfrac{2}{3}$

72. $\left|\dfrac{1 + 3x}{5}\right| > \dfrac{7}{8}$

73. $|m + 5| + 9 \le 16$

74. $|t - 7| + 3 \ge 4$

75. $25 - 2|a + 3| > 19$

76. $30 - 4|a + 2| > 12$

77. Let $f(x) = |2x - 3|$. Find all x for which $f(x) \le 4$.

78. Let $f(x) = |5x + 2|$. Find all x for which $f(x) \le 3$.

79. Let $f(x) = 2 + |3x - 4|$. Find all x for which $f(x) \ge 13$.

80. Let $f(x) = |2 - 9x|$. Find all x for which $f(x) \ge 25$.

81. Let $f(x) = 7 + |2x - 1|$. Find all x for which $f(x) < 16$.

82. Let $f(x) = 5 + |3x + 2|$. Find all x for which $f(x) < 19$.

Skill Maintenance

83. *Dinner Prices.* The Danville Volunteer Fire Department served 250 dinners. A child's dinner cost $3 and an adult's dinner cost $8. The total amount of money collected was $1410. How many of each type of dinner was served? [3.3]

84. *Agriculture.* The perimeter of a rectangular cornfield is 628 m. The length of the field is 6 m greater than the width. Find the area of the field. [3.3]

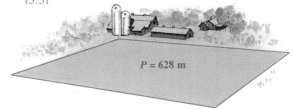

$P = 628$ m

85. Evaluate $2x^2 + 3y$ for $x = 5$ and $y = 6$. [1.1]

86. Evaluate $7a + 9b$ for $a = 8$ and $b = -3$. [1.1]

Synthesis

87. ◈ Isabel is using the following graph to solve $|x - 3| < 4$. How can you tell that a mistake has been made?

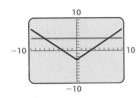

88. ◈ Explain why the inequality $|x + 7| < 1$ can be interpreted as "the distance between x and -7 is less than 1."

89. ◈ Explain in your own words why -7 is not a solution of $|x| < 5$.

90. ◈ Explain in your own words why $[6, \infty)$ is only part of the solution of $|x| \geq 6$.

91. From the definition of absolute value, $|x| = x$ only when $x \geq 0$. Thus, $|3x + 6| = 3x + 6$ only when $3x + 6 \geq 0$, which means that $x \geq -2$. Solve $|2x - 5| = 2x - 5$ using this same reasoning.

Solve.

92. $|x + 3| = x + 3$ **93.** $|7x - 2| = x + 4$

94. $|x + 2| > x$ **95.** $2 \leq |x - 1| \leq 5$

96. $|2x + 1| = 5 - x$ **97.** $|10x + 7| = 5x + 3$

98. $\left|\frac{1}{2}x - 0.4\right| = 5.4 + 2x$ **99.** $|x + 7.3| = 8.2 - x$

100. $|2x + 5| > |x - 4|$ **101.** $|x + 7| \leq \left|\frac{1}{2}x - 3\right|$

102. $2|x + 1| \geq |3.4 - x|$ **103.** $|9.3 - x| < |x - 1.5|$

Find an equivalent inequality with absolute value.

104. $-3 < x < 3$ **105.** $-5 \leq y \leq 5$

106. $x \leq -6 \text{ or } 6 \leq x$ **107.** $x < -4 \text{ or } 4 < x$

108. $x < -8 \text{ or } 2 < x$ **109.** $-5 < x < 1$

110. x is less than 2 units from 7.

111. x is more than 1 unit from 5.

112. *Motion of a Spring.* A weighted spring is bouncing up and down so that its distance d above the ground satisfies the inequality $|d - 6 \text{ ft}| \leq \frac{1}{2} \text{ ft}$ (see the figure below). Find all possible distances d.

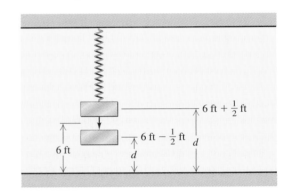

113. ◈ Is it possible for an equation in x of the form $|ax + b| = c$ to have exactly one solution? Why or why not?

4.4

Inequalities in Two Variables

- *Graphs of Linear Inequalities*
- *Systems of Linear Inequalities*

In Section 4.1, we graphed inequalities in one variable on a number line. Now we graph inequalities in two variables on a plane.

Graphs of Linear Inequalities

When the equals sign in a linear equation is replaced with an inequality sign, a **linear inequality** is formed. Solutions of linear inequalities are ordered pairs.

Example 1   Determine whether $(-3, 2)$ and $(6, -7)$ are solutions of the inequality $5x - 4y > 13$.

SOLUTION   Below, on the left, we replace x with -3 and y with 2. On the right, we replace x with 6 and y with -7.

$$
\begin{array}{c|c}
5x - 4y > 13 \\
\hline
5(-3) - 4 \cdot 2 \;?\; 13 \\
-15 - 8 \\
\hspace{1cm} -23 \;\big|\; 13 \quad \text{FALSE}
\end{array}
\qquad
\begin{array}{c|c}
5x - 4y > 13 \\
\hline
5(6) - 4(-7) \;?\; 13 \\
30 + 28 \\
\hspace{1cm} 58 \;\big|\; 13 \quad \text{TRUE}
\end{array}
$$

Since $-23 > 13$ is false, $(-3, 2)$ is not a solution.

Since $58 > 13$ is true, $(6, -7)$ is a solution.

The graph of a linear equation is a straight line. The graph of a linear inequality is a half-plane, bordered by the graph of the *related equation.* To find an inequality's related equation, we simply replace the inequality sign with an equals sign.

Example 2   Graph: $y \leq x$.

SOLUTION   We first graph the related equation $y = x$. Every solution of $y = x$ is an ordered pair, like $(3, 3)$, in which both coordinates are the same. The graph of $y = x$ is shown on the left below. Every point on the line is a solution of $y \leq x$, so the line is part of the graph of $y \leq x$.

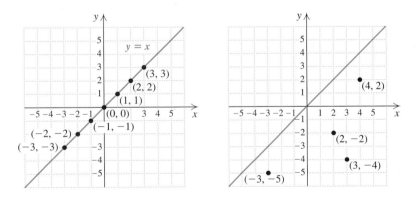

Notice that in the graph on the right above each ordered pair on the half-plane below $y = x$ contains a y-coordinate that is less than the x-coordinate. All these pairs represent solutions of $y \leq x$. We check one pair, $(4, 2)$, as follows:

$$
\begin{array}{c|c}
y \leq x \\
\hline
2 \;\big|\; 4 \quad \text{TRUE}
\end{array}
$$

It turns out that *any* point on the same side of $y = x$ as $(4, 2)$ is also a solution. Thus, if one point in a half-plane is a solution, then *all* points in that half-plane are solutions. We complete the drawing of the solution set by shading the half-plane below $y = x$. The complete solution set consists of the shaded half-plane and the line. Note too that for any inequality of the form $y \leq mx + b$ or $y < mx + b$, we shade *below* the graph of $y = mx + b$.

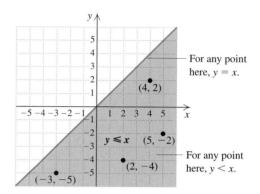

Example 3   Graph: $8x + 3y > 24$.

SOLUTION   First, we sketch the line $8x + 3y = 24$. Since the inequality sign is $>$, points on this line are not part of the graph of $8x + 3y > 24$, so the line is drawn dashed. Points representing solutions of $8x + 3y > 24$ are in either the half-plane above the line or the half-plane below the line. To determine which, we select a point that is not on the line and determine whether it is a solution of $8x + 3y > 24$. We try $(-3, 4)$ as a test point:

$$
\begin{array}{c|c}
\multicolumn{2}{c}{8x + 3y > 24} \\
\hline
8(-3) + 3 \cdot 4 \ ? \ 24 & \\
-24 + 12 & \\
-12 & 24 \quad \text{FALSE}
\end{array}
$$

Since $-12 > 24$ is *false*, $(-3, 4)$ is not a solution. Thus no point in the half-plane containing $(-3, 4)$ is a solution. The points in the other half-plane *are* solutions, so we shade that half-plane and obtain the following graph.

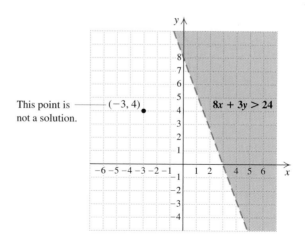

> *Steps for Graphing Linear Inequalities*
>
> 1. Replace the inequality sign with an equals sign and graph this related equation. If the inequality symbol is $<$ or $>$, draw the line dashed. If the inequality symbol is $\leq$ or $\geq$, draw the line solid.
> 2. The graph consists of a half-plane on one side or the other of the line, and, if the line is solid, the line as well. To determine which half-plane to shade, test a point not on the line. If that point is a solution of the inequality, shade the half-plane containing the point. If it is not, shade the other half-plane.

Many graphers can draw the solution set of an inequality in two variables. The procedures used vary widely. Some graphers allow you to choose to shade above or below a boundary line at the time that you enter the equation.

Example 4   Use a grapher to graph $8x + 3y > 24$ (see Example 3):

SOLUTION   First, we solve for y:

$$8x + 3y > 24$$
$$3y > -8x + 24$$
$$y > -\tfrac{8}{3}x + 8.$$

Next, we enter the related equation $y_1 = -\tfrac{8}{3}x + 8$ and move the cursor to the symbol before Y1. Since the inequality states that y is *greater than* $-\tfrac{8}{3}x + 8$, we shade the half-plane *above* the line $y = -\tfrac{8}{3}x + 8$. We press ENTER until the symbol indicates shading above a line. Now we can view the graph of the inequality.

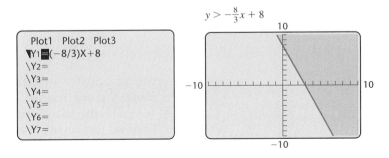

Example 5   Graph: $6x - 2y < 12$.

SOLUTION   We graph both by hand and using a grapher.

┌--- *BY HAND*

We first graph the related equation, $6x - 2y = 12$, as a dashed line. This line serves as the boundary line of the solution set of the inequality. To determine which half-plane to shade, we test a point *not* on the line. The pair $(0, 0)$ is easy to substitute:

$$\frac{6x - 2y < 12}{\begin{array}{l} 6 \cdot 0 - 2 \cdot 0 \;?\; 12 \\ \quad\quad 0 - 0 \mid \\ \quad\quad\quad 0 \mid 12 \quad \text{TRUE} \end{array}}$$

Since the inequality $0 < 12$ is *true*, the point $(0, 0)$ is a solution, as is every other point in the half-plane containing $(0, 0)$. The graph is shown below.

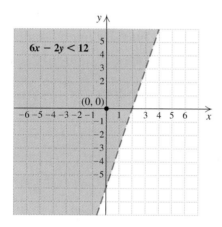

┌--- *USING A GRAPHER*

We solve the inequality for y:

$$6x - 2y < 12$$
$$-2y < 12 - 6x$$
$$y > -6 + 3x \text{ or } y > 3x - 6.$$

We enter the related equation, $y = 3x - 6$, into the grapher and shade the half-plane above the line.

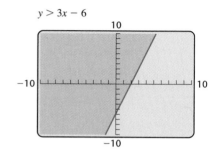

Note the grapher does not draw a dashed line, so the graph of $y > 3x - 6$ appears to be the same as the graph of $y \geq 3x - 6$.

Example 6 Graph $x > -3$ on a plane.

SOLUTION There is a missing variable in this inequality. If we graph the inequality on a line, its graph is as follows:

However, we can also write this inequality as $x + 0y > -3$ and graph it on a plane. We can use the same technique as in the examples above. First, we graph the related equation $x = -3$ in the plane, using a dashed line. Then we test some point, say, $(2, 5)$:

$$\frac{x + 0y > -3}{\begin{array}{l} 2 + 0 \cdot 5 \;?\; -3 \\ \quad\quad\quad 2 \mid -3 \quad \text{TRUE} \end{array}}$$

Since $(2, 5)$ is a solution, all points in the half-plane containing $(2, 5)$ are solutions. We shade that half-plane.

Another approach is to simply note that the solutions of $x > -3$ are all pairs with first coordinates greater than -3. Although many graphers can graph this inequality from the DRAW menu, it is simpler to graph it by hand.

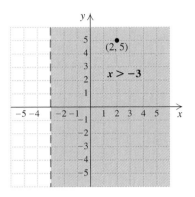

Example 7 Graph $y \le 4$ on a plane.

SOLUTION One approach is to graph $y = 4$, using a solid line, and then use $(2, -3)$ as a test point:

$$
\begin{array}{c}
0x + y \le 4 \\
\hline
0 \cdot 2 + (-3) \; ? \; 4 \\
-3 \; | \; 4 \quad \text{TRUE}
\end{array}
$$

We see that $(2, -3)$ is a solution, so all points in the half-plane containing $(2, -3)$ are solutions. Probably the best approach, however, is to simply note that the solutions of $y \le 4$ are all pairs with y-coordinates less than or equal to 4. A grapher is not needed for this example.

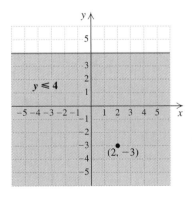

Systems of Linear Inequalities

To graph a system of equations, we graph the individual equations and then find the intersection of the individual graphs. We do the same thing for a system of inequalities, that is, we graph each inequality and find the intersection of the individual graphs.

Example 8 Graph the system

$$x + y \leq 4,$$
$$x - y < 4.$$

SOLUTION To graph $x + y \leq 4$, we graph $x + y = 4$ (or, equivalently, $y = -x + 4$) using a solid line. Since the test point $(0, 0)$ *is* a solution, we shade all points below the line red. The arrows near the ends of the line are another way of indicating the half-plane containing solutions.

Next, we graph $x - y < 4$ (or, equivalently, $y > x - 4$). We graph $x - y = 4$ using a dashed line and consider $(0, 0)$ as a test point. Again, $(0, 0)$ is a solution, so we shade that side of the line blue. The solution set of the system is the region that is shaded purple (both red and blue) and part of the line $x + y = 4$.

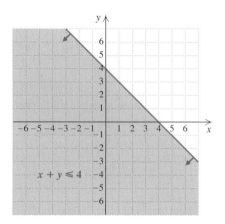 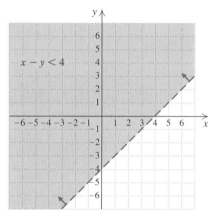 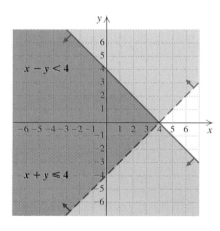

On a grapher, the intersection of the graphs of inequalities is shown by using different shading patterns.

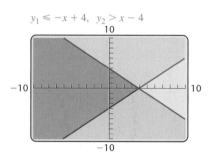

Example 9 Graph: $-2 < x \leq 3$.

SOLUTION This is a system of inequalities:

$$-2 < x,$$
$$x \leq 3.$$

We graph the equation $-2 = x$, and see that the graph of the first inequality is the half-plane to the right of the line $-2 = x$. It is shaded red.

We graph the second inequality, starting with the line $x = 3$, and find that its graph is the line and also the half-plane to its left. It is shaded blue.

The solution set of the system is the region that is the intersection of the individual graphs. Since it is shaded both blue and red, it appears to be purple. All points in this region have x-coordinates that are greater than -2 but do not exceed 3. Some graphers can generate this type of graph using the DRAW menu. Consult your user's manual for instructions.

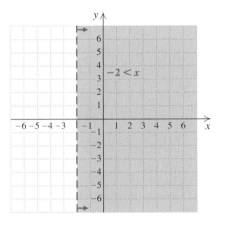

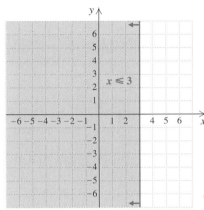

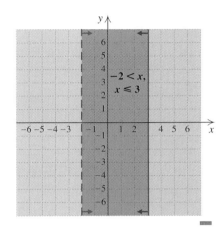

A system of inequalities may have a graph that consists of a polygon and its interior.

Example 10 Graph the system of inequalities. Find the coordinates of any vertices formed.

$$6x - 2y \le 12, \qquad (1)$$
$$y - 3 \le 0, \qquad (2)$$
$$x + y \ge 0. \qquad (3)$$

SOLUTION We graph both by hand and using a grapher.

┌-- BY HAND

We graph the lines

$$6x - 2y = 12,$$
$$y - 3 = 0,$$
and $\qquad x + y = 0$

using solid lines. The regions for each inequality are indicated by the arrows near the ends of the lines. We note where the regions overlap and shade the region of solutions purple.

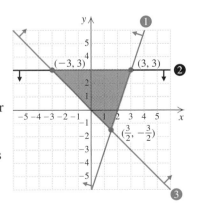

To find the vertices, we solve three different systems of equations. The system of related equations from inequalities (1) and (2) is

$$6x - 2y = 12,$$
$$y - 3 = 0.$$

Solving, we obtain the vertex (3, 3).

The system of related equations from inequalities (1) and (3) is

$$6x - 2y = 12,$$
$$x + y = 0.$$

Solving, we obtain the vertex $\left(\frac{3}{2}, -\frac{3}{2}\right)$.

The system of related equations from inequalities (2) and (3) is

$$y - 3 = 0,$$
$$x + y = 0.$$

Solving, we obtain the vertex (−3, 3).

USING A GRAPHER

First, we solve each inequality for y and obtain the equivalent system of inequalities

$$y_1 \geq (12 - 6x)/(-2),$$
$$y_2 \leq 3,$$
$$y_3 \geq -x.$$

We enter the related equations

$$y_1 = (12 - 6x)/(-2),$$
$$y_2 = 3,$$
and $$y_3 = -x$$

and shade the half-planes *above* y_1, *below* y_2, and *above* y_3. The intersection of the half-planes is the graph of the system.

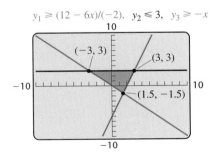

The vertices are the points of intersection of the graphs of the related equations y_1 and y_2, y_1 and y_3, and y_2 and y_3. We can find them by using INTERSECT and choosing the appropriate curves. The vertices are (3, 3), (1.5, −1.5), and (−3, 3).

4.4 Exercise Set

Determine whether each ordered pair is a solution of the given inequality.

1. $(-4, 2)$; $2x + 3y < -1$

2. $(3, -6)$; $4x + 2y \leq -2$

3. $(8, 14)$; $2y - 3x \geq 9$

4. $(7, 20)$; $3x - y > -1$

Graph on a plane.

5. $y < \frac{1}{2}x$

6. $y > 2x$

7. $y \leq x - 4$

8. $y < x + 3$

9. $y \geq x + 4$

10. $y > x - 2$

11. $x - y \geq 5$

12. $x + y < 4$

13. $2x + 3y < 6$

14. $3x + 4y \leq 12$

15. $2y - x \leq 4$

16. $2y - 3x > 6$

17. $2x - 2y \geq 8 + 2y$

18. $3x - 2 \leq 5x + y$

19. $x \leq 6$

20. $y > -3$

21. $-2 < y < 5$

22. $-4 < x < -1$

23. $-4 \leq x \leq 4$

24. $-3 \leq y \leq 4$

Graph using a grapher.

25. $y > x + 3.5$

26. $7y \leq 2x + 5$

27. $8x - 2y < 11$

28. $11x + 13y + 4 \geq 0$

Graph each system.

29. $y > x$,
$y < -x + 5$

30. $y < x$,
$y > -x + 1$

31. $y \geq x$,
$y \leq -x + 2$

32. $y \geq x$,
$y \leq -x + 4$

33. $y \leq -2$,
$x \geq -1$

34. $y \geq -3$,
$x \geq 1$

35. $x > -2$,
$y < -2x + 3$

36. $x < 3$,
$y > -3x + 2$

37. $y \leq 4$,
$y \geq -x + 2$

38. $y \geq -2$,
$y \geq x + 3$

39. $x + y \leq 3$,
$x - y \leq 4$

40. $x + y < 1$,
$x - y < 2$

41. $y + 3x > 0$,
$y + 3x < 2$

42. $y - 2x \geq 1$,
$y - 2x \leq 3$

Graph each system of inequalities. Find the coordinates of any vertices formed.

43. $y \leq 2x - 1$,
$y \geq -2x + 1$,
$x \leq 3$

44. $2y - x \leq 2$,
$y - 3x \geq -4$,
$y \geq -1$

45. $x + 2y \leq 12$,
$2x + y \leq 12$,
$x \geq 0$,
$y \geq 0$

46. $x - y \leq 2$,
$x + 2y \geq 8$,
$y \leq 4$

47. $8x + 5y \leq 40$,
$x + 2y \leq 8$,
$x \geq 0$,
$y \geq 0$

48. $4y - 3x \geq -12$,
$4y + 3x \geq -36$,
$y \leq 0$,
$x \leq 0$

49. $y - x \geq 1$,
$y - x \leq 3$,
$2 \leq x \leq 5$

50. $3x + 4y \geq 12$,
$5x + 6y \leq 30$,
$1 \leq x \leq 3$

Skill Maintenance

51. One side of a square is 5 less than a side of an equilateral triangle. If the perimeter of the square is the same as the perimeter of the triangle, what is the length of a side of the square? of a side of the triangle? [3.3]

Solve.

52. $4y - 3x = 8$,
$2x + 5y = -1$ [3.2]

53. $5(3x - 4) = -2(x + 5)$ [2.2]

54. $4(3x + 4) = 2 - x$ [2.2]

Synthesis

55. ◆ Do all systems of linear inequalities have solutions? Why or why not?

56. ◆ Explain how a system of linear inequalities could have a solution set containing exactly one pair.

57. ◆ Is $(0, 0)$ always the best test point to use when graphing inequalities? Why or why not?

58. ◆ When graphing linear inequalities, Ron makes a habit of always shading above the line when the symbol $\geq$ is used. Is this wise? Why or why not?

Graph.

59. $x + y > 8$,
$x + y \leq -2$

60. $x - 2y \le 0,$
$-2x + y \le 2,$
$x \le 2,$
$y \le 2,$
$x + y \le 4$

61. $x + y \ge 1,$
$-x + y \ge 2,$
$x \le 4,$
$y \ge 0,$
$y \le 4,$
$x \le 2$

62. Write four systems of four inequalities that describe a 2-unit by 2-unit square that has (0, 0) as one of the vertices.

63. *Luggage Size.* Unless an additional fee is paid, most major airlines will not check any luggage that is more than 62 in. long. The U.S. Postal Service will ship a package only if the sum of the package's length and girth (distance around its midsection) does not exceed 108 in. Concert Productions is ordering several 62-in. long trunks that will be both mailed and checked as luggage. Using *w* and *h* for width and height (in inches), respectively, write and graph an inequality that represents all acceptable combinations of width and height.

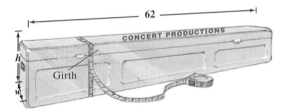

64. *Widths of a Basketball Floor.* Sizes of basketball floors vary due to building sizes and other constraints such as cost. The length *L* is to be at most 94 ft and the width *W* is to be at most 50 ft. Graph a system of inequalities that describes the possible dimensions of a basketball floor.

65. *Hockey Wins and Losses.* The Skating Stars figure that they need at least 60 points for the season in order to make the playoffs. A win is worth 2 points and a tie is worth 1 point. Graph a system of inequalities that describes the situation. (*Hint:* Let *w* = the number of wins and *t* = the number of ties.)

66. *Elevators.* Many elevators have a capacity of 1 metric ton (1000 kg). Suppose that *c* children, each weighing 35 kg, and *a* adults, each 75 kg, are on an elevator. Graph a system of inequalities that indicates when the elevator is overloaded.

Write a system of inequalities for each region shown.

67.

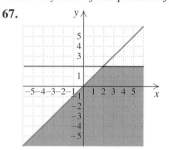

68.

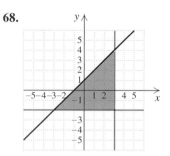

69.

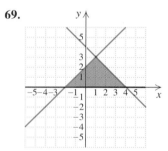

70.

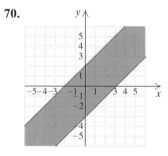

Focus: Linear inequalities

Time: 15–25 minutes

Group size: 3

Under a proposed "Rule of 85," full-time faculty in the California State Teachers Retirement System (kindergarten through community college) who are a years old with y years of service would have the option of retirement if $a + y \geq 85$.

Activity

1. Decide, as a group, how old you think someone would have to be in order to teach full-time. Express this age range as an inequality involving a.

2. Decide, as a group, the number of years someone could teach full-time before retiring. Ex-

press this answer as a compound inequality involving y.

3. Using the Rule of 85 and the answers to parts (1) and (2) above, write a system of inequalities. Then, using a scale of 5 yr per square, graph the system. To facilitate comparisons with graphs from other groups, plot a on the horizontal axis and y on the vertical axis.

4. Compare the graphs from all groups. Try to reach consensus on the graph that most clearly illustrates what the status would be of someone who would have the option of retirement under the Rule of 85.

5. If your instructor is agreeable to the idea, attempt to represent him or her with a point on your graph.

CHAPTER 4

Summary and Review

Key Terms

Inequality, p. 220
Solution, p. 220
Solution set, p. 220
Set-builder notation, p. 220
Interval notation, p. 221
Open interval, p. 221

Closed interval, p. 221
Half-open interval, p. 221
Compound inequality, p. 233
Intersection, p. 233
Conjunction, p. 233

Union, p. 236
Disjunction, p. 237
Absolute value, p. 244
Linear inequality, p. 252
Related equation, p. 253

Important Properties and Formulas

The Addition Principle for Inequalities

For any real numbers a, b, and c:

$$a < b \quad \text{is equivalent to} \quad a + c < b + c;$$
$$a > b \quad \text{is equivalent to} \quad a + c > b + c.$$

Similar statements hold for $\leq$ and $\geq$.

_____

The Multiplication Principle for Inequalities

For any real numbers a and b, and for any *positive* number c,

$$a < b \quad \text{is equivalent to} \quad ac < bc;$$
$$a > b \quad \text{is equivalent to} \quad ac > bc.$$

For any real numbers a and b, and for any *negative* number c,

$$a < b \quad \text{is equivalent to} \quad ac > bc;$$
$$a > b \quad \text{is equivalent to} \quad ac < bc.$$

Similar statements hold for $\leq$ and $\geq$.

_____

Set intersection:

$$A \cap B = \{x \mid x \text{ is in } A \text{ and } x \text{ is in } B\}$$

Set union:

$$A \cup B = \{x \mid x \text{ is in } A \text{ or in } B, \text{ or both}\}$$

For any real numbers a and b with $a < b$, $a < x$ and $x < b$ can be abbreviated $a < x < b$.

Intersection corresponds to "and"; union corresponds to "or."

_____

$$|x| = x \text{ if } x \geq 0; \qquad |x| = -x \text{ if } x < 0.$$

_____

The Absolute-Value Principles for Equations and Inequalities

For any positive number p and any algebraic expression X:

a) The solutions of $|X| = p$ are those numbers that satisfy $X = -p$ or $X = p$.

b) The solutions of $|X| < p$ are those numbers that satisfy $-p < X < p$.

c) The solutions of $|X| > p$ are those numbers that satisfy $X < -p$ or $p < X$.

If $|X| = 0$, then $X = 0$. If p is negative, then $|X| = p$ has no solution.

REVIEW EXERCISES

Graph each inequality and write the solution set using both set-builder and interval notation.

1. $x \leq -4$

2. $a + 7 \leq -14$

3. $y - 5 \geq -12$

4. $4y > -15$

5. $-0.3y < 9$

6. $-6x - 5 < 13$

7. $-\frac{1}{2}x - \frac{1}{4} > \frac{1}{2} - \frac{1}{4}x$

8. $0.3y - 7 < 2.6y + 15$

9. $-2(x - 5) \geq 6(x + 7) - 12$

10. Let $f(x) = 3x - 5$ and $g(x) = 11 - x$. Find all values of x for which $f(x) \leq g(x)$.

Solve.

11. Jessica can choose between two summer jobs. She can work as a checker in a discount store for $5.40 an hour, or she can mow lawns for $9.00 an hour. In order to mow lawns, she must buy a $450 lawn-mower. How many hours of labor will it take Jessica to make more money mowing lawns?

12. Jerry is going to invest $4500, part at 6% simple interest and the rest at 7%. What is the most he can invest at 6% and still be guaranteed $300 in interest each year?

13. Find the intersection:

$$\{1, 2, 5, 6, 9\} \cap \{1, 3, 5, 9\}.$$

14. Find the union:

$$\{1, 2, 5, 6, 9\} \cup \{1, 3, 5, 9\}.$$

Graph and write interval notation.

15. $x \le 3 \text{ and } x > -5$

16. $x \le 3 \text{ or } x > -5$

Solve and graph each solution set.

17. $-4 < x + 3 \le 5$

18. $-15 < -4x - 5 < 0$

19. $3x < -9 \text{ or } -5x < -5$

20. $2x + 5 < -17 \text{ or } -4x + 10 \le 34$

21. $2x + 7 \le -5 \text{ or } x + 7 \ge 15$

22. $f(x) < -5 \text{ or } f(x) > 5$, where $f(x) = 3 - 5x$

23. Write the domain of f using interval notation if $f(x) = \sqrt{x - 3}$.

Solve.

24. $|x| = 6$

25. $|x| \ge 3.5$

26. $|x - 2| = 7$

27. $|2x + 5| < 12$

28. $|3x - 4| \ge 15$

29. $|2x + 5| = |x - 9|$

30. $|5x + 6| = -8$

31. $\left| \dfrac{x + 4}{8} \right| \le 1$

32. $2|x - 5| - 7 > 3$

33. Let $f(x) = |3x - 5|$. Find all x for which $f(x) < 0$.

34. Graph $x - 2y \ge 6$ on a plane.

Graph each system of inequalities. Fin of any vertices formed.

35. $x + 3y > -1,$
     $x + 3y < 4$

36. $x - 3y \le 3,$
     $x + 3y \ge 9,$
     $y \le 6$

Skill Maintenance

Solve.

37. $5x - 4y = -10,$
     $4x + 2y = 5$

38. $3(x + 4) = 2(x - 5)$

39. Graph: $f(x) = -2x - 6$.

40. The perimeter of a rectangular field is 786 ft. The length is 9 ft longer than the width. Find the area of the field.

Synthesis

41. ◈ Explain in your own words why $|x| = p$ has two solutions when p is positive and no solution when p is negative.

42. ◈ Explain why the graph of the solution of a system of linear inequalities is the intersection, not the union, of the individual graphs.

43. Solve: $|2x + 5| \le |x + 3|$.

44. Classify as true or false: If $x < 3$, then $x^2 < 9$. If false, give an example showing why.

45. Just-For-Fun manufactures marbles with a 1.1-cm diameter and a ±0.03-cm manufacturing tolerance, or allowable variation in diameter. Write the tolerance as an inequality with absolute value.

CHAPTER 4 TEST

Graph each inequality and write the solution set using both set-builder and interval notation.

1. $x - 2 < 12$

2. $-0.6y < 30$

3. $-4y - 3 \ge 5$

4. $3a - 5 \le -2a + 6$

5. $4(5 - x) < 2x + 5$

6. $-8(2x + 3) + 6(4 - 5x) \ge 2(1 - 7x) - 4(4 + 6x)$

7. Let $f(x) = -5x - 1$ and $g(x) = -9x + 3$. Find all values of x for which $f(x) > g(x)$.

8. Lia can rent a van for either $40 per day with unlimited mileage or $30 per day with 100 free miles and an extra charge of 15¢ for each mile over

100. For what numbers of miles traveled would the unlimited mileage plan save Lia money?

9. A refrigeration repair company charges $40 for the first half-hour of work and $30 for each additional hour. Blue Mountain Camp has budgeted $100 to repair its walk-in cooler. For what lengths of a service call will the budget not be exceeded?

10. Find the intersection:

$$\{1, 3, 5, 7, 9\} \cap \{3, 5, 11, 13\}.$$

11. Find the union:

$$\{1, 3, 5, 7, 9\} \cup \{3, 5, 11, 13\}.$$

12. Write the domain of f using interval notation if $f(x) = \sqrt{7 - x}$.

Solve and graph each solution set.

13. $-3 < x - 2 < 4$

14. $-11 \leq -5x - 2 < 0$

15. $3x - 2 < 7$ *or* $x - 2 > 4$

16. $-3x > 12$ *or* $4x > -10$

17. $-\frac{1}{3} \leq \frac{1}{6}x - 1 < \frac{1}{4}$

18. $|x| = 9$

19. $|x| > 3$

20. $|4x - 1| < 4.5$

21. $|-5x - 3| \geq 10$

22. $|2 - 5x| = -10$

23. $g(x) < -3$ *or* $g(x) > 3$, where $g(x) = 4 - 2x$

24. Let $f(x) = |x + 10|$ and $g(x) = |x - 12|$. Find all values of x for which $f(x) = g(x)$.

Graph the system of inequalities. Find the coordinates of any vertices formed.

25. $x + y \geq 3$,
   $x - y \geq 5$

26. $2y - x \geq -7$,
   $2y + 3x \leq 15$,
   $y \leq 0$,
   $x \leq 0$

Skill Maintenance

Solve.

27. $4(x - 2) - 3(2x + 7) = 4$

28. $3x + 5y = 8$,
   $7x + 9y = -11$

29. Graph: $2x - y = -6$.

30. A disc jockey must play 15 commercials during 1 hr of a radio show. Each commercial is either 30 sec or 60 sec long. The total commercial time during that hour is 10 min. How many of each type of commercial were played?

Synthesis

Solve. Write the solution set using interval notation.

31. $|2x - 5| \leq 7$ *and* $|x - 2| \geq 2$

32. $7x < 8 - 3x < 6 + 7x$

Polynomials and Polynomial Functions 5

There are many different kinds of functions. One type that we have already studied in some detail is a *linear function*. In this chapter, we introduce *polynomial functions*. A linear function is one type of polynomial function. We will examine polynomials both algebraically and graphically.

A P P L I C A T I O N

DOG YEARS. A dog's life span is typically much shorter than that of a human. Given the following age equivalents for dogs and humans, use regression to find and graph a third-degree polynomial function h to determine the equivalent human age $h(x)$ for a dog that is x years old.

AGE OF DOG (IN YEARS)	HUMAN AGE (IN YEARS)
0.25	5
0.5	10
1	15
2	24
4	32
6	40
8	48
10	56
14	72
18	91
21	106

Source: Country, December 1992: 60

This problem appears as Example 5 in Section 5.7.

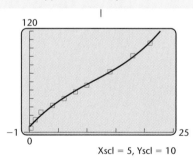

Xscl = 5, Yscl = 10

5.1

Introduction to Polynomials and Polynomial Functions

• *Algebraic Expressions and Polynomials*
• *Polynomial Functions*
• *Adding Polynomials*
• *Opposites and Subtraction*

In Chapter 2, we saw that a *linear function* is a function f that can be expressed in the form $f(x) = mx + b$. The graph of a linear function is a straight line. Its slope, or rate of change, is constant. We can think of linear functions as a "family" of functions, with the function $f(x) = x$ being the simplest such function.

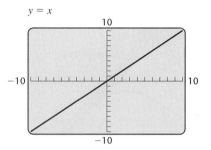
$y = x$

We have also studied some nonlinear functions. In Chapter 4, we solved equations and inequalities containing absolute-value symbols by examining graphs of absolute-value functions. An *absolute-value function* of the form $f(x) = |ax + b|$, where $a \neq 0$, has a graph similar to that of $f(x) = |x|$.

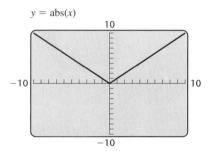
$y = \mathrm{abs}(x)$

Another nonlinear function that we considered in Chapter 4 when examining domains is a square-root function. A *square-root function* $f(x) = \sqrt{ax + b}$, where $a \neq 0$, has a graph similar to the graph of $f(x) = \sqrt{x}$.

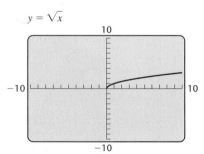
$y = \sqrt{x}$

These basic functions, along with others that we will develop throughout the text, form a **library of functions**. Knowing the general shape of the graphs of different types of functions will help in analyzing both data and graphs.

In this chapter, we look at polynomial functions. The graphs of polynomial functions have certain common characteristics.

Interactive Discovery

The following table lists some polynomial functions and some nonpolynomial functions. Graph each function and compare the graphs.

POLYNOMIAL FUNCTIONS	NONPOLYNOMIAL FUNCTIONS
$f(x) = x^2 + 3x + 5$ $f(x) = 4$ $f(x) = -0.5x^4 + 5x - 2.3$	$f(x) = \|x - 4\|$ $f(x) = 1 + \sqrt{2x - 5}$ $f(x) = \dfrac{x - 7}{2x}$

What are some characteristics of graphs of polynomial functions?

You may have noticed the following:

The graph of a polynomial function is "smooth," that is, there are no sharp corners.

The graph of a polynomial function is continuous, that is, there are no holes or breaks.

The domain of a polynomial function, unless otherwise specified, is all real numbers.

There are other characteristics of polynomial functions, but before discussing them, we need to establish some terminology.

Algebraic Expressions and Polynomials

A **term** is a number or a product of a number and a variable or variables raised to a power. Some examples of terms are

$$3, \quad x, \quad -5a, \quad xyz, \quad \tfrac{3}{13}a^2, \quad 0.2x^{-2}, \quad 9p^2q^3r^{-1}.$$

When all variables in a term are raised to whole-number powers, the term is a **monomial**. Of the expressions above, 3, x, $-5a$, xyz, and $\tfrac{3}{13}a^2$ are monomials.

 A **polynomial** is a sum of monomials. Each monomial in a polynomial is a *term* of the polynomial. A number like 5 in the term $5x^2$ is called the **coefficient** of that term, and the sum of the exponents of the variables in the term is the **degree of the term**. If a term contains no variables (is constant), it has degree 0. The only exception is the term 0, which is said to have no degree.

Polynomials, as well as terms, are described by their degree. The

leading term of a polynomial is the term of highest degree. Its coefficient is called the **leading coefficient**. The **degree of a polynomial** is the same as the degree of its leading term.

Example 1 State whether each expression is a polynomial. If it is, find the degree of each term, the degree of the polynomial, the leading term, and the leading coefficient.

a) $2x^3 + 8x^2 - 17x - 3$

b) $4x^3 + 9x^{-1}$

c) $\dfrac{3}{x^2 + 5}$

d) $6x^2 + 8x^2y^3 - 17xy - 24xy^2z^4 + 2y + 3$

SOLUTION Expressions (a) and (d) are polynomials, since each can be written as the sum of monomials. The variable in $9x^{-1}$ is raised to a negative exponent, so $9x^{-1}$ is not a monomial. Thus expression (b) is not a polynomial. Expression (c) cannot be written as the sum of monomials, so it is not a polynomial.

The degrees, leading terms, and leading coefficients of expressions (a) and (d) are shown below.

$$2x^3 + 8x^2 - 17x - 3 \qquad\qquad 6x^2 + 8x^2y^3 - 17xy - 24xy^2z^4 + 2y + 3$$

TERM	$2x^3$	$8x^2$	$-17x$	-3	$6x^2$	$8x^2y^3$	$-17xy$	$-24xy^2z^4$	$2y$	3
DEGREE	3	2	1	0	2	5	2	7	1	0
DEGREE OF POLYNOMIAL	3				7					
LEADING TERM	$2x^3$				$-24xy^2z^4$					
LEADING COEFFICIENT	2				-24					

If a polynomial contains only one variable, it is a *polynomial in one variable*. If it contains two variables, it is a *polynomial in two variables*. A polynomial in one variable is said to be **quadratic** if it is of degree 2 and **cubic** if it is of degree 3. A polynomial of degree 0 or 1 is called **linear**.

The following are some names for certain kinds of polynomials.

TYPE	DEFINITION	EXAMPLES
Monomial	A polynomial of one term	$4, -3p, 5x^2, -7a^2b^3, 0, xyz$
Binomial	A polynomial of two terms	$2x + 7, a - 3b, 5x^2 + 7y^3$
Trinomial	A polynomial of three terms	$x^2 - 7x + 12, 4a^2 + 2ab + b^2$

Polynomials in one variable are generally written so that the exponents *decrease* from left to right. This is called **descending order**. Some polynomials may be written with exponents *increasing* from left to right, which is **ascending order**. In general, if an exercise is written in one kind of order, the answer is written in that same order.

Example 2   Arrange in ascending order: $12 + 2x^3 - 7x + x^2$.

SOLUTION

$$12 + 2x^3 - 7x + x^2 = 12 - 7x + x^2 + 2x^3$$

Polynomials in several variables can be arranged with respect to the powers of one of the variables.

Example 3   Arrange in descending powers of x: $y^4 + 2 - 5x^2 + 3x^3y + 7xy^2$.

SOLUTION

$$y^4 + 2 - 5x^2 + 3x^3y + 7xy^2 = 3x^3y - 5x^2 + 7xy^2 + y^4 + 2$$

Polynomial Functions

A **polynomial function** is a function like

$$P(x) = 5x^7 + 3x^5 - 4x^2 - 5,$$

where the expression used to describe the function P is a polynomial. In this text, we concentrate on polynomial functions in one variable. The linear functions that we studied in Chapter 2 are polynomial functions, as are the quadratic functions that we will study in Chapter 8.

To evaluate a polynomial function, we find the function value at a specified value of x, just as we did in Chapter 2.

Example 4   Find $P(-5)$ for the polynomial function given by $P(x) = -x^2 + 4x - 1$.

SOLUTION   We evaluate the function using several methods discussed in Chapter 2.

Using algebraic substitution. We substitute -5 for x and carry out the operations using the rules for order of operations:

$$
\begin{aligned}
P(-5) &= -(-5)^2 + 4(-5) - 1 \\
&= -25 - 20 - 1 \qquad \text{We square the input before} \\
&\qquad\qquad\qquad\qquad \text{taking its opposite.} \\
&= -46.
\end{aligned}
$$

```
Y1(-5)
                    -46
```

Using a grapher. We let $y_1 = -x^2 + 4x - 1$. We enter the function into the grapher and evaluate Y1 using the YVARS menu.

Using a table. If $y_1 = -x^2 + 4x - 1$, we can find the value of y_1 for $x = -5$ by setting Indpnt to Ask in the TABLE SETUP.

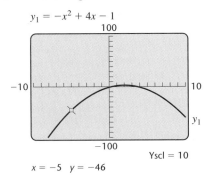

Using the graph of the function. We graph $y_1 = -x^2 + 4x - 1$ and find the value of y_1 for $x = -5$ using VALUE.

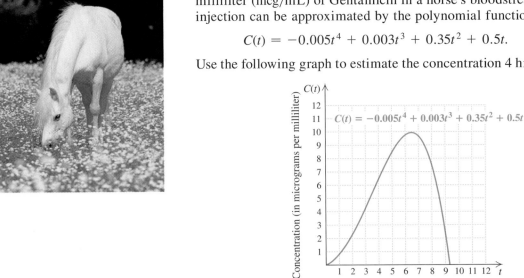

$$y_1 = -x^2 + 4x - 1$$

Yscl = 10

$x = -5 \quad y = -46$

No matter which method we use, we have $P(-5) = -46$. ▬

Sometimes an approximate function value read from a graph is sufficient.

Example 5 *Veterinary Medicine.* Gentamicin is an antibiotic frequently used by veterinarians. The concentration in micrograms per milliliter (mcg/mL) of Gentamicin in a horse's bloodstream t hours after injection can be approximated by the polynomial function given by

$$C(t) = -0.005t^4 + 0.003t^3 + 0.35t^2 + 0.5t.$$

Use the following graph to estimate the concentration 4 hr after injection.

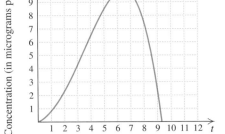

$$C(t) = -0.005t^4 + 0.003t^3 + 0.35t^2 + 0.5t$$

Time (in hours)

SOLUTION   To estimate $C(4)$, the concentration after 4 hr, we locate 4 on the horizontal axis. From there we move vertically to the graph of the function and then horizontally to the $C(t)$-axis. This locates a value of about 6.5. Thus,

$$C(4) \approx 6.5.$$

The concentration of Gentamicin 4 hr after injection is about 6.5 mcg/mL.

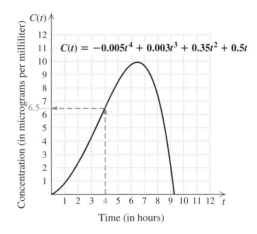

Recall from Section 2.7 that the domain of a function, when not specified, is the set of all real numbers for which the function is defined. Every polynomial function is defined for all real numbers. In other words,

The domain of a polynomial function is $(-\infty, \infty)$.

We can estimate the range of a polynomial function from its graph.

Example 6   Graph each function and estimate its range.

a) $p(x) = 2x^3 - 9x^2 + 5x - 7$ **b)** $q(x) = -5x^4 + 3x^2 + 6$

SOLUTION

a) We enter the function and graph. Since we are raising x to the third power, let's try a viewing window of $[-10, 10, -50, 50]$.

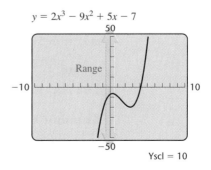

The range of p appears to be the set of all real numbers, or $(-\infty, \infty)$.

b) We first graph $q(x)$ using a viewing window of $[-10, 10, -50, 50]$. It appears that the greatest function value is about 5. We can magnify that part of the graph by changing the window dimensions to $[-5, 5, -10, 10]$.

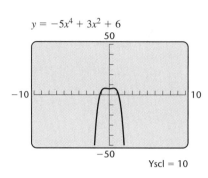

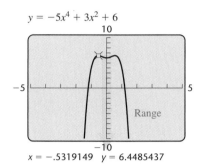

Using TRACE, we estimate that the greatest function value is 6.4. We could continue to TRACE and ZOOM to get a better estimate if we wish. The range of q is approximately $(-\infty, 6.4]$.

We can tell something about the range of a polynomial function from its degree.

Interactive Discovery

For each of the following functions, **(a)** tell whether its degree is even or odd and **(b)** estimate the range of the function.

1. $p(x) = x^3 - 7$ **2.** $p(x) = 2x^4 - 6x^3 - 2$

3. $p(x) = x^5 - 3x^4 + x^2 + 2$ **4.** $p(x) = -4x^3 + x - 3$

5. $p(x) = -6x^2 - x + 7$ **6.** $p(x) = 2x + 1.7$

Which polynomial functions have a range of $(-\infty, \infty)$?

The pattern you may have observed is true in general:

The range of a polynomial function of odd degree is $(-\infty, \infty)$.

The range of a polynomial function of even degree, n, $n \neq 0$, is $(-\infty, m]$ or $[m, \infty)$, for some real number m.

Note in Example 6(a) that the degree of $p(x) = 2x^3 - 9x^2 + 5x - 7$ is odd, and the range of p is $(-\infty, \infty)$. In Example 6(b), the degree of $q(x) = -5x^4 + 3x^2 + 6$ is even, and the range of q is approximately $(-\infty, 6.4]$.

We can tell from the graph of a polynomial function whether its degree is even or odd.

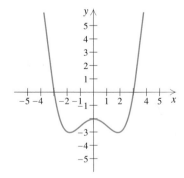

Example 7   The figure at left is the graph of a polynomial function. Determine whether the degree of the polynomial is even or odd.

SOLUTION   The range of the polynomial appears to be about $[-3, \infty)$. Thus the degree of the polynomial must be even.

Adding Polynomials

Recall from Section 1.3 that when two terms have the same variable(s) raised to the same power(s), they are **similar**, or **like**, **terms** and can be "combined" or "collected."

Example 8 Combine like terms.

a) $3x^2 - 4y + 2x^2$
b) $9x^3 + 5x - 4x^2 - 2x^3 + 5x^2$
c) $3x^2y + 5xy^2 - 3x^2y - xy^2$

SOLUTION

a) $3x^2 - 4y + 2x^2 = 3x^2 + 2x^2 - 4y$ **Rearranging terms using the commutative law for addition**

$$= (3 + 2)x^2 - 4y$$ **Using the distributive law**

$$= 5x^2 - 4y$$

b) $9x^3 + 5x - 4x^2 - 2x^3 + 5x^2 = 7x^3 + x^2 + 5x$ **We usually perform the middle steps mentally and write just the answer.**

c) $3x^2y + 5xy^2 - 3x^2y - xy^2 = 4xy^2$ —

The sum of two polynomials can be found by writing a plus sign between them and then combining like terms.

Example 9 Add: $(-3x^3 + 2x - 4) + (4x^3 + 3x^2 + 2)$.

SOLUTION

$$(-3x^3 + 2x - 4) + (4x^3 + 3x^2 + 2) = x^3 + 3x^2 + 2x - 2$$

We can check addition of polynomials in one variable using a grapher.

Interactive Discovery

In Example 9, we added polynomials to obtain the identity

$$(-3x^3 + 2x - 4) + (4x^3 + 3x^2 + 2) = x^3 + 3x^2 + 2x - 2.$$

Let

$$y_1 = (-3x^3 + 2x - 4) + (4x^3 + 3x^2 + 2)$$

and $y_2 = x^3 + 3x^2 + 2x - 2.$

If the addition is correct, how should the graphs of y_1 and y_2 compare? Check the addition by graphing y_1 and y_2.

Now let $y_3 = y_1 - y_2$. If our addition is correct, what will be the graph of y_3? Check Example 9 by graphing y_3.

As a third check, compare the values of y_1 and y_2 using a table.

Now check the addition

$$(-2x^4 + 3x^2 + 5) + (6x^4 - 2x - 8) = 4x^4 + 3x^2 - 2x.$$

Let

$$y_1 = (-2x^4 + 3x^2 + 5) + (6x^4 - 2x - 8)$$

and

$$y_2 = 4x^4 + 3x^2 - 2x.$$

First, graph y_1 and y_2 using a viewing window of $[-10, 10, -100, 100]$. Do the graphs coincide?

Now graph $y_3 = y_1 - y_2$. Is the graph of y_3 the x-axis? What advantage(s) does this check have over graphing y_1 and y_2?

Finally, check the addition using a table of values.

Graphs and tables of values provide good checks of the results of algebraic manipulation. When checking by graphing, remember that two different graphs might not appear different in some viewing windows. It is also possible for two different functions to have the same value for several entries in a table. However, when enough values of two polynomial functions are the same, a table does provide a foolproof check.

If the values of two nth-degree polynomial functions are the same for $n + 1$ or more values, then the polynomials are equivalent.

Using columns for addition is sometimes helpful. To do so, we write the polynomials one under the other, aligning like terms and leaving spaces for missing terms.

Example 10   Add: $4x^3 + 4x - 5$ and $-x^3 + 7x^2 - 2$. Check using a table of values.

SOLUTION   We have

$$\begin{array}{r} 4x^3 + 4x - 5 \\ -x^3 + 7x^2 - 2 \\ \hline 3x^3 + 7x^2 + 4x - 7 \end{array}$$

To check, we let

$$y_1 = (4x^3 + 4x - 5) + (-x^3 + 7x^2 - 2)$$

and $y_2 = 3x^3 + 7x^2 + 4x - 7.$

Because the polynomials are of degree 3, it suffices to show that $y_1 = y_2$ for four different x-values. This can be accomplished quickly by creating a table of values, as shown at left. The answer checks.

Note that if a grapher were not available, it would suffice to show, manually, that $y_1 = y_2$ for four different x-values.

X	Y1	Y2
0	−7	−7
1	7	7
2	53	53
3	149	149
4	313	313
5	563	563
6	917	917

X = 0

Opposites and Subtraction

If the sum of two polynomials is 0, the polynomials are *opposites,* or *additive inverses,* of each other. For example,

$$(3x^2 - 5x + 2) + (-3x^2 + 5x - 2) = 0,$$

so the opposite of $(3x^2 - 5x + 2)$ is $(-3x^2 + 5x - 2)$. We can say the same thing using algebraic symbolism, as follows:

$$\text{The opposite of } (3x^2 - 5x + 2) \text{ is } (-3x^2 + 5x - 2).$$

$$- \qquad (3x^2 - 5x + 2) = \quad -3x^2 + 5x - 2$$

To form the opposite of a polynomial, we can think of distributing the "$-$" sign, or multiplying each term of the polynomial by -1, and removing the parentheses. The effect is to change the sign of each term in the polynomial.

The Opposite of a Polynomial

The *opposite* of a polynomial P can be written as $-P$ or, equivalently, by replacing each term with its opposite.

Example 11 Write two equivalent expressions for the opposite of

$$7xy^2 - 6xy - 4y + 3.$$

SOLUTION

a) $-(7xy^2 - 6xy - 4y + 3)$ Writing the opposite of P as $-P$

b) $-7xy^2 + 6xy + 4y - 3$ Multiplying each term by -1 and removing parentheses

To subtract a polynomial, we add its opposite.

Example 12 Subtract: $(-5x^2 + 4) - (2x^2 + 3x - 1)$.

SOLUTION We have

$$(-5x^2 + 4) - (2x^2 + 3x - 1)$$
$$= (-5x^2 + 4) + (-2x^2 - 3x + 1) \qquad \text{Adding the opposite of the polynomial being subtracted}$$
$$= -7x^2 - 3x + 5.$$

To check, we let

$$y_1 = (-5x^2 + 4) - (2x^2 + 3x - 1),$$
$$y_2 = -7x^2 - 3x + 5,$$

and $y_3 = y_1 - y_2.$

X	Y3
0	0
1	0
2	0
3	0
4	0
5	0
6	0

X = 0

If our subtraction is correct, $y_3 = 0$, and the graph of y_3 will be the x-axis. Using a split screen, we can view both the graph and the table of values, as shown at left. It appears that the graph of y_3 is the x-axis, and this is confirmed by the table. We could also check by comparing the values of y_1 and y_2. Using either method, we find that the answer checks.

With practice, you will find that you can skip some steps, by mentally taking the opposite of each term and then combining like terms. Eventually, all you will write is the answer.

To use columns for subtraction, we mentally change the signs of the terms being subtracted.

Example 13 Subtract:

$$(4x^2y - 6x^3y^2 + x^2y^2) - (4x^2y + x^3y^2 + 3x^2y^3 - 8x^2y^2).$$

SOLUTION

Write: (Subtract)

$$4x^2y - 6x^3y^2 \qquad + x^2y^2$$
$$-(4x^2y + x^3y^2 + 3x^2y^3 - 8x^2y^2)$$

Think: (Add)

$$4x^2y - 6x^3y^2 \qquad + x^2y^2$$
$$-4x^2y - x^3y^2 - 3x^2y^3 + 8x^2y^2$$
$$\overline{\qquad -7x^3y^2 - 3x^2y^3 + 9x^2y^2}$$

Take the opposite of each term mentally and add.

5.1 Exercise Set

Determine whether each expression is a polynomial.

1. $3x - 7$

2. $-2x^5 + 9 - 7x^2$

3. $\dfrac{x^2 + x + 1}{x^3 - 7}$

4. -10

5. $\frac{1}{4}x^{10} - 8.6$

6. $3x^{-4} - x^{-1} + 13$

Determine the degree of each term and the degree of the polynomial.

7. $-7x^5 - x^3 + x^2 + 3x - 9$

8. $t^3 - 5t^2 + t + 1$

9. $a^5 + 4a^2b^4 + 6ab + 4a - 3$

10. $8p^6 + 2p^4t^4 - 7p^3t + 5p^2 - 14$

Arrange in descending order. Then find the leading term and the leading coefficient.

11. $15 - 4y^3 + 7y - 6y^2$

12. $2 - 5y + 6y^2 + 11y^3 - 18y^4$

13. $a + 5a^3 - a^7 - 19a^2 + 8a^5$

14. $a^3 - 7 + 11a^4 + a^9 - 5a^2$

Arrange in ascending powers of x.

15. $7x - 9 + 3x^4 - 5x^2$

16. $-3x^4 + 2x^3 - x + 9$

17. $-9x^3y + 3xy^3 + x^2y^2 + 2x^4$

18. $5x^2y^2 - 9xy + 8x^3y^2 - 5x^4$

Find the specified function values.

19. Find $P(4)$ and $P(0)$: $P(x) = 3x^2 - 2x + 5$.

20. Find $Q(3)$ and $Q(-1)$: $Q(x) = -4x^3 + 7x^2 - 1$.

21. Find $P(-2)$ and $P(\frac{1}{3})$: $P(y) = 8y^3 - 12y - 5$.

22. Find $Q(-3)$ and $Q(0)$:

$$Q(y) = 9y^3 + 8y^2 - 4y - 9.$$

Evaluate each polynomial for x = 4.

23. $-7x + 5$

24. $4x - 13$

25. $x^3 - 5x^2 + x$

26. $7 - x + 3x^2$

Evaluate each polynomial function for $x = -1$.

27. $f(x) = -5x^3 + 3x^2 - 4x - 3$

28. $g(x) = -4x^3 + 2x^2 + 5x - 7$

Electing Officers. For a club consisting of n people, the number of ways in which a president, vice president, and treasurer can be elected is given by the polynomial function

$$p(n) = n^3 - 3n^2 + 2n.$$

29. The Stage Right drama club has 12 members. In how many ways can a president, vice president, and treasurer be elected?

30. The Southside Rugby Club has 20 members. In how many ways can they elect a president, vice president, and treasurer?

Falling Distance. The distance $s(t)$, in feet, traveled by a body falling freely from rest in t seconds is approximated by the function

$$s(t) = 16t^2.$$

31. A paintbrush falls from a scaffold and takes 3 sec to hit the ground. How high is the scaffold?

$s(t) = 16t^2$

32. A stone is dropped from a cliff and takes 8 sec to hit the ground. How high is the cliff?

Total Revenue. An electronics firm is marketing a new kind of stereo. The firm determines that when it sells x stereos, its total revenue is

$$R(x) = 280x - 0.4x^2 \text{ dollars.}$$

33. What is the total revenue from the sale of 75 stereos?

34. What is the total revenue from the sale of 100 stereos?

Total Cost. The electronics firm determines that the total cost, in dollars, of producing x stereos is given by

$$C(x) = 5000 + 0.6x^2.$$

35. What is the total cost of producing 75 stereos?

36. What is the total cost of producing 100 stereos?

Daily Accidents. The number of daily accidents (the average number of accidents per day) involving drivers of age a is approximated by the polynomial function

$$P(a) = 0.4a^2 - 40a + 1039.$$

37. Find the number of daily accidents involving a 20-year-old driver.

38. Find the number of daily accidents involving a 25-year-old driver.

Medicine. Ibuprofen is a medication used to relieve pain. The polynomial function

$$M(t) = 0.5t^4 + 3.45t^3 - 96.65t^2 + 347.7t, \quad 0 \le t \le 6$$

can be used to estimate the number of milligrams of ibuprofen in the bloodstream t hours after 400 mg of the medication has been swallowed. Use the following graph for Exercises 39–42.

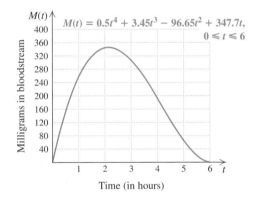

39. Estimate the number of milligrams of ibuprofen in the bloodstream 2 hr after 400 mg has been swallowed.

40. Estimate the number of milligrams of ibuprofen in the bloodstream 4 hr after 400 mg has been swallowed.

41. Approximate $M(5)$.

42. Approximate $M(3)$.

Surface Area of a Right Circular Cylinder. The surface area of a right circular cylinder is given by the polynomial

$$2\pi rh + 2\pi r^2,$$

where h is the height, r is the radius of the base, and h and r are given in the same units.

43. A 16-oz beverage can has height 6.3 in. and radius 1.2 in. Find the surface area of the can. (Use the $\boxed{\pi}$ key on a calculator.)

44. A 12-oz beverage can has height 4.7 in. and radius 1.2 in. Find the surface area of the can. (Use the $\boxed{\pi}$ key on a calculator.)

Graph each polynomial function and estimate its range.

45. $f(x) = x^2 + 2x + 1$

46. $p(x) = x^2 + x - 6$

47. $q(x) = -2x^2 + 5$

48. $g(x) = 1 - x^2$

49. $p(x) = -2x^3 + x + 5$

50. $f(x) = -x^4 + 2x^3 - 10$

51. $g(x) = x^4 + 2x^3 - 5$

52. $q(x) = x^5 + x^4 + x^3$

🔍 *The graphs in Exercises 53–56 are those of polynomial functions. Determine from the graph whether the degree of each polynomial is even or odd.*

53.

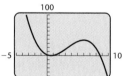

Yscl = 10

54.

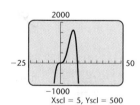

Xscl = 5, Yscl = 500

55.

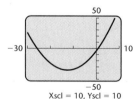

Xscl = 10, Yscl = 10

56.

500

-10 ┤├├├├├├├├┼├├├├├├├├┤ 10

-500

Yscl = 100

Combine like terms.

57. $4a + 7 - 4 + 2a^3 - 6a + 3$

58. $8x + 12 - 8 - 7x + 5x^2 + 10$

59. $3a^2b + 4b^2 - 9a^2b - 6b^2$

60. $5x^2y^2 + 4x^3 - 8x^2y^2 - 12x^3$

61. $8x^2 - 3xy + 12y^2 + x^2 - y^2 + 5xy + 4y^2$

62. $a^2 - 2ab + b^2 + 9a^2 + 5ab - 4b^2 + a^2$

Add.

63. $(5a + 6b - 3c) + (4a - 2b + 2c)$

64. $(6x - 5y + 3z) + (9x + 12y - 8z)$

65. $(x^2 + 2x - 3xy - 7) + (-3x^2 - x + 2xy + 6)$

66. $(3a^2 - 2b + ab + 6) + (-a^2 + 5b - 5ab - 2)$

67. $(2r^2 + 12r - 11) + (6r^2 - 2r + 4) + (r^2 - r - 2)$

68. $(5x^2 + 19x - 23) + (-7x^2 - 11x + 12) + (-x^2 - 9x + 8)$

69. $\left(\frac{1}{8}xy - \frac{3}{5}x^3y^2 + 4.3y^3\right) + \left(-\frac{1}{3}xy - \frac{3}{4}x^3y^2 - 2.9y^3\right)$

70. $\left(\frac{2}{3}xy + \frac{5}{6}xy^2 + 5.1x^2y\right) + \left(-\frac{4}{5}xy + \frac{3}{4}xy^2 - 3.4x^2y\right)$

Write two equivalent expressions for the opposite, or additive inverse, of each polynomial.

71. $5x^3 - 7x^2 + 3x - 6$

72. $-8y^4 - 18y^3 + 4y - 9$

73. $-12y^5 + 4ay^4 - 7by^2$

74. $7ax^3y^2 - 8by^4 - 7abx - 12ay$

Subtract.

75. $(8x - 4) - (-5x + 2)$

76. $(9y + 3) - (-4y - 2)$

77. $(-3x^2 + 2x + 9) - (x^2 + 5x - 4)$

78. $(-9y^2 + 4y + 8) - (4y^2 + 2y - 3)$

79. $(5a - 2b + c) - (3a + 2b - 2c)$

80. $(8x - 4y + z) - (4x + 6y - 3z)$

81. $(3x^2 - 2x - x^3) - (5x^2 - 8x - x^3)$

82. $(9y^2 - 14yz - 8z^2) - (12y^2 - 8yz + 4z^2)$

83. $\left(\frac{5}{8}x^4 - \frac{1}{4}x^2 - \frac{1}{2}\right) - \left(-\frac{3}{8}x^4 + \frac{3}{4}x^2 + \frac{1}{2}\right)$

84. $\left(\frac{5}{6}y^4 - \frac{1}{2}y^2 - 7.8y\right) - \left(-\frac{3}{8}y^4 + \frac{3}{4}y^2 + 3.4y\right)$

Total Profit. *Total profit is defined as total revenue minus total cost. In Exercises 85 and 86, R(x) and C(x) are the revenue and cost from the sale of x stereos.*

85. If $R(x) = 280x - 0.4x^2$ and $C(x) = 5000 + 0.6x^2$, find the profit from the sale of 70 stereos.

86. If $R(x) = 280x - 0.7x^2$ and $C(x) = 8000 + 0.5x^2$, find the profit from the sale of 100 stereos.

In Exercises 87–90, tell which of the following statements are true for the given polynomials.

a) The graphs of y_1 and y_2 coincide.
b) The graphs of y_1 and y_2 differ.
c) $y_1 - y_2 = 0$
d) $y_1 - y_2 \neq 0$

87. $y_1 = (5x^3 + 2x + 3) + (3x^3 - 1)$,
$y_2 = 8x^3 + 2x + 2$

88. $y_1 = (7x^2 - 9x + 3) - (x^3 - 9x + 3)$,
$y_2 = -x^3 + 7x^2$

89. $y_1 = (x^2 + 8x + 1) - (-x^2 + 3x + 5)$,
$y_2 = 5x - 4$

90. $y_1 = (x^4 + x^2 + 4) + (2x^3 - 3x^2 - 7)$,
$y_2 = 3x^4 - 2x^2 - 3$

Skill Maintenance

91. Graph: $f(x) = \frac{2}{3}x - 1$. [2.4]

92. Solve: $|2x - 3| = 7$. [4.3]

93. Multiply: $3(y - 2)$. [1.3]

94. Simplify: $3(x - 4) - 5(x + 16)$. [1.3]

Synthesis

95. ◆ Is the sum of two binomials always a binomial? Why or why not?

96. ◆ Ani claims that she can add any two polynomials but finds subtraction difficult. What advice would you offer her?

97. ◆ Three methods for checking with a grapher were described in the Interactive Discovery on p. 275. Give at least one advantage and one disadvantage of each method.

98. ◆ A student who is trying to graph
$$p(x) = 0.05x^4 - x^2 + 5$$
gets the following screen.

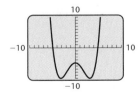

How can the student tell at a glance that a mistake has been made?

For P(x) and Q(x) as given, find the following.
$$P(x) = 13x^5 - 22x^4 - 36x^3 + 40x^2 - 16x + 75,$$
$$Q(x) = 42x^5 - 37x^4 + 50x^3 - 28x^2 + 34x + 100$$

99. $2[P(x)] + Q(x)$ **100.** $3[P(x)] - Q(x)$

101. $2[Q(x)] - 3[P(x)]$ **102.** $4[P(x)] + 3[Q(x)]$

103. *Volume of a Display.* The number of spheres in a triangular pyramid with x layers is given by the function
$$N(x) = \frac{1}{6}x^3 + \frac{1}{2}x^2 + \frac{1}{3}x.$$
The volume of a sphere of radius r is given by the function
$$V(r) = \frac{4}{3}\pi r^3.$$

Chocolate Heaven has a window display of truffles piled in a triangular pyramid formation 5 layers deep. If the diameter of each truffle is 3 cm, find the volume of chocolate in the display.

104. Find a polynomial function that gives the outside surface area of a box like this one, with an open top and dimensions as shown.

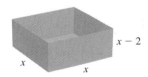

105. Develop a formula for the surface area of a right circular cylinder in which h is the height, in *centimeters*, and r is the radius, in *meters*. (See Exercises 43 and 44.)

Perform the indicated operation. Assume that the exponents are natural numbers.

106. $(2x^{2a} + 4x^a + 3) + (6x^{2a} + 3x^a + 4)$

107. $(3x^{6a} - 5x^{5a} + 4x^{3a} + 8) -$
$(2x^{6a} + 4x^{4a} + 3x^{3a} + 2x^{2a})$

108. $(2x^{5b} + 4x^{4b} + 3x^{3b} + 8) -$
$(x^{5b} + 2x^{3b} + 6x^{2b} + 9x^b + 8)$

GROUP SIZE	NUMBER OF HANDSHAKES
1	
2	
3	
4	
5	

Activity

1. All group members should shake hands with each other. Without "double counting," how many handshakes occurred?

2. Complete the table in the next column.

3. Join another group to determine the number of handshakes for a group of size 10.

4. Try to find a function of the form $H(n) = an^2 + bn$, for which $H(n)$ is the number of different handshakes that are possible in a group of n people. Make sure that $H(n)$ produces all of the values in the table above. (*Hint*: Look for a pattern in the table.)

5. Find H(class size).

5.2
Multiplication of Polynomials

- *Multiplying Monomials*
- *Multiplying Monomials and Binomials*
- *Multiplying Any Two Polynomials*
- *The Product of Two Binomials: FOIL*
- *Squares of Binomials*
- *Products of Sums and Differences*
- *Function Notation*

Just like numbers, polynomials can be multiplied. The product of two polynomials $P(x)$ and $Q(x)$ is a polynomial $R(x)$ that gives the same value as $P(x) \cdot Q(x)$ for any replacement of x.

Multiplying Monomials

To multiply monomials, we first multiply their coefficients. Then we multiply the variables using the rules for exponents and the commutative and associative laws. With practice, we can work mentally, writing only the answer.

Example 1   Multiply and simplify:

(a) $(-8x^4y^7)(5x^3y^2)$;

(b) $(-2x^2yz^5)(-6x^5y^{10}z^2)$.

SOLUTION

a) $(-8x^4y^7)(5x^3y^2) = -8 \cdot 5 \cdot x^4 \cdot x^3 \cdot y^7 \cdot y^2$ **Using the associative and commutative laws**

$$= -40x^{4+3}y^{7+2}$$ **Multiplying coefficients; adding exponents**

$$= -40x^7y^9$$

b) $(-2x^2yz^5)(-6x^5y^{10}z^2) = (-2)(-6) \cdot x^2 \cdot x^5 \cdot y \cdot y^{10} \cdot z^5 \cdot z^2$

$$= 12x^7y^{11}z^7$$ **Multiplying coefficients; adding exponents** ▬

Multiplying Monomials and Binomials

The distributive law is the basis for multiplying polynomials other than monomials. We first multiply a monomial and a binomial.

Example 2   Multiply: **(a)** $2x(3x - 5)$; **(b)** $3a^2b(a^2 - b^2)$.

SOLUTION

a) $2x \cdot (3x - 5) = 2x \cdot 3x - 2x \cdot 5$ **Using the distributive law**

$$= 6x^2 - 10x$$ **Multiplying monomials**

b) $3a^2b(a^2 - b^2) = 3a^2b \cdot a^2 - 3a^2b \cdot b^2$ **Using the distributive law**

$$= 3a^4b - 3a^2b^3$$ ▬

We can use graphs or tables to check the multiplication of polynomials in one variable. For example, we can check the product found in Example 2(a) by letting

$$y_1 = 2x(3x - 5)$$

and $y_2 = 6x^2 - 10x.$

The following table of values for y_1 and y_2 shows that the answer found is correct.

X	Y₁	Y₂
0	0	0
1	−4	−4
2	4	4
3	24	24
4	56	56
5	100	100
6	156	156
X = 0		

The distributive law is also used when multiplying two binomials. In this case, however, we begin by distributing a *binomial* rather than a monomial. With practice, some of the following steps can be combined.

Example 3   Let $P(x) = x^3 - 5$ and $Q(x) = 2x^3 + 4$. Find $P(x) \cdot Q(x)$.

SOLUTION   We have

$$P(x) \cdot Q(x) = (x^3 - 5)(2x^3 + 4)$$

$$= (x^3 - 5)2x^3 + (x^3 - 5)4 \qquad \text{``Distributing'' the } x^3 - 5$$

$$= 2x^3(x^3 - 5) + 4(x^3 - 5) \qquad \text{Using the commutative law for multiplication}$$

$$= 2x^3 \cdot x^3 - 2x^3 \cdot 5 + 4 \cdot x^3 - 4 \cdot 5 \qquad \text{Using the distributive law (twice)}$$

$$= 2x^6 - 10x^3 + 4x^3 - 20 \qquad \text{Multiplying the monomials}$$

$$= 2x^6 - 6x^3 - 20. \qquad \text{Combining like terms}$$

In this case, let's check by letting

$$y_1 = (x^3 - 5)(2x^3 + 4),$$

$$y_2 = 2x^6 - 6x^3 - 20,$$

and $y_3 = y_1 - y_2.$

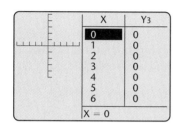

The graph and the table at left indicate that $y_3 = 0$, so the answer is correct.

Multiplying Any Two Polynomials

Repeated use of the distributive law enables us to multiply any two polynomials.

Example 4   Multiply: $(p + 2)(p^4 - 2p^3 + 3)$.

SOLUTION   By the distributive law, we have

$$(p + 2)(p^4 - 2p^3 + 3)$$

$$= (p + 2)(p^4) - (p + 2)(2p^3) + (p + 2)(3)$$

$$= p^4(p + 2) - 2p^3(p + 2) + 3(p + 2)$$

$$= p^4 \cdot p + p^4 \cdot 2 - 2p^3 \cdot p - 2p^3 \cdot 2 + 3 \cdot p + 3 \cdot 2$$

$$= p^5 + 2p^4 - 2p^4 - 4p^3 + 3p + 6$$

$$= p^5 - 4p^3 + 3p + 6. \qquad \text{Combining like terms}$$

In order for these polynomials to be equivalent, they must have the same value for at least 6 x-values, because the degree is 5. The table in the margin shows that this is true.

$y_1 = (x + 2)(x^4 - 2x^3 + 3),$
$y_2 = x^5 - 4x^3 + 3x + 6$

X	Y1	Y2
0	6	6
1	6	6
2	12	12
3	150	150
4	786	786
5	2646	2646
6	6936	6936
X = 0		

The Product of Two Polynomials

The *product* of two polynomials $P(x)$ and $Q(x)$ is found by multiplying each term of $P(x)$ by every term of $Q(x)$ and then combining like terms.

It is also possible to use columns, multiplying each term at the top by every term at the bottom, keeping like terms in columns, and leaving spaces for missing terms. Then we add.

$y_1 = (5x^3 + x - 4)(-2x^2 + 3x + 6),$
$y_2 = -10x^5 + 15x^4 + 28x^3 + 11x^2 - 6x - 24,$
$y_3 = y_1 - y_2$

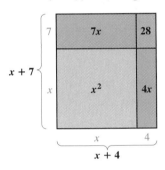

	X	Y3
	0	0
	1	0
	2	0
	3	0
	4	0
	5	0
	6	0
	X = 0	

Example 5   Multiply: $(5x^3 + x - 4)(-2x^2 + 3x + 6)$.

SOLUTION   We have

$$
\begin{array}{r}
5x^3 + x - 4 \\
-2x^2 + 3x + 6 \\
\hline
30x^3 + 6x - 24 \qquad \text{Multiplying by 6}\\
15x^4 + 3x^2 - 12x \text{Multiplying by } 3x\\
-10x^5 - 2x^3 + 8x^2 \text{Multiplying by } -2x^2\\
\hline
-10x^5 + 15x^4 + 28x^3 + 11x^2 - 6x - 24 \qquad \text{Adding}
\end{array}
$$

The graph and the table at left serve as a check.

The Product of Two Binomials: FOIL

We now consider what are called *special products*. These lead to faster ways to multiply in certain situations.

Let's find a faster special-product rule for the product of two binomials. Consider $(x + 7)(x + 4)$. We multiply each term of $(x + 7)$ by each term of $(x + 4)$.

$$(x + 7)(x + 4) = x \cdot x + x \cdot 4 + 7 \cdot x + 7 \cdot 4.$$

This multiplication illustrates a pattern that occurs whenever two binomials are multiplied:

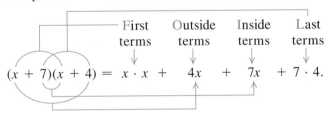

$$
\begin{array}{ccccc}
 & \text{First} & \text{Outside} & \text{Inside} & \text{Last} \\
 & \text{terms} & \text{terms} & \text{terms} & \text{terms} \\
 & \downarrow & \downarrow & \downarrow & \downarrow \\
(x + 7)(x + 4) = & x \cdot x + & 4x + & 7x & + 7 \cdot 4.
\end{array}
$$

We use the mnemonic device FOIL to remember this method for multiplying.

A visualization of
$(x + 7)(x + 4)$ using areas

$$
\begin{array}{c|c|c}
 & 7x & 28 \\
\hline
 & x^2 & 4x
\end{array}
$$

(labels: 7, $x + 7$, x on the left; x, 4, $x + 4$ along the bottom)

The FOIL Method

To multiply two binomials $A + B$ and $C + D$, multiply the First terms AC, the Outside terms AD, the Inside terms BC, and then the Last terms BD. Then combine like terms, if possible.

$$(A + B)(C + D) = AC + AD + BC + BD$$

1. Multiply First terms: AC.
2. Multiply Outside terms: AD.
3. Multiply Inside terms: BC.
4. Multiply Last terms: BD.

↓

FOIL

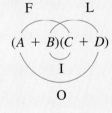

$$
\begin{array}{cc}
\text{F} & \text{L} \\
(A + B)(C + D) & \\
\text{I} & \\
\text{O} &
\end{array}
$$

Example 6 Multiply.

a) $(x + 5)(x - 8)$

b) $(2x + 3y)(x - 4y)$

c) $(5xy + 2x)(x^2 + 2xy^2)$

SOLUTION

$$\begin{array}{cccc} \text{F} & \text{O} & \text{I} & \text{L} \end{array}$$

a) $(x + 5)(x - 8) = x^2 - 8x + 5x - 40$

$\qquad\qquad\qquad = x^2 - 3x - 40$ **Combining like terms**

b) $(2x + 3y)(x - 4y) = 2x^2 - 8xy + 3xy - 12y^2$ **Using FOIL**

$\qquad\qquad\qquad\qquad = 2x^2 - 5xy - 12y^2$ **Combining like terms**

c) $(5xy + 2x)(x^2 + 2xy^2) = 5x^3y + 10x^2y^3 + 2x^3 + 4x^2y^2$

$\qquad\qquad\qquad\qquad$ **There are no like terms to combine.** ▬

Squares of Binomials

Interactive Discovery

Use a grapher to determine which of the following are identities.

1. $(x + 3)^2 = x^2 + 9$ 2. $(x + 3)^2 = x^2 + 6x + 9$

3. $(x - 3)^2 = x^2 - 6x - 9$ 4. $(x - 3)^2 = x^2 - 9$

5. $(x - 3)^2 = x^2 - 6x + 9$

A fast method for squaring binomials can be developed using FOIL:

$$(A + B)^2 = (A + B)(A + B)$$
$$= A^2 + AB + AB + B^2$$
$$= A^2 + 2AB + B^2;$$

$$(A - B)^2 = (A - B)(A - B)$$
$$= A^2 - AB - AB + B^2$$
$$= A^2 - 2AB + B^2.$$

A visualization of $(A + B)^2$ using areas

Squaring a Binomial

$$(A + B)^2 = A^2 + 2AB + B^2;$$
$$(A - B)^2 = A^2 - 2AB + B^2$$

The *square of a binomial* is the square of the first term, plus twice the product of the two terms, plus the square of the last term.

Example 7 Multiply: **(a)** $(x - 5)^2$; **(b)** $(2x + 3y)^2$; **(c)** $\left(\frac{1}{2}x - 3y^4\right)^2$.

SOLUTION

$$(A - B)^2 = A^2 - 2 \cdot A \cdot B + B^2$$

a) $(x - 5)^2 = x^2 - 2 \cdot x \cdot 5 + 5^2$

$$= x^2 - 10x + 25$$

> It can be helpful to memorize the words of the rules and say them while you are calculating.

b) $(2x + 3y)^2 = (2x)^2 + 2 \cdot 2x \cdot 3y + (3y)^2$

$$= 4x^2 + 12xy + 9y^2 \qquad \text{Raising a product to a power}$$

c) $\left(\frac{1}{2}x - 3y^4\right)^2 = \left(\frac{1}{2}x\right)^2 - 2 \cdot \frac{1}{2}x \cdot 3y^4 + (3y^4)^2$

$$= \frac{1}{4}x^2 - 3xy^4 + 9y^8 \qquad \text{Raising a product to a power; multiplying exponents}$$

Caution! Note that $(3 + 5)^2 \neq 3^2 + 5^2$ (since $64 \neq 9 + 25$). More generally,

$$(A + B)^2 \neq A^2 + B^2 \quad \text{and} \quad (A - B)^2 \neq A^2 - B^2.$$

Products of Sums and Differences

Another pattern emerges when we are multiplying a sum and a difference of the same two terms.

Interactive Discovery

Use a grapher to determine which of the following are identities.

Yes **1.** $(x + 3)(x - 3) = x^2 - 9$

2. $(x + 3)(x - 3) = x^2 + 9$

3. $(x + 3)(x - 3) = x^2 - 6x + 9$

4. $(x + 3)(x - 3) = x^2 + 6x + 9$

Using FOIL, we see that

$$\begin{array}{cccc} \text{F} & \text{O} & \text{I} & \text{L} \\ \downarrow & \downarrow & \downarrow & \downarrow \end{array}$$

$$(A + B)(A - B) = A^2 - AB + AB - B^2$$

$$= A^2 - B^2. \qquad -AB + AB = 0$$

The Product of a Sum and a Difference

$$(A + B)(A - B) = A^2 - B^2 \qquad \text{This is called a } difference \text{ of } two \text{ } squares.$$

The product of the sum and the difference of the same two terms is the square of the first term minus the square of the second term.

Example 8 Multiply.

a) $(x + 5)(x - 5)$
b) $(2xy^2 + 3x)(2xy^2 - 3x)$
c) $(0.2t - 1.4m)(0.2t + 1.4m)$
d) $\left(\frac{2}{3}n - m^3\right)\left(\frac{2}{3}n + m^3\right)$

SOLUTION

$$(A + B)(A - B) = A^2 - B^2$$

a) $(x + 5)(x - 5) = x^2 - 5^2$ Replacing A with x and B with 5

$\qquad\qquad\qquad = x^2 - 25$ Try to do problems like this mentally.

b) $(2xy^2 + 3x)(2xy^2 - 3x) = (2xy^2)^2 - (3x)^2$

$\qquad\qquad\qquad\qquad\qquad = 4x^2y^4 - 9x^2$ Raising a product to a power

c) $(0.2t - 1.4m)(0.2t + 1.4m) = (0.2t)^2 - (1.4m)^2$

$\qquad\qquad\qquad\qquad\qquad = 0.04t^2 - 1.96m^2$

d) $\left(\frac{2}{3}n - m^3\right)\left(\frac{2}{3}n + m^3\right) = \left(\frac{2}{3}n\right)^2 - (m^3)^2$

$\qquad\qquad\qquad\qquad\qquad = \frac{4}{9}n^2 - m^6$

—

Example 9 Multiply.

a) $(5y + 4 + 3x)(5y + 4 - 3x)$
b) $(3xy^2 + 4y)(-3xy^2 + 4y)$
c) $(a - 5b)(a + 5b)(a^2 - 25b^2)$

SOLUTION

a) $(5y + 4 + 3x)(5y + 4 - 3x) = (5y + 4)^2 - (3x)^2$

$\qquad\qquad\qquad\qquad\qquad = 25y^2 + 40y + 16 - 9x^2$

Note that $(5y + 4 + 3x)(5y + 4 - 3x)$ can be multiplied using columns, but not as quickly.

b) $(3xy^2 + 4y)(-3xy^2 + 4y) = (4y + 3xy^2)(4y - 3xy^2)$ Rewriting

$\qquad\qquad\qquad\qquad\qquad = (4y)^2 - (3xy^2)^2$

$\qquad\qquad\qquad\qquad\qquad = 16y^2 - 9x^2y^4$

c) $(a - 5b)(a + 5b)(a^2 - 25b^2) = (a^2 - 25b^2)(a^2 - 25b^2)$

$\qquad\qquad\qquad\qquad\qquad = (a^2 - 25b^2)^2$

$\qquad\qquad\qquad\qquad\qquad = (a^2)^2 - 2(a^2)(25b^2) + (25b^2)^2$

Squaring a binomial

$\qquad\qquad\qquad\qquad\qquad = a^4 - 50a^2b^2 + 625b^4$ —

Function Notation

Our work with multiplying can be used when evaluating functions.

Example 10 Given $f(x) = x^2 - 4x + 5$, find and simplify each of the following.

a) $f(a + 3)$

b) $f(a + h) - f(a)$

SOLUTION

a) To find $f(a + 3)$, we replace x with $a + 3$. Then we simplify:

$$f(a + 3) = (a + 3)^2 - 4(a + 3) + 5$$
$$= a^2 + 6a + 9 - 4a - 12 + 5$$
$$= a^2 + 2a + 2.$$

b) $f(a + h) - f(a) = [(a + h)^2 - 4(a + h) + 5] - [a^2 - 4a + 5]$
$$= a^2 + 2ah + h^2 - 4a - 4h + 5 - a^2 + 4a - 5$$
$$= 2ah + h^2 - 4h$$

5.2 Exercise Set

Multiply.

1. $3a^2 \cdot 7a$

2. $-5x^3 \cdot 2x$

3. $5x(-4x^2y)$

4. $-3ab^2(2a^2b^2)$

5. $(2x^3y^2)(-5x^2y^4)$

6. $(7a^2bc^4)(-8ab^3c^2)$

7. $8x(2 - x)$

8. $3a(a^2 - 4a)$

9. $5cd(3c^2d - 5cd^2)$

10. $a^2(2a^2 - 5a^3)$

11. $(2x + 5)(3x - 4)$

12. $(2a + 3b)(4a - b)$

13. $(m + 2n)(m - 3n)$

14. $(m - 5)(m + 5)$

15. $(a^2 - 2b^2)(a^2 - 3b^2)$

16. $(2m^2 - n^2)(3m^2 - 5n^2)$

17. $(x - 4)(x^2 + 4x + 16)$

18. $(y + 3)(y^2 - 3y + 9)$

19. $(a^2 + a - 1)(a^2 + 4a - 5)$

20. $(x^2 - 2x + 1)(x^2 + x + 2)$

21. $(4a^2b - 2ab + 3b^2)(ab - 2b + a)$

22. $(2x^2 + y^2 - 2xy)(x^2 - 2y^2 - xy)$

23. $\left(x - \frac{1}{2}\right)\left(x - \frac{1}{4}\right)$

24. $\left(b - \frac{1}{3}\right)\left(b - \frac{1}{3}\right)$

25. $(1.2x - 3y)(2.5x + 5y)$

26. $(40a - 0.24b)(0.3a + 10b)$

27. Let $P(x) = 3x^2 - 5$ and $Q(x) = 4x^2 - 7x + 2$. Find $P(x) \cdot Q(x)$.

28. Let $P(x) = x^2 - x + 1$ and $Q(x) = x^3 + x^2 + x + 2$. Find $P(x) \cdot Q(x)$.

Multiply.

29. $(a + 8)(a + 5)$

30. $(x + 3)(x + 2)$

31. $(y - 4)(y + 3)$

32. $(y - 1)(y + 5)$

33. $(x + 3)^2$

34. $(y - 7)^2$

35. $(x - 2y)^2$

36. $(2s + 3t)^2$

37. $(2x + 9)(x + 2)$

38. $(3b + 2)(2b - 5)$

39. $(10a - 0.12b)^2$

40. $(10p^2 + 2.3q)^2$

41. $\left(2a + \frac{1}{3}\right)^2$

42. $\left(3c - \frac{1}{2}\right)^2$

43. $(2x^3 - 3y^2)^2$

44. $(3s^2 + 4t^3)^2$

45. $(a^2b^2 + 1)^2$

46. $(x^2y - xy^2)^2$

47. Let $P(x) = 4x - 1$. Find $P(x) \cdot P(x)$.

48. Let $Q(x) = 3x^2 + 1$. Find $Q(x) \cdot Q(x)$.

Multiply.

49. $(c + 2)(c - 2)$

50. $(x - 3)(x + 3)$

51. $(2a + 1)(2a - 1)$

52. $(3 - 2x)(3 + 2x)$

53. $(x^3 + yz)(x^3 - yz)$

54. $(2a^3 + 5ab)(2a^3 - 5ab)$

55. $(-mn + m^2)(mn + m^2)$

56. $(-3b + a^2)(3b + a^2)$

57. $(x + 1)(x - 1)(x^2 + 1)$

58. $(y - 2)(y + 2)(y^2 + 4)$

59. $(a - b)(a + b)(a^2 - b^2)$

60. $(2x - y)(2x + y)(4x^2 - y^2)$

61. $(a + b + 1)(a + b - 1)$

62. $(m + n + 2)(m + n - 2)$

63. $(2x + 3y + 4)(2x + 3y - 4)$

64. $(3a - 2b + c)(3a - 2b - c)$

65. *Compounding Interest.* Suppose that P dollars is invested in a savings account at interest rate i, compounded annually, for 2 yr. The amount A in the account after 2 yr is given by

$$A = P(1 + i)^2.$$

Find an equivalent expression for A.

66. *Compounding Interest.* Suppose that P dollars is invested in a savings account at interest rate i, compounded semiannually, for 1 yr. The amount A in the account after 1 yr is given by

$$A = P\left(1 + \frac{i}{2}\right)^2.$$

Find an equivalent expression for A.

67. Given $f(x) = 5x + x^2$, find and simplify.

a) $f(t - 1)$

b) $f(a + h) - f(a)$

68. Given $f(x) = 4 + 3x - x^2$, find and simplify.

a) $f(p + 1)$

b) $f(a + h) - f(a)$

69. Given $f(x) = 2 - x^2$, find and simplify.

a) $f(2t + 1)$

b) $f(a + h) - f(a)$

70. Given $f(x) = x^2 - 3x$, find and simplify.

a) $f(3r - 2)$

b) $f(a + h) - f(a)$

71. Given $f(x) = 2x^2 + x + 5$, find and simplify.

a) $f(p - 5)$

b) $f(a + h) - f(a)$

72. Given $f(x) = -2x^2 - 5x + 7$, find and simplify.

a) $f(t + 1)$

b) $f(a + h) - f(a)$

Skill Maintenance

73. *Wages.* Takako worked a total of 17 days last month at her father's restaurant. She earned $50 a day during the week and $60 a day during the weekend. Last month Takako earned $940. How many weekdays did she work? [3.3]

74. *Geometry.* The perimeter of a triangle is 174. The lengths of the three sides are consecutive even numbers. What are the lengths of the sides of the triangle? [2.3]

75. *Manufacturing.* In a factory, there are three machines A, B, and C. When all three are running, they produce 222 suitcases per day. If A and B work but C does not, they produce 159 suitcases per day. If B and C work but A does not, they produce 147 suitcases. What is the daily production of each machine? [3.5]

76. *Value of Coins.* There are 50 dimes in a roll of dimes, 40 nickels in a roll of nickels, and 40 quarters in a roll of quarters. Kacie has 13 rolls of coins, which have a total value of $89. There are three more rolls of dimes than nickels. How many of each type of roll does she have? [3.5]

Synthesis

77. ◆ Find two binomials whose product is $x^2 - 25$ and explain how you decided on those two binomials.

78. ◆ Find two binomials whose product is $x^2 - 6x + 9$ and explain how you decided on those two binomials.

79. ◆ A student incorrectly claims that since $2x^2 \cdot 2x^2 = 4x^4$, it follows that $5x^5 \cdot 5x^5 = 25x^{25}$. What mistake is the student making?

80. ◆ We have seen that $(a - b)(a + b) = a^2 - b^2$. Explain how this result can be used to develop a fast way of multiplying $95 \cdot 105$.

81. Draw rectangles similar to those on p. 285 to show that $(x + 2)(x + 5) = x^2 + 7x + 10$.

82. Use a grapher to determine whether each of the following is an identity.

a) $(x - 1)^2 = x^2 - 1$

b) $(x - 2)(x + 3) = x^2 + x - 6$

c) $(x - 1)^3 = x^3 - 3x^2 + 3x - 1$

d) $(x + 1)^4 = x^4 + 1$

e) $(x + 1)^4 = x^4 + 4x^3 + 8x^2 + 4x + 1$

Multiply. Assume that variables in exponents represent natural numbers.

83. $(ab^{3n})^{2n}$

84. $[(-x^a y^b)^4]^a$

85. $(z^{n^2})^{n^3}(z^{4n^3})^{n^2}$

86. $(a^x b^{2y})\left(\frac{1}{2}a^{3x}b\right)^2$

87. $(a^x b^y)^{w+z}$

88. $y^3 z^n(y^{3n}z^3 - 4yz^{2n})$

89. $[(a + b)(a - b)][5 - (a + b)][5 + (a + b)]$

90. $[x + y + 1][x^2 - x(y + 1) + (y + 1)^2]$

91. $(y - 1)^6(y + 1)^6$

92. $(a - b + c - d)(a + b + c + d)$

93. $\left(\frac{2}{3}x + \frac{1}{3}y + 1\right)\left(\frac{2}{3}x - \frac{1}{3}y - 1\right)$

94. $\left(x - \frac{1}{7}\right)\left(x^2 + \frac{1}{7}x + \frac{1}{49}\right)$

95. $(4x^2 + 2xy + y^2)(4x^2 - 2xy + y^2)$

96. $(x^2 - 7x + 12)(x^2 + 7x + 12)$

97. $(x^a + y^b)(x^a - y^b)(x^{2a} + y^{2b})$

98. $(x - 1)(x^2 + x + 1)(x^3 + 1)$

99. $(x^{a-b})^{a+b}$

100. $(M^{x+y})^{x+y}$

COLLABORATIVE CORNER

Focus: Polynomial multiplication

Time: 15–20 minutes

Group size: 2

Consider the following dialogue:

Jinny: Cal, let me do a number trick with you. Think of a number between 1 and 7. I'll have you perform some manipulations to this number, you'll tell me the result, and I'll tell you your number.

Cal: Okay. I've thought of a number.

Jinny: Good. Write it down so I can't see it, double it, and then subtract x from the result.

Cal: Hey, this is algebra!

Jinny: I know. Now square your binomial and subtract x^2.

Cal: How did you know I had an x^2? I *thought* this was rigged!

Jinny: It is. Now, divide by 4 and tell me either your constant term or your x-term. I'll tell you the other term and the number you chose.

Cal: Okay. The constant term is 16.

Jinny: Then the other term is $-4x$ and the number you chose was 4.

Cal: You're right! How did you do it?

Activity

1. Each group member should follow Jinny's instructions. Then determine how Jinny determined Cal's number and the other term.

2. Suppose that, at the end, Cal told Jinny the x-term. How would Jinny have determined Cal's number and the other term?

3. Would Jinny's "trick" work with *any* real number? Why do you think she specified numbers between 1 and 7?

4. Each group member should create a new number "trick" and perform it on the other group member. Be sure to include a variable so that both members can gain practice with polynomials.

5.3
Polynomial Equations and Factoring

- *Graphical Solutions*
- *The Principle of Zero Products*
- *Factoring Out Common Factors*
- *Factoring by Grouping*
- *Factoring and Equations*

Whenever two polynomials are set equal to each other, the result is a **polynomial equation**. In this section, we learn how to solve such equations both graphically and algebraically by *factoring*.

Graphical Solutions

We can find real-number solutions of a polynomial equation by graphing both sides of the equation, much as we did in Section 2.2.

Example 1   Solve: $x^2 = 6x$.

SOLUTION   We graph $f(x) = x^2$ and $g(x) = 6x$ and find the coordinates of any points of intersection.

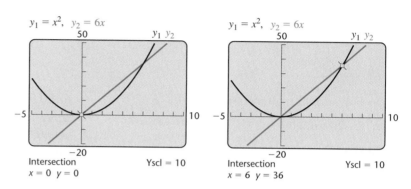

The x-coordinates of the solutions are 0 and 6. We check both in the original equation.

CHECK: *For 0:*

$$x^2 = 6x$$

$$(0)^2 \ ? \ 6(0)$$

$$0 \ | \ 0 \qquad \text{TRUE}$$

For 6:

$$x^2 = 6x$$

$$(6)^2 \ ? \ 6(6)$$

$$36 \ | \ 36 \qquad \text{TRUE}$$

Both numbers check, so the solutions are 0 and 6.

We can also solve the equation in Example 1 by rewriting the equation so that one side is 0:

$$x^2 = 6x$$

$$x^2 - 6x = 0. \qquad \textbf{Adding } -6x \textbf{ on both sides}$$

To solve $x^2 - 6x = 0$, we graph the function $g(x) = x^2 - 6x$ and look for values of x for which $g(x) = 0$. One advantage of this method is that we need not worry about setting the y-dimensions, since we are searching only the x-axis.

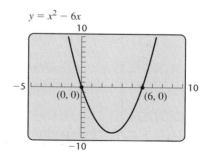

We see from the graph that $g(x) = 0$ when $x = 0$ or $x = 6$. The values 0 and 6 are called **zeros** of the function $g(x)$. They are also referred to as **roots** of the equation $g(x) = 0$.

Zeros and Roots

The x-values for which a function $f(x)$ is 0 are called the *zeros* of the function.

The x-values for which an equation such as $f(x) = 0$ is true are called the *roots* of the equation.

In this chapter, we consider only the real-number zeros of functions. The zeros of a function are the x-coordinates of any points at which the graph of the function crosses the x-axis. Thus the zeros of $g(x) = x^2 - 6x$ are 0 and 6. We can solve, or find the roots of, the equation $g(x) = 0$ by finding the zeros of the function g.

Example 2   Find the zeros of the function given by $f(x) = x^3 - 3x^2 - 4x + 12$.

SOLUTION   First, we graph the equation $y = x^3 - 3x^2 - 4x + 12$, choosing a viewing window that shows the x-intercepts of the graph. It may require trial and error to choose an appropriate viewing window. We might try the standard viewing window first, as shown on the left below. It appears that there are three zeros, but we cannot see the shape of the graph well; Ymax should be increased. We might next try a viewing window of $[-10, 10, -100, 100]$, with Yscl $= 10$, as shown in the middle below. Here we get a better idea of the overall shape of the graph, but we cannot see all three x-intercepts clearly.

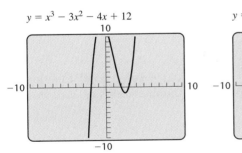

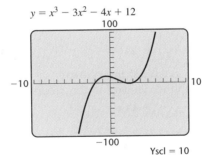

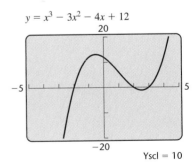

We magnify the portion of the graph close to the origin by making the window dimensions smaller. The window shown on the right above, $[-5, 5, -20, 20]$, is a good choice for viewing the zeros of this function. There are other good choices as well. The zeros of the function seem to be about -2, 2, and 3.

Many graphers have a zero option in the CALC menu that will calculate an x-intercept of a graph. You must supply a nearby x-value to the left and one to the right of each x-intercept. The function values of the boundaries supplied should have different signs.

To find the zero that appears to be about -2, we choose a Left Bound to the left of -2 on the x-axis. Next, we choose a Right Bound to the right of -2 on the x-axis. For a Guess, we choose a point close to the apparent root. We see that -2 is indeed a zero of the function f.

$y = x^3 - 3x^2 - 4x + 12$

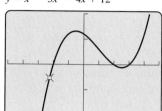

$y = x^3 - 3x^2 - 4x + 12$

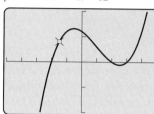

$y = x^3 - 3x^2 - 4x + 12$

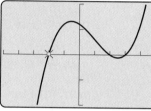

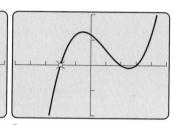

Left Bound?
$x = -2.234043$ $y = -5.186654$

Right Bound?
$x = -1.489362$ $y = 7.9991524$

Guess?
$x = -2.021277$ $y = -.4296158$

Zero
$x = -2$ $y = 0$

Using the same procedure for each of the other two zeros, we find that the zeros of the function $f(x) = x^3 - 3x^2 - 4x + 12$ are -2, 2, and 3.

We saw in the discussion after Example 1 that $g(x) = x^2 - 6x$ has 2 zeros. We also saw in Example 2 that $f(x) = x^3 - 3x^2 - 4x + 12$ has 3 zeros. We can tell something about the number of zeros of a polynomial function from its degree.

Interactive Discovery

Each of the following is a second-degree polynomial function. Use a graph to determine the *number* of real-number zeros of each function.

1. $f(x) = x^2 - 2x + 1$

2. $g(x) = x^2 + 9$

3. $h(x) = 2x^2 + 7x - 4$

Each of the following is a third-degree polynomial function. Use a graph to determine the *number* of real-number zeros of each function.

4. $f(x) = 2x^3 - 5x^2 - 3x$

5. $g(x) = x^3 - 7x^2 + 16x - 12$

6. $h(x) = x^3 - 4x^2 + 5x - 20$

Compare the degree of each function with the number of real-number zeros of that function. What conclusion can you draw?

A second-degree polynomial function will have 0, 1, or 2 real-number zeros. A third-degree polynomial function will have 1, 2, or 3 real-number zeros. This result can be generalized.

> An nth-degree polynomial function will have at most n zeros.

"Seeing" all the zeros of a polynomial function when solving an equation graphically depends on a good choice of viewing window. Many polynomial equations can be solved algebraically. One principle used in solving polynomial equations is the *principle of zero products*.

The Principle of Zero Products

When a polynomial is written as a product, we say that it is *factored*. The product is called a *factorization* of the polynomial. For example,

$$3x(x + 4)$$

is a factorization of

$$3x^2 + 12x$$

since $3x(x + 4)$ is a product and

$$3x(x + 4) = 3x^2 + 12x.$$

The polynomials $3x$ and $x + 4$ are *factors* of $3x^2 + 12x$. The zeros of a polynomial function are related to the factorization of the polynomial.

Interactive Discovery

Consider the function f given by $f(x) = 3x^2 + 12x$. If $g(x) = 3x$ and $h(x) = x + 4$, then $f(x) = g(x) \cdot h(x)$.

Graph $f(x)$ and find the zeros. Now using the same window, graph $g(x)$ and $h(x)$ and find the zero of each function. How are the zeros of $f(x)$ and the zeros of $g(x)$ and $h(x)$ related?

We see that if a polynomial function f is given by $f(x) = g(x) \cdot h(x)$, then $f(x) = 0$ at the same x-values for which $g(x) = 0$ or $h(x) = 0$.

The zeros of a polynomial function are the zeros of the functions described by the factors of the polynomial.

This result leads us to the following principle, which is used to solve factored polynomials algebraically.

The Principle of Zero Products

For any real numbers a and b:

If $ab = 0$, then $a = 0$ or $b = 0$ (or both).

If $a = 0$ or $b = 0$, then $ab = 0$.

The principle of zero products is based on a multiplication property of 0. When we multiply two or more numbers, the product is 0 if one of the factors is 0. Conversely, if a product is 0, then at least one of the factors must be 0. Note that in order to solve a polynomial equation using the principle of zero products, we must have 0 on one side of the equation and the other side must be factored.

Example 3   Solve: $(x - 3)(x + 2) = 0$.

SOLUTION   The principle of zero products says that in order for $(x - 3)(x + 2)$ to be 0, at least one factor must be 0. Thus,

$$x - 3 = 0 \quad or \quad x + 2 = 0. \qquad \text{Using the principle of zero products}$$

Each of these linear equations is then solved separately:

$$x = 3 \quad or \quad x = -2.$$

We check both algebraically and graphically, as follows:

ALGEBRAIC CHECK

For 3:

$$\frac{(x - 3)(x + 2) = 0}{(3 - 3)(3 + 2) \; ? \; 0}$$
$$0(5) \quad \Big|$$
$$0 \quad \Big| \quad 0 \quad \text{TRUE}$$

For −2:

$$\frac{(x - 3)(x + 2) = 0}{(-2 - 3)(-2 + 2) \; ? \; 0}$$
$$(-5)(0) \quad \Big|$$
$$0 \quad \Big| \quad 0 \quad \text{TRUE}$$

GRAPHICAL CHECK

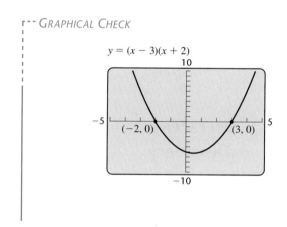

$y = (x - 3)(x + 2)$

The solutions are -2 and 3.

Example 4   Given that $f(x) = x(3x + 2)$, find all the values of a for which $f(a) = 0$.

SOLUTION   We are looking for all numbers a for which $f(a) = 0$. Since $f(a) = a(3a + 2)$, we must have

$$a(3a + 2) = 0 \qquad \text{Setting } f(a) \text{ equal to } 0$$
$$a = 0 \quad or \quad 3a + 2 = 0 \qquad \text{Using the principle of zero products}$$
$$a = 0 \quad or \qquad a = -\tfrac{2}{3}.$$

We check by evaluating $f(0)$ and $f\left(-\tfrac{2}{3}\right)$. One way to check with a grapher is to enter $y_1 = x(3x + 2)$ and calculate Y1(0) and Y1(−2/3).

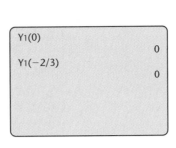

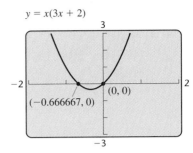

$y = x(3x + 2)$

We can also graph $f(x) = x(3x + 2)$ and observe that its zeros are approximately $-\frac{2}{3}$ and 0. Thus, to have $f(a) = 0$, we must have $a = 0$ or $a = \frac{2}{3}$.

If the polynomial in an equation of the form $p(x) = 0$ is not in factored form, we must factor it before we can use the principle of zero products to solve the equation. To **factor** an expression means to write an equivalent expression that is a product.

Factoring Out Common Factors

When factoring, we first look for factors common to every term in an expression and then use the distributive law.

Example 5 Factor out a common factor: $4x^2 - 8$.

SOLUTION

$$4x^2 - 8 = 4 \cdot x^2 - 4 \cdot 2 \qquad \textbf{Noting that 4 is a common factor}$$
$$= 4(x^2 - 2) \qquad \textbf{Using the distributive law}$$

In some cases, there is more than one common factor. In Example 6(a) below, for instance, 5 is a common factor, x^3 is a common factor, and $5x^3$ is a common factor. If there is more than one common factor, we factor out the *largest common factor*, that is, the factor that has the largest coefficient and the highest degree. In Example 6(a), the largest common factor is $5x^3$.

Example 6 Factor out a common factor.

a) $5x^4 + 20x^3$ **b)** $12x^2y - 20x^3y$ **c)** $10p^6q^2 - 4p^5q^3 - 2p^4q^4$

SOLUTION

a) $5x^4 + 20x^3 = 5x^3(x + 4)$ **Try to write your answer directly. Multiply mentally to check your answer.**

We can always check a factorization by multiplying. As another check, we can compare graphs or a table of values.

ALGEBRAIC CHECK

$$5x^3(x + 4) = 5x^3 \cdot x + 5x^3 \cdot 4$$
$$= 5x^4 + 20x^3$$

GRAPHICAL CHECK

$y_1 = 5x^4 + 20x^3,\ \ y_2 = (5x^3)(x + 4),$
$y_3 = y_1 - y_2$

X	Y₃
0	0
1	0
2	0
3	0
4	0
5	0
6	0

X = 0

Since $y_3 = 0$, our multiplication is correct.

b) $12x^2y - 20x^3y = 4x^2y(3 - 5x)$

Check: $4x^2y \cdot 3 = 12x^2y$ and $4x^2y(-5x) = -20x^3y$,
so $4x^2y(3 - 5x) = 12x^2y - 20x^3y$.

c) $10p^6q^2 - 4p^5q^3 - 2p^4q^4 = 2p^4q^2(5p^2 - 2pq - q^2)$

The check is left to the student.

The polynomials in Examples 5 and 6 cannot be factored further unless (in the cases of Examples 5 and 6c) irrational numbers or (in the case of Example 6b) fractions are used. In both examples, we have **factored completely** over the set of integers. The factors used are said to be **prime polynomials** over the set of integers.

When a factor contains more than one term, it is usually desirable for the leading coefficient to be positive. To achieve this may require factoring out a common factor with a negative coefficient.

Example 7   Factor out a common factor with a negative coefficient.

a) $-4x - 24$ **b)** $-2x^3 + 6x^2 - 10x$

SOLUTION

a) $-4x - 24 = -4(x + 6)$
b) $-2x^3 + 6x^2 - 10x = -2x(x^2 - 3x + 5)$

Example 8   *Height of a Thrown Object.* Suppose that a baseball is thrown upward with an initial velocity of 64 ft/sec. Its height in feet, $h(t)$, after t seconds is given by

$$h(t) = -16t^2 + 64t.$$

Find an equivalent expression for $h(t)$ by factoring out a common factor.

SOLUTION   We factor out $-16t$ as follows:

$$h(t) = -16t^2 + 64t = -16t(t - 4).$$

$h(t) = -16t^2 + 64t$

$y_1 = -16x^2 + 64x,$
$y_2 = (-16x)(x - 4)$

X	Y1	Y2
0	0	0
1	48	48
2	64	64
3	48	48
4	0	0
5	-80	-80
6	-192	-192

X = 0

In Example 8, if we evaluate $-16t^2 + 64t$ and $-16t(t - 4)$ using any value for t, the results should match. A quick partial check of any factorization is to evaluate the factorization and the original polynomial for one or two convenient replacements. Recall that two nth-degree polynomials are equivalent if their values are the same for $(n + 1)$ x-values. Thus a check of Example 8 by evaluation becomes foolproof if three replacements are used. The table shown at left confirms that the factorization is correct.

Some polynomials with four terms can be factored by *grouping* the terms.

Factoring by Grouping

The largest common factor is sometimes a binomial.

Example 9   Factor: $(a - b)(x + 5) + (a - b)(x - y^2)$.

SOLUTION   Here the largest common factor is the binomial $a - b$:

$$(a - b)(x + 5) + (a - b)(x - y^2) = (a - b)[(x + 5) + (x - y^2)]$$
$$= (a - b)[2x + 5 - y^2]. \quad \blacksquare$$

Often, in order to find a common binomial factor, we must regroup into two groups of two terms each.

$y_1 = x^3 + 3x^2 + 4x + 12,$
$y_2 = (x + 3)(x^2 + 4)$

X	Y1	Y2
0	12	12
1	20	20
2	40	40
3	78	78
4	140	140
5	232	232
6	360	360

X = 0

Example 10   Factor: **(a)** $t^3 + 3t^2 + 4t + 12$; **(b)** $4x^3 - 15 + 20x^2 - 3x$.

SOLUTION

a) $t^3 + 3t^2 + 4t + 12 = (t^3 + 3t^2) + (4t + 12)$ Grouping

$\qquad\qquad\qquad\qquad\quad = t^2(t + 3) + 4(t + 3)$ Factoring out common factors

$\qquad\qquad\qquad\qquad\quad = (t + 3)(t^2 + 4)$ Factoring out $t + 3$

A check by comparing values is shown in the table at left.

b) When we try grouping $4x^3 - 15 + 20x^2 - 3x$ as

$$(4x^3 - 15) + (20x^2 - 3x),$$

we are unable to factor $4x^3 - 15$. When this happens, we can rearrange the polynomial and try a different grouping:

$4x^3 - 15 + 20x^2 - 3x = 4x^3 + 20x^2 - 3x - 15$ Using the commutative law to rearrange the terms

$y_1 = 4x^3 - 15 + 20x^2 - 3x,$
$y_2 = (x + 5)(4x^2 - 3),$
$y_3 = y_2 - y_1$

X	Y3
0	0
1	0
2	0
3	0
4	0
5	0
6	0

X = 0

$\qquad\qquad\qquad\quad = 4x^2(x + 5) - 3(x + 5)$

$\qquad\qquad\qquad\quad = (x + 5)(4x^2 - 3).$

We check by graphing the difference of the polynomial and its factorization as shown at left. $\quad \blacksquare$

In Section 1.3, we saw that the expressions $b - a$ and $-(a - b)$ or $-1(a - b)$ are equivalent. Remembering this can help when we need to reverse a subtraction.

Example 11   Factor: $ax - bx + by - ay$.

SOLUTION   We have

$$
\begin{aligned}
ax - bx + by - ay &= (ax - bx) + (by - ay) &&\text{Grouping}\\
&= x(a - b) + y(b - a) &&\text{Factoring each binomial}\\
&= x(a - b) + y(-1)(a - b) &&\text{Factoring out } -1 \text{ to reverse } b - a\\
&= x(a - b) - y(a - b) &&\text{Simplifying}\\
&= (a - b)(x - y). &&\text{Factoring out } a - b
\end{aligned}
$$

We can always check our factoring by multiplying:

CHECK: $(a - b)(x - y) = ax - ay - bx + by = ax - bx + by - ay.$

Some polynomials with four terms, like $x^3 + x^2 + 3x - 3$, are prime. Not only is there no common monomial factor, but no matter how we group terms, there is no common binomial factor:

$$x^3 + x^2 + 3x - 3 = x^2(x + 1) + 3(x - 1);$$ No common factor
$$x^3 + 3x + x^2 - 3 = x(x^2 + 3) + (x^2 - 3);$$ No common factor
$$x^3 - 3 + x^2 + 3x = (x^3 - 3) + x(x + 3).$$ No common factor

Factoring and Equations

Factoring can help us solve polynomial equations.

Example 12   Solve: $6x^2 = 30x$.

SOLUTION   We can use the principle of zero products if there is a 0 on one side of the equation and the other side is in factored form:

$$
\begin{aligned}
6x^2 &= 30x\\
6x^2 - 30x &= 0 &&\text{Subtracting 30x. One side is now 0.}\\
6x(x - 5) &= 0 &&\text{Factoring}\\
6x = 0 \quad &or \quad x - 5 = 0 &&\text{Using the principle of zero products}\\
x = 0 \quad &or \qquad\quad x = 5.
\end{aligned}
$$

We check by substitution or graphically, as shown in the figure at left. The solutions are 0 and 5.

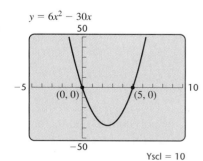

$y = 6x^2 - 30x$

50

−5   10

(0, 0)   (5, 0)

−50

Yscl = 10

Caution! We can *factor expressions*, such as $6x^2 - 30x$. We can *solve equations*, such as $6x^2 = 30x$. Do not make the mistake of attempting to solve an expression.

The principle of zero products can be used to show that an *n*th-degree polynomial function can have at most *n* zeros. In Example 12, we wrote a quadratic polynomial as a product of two linear factors. Each linear factor corresponded to one zero of the polynomial function. In general, a polynomial function of degree *n* can have at most *n* linear factors, so a polynomial function of degree *n* can have at most *n* zeros. Thus, when solving an *n*th-degree polynomial equation, we need not look for more than *n* zeros.

To Use the Principle of Zero Products:

1. Obtain a 0 on one side of the equation using the addition principle, if necessary.
2. Factor the expression on the nonzero side of the equation.
3. Set each factor that is not a constant equal to 0.
4. Solve the resulting equations.

5.3 | *Exercise Set*

1. Use the following graph to solve $x^2 - 2x - 15 = 0$.

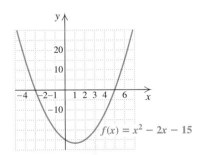

2. Use the following graph to solve $x^2 + 7x + 10 = 0$.

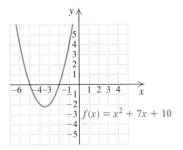

3. 🔍 Use the following graph to solve $x^2 + 2x = 3$.

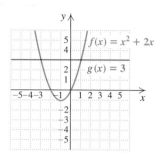

4. 🔍 Use the following graph to solve $x^2 = 4$.

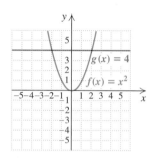

5. 🔍 Use the following graph to find the zeros of the function given by $f(x) = x^2 + 2x - 8$.

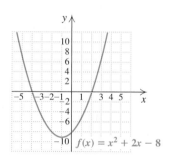

6. 🔍 Use the following graph to find the zeros of the function given by $f(x) = x^2 - 2x + 1$.

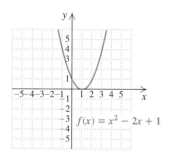

Solve using a grapher.

7. $x^2 = 5x$

8. $2x^2 = 20x$

9. $4x = x^2 + 3$

10. $x^2 = 1$

11. $x^2 + 150 = 25x$

12. $2x^2 + 25 = 51x$

13. $x^3 - 3x^2 + 2x = 0$

14. $x^3 + 2x^2 = x + 2$

15. $x^3 - 3x^2 - 198x + 1080 = 0$

16. $2x^3 + 25x^2 - 282x + 360 = 0$

17. $21x^2 + 2x - 3 = 0$

18. $66x^2 - 49x - 5 = 0$

Find the zeros of each function.

19. $f(x) = x^2 - 4x + 45$

20. $g(x) = x^2 + x - 20$

21. $p(x) = 2x^2 - 13x - 7$

22. $f(x) = 6x^2 + 17x + 6$

23. $f(x) = x^3 - 2x^2 - 3x$

24. $r(x) = 3x^3 - 12x$

🔍 *Match each graph to the corresponding function in Exercises 25–28.*

I

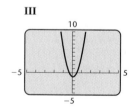

II

Yscl = 10

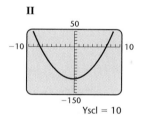

III

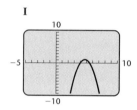

IV

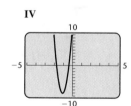

25. $f(x) = (2x - 1)(3x + 1)$

26. $f(x) = (2x + 15)(x - 7)$

27. $f(x) = (4 - x)(2x - 11)$

28. $f(x) = (5x + 2)(4x + 7)$

Tell whether each of the following is an expression or an equation.

29. $x^2 + 6x + 9$

30. $x^3 = x^2 - x + 3$

31. $3x^2 = 3x$

32. $x^4 + 3x^3 + x^2$

33. $2x^3 + x^2 = 0$

34. $5x^4 + 5x$

Factor.

35. $8t^2 + 2t$

36. $6y^2 + 3y$

37. $y^3 + 9y^2$

38. $x^3 + 8x^2$

39. $5x^2 - 15x^4$

40. $8y^2 + 4y^4$

41. $4x^2y - 12xy^2$

42. $5x^2y^3 + 15x^3y^2$

43. $3y^2 - 3y - 9$

44. $5x^2 - 5x + 15$

45. $6ab - 4ad + 12ac$

46. $8xy + 10xz - 14xw$

47. $9x^3y^6z^2 - 12x^4y^4z^4 + 15x^2y^5z^3$

48. $14a^4b^3c^5 + 21a^3b^5c^4 - 35a^4b^4c^3$

Factor out a factor with a negative coefficient.

49. $-5x + 15$

50. $-5x - 40$

51. $-6y - 72$

52. $-8t + 72$

53. $-2x^2 + 4x - 12$

54. $-2x^2 + 12x + 40$

55. $-3y^3 + 12y^2 - 15y$

56. $-4m^4 - 32m^3 + 64m$

57. $-x^2 + 5x - 9$

58. $-p^3 - 4p^2 + 11$

59. *Height of a Rocket.* A model rocket is launched upward with an initial velocity of 96 ft/sec. Its height in feet, $h(t)$, after t seconds is given by

$$h(t) = -16t^2 + 96t.$$

a) Find an equivalent expression for $h(t)$ by factoring out a common factor with a negative coefficient.

b) Check your factoring by evaluating both expressions for $h(t)$ at $t = 2$.

60. *Height of a Baseball.* A baseball is popped up with an upward velocity of 72 ft/sec. Its height in feet, $h(t)$, after t seconds is given by

$$h(t) = -16t^2 + 72t.$$

a) Find an equivalent expression for $h(t)$ by factoring out a common factor with a negative coefficient.

b) Perform a partial check of part (a) by evaluating both expressions for $h(t)$ at $t = 2$.

61. *Counting Spheres in a Pile.* The number N of spheres in a triangular pile like the one shown below is a polynomial function given by

$$N(x) = \tfrac{1}{6}x^3 + \tfrac{1}{2}x^2 + \tfrac{1}{3}x,$$

where x is the number of layers and $N(x)$ is the number of spheres. Find an equivalent expression for $N(x)$ by factoring out a common factor.

62. *Number of Games in a League.* If there are n teams in a league and each team plays every other team once, we can find the total number of games played by using the polynomial function $f(n) = \tfrac{1}{2}n^2 - \tfrac{1}{2}n$. Find an equivalent expression for $f(n)$ by factoring out a common factor.

63. *Surface Area of a Silo.* A silo is a structure that is shaped like a right circular cylinder with a half sphere on top. The surface area of a silo of height h and radius r (including the area of the base) is given by the polynomial $2\pi rh + \pi r^2$. (Note that h is the height of the silo.) Find an equivalent expression by factoring out a common factor.

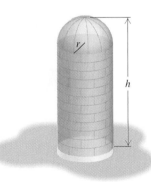

64. *Number of Diagonals.* The number of diagonals of a polygon having n sides is given by the polynomial function

$$P(n) = \tfrac{1}{2}n^2 - \tfrac{3}{2}n.$$

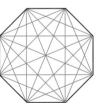

Find an equivalent expression for $P(n)$ by factoring out a common factor.

65. *Total Revenue.* Household Sound is marketing a new kind of stereo. The firm determines that when it sells x stereos, the total revenue R is given by the polynomial function

$$R(x) = 280x - 0.4x^2 \text{ dollars.}$$

Find an equivalent expression for $R(x)$ by factoring out $0.4x$.

66. *Total Cost.* Household Sound determines that the total cost C of producing x stereos is given by the polynomial function

$$C(x) = 0.18x + 0.6x^2.$$

Find an equivalent expression for $C(x)$ by factoring out $0.6x$.

Factor.

67. $a(b - 2) + c(b - 2)$

68. $a(x^2 - 3) - 2(x^2 - 3)$

69. $(x + 7)(x - 1) + (x + 7)(x - 2)$

70. $(a + 5)(a - 2) + (a + 5)(a + 1)$

71. $a^2(x - y) + 5(y - x)$

72. $3x^2(x - 6) + 2(6 - x)$

73. $ac + ad + bc + bd$

74. $xy + xz + wy + wz$

75. $b^3 - b^2 + 2b - 2$

76. $y^3 - y^2 + 3y - 3$

77. $a^3 - 3a^2 + 6 - 2a$

78. $t^3 + 6t^2 - 2t - 12$

79. $24x^3 - 36x^2 + 72x$

80. $12a^4 - 21a^3 - 9a^2$

81. $x^6 - x^5 - x^3 + x^4$

82. $y^4 - y^3 - y + y^2$

83. $2y^4 + 6y^2 + 5y^2 + 15$

84. $2xy - x^2y - 6 + 3x$

Solve using the principle of zero products.

85. $x(x + 1) = 0$

86. $5x(x - 2) = 0$

87. $x^2 - 3x = 0$

88. $2x^2 + 8x = 0$

89. $-5x^2 = 15x$

90. $2x - 4x^2 = 0$

91. $12x^4 + 4x^3 = 0$

92. $21x^3 = 7x^2$

Skill Maintenance

Let $f(x) = 3x - 5$.

93. Draw the graph of f. [2.4]

94. Find $f(2a + 1)$. [2.1]

95. *Mixing Rice.* Countryside Rice is 90% white rice and 10% wild rice. Mystic Rice is 50% wild rice. How much of each type should be used to create a 25-lb batch of rice that is 35% wild rice? [3.3]

96. *Geometry.* The first angle of a triangle is four times the second angle. The third angle is 75°. What are the measures of the first two angles? [2.3]

Synthesis

97. ◈ Under what conditions would it be easier to evaluate a polynomial *after* it has been factored?

98. ◈ Checking the factorization of a second-degree polynomial by making a single replacement is only a *partial* check. Write an *incorrect* factorization and explain how evaluating both the polynomial and the factorization might not catch the mistake.

99. ◈ Use Exercise 17 to explain a disadvantage of solving equations graphically.

100. ◈ Mary Louise is attempting to solve $x^3 + 20x^2 + 4x + 80 = 0$ using a grapher. Unfortunately, when she graphs $y_1 = x^3 + 20x^2 + 4x + 80$ in a standard $[-10, 10, -10, 10]$ window, she sees no graph at all, let alone any x-intercept. Can this problem be solved graphically? If so, how? If not, why not?

101. Use the results of Exercise 1 to factor $x^2 - 2x - 15$.

102. Use the results of Exercise 2 to factor $x^2 + 7x + 10$.

103. Use the results of Exercise 5 to factor $x^2 + 2x - 8$.

104. Use the results of Exercise 6 to factor $x^2 - 2x + 1$.

Complete each of the following.

105. $x^5y^4 + \underline{\hspace{1cm}} = x^3y(\underline{\hspace{1cm}} + xy^5)$

106. $a^3b^7 - \underline{\hspace{1cm}} = \underline{\hspace{1cm}} (ab^4 - c^2)$

Factor.

107. $rx^2 - rx + 5r + sx^2 - sx + 5s$

108. $3a^2 + 6a + 30 + 7a^2b + 14ab + 70b$

109. $5x^2 - x^2y + 10x - 2xy + 15xz - 3xyz$

110. $a^4x^4 + a^4x^2 + 5a^4 + a^2x^4 + a^2x^2 + 5a^2 + 5x^4 + 5x^2 + 25$
(*Hint:* Use three groups of three.)

Factor. Assume that all exponents are natural numbers.

111. $2x^{3a} + 8x^a + 4x^{2a}$

112. $3a^{n+1} + 6a^n - 15a^{n+2}$

113. $4x^{a+b} + 7x^{a-b}$

114. $7y^{2a+b} - 5y^{a+b} + 3y^{a+2b}$

5.4
Equations Containing Trinomials of the Type $x^2 + bx + c$

- *Factoring Trinomials of the Type $x^2 + bx + c$*
- *Equations Containing Trinomials*
- *Zeros and Factoring*

In this section, we expand our list of the types of polynomials that we can factor so that we can solve a wider variety of polynomial equations. We begin by factoring trinomials of the type $x^2 + bx + c$ and then solve equations containing such trinomials. We then see how to use zeros of a function to factor a polynomial.

Factoring Trinomials of the Type $x^2 + bx + c$

When trying to factor trinomials of the type $x^2 + bx + c$, we can use a trial-and-error procedure.

CONSTANT TERM POSITIVE Recall the FOIL method of multiplying two binomials:

$$
\begin{array}{cccc}
\text{F} & \text{O} & \text{I} & \text{L} \\
\end{array}
$$
$$(x + 3)(x + 5) = x^2 + \underbrace{5x + 3x} + 15$$
$$= x^2 + \quad 8x \quad + 15.$$

The product is a trinomial in which the leading coefficient is 1. To factor $x^2 + 8x + 15$, we think of FOIL: The first term, x^2, is the product of the First terms of two binomial factors, so the first term in each binomial will be x. The challenge is to find two numbers p and q such that

$$x^2 + 8x + 15 = (x + p)(x + q).$$

Note that the Outer and Inner products, qx and px, can be written as $(p + q)x$. The Last product, pq, will be a constant. Thus the numbers p and q must be selected so that their product is 15 and their sum is 8. In this case, we know from above that these numbers are 3 and 5. The factorization is

$$(x + 3)(x + 5), \quad \text{or} \quad (x + 5)(x + 3). \qquad \text{Using a commutative law}$$

In general,

$$(x + p)(x + q) = x^2 + (p + q)x + pq.$$

To factor $x^2 + (p + q)x + pq$, we use FOIL in reverse:

$$x^2 + (p + q)x + pq = (x + p)(x + q).$$

Example 1 Factor: $x^2 + 9x + 8$.

SOLUTION We think of FOIL in reverse. The first term of each factor is x.

We are looking for numbers p and q such that

$$x^2 + 9x + 8 = (x + p)(x + q) = x^2 + (p + q)x + pq.$$

Thus we look for factors of 8 whose sum is 9.

PAIR OF FACTORS	SUM OF FACTORS
2, 4	6
1, 8	9 ←

The numbers we need are 1 and 8.

The factorization is thus $(x + 1)(x + 8)$. We can check by multiplying to see if the product is the original trinomial.

When factoring trinomials with a leading coefficient of 1, it suffices to consider all pairs of factors along with their sums, as we did above. At times, however, you may be tempted to form factors without calculating any sums. It is essential that you check any attempt made in this manner! For example, if we attempt the factorization

$$x^2 + 9x + 8 \stackrel{?}{=} (x + 2)(x + 4),$$

a check reveals that $(x + 2)(x + 4) = x^2 + 6x + 8 \neq x^2 + 9x + 8$. This type of trial-and-error procedure becomes easier to use with time. As you gain experience, you will find that many trials can be performed mentally.

When the constant term of a trinomial is positive, the constant terms in the binomial factors both have the same sign. This ensures a positive product. The sign used is that of the trinomial's middle term.

Example 2 Factor: $t^2 - 9t + 20$.

SOLUTION Since the constant term is positive and the coefficient of the middle term is negative, we look for a factorization of 20 in which both factors are negative. Their sum must be -9.

$y_1 = x^2 - 9x + 20,$
$y_2 = (x - 4)(x - 5)$

X	Y1	Y2
0	20	20
1	12	12
2	6	6
3	2	2
4	0	0
5	0	0
6	2	2

X = 0

PAIR OF FACTORS	SUM OF FACTORS
$-1, -20$	-21
$-2, -10$	-12
$-4, -5$	-9 ←

The numbers we need are -4 and -5.

The factorization is $(t - 4)(t - 5)$. We check by comparing values using a table, as shown at left.

CONSTANT TERM NEGATIVE When the constant term of a trinomial is negative, we look for one negative factor and one positive factor. The sum of the factors must still be the coefficient of the middle term.

Example 3   Factor: $x^3 - x^2 - 30x$.

SOLUTION   *Always* look first for a common factor! This time there is one, x. We factor it out:

$$x^3 - x^2 - 30x = x(x^2 - x - 30).$$

Now we consider $x^2 - x - 30$. We need a factorization of -30 in which one factor is positive, the other factor is negative, and the sum of the factors is -1. Since the sum is to be negative, the negative factor must be further from 0 than the positive factor is. Thus we need consider only pairs of factors in which the negative term has the larger absolute value.

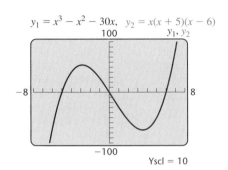

$y_1 = x^3 - x^2 - 30x$, $y_2 = x(x + 5)(x - 6)$

Yscl = 10

PAIR OF FACTORS	SUM OF FACTORS
1, −30	−29
3, −10	−7
5, −6	−1 ←

The numbers we need are 5 and −6.

The factorization of $x^2 - x - 30$ is $(x + 5)(x - 6)$. *Don't forget to include the factor that was factored out earlier!* In this case, the factorization of the original trinomial is $x(x + 5)(x - 6)$. We check by graphing $y_1 = x^3 - x^2 - 30x$ and $y_2 = x(x + 5)(x - 6)$ (see the figure at left).

Some polynomials are not factorable using integers.

Example 4   Factor: $x^2 - x - 7$.

SOLUTION   There are no factors of -7 that add to -1. Thus this trinomial is *not* factorable into binomials with integer coefficients. The polynomial is *prime*.

Tips for Factoring $x^2 + bx + c$

1. If necessary, rewrite the trinomial in descending order. Search for factors of c that add up to b. Remember the following:

   ■ If c is positive, the signs of the factors of c are the same as the sign of b.

   ■ If c is negative, one factor of c is positive and the other is negative.

   ■ If the sum of the two factors of c is the opposite of b, changing the signs of both factors will give the desired factors whose sum is b.

2. Check the result by multiplying the binomials. If the trinomial is in one variable, you can also check using a grapher.

Sometimes a trinomial like $x^6 + 2x^3 - 15$ can be factored if we first think of it as $(x^3)^2 + 2x^3 - 15$. To do this, we make a substitution (perhaps just mentally) in which $u = x^3$. The trinomial then becomes

$$u^2 + 2u - 15.$$

We try factoring this trinomial and if a factorization is found, we replace all occurrences of u with x^3. Since

$$u^2 + 2u - 15 = (u - 3)(u + 5),$$

the original polynomial can be factored as

$$x^6 + 2x^3 - 15 = (x^3 - 3)(x^3 + 5).$$

Equations Containing Trinomials

We can now use our new factoring skills to solve some polynomial equations.

Example 5   Solve each of the following.

a) $x^2 + 9x + 8 = 0$

b) $(t - 10)(t + 1) = -30$

c) $x^3 = x^2 + 30x$

> **Caution!** These are equations that we can try to solve. Do not try to solve an expression!

SOLUTION

a) We solve using two methods.

┌ ─ ─ *ALGEBRAIC SOLUTION*

We use the principle of zero products:

$x^2 + 9x + 8 = 0$

$(x + 1)(x + 8) = 0$ **Using the factorization from Example 1**

$x + 1 = 0$ *or* $x + 8 = 0$ **Using the principle of zero products**

$x = -1$ *or* $x = -8$.   **Solving each equation for x**

The solutions are -1 and -8.

┌ ─ ─ *GRAPHICAL SOLUTION*

The real-number solutions of $x^2 + 9x + 8 = 0$ are the x-intercepts of the graph of $f(x) = x^2 + 9x + 8$.

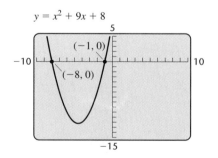

$y = x^2 + 9x + 8$

We find that -1 and -8 are solutions.

CHECK:

For -1:

$$\frac{x^2 + 9x + 8 = 0}{(-1)^2 + 9(-1) + 8 \ ? \ 0}$$
$$1 - 9 + 8 \ \Big| $$
$$0 \ \Big| \ 0 \quad \text{TRUE}$$

For -8:

$$\frac{x^2 + 9x + 8 = 0}{(-8)^2 + 9(-8) + 8 \ ? \ 0}$$
$$64 - 72 + 8 \ \Big| $$
$$0 \ \Big| \ 0 \quad \text{TRUE}$$

The solutions are -1 and -8.

b) We solve $(t - 10)(t + 1) = -30$ using both the principle of zero products and a grapher.

ALGEBRAIC SOLUTION

Note that the left side of the equation is factored. It may be tempting to set both factors equal to -30 and solve, but this is *not* correct. The principle of zero products requires 0 on one side of the equation. We begin by multiplying the left side.

$(t - 10)(t + 1) = -30$

$\quad t^2 - 9t - 10 = -30$ **Multiplying**

$\quad t^2 - 9t + 20 = 0$ **Adding 30 on both sides to get 0 on one side**

$\quad (t - 4)(t - 5) = 0$ **Using the factorization from Example 2**

$\quad t - 4 = 0 \quad or \quad t - 5 = 0$ **Using the principle of zero products**

$\quad\quad\quad t = 4 \quad or \quad\quad\quad t = 5$

The solutions are 4 and 5.

GRAPHICAL SOLUTION

There is no need to multiply. We let $y_1 = (x - 10)(x + 1)$ and $y_2 = -30$ and look for any points of intersection of the graphs. The x-coordinates of these points are the solutions of the equation.

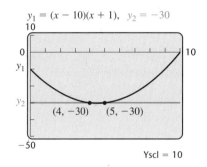

We find that 4 and 5 are solutions.

Both numbers check in the original equation. As a further check, note that we obtained the same solutions using two methods. The solutions are 4 and 5.

c) Again, we solve using two methods.

ALGEBRAIC SOLUTION

In order to solve $x^3 = x^2 + 30x$ using the principle of zero products, we must rewrite the equation with 0 on one side.

We have

$$x^3 = x^2 + 30x$$

$$x^3 - x^2 - 30x = 0 \quad \text{Getting 0 on one side}$$

$$x(x - 6)(x + 5) = 0 \quad \text{Using the factorization from Example 3}$$

$$x = 0 \quad or \quad x - 6 = 0 \quad or \quad x + 5 = 0 \quad \text{Using the principle of zero products}$$

$$x = 0 \quad or \quad\quad\quad x = 6 \quad or \quad\quad\quad x = -5.$$

The solutions are 0, 6, and -5.

GRAPHICAL SOLUTION

We let $y_1 = x^3$ and $y_2 = x^2 + 30x$ and look for points of intersection of the graphs.

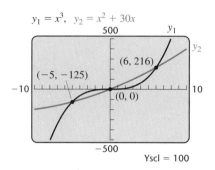

The viewing window $[-10, 10, -500, 500]$ shows three points of intersection. Since the polynomial equation is of degree 3, we know there will be no more than 3 solutions. The x-coordinates of the points of intersection are -5, 0, and 6.

We check by substituting -5, 0, and 6 into the original equation. The solutions are -5, 0, and 6.

Zeros and Factoring

Note in Example 5(a) the relationship between the factors of $x^2 + 9x + 8$ and the solutions of the equation $x^2 + 9x + 8 = 0$. The graphs of $f(x) = x^2 + 9x + 8$, $g(x) = x + 1$, and $h(x) = x + 8$ are shown below. Note that -8 is a zero of both $f(x)$ and $h(x)$ and -1 is a zero of both $f(x)$ and $g(x)$. We can use the principle of zero products "in reverse" to factor a polynomial and to write a function with given zeros.

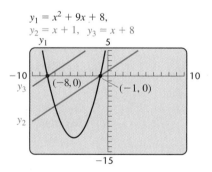

Example 6   Factor: $x^2 - 13x - 608$.

SOLUTION   The factorization of the trinomial $x^2 - 13x - 608$ will be in the form

$$x^2 - 13x - 608 = (x + p)(x + q).$$

We could use trial and error to find p and q, but there are many factors of 608. Thus we factor the trinomial using the principle of zero products.

Note that the roots of the equation

$$x^2 - 13x - 608 = 0$$

are the zeros of the function $f(x) = x^2 - 13x - 608$. We graph the function and find the zeros.

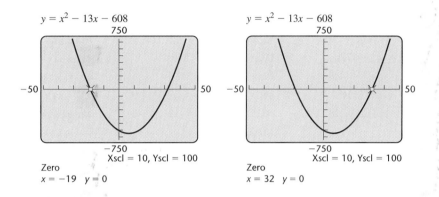

The zeros are -19 and 32. Each zero is a root of $x^2 - 13x - 608 = 0$. We also know that -19 is a zero of the linear function $g(x) = x + 19$ and that 32 is a zero of the linear function $h(x) = x - 32$. This suggests that $x + 19$ and $x - 32$ are factors of $x^2 - 13x - 608$. Using the principle of zero products in reverse, we now have

$$x^2 - 13x - 608 = (x + 19)(x - 32).$$

Multiplication indicates that the factorization is correct.   ▬

We can also use the principle of zero products to write a function whose zeros are given.

Example 7   Write a polynomial function $f(x)$ whose zeros are -1, 0, and 3.

SOLUTION   Each zero of the polynomial function f is a zero of a linear factor of the polynomial. We write a linear function for each given zero:

$$-1 \text{ is a zero of } g(x) = x + 1;$$
$$0 \text{ is a zero of } h(x) = x;$$

and $3 \text{ is a zero of } k(x) = x - 3.$

Thus a polynomial function with zeros -1, 0, and 3 is

$$f(x) = (x + 1) \cdot x \cdot (x - 3);$$

multiplying gives us

$$f(x) = x^3 - 2x^2 - 3x.$$ ▬

5.4 | Exercise Set

Factor.

1. $x^2 + 8x + 12$

2. $x^2 + 6x + 5$

3. $t^2 - 8t + 15$

4. $y^2 - 12y + 27$

5. $x^2 - 27 - 6x$

6. $t^2 - 15 - 2t$

7. $2n^2 - 20n + 50$

8. $2a^2 - 16a + 32$

9. $a^3 + a^2 - 72a$

10. $x^3 + 3x^2 - 54x$

11. $14x + x^2 + 45$

12. $12y + y^2 + 32$

13. $3x + x^2 - 10$

14. $x + x^2 - 6$

15. $3x^2 + 15x + 18$

16. $5y^2 + 40y + 35$

17. $56 + x - x^2$

18. $32 + 4y - y^2$

19. $32y + 4y^2 - y^3$

20. $56x + x^2 - x^3$

21. $x^4 + 11x^2 - 80$

22. $y^4 + 5y^2 - 84$

23. $x^2 + 12x + 13$

24. $x^2 - 3x + 7$

25. $p^2 - 5pq - 24q^2$

26. $x^2 + 12xy + 27y^2$

27. $y^2 + 8yz + 16z^2$

28. $x^2 - 14xy + 49y^2$

29. $p^4 + 80p^2 + 79$

30. $x^4 + 50x^2 + 49$

31. Use the results of Exercise 1 to solve $x^2 + 8x + 12 = 0$.

32. Use the results of Exercise 2 to solve $x^2 + 6x + 5 = 0$.

33. Use the results of Exercise 7 to solve $2n^2 + 50 = 20n$.

34. Use the results of Exercise 18 to solve $32 + 4y = y^2$.

35. Use the results of Exercise 9 to solve $a^3 + a^2 = 72a$.

36. Use the results of Exercise 20 to solve $56x + x^2 = x^3$.

In Exercises 37–40, use the graph to solve the given equation. Check by substituting into the equation.

37. $x^2 + 4x - 5 = 0$

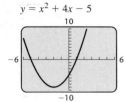
$y = x^2 + 4x - 5$

38. $x^2 - x - 6 = 0$

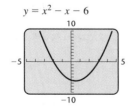
$y = x^2 - x - 6$

39. $x^2 + x - 6 = 0$

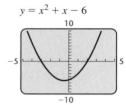
$y = x^2 + x - 6$

40. $x^2 + 8x + 15 = 0$

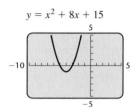
$y = x^2 + 8x + 15$

Find the zeros of each function.

41. $f(x) = x^2 - 4x + 45$

42. $f(x) = x^2 + x - 20$

43. $r(x) = x^3 - 2x^2 - 3x$

44. $g(x) = 3x^2 + 21x + 30$

Solve.

45. $x^2 + 4x = 45$

46. $t^2 - 3t = 28$

47. $x^2 - 9x = 0$

48. $a^2 + 18a = 0$

49. $a^3 - 3a^2 = 40a$

50. $x^3 - 2x^2 = 63x$

51. $(x - 3)(x + 2) = 14$

52. $(z + 4)(z - 2) = -5$

53. $35 - x^2 = 2x$

54. $40 - x^2 + 3x = 0$

In Exercises 55 and 56, use the graph to factor the given polynomial.

55. $x^2 + 10x - 264$

56. $x^2 + 16x - 336$

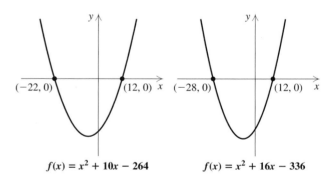
$f(x) = x^2 + 10x - 264$ $f(x) = x^2 + 16x - 336$

In Exercises 57–60, use a graph to help factor each polynomial.

57. $x^2 + 40x + 384$

58. $x^2 - 13x - 300$

59. $x^2 + 26x - 2432$

60. $x^2 - 46x + 504$

Write a polynomial function that has the given zeros. Answers may vary.

61. $-1, 2$

62. $2, 5$

63. $-7, -10$

64. $8, -3$

65. $0, 1, 2$

66. $-3, 0, 5$

Skill Maintenance

67. Graph: $f(x) = -\frac{3}{4}x + 6$. [2.4]

68. *Exam Scores.* There are 75 questions on a college entrance examination. Two points are awarded for each correct answer, and one-half point is deducted for each incorrect answer. A score of 100 indicates how many correct and how many incorrect answers, assuming that all questions have been answered? [3.3]

69. *Perimeter.* A pentagon with all five sides the same size has the same perimeter as an octagon in which all eight sides are the same size. One side of the pentagon is 2 less than three times the length of one side of the octagon. Find the perimeters. [3.3]

70. Solve: $|x - 7| < 3$. [4.3]

Synthesis

71. ◈ Explain how to conclude that $x^2 + 5x + 200$ is a prime polynomial without performing any trials.

72. ◈ Explain how to conclude that $x^2 - 59x + 6$ is a prime polynomial without performing any trials.

73. ◈ Explain how the following graph of
$$y = x^2 + 3x - 2 - (x - 2)(x + 1)$$
can be used to show that
$$x^2 + 3x - 2 \neq (x - 2)(x + 1).$$

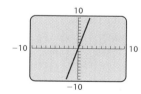

74. ◈ Explain why Example 7 reads "Write *a* polynomial function $f(x)$" and not "Write *the* polynomial function $f(x)$."

75. Use the following graph of $f(x) = x^2 - 2x - 3$ to solve $x^2 - 2x - 3 = 0$ and to solve $x^2 - 2x - 3 < 5$.

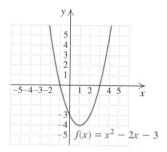

76. Use the following graph of $g(x) = -x^2 - 2x + 3$ to solve $-x^2 - 2x + 3 = 0$ and to solve $-x^2 - 2x + 3 \geq -5$.

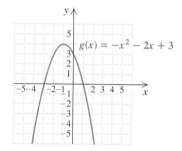

77. Find a polynomial function f for which $f(2) = 0$, $f(-1) = 0$, $f(3) = 0$, and $f(0) = 30$.

78. Find a polynomial function g for which $g(-3) = 0$, $g(1) = 0$, $g(5) = 0$, and $g(0) = 45$.

In Exercises 79–81, use a grapher to find any solutions that exist accurate to two decimal places.

79. $-x^2 + 13.80x = 47.61$

80. $-x^2 + 3.63x + 34.34 = x^2$

81. $x^3 - 3.48x^2 + x = 3.48$

Factor. Assume that variables in exponents represent positive integers.

82. $x^2 - \frac{4}{25} + \frac{3}{5}x$

83. $y^2 - \frac{8}{49} + \frac{2}{7}y$

84. $y^2 + 0.4y - 0.05$

85. $x^{2a} + 5x^a - 24$

86. $(x + 3)^2 - 2(x + 3) - 35$

87. Find all integers m for which $x^2 + mx + 75$ can be factored.

88. Find all integers q for which $x^2 + qx - 32$ can be factored.

89. One of the factors of $x^2 - 345x - 7300$ is $x + 20$. Find the other factor.

5.5

Equations Containing Trinomials of the Type $ax^2 + bx + c$, $a \neq 1$

- *Factoring and Equations*
- *Factoring and Functions*

Factoring and Equations

Now we look at trinomials in which the leading coefficient is not 1. We consider two methods. Use what works best for you or what your instructor chooses for you.

METHOD 1: THE FOIL METHOD   We first consider the **FOIL method** for factoring trinomials of the type

$$ax^2 + bx + c, \quad \text{where } a \neq 1.$$

Consider the following multiplication.

$$
\begin{array}{c}
\quad\ \text{F} \qquad \text{O} \quad\ \text{I} \quad\ \text{L} \\
\quad\ \downarrow \qquad \downarrow \quad\ \downarrow \quad\ \downarrow \\
(3x + 2)(4x + 5) = 12x^2 + \underbrace{15x + 8x} + 10 \\
\qquad\qquad\qquad\ \downarrow \qquad\quad \downarrow \qquad\ \downarrow \\
\qquad\quad\ = 12x^2 + \quad 23x \quad\ + 10
\end{array}
$$

To factor $12x^2 + 23x + 10$, we must reverse what we just did. We look for two binomials whose product is this trinomial. The product of the First terms must be $12x^2$. The product of the Outside terms plus the product of the Inside terms must be $23x$. The product of the Last terms must be 10. We know from the preceding discussion that the answer is

$$(3x + 2)(4x + 5).$$

In general, however, finding such an answer involves trial and error. We use the following method.

To Factor $ax^2 + bx + c$ Using FOIL:

1. Factor out the largest common factor, if one exists. Here we assume none does.

2. Find two **First** terms whose product is ax^2:

$$(\ \square\ x + \quad)(\ \square\ x + \quad) = ax^2 + bx + c.$$
$$\underbrace{\qquad\qquad\qquad\qquad}_{\text{FOIL}}$$

3. Find two **Last** terms whose product is c:

$$(\ x + \square\ )(\ x + \square\ ) = ax^2 + bx + c.$$
$$\underbrace{\qquad\qquad\qquad\qquad}_{\text{FOIL}}$$

4. Repeat steps (2) and (3) until a combination is found for which the sum of the **Outer** and **Inner** products is bx:

$$(\ \square\ x + \square\ )(\ \square\ x + \square\ ) = ax^2 + bx + c.$$
$$\qquad\qquad\qquad\qquad\qquad \text{FOIL}$$

Example 1 Factor: $3x^2 + 10x - 8$.

SOLUTION

1. First, note that there is no common factor (other than 1 or -1).

2. Next, factor the first term, $3x^2$. The only possibility for factors is $3x \cdot x$. Thus, if a factorization exists, it must be of the form

   $$(3x + \quad)(x + \quad).$$

   We need to find the right numbers for the blanks.

3. Note that the constant term, -8, can be factored as $(-8)(1)$, $8(-1)$, $(-2)4$, and $2(-4)$, as well as $(1)(-8)$, $(-1)8$, $(-4)2$, and $4(-2)$.

4. Find a pair of factors for which the sum of the products (the "outside" and "inside" parts of FOIL) is the middle term, $10x$. Each possibility should be checked by multiplying:

   $$(3x - 8)(x + 1) = 3x^2 - 5x - 8.$$

This gives a middle term with a negative coefficient. Since a positive coefficient is needed, a second possibility must be tried:

$$(3x + 8)(x - 1) = 3x^2 + 5x - 8.$$

Note that changing the signs of the two constant terms changes only the sign of the middle term. We try again:

$$(3x - 2)(x + 4) = 3x^2 + 10x - 8. \qquad \text{This is what we wanted.}$$

Thus the desired factorization is $(3x - 2)(x + 4)$. The table at left serves as a check. ▬

$y_1 = 3x^2 + 10x - 8,$
$y_2 = (3x - 2)(x + 4)$

X	Y₁	Y₂
0	−8	−8
1	5	5
2	24	24
3	49	49
4	80	80
5	117	117
6	160	160
X = 0		

Example 2 Factor: $6x^6 - 19x^5 + 10x^4$.

SOLUTION

1. First, factor out the common factor x^4:

   $$x^4(6x^2 - 19x + 10).$$

2. Note that $6x^2 = 6x \cdot x$ and $6x^2 = 3x \cdot 2x$. Thus, $6x^2 - 19x + 10$ may factor into

   $$(3x + \quad)(2x + \quad) \quad \text{or} \quad (6x + \quad)(x + \quad).$$

3. We factor the last term, 10. The possibilities are $10 \cdot 1$, $(-10)(-1)$, $5 \cdot 2$, and $(-5)(-2)$, as well as $1 \cdot 10$, $(-1)(-10)$, $2 \cdot 5$, and $(-2)(-5)$.

4. There are 8 possibilities for *each* factorization in step (2). We need factors for which the sum of the products (the "outer" and "inner" parts of FOIL) is the middle term, $-19x$. Since the x-coefficient is negative, we consider pairs of negative factors. Each possible factorization must be checked by multiplying:

   $$(3x - 10)(2x - 1) = 6x^2 - 23x + 10.$$

We try again:

$$(3x - 5)(2x - 2) = 6x^2 - 16x + 10.$$

Actually this last attempt could have been rejected by simply noting that $2x - 2$ has a common factor, 2. Since the *largest* common factor was removed in step (1), no other common factors can exist. We try again, reversing the -5 and -2:

$$(3x - 2)(2x - 5) = 6x^2 - 19x + 10. \qquad \text{This is what we}$$
wanted.

The factorization of $6x^2 - 19x + 10$ is $(3x - 2)(2x - 5)$. But do not forget the common factor! We must include it to get the complete factorization of the original trinomial:

$$6x^6 - 19x^5 + 10x^4 = x^4(3x - 2)(2x - 5).$$

The graphs of the original polynomial and the factorization coincide, as shown in the figure below.

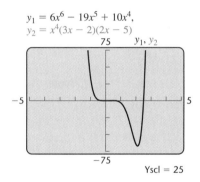

$y_1 = 6x^6 - 19x^5 + 10x^4,$
$y_2 = x^4(3x - 2)(2x - 5)$

Tips for Factoring with FOIL

1. If the largest common factor has been factored out of the original trinomial, then no binomial factor can have a common factor (other than 1 or -1).

2. If a and c are both positive, then the signs in the factors will be the same as the sign of b.

3. When a possible factoring produces the opposite of the desired middle term, reverse the signs of the constants in the factors.

4. Be systematic about your trials. Keep track of those possibilities that you have tried and those that you have not.

Keep in mind that this method of factoring involves trial and error. With practice, you will find yourself making fewer and better guesses.

METHOD 2: THE GROUPING METHOD   The second method for factoring trinomials of the type $ax^2 + bx + c$, $a \neq 1$, is known as the *grouping method*. It involves not only trial and error and FOIL but also factoring by grouping. We know that

$$x^2 + 7x + 10 = x^2 + 2x + 5x + 10$$
$$= x(x + 2) + 5(x + 2)$$
$$= (x + 2)(x + 5),$$

but what if the leading coefficient is not 1? Consider $6x^2 + 23x + 20$. The method is similar to what we just did with $x^2 + 7x + 10$, but we need two more steps. First, multiply the leading coefficient, 6, and the constant, 20, to get 120. Then find a factorization of 120 in which the sum of the factors is the coefficient of the middle term: 23. The middle term is then split into a sum or difference using these factors.

$6x^2 + 23x + 20$

(1) Multiply 6 and 20: $6 \cdot 20 = 120$.

(2) Factor 120: $120 = 8 \cdot 15$, and $8 + 15 = 23$.

(3) Split the middle term: $23x = 8x + 15x$.

(4) Factor by grouping.

We factor by grouping as follows:

$$6x^2 + 23x + 20 = 6x^2 + 8x + 15x + 20$$
$$= 2x(3x + 4) + 5(3x + 4)$$
$$= (3x + 4)(2x + 5).$$

Factoring by grouping

To Factor $ax^2 + bx + c$ Using Grouping:

1. Make sure that any common factors have been factored out.

2. Multiply the leading coefficient a and the constant c.

3. Try to factor the product ac so that the sum of the factors is b. That is, find integers p and q so that $pq = ac$ and $p + q = b$.

4. Split the middle term. That is, write bx as $px + qx$.

5. Factor by grouping.

Example 3 Factor: $3x^2 + 10x - 8$.

SOLUTION

1. First, look for a common factor. There is none (other than 1 or -1).

2. Multiply the leading coefficient and the constant, 3 and -8:

$$3(-8) = -24.$$

3. Try to factor -24 so that the sum of the factors is 10:

$$-24 = 12(-2) \quad \text{and} \quad 12 + (-2) = 10.$$

4. Split $10x$ using the results of step (3):

$$10x = 12x - 2x.$$

5. Finally, factor by grouping:

$$3x^2 + 10x - 8 = 3x^2 + 12x - 2x - 8$$
$$= 3x(x + 4) - 2(x + 4)$$
$$= (x + 4)(3x - 2).$$

Factoring by grouping

We check the solution both algebraically and graphically.

┌─ ALGEBRAIC CHECK

$(x + 4)(3x - 2) = 3x^2 - 2x + 12x - 8$
$= 3x^2 + 10x - 8$

┌─ GRAPHICAL CHECK

$y_1 = 3x^2 + 10x - 8$,
$y_2 = (x + 4)(3x - 2)$

X	Y₁	Y₂
0	−8	−8
1	5	5
2	24	24
3	49	49
4	80	80
5	117	117
6	160	160

X = 0

We now use our new factoring skill to solve a polynomial equation. We factor a polynomial *expression* and use the principle of zero products to solve the *equation*.

Example 4 Solve: $6x^6 - 19x^5 + 10x^4 = 0$.

SOLUTION We note at the outset that the polynomial is of degree 6, so there will be at most 6 solutions of the equation.

┌─ ALGEBRAIC SOLUTION

We solve as follows:

$$6x^6 - 19x^5 + 10x^4 = 0$$
$$x^4(3x - 2)(2x - 5) = 0$$

Factoring as in Example 2

$x^4 = 0$ *or* $3x - 2 = 0$ *or* $2x - 5 = 0$

Using the principle of zero products

$x = 0$ *or* $x = \frac{2}{3}$ *or* $x = \frac{5}{2}$.

Solving for x; $x^4 = x \cdot x \cdot x \cdot x = 0$ when $x = 0$

The solutions are 0, $\frac{2}{3}$, and $\frac{5}{2}$.

┌─ GRAPHICAL SOLUTION

We find the *x*-intercepts of the function

$$f(x) = 6x^6 - 19x^5 + 10x^4.$$

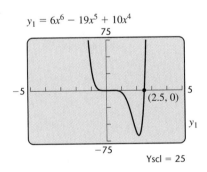

$y_1 = 6x^6 - 19x^5 + 10x^4$

(2.5, 0)

y_1

Yscl = 25

Using the viewing window $[-5, 5, -75, 75]$, we see that 2.5 is a zero of the function. From this window, we cannot tell how many times the graph of the function intersects the *x*-axis between -1 and 1. We magnify that portion of the *x*-axis.

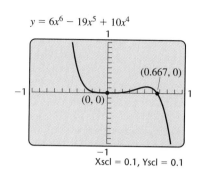

$$y = 6x^6 - 19x^5 + 10x^4$$

The x-coordinates of the x-intercepts are 0, 0.667, and 2.5.

Since $\frac{5}{2} = 2.5$ and $\frac{2}{3} \approx 0.667$, we obtain the same solutions using both methods. The solutions are 0, $\frac{2}{3}$, and $\frac{5}{2}$.

Example 4 illustrates two disadvantages of solving equations graphically: (1) It can be easy to "miss" solutions in some viewing windows and (2) solutions may be given as approximations.

Factoring and Functions

Our work with factoring can help us when we are working with functions.

Example 5   Given that $f(x) = 3x^2 - 4x$, find all values of a for which $f(a) = 4$.

SOLUTION   We want all numbers a for which $f(a) = 4$. Since $f(a) = 3a^2 - 4a$, we must have

$$3a^2 - 4a = 4 \qquad \text{Setting } f(a) \text{ equal to 4}$$
$$3a^2 - 4a - 4 = 0 \qquad \text{Getting 0 on one side}$$
$$(3a + 2)(a - 2) = 0 \qquad \text{Factoring}$$
$$3a + 2 = 0 \quad or \quad a - 2 = 0$$
$$a = -\tfrac{2}{3} \quad or \qquad a = 2.$$

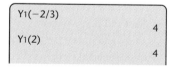

To check using a grapher, we enter $y_1 = 3x^2 - 4x$ and calculate Y1(−2/3) and Y2(2). To have $f(a) = 4$, we must have $a = -\frac{2}{3}$ or $a = 2$.

Example 6   Find the domain of F if $F(x) = \dfrac{x - 2}{x^2 + 2x - 15}$.

SOLUTION   The domain of F is the set of all values for which

$$\frac{x - 2}{x^2 + 2x - 15}$$

is a real number. Since division by 0 is undefined, $F(x)$ cannot be calculated for any x-value for which the denominator, $x^2 + 2x - 15$, is 0. To make sure these values are *excluded,* we solve:

$$x^2 + 2x - 15 = 0 \qquad \text{Setting the denominator equal to 0}$$
$$(x - 3)(x + 5) = 0 \qquad \text{Factoring}$$
$$x - 3 = 0 \quad or \quad x + 5 = 0$$
$$x = 3 \quad or \qquad x = -5. \qquad \text{These are the values}$$
$$\text{to } \textit{exclude.}$$

We can check to see whether either 3 or -5 is in the domain of F using a grapher, as shown at left. We enter $y = (x - 2)/(x^2 + 2x - 15)$ and set up a table with Indpnt set to Ask. When the x-values of 3 and -5 are supplied, the grapher should return an error message, indicating that those numbers are not in the domain of the function. There will be a function value given for any other value of x.

The domain of F is $\{x \mid x$ is a real number and $x \neq -5$ and $x \neq 3\}$.

X	Y1
3	ERROR
−5	ERROR
0	.13333
.5	.10909
8	.09231
127	.00764
−3	.41667
X = 3	

5.5 Exercise Set

Factor.

1. $6x^2 - 5x - 25$
2. $3x^2 - 16x - 12$
3. $10y^3 - 12y - 7y^2$
4. $6x^3 - 15x - x^2$
5. $24a^2 - 14a + 2$
6. $3a^2 - 10a + 8$
7. $35y^2 + 34y + 8$
8. $9a^2 + 18a + 8$
9. $4t + 10t^2 - 6$
10. $8x + 30x^2 - 6$
11. $8x^2 - 16 - 28x$
12. $18x^2 - 24 - 6x$
13. $14x^4 - 19x^3 - 3x^2$
14. $70x^4 - 68x^3 + 16x^2$
15. $12a^2 - 4a - 16$
16. $12a^2 - 14a - 20$
17. $9x^2 + 15x + 4$
18. $6y^2 - y - 2$
19. $8 - 6z - 9z^2$
20. $3 + 35a - 12a^2$
21. $18xy^3 + 3xy^2 - 10xy$
22. $3x^3y^2 - 5x^2y^2 - 2xy^2$
23. $24x^2 - 2 - 47x$
24. $15z^2 - 10 - 47z$
25. $63x^3 + 111x^2 + 36x$
26. $50t^3 + 115t^2 + 60t$
27. $24x^4 + 2x^2 - 15$
28. $40y^4 + 4y^2 - 12$

29. Use the results of Exercise 1 to solve
$6x^2 - 5x - 25 = 0$.

30. Use the results of Exercise 2 to solve
$3x^2 - 16x - 12 = 0$.

31. Use the results of Exercise 19 to solve
$9z^2 + 6z = 8$.

32. Use the results of Exercise 20 to solve
$3 + 35a = 12a^2$.

33. Use the results of Exercise 25 to solve
$63x^3 + 111x^2 + 36x = 0$.

34. Use the results of Exercise 26 to solve
$50t^3 + 115t^2 + 60t = 0$.

Solve.

35. $3x^2 - 8x + 4 = 0$
36. $9x^2 - 15x + 4 = 0$
37. $4t^3 + 11t^2 + 6t = 0$
38. $8n^3 + 10n^2 + 3n = 0$
39. $6x^2 = 13x + 5$
40. $40x^2 + 43x = 6$
41. $x(5 + 12x) = 28$
42. $a(1 + 21a) = 10$

43. Find the zeros of the function given by
$f(x) = 2x^2 - 13x - 7$.

44. Find the zeros of the function given by
$g(x) = 6x^2 + 13x + 6$.

45. Let $f(x) = x^2 + 12x + 40$. Find a such that
$f(a) = 8$.

46. Let $f(x) = x^2 + 14x + 50$. Find a such that
$f(a) = 5$.

47. Let $g(x) = 2x^2 + 5x$. Find a such that $g(a) = 12$.

48. Let $g(x) = 2x^2 - 15x$. Find a such that $g(a) = -7$.

49. Let $h(x) = 12x + x^2$. Find a such that $h(a) = -27$.

50. Let $h(x) = 4x - x^2$. Find a such that $h(a) = -32$.

Find the domain of the function f given by each of the following.

51. $f(x) = \dfrac{3}{x^2 - 4x - 5}$
52. $f(x) = \dfrac{2}{x^2 - 7x + 6}$

53. $f(x) = \dfrac{x - 5}{9x - 18x^2}$
54. $f(x) = \dfrac{1 + x}{3x - 15x^2}$

55. $f(x) = \dfrac{7}{5x^3 - 35x^2 + 50x}$

56. $f(x) = \dfrac{3}{2x^3 - 2x^2 - 12x}$

Skill Maintenance

57. The width of a rectangle is 7 ft less than its length. If the width is increased by 2 ft, the perimeter is then 66 ft. What is the area of the original rectangle? [3.3]

58. If $f(x) = 7 - x^2$, find $f(-3)$. [2.1]

59. Find the slope and the y-intercept of the line given by $4x - 3y = 8$. [2.4]

Solve. [4.3]

60. $|x| = 27$

61. $|5x - 6| \le 39$

62. $|5x - 6| > 39$

Synthesis

63. ◈ Describe in your own words an approach that can be used to factor any "nonprime" trinomial of the form $ax^2 + bx + c$.

64. ◈ Suppose that $(rx + p)(sx - q) = ax^2 - bx + c$ is true. Explain how this can be used to factor $ax^2 + bx + c$.

65. ◈ Emily has factored a polynomial as $(a - b)(x - y)$, while Jorge has factored the same polynomial as $(b - a)(y - x)$. Who is correct, and why?

66. ◈ Austin says that the domain of the function

$$F(x) = \frac{x + 3}{3x^2 - x - 2}$$

is $\left\{-\frac{2}{3}, 1\right\}$. Is he correct? Why or why not?

Use a graph to help factor each polynomial.

67. $4x^2 + 120x + 675$ **68.** $4x^2 + 164x + 1197$

69. $3x^3 + 150x^2 - 3672x$ **70.** $5x^4 + 20x^3 - 1600x^2$

Solve.

71. $(8x + 11)(12x^2 - 5x - 2) = 0$

72. $(x + 1)^3 = (x - 1)^3 + 26$

73. $(x - 2)^3 = x^3 - 2$

Factor. Assume that variables in exponents represent positive integers.

74. $6(x - 7)^2 + 13(x - 7) - 5$

75. $2a^4b^6 - 3a^2b^3 - 20$

76. $5x^8y^6 + 35x^4y^3 + 60$

77. $4x^{2a} - 4x^a - 3$

5.6

Equations Containing Perfect-Square Trinomials and Differences of Squares

- *Factoring Perfect-Square Trinomials*
- *Factoring Differences of Squares*
- *Solving Equations*

We now introduce a faster way to factor trinomials that are squares of binomials. A faster way to factor differences of squares is also developed. These factoring methods enable us to solve more types of polynomial equations using the principle of zero products.

Factoring Perfect-Square Trinomials

Consider the trinomial

$$x^2 + 6x + 9.$$

To factor it, we can look for factors of 9 that add to 6. These factors are 3 and 3 and the factorization is

$$x^2 + 6x + 9 = (x + 3)(x + 3) = (x + 3)^2.$$

Note that the result is the square of a binomial. Because of this, we call $x^2 + 6x + 9$ a **perfect-square trinomial**. Although trial and error can be used to factor any perfect-square trinomial, a faster procedure can be developed. In order to do so, we must first recognize when a trinomial is a perfect square.

To Recognize a Perfect-Square Trinomial:

A perfect-square trinomial is of the form

$$A^2 + 2AB + B^2 \quad \text{or} \quad A^2 - 2AB + B^2.$$

Note that:

- Two of the terms must be squares, such as A^2 and B^2.
- There must be no minus sign before A^2 or B^2.
- The remaining term is twice the product of A and B, $2AB$, or its opposite, $-2AB$.

For example, $x^2 + 10x + 25$ and $100a^2 + 81 - 180a$ are perfect-square trinomials, but $4x + 16 + 3x^2$ is not, because in the last case, only one term, 16, is square.

To factor a perfect-square trinomial, we use the same equations that we used for squaring a binomial. The pattern learned in Section 5.2 is reversed.

Factoring a Perfect-Square Trinomial

$$A^2 + 2AB + B^2 = (A + B)^2;$$
$$A^2 - 2AB + B^2 = (A - B)^2$$

Example 1   Factor.

a) $x^2 - 10x + 25$

b) $16y^2 + 49 + 56y$

c) $-20xy + 4y^2 + 25x^2$

> **Caution!** These are expressions that can be factored. Do not try to solve an expression.

SOLUTION

a) $x^2 - 10x + 25 = (x - 5)^2$

Note the sign!

We find the square terms and write the quantities that were squared with a minus sign between them.

b) $16y^2 + 49 + 56y = 16y^2 + 56y + 49$ Using a commutative law

$$= (4y + 7)^2$$

We find the square terms and write the quantities that were squared with a plus sign between them.

c) $-20xy + 4y^2 + 25x^2 = 4y^2 - 20xy + 25x^2$ Writing descending order with respect to y

$$= (2y - 5x)^2$$

This square can also be expressed as

$$25x^2 - 20xy + 4y^2 = (5x - 2y)^2.$$

As always, any factorization can be checked by multiplying:

$$(5x - 2y)^2 = (5x - 2y)(5x - 2y) = 25x^2 - 20xy + 4y^2. \quad \blacksquare$$

When factoring, always look first for a factor common to all the terms.

Example 2 Factor: **(a)** $2x^2 - 12xy + 18y^2$; **(b)** $-4y^2 - 144y^8 + 48y^5$.

SOLUTION

a) We first look for a common factor. This time, there is a common factor, 2.

$$2x^2 - 12xy + 18y^2 = 2(x^2 - 6xy + 9y^2) \quad \text{Factoring out the 2}$$
$$= 2(x - 3y)^2 \quad \text{Factoring the perfect-square trinomial}$$

b) $-4y^2 - 144y^8 + 48y^5 = -4y^2(1 + 36y^6 - 12y^3)$ Factoring out the common factor

$$= -4y^2(36y^6 - 12y^3 + 1)$$ Changing order. Note that $(y^3)^2 = y^6$.

$$= -4y^2(6y^3 - 1)^2$$ Factoring the perfect-square trinomial $\blacksquare$

Factoring Differences of Squares

In Section 5.2, we saw that the product of the sum and the difference of the same two terms is a difference of two squares:

$$(A + B)(A - B) = A^2 - B^2.$$

When an expression like $x^2 - 9$ is recognized as a difference of two squares, we can reverse this pattern.

Factoring a Difference of Two Squares

$$A^2 - B^2 = (A + B)(A - B)$$

To factor a difference of two squares, write the product of the sum and the difference of the two quantities being squared.

Example 3   Factor:

a) $x^2 - 9$;

b) $25y^6 - 49x^2$

SOLUTION

a) $x^2 - 9 = x^2 - 3^2 = (x + 3)(x - 3)$

$$A^2 \quad - \quad B^2 \quad = (A \quad + \quad B)(A \quad - \quad B)$$

b) $25y^6 - 49x^2 = (5y^3)^2 - (7x)^2 = (5y^3 + 7x)(5y^3 - 7x)$

As always, the first step in factoring is to look for common factors.

Example 4   Factor:

a) $5 - 5x^2y^6$;

b) $16x^4y - 81y$.

SOLUTION

a) $5 - 5x^2y^6 = 5(1 - x^2y^6)$ Factoring out the common factor

$\qquad = 5[1^2 - (xy^3)^2]$ Rewriting x^2y^6 as a quantity squared

$\qquad = 5(1 + xy^3)(1 - xy^3)$ Factoring the difference of squares

b) $16x^4y - 81y = y(16x^4 - 81)$ Factoring out the common factor

$\qquad = y[(4x^2)^2 - 9^2]$

$\qquad = y(4x^2 + 9)(4x^2 - 9)$ Factoring the difference of squares

$\qquad = y(4x^2 + 9)(2x + 3)(2x - 3)$ Factoring $4x^2 - 9$, which is *also* a difference of squares

In Example 4(b), it is tempting to try to factor $(4x^2 + 9)$. Note that it is a sum of two squares. Apart from possibly removing a common factor, it is impossible to factor a sum of squares using real numbers. Note also in Example 4(b) that $4x^2 - 9$ *could* be factored further. Whenever a factor itself can be factored, be sure to do so. That way you will be factoring *completely*.

Example 5   Factor:

$$x^3 + 3x^2 - 4x - 12.$$

SOLUTION

$$x^3 + 3x^2 - 4x - 12 = x^2(x + 3) - 4(x + 3)$$

Factoring by grouping

$$= (x + 3)(x^2 - 4)$$

Factoring out $x + 3$

$$= (x + 3)(x + 2)(x - 2)$$

Factoring $x^2 - 4$ to factor completely

Solving Equations

We can now solve polynomial equations involving differences of squares and perfect-square trinomials.

Example 6   Solve: $x^3 + 3x^2 = 4x + 12$.

SOLUTION   We first note that the equation is a third-degree polynomial equation. Thus it will have 3 or fewer solutions.

ALGEBRAIC SOLUTION

We have

$$x^3 + 3x^2 = 4x + 12$$
$$x^3 + 3x^2 - 4x - 12 = 0 \quad \text{Getting 0 on one side}$$
$$(x + 3)(x + 2)(x - 2) = 0$$

Factoring; using the results of Example 5

$$x + 3 = 0 \quad or \quad x + 2 = 0 \quad or \quad x - 2 = 0$$

Using the principle of zero products

$$x = -3 \quad or \qquad x = -2 \quad or \qquad x = 2.$$

The solutions are -3, -2, and 2.

GRAPHICAL SOLUTION

We let $f(x) = x^3 + 3x^2$ and $g(x) = 4x + 12$ and look for any points of intersection of the graphs of f and g.

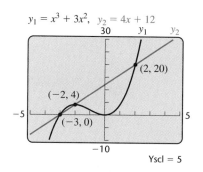

$y_1 = x^3 + 3x^2, \quad y_2 = 4x + 12$

$(2, 20)$

$(-2, 4)$

$(-3, 0)$

$Yscl = 5$

The x-coordinates of the points of intersection are -3, -2, and 2.

As a partial check, note that both methods yield the same solutions. Substituting -3, -2, and 2 into the original equation, we see that all three numbers check. The solutions are -3, -2, and 2.

Example 7 Find the zeros of the function given by $f(x) = x^4 - 8x^2 + 16$.

SOLUTION   The function is a fourth-degree polynomial function, so there will be at most 4 zeros. To find the zeros of the function, we find the roots of the equation $x^4 - 8x^2 + 16 = 0$.

┌--- ALGEBRAIC SOLUTION

We have

$$x^4 - 8x^2 + 16 = 0$$

$$(x^2)^2 - 2 \cdot 4 \cdot x^2 + 4^2 = 0$$

Recognizing a perfect-square trinomial

$$(x^2 - 4)^2 = 0$$

Factoring a perfect-square trinomial

$$(x + 2)(x - 2)(x + 2)(x - 2) = 0$$

Factoring the differences of squares

$$x + 2 = 0 \quad or \ x - 2 = 0 \ or \ x + 2 = 0 \quad or \ x - 2 = 0$$

Using the principle of zero products

$$x = -2 \ or \qquad x = 2 \ or \qquad x = -2 \ or \qquad x = 2.$$

We have -2 and 2 as solutions.

┌--- GRAPHICAL SOLUTION

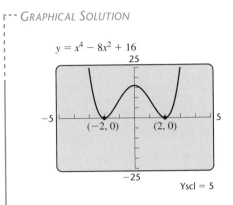

$y = x^4 - 8x^2 + 16$

Yscl = 5

The zeros are -2 and 2.

The numbers -2 and 2 check in the original equation. The solutions are -2 and 2.

Note in Example 7 that the factors $(x + 2)$ and $(x - 2)$ each appear twice in the factorization. When a factor appears two or more times, we say that we have a **repeated root**. If the factor occurs twice, we say that we have a **double root**, or a **root of multiplicity two**.

In this chapter, we have emphasized the relationship between factoring and equation solving. There are many polynomial equations, however, that cannot be solved by factoring. Nonetheless, we can find approximations of any real-number solutions using a grapher.

Example 8 Solve: $x^3 - 6x^2 + 5x + 1 = 0$.

SOLUTION   The polynomial $x^3 - 6x^2 + 5x + 1$ is prime; it cannot be factored using rational coefficients. We solve the equation by graphing $f(x) = x^3 - 6x^2 + 5x + 1$ and finding the zeros of the function. Since the polynomial is of degree 3, we need not look for more than 3 zeros.

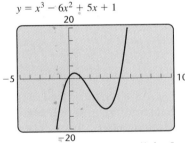

$y = x^3 - 6x^2 + 5x + 1$

Yscl = 5

X	Y₁	
−.166	9E−5	
1.217	9.5E−4	
4.949	.00328	

X = −.166

There are three x-intercepts of the graph. We use the ZERO feature to find each root, or zero. Since the solutions are irrational, the grapher will give approximations. Rounding each to the nearest thousandth, we have the solutions -0.166, 1.217, and 4.949. We check the solutions using a table. Note that the function values are close to 0 for each approximate solution.

We have considered both algebraic and graphical methods of solving polynomial equations. It is important to understand and be able to use both methods. Some of the advantages and disadvantages of each method are given in the following table.

	ADVANTAGES	DISADVANTAGES
ALGEBRAIC METHOD	■ Can find exact answers ■ Works well when the polynomial is in factored form or can be readily factored ■ Can be used to find solutions that are not real numbers (see Chapter 8)	■ Cannot be used if the polynomial is not factorable ■ Can be difficult to use if factorization is not readily apparent
GRAPHICAL METHOD	■ Does not require the polynomial to be factored ■ Can visualize solutions	■ Easy to miss solutions if an appropriate viewing window is not chosen ■ Gives approximations of solutions ■ For most graphers, only real-number solutions can be found.

5.6 Exercise Set

Factor.

1. $x^2 + 8x + 16$

2. $t^2 + 6t + 9$

3. $a^2 - 16a + 64$

4. $a^2 - 14a + 49$

5. $2a^2 + 8a + 8$

6. $4a^2 - 16a + 16$

7. $y^2 + 36 - 12y$

8. $y^2 + 36 + 12y$

9. $24a^2 + a^3 + 144a$

10. $-18y^2 + y^3 + 81y$

11. $32x^2 + 48x + 18$

12. $2x^2 - 40x + 200$

13. $64 + 25a^2 - 80a$

14. $1 - 8d + 16d^2$

15. $0.25x^2 + 0.30x + 0.09$

16. $0.04x^2 - 0.28x + 0.49$

17. $p^2 - 2pq + q^2$

18. $m^2 + 2mn + n^2$

19. $a^4 - 10a^2 + 25$

20. $n^4 + 8n^2 + 16$

21. $25a^2 - 30ab + 9b^2$

22. $49p^2 - 84pq + 36q^2$

23. $t^8 + 2t^4s^4 + s^8$

24. $a^4 + 2a^2b^2 + b^4$

25. $x^2 - 16$

26. $y^2 - 100$

27. $p^2 - 49$

28. $m^2 - 64$

29. $a^2b^2 - 81$

30. $p^2q^2 - 25$

31. $6x^2 - 6y^2$

32. $8x^2 - 8y^2$

33. $7xy^4 - 7xz^4$

34. $25ab^4 - 25az^4$

35. $4a^3 - 49a$

36. $9x^4 - 25x^2$

37. $3x^8 - 3y^8$

38. $9a^4 - a^2b^2$

39. $9a^4 - 25a^2b^4$

40. $16x^6 - 121x^2y^4$

41. $\frac{1}{25} - x^2$

42. $\frac{1}{16} - y^2$

43. $(a + b)^2 - 9$

44. $(p + q)^2 - 25$

45. $36 - (x + y)^2$

46. $49 - (a + b)^2$

47. $m^3 - 7m^2 - 4m + 28$

48. $x^3 + 8x^2 - x - 8$

49. $a^3 - ab^2 - 2a^2 + 2b^2$

50. $p^2q - 25q + 3p^2 - 75$

51. Use the results of Exercise 1 to solve $x^2 + 8x + 16 = 0$.

52. Use the results of Exercise 4 to solve $a^2 - 14a + 49 = 0$.

53. Use the results of Exercise 25 to solve $x^2 = 16$.

54. Use the results of Exercise 26 to solve $y^2 = 100$.

Solve. Round any irrational solutions to the nearest thousandth.

55. $a^2 + 1 = 2a$

56. $r^2 + 16 = 8r$

57. $2x^2 - 24x + 72 = 0$

58. $-t^2 - 16t - 64 = 0$

59. $x^2 - 9 = 0$

60. $r^2 - 64 = 0$

61. $a^2 = \frac{1}{25}$

62. $x^2 = \frac{1}{100}$

63. $t^4 - 26t^2 + 25 = 0$

64. $t^4 - 13t^2 + 36 = 0$

65. $x^3 + 3 = 3x^2 + x$

66. $x^3 + x^2 = 16x + 16$

67. $x^2 - 3x - 7 = 0$

68. $x^2 - 5x + 1 = 0$

69. $2x^2 + 8x + 1 = 0$

70. $3x^2 + x - 1 = 0$

71. $x^3 + 3x^2 + x - 1 = 0$

72. $x^3 + x^2 + x - 1 = 0$

73. Let $f(x) = x^2 - 12x$. Find a such that $f(a) = -36$.

74. Let $g(x) = x^2$. Find a such that $g(a) = 144$.

Skill Maintenance

Solve.

75. $x - y + z = 6$,
$2x + y - z = 0$,
$x + 2y + z = 3$ [3.4]

76. $|5 - 7x| \geq 9$ [4.3]

77. $|5 - 7x| \leq 9$ [4.3]

78. $5 - 7x > -9 + 12x$ [4.1]

Synthesis

79. ◆ Describe a procedure that could be used to find a polynomial with four terms that can be factored as a difference of two squares.

80. ◆ Are the product and power rules for exponents (see Section 1.4) important when factoring differences of squares? Why or why not?

81. ◆ Under what conditions can a sum of two squares be factored?

82. ◆ Without finding the entire factorization, determine the number of factors of $x^{256} - 1$. Explain how you arrived at your answer.

83. *Volume of Carpeting.* The volume of a carpet that is rolled up can be estimated by the polynomial $\pi R^2 h - \pi r^2 h$.

a) Factor the polynomial.
b) Use both the original and the factored forms to find the volume of a roll for which $R = 50$ cm, $r = 10$ cm, and $h = 4$ m.

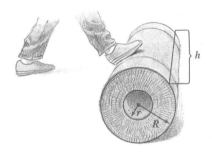

84. Use a grapher to show that
$$(x^2 - 3x + 2)^4 = x^8 + 81x^4 + 16$$
is *not* an identity.

Factor. Assume that variables in exponents represent positive integers.

85. $x^{2a} - y^2$

86. $x^{4a} - y^{2b}$

87. $9x^{2n} - 6x^n + 1$

88. $c^{2w+1} + 2c^{w+1} + c$

89. $5c^{100} - 80d^{100}$

Although sums of squares cannot be factored if there is no common factor, sums and differences of cubes can be factored using the following formulas:
$$A^3 + B^3 = (A + B)(A^2 - AB + B^2),$$
$$A^3 - B^3 = (A - B)(A^2 + AB + B^2).$$

Factor each of the following.

90. $t^3 - 8$

91. $x^3 + 64$

92. $x^3 + 27$

93. $z^3 - 1$

94. $8a^3 + 1000$

95. $54x^3 + 2$

96. $rs^3 + 64r$

97. $ab^3 + 125a$

98. $2y^3 - 54z^3$

99. $5x^3 - 40z^3$

100. $y^3 + 0.125$

101. $x^3 + 0.001$

102. $125c^6 - 8d^6$

103. $64x^6 - 8t^6$

104. $3z^5 - 3z^2$

105. $2y^4 - 128y$

5.7
Applications of Polynomial Equations

- *Problem Solving*
- *Fitting Polynomial Functions to Data*

Polynomial equations and functions occur frequently in applications. We can use polynomial equations to solve problems and to analyze data.

Problem Solving

Some problems can be translated to polynomial equations, which we are now able to solve. The problem-solving process is the same as that used for other kinds of problems.

Example 1 *Fireworks Displays.* Fireworks are typically launched from a mortar with an upward velocity (initial speed) of about 64 ft/sec. The height $h(t)$, in feet, of a "weeping willow" display, t seconds after having been launched from an 80-ft-high rooftop, is given by

$$h(t) = -16t^2 + 64t + 80.$$

After how long will the cardboard shell from the fireworks reach the ground?

SOLUTION

1. **Familiarize.** We make a drawing and label it, using the information provided. If we wanted to, we could evaluate $h(t)$ for a few values of t. Note that t cannot be negative, since it represents time from launch.

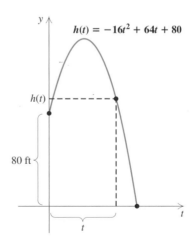

2. **Translate.** The relevant function has been provided. Since we are asked to determine how long it will take for the shell to hit the ground, we are interested in the value of t for which $h(t) = 0$:

$$-16t^2 + 64t + 80 = 0.$$

3. **Carry out.** We solve both algebraically and graphically, as follows.

We solve by factoring:

$$-16t^2 + 64t + 80 = 0$$

$$\left.\begin{array}{l} -16(t^2 - 4t - 5) = 0 \\ -16(t - 5)(t + 1) = 0 \end{array}\right\} \text{ Factoring}$$

$$t - 5 = 0 \quad or \quad t + 1 = 0$$
$$t = 5 \quad or \quad t = -1.$$

The solutions appear to be 5 and -1.

We graph the function and find its zeros. Using the interval $[-10, 10, -10, 150]$, we can see both x-intercepts and the shape of the curve.

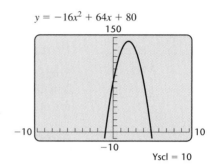

We use the ZERO option on the CALC menu to find the zeros of the function. They are -1 and 5.

4. **Check.** Since t cannot be negative, we need check only 5:

$$h(5) = -16 \cdot 5^2 + 64 \cdot 5 + 80 = 0.$$

The number 5 checks. Note that another check is provided by the fact that both the algebraic and graphical approaches yielded the same solution.

5. **State.** The cardboard shell will hit the ground about 5 sec after the fireworks display has been launched.

The following problem involves the **Pythagorean theorem**, which relates the lengths of the sides of a right triangle. A **right triangle** has a 90°, or right, angle, which is denoted by the symbol ⌐ or ⌐. The longest side, opposite the 90° angle, is called the **hypotenuse**. The other sides, called **legs**, form the two sides of the right angle.

The Pythagorean Theorem

The sum of the squares of the legs of a right triangle is equal to the square of the hypotenuse:

$$a^2 + b^2 = c^2.$$

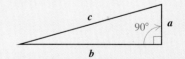

Example 2 *Carpentry.* In order to build a deck at a right angle to their house, Lucinda and Felipe decide to plant a stake in the ground a precise distance from the back wall of their house. This stake will combine with

two marks on the house to form a right triangle. From a course in geometry, Lucinda remembers that there are three consecutive integers that can work as sides of a right triangle. Find the measurements of that triangle.

SOLUTION

1. **Familiarize.** Recall that x, $x + 1$, and $x + 2$ can be used to represent three unknown consecutive integers. Since $x + 2$ is the largest number, it must represent the hypotenuse. The legs serve as the sides of the right angle, so one leg must be formed by the marks on the house. We make a drawing in which

   $x =$ the distance between the marks on the house,

   $x + 1 =$ the length of the other leg,

   and

   $x + 2 =$ the length of the hypotenuse.

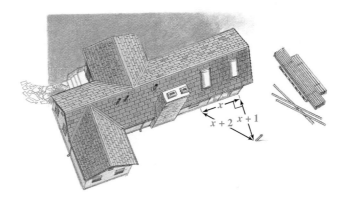

2. **Translate.** Applying the Pythagorean theorem, we translate as follows:

$$a^2 + b^2 = c^2$$
$$x^2 + (x + 1)^2 = (x + 2)^2.$$

3. **Carry out.** We solve both algebraically and graphically, as follows.

ALGEBRAIC SOLUTION

We solve the equation as follows:

$$x^2 + (x^2 + 2x + 1) = x^2 + 4x + 4 \qquad \text{Squaring the binomials}$$
$$2x^2 + 2x + 1 = x^2 + 4x + 4 \qquad \text{Combining like terms}$$
$$x^2 - 2x - 3 = 0 \qquad \text{Adding } -x^2 \text{ and } -4x \text{ and } -4 \text{ on both sides}$$
$$(x - 3)(x + 1) = 0 \qquad \text{Factoring}$$
$$x - 3 = 0 \quad or \quad x + 1 = 0 \qquad \text{Using the principle of zero products}$$
$$x = 3 \quad or \qquad x = -1.$$

The possible solutions are 3 and -1.

GRAPHICAL SOLUTION

We graph $y_1 = x^2 + (x + 1)^2$ and $y_2 = (x + 2)^2$ and look for the points of intersection of the graphs. A standard $[-10, 10, -10, 10]$ viewing window shows one point of intersection.

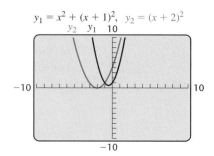

To see if the graphs intersect at another point, we use a larger viewing window. The window $[-10, 10, -10, 50]$ shows that there is indeed another point of intersection.

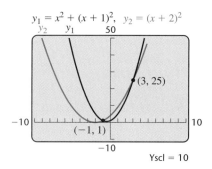

The points of intersection are $(-1, 1)$ and $(3, 25)$, so the possible solutions are -1 and 3.

4. Check. The integer -1 cannot be a length of a side because it is negative. When $x = 3$, $x + 1 = 4$, and $x + 2 = 5$. Since $3^2 + 4^2 = 5^2$, the lengths 3, 4, and 5 determine a right triangle. Thus, 3, 4, and 5 check.

5. State. Lucinda and Felipe should use a triangle with sides having a ratio of $3:4:5$. Thus, if the marks on the house are 3 ft apart, they should locate the stake at the point in the yard that is precisely 4 ft from one mark and 5 ft from the other mark.

Example 3 *Display of a Stamp.* A valuable stamp is 4 cm wide and 5 cm long. The stamp is to be mounted on a rectangular sheet of paper that is $5\frac{1}{2}$ times the area of the stamp. Determine the dimensions of the paper that will ensure a uniform border around the stamp.

SOLUTION

1. Familiarize. We make a drawing and label it, using x to represent the width of the border, in centimeters. Since the border extends uniformly around the entire stamp, the length of the sheet of paper must be $5 + 2x$ and the width must be $4 + 2x$.

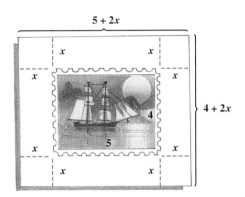

2. Translate. We rephrase the information given and translate as follows:

$$\underbrace{\text{Area of sheet}}_{(5 + 2x)(4 + 2x)} \;\; \underbrace{\text{is}}_{=} \;\; \underbrace{5\tfrac{1}{2} \text{ times}}_{5\tfrac{1}{2} \cdot} \;\; \underbrace{\text{Area of stamp.}}_{5 \cdot 4}$$

3. Carry out. We solve both algebraically and graphically, as follows.

ALGEBRAIC SOLUTION

We solve the equation:

$(5 + 2x)(4 + 2x) = 5\tfrac{1}{2} \cdot 5 \cdot 4$

$20 + 10x + 8x + 4x^2 = 110$ **Multiplying**

$4x^2 + 18x - 90 = 0$ **Finding standard form**

$2x^2 + 9x - 45 = 0$ **Multiplying by $\tfrac{1}{2}$ on both sides**

$(2x + 15)(x - 3) = 0$ **Factoring**

$2x + 15 = 0 \quad or \quad x - 3 = 0$ **Principle of zero products**

$x = -7\tfrac{1}{2} \quad or \quad x = 3.$

The possible solutions are $-7\tfrac{1}{2}$ and 3.

GRAPHICAL SOLUTION

We graph $y_1 = (5 + 2x)(4 + 2x)$ and $y_2 = 5.5(5)(4)$ and look for the points of intersection of the graphs. Since we have $y_2 = 110$, we use a viewing window with Ymax > 110.

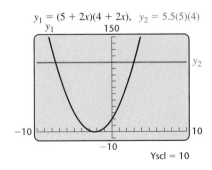

Using INTERSECT, we find the points of intersection $(-7.5, 110)$ and $(3, 110)$. The possible solutions are -7.5 and 3.

4. **Check.** We check 3 in the original problem. (Note that $-7\frac{1}{2}$ is not a solution because measurements cannot be negative.) If the border is 3 cm wide, the paper will have a length of $5 + 2 \cdot 3$, or 11 cm, and a width of $4 + 2 \cdot 3$, or 10 cm. The area of the paper is thus $11 \cdot 10$, or 110 cm^2. Since the area of the stamp is 20 cm^2 and 110 cm^2 is $5\frac{1}{2}$ times 20 cm^2, the number 3 checks. Note too that both the algebraic and graphical approaches yielded the same solution.

5. **State.** The sheet of paper should be 11 cm long and 10 cm wide.

Fitting Polynomial Functions to Data

In Chapters 2–4, we modeled data using linear functions. There are many situations, however, that cannot be modeled using a linear function.

Example 4 Determine whether each of the following sets of data can be modeled using a linear function.

a) *Media Usage by Eighteen-Year Olds*

YEAR*	MEDIA USAGE (IN HOURS PER DAY)
1990	8.944
1991	8.917
1992	9.114
1993	9.040
1994	9.314
1995	9.402
1996	9.465
1997	9.552
1998	9.659
1999	9.747

*Data from 1995 on are projected.
Source: Veronis, Suhler, & Associates, Inc., New York, New York

b) *Growth of World Wide Web Sites*

YEAR	NUMBER OF WEB SITES (IN MILLIONS)
1995	8
1996	11
1997	27
1998	50
1999	78
2000	140

Source: International Data Corporation, 1996

c) *Age Equivalents for Dogs and Humans*

AGE OF DOG (IN YEARS)	HUMAN AGE (IN YEARS)
0.25	5
0.5	10
1	15
2	24
4	32
6	40
8	48
10	56
14	72
18	91
21	106

Source: Country, December 1992: 60

SOLUTION   We plot each set of data and determine visually whether the data can be modeled closely by a linear function.

a) We let x = the number of years since 1990 and y = media usage, in number of hours per day. Next, we enter the data using the STAT menu and graph. Although not all the points lie on a straight line, the data points approximate a straight line. We illustrate this by sketching a line through the points. This situation can be modeled using a linear function.

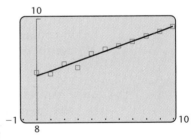

b) We let x = the number of years since 1995 and y = the number of web sites, in millions. The graph of the data shows us that the rate of change in the number of web sites per year is not constant. This situation cannot be modeled by a linear function.

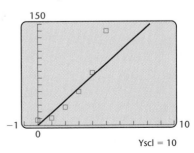

Yscl = 10

c) We let x = dog age, in years, and y = human age, in years. We graph the data and sketch a line through the points. This situation *could* be modeled using a linear function. However, note from the graph that dogs mature more rapidly during the first year than in later years. The *rate of change* in age equivalence from dog to human is not the same throughout a dog's life. A nonlinear function may be a *better* model for this relationship.

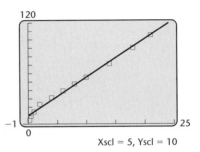

Xscl = 5, Yscl = 10

Many graphers have the capability to fit nonlinear polynomial equations to data using regression.

Example 5 *Dog Years.* The lifespan of a dog is typically much shorter than that of a human. Refer to the data in Example 4(c) to do the following.

a) Use regression to find and graph a polynomial function h of degree 3 that can be used to determine the equivalent human age $h(x)$ for a dog that is x years old.

b) Estimate the equivalent human age for a dog that is 24 yr old.

c) A common retirement age for humans is 65. Estimate the equivalent age for a dog.

SOLUTION

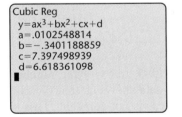

a) First, we enter the data, with the age of a dog in list L1 and the equivalent human age in L2. A polynomial in one variable of degree 3 is called a *cubic*. The CubicReg option in the STAT CALC menu uses regression to fit a cubic equation to a set of data, as shown at left.

The coefficients are given for a cubic function of the form $h(x) = ax^3 + bx^2 + cx + d$. The graph of the equation is shown below, along with the graph of the data.

$$y = 0.0102548814x^3 - 0.3401188859x^2 + 7.397498939x + 6.618361098$$

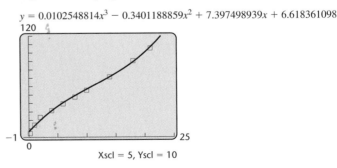

Xscl = 5, Yscl = 10

b) The equivalent human age of a 24-yr-old dog is given by $h(24)$. Using the VALUE option in the CALC menu, we find that $h(24) \approx 130$.

c) To find the dog age that is equivalent to a human age of 65, we must solve the equation

$$h(x) = 65.$$

We graph $y_1 = h(x)$ and $y_2 = 65$ and determine the coordinates of any points of intersection.

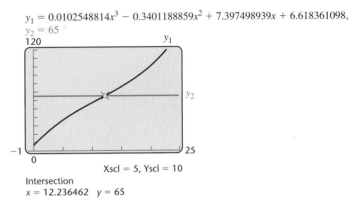

$y_1 = 0.0102548814x^3 - 0.3401188859x^2 + 7.397498939x + 6.618361098,$
$y_2 = 65$

Xscl = 5, Yscl = 10

Intersection
$x = 12.236462$ $y = 65$

The graphs intersect at approximately $(12.2, 65)$. A human age of 65 yr is equivalent to a dog age of 12.2 yr.

In Example 5, we chose to use a cubic function because neither a quadratic function nor a linear function described the rapid early maturing shown by the data. The best type of model to use for a situation is often difficult to determine.

5.7 Exercise Set

Solve.

1. The square of a number plus the number is 156. What is the number?

2. The square of a number plus the number is 132. What is the number?

3. A book is 5 cm longer than it is wide. Find the length and the width if the area is 84 cm^2.

4. An envelope is 4 cm longer than it is wide. The area is 96 cm^2. Find the length and the width.

5. *Geometry.* If each of the sides of a square is lengthened by 6 cm, the area becomes 144 cm^2. Find the length of a side of the original square.

6. *Geometry.* If each of the sides of a square is lengthened by 4 m, the area becomes 49 m^2. Find the length of a side of the original square.

7. *Framing a Picture.* A picture frame measures 12 cm by 20 cm, and 84 cm^2 of picture shows. Find the width of the frame.

8. *Framing a Picture.* A picture frame measures 14 cm by 20 cm, and 160 cm^2 of picture shows. Find the width of the frame.

9. *Landscaping.* A rectangular garden is 30 ft by 40 ft. Part of the garden is removed in order to install a walkway of uniform width around it. The area of the new garden is one-half the area of the old garden. How wide is the walkway?

10. *Landscaping.* A rectangular lawn measures 60 ft by 80 ft. Part of the lawn is torn up to install a sidewalk of uniform width around it. The area of the new lawn is 2400 ft². How wide is the sidewalk?

11. *Tent Design.* The triangular entrance to a tent is 2 ft taller than it is wide. The area of the entrance is 12 ft². Find the height and the base.

12. Three consecutive even integers are such that the square of the first plus the square of the third is 136. Find the three integers.

13. Three consecutive even integers are such that the square of the third is 76 more than the square of the second. Find the three integers.

14. *Antenna Wires.* A wire is stretched from the ground to the top of an antenna tower, as shown. The wire is 20 ft long. The height of the tower is 4 ft greater than the distance d from the tower's base to the end of the wire. Find the distance d and the height of the tower.

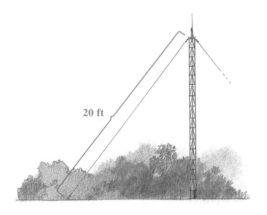

20 ft

15. *Parking Lot Design.* A rectangular parking lot is 50 ft longer than it is wide. Determine the dimensions of the parking lot if it measures 250 ft diagonally.

16. *Sailing.* A triangular sail is 9 m taller than it is wide. The area is 56 m². Find the height and the base of the sail.

Area = 56m²

17. *Ladder Location.* The foot of an extension ladder is 9 ft from a wall. The height that the ladder reaches on the wall and the length of the ladder are consecutive integers. How long is the ladder?

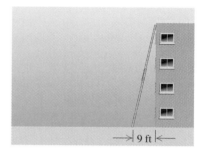

9 ft

18. *Ladder Location.* The foot of an extension ladder is 10 ft from a wall. The ladder is 2 ft longer than the height that it reaches on the wall. How far up the wall does the ladder reach?

19. *Garden Design.* Ignacio is planning a garden that is 25 m longer than it is wide. The garden will have an area of 7500 m². What will its dimensions be?

20. *Garden Design.* A flower bed is to be 3 m longer than it is wide. The flower bed will have an area of 108 m². What will its dimensions be?

21. *Fireworks.* Suppose that a bottle rocket is launched upward with an initial velocity of 96 ft/sec from a height of 880 ft. Its height in feet, $h(t)$, after t seconds is given by

$$h(t) = -16t^2 + 96t + 880.$$

After how long will the rocket reach the ground?

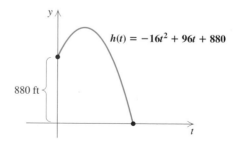

$h(t) = -16t^2 + 96t + 880$

880 ft

22. *Safety Flares.* Suppose that a flare is launched upward with an initial velocity of 80 ft/sec from a height of 224 ft. Its height in feet, $h(t)$, after t seconds is given by

$$h(t) = -16t^2 + 80t + 224.$$

After how long will the flare reach the ground?

23. *Camcorder Production.* Suppose that the cost of making x video cameras is $C(x) = \frac{1}{9}x^2 + 2x + 1$, where $C(x)$ is in thousands of dollars. If the revenue from the sale of x video cameras is given by $R(x) = \frac{5}{36}x^2 + 2x$, where $R(x)$ is in thousands of dollars, how many cameras must be sold in order for the firm to break even?

24. *Cabinet Making.* Dovetail Woodworking determines that the revenue R, in thousands of dollars, from the sale of x sets of cabinets is given by $R(x) = 2x^2 + x$. If the cost C, in thousands of dollars, of producing x sets of cabinets is given by $C(x) = x^2 - 2x + 10$, how many sets must be produced and sold in order for the company to break even?

Determine whether each set of data can be modeled by a linear function.

25. *Number of Farms in the United States*

Year	Number of Farms (in millions)
1850	2.1
1870	3.2
1890	5
1910	6.7
1930	6.6
1950	5.8
1970	3.5
1985	2.293
1990	2.146
1996	2.063

Source: The Macmillan Visual Almanac, 1996; Statistical Abstract of the United States, 1997

26. *Health-Care Payments*

Year	Health-Care Payments (in billions)
1990	$175.9
1991	204.4
1992	245.1
1993	268.7
1994	328.9
1995	344.3
1996	377.1
1997	405.7

Source: Adams Media Research, Carmel, CA

27. *Amusement Park Attendance in the United States*

Year	Attendance at Amusement Parks (in millions)
1986	97
1987	115
1988	129
1989	124
1990	122
1991	120
1992	131

Source: The Macmillan Visual Almanac, 1996

28. *Leisure Time for U.S. Workers*

Year	Leisure Time (in hours per week)
1973	26.2
1975	24.3
1980	19.2
1984	18.1
1987	16.6
1989	18.8
1993	18.8
1994	19.5
1995	19.2
1997	19.5

Source: The Wall Street Journal Almanac, 1998

29. *Computer Usage in Schools*

YEAR	RATIO OF STUDENTS TO COMPUTERS
1993	12
1994	10.8
1995	9.1
1997	7.3

Source: The Wall Street Journal Almanac, 1998

30. *Milk Price Index*

MONTH AND YEAR	PRICE PAID TO FARMER (BASED ON 100 IN 1982)
July 1996	114
September 1996	120
November 1996	111
February 1997	96
April 1997	98

Source: U.S. Department of Agriculture

31. *Number of Farms.* The table in Exercise 25 lists the number of U.S. farms for various years.

a) Use regression to find a polynomial function f of degree 4 (quartic) that can be used to estimate the number of farms $f(x)$, in millions, x years after 1850.

b) Estimate the number of farms in 1980.

c) Estimate the year or years in which there were 4 million farms.

32. *Amusement Park Attendance.* The table in Exercise 27 lists the attendance at amusement parks for various years.

a) Use regression to find a polynomial function p of degree 3 (cubic) that can be used to estimate the attendance at amusement parks $p(x)$, in millions, x years after 1986.

b) Estimate the attendance at amusement parks in 1995.

c) Estimate the year or years in which the attendance at amusement parks was 250 million.

33. *World Wide Web Sites.* The table in Example 4(b) lists the number of web sites for various years.

a) Use regression to find a polynomial function w of degree 2 (quadratic) that can be used to estimate the number of web sites $w(x)$, in millions, x years after 1995.

b) Estimate the number of web sites in 2010.

c) Estimate the year or years in which there will be 200 million web sites.

34. *Leisure Time.* The table in Exercise 28 lists the median number of hours of leisure per week for U.S. workers for various years.

a) Use regression to find a polynomial function l of degree 4 (quartic) that can be used to estimate the leisure time $l(x)$, in number of hours per week, x years after 1973.

b) Estimate the number of leisure hours per week in 2000.

c) Estimate the year or years in which the median number of hours of leisure per week was 19.0.

Skill Maintenance

35. *Driving.* At noon, two cars start from the same location and travel in opposite directions at different speeds. After 7 hr, they are 651 mi apart. If one car is traveling 15 mph slower than the other car, what are their respective speeds? [3.3]

36. *Television Sales.* At the beginning of the month, J.C.'s Appliances had 150 televisions in stock. During the month, they sold 45% of their conventional televisions and 60% of their surround-sound televisions. If they sold a total of 78 televisions, how many of each type did they sell? [3.3]

Solve.

37. $2x - 14 + 9x > -8x + 16 + 10x$ [4.1]

38. $x + y = 0,$
$z - y = -2,$
$x - z = 6$ [3.4]

Synthesis

39. ◆ Sandra says that she will never miss a point of intersection when solving graphically because she always uses a $[-100, 100, -100, 100]$ window. Is she correct? Why or why not?

40. ◆ Tyler disregards any negative solutions that he finds when solving applied problems. Is this approach correct? Why or why not?

41. ◆ Write a chart of the population of two imaginary cities. Devise the numbers in such a way that one city has linear growth and the other has nonlinear growth.

42. ◆ Explain how you might decide what kind of function would fit a particular set of data.

43. *Box Construction.* A rectangular piece of tin is twice as long as it is wide. Squares 2 cm on a side are cut out of each corner, and the ends are turned up to make a box whose volume is 480 cm³. What are the dimensions of the piece of tin?

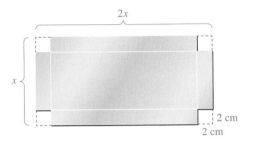

44. *Navigation.* A tugboat and a freighter leave the same port at the same time at right angles. The freighter travels 7 km/h slower than the tugboat. After 4 hr, they are 68 km apart. Find the speed of each boat.

45. *Skydiving.* During the first 13 sec of a jump, a skydiver falls approximately $11.12t^2$ feet in t seconds. A small heavy object (with less wind resistance) falls about $15.4t^2$ feet in t seconds. Suppose that a skydiver jumps from 30,000 ft, and 1 sec later a camera falls out of the airplane. How long will it take the camera to catch up to the skydiver?

46. The sum of two numbers is 17, and the sum of their squares is 205. Find the numbers.

COLLABORATIVE CORNER

Focus: Polynomial functions and data analysis

Time: 20 minutes

Group size: 3

A country's average life expectancy and infant mortality rates are related to the diet of its citizens.

The following table shows daily caloric intake, life expectancy, and infant mortality rates for several countries.

COUNTRY	DAILY CALORIC INTAKE, c (IN 1992)	LIFE EXPECTANCY, L (PROJECTED IN 2000)	INFANT MORTALITY RATE, M (IN NUMBER OF DEATHS PER 1000 BIRTHS)
Argentina	2880	72.3	26.1
Bolivia	2100	62.0	60.2
Canada	3482	80.0	5.5
Dominican Republic	2359	70.4	40.8
Germany	3443	76.7	22.2
Haiti	1707	50.2	98.4
Mexico	3181	75.0	20.7
United States	3671	76.3	6.2

Source: The Universal Almanac; Statistical Abstract of the United States

(continued)

Activity

1. Consider life expectancy as a function of daily caloric intake.

   a) Use regression to fit a linear function, a quadratic function, and a cubic function to the data, with each group member modeling one type of function.

   b) Use each of the three functions to estimate the life expectancy for a country with a daily caloric intake of 1500 calories; of 4500 calories.

   c) Decide as a group which of the three functions best models the situation.

   d) Is there an "optimal" daily caloric intake; that is, one for which the life expectancy is greatest?

2. Consider infant mortality rate as a function of daily caloric intake.

   a) Use regression to fit a linear function, a quadratic function, and a cubic function to the data, with each group member modeling one type of function.

   b) Use each of the three functions to estimate the infant mortality rate for a country with a daily caloric intake of 1500 calories; of 4500 calories.

   c) Decide as a group which of the three functions best models the situation.

   d) Is there an "optimal" daily caloric intake; that is, one for which the infant mortality rate is lowest?

CHAPTER 5

Summary and Review

Key Terms

Linear function, p.268
Absolute-value function, p. 268
Square-root function, p. 268
Library of functions, p. 269
Term, p. 269
Monomial, p. 269
Polynomial, p. 269
Coefficient, p. 269
Degree, p. 269
Leading term, p. 270
Leading coefficient, p. 270
Quadratic, p. 270
Cubic, p. 270

Linear, p. 270
Binomial, p. 270
Trinomial, p. 270
Descending order, p. 271
Ascending order, p. 271
Polynomial function, p. 271
Similar, or like, terms, p. 275
FOIL, p. 285
Square of a binomial, p. 286
Difference of two squares, p. 287
Polynomial equation, p. 291
Zero, p. 293
Root, p. 293

Factor, p. 295
Factored completely, p. 298
Prime polynomial, p. 298
Factoring by grouping, p. 299
Perfect-square trinomial, p. 322
Repeated root, p. 326
Double root, p. 326
Root of multiplicity two, p. 326
Pythagorean theorem, p. 330
Right triangle, p. 330
Hypotenuse, p. 330
Leg, p. 330

Important Properties and Formulas

Factoring Formulas

$A^2 + 2AB + B^2 = (A + B)^2;$

$A^2 - 2AB + B^2 = (A - B)^2;$

$A^2 - B^2 = (A + B)(A - B)$

To Factor $ax^2 + bx + c$ Using FOIL

1. Factor out the largest common factor, if one exists. Here we assume none does.
2. Find two **First** terms whose product is ax^2:

$$(\quad x + \quad)(\quad x + \quad) = ax^2 + bx + c.$$
$$\text{FOIL}$$

3. Find two **Last** terms whose product is c:

$$(\quad x + \quad)(\quad x + \quad) = ax^2 + bx + c.$$
$$\text{FOIL}$$

4. Repeat steps (2) and (3) until a combination is found for which the sum of the **O**uter and **I**nner products is bx:

$$(\quad x + \quad)(\quad x + \quad) = ax^2 + bx + c.$$
$$\text{I}$$
$$\text{FOIL}$$
$$\text{O}$$

To Factor $ax^2 + bx + c$ Using Grouping

1. Make sure that any common factors have been factored out.
2. Multiply the leading coefficient a and the constant c.
3. Try to factor the product ac so that the sum of the factors is b. That is, find integers p and q so that $pq = ac$ and $p + q = b$.

4. Split the middle term. That is, write bx as $px + qx$.
5. Factor by grouping.

To Factor a Polynomial

A. Always factor out the largest common factor.

B. Look at the number of terms.

Two terms: Try factoring as a difference of squares. Do *not* try to factor a *sum* of squares.

Three terms: Try factoring as a perfect-square trinomial. Next, try trial and error, using the FOIL method or the grouping method.

Four or more terms: Try factoring by grouping and factoring out a common binomial factor. Next, try grouping into a difference of squares, one of which is a trinomial.

C. Always *factor completely*. If a factor with more than one term can itself be factored further, do so.

The Principle of Zero Products

For any real numbers a and b:

If $ab = 0$, then $a = 0$ or $b = 0$ (or both).
If $a = 0$ or $b = 0$, then $ab = 0$.

REVIEW EXERCISES

1. Given the polynomial
$$2xy^6 - 7x^8y^3 + 2x^3 - 3,$$
determine the degree of each term and the degree of the polynomial.

2. Given the polynomial
$$4x - 5x^3 + 2x^2 - 7,$$
arrange in descending order and determine the leading term and the leading coefficient.

3. Arrange in ascending powers of x:
$$3x^6y - 7x^8y^3 + 2x^3 - 3x^2.$$

4. Find $P(0)$ and $P(-1)$:
$$P(x) = x^3 - x^2 + 4x.$$

5. Evaluate the polynomial function for $x = -2$:
$$P(x) = 4 - 2x - x^2.$$

6. Let $f(x) = x^2 - x - 6$.
   a) Find the domain of f.
   b) Estimate the range of f.

Combine like terms.

7. $4x^2y - 3xy^2 - 5x^2y + xy^2$

8. $3ab - 10 + 5ab^2 - 2ab + 7ab^2 + 14$

Add.

9. $(-6x^3 - 4x^2 + 3x + 1) + (5x^3 + 2x + 6x^2 + 1)$

10. $(-9xy^2 - xy + 6x^2y) + (-5x^2y - xy + 4xy^2) + (12x^2y - 3xy^2 + 6xy)$

Subtract.

11. $(3x - 5) - (-6x + 2)$

12. $(4a - b + 3c) - (6a - 7b - 4c)$

13. $(8x^2 - 4xy + y^2) - (2x^2 + 3xy - 2y^2)$

Multiply.

14. $(3x^2y)(-6xy^3)$

15. $(x^4 - 2x^2 + 3)(x^4 + x^2 - 1)$

16. $(4ab + 3c)(2ab - c)$

17. $(2x + 5y)(2x - 5y)$

18. $(2x - 5y)^2$

19. $(x + 3)(2x - 1)$

20. $(x - 5)(x^2 + 5x + 25)$

21. $\left(x - \frac{1}{3}\right)\left(x - \frac{1}{6}\right)$

22. Given $f(x) = x^2 + 2x + 1$, find and simplify $f(t + 1)$.

Factor.

23. $x^2 - 50x + 576$

24. $9y^4 - 3y^2$

25. $15x^4 - 18x^3 + 21x^2 - 9x$

26. $a^2 - 12a + 27$

27. $3m^2 + 14m + 8$

28. $25x^2 + 20x + 4$

29. $4y^2 - 16$

30. $5x^2 + x^3 - 14x$

31. $ax + 2bx - ay - 2by$

32. $3y^3 + 6y^2 - 5y - 10$

33. $a^4 - 81$

34. $4x^4 + 4x^2 + 20$

35. $y^5 + y$

36. $2z^8 - 16z^6$

37. $6t^2 + 17pt + 5p^2$

38. $36x^2 - 120x + 100$

39. $x^3 + 2x^2 - 9x - 18$

40. The graph shown below is that of a polynomial function $p(x)$. Using the graph, do the following.
   a) State whether the degree of the polynomial is even or odd.
   b) List the zeros of p.

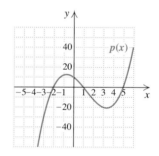

41. Find the zeros of the function given by
$$f(x) = x^2 - 11x + 28.$$

Solve. Where appropriate, round answers to the nearest thousandth.

42. $x^2 - 20x = -100$

43. $6b^2 - 13b + 6 = 0$

44. $8t^2 = 14t$

45. $r^2 = 16$

46. $x^3 - 5x^2 - 16x + 80 = 0$

47. $x^2 + 180 = 27x$

48. Let $f(x) = x^2 - 7x - 40$. Find a such that $f(a) = 4$.

49. Find the domain of the function f given by
$$f(x) = \frac{x - 3}{3x^2 + 19x - 14}.$$

50. The area of a square is 5 more than four times the length of a side. What is the length of a side of the square?

51. The sum of the squares of three consecutive odd numbers is 83. Find the numbers.

52. A photograph is 3 in. longer than it is wide. When a 2-in. border is placed around the photograph, the total area of the photograph and the border is 108 in². Find the dimensions of the photograph.

53. *Movie Revenue.* The following table shows the weekly U.S. revenue, in millions of dollars, from the movie *Air Force One.*

NUMBER OF WEEKS SINCE MOVIE RELEASED	MOVIE REVENUE (IN MILLIONS)
1	$38
2	42
3	37
4	26
5	22
6	23
7	28
8	29

Source: Exhibitor Relations Co., Inc.

a) Use regression to find a polynomial function m of degree 4 (quartic) that can be used to estimate the weekly revenue $m(x)$, in millions of dollars, x weeks after the movie was released.

b) Estimate the weekly revenue 9 weeks after the movie was released.

c) During what week or weeks was the movie revenue $16 million?

Skill Maintenance

Solve.

54. $3x + 2y + z = 3,$
$2x - y + 2z = 16,$
$x + y - z = -9$

55. $-19x + 10 + 15x > 2x - 4 - 12x$

56. $|10 - 3x| \leq 14$

57. Graph: $f(x) = \frac{2}{3}x + 2.$

58. There are 70 questions on a test. The questions are either multiple-choice, true–false, or fill-in. There are twice as many true–false as fill-in and 5 more multiple-choice than true–false. How many of each type of question are there on the test?

Synthesis

59. ◈ Explain how to find the zeros of a polynomial function from its graph.

60. ◈ Explain in your own words why there must be a 0 on one side of an equation before you can use the principle of zero products.

Factor.

61. $2x^3 - 2y^3$ **62.** $a^3b^3 + c^3d^3$

Multiply.

63. $[a - (b - 1)][(b - 1)^2 + a(b - 1) + a^2]$

64. $(z^{n^2})^{n^3}(z^{4n^3})^{n^2}$

CHAPTER 5 TEST

Given the polynomial $3y^3 - 4y + 5y^4 - 2.$

1. Determine the degree of the polynomial.

2. Arrange in descending order.

3. Determine the leading term of the polynomial $8a - 2 + a^2 - 4a^3.$

4. Given $P(x) = 2x^3 + 3x^2 - x + 4$, find $P(0)$ and $P(-2).$

5. Given $P(x) = x^2 - 5x$, find and simplify
$$P(a + h) - P(a).$$

6. Combine like terms:
$$5xy - 2xy^2 - 2xy + 5xy^2.$$

Add.

7. $(-6x^3 + 3x^2 - 4y) + (3x^3 - 2y - 7y^2)$

8. $(5m^3 - 4m^2n - 6mn^2 - 3n^3) + (9mn^2 - 4n^3 + 2m^3 + 6m^2n)$

Subtract.

9. $(9a - 4b) - (3a + 4b)$

10. $(6y^2 - 2y - 5y^3) - (4y^2 - 7y - 6y^3)$

Multiply.

11. $(-4x^2y)(-16xy^2)$

12. $(6a - 5b)(2a + b)$

13. $(x - y)(x^2 - xy - y^2)$

14. $(3m^2 + 4m - 2)(-m^2 - 3m + 5)$

15. $(4y - 9)^2$

16. $(x - 2y)(x + 2y)$

Factor.

17. $15x^2 - 5x^4$ **18.** $y^3 + 5y^2 - 4y - 20$

19. $p^2 - 12p - 28$ **20.** $12m^2 + 20m + 3$

21. $9y^2 - 25$

22. $x^2 - 7x - 330$

23. $9x^2 + 25 - 30x$

24. $x^8 - y^8$

25. $24x^2 - 46x + 10$

26. $20a^2 - 5b^2$

27. $4y^4x + 36yx^2 + 8y^2x^3 - 16xy$

28. The graph shown below is that of a polynomial function $p(x)$. Using the graph, do the following.

   a) Determine the zeros of p.

   b) State whether the degree of the polynomial is even or odd.

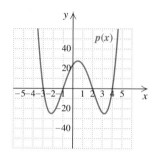

29. Let $f(x) = 2x^2 - 11x - 40$.

   a) Find the domain of f.

   b) Find the zeros of f.

   c) Estimate the range of f.

Solve. Where appropriate, round solutions to the nearest thousandth.

30. $x^2 - 18 = 3x$ **31.** $5t^2 = 125$

32. $2x^2 + 21 = -17x$ **33.** $9x^2 + 3x = 0$

34. $x^2 + 81 = 18x$ **35.** $x^2(x + 1) = 8x$

36. Let $f(x) = 3x^2 - 15x + 11$. Find a such that $f(a) = 11$.

37. Find the domain of the function f given by

$$f(x) = \frac{3 - x}{x^2 + 2x + 1}.$$

38. A photograph is 3 cm longer than it is wide. Its area is 40 cm^2. Find its length and its width.

39. *Birth Rate.* The following table lists the average number of live births to women who are x years old.

AGE	AVERAGE NUMBER OF LIVE BIRTHS PER 1000 WOMEN
16	34
18.5	86.5
22	111.1
27	113.9
32	84.5
37	35.4
42	6.8

Source: Centers for Disease Control and Prevention

   a) Use regression to find a polynomial function $b(x)$ of degree 3 (cubic) that can be used to estimate the average number of live births per 1000 women who are x years old.

   b) Estimate the average number of live births per 1000 20-yr-old women.

   c) For what age or ages are there approximately 100 live births per 1000 women?

Skill Maintenance

Solve.

40. $|3x + 8| < 10$

41. $-3x + 4 - 5x > 8 - 9x - 3$

42. $2x - y + z = 9,$
$x - y + z = 4,$
$x + 2y - z = 5$

43. Find the slope and the y-intercept of the line given by

$$x + 3y = 8.$$

Synthesis

44. a) Multiply: $(x^2 + x + 1)(x^3 - x^2 + 1)$.

   b) Factor: $x^5 + x + 1$.

45. Factor: $6x^{2n} - 7x^n - 20$.

Rational Equations and Functions 6

APPLICATION

COMMUTER TRAVEL. The number of trips made to a city decreases as the distance from the city increases. Graph the data and determine whether it can be modeled by a function of the type $f(x) = k/x$.

MILES FROM CITY	NUMBER OF TRIPS TO CITY PER DAY PER RESIDENTIAL ACRE
2	130
6	120
8	60
10	45
14	40
18	25
20	20
22	19
24	20
26	19
28	20

Source: Kolars, John F. and John D. Nyusten, *Geography.* (New York: McGraw-Hill, 1974).

This problem appears as Example 5 in Section 6.7.

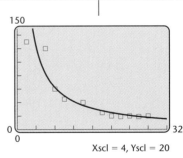

150

0 · 0 · 32

Xscl = 4, Yscl = 20

A *rational expression* is an expression that indicates division, as the fractional symbols in arithmetic do. In this chapter, we add, subtract, multiply, and divide rational expressions, and use them in equations and functions. We then use rational expressions to solve problems that we could not have solved before.

6.1

Rational Expressions and Functions: Multiplying and Dividing

- *Rational Functions*
- *Multiplying and Simplifying Rational Expressions*
- *Dividing and Simplifying*

An expression that consists of a polynomial divided by a nonzero polynomial is called a **rational expression**. The following are examples of rational expressions:

$$\frac{7}{8}, \quad \frac{a}{b}, \quad \frac{8}{y+5}, \quad \frac{x^2 + 7xy - 4}{x^3 - y^3}, \quad \frac{1 + z^3}{1 - z^6}.$$

Rational Functions

Like polynomials, certain rational expressions are used to describe functions. Such functions are called **rational functions**.

Example 1   The function given by

$$T(t) = \frac{t^2 + 5t}{2t + 5}$$

gives the time required for two machines, working together, to complete a job that the first machine could do alone in t hours and the other machine could do in $t + 5$ hours. How long will the two machines, working together, require for the job if the first machine alone would take **(a)** 1 hr? **(b)** 5 hr?

SOLUTION

a) $T(1) = \dfrac{1^2 + 5 \cdot 1}{2 \cdot 1 + 5} = \dfrac{1 + 5}{2 + 5} = \dfrac{6}{7}$ hr

b) $T(5) = \dfrac{5^2 + 5 \cdot 5}{2 \cdot 5 + 5} = \dfrac{25 + 25}{10 + 5} = \dfrac{50}{15} = \dfrac{10}{3}$ hr

Recall that the domain of a function is the set of all inputs for which the function is defined. The domain of a rational function will exclude all values of x that make a denominator 0. For instance, the domain of the function in Example 1 does not include the number $-\frac{5}{2}$, because

$$T = \left(-\frac{5}{2}\right) = \frac{-25/4}{0}$$

and division by 0 is undefined.

How is the domain of a rational function related to its graph?

Interactive Discovery

Consider the function given by

$$f(x) = \frac{2}{x - 3}.$$

What number is not in the domain of the function? Let $y = 2/(x - 3)$.

Graph the function and trace along the graph, starting at $x = 0$ and moving to the right. What happens to the function values as x gets closer to 3? What happens to the function values for $x > 3$? What happens when you attempt to find the value of the function for $x = 3$?

Although we will not consider graphs of rational functions in detail in this text, we will make some general observations.

If a is not in the domain of f, the graph of f does not cross the line $x = a$.

Consider the graph of the function in Example 1,

$$T(t) = \frac{t^2 + 5t}{2t + 5}.$$

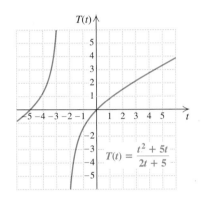

In Section 5.5, we found the domain of a rational function by determining when the denominator is 0. Since $2t + 5 = 0$ when $t = -\frac{5}{2}$, the domain of T is $\left(-\infty, -\frac{5}{2}\right) \cup \left(-\frac{5}{2}, \infty\right)$. Note that the graph consists of two unconnected "branches." It does not touch the vertical line $x = -\frac{5}{2}$. The line $x = -\frac{5}{2}$ is a *vertical asymptote*, that is, the graph of T gets very close to the line but never touches or crosses it.

Now consider the graph of

$$T(t) = \frac{t^2 + 5t}{2t + 5}$$

as drawn using a grapher. The vertical line that appears on the grapher screen on the left below is not part of the graph, nor should it be considered the vertical asymptote. Since a grapher graphs an equation by plotting points and connecting them, the vertical line is actually connecting a point just to the left of the line $x = -\frac{5}{2}$ with a point just to the right of the line $x = -\frac{5}{2}$. Such a vertical line may or may not appear for values not in the domain of a function.

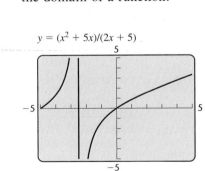

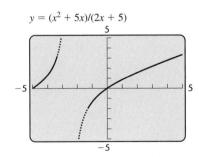

If we tell the grapher not to connect the points that it plots, the vertical line will not appear. Changing from CONNECTED mode to DOT mode gives us the graph on the right above.

For the graph of $T(t) = (t^2 + 5t)/(2t + 5)$, the vertical asymptote, $x = -\frac{5}{2}$, corresponds to the input that makes the denominator 0, that is, $-\frac{5}{2}$. This is true in general only if a function is described by a rational expression in *simplified form*.

Multiplying and Simplifying Rational Expressions

To simplify rational expressions, we factor them, which is the reverse of multiplying.

> **Products of Rational Expressions**
>
> To multiply two rational expressions, multiply numerators and multiply denominators:
>
> $$\frac{A}{B} \cdot \frac{C}{D} = \frac{AC}{BD}, \quad \text{where } B \neq 0, D \neq 0.$$

Example 2   Multiply: $\dfrac{x + 1}{y - 3} \cdot \dfrac{x^2}{y + 1}$.

SOLUTION

$$\frac{x + 1}{y - 3} \cdot \frac{x^2}{y + 1} = \frac{(x + 1)x^2}{(y - 3)(y + 1)} \qquad \text{Multiplying the numerators and multiplying the denominators}$$

Recall from arithmetic that multiplication by 1 can be used to find equivalent expressions:

$$\frac{3}{5} = \frac{3}{5} \cdot \frac{2}{2} \qquad \text{Multiplying by } \tfrac{2}{2}, \text{ which is 1}$$

$$= \frac{6}{10}.$$

In a similar manner, multiplication by 1 can be used to find equivalent rational expressions in algebra:

$$\frac{x - 5}{x + 2} = \frac{x - 5}{x + 2} \cdot \frac{x + 3}{x + 3}$$

$$= \frac{(x - 5)(x + 3)}{(x + 2)(x + 3)}. \qquad \text{Multiplying by } \frac{x + 3}{x + 3}, \text{ which, provided } x \neq -3, \text{ is 1}$$

The expressions

$$\frac{x - 5}{x + 2} \quad \text{and} \quad \frac{(x - 5)(x + 3)}{(x + 2)(x + 3)}$$

are equivalent: So long as x is replaced with a number that can be used in either expression, both expressions will represent the same value.

Example 3 Multiply to find an equivalent expression:

$$\frac{4-x}{y-x} \cdot \frac{-1}{-1}.$$

SOLUTION   We have

$$\frac{4-x}{y-x} \cdot \frac{-1}{-1} = \frac{-4+x}{-y+x} = \frac{x-4}{x-y}. \qquad \text{Multiplication by } -1 \\ \text{reverses subtraction.}$$

Multiplying by $-1/-1$ is the same as multiplying by 1, so

$$\frac{4-x}{y-x} \quad \text{is equivalent to} \quad \frac{x-4}{x-y}.$$

As in arithmetic, rational expressions are *simplified* by "removing" a factor equal to 1. This reverses the process shown above:

$$\frac{6}{10} = \frac{3 \cdot 2}{5 \cdot 2} = \frac{3}{5} \cdot \frac{2}{2} = \frac{3}{5}. \qquad \text{We "removed" the factor that} \\ \text{equals 1: } \frac{2}{2} = 1.$$

Similarly,

$$\frac{(x-5)(x+3)}{(x+2)(x+3)} = \frac{x-5}{x+2} \cdot \frac{x+3}{x+3} = \frac{x-5}{x+2}. \qquad \text{We "removed" the factor} \\ \text{that equals 1: } \frac{x+3}{x+3} = 1.$$

A rational expression is said to be **simplified** when no factors equal to 1 can be removed. When simplifying, it is important to look for the *greatest* common factor of both the numerator and the denominator.

Example 4 Simplify: $\dfrac{7x^2 + 21x}{14x}$.

SOLUTION   We first factor the numerator and the denominator, looking for the largest factor common to both. Once the greatest common factor is identified, we use it to write 1 and simplify:

$$\frac{7x^2 + 21x}{14x} = \frac{7x(x+3)}{7 \cdot 2 \cdot x} \qquad \text{Factoring. The greatest common} \\ \text{factor is } 7x.$$

$$= \left(\frac{7x}{7x}\right) \cdot \frac{x+3}{2} \qquad \text{Factoring the rational expression}$$

$$= 1 \cdot \frac{x+3}{2} \qquad \frac{7x}{7x} = 1$$

$$= \frac{x+3}{2}. \qquad \text{Removing the factor 1}$$

As a check, we let $y_1 = (7x^2 + 21x)/(14x)$ and $y_2 = (x+3)/2$. Let's compare values in a table. Note that the y-values are the same for any given x-value except 0. We exclude 0 because although $(x+3)/2$ is defined when $x = 0$, $(7x^2 + 21x)/(14x)$ is not. Both expressions represent the same value when x is replaced with a number that can be used in *either* expression, so they are equivalent.

X	Y₁	Y₂
−3	0	0
−2	.5	.5
−1	1	1
0	ERROR	1.5
1	2	2
2	2.5	2.5
3	3	3

X = −3

In Example 4, we found that

$$\frac{7x^2 + 21x}{14x} = \frac{x + 3}{2}.$$

These expressions are equivalent; their values are equal for every value of x except $x = 0$, for which the expression on the left is not defined. Special care must be taken, however, when simplifying a rational expression within a function. If

$$f(x) = \frac{7x^2 + 21x}{14x}$$

and

$$g(x) = \frac{x + 3}{2},$$

 we cannot say that $f = g$, because the domains of the functions are not the same. If it is necessary to simplify a rational expression that defines a function, the domain must be carefully specified.

Example 5   Given the function $f(x) = \dfrac{x^2 - 4}{2x^2 - 3x - 2}$.

a) Simplify the rational expression.

b) Find the domain of f.

c) Find the domain of the function given by the simplified rational expression.

SOLUTION

a) We have

$$\frac{x^2 - 4}{2x^2 - 3x - 2} = \frac{(x - 2)(x + 2)}{(2x + 1)(x - 2)} \qquad \text{Factoring the numerator and the denominator}$$

$$= \frac{x - 2}{x - 2} \cdot \frac{x + 2}{2x + 1} \qquad \text{Factoring the rational expression; } \frac{x - 2}{x - 2} = 1$$

$$= \frac{x + 2}{2x + 1}. \qquad \text{Removing a factor equal to 1}$$

The simplified form of the expression is $\dfrac{x + 2}{2x + 1}$.

b) To find the domain of f, we set the denominator equal to 0 and solve:

$$2x^2 - 3x - 2 = 0$$

$$(2x + 1)(x - 2) = 0 \qquad \text{Factoring the denominator}$$

$$2x + 1 = 0 \quad \text{or} \quad x - 2 = 0 \qquad \text{Using the principle of zero products}$$

$$x = -\tfrac{1}{2} \quad \text{or} \qquad x = 2.$$

The numbers $-\tfrac{1}{2}$ and 2 are not in the domain of the function. Thus the domain of f is $\left\{x \mid x \text{ is a real number } and \ x \neq -\tfrac{1}{2} \text{ and } x \neq 2\right\}$.

c) The function given by the simplified form of the rational expression is

$$g(x) = \frac{x + 2}{2x + 1}.$$

The function is undefined when

$$2x + 1 = 0$$

or $x = -\frac{1}{2}.$

The domain of g is $\left\{ x \mid x \text{ is a real number } and \ x \neq -\frac{1}{2} \right\}.$

In Example 5, we cannot say that

$$f(x) = \frac{x + 2}{2x + 1}$$

unless we specify that 2 is not in the domain of f.

Interactive Discovery

Let

$$f(x) = \frac{x^2 - 4}{2x^2 - 3x - 2} \quad \text{and} \quad g(x) = \frac{x + 2}{2x + 1}.$$

As we saw in Example 5, the rational expressions describing these functions are equivalent but the domains of the functions are different. Graph $f(x)$. How many vertical asymptotes does the graph appear to have? What is the vertical asymptote? Graph $g(x)$ and determine the vertical asymptote.

> **If a function $f(x)$ is described by a *simplified* rational expression, and a is a number that makes the denominator 0, then $x = a$ is a vertical asymptote of the graph of $f(x)$.**

Example 6   Determine the vertical asymptotes of the graph of

$$f(x) = \frac{9x^2 + 6x - 3}{12x^2 - 12}.$$

SOLUTION   We first simplify the rational expression describing the function:

$$\frac{9x^2 + 6x - 3}{12x^2 - 12} = \frac{3(x + 1)(3x - 1)}{3 \cdot 4(x + 1)(x - 1)} \qquad \text{Factoring the numerator and the denominator}$$

$$= \frac{3(x + 1)}{3(x + 1)} \cdot \frac{3x - 1}{4(x - 1)} \qquad \text{Factoring the rational expression}$$

$$= \frac{3x - 1}{4(x - 1)}. \qquad \text{Removing a factor equal to 1}$$

The denominator of the simplified expression, $4(x - 1)$, is 0 when $x = 1$. Thus, $x = 1$ is a vertical asymptote of the graph. Although the domain of the function also excludes −1, there is no asymptote at $x = -1$, only a "hole." This hole should be indicated on a hand-drawn graph but it will not be obvious on a grapher screen. A table will show that −1 is not in the domain of this function.

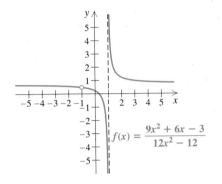

$$f(x) = \frac{9x^2 + 6x - 3}{12x^2 - 12}$$

$y = (9x^2 + 6x - 3)/(12x^2 - 12)$

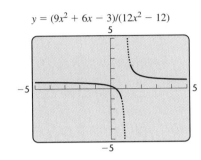

CANCELING   Canceling is a shortcut that you may have used for removing a factor equal to 1 when working with fractional notation or rational expressions. With great concern, we mention it here as a possible way to speed up your work. Canceling *can* be done to remove factors equal to 1 in products. It *cannot* be done in sums or when adding expressions together. Our concern is that canceling be done with care and understanding. The simplification in Example 6 might have been done faster as follows:

$$\frac{9x^2 + 6x - 3}{12x^2 - 12} = \frac{3(x + 1)(3x - 1)}{3 \cdot 4(x + 1)(x - 1)}$$ When a factor that equals 1 is noted, it is "canceled" as shown.

$$= \frac{3x - 1}{4(x - 1)}.$$ Removing a factor equal to 1: $\dfrac{3(x + 1)}{3(x + 1)} = 1$

Caution!   Canceling is often performed incorrectly:

$$\frac{x + 3}{x} = 3, \quad \frac{4x + 3}{2} = 2x + 3, \quad \frac{5}{5 + x} = \frac{1}{x}.$$ To check that these are not equivalent, substitute a number for x.

Wrong!   Wrong!   Wrong!

In each of these situations, the expressions canceled are *not* factors that equal 1. Factors are parts of products. For example, in $x \cdot 3$, the x and the 3 are factors, but in $x + 3$, the x and the 3 are *not* factors, but terms. If you can't factor, you can't cancel! If in doubt, don't cancel!

After multiplying, we usually simplify, if possible. That is one reason why we leave the numerator and the denominator in factored form. Even so, we might need to factor them further in order to simplify.

Example 7 Multiply. Then simplify by removing a factor equal to 1.

$$\frac{x + 2}{x - 3} \cdot \frac{x^2 - 4}{x^2 + x - 2}$$

SOLUTION We have

$$\frac{x + 2}{x - 3} \cdot \frac{x^2 - 4}{x^2 + x - 2} = \frac{(x + 2)(x^2 - 4)}{(x - 3)(x^2 + x - 2)}$$

Multiplying the numerators and also the denominators

$$= \frac{(x + 2)(x - 2)(x + 2)}{(x - 3)(x + 2)(x - 1)}$$

Factoring the numerator and the denominator and identifying common factors

$$= \frac{(x + 2)(x + 2)(x - 2)}{(x - 3)(x + 2)(x - 1)}$$

Removing a factor equal to 1: $\frac{x + 2}{x + 2} = 1$

$$= \frac{(x + 2)(x - 2)}{(x - 3)(x - 1)}.$$

Simplifying

$y_1 = ((x + 2)/(x - 3))*$
$((x^2 - 4)/(x^2 + x - 2)),$
$y_2 = ((x + 2)(x - 2))/$
$((x - 3)(x - 1))$

X	Y₁	Y₂
−2	ERROR	0
−1	−.375	−.375
0	−1.333	−1.333
1	ERROR	ERROR
2	0	0
3	ERROR	ERROR
4	4	4

X = −2

To check, we can compare tables of values, as shown in the table at left. For purposes of our later work, we generally do not multiply out the numerator and the denominator.

Dividing and Simplifying

Two expressions are reciprocals of each other if their product is 1. As in arithmetic, to find the reciprocal of a rational expression, we interchange numerator and denominator.

The reciprocal of $\dfrac{x}{x^2 + 3}$ is $\dfrac{x^2 + 3}{x}$.

The reciprocal of $y - 8$ is $\dfrac{1}{y - 8}$.

Quotients of Rational Expressions

For any rational expressions A/B and C/D, with $B, C, D \neq 0$,

$$\frac{A}{B} \div \frac{C}{D} = \frac{A}{B} \cdot \frac{D}{C}.$$

(To divide two rational expressions, multiply by the reciprocal of the divisor. We often say that we "*invert* and multiply.")

Example 8 Divide. Simplify by removing a factor equal to 1 if possible.

a) $\dfrac{x - 2}{x + 1} \div \dfrac{x + 5}{x - 3}$

b) $\dfrac{a^2 - 1}{a - 1} \div \dfrac{a^2 - 2a + 1}{a + 1}$

SOLUTION

a) $\dfrac{x-2}{x+1} \div \dfrac{x+5}{x-3} = \dfrac{x-2}{x+1} \cdot \dfrac{x-3}{x+5}$ **Multiplying by the reciprocal**

$\qquad\qquad\qquad = \dfrac{(x-2)(x-3)}{(x+1)(x+5)}$ **Multiplying the numerators and the denominators**

b) $\dfrac{a^2-1}{a-1} \div \dfrac{a^2-2a+1}{a+1} = \dfrac{a^2-1}{a-1} \cdot \dfrac{a+1}{a^2-2a+1}$ **Multiplying by the reciprocal**

$\qquad\qquad\qquad = \dfrac{(a^2-1)(a+1)}{(a-1)(a^2-2a+1)}$ **Multiplying the numerators and the denominators**

$\qquad\qquad\qquad = \dfrac{(a+1)(a-1)(a+1)}{(a-1)(a-1)(a-1)}$ **Factoring the numerator and the denominator**

$\qquad\qquad\qquad = \dfrac{(a+1)(a\!-\!1)(a+1)}{(a-1)(a\!-\!1)(a-1)}$ **Removing a factor equal to 1:** $\dfrac{a-1}{a-1} = 1$

$\qquad\qquad\qquad = \dfrac{(a+1)(a+1)}{(a-1)(a-1)}$ **Simplifying**

> The procedures covered in this chapter are by their nature rather long. It may help to write out lots of steps as you do the problems. If you have difficulty, consider taking a clean sheet of paper and starting over. Don't squeeze your work into a small amount of space. When using lined paper, consider using two spaces at a time, with the paper's line locating the fraction bar.

6.1   Exercise Set

For each rational function, find the function values indicated, provided the value exists.

1. $r(y) = \dfrac{3y^3 - 2y}{y - 5};$ $r(0), r(4), r(5)$

2. $f(r) = \dfrac{5r^2 - r^3}{r - 4};$ $f(2), f(5), f(-1)$

3. $g(x) = \dfrac{2x^3 - 9}{x^2 - 4x + 4};$ $g(0), g(2), g(-1)$

4. $r(t) = \dfrac{t^2 - 5t + 4}{t^2 - 9};$ $r(1), r(2), r(-3)$

Multiply to obtain equivalent expressions. Do not simplify. Assume that all denominators are nonzero.

5. $\dfrac{5x}{5x} \cdot \dfrac{x-3}{x+2}$

6. $\dfrac{3-a^2}{a-7} \cdot \dfrac{-1}{-1}$

7. $\dfrac{t-2}{t+3} \cdot \dfrac{-1}{-1}$

8. $\dfrac{x-4}{x+5} \cdot \dfrac{x-5}{x-5}$

Simplify by removing a factor equal to 1.

9. $\dfrac{15x}{5x^2}$

10. $\dfrac{7a^3}{21a}$

11. $\dfrac{18t^3}{27t^7}$

12. $\dfrac{8y^5}{4y^9}$

13. $\dfrac{2a - 10}{2}$

14. $\dfrac{3a + 12}{3}$

15. $\dfrac{15}{25a - 30}$

16. $\dfrac{21}{6x - 9}$

17. $\dfrac{5x + 20}{x^2 + 4x}$

18. $\dfrac{3x + 21}{x^2 + 7x}$

19. $\dfrac{3a - 1}{2 - 6a}$

20. $\dfrac{6 - 5a}{10a - 12}$

21. $\dfrac{8t - 16}{t^2 - 4}$

22. $\dfrac{t^2 - 9}{5t + 15}$

23. $\dfrac{2t - 1}{1 - 4t^2}$

24. $\dfrac{3a - 2}{4 - 9a^2}$

25. $\dfrac{a^2 - 25}{a^2 + 10a + 25}$

26. $\dfrac{a^2 - 16}{a^2 - 8a + 16}$

27. $\dfrac{x^2 + 9x + 8}{x^2 - 3x - 4}$

28. $\dfrac{t^2 - 8t - 9}{t^2 + 5t + 4}$

29. $\dfrac{16 - t^2}{t^2 - 8t + 16}$

30. $\dfrac{25 - p^2}{p^2 + 10p + 25}$

Determine the vertical asymptotes of the graph of each function.

31. $f(x) = \dfrac{3x - 12}{3x + 15}$

32. $f(x) = \dfrac{4x - 20}{4x + 12}$

33. $g(x) = \dfrac{12 - 6x}{5x - 10}$

34. $r(x) = \dfrac{21 - 7x}{3x - 9}$

35. $t(x) = \dfrac{x^3 + 3x^2}{x^2 + 6x + 9}$

36. $g(x) = \dfrac{x^2 - 4}{2x^2 - 5x + 2}$

37. $f(x) = \dfrac{x^2 - x - 6}{x^2 - 6x + 8}$

38. $f(x) = \dfrac{x^2 + 2x + 1}{x^2 - 2x + 1}$

In Exercises 39–44, match each function with one of the following graphs.

a)

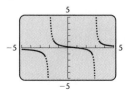

b)

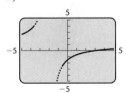

c)

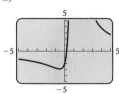

d)

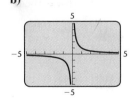

e)

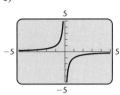

f)

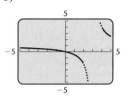

39. $h(x) = \dfrac{1}{x}$

40. $q(x) = -\dfrac{1}{x}$

41. $f(x) = \dfrac{x}{x - 3}$

42. $g(x) = \dfrac{x - 3}{x + 2}$

43. $r(x) = \dfrac{4x - 2}{x^2 - 2x + 1}$

44. $t(x) = \dfrac{x - 1}{x^2 - x - 6}$

Multiply and simplify.

45. $\dfrac{5a^3}{3b} \cdot \dfrac{7b^3}{10a^7}$

46. $\dfrac{25a}{9b^8} \cdot \dfrac{3b^5}{5a^2}$

47. $\dfrac{3x - 6}{5x} \cdot \dfrac{x^3}{5x - 10}$

48. $\dfrac{5t^3}{4t - 8} \cdot \dfrac{6t - 12}{10t}$

49. $\dfrac{y^2 - 16}{2y + 6} \cdot \dfrac{y + 3}{y - 4}$

50. $\dfrac{m^2 - n^2}{4m + 4n} \cdot \dfrac{m + n}{m - n}$

51. $\dfrac{x^2 - 16}{x^2} \cdot \dfrac{x^2 - 4x}{x^2 - x - 12}$

52. $\dfrac{y^2 + 10y + 25}{y^2 - 9} \cdot \dfrac{y^2 + 3y}{y + 5}$

53. $\dfrac{7a - 14}{4 - a^2} \cdot \dfrac{5a^2 + 6a + 1}{35a + 7}$

54. $\dfrac{a^2 - 1}{2 - 5a} \cdot \dfrac{15a - 6}{a^2 + 5a - 6}$

55. $\dfrac{6 - 2t}{t^2 + 4t + 4} \cdot \dfrac{t^3 + 2t^2}{t^8 - 9t^6}$

56. $\dfrac{x^2 - 6x + 9}{12 - 4x} \cdot \dfrac{x^6 - 9x^4}{x^3 - 3x^2}$

57. $\dfrac{x^2 - 2x - 35}{2x^3 - 3x^2} \cdot \dfrac{4x^3 - 9x}{7x - 49}$

58. $\dfrac{y^2 - 10y + 9}{y^2 - 1} \cdot \dfrac{y + 4}{y^2 - 5y - 36}$

Divide and simplify.

59. $\dfrac{9x^5}{8y^2} \div \dfrac{3x}{16y^9}$

60. $\dfrac{16a^7}{3b^5} \div \dfrac{8a^3}{6b}$

61. $\dfrac{6x + 12}{x^8} \div \dfrac{x + 2}{x^3}$

62. $\dfrac{3y + 15}{y^7} \div \dfrac{y + 5}{y^2}$

63. $\dfrac{x^2 - 4}{x^3} \div \dfrac{x^5 - 2x^4}{x + 4}$

64. $\dfrac{y^2 - 9}{y^2} \div \dfrac{y^5 + 3y^4}{y + 2}$

65. $\dfrac{25x^2 - 4}{x^2 - 9} \div \dfrac{2 - 5x}{x + 3}$

66. $\dfrac{4a^2 - 1}{a^2 - 4} \div \dfrac{2a - 1}{2 - a}$

67. $\dfrac{5y - 5x}{15y^3} \div \dfrac{x^2 - y^2}{3x + 3y}$

68. $\dfrac{x^2 - y^2}{4x + 4y} \div \dfrac{3y - 3x}{12x^2}$

69. $\dfrac{x^2 - 16}{x^2 - 10x + 25} \div \dfrac{3x - 12}{x^2 - 3x - 10}$

70. $\dfrac{y^2 - 36}{y^2 - 8y + 16} \div \dfrac{3y - 18}{y^2 - y - 12}$

71. $\dfrac{y^3 + 3y}{y^2 - 9} \div \dfrac{y^2 + 5y - 14}{y^2 + 4y - 21}$

72. $\dfrac{a^3 + 4a}{a^2 - 16} \div \dfrac{a^2 + 8a + 15}{a^2 + a - 20}$

Skill Maintenance

73. Solve by substitution: [3.2]
$$3x + y = 13,$$
$$x = y + 1.$$

74. For $f(x) = x^2 - x$, find all values of a for which $f(a) = 6$. [5.4]

75. Solve: $\frac{2}{3}(3x - 4) = 8$. [2.2]

76. A concert committee needs to take in \$4000 from ticket sales in order to break even. If a total of 400 tickets is to be sold at full price and 200 tickets sold at half price, how should the tickets be priced? [3.3]

Synthesis

77. ◈ To check Example 4, Kara lets
$$y_1 = \dfrac{7x^2 + 21x}{14x} \quad \text{and} \quad y_2 = \dfrac{x + 3}{2}.$$
Since the graphs of y_1 and y_2 appear to be identical, Kara believes that the domains of the functions described by y_1 and y_2 are the same, $\mathbb{R}$. How could you convince Kara otherwise?

78. ◈ Below are unlabeled graphs of $f(x) = x + 2$ and $g(x) = (x^2 - 4)/(x - 2)$. How could you determine which graph represents f and which graph represents g?

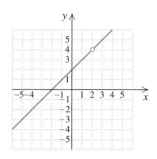
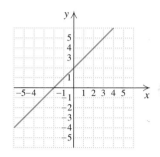

79. ◈ Nancy *incorrectly* simplifies $\dfrac{x + 2}{x}$ as
$$\dfrac{x + 2}{x} = \dfrac{\not x + 2}{\not x} = 1 + 2 = 3.$$
She insists that this is correct because it checks when x is replaced with 1. Explain her misconception.

80. ◈ Tony *incorrectly* argues that since
$$\dfrac{a^2 - 4}{a - 2} = \dfrac{a^2}{a} + \dfrac{-4}{-2} = a + 2,$$
it follows that
$$\dfrac{x^2 + 9}{x + 1} = \dfrac{x^2}{x} + \dfrac{9}{1} = x + 9.$$
Explain his misconception.

81. Graph the function given by
$$f(x) = \dfrac{x^2 - 9}{x - 3}.$$
(*Hint:* Determine the domain of f and simplify.)

82. Let
$$g(x) = \dfrac{2x + 3}{4x - 1}.$$
Determine each of the following.

a) $g(x + h)$
b) $g(2x - 2) \cdot g(x)$
c) $g\!\left(\frac{1}{2}x + 1\right) \cdot g(x)$

Perform the indicated operations and simplify.

83. $\left[\dfrac{d^2 - d}{d^2 - 6d + 8} \cdot \dfrac{d - 2}{d^2 + 5d}\right] \div \dfrac{5d}{d^2 - 9d + 20}$

84. $\left[\dfrac{r^2 - 4s^2}{r + 2s} \div (r + 2s)\right] \cdot \dfrac{2s}{r - 2s}$

Simplify.

85. $\dfrac{m^2 - t^2}{m^2 + t^2 + m + t + 2mt}$

86. $\dfrac{a^3 - 2a^2 + 2a - 4}{a^3 - 2a^2 - 3a + 6}$

87. $\dfrac{x^3 + x^2 - y^3 - y^2}{x^2 - 2xy + y^2}$

88. Let

$$f(x) = \frac{4}{x^2 - 1} \quad \text{and} \quad g(x) = \frac{4x^2 + 8x + 4}{x^3 - 1}.$$

Find each of the following.

a) $(f \cdot g)(x)$
b) $(f / g)(x)$
c) $(g / f)(x)$

Determine the domain and the range of each function from its graph.

89.

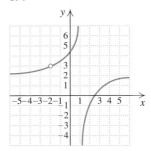

90.

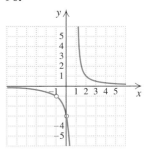

91.

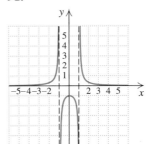

COLLABORATIVE CORNER

Focus: Graphs of rational functions

Time: 10–15 minutes

Group size: 3–5

Materials: Graphing calculators

Activity

1. Each group member should privately choose one of the following functions and graph it using a calculator.

$$f(x) = \frac{x + 1}{x^2 - x - 6}$$

$$g(x) = 2 + \frac{x + 5}{x - 2}$$

$$h(x) = \frac{4}{x^2}$$

$$r(x) = \frac{2}{x^2 + 2x + 1}$$

$$q(x) = \frac{3}{x^3 + x^2 - 12x}$$

2. When everyone has finished graphing the function, the calculators should be exchanged among the group. Using TRACE and/or a table, each group member should decide which function the other member chose.

6.2
Rational Expressions and Functions: Adding and Subtracting

- *When Denominators Are the Same*
- *When Denominators Are Different*

Rational expressions are added in much the same way as the fractions of arithmetic.

When Denominators Are the Same

Addition with Like Denominators

To add or subtract when denominators are the same, add or subtract the numerators and keep the same denominator.

$$\frac{A}{C} + \frac{B}{C} = \frac{A + B}{C} \quad \text{and} \quad \frac{A}{C} - \frac{B}{C} = \frac{A - B}{C}, \quad \text{where } C \neq 0.$$

X	Y₁	Y₂
−2	−2.5	−2.5
−1	−6	−6
0	ERROR	ERROR
1	8	8
2	4.5	4.5
3	3.3333	3.3333
4	2.75	2.75

X = −2

Example 1   Add: $\dfrac{3 + x}{x} + \dfrac{4}{x}$.

SOLUTION   We have

$$\frac{3 + x}{x} + \frac{4}{x} = \frac{7 + x}{x}.$$ **This expression cannot be simplified further because x is not a factor of $7 + x$.**

To check, we let $y_1 = (3 + x)/x + 4/x$ and $y_2 = (7 + x)/x$. The table at left shows that $y_1 = y_2$ for all x not equal to 0. ▬

Example 2   Add: $\dfrac{4x^2 - 5x}{x^2 - 1} + \dfrac{2x - 1}{x^2 - 1}$.

SOLUTION

$$\frac{4x^2 - 5x}{x^2 - 1} + \frac{2x - 1}{x^2 - 1} = \frac{4x^2 - 3x - 1}{x^2 - 1}$$ **Adding the numerators and combining like terms. The denominator is unchanged.**

$$= \frac{(x - 1)(4x + 1)}{(x - 1)(x + 1)}$$ **Factoring the numerator and the denominator and looking for common factors**

$$= \frac{(x - 1)(4x + 1)}{(x - 1)(x + 1)}$$ **Removing a factor equal to 1: $\dfrac{x - 1}{x - 1} = 1$**

$$= \frac{4x + 1}{x + 1}$$ **Simplifying** ▬

Note from Example 2 that the expressions

$$\frac{4x^2 - 5x}{x^2 - 1} + \frac{2x - 1}{x^2 - 1} \quad \text{and} \quad \frac{4x + 1}{x + 1}$$

are equivalent; they have the same value for all values of x except -1 and 1. However, the functions

$$f(x) = \frac{4x^2 - 5x}{x^2 - 1} + \frac{2x - 1}{x^2 - 1} \quad \text{and} \quad g(x) = \frac{4x + 1}{x + 1}$$

are not equal because their domains are not the same.

Recall from Chapter 1 that a fraction bar is a grouping symbol. Thus, when a numerator containing a polynomial is subtracted, care must be taken to subtract, or change the sign of, *each* term in that polynomial.

Example 3 If

$$f(x) = \frac{4x + 5}{x + 3} - \frac{x - 2}{x + 3},$$

find a simplified form of $f(x)$.

SOLUTION We have

$$f(x) = \frac{4x + 5}{x + 3} - \frac{x - 2}{x + 3}$$

$$= \frac{4x + 5 - (x - 2)}{x + 3}$$

The parentheses remind us to subtract *both* terms.

$$= \frac{4x + 5 - x + 2}{x + 3}$$

$$= \frac{3x + 7}{x + 3}.$$

As a check, we let $y_1 = (4x + 5)/(x + 3) - (x - 2)/(x + 3)$ and $y_2 = (3x + 7)/(x + 3)$ and graph $y_3 = y_1 - y_2$. Since $y_3 = 0$ for all x except -3, the rational expressions are equivalent. Also note that the domain of the function is the same for both expressions. ▬

When Denominators Are Different

In order to add rational expressions such as

$$\frac{7}{12xy^2} + \frac{8}{15x^3y} \quad \text{or} \quad \frac{x}{x^2 - y^2} + \frac{y}{x^2 - 4xy + 3y^2},$$

we must first find common denominators. As in arithmetic, our work is easier when we use the *least common multiple* (LCM) of the denominators.

Least Common Multiple

To find the least common multiple (LCM) of two or more expressions, find the prime factorization of each expression and form a product that uses each factor the greatest number of times that it occurs in any one prime factorization.

Example 4   Find the least common multiple of each pair of polynomials.

a) $21x$ and $3x^2$

b) $x^2 + x - 12$ and $x^2 - 16$

SOLUTION

a) We write the prime factorizations of $21x$ and $3x^2$:

$$21x = 3 \cdot 7 \cdot x \quad \text{and} \quad 3x^2 = 3 \cdot x \cdot x.$$

The factors 3, 7, and x must appear in the LCM if $21x$ is to be a factor of the LCM. Note that $3x^2$, the other polynomial, is not a factor of $3 \cdot 7 \cdot x$. This is because the prime factors of $3x^2$—namely, 3, x, and x—do not all appear in $3 \cdot 7 \cdot x$. By including a second factor of x, we form a product that contains both $21x$ and $3x^2$ as factors:

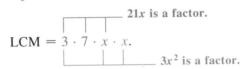

Note that each factor (3, 7, and x) is used the greatest number of times that it occurs as a factor of either $21x$ or $3x^2$. The LCM is $3 \cdot 7 \cdot x \cdot x$, or $21x^2$.

b) We factor both expressions:

$$x^2 + x - 12 = (x - 3)(x + 4),$$
$$x^2 - 16 = (x + 4)(x - 4).$$

The LCM must contain each polynomial as a factor. By multiplying the factors of $x^2 + x - 12$ by $x - 4$, we form a product that contains both $x^2 + x - 12$ and $x^2 - 16$ as factors:

$$
\begin{array}{l}
\qquad\qquad\qquad x^2 + x - 12 \text{ is a factor.}\\
\text{LCM} = (x - 3)\,(x + 4)\,(x - 4).\\
\qquad\qquad\qquad\qquad x^2 - 16 \text{ is a factor.}
\end{array}
$$

Before adding or subtracting rational expressions with unlike denominators, we determine the *least common denominator,* or LCD, by finding the LCM of the denominators. Then we multiply each rational expression by a form of 1, as needed, to form an equivalent expression that has the LCD.

Example 5   Add: $\dfrac{2}{21x} + \dfrac{5}{3x^2}$.

SOLUTION   In Example 4(a), we found that the LCD is $3 \cdot 7 \cdot x \cdot x$, or $21x^2$. We now multiply each rational expression by 1, using expressions for 1 that give us the LCD in each expression. In this case, we use x/x

and 7/7:

$$\frac{2}{21x} \cdot \frac{x}{x} + \frac{5}{3x^2} \cdot \frac{7}{7} = \frac{2x}{21x^2} + \frac{35}{21x^2} \qquad \text{We now have a common denominator.}$$

$$= \frac{2x + 35}{21x^2}. \qquad \text{This expression cannot be simplified.}$$

Note that multiplying by x/x in the first expression gave us the LCD, $21x^2$. Similarly, multiplying by 7/7 in the second expression also gave us the denominator $21x^2$. ▬

Example 6 Add: $\dfrac{x^2}{x^2 + 2xy + y^2} + \dfrac{2x - 2y}{x^2 - y^2}$.

SOLUTION Before looking for an LCD, we factor each denominator:

$$\frac{x^2}{x^2 + 2xy + y^2} + \frac{2x - 2y}{x^2 - y^2} = \frac{x^2}{(x + y)(x + y)} + \frac{2x - 2y}{(x + y)(x - y)}.$$

Although we need not *always* factor the numerators, we do so if it enables us to simplify. In this case, the rightmost rational expression can be simplified:

$$\frac{x^2}{(x + y)(x + y)} + \frac{2(x - y)}{(x + y)(x - y)} = \frac{x^2}{(x + y)(x + y)} + \frac{2}{x + y}.$$

Factoring and removing a factor equal to 1: $\dfrac{x - y}{x - y} = 1$

Note that the LCM of $(x + y)(x + y)$ and $(x + y)$ is $(x + y)(x + y)$. To get the LCD in the second expression, we multiply by 1, using $(x + y)/(x + y)$. Then we add and, if possible, simplify.

$$\frac{x^2}{(x + y)(x + y)} + \frac{2}{x + y} = \frac{x^2}{(x + y)(x + y)} + \frac{2}{x + y} \cdot \frac{x + y}{x + y}$$

$$= \frac{x^2}{(x + y)(x + y)} + \frac{2x + 2y}{(x + y)(x + y)}$$
$$\text{We now have the LCD.}$$

$$= \frac{x^2 + 2x + 2y}{(x + y)(x + y)} \qquad \begin{array}{l}\text{Since the numerator can-}\\\text{not be factored, we can-}\\\text{not simplify further.}\end{array}$$
▬

Example 7 Subtract: $\dfrac{2y + 1}{y^2 - 7y + 6} - \dfrac{y + 3}{y^2 - 5y - 6}$.

SOLUTION

$$\frac{2y + 1}{y^2 - 7y + 6} - \frac{y + 3}{y^2 - 5y - 6} = \frac{2y + 1}{(y - 6)(y - 1)} - \frac{y + 3}{(y - 6)(y + 1)} \qquad \begin{array}{l} \text{The LCD is} \\ (y - 6)(y - 1)(y + 1). \end{array}$$

$$= \frac{2y + 1}{(y - 6)(y - 1)} \cdot \frac{y + 1}{y + 1} - \frac{y + 3}{(y - 6)(y + 1)} \cdot \frac{y - 1}{y - 1}$$

$$\underbrace{}_{\begin{array}{c}\text{Multiplying by 1 to get the}\\ \text{LCD in each expression}\end{array}}$$

$$= \frac{(2y + 1)(y + 1) - (y + 3)(y - 1)}{(y - 6)(y - 1)(y + 1)}$$

$$= \frac{2y^2 + 3y + 1 - (y^2 + 2y - 3)}{(y - 6)(y - 1)(y + 1)} \qquad \begin{array}{l}\text{The parentheses}\\ \text{are important.}\end{array}$$

$$= \frac{2y^2 + 3y + 1 - y^2 - 2y + 3}{(y - 6)(y - 1)(y + 1)}$$

$$= \frac{y^2 + y + 4}{(y - 6)(y - 1)(y + 1)} \qquad \begin{array}{l}\text{We leave the}\\ \text{denominator in}\\ \text{factored form.}\end{array}$$

Example 8 Add: $\dfrac{3}{8a} + \dfrac{1}{-8a}$.

SOLUTION

$$\frac{3}{8a} + \frac{1}{-8a} = \frac{3}{8a} + \frac{-1}{-1} \cdot \frac{1}{-8a}$$

> **When denominators are opposites, we multiply one rational expression by $-1/-1$ to get the LCD.**

$$= \frac{3}{8a} + \frac{-1}{8a} = \frac{2}{8a}$$

$$= \frac{2 \cdot 1}{2 \cdot 4a} = \frac{1}{4a} \qquad \begin{array}{l}\text{Simplifying by removing a factor equal}\\ \text{to 1: } \dfrac{2}{2} = 1\end{array}$$

Example 9 Subtract: $\dfrac{5x}{x - 2y} - \dfrac{3y - 7}{2y - x}$.

SOLUTION

$$\frac{5x}{x - 2y} - \frac{3y - 7}{2y - x} = \frac{5x}{x - 2y} - \frac{-1}{-1} \cdot \frac{3y - 7}{2y - x} \qquad \begin{array}{l}\text{Note that } x - 2y \text{ and}\\ 2y - x \text{ are opposites.}\end{array}$$

$$= \frac{5x}{x - 2y} - \frac{7 - 3y}{x - 2y} \qquad \begin{array}{l}\text{Performing the multiplication.}\\ \textit{Note: } -1(2y - x) = -2y + x\\ = x - 2y.\end{array}$$

$$= \frac{5x - (7 - 3y)}{x - 2y} \qquad \begin{array}{l}\text{Subtracting. The parentheses}\\ \text{are important.}\end{array}$$

$$= \frac{5x - 7 + 3y}{x - 2y}$$

In Example 9, you may have noticed that when $3y - 7$ is multiplied by -1 and subtracted, the result is $-7 + 3y$, which is equivalent to the original $3y - 7$. Thus, instead of multiplying the numerator by -1 and then subtracting, we could have simply *added* $3y - 7$ to $5x$, as in the following:

$$\frac{5x}{x - 2y} - \frac{3y - 7}{2y - x} = \frac{5x}{x - 2y} + (-1) \cdot \frac{3y - 7}{2y - x} \qquad \text{Rewriting subtraction as addition}$$

$$= \frac{5x}{x - 2y} + \frac{1}{-1} \cdot \frac{3y - 7}{2y - x} \qquad \text{Writing } -1 \text{ as } \frac{1}{-1}$$

$$= \frac{5x}{x - 2y} + \frac{3y - 7}{x - 2y} \qquad \text{The opposite of } 2y - x \text{ is } x - 2y.$$

$$= \frac{5x + 3y - 7}{x - 2y}. \qquad \text{This checks with the answer to Example 9.}$$

Example 10   Perform the indicated operations and simplify:

$$\frac{2x}{x^2 - 4} + \frac{5}{2 - x} - \frac{1}{2 + x}.$$

SOLUTION   We have

$$\frac{2x}{x^2 - 4} + \frac{5}{2 - x} - \frac{1}{2 + x}$$

$$= \frac{2x}{(x - 2)(x + 2)} + \frac{5}{2 - x} - \frac{1}{2 + x} \qquad \text{Factoring}$$

$$= \frac{2x}{(x - 2)(x + 2)} + \frac{-1}{-1} \cdot \frac{5}{(2 - x)} - \frac{1}{x + 2} \qquad \begin{array}{l} \text{Multiplying by } \frac{-1}{-1} \\ \text{since } 2 - x \text{ is the} \\ \text{opposite of } x - 2 \end{array}$$

$$= \frac{2x}{(x - 2)(x + 2)} + \frac{-5}{x - 2} - \frac{1}{x + 2} \qquad \text{The LCD is } (x - 2)(x + 2).$$

$$= \frac{2x}{(x - 2)(x + 2)} + \frac{-5}{x - 2} \cdot \frac{x + 2}{x + 2} - \frac{1}{x + 2} \cdot \frac{x - 2}{x - 2} \qquad \begin{array}{l} \text{Multiplying} \\ \text{by 1 to get} \\ \text{the LCD} \end{array}$$

$$= \frac{2x - 5(x + 2) - (x - 2)}{(x - 2)(x + 2)} = \frac{2x - 5x - 10 - x + 2}{(x - 2)(x + 2)}$$

$$= \frac{-4x - 8}{(x - 2)(x + 2)} = \frac{-4(x + 2)}{(x - 2)(x + 2)}$$

$$= \frac{-4(x + 2)}{(x - 2)(x + 2)} \qquad \text{Removing a factor equal to 1: } \frac{x + 2}{x + 2} = 1$$

$$= \frac{-4}{x - 2}, \text{ or } -\frac{4}{x - 2}.$$

Another correct form of the answer is $\dfrac{4}{2 - x}$. It is found by writing $-\dfrac{4}{x - 2}$ as $\dfrac{4}{-(x - 2)}$ and removing parentheses.

We can check using a table of values, as shown below. The table indicates that the values of both the original and the simplified expressions are the same for all *x*-values for which both expressions are defined.

$$y_1 = 2x/(x^2 - 4) + 5/(2 - x) - 1/(2 + x),$$
$$y_2 = -4/(x - 2)$$

X	Y₁	Y₂
−3	.8	.8
−2	ERROR	1
−1	1.3333	1.3333
0	2	2
1	4	4
2	ERROR	ERROR
3	−4	−4

X = −3

6.2 Exercise Set

Perform the indicated operations. Simplify when possible.

1. $\dfrac{4}{3a} + \dfrac{8}{3a}$

2. $\dfrac{3}{2y} + \dfrac{5}{2y}$

3. $\dfrac{3}{4a^2b} - \dfrac{7}{4a^2b}$

4. $\dfrac{5}{3m^2n^2} - \dfrac{4}{3m^2n^2}$

5. $\dfrac{a - 5b}{a + b} + \dfrac{a + 7b}{a + b}$

6. $\dfrac{x - 3y}{x + y} + \dfrac{x + 5y}{x + y}$

7. $\dfrac{4y + 2}{y - 2} - \dfrac{y - 3}{y - 2}$

8. $\dfrac{3t + 2}{t - 4} - \dfrac{t - 2}{t - 4}$

9. $\dfrac{3a - 2}{a^2 - 25} - \dfrac{4a - 7}{a^2 - 25}$

10. $\dfrac{2a - 5}{a^2 - 9} - \dfrac{3a - 8}{a^2 - 9}$

11. $\dfrac{a^2}{a - b} + \dfrac{b^2}{b - a}$

12. $\dfrac{s^2}{r - s} + \dfrac{r^2}{s - r}$

13. $\dfrac{3}{x} - \dfrac{8}{-x}$

14. $\dfrac{2}{a} - \dfrac{5}{-a}$

15. $\dfrac{x - 7}{x^2 - 16} - \dfrac{x - 1}{16 - x^2}$

16. $\dfrac{y - 4}{y^2 - 25} - \dfrac{9 - 2y}{25 - y^2}$

17. $\dfrac{t^2 + 3}{t^4 - 16} + \dfrac{7}{16 - t^4}$

18. $\dfrac{y^2 - 5}{y^4 - 81} + \dfrac{4}{81 - y^4}$

19. $\dfrac{a + 2}{a - 4} + \dfrac{a - 2}{a + 3}$

20. $\dfrac{a + 3}{a - 5} + \dfrac{a - 2}{a + 4}$

21. $2 + \dfrac{x - 3}{x + 1}$

22. $3 + \dfrac{y + 2}{y - 5}$

23. $\dfrac{4xy}{x^2 - y^2} + \dfrac{x - y}{x + y}$

24. $\dfrac{5ab}{a^2 - b^2} + \dfrac{a + b}{a - b}$

25. $\dfrac{8}{2x^2 - 7x + 5} + \dfrac{3x + 2}{2x^2 - x - 10}$

26. $\dfrac{7}{3y^2 + y - 4} + \dfrac{9y + 2}{3y^2 - 2y - 8}$

27. $\dfrac{4}{x + 1} + \dfrac{x + 2}{x^2 - 1} + \dfrac{3}{x - 1}$

28. $\dfrac{-2}{y + 2} + \dfrac{5}{y - 2} + \dfrac{y + 3}{y^2 - 4}$

29. $\dfrac{x + 6}{5x + 10} - \dfrac{x - 2}{4x + 8}$

30. $\dfrac{a + 3}{5a + 25} - \dfrac{a - 1}{3a + 15}$

31. $\dfrac{x}{x^2 + 9x + 20} - \dfrac{4}{x^2 + 7x + 12}$

32. $\dfrac{x}{x^2 + 11x + 30} - \dfrac{5}{x^2 + 9x + 20}$

33. $\dfrac{3y}{y^2 - 7y + 10} - \dfrac{2y}{y^2 - 8y + 15}$

34. $\dfrac{5x}{x^2 - 6x + 8} - \dfrac{3x}{x^2 - x - 12}$

35. $\dfrac{2x + 1}{x - y} + \dfrac{5x^2 - 5xy}{x^2 - 2xy + y^2}$

36. $\dfrac{2 - 3a}{a - b} + \dfrac{3a^2 + 3ab}{a^2 - b^2}$

37. $\dfrac{a - 3}{a^2 - 16} - \dfrac{3a - 2}{a^2 + 2a - 24}$

38. $\dfrac{t+4}{t^2-9} - \dfrac{3t-1}{t^2+2t-3}$

39. $3 + \dfrac{t}{t+2} - \dfrac{2}{t^2-4}$

40. $2 + \dfrac{t}{t-3} - \dfrac{3}{t^2-9}$

41. $\dfrac{2}{y+3} - \dfrac{y}{y-1} + \dfrac{y^2+2}{y^2+2y-3}$

42. $\dfrac{1}{x+1} - \dfrac{x}{x-2} + \dfrac{x^2+2}{x^2-x-2}$

43. $\dfrac{5y}{1-2y} - \dfrac{2y}{2y+1} + \dfrac{3}{4y^2-1}$

44. $\dfrac{4x}{x^2-1} + \dfrac{3x}{1-x} - \dfrac{4}{x-1}$

45. $\dfrac{2}{x^2-5x+6} - \dfrac{4}{x^2-2x-3} + \dfrac{2}{x^2+4x+3}$

46. $\dfrac{1}{t^2+5t+6} - \dfrac{2}{t^2+3t+2} + \dfrac{1}{t^2-3t-4}$

Skill Maintenance

47. Simplify. Use only positive exponents in your answer. [1.4]

$$\dfrac{15x^{-7}y^{12}z^4}{35x^{-2}y^6z^{-3}}$$

48. Find an equation for the line that passes through the point $(-2, 3)$ and is perpendicular to the line $f(x) = -\frac{4}{5}x + 7$. [2.6]

49. *Value of Coins.* There are 50 dimes in a roll of dimes, 40 nickels in a roll of nickels, and 40 quarters in a roll of quarters. Robert has 12 rolls of coins with a total value of $70.00. If he has 3 more rolls of nickels than dimes, how many of each roll of coins does he have? [3.5]

50. *Audiotapes.* Anna wants to buy tapes for her work at the campus radio station. She needs some 30-min tapes and some 60-min tapes. If she buys 12 tapes with a total recording time of 10 hr, how many tapes of each length did she buy? [3.3]

Synthesis

51. ◆ When two rational expressions are added or subtracted, should the numerator of the sum or difference be factored? Why or why not?

52. ◆ Janine found that the sum of two rational expressions was $(3-x)/(x-5)$. The answer given at the back of the book is $(x-3)/(5-x)$. Is Janine's answer incorrect? Why or why not?

53. ◆ Many students make the mistake of always multiplying denominators when looking for a common denominator. Use Example 7 to explain why this approach can yield results that are more difficult to simplify.

54. ◆ Is the sum of two rational expressions always a rational expression? Why or why not?

55. *Music.* To duplicate a common African polyrhythm, a drummer needs to play sextuplets (6 beats per measure) on a tom-tom while simultaneously playing quarter notes (4 beats per measure) on a bass drum. Into how many equally sized parts must a measure be divided, in order to precisely execute this rhythm?

56. *Astronomy.* Earth, Jupiter, Saturn, and Uranus all revolve around the sun. Earth takes 1 yr, Jupiter 12 yr, Saturn 30 yr, and Uranus 84 yr. How frequently do these four planets line up with each other?

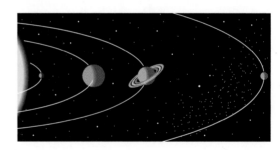

Find the LCM.

57. $x^8 - x^4, \quad x^5 - x^2, \quad x^5 - x^3, \quad x^5 + x^2$

58. $2a^3 + 2a^2b + 2ab^2, \quad a^6 - b^6,$
$2b^2 + ab - 3a^2, \quad 2a^2b + 4ab^2 + 2b^3$

59. The LCM of two expressions is $8a^4b^7$. One of the expressions is $2a^3b^7$. List all the possibilities for the other expression.

If

$$f(x) = \dfrac{x^3}{x^2-4} \quad and \quad g(x) = \dfrac{x^2}{x^2+3x-10},$$

find each of the following.

60. $(f+g)(x)$

61. $(f-g)(x)$

62. $(f \cdot g)(x)$

63. $(f/g)(x)$

64. The domain of $f + g$

65. The domain of f / g

Perform the indicated operations and simplify.

66. $5(x - 3)^{-1} + 4(x + 3)^{-1} - 2(x + 3)^{-2}$

67. $4(y - 1)(2y - 5)^{-1} + 5(2y + 3)(5 - 2y)^{-1} +$
$(y - 4)(2y - 5)^{-1}$

68. $\dfrac{x + 4}{6x^2 - 20x} \cdot \left(\dfrac{x}{x^2 - x - 20} + \dfrac{2}{x + 4} \right)$

69. $\dfrac{8t^5}{2t^2 - 10t + 12} \div \left(\dfrac{2t}{t^2 - 8t + 15} - \dfrac{3t}{t^2 - 7t + 10} \right)$

Determine the domain and estimate the range of each function.

70. $f(x) = 2 + \dfrac{x - 3}{x + 1}$

71. $g(x) = \dfrac{2}{(x + 1)^2} + 5$

72. $r(x) = \dfrac{1}{x^2} + \dfrac{1}{(x - 1)^2}$

6.3

Complex Rational Expressions

- *Multiplying by 1*
- *Dividing Two Rational Expressions*

A **complex rational expression** is a rational expression that contains rational expressions within its numerator and/or its denominator. Here are some examples:

$$\dfrac{x + \dfrac{5}{x}}{4x}, \quad \dfrac{\dfrac{x - y}{x + y}}{\dfrac{2x - y}{3x + y}}, \quad \dfrac{\dfrac{2}{3}}{\dfrac{4}{5}}, \quad \dfrac{\dfrac{3x}{5} - \dfrac{2}{x}}{\dfrac{4x}{3} + \dfrac{7}{6x}}.$$

The rational expressions within each complex rational expression are red.

Two methods are used to simplify complex rational expressions.

Method 1: Multiplying by 1

One method of simplifying a complex rational expression is to multiply the entire expression by 1. To write 1, we use the LCD of the expressions within the complex rational expression.

Example 1   Simplify:

$$\dfrac{\dfrac{1}{x^2 y} - \dfrac{1}{y}}{\dfrac{1}{xy^2} - \dfrac{1}{y^2}}.$$

SOLUTION   The denominators within the complex rational expression are x^2y, y, xy^2, and y^2. Thus the LCD is x^2y^2. We multiply by 1, using x^2y^2/x^2y^2:

$$\dfrac{\dfrac{1}{x^2y} - \dfrac{1}{y}}{\dfrac{1}{xy^2} - \dfrac{1}{y^2}} = \dfrac{\dfrac{1}{x^2y} - \dfrac{1}{y}}{\dfrac{1}{xy^2} - \dfrac{1}{y^2}} \cdot \dfrac{x^2y^2}{x^2y^2}$$

Multiplying by 1, using the LCD

$$= \dfrac{\left(\dfrac{1}{x^2y} - \dfrac{1}{y}\right)x^2y^2}{\left(\dfrac{1}{xy^2} - \dfrac{1}{y^2}\right)x^2y^2}$$

Multiplying the numerator and the denominator. Remember to use parentheses.

$$= \dfrac{\dfrac{1}{x^2y} \cdot x^2y^2 - \dfrac{1}{y} \cdot x^2y^2}{\dfrac{1}{xy^2} \cdot x^2y^2 - \dfrac{1}{y^2} \cdot x^2y^2}$$

Using the distributive law to carry out the multiplications

$$= \dfrac{\dfrac{x^2y}{x^2y} \cdot y - \dfrac{y}{y} \cdot x^2y}{\dfrac{xy^2}{xy^2} \cdot x - \dfrac{y^2}{y^2} \cdot x^2}$$

Removing factors that equal 1. Study this carefully.

$$= \dfrac{y - x^2y}{x - x^2}$$

Simplifying

$$= \dfrac{y(1 - x^2)}{x(1 - x)}$$

Factoring

$$= \dfrac{y(1 + x)(1 - x)}{x(1 - x)}$$

Factoring further and identifying a factor that equals 1

$$= \dfrac{y(1 + x)}{x}.$$

Simplifying

Using Multiplication by 1 to Simplify a Complex Rational Expression

1. Find the LCD of all expressions *within* the complex rational expression.
2. Multiply the complex rational expression by 1, using the LCD to form the expression for 1.
3. Distribute and simplify so that the numerator and the denominator of the complex rational expression are simply polynomials.
4. Factor and simplify, if possible.

Note that using the LCD to form 1 clears the numerator and the denominator of the complex rational expression of all rational expressions.

Example 2 Simplify:

$$\frac{\dfrac{3}{2x - 2} - \dfrac{1}{x + 1}}{\dfrac{1}{x - 1} + \dfrac{x}{x^2 - 1}}.$$

SOLUTION   In this case, to find the LCD, we must first factor:

$$\frac{\dfrac{3}{2x - 2} - \dfrac{1}{x + 1}}{\dfrac{1}{x - 1} + \dfrac{x}{x^2 - 1}} = \frac{\dfrac{3}{2(x - 1)} - \dfrac{1}{x + 1}}{\dfrac{1}{x - 1} + \dfrac{x}{(x - 1)(x + 1)}}$$

The LCD is
$2(x - 1)(x + 1)$.

$$= \frac{\dfrac{3}{2(x - 1)} - \dfrac{1}{x + 1}}{\dfrac{1}{x - 1} + \dfrac{x}{(x - 1)(x + 1)}} \cdot \frac{2(x - 1)(x + 1)}{2(x - 1)(x + 1)}$$

Multiplying by 1, using the LCD

$$= \frac{\dfrac{3}{2(x - 1)} \cdot 2(x - 1)(x + 1) - \dfrac{1}{x + 1} \cdot 2(x - 1)(x + 1)}{\dfrac{1}{x - 1} \cdot 2(x - 1)(x + 1) + \dfrac{x}{(x - 1)(x + 1)} \cdot 2(x - 1)(x + 1)}$$

Using the distributive law

$$= \frac{\dfrac{2(x - 1)}{2(x - 1)} \cdot 3(x + 1) - \dfrac{x + 1}{x + 1} \cdot 2(x - 1)}{\dfrac{x - 1}{x - 1} \cdot 2(x + 1) + \dfrac{(x - 1)(x + 1)}{(x - 1)(x + 1)} \cdot 2x}$$

Removing factors that equal 1

$$= \frac{3(x + 1) - 2(x - 1)}{2(x + 1) + 2x}$$

Simplifying

$$= \frac{3x + 3 - 2x + 2}{2x + 2 + 2x}$$

Using the distributive law

$$= \frac{x + 5}{4x + 2}.$$

Method 2: Dividing Two Rational Expressions

Another method for simplifying complex rational expressions involves first adding or subtracting, as necessary, to get one rational expression in the numerator and one rational expression in the denominator. The problem is thereby simplified to one involving the division of two rational expressions.

Example 3   Simplify:

$$\frac{\dfrac{3}{x} - \dfrac{2}{x^2}}{\dfrac{3}{x-2} + \dfrac{1}{x^2}}.$$

SOLUTION   We have

$$\frac{\dfrac{3}{x} - \dfrac{2}{x^2}}{\dfrac{3}{x-2} + \dfrac{1}{x^2}} = \frac{\dfrac{3}{x} \cdot \dfrac{x}{x} - \dfrac{2}{x^2}}{\dfrac{3}{x-2} \cdot \dfrac{x^2}{x^2} + \dfrac{1}{x^2} \cdot \dfrac{x-2}{x-2}}$$

$\left.\begin{array}{l}\end{array}\right\}$ Multiplying 3/x by 1 to get x^2 as a common denominator

$\left.\begin{array}{l}\end{array}\right\}$ Multiplying by 1, twice, to get $x^2(x-2)$ as a common denominator

$$= \frac{\dfrac{3x}{x^2} - \dfrac{2}{x^2}}{\dfrac{3x^2}{(x-2)x^2} + \dfrac{x-2}{x^2(x-2)}}$$

There is now a common denominator in the numerator and a common denominator in the denominator of the complex rational expression.

$$= \frac{\dfrac{3x-2}{x^2}}{\dfrac{3x^2 + x - 2}{(x-2)x^2}}$$

Subtracting in the numerator and adding in the denominator. We now have one rational expression divided by another rational expression.

$$= \frac{3x-2}{x^2} \div \frac{3x^2 + x - 2}{(x-2)x^2}$$

Rewriting with a division symbol

$$= \frac{3x-2}{x^2} \cdot \frac{(x-2)x^2}{3x^2 + x - 2}$$

To divide, multiply by the reciprocal of the divisor.

$$= \frac{(3x-2)(x-2)x^2}{x^2(3x-2)(x+1)}$$

Factoring and removing a factor equal to 1: $\dfrac{x^2(3x-2)}{x^2(3x-2)} = 1$

$$= \frac{x-2}{x+1}.$$

To check, we let $y_1 = (3/x - 2/x^2)/(3/(x-2) + 1/x^2)$ and $y_2 = (x-2)/(x+1)$. A table of values, like that shown at left, indicates that the values of y_1 and y_2 are the same for all x-values for which both expressions are defined.

$y_1 = (3/x - 2/x^2)/(3/(x-2) + 1/x^2)$,
$y_2 = (x-2)/(x+1)$

X	Y1	Y2
−3	2.5	2.5
−2	4	4
−1	ERROR	ERROR
0	ERROR	−2
1	−.5	−.5
2	ERROR	0
3	.25	.25
X = −3		

Using Division to Simplify a Complex Rational Expression

1. Add or subtract, as necessary, to get one rational expression in the numerator.

2. Add or subtract, as necessary, to get one rational expression in the denominator.

3. Perform the indicated division (invert the divisor and multiply).

4. Simplify, if possible, by removing any factors that equal 1.

Example 4   Simplify:

$$\frac{1 + \dfrac{2}{x}}{1 - \dfrac{4}{x^2}}.$$

SOLUTION   We have

$$\frac{1 + \dfrac{2}{x}}{1 - \dfrac{4}{x^2}} = \frac{\dfrac{x}{x} + \dfrac{2}{x}}{\dfrac{x^2}{x^2} - \dfrac{4}{x^2}} \left. \begin{array}{c} \\ \end{array} \right\} \quad \text{Finding a common denominator}$$
$$\hspace{5cm} \left. \begin{array}{c} \\ \end{array} \right\} \quad \text{Finding a common denominator}$$

$$= \frac{\dfrac{x + 2}{x}}{\dfrac{x^2 - 4}{x^2}} \quad \begin{array}{l} \text{Adding in the numerator} \\[1em] \text{Subtracting in the denominator} \end{array}$$

$$= \frac{x + 2}{x} \cdot \frac{x^2}{x^2 - 4} \quad \begin{array}{l} \text{Multiplying by the reciprocal} \\ \text{of the divisor} \end{array}$$

$$= \frac{(x + 2) \cdot x^2}{x(x + 2)(x - 2)} \quad \text{Factoring}$$

$$= \frac{\cancel{(x + 2)}x \cdot x}{x\cancel{(x + 2)}(x - 2)} \quad \text{Removing a factor equal to 1: } \frac{(x + 2)x}{(x + 2)x} = 1$$

$$= \frac{x}{x - 2}. \quad \text{Simplifying}$$

We could again check by comparing graphs or tables; however, entering a complex rational expression can be time-consuming.

As a quick partial check, we select a convenient value for x—say, 1:

$$\frac{1 + \dfrac{2}{1}}{1 - \dfrac{4}{1^2}} = \frac{1 + 2}{1 - 4} = \frac{3}{-3} = -1 \quad \begin{array}{l} \text{We evaluated the original} \\ \text{expression for } x = 1. \end{array}$$

and

$$\frac{1}{1 - 2} = \frac{1}{-1} = -1. \quad \begin{array}{l} \text{We evaluated the simplified} \\ \text{result for } x = 1. \end{array}$$

Since both expressions yield the same result, our simplification is probably correct. More evaluation would provide a more definitive check.   ▬

If negative exponents occur, we first find an equivalent expression using positive exponents and then proceed as in the examples above.

Example 5 Simplify:

$$\frac{a^{-1} + b^{-1}}{a^{-2} - b^{-2}}.$$

SOLUTION

$$\frac{a^{-1} + b^{-1}}{a^{-2} - b^{-2}} = \frac{\dfrac{1}{a} + \dfrac{1}{b}}{\dfrac{1}{a^2} - \dfrac{1}{b^2}} \qquad$$ Rewriting with positive exponents. We continue, using method 2.

$$= \frac{\dfrac{1}{a} \cdot \dfrac{b}{b} + \dfrac{1}{b} \cdot \dfrac{a}{a}}{\dfrac{1}{a^2} \cdot \dfrac{b^2}{b^2} - \dfrac{1}{b^2} \cdot \dfrac{a^2}{a^2}} \qquad$$ Finding a common denominator

Finding a common denominator

$$= \frac{\dfrac{b}{ab} + \dfrac{a}{ab}}{\dfrac{b^2}{a^2b^2} - \dfrac{a^2}{a^2b^2}}$$

$$= \frac{\dfrac{b + a}{ab}}{\dfrac{b^2 - a^2}{a^2b^2}} \qquad$$ Adding in the numerator

Subtracting in the denominator

$$= \frac{b + a}{ab} \cdot \frac{a^2b^2}{b^2 - a^2} \qquad$$ Multiplying by the reciprocal of the divisor

$$= \frac{(b + a) \cdot ab \cdot ab}{ab(b + a)(b - a)} \qquad$$ Factoring and looking for common factors

$$= \frac{\cancel{(b + a)} \cdot \cancel{ab} \cdot ab}{\cancel{ab}\cancel{(b + a)}(b - a)} \qquad$$ Removing a factor equal to 1: $\dfrac{(b + a)ab}{(b + a)ab} = 1$

$$= \frac{ab}{b - a}$$

It is difficult to say which method is better to use. For expressions like

$$\frac{\dfrac{3x + 1}{x - 5}}{\dfrac{2 - x}{x + 3}}$$

or

$$\frac{\dfrac{3}{x} - \dfrac{2}{x}}{\dfrac{1}{x + 1} + \dfrac{5}{x + 1}},$$

the second method is probably easier to use since it is little or no work to write the expression as a quotient of two rational expressions.

On the other hand, expressions like

$$\frac{\dfrac{3}{a^2b} - \dfrac{4}{bc^3}}{\dfrac{1}{b^3c} + \dfrac{2}{ac^4}} \quad \text{or} \quad \frac{\dfrac{5}{a^2 - b^2} + \dfrac{2}{a^2 + 2ab + b^2}}{\dfrac{1}{a - b} + \dfrac{4}{a + b}}$$

require fewer steps if we use the first method. Either method works for any complex rational expression.

6.3 Exercise Set

Simplify. If possible, use a second method or evaluation as a check.

1. $\dfrac{5 + \dfrac{1}{a}}{\dfrac{1}{a} - 2}$

2. $\dfrac{\dfrac{1}{y} + 7}{\dfrac{1}{y} - 5}$

3. $\dfrac{x - x^{-1}}{x + x^{-1}}$

4. $\dfrac{y + y^{-1}}{y - y^{-1}}$

5. $\dfrac{\dfrac{3}{x} + \dfrac{4}{y}}{\dfrac{4}{x} - \dfrac{3}{y}}$

6. $\dfrac{\dfrac{5}{z} + \dfrac{2}{y}}{\dfrac{4}{z} - \dfrac{1}{y}}$

7. $\dfrac{\dfrac{x^2 - y^2}{xy}}{\dfrac{x - y}{y}}$

8. $\dfrac{\dfrac{a^2 - b^2}{ab}}{\dfrac{a - b}{b}}$

9. $\dfrac{\dfrac{3x}{y} - x}{2y - \dfrac{y}{x}}$

10. $\dfrac{1 - \dfrac{2}{3x}}{x - \dfrac{4}{9x}}$

11. $\dfrac{a^{-1} + b^{-1}}{\dfrac{a^2 - b^2}{ab}}$

12. $\dfrac{x^{-1} + y^{-1}}{\dfrac{x^2 - y^2}{xy}}$

13. $\dfrac{\dfrac{1}{x + h} - \dfrac{1}{x}}{h}$

14. $\dfrac{\dfrac{1}{a - h} - \dfrac{1}{a}}{h}$

15. $\dfrac{\dfrac{a^2 - 4}{a^2 + 3a + 2}}{\dfrac{a^2 - 5a - 6}{a^2 - 6a - 7}}$

16. $\dfrac{\dfrac{x^2 - x - 12}{x^2 - 2x - 15}}{\dfrac{x^2 + 8x + 12}{x^2 - 5x - 14}}$

17. $\dfrac{\dfrac{2}{y - 3} + \dfrac{1}{y + 1}}{\dfrac{3}{y + 1} + \dfrac{4}{y - 3}}$

18. $\dfrac{\dfrac{1}{x - 2} + \dfrac{3}{x - 1}}{\dfrac{2}{x - 1} + \dfrac{5}{x - 2}}$

19. $\dfrac{a(a + 3)^{-1} - 2(a - 1)^{-1}}{a(a + 3)^{-1} - (a - 1)^{-1}}$

20. $\dfrac{a(a + 2)^{-1} - 3(a - 3)^{-1}}{a(a + 2)^{-1} - (a - 3)^{-1}}$

21. $\dfrac{\dfrac{x}{x^2 + 3x - 4} - \dfrac{1}{x^2 + 3x - 4}}{\dfrac{x}{x^2 + 6x + 8} + \dfrac{3}{x^2 + 6x + 8}}$

22. $\dfrac{\dfrac{x}{x^2 + 5x - 6} + \dfrac{6}{x^2 + 5x - 6}}{\dfrac{x}{x^2 - 5x + 4} - \dfrac{2}{x^2 - 5x + 4}}$

23. $\dfrac{\dfrac{3}{a^2 - 9} + \dfrac{2}{a + 3}}{\dfrac{4}{a^2 - 9} + \dfrac{1}{a + 3}}$

24. $\dfrac{\dfrac{2}{a^2 - 1} + \dfrac{1}{a + 1}}{\dfrac{3}{a^2 - 1} + \dfrac{2}{a - 1}}$

25. $\dfrac{\dfrac{4}{x^2 - 1} - \dfrac{3}{x + 1}}{\dfrac{5}{x^2 - 1} - \dfrac{2}{x - 1}}$

26. $\dfrac{\dfrac{5}{x^2 - 4} - \dfrac{3}{x - 2}}{\dfrac{4}{x^2 - 4} - \dfrac{2}{x + 2}}$

27. $\dfrac{\dfrac{y}{y^2-1}+\dfrac{3}{1-y^2}}{\dfrac{y^2}{y^2-1}+\dfrac{9}{1-y^2}}$

28. $\dfrac{\dfrac{y}{y^2-4}+\dfrac{5}{4-y^2}}{\dfrac{y^2}{y^2-4}+\dfrac{25}{4-y^2}}$

29. $\dfrac{\dfrac{y^2}{y^2-9}-\dfrac{y}{y+3}}{\dfrac{y}{y^2-9}-\dfrac{1}{y-3}}$

30. $\dfrac{\dfrac{y^2}{y^2-25}-\dfrac{y}{y-5}}{\dfrac{y}{y^2-25}-\dfrac{1}{y+5}}$

31. $\dfrac{\dfrac{a}{a+3}+\dfrac{4}{5a}}{\dfrac{a}{2a+6}+\dfrac{3}{a}}$

32. $\dfrac{\dfrac{a}{a+2}+\dfrac{5}{a}}{\dfrac{a}{2a+4}+\dfrac{1}{3a}}$

33. $\dfrac{\dfrac{1}{x^2-3x+2}+\dfrac{1}{x^2-4}}{\dfrac{1}{x^2+4x+4}+\dfrac{1}{x^2-4}}$

34. $\dfrac{\dfrac{1}{x^2+3x+2}+\dfrac{1}{x^2-1}}{\dfrac{1}{x^2-1}+\dfrac{1}{x^2-4x+3}}$

35. $\dfrac{\dfrac{3}{x^2+2x-3}-\dfrac{1}{x^2-3x-10}}{\dfrac{3}{x^2-6x+5}-\dfrac{1}{x^2+5x+6}}$

36. $\dfrac{\dfrac{1}{a^2+7a+12}+\dfrac{1}{a^2+a-6}}{\dfrac{1}{a^2+2a-8}+\dfrac{1}{a^2+5a+4}}$

Skill Maintenance

37. If $f(x)=x^2-3$, find $f(-5)$. [2.1]

38. Solve for y: $\dfrac{a}{x+y}=b$. [2.3]

39. Graph: $f(x)=-3x+7$. [2.4]

40. Solve: $|2x-5|=7$. [4.3]

41. *Earnings.* Antonio received $28 in tips on Monday, $22 in tips on Tuesday, and $36 in tips on Wednesday. How much will Antonio need to receive in tips on Thursday if his average for the four days is to be $30? [2.3]

42. *Framing.* Glenn has two rectangular frames. The first frame is 3 cm shorter, and 4 cm narrower, than the second frame. If the perimeter of the second frame is 1 cm less than twice the perimeter of the first, what is the perimeter of each frame? [3.3]

Synthesis

43. ◆ To simplify a complex rational expression in which the sum of two fractions is divided by the difference of the same two fractions, which method is easiest? Why?

44. ◆ In arithmetic, we are taught that

$$\dfrac{a}{b}\div\dfrac{c}{d}=\dfrac{a}{b}\cdot\dfrac{d}{c}$$

(to divide by a fraction, we invert and multiply). Use method 1 to explain *why* we do this.

45. ◆ Use algebra to determine the domain of the function given by

$$f(x)=\dfrac{\dfrac{1}{x-2}}{\dfrac{x}{x-2}-\dfrac{5}{x-2}}.$$

Then explain how a grapher could be used to check your answer.

Simplify.

46. $\dfrac{5x^{-2}+10x^{-1}y^{-1}+5y^{-2}}{3x^{-2}-3y^{-2}}$

47. $(a^2-ab+b^2)^{-1}(a^2b^{-1}+b^2a^{-1})\times$
$(a^{-2}-b^{-2})(a^{-2}+2a^{-1}b^{-1}+b^{-2})^{-1}$

48. *Astronomy.* When two galaxies are moving in opposite directions at velocities v_1 and v_2, an observer in one of the galaxies would see the other galaxy receding at speed

$$\dfrac{v_1+v_2}{1+\dfrac{v_1v_2}{c^2}},$$

where c is the speed of light. Determine the observed speed if v_1 and v_2 are both one-fourth the speed of light.

Find the reciprocal of the expression shown. Then simplify, if possible.

49. $1+\dfrac{1}{1+\dfrac{1}{1+\dfrac{1}{x}}}$

50. $x^2 + x + 1 + \dfrac{1}{x} + \dfrac{1}{x^2}$

51. For $f(x) = \dfrac{1}{1+x}$, find $f(f(a))$.

52. For $g(x) = \dfrac{x+1}{x-2}$, find $g(g(a))$.

Find and simplify

$$\dfrac{f(x+h) - f(x)}{h}$$

for each rational function f in Exercises 53–56.

53. $f(x) = \dfrac{3}{x^2}$ **54.** $f(x) = \dfrac{5}{x}$

55. $f(x) = \dfrac{1}{1-x}$

56. $f(x) = \dfrac{x}{1+x}$

57. If

$$F(x) = \dfrac{3 + \dfrac{1}{x}}{2 - \dfrac{8}{x^2}},$$

find the domain of F.

58. If

$$G(x) = \dfrac{x - \dfrac{1}{x^2 - 1}}{\dfrac{1}{9} - \dfrac{1}{x^2 - 16}},$$

find the domain of G.

59. Let

$$f(x) = \left[\dfrac{\dfrac{x+3}{x-3} + 1}{\dfrac{x+3}{x-3} - 1}\right]^4.$$

Find a simplified form of $f(x)$ and specify the domain of f.

6.4 Rational Equations

- *Solving Rational Equations*
- *Rational Functions*

Solving Rational Equations

In Sections 6.1–6.3, we learned how to *simplify expressions*. We now learn to *solve* a new type of *equation*. A **rational equation** is an equation that contains one or more rational expressions. Here are some examples:

$$\frac{2}{3} - \frac{5}{6} = \frac{1}{t}, \qquad \frac{a-1}{a-5} = \frac{4}{a^2 - 25}, \qquad x^3 + \frac{6}{x} = 5.$$

As you will see in Section 6.5, equations of this type occur frequently in applications. To solve rational equations, recall that one way to *clear fractions* is to multiply by the LCD.

> **To Solve a Rational Equation:**
>
> Multiply on both sides by the LCD. This is called *clearing fractions* and produces an equation similar to those we have already solved.

Recall that division by 0 is undefined. Note, too, that variables generally appear in at least one denominator of a rational equation. Thus certain numbers can often be ruled out as possible solutions before we ever attempt to solve a given rational equation.

Example 1   Solve: $\dfrac{x+4}{3x} + \dfrac{x+8}{5x} = 2$.

ALGEBRAIC SOLUTION

Because the left side of this equation is undefined when x is 0, we state at the outset that $x \neq 0$. Next, we multiply both sides of the equation by the LCD, $3 \cdot 5 \cdot x$:

$$3 \cdot 5 \cdot x\left(\frac{x+4}{3x} + \frac{x+8}{5x}\right) = 3 \cdot 5 \cdot x \cdot 2$$

Multiplying by the LCD to clear fractions

$$3 \cdot 5 \cdot x \cdot \frac{x+4}{3x} + 3 \cdot 5 \cdot x \cdot \frac{x+8}{5x} = 3 \cdot 5 \cdot x \cdot 2$$

Using the distributive law

$$\frac{3 \cdot 5 \cdot x \cdot (x+4)}{3x} + \frac{3 \cdot 5 \cdot x \cdot (x+8)}{5x} = 30x$$

Locating factors equal to 1

$$5(x+4) + 3(x+8) = 30x$$

Removing factors equal to 1: $\dfrac{3x}{3x} = 1;\ \dfrac{5x}{5x} = 1$

$$5x + 20 + 3x + 24 = 30x \qquad \text{Using the distributive law}$$

$$8x + 44 = 30x$$

$$44 = 22x$$

$$2 = x. \qquad \text{This should check since } x \neq 0.$$

The solution is 2.

GRAPHICAL SOLUTION

We let $y_1 = (x+4)/(3x) + (x+8)/(5x)$ and $y_2 = 2$. The solutions of the equation will be the x-coordinates of any points of intersection.

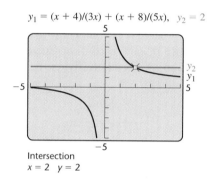

$y_1 = (x+4)/(3x) + (x+8)/(5x),\ \ y_2 = 2$

Intersection
$x = 2\ \ y = 2$

It appears from the graph that there is one point of intersection, $(2, 2)$. The solution is 2.

CHECK:
$$\frac{x+4}{3x} + \frac{x+8}{5x} = 2$$

$$\frac{2+4}{3\cdot 2} + \frac{2+8}{5\cdot 2} \ ?\ 2$$

$$\frac{6}{6} + \frac{10}{10}$$

$$2 \ \bigm|\ 2 \quad \text{TRUE}$$

As a further check, we see that since both approaches yield the same solution, the number 2 is the solution.

Note that when we clear fractions, all denominators "disappear." This leaves an equation without rational expressions, which we know how to solve.

Example 2 Solve: $\dfrac{x-1}{x-5} = \dfrac{4}{x-5}$.

ALGEBRAIC SOLUTION

To ensure that neither denominator is 0, we state at the outset the restriction that $x \neq 5$. Then we proceed as before, multiplying by the LCD, $x - 5$, on both sides:

$$(x-5) \cdot \dfrac{x-1}{x-5} = (x-5) \cdot \dfrac{4}{x-5}$$

$$x - 1 = 4$$

$$x = 5. \qquad \text{But recall that } x \neq 5.$$

In this case, it is important to remember that, because of our restriction above, 5 cannot be a solution. A check confirms the necessity of that restriction.

CHECK:
$$\dfrac{x-1}{x-5} = \dfrac{4}{x-5}$$

$$\dfrac{5-1}{5-5} \overset{?}{=} \dfrac{4}{5-5}$$

$$\dfrac{4}{0} \quad \Big| \quad \dfrac{4}{0} \qquad \text{Division by 0 is undefined.}$$

GRAPHICAL SOLUTION

We graph $y_1 = (x-1)/(x-5)$ and $y_2 = 4/(x-5)$ and look for any points of intersection of the graphs.

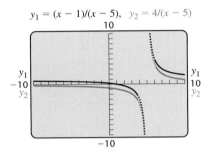

$$y_1 = (x-1)/(x-5), \quad y_2 = 4/(x-5)$$

It appears that the graphs may intersect around $x = 5$. If we use INTERSECT and make a guess close to $x = 5$, the calculator returns an error message. This in itself does not tell us that there is no point of intersection, because we do not see all the graph in the viewing window. In order to state positively that there is no solution, we must solve the equation with algebra.

This equation has no solution. ▬

To help see why 5 is not a solution to Example 2, consider the fact that the multiplication principle for equations requires that we multiply by a *nonzero* number on both sides if we are to form an equivalent equation. When both sides of an equation are multiplied by an expression containing variables, it is possible that certain replacements will make that expression equal to 0. Thus it is safe to say that *if* a solution of

$$\dfrac{x-1}{x-5} = \dfrac{4}{x-5}$$

exists, then it is also a solution of $x - 1 = 4$. We *cannot* conclude that every solution of $x - 1 = 4$ is a solution of the original equation. Thus it is important to check every possible solution in the original equation.

Example 2 illustrates a disadvantage of graphical methods of solving equations. It can be difficult to tell whether two graphs actually intersect. Also, unless you already know the shape of the graph, it is easy to "miss" seeing a solution. This is especially true for rational equations, where there are usually two or more branches of the graph. Solving by graphing does make a good check, however, of an algebraic solution.

Example 3   Solve: $\dfrac{x^2}{x-3} = \dfrac{9}{x-3}$.

SOLUTION   Note that $x \neq 3$. Since the LCD is $x - 3$, we multiply by $x - 3$ on both sides:

$$(x - 3) \cdot \frac{x^2}{x - 3} = (x - 3) \cdot \frac{9}{x - 3}$$

$$x^2 = 9 \qquad \text{Simplifying}$$

$$x^2 - 9 = 0 \qquad \text{Getting 0 on one side}$$

$$(x - 3)(x + 3) = 0 \qquad \text{Factoring}$$

$$x = 3 \quad \text{or} \quad x = -3. \qquad \text{Using the principle of zero products}$$

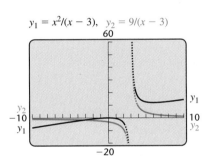

$y_1 = x^2/(x - 3), \quad y_2 = 9/(x - 3)$

Yscl = 10

Although 3 is a solution of $x^2 = 9$, it must be rejected as a solution of the rational equation because it makes the denominators 0. The graphs of $y_1 = x^2/(x - 3)$ and $y_2 = 9/(x - 3)$ are shown in the figure at left. They clearly intersect at $x = -3$, but $x = 3$ is a vertical asymptote, and the graphs cannot intersect at this x-value. You should perform a check to confirm that -3 *is* a solution.

Example 4   Solve: $\dfrac{2}{x+5} + \dfrac{1}{x-5} = \dfrac{16}{x^2-25}$.

SOLUTION   To find all restrictions and to assist in finding the LCD, we factor:

$$\frac{2}{x + 5} + \frac{1}{x - 5} = \frac{16}{(x + 5)(x - 5)}. \qquad \text{Factoring } x^2 - 25$$

Note that $x \neq -5$ and $x \neq 5$. We multiply by the LCD, $(x + 5)(x - 5)$, and then use the distributive law:

$$(x + 5)(x - 5)\left(\frac{2}{x + 5} + \frac{1}{x - 5}\right) = (x + 5)(x - 5) \cdot \frac{16}{(x + 5)(x - 5)}$$

$$(x + 5)(x - 5)\frac{2}{x + 5} + (x + 5)(x - 5)\frac{1}{x - 5}$$

$$= (x + 5)(x - 5) \cdot \frac{16}{(x + 5)(x - 5)}$$

$$2(x - 5) + (x + 5) = 16$$

$$2x - 10 + x + 5 = 16$$

$$3x - 5 = 16$$

$$3x = 21$$

$$x = 7.$$

The check is left to the student. The solution is 7.

Rational Functions

Example 5   Let $f(x) = x + \dfrac{6}{x}$. Find all values of a for which $f(a) = 5$.

SOLUTION   Since $f(a) = a + \dfrac{6}{a}$, the problem asks that we find all values of a for which

$$a + \frac{6}{a} = 5.$$

To solve for a, we first note that $a \neq 0$. Next, we multiply by the LCD, a, on both sides:

$$a\left(a + \frac{6}{a}\right) = 5 \cdot a \qquad \text{Multiplying on both sides by } a. \text{ Parentheses are important.}$$

$$a \cdot a + a \cdot \frac{6}{a} = 5a \qquad \text{Using the distributive law}$$

$$a^2 + 6 = 5a \qquad \text{Simplifying}$$

$$a^2 - 5a + 6 = 0 \qquad \text{Getting 0 on one side}$$

$$(a - 3)(a - 2) = 0 \qquad \text{Factoring}$$

$$a = 3 \quad or \quad a = 2. \qquad \text{Using the principle of zero products}$$

CHECK: $f(3) = 3 + \dfrac{6}{3} = 3 + 2 = 5;$

$f(2) = 2 + \dfrac{6}{2} = 2 + 3 = 5.$

The solutions are 2 and 3. For $a = 2$ or $a = 3$, we have $f(a) = 5$.

The graph at left shows the points of intersection of the graph of $f(x)$ and the line $y = 5$. It also shows that we have two solutions of $f(x) = 5$, 2 and 3. For $a = 2$ or $a = 3$, we have $f(a) = 5$.   ▬

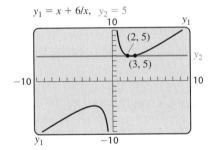

$y_1 = x + 6/x, \quad y_2 = 5$

6.4 Exercise Set

Solve.

1. $\dfrac{1}{3} + \dfrac{4}{5} = \dfrac{x}{9}$

2. $\dfrac{7}{8} + \dfrac{2}{5} = \dfrac{x}{20}$

3. $\dfrac{x}{3} - \dfrac{x}{4} = 12$

4. $\dfrac{t}{5} - \dfrac{t}{3} = 15$

5. $\dfrac{5}{8} - \dfrac{1}{a} = \dfrac{2}{5}$

6. $\dfrac{1}{3} - \dfrac{1}{x} = \dfrac{5}{6}$

7. $\dfrac{2}{3} - \dfrac{1}{5} = \dfrac{7}{3x}$

8. $\dfrac{1}{2} - \dfrac{2}{7} = \dfrac{3}{2x}$

9. $\dfrac{2}{6} + \dfrac{1}{2x} = \dfrac{1}{3}$

10. $\dfrac{12}{15} - \dfrac{1}{3x} = \dfrac{4}{5}$

11. $a + \dfrac{4}{a} = -5$

12. $\dfrac{4}{3n} - \dfrac{3}{n} = \dfrac{10}{3}$

13. $\dfrac{p-1}{p-3} = \dfrac{2}{p-3}$

14. $\dfrac{x-2}{x-4} = \dfrac{2}{x-4}$

15. $\dfrac{3}{x-2} = \dfrac{5}{x+4}$

16. $\dfrac{5}{4t} = \dfrac{7}{5t-2}$

17. $\dfrac{x^2-1}{x+2} = \dfrac{3}{x+2}$

18. $\dfrac{x^2+4}{x-1} = \dfrac{5}{x-1}$

19. $\dfrac{4}{a-7} = \dfrac{-2a}{a+3}$

20. $\dfrac{6}{a+1} = \dfrac{a}{a-1}$

21. $\dfrac{50}{t-2} - \dfrac{16}{t} = \dfrac{30}{t}$

22. $\dfrac{60}{t-5} - \dfrac{18}{t} = \dfrac{40}{t}$

23. $\dfrac{3}{x} + \dfrac{x}{x+2} = \dfrac{4}{x^2+2x}$

24. $\dfrac{x}{x+1} + \dfrac{5}{x} = \dfrac{1}{x^2+x}$

In Exercises 25–30, a rational function f is given. Find all values of a for which f(a) is the indicated value.

25. $f(x) = 2x - \dfrac{15}{x};\quad f(a) = 1$

26. $f(x) = 2x - \dfrac{6}{x};\quad f(a) = 1$

27. $f(x) = \dfrac{x-5}{x+1};\quad f(a) = \dfrac{3}{5}$

28. $f(x) = \dfrac{x-3}{x+2};\quad f(a) = \dfrac{1}{5}$

29. $f(x) = \dfrac{12}{x} - \dfrac{12}{2x};\quad f(a) = 8$

30. $f(x) = \dfrac{6}{x} - \dfrac{6}{2x};\quad f(a) = 5$

Solve.

31. $\dfrac{5}{x+2} - \dfrac{3}{x-2} = \dfrac{2x}{4-x^2}$

32. $\dfrac{t+3}{t+2} - \dfrac{t}{t^2-4} = \dfrac{t}{t-2}$

33. $\dfrac{2}{a+4} + \dfrac{2a-1}{a^2+2a-8} = \dfrac{1}{a-2}$

34. $\dfrac{3}{x^2-6x+9} + \dfrac{x-2}{3x-9} = \dfrac{x}{2x-6}$

35. $\dfrac{2}{x+3} - \dfrac{3x+5}{x^2+4x+3} = \dfrac{5}{x+1}$

36. $\dfrac{3-2t}{t+1} - \dfrac{10}{t^2-1} = \dfrac{2t+3}{1-t}$

37. $\dfrac{x-1}{x^2-2x-3} + \dfrac{x+2}{x^2-9} = \dfrac{2x+5}{x^2+4x+3}$

38. $\dfrac{2x+1}{x^2-3x-10} + \dfrac{x-1}{x^2-4} = \dfrac{3x-1}{x^2-7x+10}$

39. $\dfrac{3}{x^2-x-12} + \dfrac{1}{x^2+x-6} = \dfrac{4}{x^2+3x-10}$

40. $\dfrac{3}{x^2-2x-3} - \dfrac{1}{x^2-1} = \dfrac{2}{x^2-8x+7}$

Skill Maintenance

41. Factor completely: $81x^4 - y^4$. [5.6]

42. Determine whether each of the following systems is consistent or inconsistent. [3.2]

  a) $2x - 3y = 4,$
  $4x - 6y = 7$

  b) $x + 3y = 2,$
  $2x - 3y = 1$

43. Solve: $|x-2| > 3$. [4.3]

44. *Test Questions.* There are 70 questions on a test. The questions are either multiple-choice, true–false, or fill-in. There are twice as many true–false as fill-in and 5 fewer multiple-choice than true–false. How many of each type of question are there on the test? [3.5]

45. Find two consecutive even numbers whose product is 288. [5.7]

46. Simplify: $(a^2b^3)^5/(a^{-4}b^2)^3$. [1.4]

Synthesis

47. ◆ Is the following statement true or false: "For any real numbers a, b, and c, if $ac = bc$, then $a = b$"? Explain why you answered as you did.

48. ◆ When checking a possible solution of a rational equation, is it sufficient to check that the "solution" does not make any denominator equal to 0? Why or why not?

49. ◈ Karin and Kyle are working together to solve the equation

$$\frac{x}{x^2 - x - 6} = 2x - 1.$$

Karin obtains the first graph below using the DOT mode of a grapher, while Kyle obtains the second graph below using the CONNECTED mode. Which approach is the better method of showing the solutions? Why?

$y_1 = x/(x^2 - x - 6), \quad y_2 = 2x - 1$

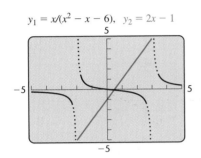

$y_1 = x/(x^2 - x - 6), \quad y_2 = 2x - 1$

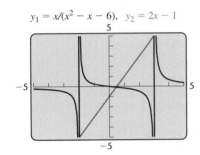

50. ◈ Explain how one can easily produce rational equations for which no solution exists. (*Hint:* Examine Example 2.)

For each pair of functions f and g, find all values of a for which $f(a) = g(a)$.

51. $f(x) = \dfrac{2 + \dfrac{x}{2}}{2 - \dfrac{x}{4}}, \quad g(x) = \dfrac{2}{\dfrac{x}{4} - 2}$

52. $f(x) = \dfrac{x - \dfrac{3}{2}}{x + \dfrac{2}{3}}, \quad g(x) = \dfrac{x + \dfrac{1}{2}}{x - \dfrac{2}{3}}$

53. $f(x) = \dfrac{1}{1 + x} + \dfrac{x}{1 - x}, \quad g(x) = \dfrac{1}{1 - x} - \dfrac{x}{1 + x}$

54. $f(x) = \dfrac{x + 3}{x + 2} - \dfrac{x + 4}{x + 3}, \quad g(x) = \dfrac{x + 5}{x + 4} - \dfrac{x + 6}{x + 5}$

55. $f(x) = \dfrac{0.793}{x} + 18.15, \quad g(x) = \dfrac{6.034}{x} - 43.17$

56. $f(x) = \dfrac{2.315}{x} - \dfrac{12.6}{17.4}, \quad g(x) = \dfrac{6.71}{x} + 0.763$

Recall that identities are true for any possible replacement of the variable(s). Determine whether each of the following equations is an identity.

57. $\dfrac{x^2 + 6x - 16}{x - 2} = x + 8, \; x \neq 2$

58. $\dfrac{x^3 + 8}{x^2 - 4} = \dfrac{x^2 - 2x + 4}{x - 2}, \; x \neq -2, x \neq 2$

6.5

Applications and Models Using Rational Functions

• *Problems Involving Work*
• *Problems Involving Motion*

Now that we are able to solve rational equations, we can solve certain problems that we could not have handled before. The five problem-solving steps remain the same.

Problems Involving Work

Example 1   Lon can mow a lawn in 4 hr. Penny can mow the same lawn in 5 hr. How long would it take both of them, working together, to mow the lawn?

SOLUTION

1. Familiarize. We familiarize ourselves with the problem by considering two *incorrect* ways of translating the problem to mathematical language.

a) One *incorrect* way to translate the problem is to add the two times:

4 hr + 5 hr = 9 hr.

Let's think about this. Lon can do the job *alone* in 4 hr. If Lon and Penny work together, whatever time it takes them must be *less* than 4 hr.

b) Another *incorrect* approach is to assume that each person mows half the lawn. Were this the case,

Lon would mow $\frac{1}{2}$ the lawn in $\frac{1}{2}$(4 hr), or 2 hr

and

Penny would mow $\frac{1}{2}$ the lawn in $\frac{1}{2}$(5 hr), or $2\frac{1}{2}$ hr.

But time would be wasted since Lon would finish $\frac{1}{2}$ hr before Penny. Were Lon to help Penny after completing his half, the entire job would take between 2 and $2\frac{1}{2}$ hr. This information provides a partial check on any answer we get—the answer should be between 2 and $2\frac{1}{2}$ hr.

Let's consider how much of the job each person completes in 1 hr, 2 hr, 3 hr, and so on. Since Lon takes 4 hr to mow the entire lawn, in 1 hr he mows $\frac{1}{4}$ of the lawn. Since Penny takes 5 hr to mow the entire lawn, in 1 hr she mows $\frac{1}{5}$ of the lawn. Together, Lon and Penny mow

$\frac{1}{4} + \frac{1}{5}$ of the lawn in 1 hr.

In 2 hr, Lon mows $\frac{1}{4} \cdot 2$ of the lawn and Penny mows $\frac{1}{5} \cdot 2$ of the lawn. Together, they mow

$\frac{1}{4} \cdot 2 + \frac{1}{5} \cdot 2$ of the lawn in 2 hr.

Continuing this pattern, we can form a table like the following one.

	FRACTION OF THE LAWN MOWED		
TIME	BY LON	BY PENNY	TOGETHER
1 hr	$\frac{1}{4}$	$\frac{1}{5}$	$\frac{1}{4} + \frac{1}{5}$, or $\frac{9}{20}$
2 hr	$\frac{1}{4} \cdot 2$	$\frac{1}{5} \cdot 2$	$\frac{1}{4} \cdot 2 + \frac{1}{5} \cdot 2$, or $\frac{9}{10}$
3 hr	$\frac{1}{4} \cdot 3$	$\frac{1}{5} \cdot 3$	$\frac{1}{4} \cdot 3 + \frac{1}{5} \cdot 3$, or $1\frac{7}{20}$
t hr	$\frac{1}{4} \cdot t$	$\frac{1}{5} \cdot t$	$\frac{1}{4} \cdot t + \frac{1}{5} \cdot t$

From the table, we note that what is needed is the number of hours t required for Lon and Penny to mow exactly one lawn.

2. Translate.   From the table, we see that t must be some number for which

$$\frac{1}{4} \cdot t + \frac{1}{5} \cdot t = 1,$$

Portion of work done by Lon in t hr

Portion of work done by Penny in t hr

or

$$\frac{t}{4} + \frac{t}{5} = 1,$$

where 1 represents the idea that one entire job is completed in t hours.

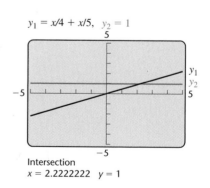

3. Carry out.   We solve the equation

$$\frac{t}{4} + \frac{t}{5} = 1$$

both algebraically and graphically.

ALGEBRAIC SOLUTION

We have

$$\frac{t}{4} + \frac{t}{5} = 1$$

$$20\left(\frac{t}{4} + \frac{t}{5}\right) = 20 \cdot 1 \qquad \text{Multiplying by the LCD}$$

$$20 \cdot \frac{t}{4} + 20 \cdot \frac{t}{5} = 20 \qquad \text{Using the distributive law}$$

$$5t + 4t = 20$$

$$9t = 20$$

$$t = \frac{20}{9}, \text{ or } 2\frac{2}{9} \text{ hr.}$$

The solution is $2\frac{2}{9}$ hr.

GRAPHICAL SOLUTION

We graph $y_1 = x/4 + x/5$ and $y_2 = 1$.

$y_1 = x/4 + x/5, \quad y_2 = 1$

Intersection
$x = 2.2222222 \quad y = 1$

Since the graphs intersect at the point (2.2222222, 1), the solution is about 2.22.

4. **Check.** We note first that both approaches yield the same solution, since $2\frac{2}{9} \approx 2.22$. In $\frac{20}{9}$ hr, Lon mows $\frac{1}{4} \cdot \frac{20}{9}$, or $\frac{5}{9}$ of the lawn and Penny mows $\frac{1}{5} \cdot \frac{20}{9}$, or $\frac{4}{9}$ of the lawn. Together, they mow $\frac{5}{9} + \frac{4}{9}$, or 1 lawn. The fact that our solution is between 2 and $2\frac{1}{2}$ hr (see step 1 above) is also a check.

5. **State.** It will take $2\frac{2}{9}$ hr for Lon and Penny, working together, to mow the lawn.

Example 2 It takes Red 9 hr more than it does Hannah to construct a stone wall. Working together, they can build the wall in 20 hr. How long would it take each, working alone, to build the wall?

SOLUTION

1. **Familiarize.** Unlike Example 1, this problem does not provide us with the times required by the individuals to do the job alone. Let's have $h =$ the amount of time it would take Hannah working alone and $h + 9 =$ the amount of time it would take Red working alone.

2. **Translate.** Using the same reasoning as in Example 1, we see that Hannah can build $\frac{1}{h}$ of a wall in 1 hr and Red can build $\frac{1}{h + 9}$ of a wall in 1 hr. In 20 hr, Hannah builds $\left(\frac{1}{h}\right)20$, or $\frac{20}{h}$ of the wall and Red builds $\left(\frac{1}{h + 9}\right)20$, or $\frac{20}{h + 9}$ of the wall. Since Hannah and Red complete 1 entire wall in 20 hr, we have

$$\underbrace{\frac{20}{h}}_{\substack{\text{Hannah's portion} \\ \text{of the work}}} + \underbrace{\frac{20}{h + 9}}_{\substack{\text{Red's portion} \\ \text{of the work}}} = 1.$$

3. **Carry out.** We solve the equation

$$\frac{20}{h} + \frac{20}{h + 9} = 1$$

both algebraically and graphically.

┌ ─ ─ *ALGEBRAIC SOLUTION*

We have

$$\frac{20}{h} + \frac{20}{h + 9} = 1$$

$$h(h + 9)\left(\frac{20}{h} + \frac{20}{h + 9}\right) = h(h + 9)1$$

Multiplying by the LCD

$$(h + 9)20 + h \cdot 20 = h(h + 9)$$

Distributing and simplifying

$$40h + 180 = h^2 + 9h$$

$$0 = h^2 - 31h - 180$$

Getting 0 on one side

$$0 = (h - 36)(h + 5)$$

Factoring

$$h - 36 = 0 \quad or \quad h + 5 = 0$$

Principle of zero products

$$h = 36 \quad or \quad h = -5.$$

The possible solutions are 36 and −5.

┌ ─ ─ *GRAPHICAL SOLUTION*

We let $y_1 = 20/x + 20/(x + 9)$ and $y_2 = 1$. We graph y_1 and y_2 and find the points of intersection. If we use the window $[-8, 8, -10, 10]$, we see that there is a point of intersection near the x-value −5. It also appears that the graphs may intersect at a point off the screen to the right. Thus we adjust the window to $[-8, 50, 0, 5]$ and see that there is indeed another point of intersection.

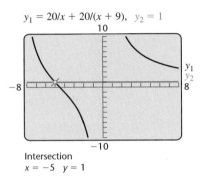

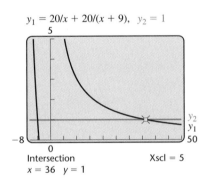

We have $x = 36$ or $x = -5$.

4. Check. Since negative time has no meaning in the problem, −5 is not a solution to the original problem. The number 36 checks since, if Hannah takes 36 hr alone and Red takes $36 + 9 = 45$ hr alone, in 20 hr they would have completed

$$\frac{20}{36} + \frac{20}{45} = \frac{5}{9} + \frac{4}{9} = 1 \text{ wall.}$$

5. State. It would take Hannah 36 hr to build the wall alone, and Red 45 hr.

The equations used in Examples 1 and 2 can be generalized, as follows.

> **Modeling Work Problems**
>
> If
>
> > a = the time needed for A to complete the work alone,
> >
> > b = the time needed for B to complete the work alone, and
> >
> > t = the time needed for A and B to complete the work together,
>
> then
>
> $$\frac{t}{a} + \frac{t}{b} = 1.$$

Problems Involving Motion

Problems dealing with distance, rate (or speed), and time are called **motion problems**. To translate them, we use either the basic motion formula, $d = rt$, or the formulas $r = d/t$ or $t = d/r$, which can be derived from $d = rt$.

Example 3 A racer is bicycling 15 km/h faster than a person on a mountain bike. In the time it takes the racer to travel 80 km, the person on the mountain bike has gone 50 km. Find the speed of each bicyclist.

SOLUTION

1. **Familiarize.** Let's guess that the person on the mountain bike is going 10 km/h. The racer would then be traveling $10 + 15$, or 25 km/h. At 25 km/h, the racer will travel 80 km in $\frac{80}{25} = 3.2$ hr. Going 10 km/h, the mountain bike will cover 50 km in $\frac{50}{10} = 5$ hr. Since $3.2 \neq 5$, our guess was wrong, but we can see that if r = the rate, in kilometers per hour, of the slower bike, then the rate of the racer = $r + 15$.

   Drawing a sketch and constructing a table can be helpful.

	DISTANCE	SPEED	TIME
MOUNTAIN BIKE	50	r	t
RACING BIKE	80	$r + 15$	t

2. Translate. By looking at how we checked our guess, we see that in the column of the table labeled "Time," the t's can be replaced with expressions involving r, using the formula *Time = Distance/Rate*, as follows.

	DISTANCE	SPEED	TIME
MOUNTAIN BIKE	50	r	$50/r$
RACING BIKE	80	$r + 15$	$80/(r + 15)$

Since we are told that the times must be the same, we can write an equation:

$$\frac{50}{r} = \frac{80}{r + 15}.$$

3. Carry out. We solve the equation

$$\frac{50}{r} = \frac{80}{r + 15}$$

both algebraically and graphically.

┌- - *ALGEBRAIC SOLUTION*

We have

$$\frac{50}{r} = \frac{80}{r + 15}$$

$$r(r + 15)\frac{50}{r} = r(r + 15)\frac{80}{r + 15} \qquad \text{Multiplying by the LCD}$$

$$50r + 750 = 80r \qquad \text{Simplifying}$$

$$750 = 30r$$

$$25 = r.$$

The solution is 25.

┌- - *GRAPHICAL SOLUTION*

We graph the equations $y_1 = 50/x$ and $y_2 = 80/(x + 15)$, and look for a point of intersection. By adjusting the window dimensions and using the INTERSECT feature, we see that the graphs intersect at the point (25, 2).

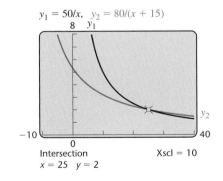

$y_1 = 50/x, \quad y_2 = 80/(x + 15)$

The solution is 25.

4. **Check.** If our answer checks, the mountain bike is going 25 km/h and the racing bike is going $25 + 15 = 40$ km/h.

   Traveling 80 km at 40 km/h, the racer is riding for $\frac{80}{40} = 2$ hr. Traveling 50 km at 25 km/h, the person on the mountain bike is riding for $\frac{50}{25} = 2$ hr. Our answer checks since the two times are the same.

5. **State.** The speed of the racer is 40 km/h, and the speed of the person on the mountain bike is 25 km/h.   ▬

In the following example, although the distance is the same in both directions, the key to the translation lies in an additional piece of given information.

Example 4   Sandy's tugboat goes 10 mph in still water. It travels 24 mi upstream and 24 mi back in a total time of 5 hr. What is the speed of the current?

SOLUTION

1. **Familiarize.**   Let's guess that the speed of the current is 4 mph. The tugboat would then be moving $10 - 4 = 6$ mph upstream and $10 + 4 = 14$ mph downstream. The tugboat would require $\frac{24}{6} = 4$ hr to travel 24 mi upstream and $\frac{24}{14} = 1\frac{5}{7}$ hr to travel 24 mi downstream. Since the total time, $4 + 1\frac{5}{7} = 5\frac{5}{7}$ hr, is not the 5 hr mentioned in the problem, we know that our guess is wrong.

   Suppose that the current's speed $= c$ mph. The tugboat would then travel $10 - c$ mph when going upstream and $10 + c$ mph when going downstream.

   A drawing and a table can help display the information.

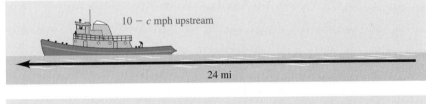

10 − c mph upstream

24 mi

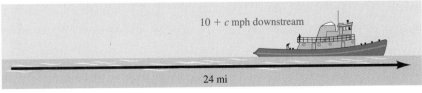

10 + c mph downstream

24 mi

	DISTANCE	SPEED	TIME
UPSTREAM	24	$10 - c$	t_1
DOWNSTREAM	24	$10 + c$	t_2

2. Translate. From examining our guess, we see that the time traveled can be represented using the formula *Time = Distance/Rate*:

	DISTANCE	SPEED	TIME
UPSTREAM	24	$10 - c$	$24/(10 - c)$
DOWNSTREAM	24	$10 + c$	$24/(10 + c)$

Since the total time upstream and back is 5 hr, we use the last column of the table to form an equation:

$$\frac{24}{10 - c} + \frac{24}{10 + c} = 5.$$

3. Carry out. We solve the equation

$$\frac{24}{10 - c} + \frac{24}{10 + c} = 5$$

both algebraically and graphically.

ALGEBRAIC SOLUTION

We have

$$\frac{24}{10 - c} + \frac{24}{10 + c} = 5$$

$$(10 - c)(10 + c)\left[\frac{24}{10 - c} + \frac{24}{10 + c}\right] = (10 - c)(10 + c)5$$

Multiplying by the LCD

$$24(10 + c) + 24(10 - c) = (100 - c^2)5$$

$$480 = 500 - 5c^2$$

Simplifying

$$5c^2 - 20 = 0$$

$$5(c^2 - 4) = 0$$

$$5(c - 2)(c + 2) = 0$$

$$c = 2 \quad or \quad c = -2.$$

The possible solutions are 2 and -2.

GRAPHICAL SOLUTION

We graph

$y_1 = 24/(10 - x) + 24/(10 + x)$ and $y_2 = 5$

and see that there are two points of intersection.

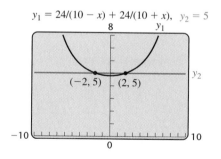

$y_1 = 24/(10 - x) + 24/(10 + x), \quad y_2 = 5$

The solutions are -2 and 2.

4. Check. Since speed cannot be negative in this problem, -2 cannot be a solution. You should confirm that 2 checks in the original problem.

5. State. The speed of the current is 2 mph.

6.5 Exercise Set

Solve.

1. The reciprocal of 3, plus the reciprocal of 6, is the reciprocal of what number?

2. The reciprocal of 5, plus the reciprocal of 7, is the reciprocal of what number?

3. The sum of a number and 6 times its reciprocal is -5. Find the number.

4. The sum of a number and 21 times its reciprocal is -10. Find the number.

5. The reciprocal of the product of two consecutive integers is $\frac{1}{42}$. Find the two integers.

6. The reciprocal of the product of two consecutive integers is $\frac{1}{72}$. Find the two integers.

7. *Painting.* Otto can paint a room in 4 hr. Sally can paint the same room in 3 hr. Working together, how long will it take them to paint the room?

8. *Mail Order.* Zoe, an experienced shipping clerk, can fill a certain order in 5 hr. Willy, a new clerk, needs 9 hr to complete the same job. Working together, how long will it take them to fill the order?

9. *Filling a Pool.* A swimming pool can be filled in 12 hr if water enters through a pipe alone or in 30 hr if water enters through a hose alone. If water is entering through both the pipe and the hose, how long will it take to fill the pool?

10. *Filling a Tank.* A tank can be filled in 18 hr by pipe A alone and in 22 hr by pipe B alone. How long will it take to fill the tank if both pipes are working?

11. *Printing.* Pronto Press can print an order of booklets in 4.5 hr. Red Dot Printers can do the same job in 5.5 hr. How long will it take if both presses are used?

12. *Wood Cutting.* Damon can clear a lot in 5.5 hr. His partner, Tyron, can complete the same job in 7.5 hr. How long will it take them to clear the lot working together?

13. *Sanding.* Mavis can sand the living room floor in 3 hr. When she works together with Henri, the job takes 2 hr. How long would it take Henri, working by himself, to sand the floor?

14. *Cutting Firewood.* Jake can cut and split a cord of firewood in 6 fewer hr than Skyler can. When they work together, it takes them 4 hr. How long would it take each of them to do the job alone?

15. *Painting.* Sara takes 3 hr longer to paint a floor than it takes Kate. When they work together, it takes them 2 hr. How long would each take to do the job alone?

16. *Painting.* Claudia can paint a neighbor's house four times as fast as Jan can. The year they worked together it took them 8 days. How long would it take each to paint the house alone?

17. *Newspaper Delivery.* Zsuzanna can deliver papers three times as fast as Stan can. If they work together, it takes them 1 hr. How long would it take each to deliver the papers alone?

18. *Waxing a Car.* Rosita can wax her car in 2 hr. When she works together with Helga, they can wax the car in 45 min. How long would it take Helga, working by herself, to wax the car?

19. *Sorting Recyclables.* Together, it takes John and Deb 2 hr 55 min to sort recyclables. Alone, John would require 2 more hr than Deb. How long would it take Deb to do the job alone? (*Hint:* Convert minutes to hours or hours to minutes.)

20. *Paving.* Together, Larry and Mo require 4 hr 48 min to pave a driveway. Alone, Larry would require 4 hr more than Mo. How long would it take Mo to do the job alone? (*Hint:* Convert minutes to hours.)

21. *Copying.* A new photocopier works twice as fast as an old one. When the machines work together, a university can produce all its staff manuals in 15 hr. Find the time it would take each machine, working alone, to complete the same job.

22. *Completing a Puzzle.* Working together, Hans and Gina can complete a jigsaw puzzle in 1.5 hr. Hans takes 4 hr longer than Gina does when working alone. How long would it take Gina alone to complete the puzzle?

23. *Boating.* The current in the Lazy River moves at a rate of 4 mph. Ken's dinghy motors 6 mi upstream in the same time it takes to motor 12 mi downstream. What is the speed of the dinghy in still water?

24. *Kayaking.* The speed of the current in Catamount Creek is 3 mph. Ahmad's kayak can travel 4 mi upstream in the same time it takes to travel 10 mi downstream. What is the speed of Ahmad's kayak in still water?

25. *Moving Sidewalks.* The moving sidewalk at O'Hare Airport in Chicago moves 1.8 ft/sec. Walking on the moving sidewalk, Camille travels 105 ft forward in the time it takes to travel 51 ft in the opposite direction. How fast would Camille be walking on a nonmoving sidewalk?

26. *Moving Sidewalks.* Newark Airport's moving sidewalk moves at a speed of 1.7 ft/sec. Walking on the moving sidewalk, Benny can travel 120 ft forward in the same time it takes to travel 52 ft in the opposite direction. How fast would Benny be walking on a nonmoving sidewalk?

27. *Train Speed.* The speed of the A&M freight train is 14 mph less than the speed of the A&M passenger train. The passenger train travels 400 mi in the same time that the freight train travels 330 mi. Find the speed of each train.

28. *Walking.* Rosanna walks 2 mph slower than Simone. In the time it takes Simone to walk 8 mi, Rosanna walks 5 mi. Find the speed of each person.

29. *Bus Travel.* A local bus travels 7 mph slower than the express. The express travels 90 mi in the time it takes the local to travel 75 mi. Find the speed of each bus.

30. *Train Speed.* The A train goes 12 mph slower than the B train. The A train travels 230 mi in the same time that the B train travels 290 mi. Find the speed of each train.

31. *Boating.* Suzie has a boat that travels 15 km/h in still water. She motors 140 km downstream in the same time it takes to travel 35 km upstream. What is the speed of the river?

32. *Boating.* A paddleboat travels 2 km/h in still water. The boat is paddled 4 km downstream in the same time it takes to go 1 km upstream. What is the speed of the river?

33. *Moped Speed.* Jaime's moped travels 8 km/h faster than Mara's. Jaime travels 69 km in the same time that Mara travels 45 km. Find the speed of each person's moped.

34. *Shipping.* A barge moves 7 km/h in still water. It travels 45 km upriver and 45 km downriver in a total time of 14 hr. What is the speed of the current?

35. *Swimming.* Al swims 55 m per minute in still water. He swims 150 m upstream and 150 m downstream in a total time of 5.5 min. What is the speed of the current?

36. *Air Travel.* A plane travels 100 mph in still air. It travels 240 mi into the wind and 240 mi with the wind in a total time of 5 hr. Find the wind speed.

37. *Travel by Van.* Cecilia's van covered 120 mi at a certain speed. Had Cecilia driven 10 mph faster, the trip would have been 2 hr shorter. How fast did Cecilia drive?

38. *Boating.* Carlos' Boston Whaler cruised 45 mi upstream and 45 mi back in a total of 8 hr. The speed of the river is 3 mph. Find the speed of the boat in still water.

Skill Maintenance

39. If

$$f(x) = \frac{x - 5}{x^2 - 4x - 5},$$

determine the domain of *f*. [5.5]

40. Solve: $|x - 2| = 9$. [4.3]

41. *Catalog Sales.* Sales from catalogs have increased steadily from 1990 to 1999, as shown in the graph below. [2.6], [5.7]

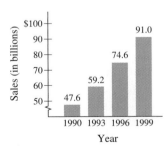

Source: Direct Marketing Association

a) Use regression to find a linear function $c(x)$ that can be used to predict catalog sales x years after 1990.

b) Use the function from part (a) to predict the year in which catalog sales will be $110 billion.

42. 240 is 16% of what number? [2.3]

Synthesis

43. ◈ Two steamrollers are paving a parking lot. Working together, will the two steamrollers take less than half as long as the slower steamroller would working alone? Why or why not?

44. ◈ Two fuel lines are filling a freighter with oil. Will the faster fuel line take more or less than twice as long to fill the freighter by itself? Why?

45. ◈ Write a work problem for a classmate to solve. Devise the problem so that the solution is "Liane and Michele will take 4 hr to complete the job, working together."

46. ◈ Write a work problem for a classmate to solve. Devise the problem so that the solution is "Jen takes 5 hr and Pablo takes 6 hr to complete the job alone."

47. *Filling a Tub.* A tub can be filled in 10 min and drained in 8 min. How long will it take to empty a full tub if the water is left on?

48. *Filling a Tank.* A tank can be filled in 9 hr and drained in 11 hr. How long will it take to fill the tank if the drain is left open?

49. *Escalators.* Together, a 100-cm-wide escalator and a 60-cm-wide escalator can empty a 1575-person auditorium in 14 min (*Source: McGraw-Hill Encyclopedia of Science and Technology*). The wider escalator moves twice as many people as the narrower one. How many people per hour does the 60-cm-wide escalator move?

50. *Aviation.* A Coast Guard plane has enough fuel to fly for 6 hr, and its speed in still air is 240 mph. The plane departs with a 40-mph tailwind and returns to the same airport flying into the same wind. How far can the plane travel under these conditions?

51. *Boating.* Shoreline Travel operates a 3-hr paddleboat cruise on the Missouri River. If the speed of the boat in still water is 12 mph, how far upriver can the pilot travel against a 5-mph current before it is time to turn around?

52. *Boating.* The speed of a motor boat in still water is three times the speed of a river's current. A trip up the river and back takes 10 hr, and the total distance of the trip is 100 km. Find the speed of the current.

53. *Travel by Car.* Melissa drives to work at 50 mph and arrives 1 min late. She drives to work at 60 mph and arrives 5 min early. How far does Melissa live from work?

54. At what time after 4:00 will the minute hand and the hour hand of a clock first be in the same position?

55. At what time after 10:30 will the hands of a clock first be perpendicular?

Average speed is defined as total distance divided by total time.

56. Lenore drove 200 km. For the first 100 km of the trip, she drove at a speed of 40 km/h. For the second half of the trip, she traveled at a speed of 60 km/h. What was the average speed for the entire trip? (It was *not* 50 km/h.)

57. For the first 50 mi of a 100-mi trip, Chip drove 40 mph. What speed would he have to travel for the last half of the trip so that the average speed for the entire trip would be 45 mph?

COLLABORATIVE CORNER

Focus: Testing a mathematical model

Time: 20–30 minutes

Group size: 2–3

Materials: An empty 1-gal plastic jug, a kitchen or laboratory sink, a stopwatch or a watch capable of measuring seconds, an inexpensive pen or pair of scissors or a nail or knife for poking holes in plastic.

Problems like Exercises 47 and 48 can be solved algebraically and then checked at home or in a laboratory.

Activity

1. Turn the water faucet on full force. While one group member fills the empty jug with water, the other group member(s) should record how many seconds this takes.

2. After carefully poking several holes in the bottom of the jug, record how many seconds it takes the full jug to empty.

3. Using the information found in parts (1) and (2) above, use algebra to predict how long it will take to fill the punctured jug.

4. Test your prediction by again turning the water on full force and timing how long it takes for the pierced jug to be filled.

5. How accurate was your prediction? How might your prediction have been made more accurate?

6.6
Division of Polynomials

- *Divisor a Monomial*
- *Divisor a Polynomial*
- *Synthetic Division*

A rational expression indicates division. Division of polynomials, like division of real numbers, relies on our multiplication and subtraction skills.

Divisor a Monomial

To divide a monomial by a monomial, we can subtract exponents when bases are the same (see Section 1.4). For example,

$$\frac{45x^{10}}{3x^4} = 15x^{10-4} = 15x^6, \qquad \frac{48a^2b^5}{-3ab^2} = \frac{48}{-3}a^{2-1}b^{5-2} = -16ab^3.$$

To divide a polynomial by a monomial, the division is regarded as a sum of quotients of monomials. This uses the fact that since

$$\frac{A}{C} + \frac{B}{C} = \frac{A + B}{C}, \quad \text{we know that} \quad \frac{A + B}{C} = \frac{A}{C} + \frac{B}{C}.$$

Example 1 Divide $12x^3 + 8x^2 + x + 4$ by $4x$.

SOLUTION

$$(12x^3 + 8x^2 + x + 4) \div (4x) = \frac{12x^3 + 8x^2 + x + 4}{4x} \qquad \begin{array}{l}\textbf{Writing a}\\ \textbf{fractional}\\ \textbf{expression}\end{array}$$

$$= \frac{12x^3}{4x} + \frac{8x^2}{4x} + \frac{x}{4x} + \frac{4}{4x} \qquad \begin{array}{l}\textbf{Writing as}\\ \textbf{a sum of}\\ \textbf{quotients}\end{array}$$

$$= 3x^2 + 2x + \frac{1}{4} + \frac{1}{x} \qquad \begin{array}{l}\textbf{Performing}\\ \textbf{the four}\\ \textbf{indicated}\\ \textbf{divisions}\end{array}$$

Example 2 Divide: $(8x^4y^5 - 3x^3y^4 + 5x^2y^3) \div x^2y^3$.

SOLUTION

$$\frac{8x^4y^5 - 3x^3y^4 + 5x^2y^3}{x^2y^3} = \frac{8x^4y^5}{x^2y^3} - \frac{3x^3y^4}{x^2y^3} + \frac{5x^2y^3}{x^2y^3} \qquad \begin{array}{l}\textbf{Try to perform}\\ \textbf{this step mentally.}\end{array}$$

$$= 8x^2y^2 - 3xy + 5$$

Division by a Monomial

To divide a polynomial by a monomial, divide each term of the polynomial by the monomial.

Divisor a Polynomial

When the divisor has more than one term, we use a procedure very similar to long division in arithmetic.

Example 3 Divide $2x^2 - 7x - 15$ by $x - 5$.

SOLUTION We have

$$
\begin{array}{r}
2x \\
x - 5 \overline{)2x^2 - 7x - 15} \\
\end{array}
$$
   Divide $2x^2$ by x: $2x^2/x = 2x$.

$-(2x^2 - 10x)$ ⟵ Multiply $x - 5$ by $2x$.

$3x - 15.$ ⟵ Subtracting: $2x^2 - 7x - 15 - (2x^2 - 10x)$
$= 2x^2 - 7x - 15 - 2x^2 + 10x$
$= 3x - 15$

We now divide $3x - 15$ by $x - 5$:

$$
\begin{array}{r}
2x \;+\; 3 \\[2pt]
x - 5\overline{)2x^2 - 7x - 15} \\
\underline{2x^2 - 10x} \\
3x - 15 \\
\underline{-(3x - 15)} \\
0.
\end{array}
$$

Divide $3x$ by x: $3x/x = 3$.
Multiply $x - 5$ by 3.
Subtract.

The quotient is $2x + 3$.

CHECK: $(x - 5)(2x + 3) = 2x^2 - 7x - 15$. The answer checks. ▬

Should a nonzero remainder occur, when do we stop dividing? We continue until the degree of the remainder is less than the degree of the divisor.

Example 4   Divide $x^2 + 5x + 8$ by $x + 3$.

SOLUTION   We have

$$
\begin{array}{r}
x \\
x + 3\overline{)x^2 + 5x + 8} \\
\underline{x^2 + 3x} \\
2x + 8.
\end{array}
$$

Divide the first term of the dividend by the first term of the divisor: $x^2/x = x$.
Multiply x above by $x + 3$.
Subtract.

The subtraction we have done is $(x^2 + 5x + 8) - (x^2 + 3x)$. *Remember:* To subtract, add the opposite (change the sign of every term, then add).
We now repeat the process:

$$
\begin{array}{r}
x \;+\; 2 \\
x + 3\overline{)x^2 + 5x + 8} \\
\underline{x^2 + 3x} \\
2x + 8 \\
\underline{2x + 6} \\
2.
\end{array}
$$

Divide the first term by the first term: $2x/x = 2$.
Multiply 2 by $x + 3$.
Subtract: $(2x + 8) - (2x + 6)$.

The quotient is $x + 2$, with remainder 2. Note that the degree of the remainder is 0 and the degree of the divisor, $x + 3$, is 1. Since $0 < 1$, the process stops.

CHECK: $(x + 3)(x + 2) + 2 = x^2 + 5x + 6 + 2$ Add the remainder to the product.

$$= x^2 + 5x + 8$$

We can write our answer as $x + 2$, R 2, or we can write our answer as

$$\text{Quotient} + \frac{\text{Remainder}}{\text{Divisor}}$$

$$x + 2 \quad + \left(\frac{2}{x + 3} \right),$$

which is how answers for the problems in this section are listed at the back of the book. Answers given in this form can be checked by multiplying:

$$(x + 3)\left[(x + 2) + \frac{2}{x + 3} \right] = (x + 3)(x + 2) + (x + 3)\frac{2}{x + 3}$$

Using the distributive law

$$= x^2 + 5x + 6 + 2$$

$$= x^2 + 5x + 8. \qquad \textbf{This was the dividend above.}$$

When the remainder is 0, as in Example 3, the results of the division can be used to factor the dividend. As we noted in the check, if $(2x^2 - 7x - 15) \div (x - 5) = 2x + 3$, then $(2x^2 - 7x - 15) = (x - 5)(2x + 3)$.

Example 5 Divide $2a^4 - 5a^3 + 5a - 2$ by $a^2 - 1$ and use the results to factor the dividend.

SOLUTION   There is no a^2-term in the dividend, so we write it in as $0a^2$ or leave a space for it. Note that we write like terms in the same column.

$$
\begin{array}{r}
2a^2 - 5a + 2 \\
a^2 - 1\overline{)2a^4 - 5a^3 + 0a^2 + 5a - 2} \\
\underline{2a^4 - 2a^2} \\
-5a^3 + 2a^2 + 5a - 2 \\
\underline{-5a^3 + 5a} \\
2a^2 - 2 \\
\underline{2a^2 - 2} \\
0
\end{array}
$$

Divide the first term of the dividend by the first term of the divisor: $2a^4/a^2 = 2a^2$.

← **Multiplying:** $2a^2(a^2 - 1) = 2a^4 - 2a^2$

← **Subtracting:** $(2a^4 - 5a^3 + 5a - 2) - (2a^4 - 2a^2) = -5a^3 + 2a^2 + 5a - 2$

← **Subtracting**

The remainder is 0, so

$$2a^4 - 5a^3 + 5a - 2 = (a^2 - 1)(2a^2 - 5a + 2)$$

$$= (a + 1)(a - 1)(2a - 1)(a - 2). \qquad \textbf{Factoring completely}$$

You may have noticed that in each example all polynomials were written in descending order. When this is not the case, we rearrange terms before dividing.

Tips for Dividing Polynomials

1. Arrange polynomials in descending order.

2. If there are missing terms in the dividend, either write them with 0 coefficients or leave space for them.

3. Continue the long-division process until the degree of the remainder is less than the degree of the divisor.

Example 6 Let $f(x) = 8 + 125x^3$ and $g(x) = 5x + 2$. If $F(x) = (f/g)(x)$, find an expression for $F(x)$.

SOLUTION Recall that $(f/g)(x) = f(x)/g(x)$. Thus,

$$F(x) = \frac{8 + 125x^3}{5x + 2}$$

and

$$
\begin{array}{r}
25x^2 - 10x + 4 \\
5x + 2 \overline{)\,125x^3 + 8} \\
\underline{125x^3 + 50x^2 } \\
-50x^2 + 8 \\
\underline{-50x^2 - 20x } \\
20x + 8 \\
\underline{20x + 8} \\
0.
\end{array}
$$

Writing in descending order and leaving space for the missing terms.

Subtracting: $(125x^3 + 8) - (125x^3 + 50x^2) = -50x^2 + 8$

Subtracting

Note that, because $F(x) = f(x)/g(x)$, $g(x)$ cannot be 0. Since $g(x)$ is 0 for $x = -\frac{2}{5}$ (check this), we have

$$F(x) = 25x^2 - 10x + 4, \quad \text{provided } x \neq -\tfrac{2}{5}.$$

That is, the domain of F is $\left\{x \mid x \text{ is a real number } and\ x \neq -\tfrac{2}{5}\right\}$. ▬

Note that Example 6 shows that $8 + 125x^3$ can be factored:

$$8 + 125x^3 = (5x + 2)(25x^2 - 10x + 4).$$

This polynomial cannot be factored further with real-number coefficients.

Synthetic Division

To divide a polynomial by a binomial of the type $x - a$, we can streamline the general procedure to develop a process called *synthetic division*.

Compare the following. When a polynomial is written in descending order, the coefficients provide the essential information:

$$
\begin{array}{r}
4x^2 + 5x + 11 \\
x-2\overline{)4x^3 - 3x^2 + x + 7} \\
\underline{4x^3 - 8x^2} \\
5x^2 + x + 7 \\
\underline{5x^2 - 10x} \\
11x + 7 \\
\underline{11x - 22} \\
29
\end{array}
\qquad
\begin{array}{r}
4 + 5 + 11 \\
1-2\overline{)4 - 3 + 1 + 7} \\
\underline{4 - 8} \\
5 + 1 + 7 \\
\underline{5 - 10} \\
11 + 7 \\
\underline{11 - 22} \\
29
\end{array}
$$

Because the coefficient of x is 1 in the divisor, each time we multiply the divisor by a term in the answer, the leading coefficient of that product duplicates a coefficient in the answer. The process of synthetic division eliminates the duplication. To simplify the process further, we reverse the sign of the constant in the divisor and add rather than subtract.

Example 7   Use synthetic division to divide: $(x^3 + 6x^2 - x - 30) \div (x - 2)$.

SOLUTION

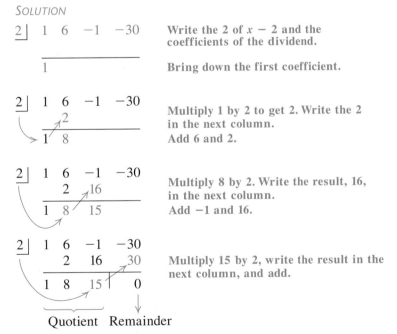

The last number, 0, is the remainder. The other numbers are the coefficients of the quotient. Note that the degree of the quotient is 1 less than the degree of the dividend.

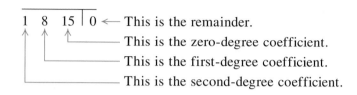

$$1 \quad 8 \quad 15 \mid 0 \longleftarrow \text{This is the remainder.}$$

This is the zero-degree coefficient.

This is the first-degree coefficient.

This is the second-degree coefficient.

The answer is $x^2 + 8x + 15$ with R 0, or just $x^2 + 8x + 15$. A table of values, like the one at left, can be used to show that $(x^3 + 6x^2 - x - 30) \div (x - 2) = x^2 + 8x + 15$ for all x-values for which both expressions are defined.

Remember that in order for this method to work, the divisor must be of the form $x - a$, that is, a variable minus a constant. The coefficient of the variable must be 1.

Example 8   Use synthetic division to divide each of the following.

a) $(2x^3 + 7x^2 - 5) \div (x + 3)$ **b)** $(10x^2 - 13x + 3x^3 - 20) \div (4 + x)$

SOLUTION

a) $(2x^3 + 7x^2 - 5) \div (x + 3)$

The dividend has no x-term, so we must write a 0 for its coefficient of x. Note that $x + 3 = x - (-3)$, so we write -3 inside the $\rfloor$.

$$\begin{array}{r|rrrr} -3 & 2 & 7 & 0 & -5 \\ & & -6 & -3 & 9 \\ \hline & 2 & 1 & -3 & \mid 4 \end{array}$$

The answer is $2x^2 + x - 3$, with R 4, or $2x^2 + x - 3 + \dfrac{4}{x + 3}$.

b) We first rewrite $(10x^2 - 13x + 3x^3 - 20) \div (4 + x)$ in descending order:

$$(3x^3 + 10x^2 - 13x - 20) \div (x + 4).$$

Next, we use synthetic division. Note that $x + 4 = x - (-4)$.

$$\begin{array}{r|rrrr} -4 & 3 & 10 & -13 & -20 \\ & & -12 & 8 & 20 \\ \hline & 3 & -2 & -5 & \mid 0 \end{array}$$

The answer is $3x^2 - 2x - 5$.

Note that Example 8 indicates that

$$3x^3 + 10x^2 - 13x - 20 = (x + 4)(3x^2 - 2x - 5).$$

Continuing to factor gives us

$$3x^3 + 10x^2 - 13x - 20 = (x + 4)(3x - 5)(x + 1).$$

6.6 Exercise Set

Divide and check.

1. $\dfrac{24x^6 + 18x^5 - 36x^2}{6x^2}$

2. $\dfrac{30y^8 - 15y^6 + 40y^4}{5y^4}$

3. $\dfrac{28a^3 + 7a^2 - 3a - 14}{7a}$

4. $\dfrac{-40x^3 + 20x^2 - 3x + 7}{5x}$

5. $(26y^3 - 9y^2 - 8y) \div (2y^2)$

6. $(6a^4 + 9a^2 - 8) \div (2a)$

7. $(18x^7 - 27x^4 - 3x^2) \div (-3x^2)$

8. $(36y^6 - 18y^4 - 12y^2) \div (-6y)$

9. $(6p^2q^2 - 9p^2q + 12pq^2) \div (-3pq)$

10. $(16y^4z^2 - 8y^6z^4 + 12y^8z^3) \div (4y^4z)$

11. $(x^2 + 10x + 21) \div (x + 3)$

12. $(y^2 - 8y + 16) \div (y - 4)$

13. $(a^2 - 8a - 16) \div (a + 4)$

14. $(y^2 - 10y - 25) \div (y - 5)$

15. $(y^2 - 25) \div (y + 5)$

16. $(a^2 - 81) \div (a - 9)$

17. $(y^3 - 4y^2 + 3y - 6) \div (y - 2)$

18. $(2x^3 + 3x^2 - x - 3) \div (x + 2)$

19. $(a^3 - a + 12) \div (a - 4)$

20. $(x^3 - x + 6) \div (x + 2)$

21. $(10y^3 + 6y^2 - 9y + 10) \div (5y - 2)$

22. $(6x^3 - 11x^2 + 11x - 2) \div (2x - 3)$

23. $(2x^4 - x^3 - 5x^2 + x - 6) \div (x^2 + 2)$

24. $(3x^4 + 2x^3 - 11x^2 - 2x + 5) \div (x^2 - 2)$

For Exercises 25–30, f(x) and g(x) are as given. Find a simplified expression for F(x) if F(x) = (f / g)(x). (See Example 6.)

25. $f(x) = 64x^3 - 8, \quad g(x) = 4x - 2$

26. $f(x) = 8x^3 + 27, \quad g(x) = 2x + 3$

27. $f(x) = 6x^2 - 11x - 10, \quad g(x) = 3x + 2$

28. $f(x) = 8x^2 - 22x - 21, \quad g(x) = 2x - 7$

29. $f(x) = x^4 - 3x^2 - 54, \quad g(x) = x^2 - 9$

30. $f(x) = x^4 - 24x^2 - 25, \quad g(x) = x^2 - 25$

Use synthetic division to divide.

31. $(x^3 - 2x^2 + 2x - 5) \div (x - 1)$

32. $(x^3 - 2x^2 + 2x - 5) \div (x + 1)$

33. $(x^3 - 7x^2 - 13x + 3) \div (x - 2)$

34. $(x^3 - 7x^2 - 13x + 3) \div (x + 2)$

35. $(3x^3 + 7x^2 - 4x + 3) \div (x + 3)$

36. $(3x^3 + 7x^2 - 4x + 3) \div (x - 3)$

37. $(y^3 - 3y + 10) \div (y - 2)$

38. $(x^3 - 2x^2 + 8) \div (x + 2)$

39. $(x^5 - 32) \div (x - 2)$

40. $(y^5 - 1) \div (y - 1)$

41. $(3x^3 + 1 - x + 7x^2) \div \left(x + \frac{1}{3}\right)$

42. $(8x^3 - 1 + 7x - 6x^2) \div \left(x - \frac{1}{2}\right)$

Divide. Use the results of the division to factor the dividend.

43. $(x^3 + 3x^2 - 13x - 15) \div (x + 5)$

44. $(x^3 + 6x^2 + 11x + 6) \div (x + 1)$

45. $(2x^3 - 7x^2 - 17x + 10) \div (2x - 1)$

46. $(6x^3 + 7x^2 - 7x - 6) \div (2x + 3)$

47. $(x^4 + 4x^3 - x^2 - 16x - 12) \div (x^2 - 4)$

48. $(2x^4 - x^3 - 19x^2 + 9x + 9) \div (x^2 - 9)$

Skill Maintenance

Solve.

49. $x^2 - 5x = 0$ [5.3]

50. $25y^2 = 64$ [5.5]

51. Find three consecutive positive integers such that the product of the first and second integers is 26 less than the product of the second and third integers. [2.3]

52. If $f(x) = 2x^3$, find $f(-3a)$. [2.1]

Solve.

53. $|2x - 3| = 7$ [4.3]

54. $|3x - 1| < 8$ [4.3]

Synthesis

55. ◆ Do addition, subtraction, and multiplication of polynomials always result in a polynomial? Does division? Why or why not?

56. ◆ Explain how you could construct a polynomial of degree 4 that has a remainder of 3 when divided by $x + 1$.

57. ◆ Can the quotient of two sums always be rewritten as a sum of two quotients? Why or why not?

58. ◆ What adjustments must be made if synthetic division is to be used to divide a polynomial by a binomial of the form $ax + b$, with $a > 1$?

Divide.

59. $(x^4 - x^3y + x^2y^2 + 2x^2y - 2xy^2 + 2y^3) \div$
$(x^2 - xy + y^2)$

60. $(4a^3b + 5a^2b^2 + a^4 + 2ab^3) \div (a^2 + 2b^2 + 3ab)$

61. $(a^7 + b^7) \div (a + b)$

62. Find k such that when $x^3 - kx^2 + 3x + 7k$ is divided by $x + 2$, the remainder is 0.

63. Find k such that when $x^2 - 3x + 2k$ is divided by $x + 2$, the remainder is 7.

64. ◆ Jamaladeen incorrectly states that
$$(x^3 + 9x^2 - 6) \div (x^2 - 1) = x + 9 + \frac{x + 4}{x^2 - 1}.$$
Without performing any long division, how could you show Jamaladeen that his division cannot possibly be correct?

65. Let $f(x) = 6x^3 - 13x^2 - 79x + 140$. Find $f(4)$ and then solve the equation $f(x) = 0$.

66. Let $f(x) = 4x^3 + 16x^2 - 3x - 45$. Find $f(-3)$ and then solve the equation $f(x) = 0$.

Nested Evaluation.   One way to evaluate a polynomial function like $P(x) = 3x^4 - 5x^3 + 4x^2 - 1$ is to successively factor out x as shown:
$$P(x) = x(x(x(3x - 5) + 4) + 0) - 1.$$

Computations are then performed using this "nested" form of $P(x)$.

67. Use nested evaluation to find $f(4)$ in Exercise 65. Note the similarities to the calculations performed with synthetic division.

68. Use nested evaluation to find $f(-3)$ in Exercise 66. Note the similarities to the calculations performed with synthetic division.

6.7
Formulas and Models

- *Formulas*
- *Models*

Formulas

Formulas occur frequently as mathematical models. Many formulas contain rational expressions, and to solve such formulas for a specified letter, we proceed as when solving rational equations.

Example 1 *Optics.* The formula $f = L/d$ tells how to calculate a camera's "f-stop." In this formula, f is the f-stop, L is the focal length (approximately the distance from the lens to the film), and d is the diameter of the lens. Solve for d.

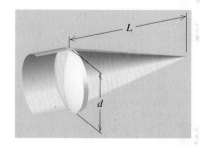

SOLUTION We solve this equation as we did those in Section 6.4:

$$f = \frac{L}{d}$$

$$d \cdot f = d \cdot \frac{L}{d} \qquad \text{**Multiplying by the LCD to clear fractions**}$$

$$df = L$$

$$df \cdot \frac{1}{f} = L \cdot \frac{1}{f} \qquad \text{**Multiplying by $\frac{1}{f}$ on both sides**}$$

$$d = \frac{L}{f}. \qquad \text{**Simplifying**}$$

The formula $d = L/f$ can now be used to determine the diameter of a lens if we know the focal length and the f-stop. ▬

Example 2 *Astronomy.* The formula

$$L = \frac{dR}{D - d},$$

where D is the diameter of the sun, d is the diameter of Earth, R is Earth's distance from the sun, and L is some fixed distance, is used in calculating when lunar eclipses occur. Solve for D.

SOLUTION We first clear fractions by multiplying by the LCD, which is $D - d$:

$$(D - d)L = (D - d)\frac{dR}{D - d}$$

$$(D - d)L = dR.$$

We do *not* multiply the factors on the left since we wish to get D all alone. Instead we multiply by $1/L$ on both sides and then add d:

$$D - d = \frac{dR}{L} \qquad \text{**Multiplying by $\frac{1}{L}$ on both sides**}$$

$$D = \frac{dR}{L} + d. \qquad \text{**Adding d on both sides**}$$

We now have D all alone on one side of the equation. Since D does not appear on the other side, we have solved the formula for D. ▬

Example 3 *Acoustics (the Doppler Effect).* The formula

$$f = \frac{sg}{s + v}$$

is used to determine the frequency f of a sound that is moving at velocity v toward a listener who hears the sound as frequency g. Here s is the speed of sound in a particular medium. Solve for s.

SOLUTION We first clear fractions by multiplying by the LCD, $s + v$:

$$f \cdot (s + v) = \frac{sg}{s + v}(s + v)$$

$$fs + fv = sg. \qquad \text{Here, because } s \textit{ does} \text{ appear on both sides, we do distribute on the left side.}$$

Next, we must get all terms containing s on one side:

$$fv = sg - fs \qquad \text{Adding } -fs \text{ on both sides}$$

$$fv = s(g - f) \qquad \text{Factoring out } s$$

$$\frac{fv}{g - f} = s. \qquad \text{Multiplying by } \frac{1}{g - f} \text{ on both sides}$$

Since s is isolated on one side, we have solved for s. This last equation can be used to determine the speed of sound whenever f, v, and g are known.

Example 4 *Resistance.* The formula

$$\frac{1}{R} = \frac{1}{r_1} + \frac{1}{r_2}$$

relates the resistance R of two resistors r_1 and r_2 connected in parallel.

a) Solve for r_2.

b) Find r_2 when $R = 3.75$ ohms and $r_1 = 6$ ohms.

SOLUTION

a) To solve for r_2, we multiply by the LCD, Rr_1r_2:

$$\frac{1}{R} = \frac{1}{r_1} + \frac{1}{r_2}$$

$$Rr_1r_2 \cdot \frac{1}{R} = Rr_1r_2 \cdot \left[\frac{1}{r_1} + \frac{1}{r_2}\right] \qquad \begin{array}{l}\text{Multiplying by the LCD}\\ \text{on both sides}\end{array}$$

$$Rr_1r_2 \cdot \frac{1}{R} = Rr_1r_2 \cdot \frac{1}{r_1} + Rr_1r_2 \cdot \frac{1}{r_2} \qquad \begin{array}{l}\text{Using the}\\ \text{distributive law}\end{array}$$

$$\left.\begin{array}{c}\dfrac{Rr_1r_2}{R} = \dfrac{Rr_1r_2}{r_1} + \dfrac{Rr_1r_2}{r_2}\\[2mm] r_1r_2 = Rr_2 + Rr_1.\end{array}\right\} \qquad \begin{array}{l}\text{Simplifying by removing}\\ \text{factors of 1:}\\ \dfrac{R}{R} = 1; \dfrac{r_1}{r_1} = 1; \dfrac{r_2}{r_2} = 1\end{array}$$

You might be tempted at this point to multiply by $1/r_1$ to get r_2 alone on the left, but note that r_2 also appears on the right. We must get all the terms involving r_2 on the *same side* of the equation:

$$r_1 r_2 - R r_2 = R r_1 \qquad \text{Adding } -Rr_2 \text{ on both sides}$$
$$r_2(r_1 - R) = R r_1 \qquad \text{Factoring out } r_2$$
$$r_2 = \frac{R r_1}{r_1 - R}. \qquad \text{Multiplying by } \frac{1}{r_1 - R} \text{ on both sides}$$

b) We use the equation for r_2, replacing R with 3.75 and r_1 with 6:

$$r_2 = \frac{(3.75)6}{6 - 3.75} = 10.$$

The second resistor has a resistance of 10 ohms.

To Solve a Rational Equation for a Specific Unknown:

1. If necessary, clear fractions.
2. Multiply, as needed, to remove parentheses.
3. Get all terms with the unknown alone on one side of the equation.
4. Factor out the unknown if it appears in more than one term.
5. Use the multiplication principle to get the unknown alone on one side.

Models

Applications from many fields can be modeled by rational functions. Many graphers, however, do not have the capability to fit a rational function to a set of data using regression.

Graphs of rational functions vary widely in shape. Functions of the type $f(x) = k/x$, $k > 0$, will have a graph similar to that of the function given by $g(x) = 1/x$.

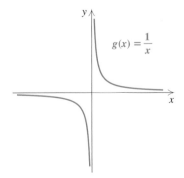

You can recognize graphed data that could be modeled by this kind of function. For $y = k/x$, $k > 0$, if the domain is a set of positive numbers, then only the branch in the first quadrant need be considered.

Example 5 *Commuter Travel.* The number of trips made to a city decreases as the distance from the city increases. Can the data in the following table be modeled by a function of the type $f(x) = k/x$?

MILES FROM CITY	NUMBER OF TRIPS TO CITY PER DAY PER RESIDENTIAL ACRE
2	130
6	120
8	60
10	45
14	40
18	25
20	20
22	19
24	20
26	19
28	20

Source: Kolars, John F., and John D. Nyusten, *Geography.*
(New York: McGraw-Hill, 1974).

SOLUTION   We enter the data and plot the points. It does appear as though the data follow the general shape of the graph of $f(x) = k/x$. We can see this by sketching a curve through the points. However, we will not attempt to fit such a model to the data here.

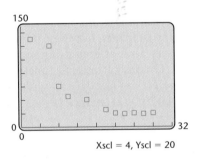

Xscl = 4, Yscl = 20

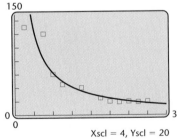

Xscl = 4, Yscl = 20

6.7 | Exercise Set

Solve the formula for the specified letter.

1. $\dfrac{W_1}{W_2} = \dfrac{d_1}{d_2}$; W_1

2. $\dfrac{W_1}{W_2} = \dfrac{d_1}{d_2}$; d_1

3. $s = \dfrac{(v_1 + v_2)t}{2}$; v_1

4. $s = \dfrac{(v_1 + v_2)t}{2}$; t

5. $\dfrac{1}{R} = \dfrac{1}{r_1} + \dfrac{1}{r_2}$; R

6. $\dfrac{1}{R} = \dfrac{1}{r_1} + \dfrac{1}{r_2}$; r_1

7. $R = \dfrac{gs}{g + s}$; g

8. $K = \dfrac{rt}{r - t}$; t

9. $I = \dfrac{2V}{R + 2r}$; R

10. $I = \dfrac{2V}{R + 2r}$; r

11. $\dfrac{1}{p} + \dfrac{1}{q} = \dfrac{1}{f}$; p

12. $\dfrac{1}{p} + \dfrac{1}{q} = \dfrac{1}{f}$; q

13. $I = \dfrac{nE}{R + nr}$; n

14. $I = \dfrac{nE}{R + nr}$; r

15. $S = \dfrac{H}{m(t_1 - t_2)}$; t_1

16. $S = \dfrac{H}{m(t_1 - t_2)}$; H

17. $\dfrac{E}{e} = \dfrac{R + r}{r}$; r

18. $\dfrac{E}{e} = \dfrac{R + r}{r}$; e

19. $S = \dfrac{a}{1 - r}$; r

20. $S = \dfrac{a - ar^n}{1 - r}$; a

Solve.

21. *Interest.* The formula

$$P = \dfrac{A}{1 + r}$$

is used to determine what principal P should be invested for 1 yr at $(100 \cdot r)\%$ simple interest in order to have A dollars after a year. Solve for r.

22. *Average Speed.* The formula

$$v = \dfrac{d_2 - d_1}{t_2 - t_1}$$

gives an object's average speed v when that object has traveled d_1 miles in t_1 hours and d_2 miles in t_2 hours. Solve for t_2.

23. *Average Speed.* At what time will Enid's Taurus, averaging a speed of 60 mph, reach Philadelphia if Enid leaves New York at 2:00 A.M. and New York is 105 mi from Philadelphia? (See Exercise 22.)

24. *Interest.* At what yearly interest rate should Bernie invest $1600 if he wants it to grow to $1712 after 1 yr? (See Exercise 21.)

25. *Earned Run Average.* The formula

$$A = 9 \cdot \dfrac{R}{I}$$

gives a pitcher's *earned run average*, where A is the earned run average, R is the number of earned runs, and I is the number of innings pitched. How many earned runs were given up if a pitcher's earned run average is 2.4 after 45 innings?

26. *Resistance.* Two resistors are connected in parallel. Their resistances are, respectively, 8 ohms and 15 ohms. What is the resistance of the combination? (See Example 4.)

27. *Resistance.* A resistor has a resistance of 50 ohms. What size resistor should be put with it, in parallel, in order to obtain a resistance of 5 ohms? (See Example 4.)

28. *Work Rate.* The formula

$$\dfrac{1}{t} = \dfrac{1}{a} + \dfrac{1}{b}$$

gives the total time t required for two workers to complete a job, if the workers' individual times are a and b. Solve for t.

29. *Area of a Trapezoid.* The area of a certain trapezoid is 25 cm². Its height is 5 cm and the length of one base is 4 cm. Find the length of the other base.

30. *Taxable Interest.* The formula

$$I_t = \dfrac{I_f}{1 - T}$$

gives the *taxable interest rate* I_t equivalent to the *tax-free interest rate* I_f for a person in the $(100 \cdot T)\%$ tax bracket. Solve for T.

31. *Escape Velocity.* The formula

$$\frac{V^2}{R^2} = \frac{2g}{R + h}$$

is used to find a satellite's *escape velocity V*, where *R* is a planet's radius, *h* is the satellite's height above the planet, and *g* is the planet's acceleration due to gravity. Solve for *h*.

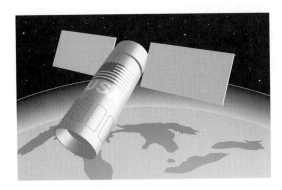

32. *Astronomy.* Solve the formula of Example 2 for *d*, the diameter of Earth.

33. *Semester Average.* The formula

$$A = \frac{2Tt + Qq}{2T + Q}$$

gives a student's average *A* after *T* tests and *Q* quizzes, where each test counts as 2 quizzes, *t* is the test average, and *q* is the quiz average. Solve for *Q*.

34. *Average Acceleration.* The formula

$$a = \frac{v_2 - v_1}{t_2 - t_1}$$

gives a vehicle's *average acceleration* when its velocity changes from v_1 at time t_1 to v_2 at time t_2. Solve for t_1.

35. A line passing through two points has a slope of $-\frac{2}{5}$. If the coordinates of the points are $(x_1, 2)$ and $(2x_1, 8)$, find the coordinates of both points.

36. Nancy has a test average of 79 and a quiz average of 90. Her grade is calculated by using the formula found in Exercise 33. If Nancy's overall average is 84 and 5 quizzes were taken, how many tests were taken?

For Exercises 37–40, determine whether the situation can be modeled by a function of the form $f(x) = k/x$.

37. *Painting a House.* The number of hours it takes to paint a particular house decreases as the number of people painting increases.

NUMBER OF PEOPLE PAINTING	NUMBER OF HOURS REQUIRED TO PAINT THE HOUSE
1	25
2	15
3	10
4	8
5	7

38. *Real Estate.* The number of houses sold increases as the interest rates decrease.

INTEREST RATE	NUMBER OF HOUSES SOLD IN A COUNTY PER MONTH
0.065	45
0.0675	38
0.07	35
0.0725	31
0.075	28
0.0775	25
0.08	22

39. *Value of an Automobile.* The value of a car decreases as its age increases.

AGE OF CAR (IN YEARS)	VALUE OF CAR
1	$15,000
2	14,000
4	11,000
6	8,000
10	4,000

40. *Perceived Height.* The height that a pole appears to be decreases as the distance from the pole increases.

DISTANCE FROM POLE (IN FEET)	PERCEIVED HEIGHT OF POLE (IN FEET)
5	4
10	2
15	$1\frac{1}{3}$
20	1
40	0.5

Skill Maintenance

41. Graph on a plane: $6x - y < 6$. [4.4]

42. If $f(x) = x^3 - x$, find $f(2a)$. [2.1]

43. Factor: $x^2 - 77x - 2940$. [5.4]

44. Solve: $6x^2 = 11x + 35$. [5.5]

Synthesis

45. ◈ Solve both Exercise 9 on p. 380 and Exercise 11 on p 407. In what ways are these exercises similar? How are they different?

46. *Escape Velocity.* A satellite's escape velocity is 6.5 mi/sec, the radius of Earth is 3960 mi, and the acceleration due to gravity is 32.2 ft/sec². How far is the satellite from the surface of Earth? (See Exercise 31.)

47. The *harmonic mean* of two numbers a and b is a number M such that the reciprocal of M is the average of the reciprocals of a and b. Find a formula for the harmonic mean.

48. Solve for x:

$$x^2\left(1 - \frac{2pq}{x}\right) = \frac{2p^2q^3 - pq^2x}{-q}.$$

49. *Average Acceleration.* The formula

$$a = \frac{\dfrac{d_4 - d_3}{t_4 - t_3} - \dfrac{d_2 - d_1}{t_2 - t_1}}{t_4 - t_2}$$

can be used to approximate average acceleration where the d's are distances and the t's are the corresponding times. Solve for t_1.

COLLABORATIVE CORNER

Focus: Application of rational expressions

Time: 20–30 minutes

Group size: 2

Materials: Graph paper

Ginny is riding in the American Diabetes Association Tour de Cure 40-mi Bikeathon. For the first 20 mi, the route is primarily downhill so Ginny averages 18 mph. Over the final 20 mi, however, Ginny can average only 12 mph. Vince is also riding in the Bikeathon, but because his 40-mi route is flat, he averages a steady 15 mph over the entire route.

Activity

1. Assume that both Ginny and Vince begin riding at 9 A.M., and draw a graph representing the situation.

2. With one group member playing the role of Ginny and the other playing the role of Vince, determine when each biker will cross the finish line.

3. *Question for Ginny:* If you knew in advance the speed at which Vince was biking, how could you have adjusted your speeds so that you would finish together?
   Question for Vince: If you knew in advance the speed at which Ginny was biking, how could you have adjusted your speeds so that you would finish together?

4. Assume now that both routes are only 20 mi long and that Ginny's speed drops from 18 mph to 12 mph at the 10-mi mark while Vince continues to ride at a steady 15 mph. Switch roles and again perform parts (2) and (3) above.

5. Suppose that Ginny rode the first half of a 10-mi route at 18 mph and the other half at 12 mph. What would be her average speed? What can you conclude, as a group, about Ginny's average speed?

CHAPTER

6 Summary and Review

Key Terms

Rational expression, p. 348
Rational function, p. 348
Simplified, p. 351
Least common multiple,
  LCM, p. 361

Least common denominator,
  LCD, p. 362
Complex rational expression,
  p. 368

Rational equation, p. 376
Clear fractions, p. 376
Motion problem, p. 387
Synthetic division, p. 398

Important Properties and Formulas

Addition: $\dfrac{A}{C} + \dfrac{B}{C} = \dfrac{A + B}{C}$

Subtraction: $\dfrac{A}{C} - \dfrac{B}{C} = \dfrac{A - B}{C}$

Multiplication: $\dfrac{A}{B} \cdot \dfrac{C}{D} = \dfrac{AC}{BD}$

Division: $\dfrac{A}{B} \div \dfrac{C}{D} = \dfrac{A}{B} \cdot \dfrac{D}{C}$

To find the least common multiple, LCM, use each factor the greatest number of times that it occurs in any one prime factorization.

Simplifying Complex Rational Expressions

I: By using multiplication by 1

1. Find the LCD of all expressions *within* the complex rational expression.
2. Multiply the complex rational expression by 1, using the LCD to form the expression for 1.
3. Distribute and simplify so that the numerator and the denominator of the complex rational expression are polynomials.

4. Factor and simplify, if possible.

II: By using division

1. Add or subtract, as necessary, to get one rational expression in the numerator.
2. Add or subtract, as necessary, to get one rational expression in the denominator.
3. Perform the indicated division (invert the divisor and multiply).
4. Simplify, if possible, by removing any factors equal to 1.

Modeling Work Problems

If

  $a =$ the time needed for A to complete the work alone,

  $b =$ the time needed for B to complete the work alone, and

  $t =$ the time needed for A and B to complete the work together,

then

$$\frac{t}{a} + \frac{t}{b} = 1.$$

$$f(x) = \frac{1}{x}$$

Domain of f: $(-\infty, 0) \cup (0, \infty)$
Range of f: $(-\infty, 0) \cup (0, \infty)$

REVIEW EXERCISES

1. If

$$f(t) = \frac{t^2 - 3t + 2}{t^2 - 9},$$

find the following function values.

a) $f(0)$ **b)** $f(-1)$ **c)** $f(2)$

Find the LCD.

2. $\dfrac{7}{6x^3}, \ \dfrac{y}{16x^2}$

3. $\dfrac{x + 8}{x^2 + x - 20}, \ \dfrac{x}{x^2 + 3x - 10}$

4. Determine the vertical asymptotes of the graph of the function given by

$$f(x) = \frac{6x^2 - 4x}{3x^2 + x - 2}.$$

Perform the indicated operations and simplify when possible.

5. $\dfrac{x^2}{x - 3} - \dfrac{9}{x - 3}$

6. $\dfrac{4x - 2}{x^2 - 5x + 4} - \dfrac{3x + 2}{x^2 - 5x + 4}$

7. $\dfrac{3a^2b^3}{5c^3d^2} \cdot \dfrac{15c^9d^4}{9a^7b}$

8. $\dfrac{5}{6m^2n^3p} + \dfrac{7}{9mn^4p^2}$

9. $\dfrac{y^2 - 64}{2y + 10} \cdot \dfrac{y + 5}{y + 8}$

10. $\dfrac{9a^2 - 1}{a^2 - 9} \div \dfrac{3a + 1}{a + 3}$

11. $\dfrac{x}{x^2 + 5x + 6} - \dfrac{2}{x^2 + 3x + 2}$

12. $\dfrac{2x^2}{x - y} + \dfrac{2y^2}{y - x}$

13. $\dfrac{3}{y + 4} - \dfrac{y}{y - 1} + \dfrac{y^2 + 3}{y^2 + 3y - 4}$

Simplify.

14. $\dfrac{\dfrac{5}{x} - 5}{\dfrac{7}{x} - 7}$

15. $\dfrac{\dfrac{2}{a} + \dfrac{2}{b}}{\dfrac{4}{a^2} - \dfrac{4}{b^2}}$

16. $\dfrac{\dfrac{y^2 + 4y - 77}{y^2 - 10y + 25}}{\dfrac{y^2 - 5y - 14}{y^2 - 25}}$

17. $\dfrac{\dfrac{5}{x^2 - 9} - \dfrac{3}{x + 3}}{\dfrac{4}{x^2 + 6x + 9} + \dfrac{2}{x - 3}}$

Solve.

18. $\dfrac{6}{x} + \dfrac{4}{x} = 5$

19. $\dfrac{5}{3x + 2} = \dfrac{3}{2x}$

20. $\dfrac{4x}{x + 1} + \dfrac{4}{x} + 9 = \dfrac{4}{x^2 + x}$

21. If

$$f(x) = \frac{2}{x - 1} + \frac{2}{x + 2},$$

find all a for which $f(a) = 1$.

Solve.

22. Kim can paint a garage in 12 hr. Kelly can paint the same garage in 9 hr. How long would it take them, working together, to paint the garage?

23. The Gold River's current is 6 mph. A boat travels 50 mi downstream in the same time that it takes to travel 30 mi upstream. What is the speed of the boat in still water?

24. A car and a motorcycle leave a rest area at the same time, with the car traveling 8 mph faster than the motorcycle. The car then travels 105 mi in the time that it takes the motorcycle to travel 93 mi. Find the speed of each vehicle.

Divide.

25. $(20r^2s^3 + 15r^2s^2 - 10r^3s^3) \div (5r^2s)$

26. $(y^3 + 125) \div (y + 5)$

27. $(4x^3 + 3x^2 - 5x - 2) \div (x^2 + 1)$

28. Divide using synthetic division. Use the results to factor the dividend.
$$(x^3 - 7x^2 + 36) \div (x - 3)$$

Solve.

29. $R = \dfrac{gs}{g + s}$, for s **30.** $S = \dfrac{H}{m(t_1 - t_2)}$, for m

31. $\dfrac{1}{ac} = \dfrac{2}{ab} - \dfrac{3}{bc}$, for c **32.** $T = \dfrac{A}{v(t_2 - t_1)}$, for t_1

33. *Serving Size.* The size of one serving of breakfast cereal decreases as the number of servings obtained from a box increases. Determine whether the data for this situation, given in the following table, can be modeled by a function of the form $f(x) = k/x$.

NUMBER OF SERVINGS	SIZE OF SERVING (IN OUNCES)
1	24
4	6
6	4
12	2
16	1.5

Skill Maintenance

Graph on a plane.

34. $y - 2x \geq 4$

35. $x > -3$

36. Factor: $x^4 - 6x^2 + 9$.

37. Factor: $6x^2 + 29x - 42$.

38. Solve: $42 = 29x + 6x^2$.

39. If $f(x) = 7x^2 - 6x$, find $f(-2)$.

Synthesis

40. ◆ Discuss at least three different uses of the LCD studied in this chapter.

41. ◆ Explain the difference between a rational expression and a rational equation.

Solve.

42. $\dfrac{5}{x - 13} - \dfrac{5}{x} = \dfrac{65}{x^2 - 13x}$

43. $\dfrac{\dfrac{x}{x^2 - 25} + \dfrac{2}{x - 5}}{\dfrac{3}{x - 5} - \dfrac{4}{x^2 - 10x + 25}} = 1$

44. One summer, Anna mowed 4 lawns for every 3 lawns mowed by her brother Franz. Together, they mowed 98 lawns. How many lawns did each mow?

CHAPTER 6 TEST

Simplify.

1. $\dfrac{t - 1}{t + 3} \cdot \dfrac{3t + 9}{4t^2 - 4}$

2. $\dfrac{4x + 12}{x^2 - x} \div \dfrac{x^2 - 9}{8x^2}$

3. Find the LCD:

$$\dfrac{3x}{x^2 + 8x - 33}, \quad \dfrac{x + 1}{x^2 - 12x + 27}.$$

4. Determine the vertical asymptotes of the function given by

$$r(x) = \dfrac{x^2 - 1}{x^3 - x^2 - 2x}.$$

Perform the indicated operation and simplify when possible.

5. $\dfrac{25x}{x + 5} + \dfrac{x^3}{x + 5}$

6. $\dfrac{3a^2}{a - b} - \dfrac{3b^2 - 6ab}{b - a}$

7. $\dfrac{4ab}{a^2 - b^2} + \dfrac{a^2 + b^2}{a + b}$

8. $\dfrac{4}{x + 3} - \dfrac{x - 2}{x^2 + 2x - 3}$

9. $\dfrac{4}{y + 3} - \dfrac{y}{y - 2} + \dfrac{y^2 + 4}{y^2 + y - 6}$

Simplify.

10. $\dfrac{\dfrac{2}{a} + \dfrac{3}{b}}{\dfrac{5}{ab} + \dfrac{1}{a^2}}$

11. $\dfrac{\dfrac{x^2 - 5x - 36}{x^2 - 36}}{\dfrac{x^2 + x - 12}{x^2 - 12x + 36}}$

Solve.

12. $\dfrac{4}{2x - 5} = \dfrac{6}{5x + 3}$

13. $\dfrac{t + 11}{t^2 - t - 12} + \dfrac{1}{t - 4} = \dfrac{4}{t + 3}$

In Exercises 14 and 15, let $f(x) = \dfrac{x + 3}{x - 1}.$

14. Find $f(2)$ and $f(-3)$.

15. Find all a for which $f(a) = 7$.

16. Kyla can cut and split a cord of wood in 3.5 hr. Ronson can cut and split a cord of wood in 4.5 hr. How long will it take them, working together, to cut and split a cord of wood?

Divide.

17. $(16ab^3c - 10ab^2c^2 + 12a^2b^2c) \div (4a^2b)$

18. $(y^2 - 20y + 64) \div (y - 6)$

19. Divide using synthetic division:
$$(x^3 + 5x^2 + 4x - 7) \div (x - 4).$$

20. Divide $x^4 + x^3 - 11x^2 - 9x + 18$ by $x^2 - 9$ and use the results of the division to factor the dividend.

21. Solve $A = \dfrac{h(b_1 + b_2)}{2}$ for b_1.

22. The product of the reciprocals of two consecutive integers is $\frac{1}{30}$. Find the integers.

23. Dodi bicycles 12 mph with no wind. Against the wind, Dodi bikes 8 mi in the same time that it takes to bike 14 mi with the wind. What is the speed of the wind?

Skill Maintenance

24. If $f(x) = x^2 - 3$, find $f(a + 1)$.

25. Factor: $16t^2 - 24t - 72$.

26. Solve: $16t^2 = 24t + 72$.

27. Graph on a plane: $2x + 5y > 10$.

Synthesis

28. Let
$$f(x) = \dfrac{1}{x + 3} + \dfrac{5}{x - 2}.$$
Find all a for which $f(a) = f(a + 5)$.

29. Solve: $\dfrac{6}{x - 15} - \dfrac{6}{x} = \dfrac{90}{x^2 - 15x}$.

30. Find the x- and y-intercepts for the function given by
$$f(x) = \dfrac{\dfrac{5}{x + 4} - \dfrac{3}{x - 2}}{\dfrac{2}{x - 3} + \dfrac{1}{x + 4}}.$$

CUMULATIVE REVIEW 1–6

1. Evaluate

$$\frac{2x - y^2}{x + y}$$

for $x = 3$ and $y = -4$.

2. Convert to scientific notation: 5,760,000,000.

3. Determine the slope and the y-intercept for the line given by $7x - 4y = 12$.

4. Find an equation for the line that passes through the points $(-1, 7)$ and $(2, -3)$.

5. Solve the system

$$5x - 2y = -23,$$
$$3x + 4y = 7.$$

6. Solve the system

$$-3x + 4y + z = -5,$$
$$x - 3y - z = 6,$$
$$2x + 3y + 5z = -8.$$

7. Luigi's Pizzeria donated 45 pizzas for a charity event. Small pizzas sold for $7.00 each and large pizzas for $10.00 each. The total amount of funds raised from the sale of the pizzas was $402. How many of each size pizza were donated?

8. The sum of three numbers is 20. The first number is 3 less than twice the third number. The second number minus the third number is -7. What are the numbers?

9. If

$$f(x) = \frac{x - 2}{x - 5},$$

find **(a)** $f(3)$ and **(b)** the domain of f.

Solve.

10. $8x = 1 + 16x^2$

11. $625 = 49y^2$

12. $20 > 2 - 6x$

13. $\frac{1}{3}x - \frac{1}{5} \geq \frac{1}{5}x - \frac{1}{3}$

14. $-8 < x + 2 < 15$

15. $3x - 2 < -6 \ or \ x + 3 > 9$

16. $|x| > 6.4$

17. $|4x - 1| \leq 14$

18. $\frac{2}{n} - \frac{7}{n} = 3$

19. $\frac{6}{x - 5} = \frac{2}{2x}$

20. $\frac{3x}{x - 2} - \frac{6}{x + 2} = \frac{24}{x^2 - 4}$

21. $\frac{3x^2}{x + 2} + \frac{5x - 22}{x - 2} = \frac{-48}{x^2 - 4}$

22. Let $f(x) = |3x - 5|$. Find all values of x for which $f(x) = 2$.

23. Write the domain of f using interval notation if $f(x) = \sqrt{x - 7}$.

24. Solve $5m - 3n = 4m + 12$ for n.

25. Solve $P = \frac{3a}{a + b}$ for a.

Graph on a plane.

26. $4x \geq 5y + 20$

27. $y < -2$

Perform the indicated operations and simplify.

28. $(2x^2 - 3x + 1) + (6x - 3x^3 + 7x^2 - 4)$

29. $(5x^3y^2)(-3xy^2)$

30. $(3a + b - 2c) - (-4b + 3c - 2a)$

31. $(5x^2 - 2x + 1)(3x^2 + x - 2)$

32. $(2x^2 - y)^2$

33. $(2x^2 - y)(2x^2 + y)$

34. $(-5m^3n^2 - 3mn^3) + (-4m^2n^2 + 4m^3n^2) - (2mn^3 - 3m^2n^2)$

35. $\frac{y^2 - 36}{2y + 8} \cdot \frac{y + 4}{y + 6}$

36. $\frac{x^4 - 1}{x^2 - x - 2} \div \frac{x^2 + 1}{x - 2}$

37. $\frac{5ab}{a^2 - b^2} + \frac{a + b}{a - b}$

38. $\frac{2}{m + 1} + \frac{3}{m - 5} - \frac{m^2 - 1}{m^2 - 4m - 5}$

39. $y - \frac{2}{3y}$

40. Simplify:

$$\frac{\dfrac{1}{x} - \dfrac{1}{y}}{x + y}.$$

41. Divide: $(9x^3 + 5x^2 + 2) \div (x + 2)$.

Factor.

42. $4x^3 + 18x^2$

43. $x^2 + 8x - 84$

44. $16y^2 - 81$

45. $x^2 - 26 + 11x$

46. $t^2 - 16t + 64$

47. $x^6 - x^2$

48. $x^2 - 25x - 2100$

49. $20x^2 + 7x - 3$

50. $3x^2 - 17x - 28$

51. $x^5 - x^3y + x^2y - y^2$

52. Let $f(x) = x^2 - 4$ and $g(x) = x^2 - 7x + 10$. Find the domain of f / g.

Solve.

53. Ed's tractor can plow a field in 3 hr. Nell's tractor can plow the same field in 1.5 hr. Working together, how long would it take them to plow the field?

54. The length of a rectangle is 3 ft longer than the width. The area is 54 ft². Find the perimeter of the rectangle.

55. The sum of the squares of three consecutive even integers is equal to 8 more than three times the square of the second number. Find the integers.

Estimate the domain and the range of each function from its graph.

56.

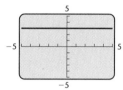

57.

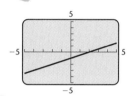

58.

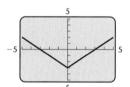

59.

In Exercises 60–63, match each equation with one of the following graphs.

a)

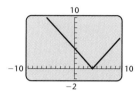

b)

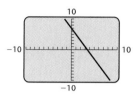

c)

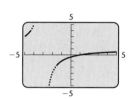

d)

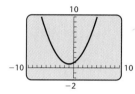

60. $y = -2x + 7$

61. $y = |x - 4|$

62. $y = x^2 + x + 1$

63. $y = \dfrac{x - 1}{x + 3}$

Synthesis

64. Multiply: $(x - 4)^3$.

65. Find all roots for $f(x) = x^4 - 34x^2 + 225$.

Solve.

66. $4 \le |3 - x| \le 6$

67. $\dfrac{18}{x - 9} + \dfrac{10}{x + 5} = \dfrac{28x}{x^2 - 4x - 45}$

68. $16x^3 = x$

Exponents and Radical Functions 7

I n this chapter, we learn about square roots, cube roots, fourth roots, and so on. These roots are studied in connection with the manipulation of radical expressions and the solution of real-world applications. Fractional exponents are also discussed and are used to ease much of our work with radicals. The chapter closes with an examination of the complex-number system.

APPLICATION

AIR-SHOW ATTENDANCE. The annual attendance at air shows in the United States has grown during recent years, as shown in the following table. Determine whether a radical function can be used to model this situation.

YEAR	ANNUAL AIR-SHOW ATTENDANCE (IN MILLIONS)
1987	14.1
1988	18.3
1989	22.0
1990	23.1
1991	23.9
1992	24.4

Source: The Macmillan Visual Almanac, 1996

We graph the points. A curve sketched through the points appears to be in the shape of the graph of a radical function.

This problem appears as Example 12 in Section 7.1.

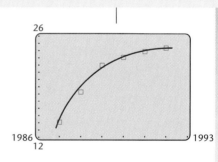

417

7.1
Radical Expressions, Functions, and Models

- *Square Roots and Square-Root Functions*
- *Expressions of the Form $\sqrt{a^2}$*
- *Cube Roots*
- *Odd and Even nth Roots*
- *Models Using Radical Functions*

In this section, we consider roots, such as square roots and cube roots. We look at the symbolism that is used and ways in which symbols can be manipulated to get equivalent expressions. All of this will be important in problem solving.

Square Roots and Square-Root Functions

When a number is raised to the second power, the number is squared. Often we need to know what number was squared in order to produce some value a. If such a number can be found, we call that number a *square root* of a.

Square Root

The number c is a *square root* of a if $c^2 = a$.

For example,

5 is a square root of 25 because $5^2 = 25$;

-5 is a square root of 25 because $(-5)^2 = 25$;

-4 does not have a real-number square root because there is no real number c such that $c^2 = -4$.

Later in this chapter, we will see that there is a number system, different from the real-number system, in which negative numbers do have square roots. Note that every positive number has two square roots, whereas 0 has only itself as a square root.

Example 1   Find the two square roots of 64.

SOLUTION   The square roots are 8 and -8, because $8^2 = 64$ and $(-8)^2 = 64$.   ▬

Principal Square Root

The *principal square root* of a nonnegative number is its nonnegative square root. The symbol $\sqrt{a}$ represents the principal square root of a and is read "radical a," "the square root of a," or simply "root a." The negative square root of a is written $-\sqrt{a}$.

Example 2 Simplify each of the following.

a) $\sqrt{25}$ **b)** $\sqrt{\dfrac{25}{64}}$ **c)** $-\sqrt{64}$ **d)** $\sqrt{0.0049}$

SOLUTION

a) $\sqrt{25} = 5$ $\sqrt{}$ indicates the principal square root.

b) $\sqrt{\dfrac{25}{64}} = \dfrac{5}{8}$ Since $\left(\dfrac{5}{8}\right)^2 = \dfrac{25}{64}$

c) $-\sqrt{64} = -8$ Since $\sqrt{64} = 8$, $-\sqrt{64} = -8$.

d) $\sqrt{0.0049} = 0.07$ $(0.07)(0.07) = 0.0049$

Radical Notation

The symbol $\sqrt{}$ is called a *radical sign* and the expression written under the radical sign is called the *radicand.* An expression written with a radical sign is called a *radical expression.*

The following are radical expressions:

$$\sqrt{5}, \qquad \sqrt{a}, \qquad -\sqrt{5x}, \qquad \sqrt{\dfrac{c^2 + 7}{\sqrt{x}}}.$$

Years ago, the symbol $\sqrt{x}$ was used to represent both square roots of x. That usage has almost completely disappeared, because today we pay a lot more attention to functions. Recall that functions must have exactly one output for each member of the domain.

Since each nonnegative real number x has exactly one principal square root, $\sqrt{x}$, there is a square-root function, given by

$$f(x) = \sqrt{x}.$$

The domain of the square-root function is $[0, \infty)$. We can draw its graph by selecting convenient values for x and calculating the corresponding outputs. Once these ordered pairs have been graphed, a smooth curve can be drawn.

x	$\sqrt{x}$	$(x, f(x))$
0	0	$(0, 0)$
1	1	$(1, 1)$
4	2	$(4, 2)$
9	3	$(9, 3)$

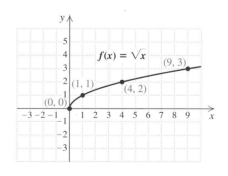

From the graph, we can see that the range of f appears to be $[0, \infty)$.

There is a second square-root function, given by

$$g(x) = -\sqrt{x}.$$

x	$-\sqrt{x}$	$(x, g(x))$
0	0	$(0, 0)$
1	-1	$(1, -1)$
4	-2	$(4, -2)$
9	-3	$(9, -3)$

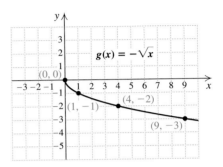

The graph shows that the domain of g is $[0, \infty)$, and the range is $(-\infty, 0]$.

The square roots of numbers that are not perfect squares can be approximated using a calculator. For example,

$$\sqrt{5} \approx 2.23606798.$$

The square root found is an approximation. The exact value of $\sqrt{5}$ is not given by any repeating or terminating decimal. The same is true for the square root of any whole number that is not a perfect square. We discussed such *irrational numbers* in Chapter 1.

Example 3 For each function, find the indicated function value.

a) $f(x) = \sqrt{3x - 2}$; $f(1)$

b) $g(z) = -\sqrt{6z + 4}$; $g(3)$

c) $q(x) = \sqrt{4 - x}$; $q(5)$

SOLUTION

a) $f(1) = \sqrt{3 \cdot 1 - 2}$ Substituting

$ = \sqrt{1} = 1$ **Simplifying and taking the square root**

b) $g(3) = -\sqrt{6 \cdot 3 + 4}$ Substituting

$ = -\sqrt{22}$ **Simplifying**

$ \approx -4.69041576$ **Using a calculator to approximate $\sqrt{22}$**

c) $q(5) = \sqrt{4 - 5}$

$ = \sqrt{-1}$ **Does not exist as a real number**

In Example 3(c), we saw that 5 is not in the domain of the function q. When a number a is not in the domain of a function f, there will be no point on the graph of f corresponding to a.

Interactive Discovery

Graph each of the following functions. By examining the graphs and tables, estimate the domain and the range of each function. When entering the functions, be sure to enclose each radicand in parentheses.

1. $g(x) = -\sqrt{x}$ **2.** $q(x) = \sqrt{-x}$

3. $t(x) = \sqrt{x - 2}$ **4.** $h(x) = \sqrt{2 - x}$

5. $f(x) = \sqrt{x^2 + 1}$

For a function f given by an expression involving square roots, any replacement for x that will make a radicand negative is not in the domain of f. Thus, to find the domain of f, we find all values of x for which the radicand is nonnegative.

Example 4 Find the domain of each of the following functions. Check by graphing the function. Then, from the graph, estimate the range of the function.

a) $q(x) = \sqrt{-x}$

b) $t(x) = \sqrt{2x - 5} - 3$

c) $f(x) = \sqrt{x^2 + 1}$

SOLUTION

a) We find all values of x for which the radicand is nonnegative:

$$-x \geq 0$$

$$x \leq 0. \qquad \text{Multiplying by } -1\text{; reversing the direction of the inequality}$$

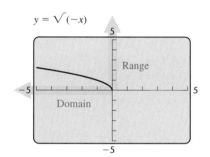

$y = \sqrt{(-x)}$

The domain is $\{x \mid x \leq 0\}$, or $(-\infty, 0]$, as indicated on the graph at left by the shading on the x-axis. The range appears to be $[0, \infty)$, as indicated by the shading on the y-axis. This can be confirmed by examining a table of values.

b) We have

$$2x - 5 \geq 0$$

$$2x \geq 5 \qquad \text{Adding 5}$$

$$x \geq \tfrac{5}{2}. \qquad \text{Dividing by 2}$$

The domain is $\left\{x \mid x \geq \tfrac{5}{2}\right\}$, or $\left[\tfrac{5}{2}, \infty\right)$, as indicated on the graph by the shading on the x-axis. The range appears to be $[-3, \infty)$, as indicated by the shading on the y-axis.

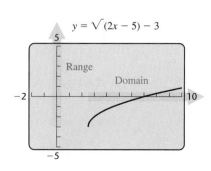

$y = \sqrt{(2x - 5)} - 3$

c) The radicand in the expression for $f(x)$ is $x^2 + 1$. We must have

$$x^2 + 1 \geq 0$$
$$x^2 \geq -1.$$

Since x^2 is nonnegative for all real numbers x, the inequality is true for all real numbers. The domain is $(-\infty, \infty)$. The range appears to be $[1, \infty)$.

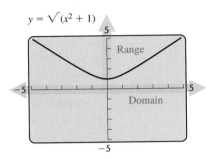

Expressions of the Form $\sqrt{a^2}$

Interactive Discovery

Use graphs to determine which of the following are identities. Be sure to enclose the entire radicand in parentheses.

1. $\sqrt{x^2} = x$

2. $\sqrt{x^2} = -x$

3. $\sqrt{x^2} = |x|$

4. $\sqrt{(x + 3)^2} = x + 3$

5. $\sqrt{(x + 3)^2} = |x + 3|$

In general, we cannot say that $\sqrt{x^2} = x$. However, we can simplify $\sqrt{x^2}$ using absolute value.

Simplifying $\sqrt{a^2}$

For any real number a,

$$\sqrt{a^2} = |a|.$$

(The principal square root of a^2 is the absolute value of a.)

When a radicand consists of a perfect square, like $25x^2$ or $(m - 3)^2$, absolute-value signs are needed when simplifying. We use absolute-value signs unless we know that the quantities being squared are nonnegative.

Example 5   Simplify each expression. Assume that the variable can represent any real number.

a) $\sqrt{(x + 1)^2}$

b) $\sqrt{x^2 - 8x + 16}$

c) $\sqrt{a^8}$

d) $\sqrt{t^6}$

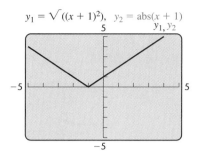

$y_1 = \sqrt{((x+1)^2)}, \quad y_2 = \text{abs}(x+1)$

SOLUTION

a) $\sqrt{(x+1)^2} = |x+1|$ Since $x + 1$ might be negative (for example, if $x = -3$), absolute-value notation is necessary.

The graph at left confirms the identity.

b) $\sqrt{x^2 - 8x + 16} = \sqrt{(x-4)^2} = |x-4|$ Since $x - 4$ might be negative, absolute-value notation is necessary.

$(x-4)(x-4)$

c) Note that $(a^4)^2 = a^8$ and that a^4 is never negative. Thus,

$$\sqrt{a^8} = a^4.$$ Absolute-value notation is unnecessary here.

d) Note that $(t^3)^2 = t^6$. Thus,

$$\sqrt{t^6} = |t^3|.$$ Since t^3 might be negative, absolute-value notation is necessary.   ▬

For many applications, radicands are nonnegative, and absolute-value signs are not needed when simplifying. If functions are involved, this corresponds to restricting the domains.

Example 6   Simplify each expression. Assume that no radicands were formed by raising negative quantities to even powers.

a) $\sqrt{t^2}$ **b)** $\sqrt{a^{10}}$ **c)** $\sqrt{9x^2 + 6x + 1}$

SOLUTION

a) $\sqrt{t^2} = t$ We are assuming that t is nonnegative, so no absolute-value notation is necessary. When t *is* negative, $\sqrt{t^2} \neq t$.

In function notation, if

$$f(t) = \sqrt{t^2}, \ t \geq 0,$$

and

$$g(t) = t,$$

then f and g are equivalent.

b) $\sqrt{a^{10}} = a^5$ Assuming that a^5 is nonnegative. Note that $(a^5)^2 = a^{10}$.

c) $\sqrt{9x^2 + 6x + 1} = \sqrt{(3x+1)^2} = 3x+1$ Assuming that $3x + 1$ is nonnegative   ▬

$(3x+1)(3x+1)$

Cube Roots

We often need to know what number was cubed in order to produce a certain value. When such a number is found, we say that we have found a *cube root*.

Cube Root

The number c is the *cube root* of a if $c^3 = a$. In symbols, we write $\sqrt[3]{a}$ to denote the cube root of a.

For example,

$$2 \text{ is the cube root of } 8 \text{ because } 2^3 = 2 \cdot 2 \cdot 2 = 8;$$
$$-4 \text{ is the cube root of } -64 \text{ because } (-4)^3 = (-4)(-4)(-4) = -64.$$

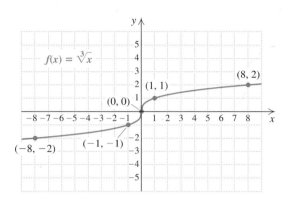

In the real-number system, every number has exactly one cube root. Thus the domain of $f(x) = \sqrt[3]{x}$ is $(-\infty, \infty)$. The cube root of a positive number is positive, the cube root of a negative number is negative, and the cube root of 0 is 0.

Interactive Discovery

Use graphs or tables to determine which of the following are identities. Be sure to enclose the entire radicand in parentheses.

1. $\sqrt[3]{x^3} = x$ **2.** $\sqrt[3]{x^3} = -x$

3. $\sqrt[3]{x^3} = |x|$ **4.** $\sqrt[3]{(x-1)^3} = x - 1$

5. $\sqrt[3]{(x-1)^3} = |x - 1|$

We see that although $\sqrt{x^2} = |x|$, $\sqrt[3]{x^3} \neq |x|$. In fact, $\sqrt[3]{x^3} = x$. **In simplifying expressions involving cube roots, we do not use absolute-value signs.**

Example 7 For each function, find the indicated function value.

a) $f(t) = \sqrt[3]{t};\ f(125)$ **b)** $g(x) = \sqrt[3]{x - 3};\ g(-24)$

SOLUTION

a) $f(125) = \sqrt[3]{125} = 5$ Since $5 \cdot 5 \cdot 5 = 125$

b) $g(-24) = \sqrt[3]{-24 - 3}$
$$= \sqrt[3]{-27}$$
$$= -3 \quad \text{Since } (-3)(-3)(-3) = -27$$

Example 8 Simplify: $\sqrt[3]{-8y^3}$.

SOLUTION

$$\sqrt[3]{-8y^3} = -2y \quad \text{Since } (-2y)(-2y)(-2y) = -8y^3$$

Odd and Even nth Roots

The fifth root of a number a is the number c for which $c^5 = a$. There are also 6th roots, 7th roots, and so on. We write $\sqrt[n]{a}$ for the nth root. The number n is called the *index* (plural, *indices*). When the index is 2, we do not write it.

If n is odd, we say that we are taking an *odd root*. Every number has just one real root when n is odd. Odd roots of positive numbers are positive and odd roots of negative numbers are negative. Absolute-value signs are not used when finding odd roots.

Example 9 Find each of the following:

a) $\sqrt[5]{32}$;
b) $\sqrt[5]{-32}$;
c) $-\sqrt[5]{32}$;
d) $-\sqrt[5]{-32}$;
e) $\sqrt[7]{x^7}$;
f) $\sqrt[9]{(x-1)^9}$.

SOLUTION

a) $\sqrt[5]{32} = 2$ Since $2^5 = 32$
b) $\sqrt[5]{-32} = -2$ Since $(-2)^5 = -32$
c) $-\sqrt[5]{32} = -2$ Taking the opposite of $\sqrt[5]{32}$
d) $-\sqrt[5]{-32} = -(-2) = 2$ Taking the opposite of $\sqrt[5]{-32}$
e) $\sqrt[7]{x^7} = x$
f) $\sqrt[9]{(x-1)^9} = x - 1$

When the index n in $\sqrt[n]{a}$ is an even number, we say that we are taking an *even root*. Every positive real number has two real nth roots when n is even. One root is positive and one is negative. Negative numbers do not have real nth roots when n is even.

When n is even and a is positive, the notation $\sqrt[n]{a}$ indicates the positive nth root. Thus, when we are finding even nth roots, absolute-value signs are often necessary.

Example 10 Simplify each expression. Assume that variables can represent any real number.

a) $\sqrt[4]{16}$
b) $-\sqrt[4]{16}$
c) $\sqrt[4]{-16}$
d) $\sqrt[4]{81x^4}$
e) $\sqrt[6]{(y+7)^6}$

SOLUTION

a) $\sqrt[4]{16} = 2$ Since $2^4 = 16$
b) $-\sqrt[4]{16} = -2$ Taking the opposite of $\sqrt[4]{16}$
c) $\sqrt[4]{-16}$ cannot be simplified.   No real-number even root exists.
d) $\sqrt[4]{81x^4} = 3|x|$ Using absolute-value notation since x could represent a negative number
e) $\sqrt[6]{(y+7)^6} = |y+7|$ Using absolute-value notation since $y+7$ could be negative

---

Simplifying $\sqrt[n]{a^n}$

For any real number a:

a) $\sqrt[n]{a^n} = |a|$ when n is even. Unless a is known to be nonnegative, absolute-value notation is needed when n is even.

b) $\sqrt[n]{a^n} = a$ when n is odd. Absolute-value notation is not used when n is odd.

Note that $\sqrt[n]{0} = 0$ for any index n.

A *radical function* is a function that can be described by a radical expression. If a function is given by a radical expression with an odd index, the domain is the set of all real numbers. If a function is given by a radical expression with an even index, the domain is the set of replacements for which the radicand is nonnegative.

Example 11   Determine the domain of $g(x) = \sqrt[6]{7 - 3x}$.

SOLUTION   Since the index is even, the radicand, $7 - 3x$, must be nonnegative. We solve the inequality:

$$7 - 3x \geq 0 \qquad \text{We cannot find the 6th root of a negative number.}$$
$$-3x \geq -7$$
$$x \leq \tfrac{7}{3}. \qquad \text{Multiplying by } -\tfrac{1}{3} \text{ on both sides and reversing the inequality}$$

Thus,

$$\text{Domain of } g = \left\{ x \,\middle|\, x \leq \tfrac{7}{3} \right\}$$
$$= \left(-\infty, \tfrac{7}{3} \right].$$

Models Using Radical Functions

Some situations can be modeled using radical functions. The graphs of radical functions can have many different shapes. However, radical functions of the form

$$r(x) = \sqrt{ax + b}$$

will have the general shape of the graph of $f(x) = \sqrt{x}$.

We can determine whether a radical function might fit a set of data by plotting the points.

Example 12   *Air-Show Attendance.* The annual attendance at air shows in the United States has grown during recent years, as shown in the following table. Determine whether a radical function can be used to model this situation.

YEAR	ANNUAL AIR-SHOW ATTENDANCE (IN MILLIONS)
1987	14.1
1988	18.3
1989	22.0
1990	23.1
1991	23.9
1992	24.4

Source: The Macmillan Visual Almanac, 1996

SOLUTION We enter the data and graph the points.

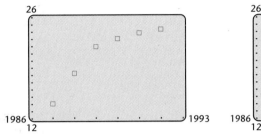 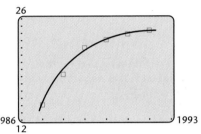

If we sketch a curve through the points, we see that it is the general shape of the graph of $f(x) = \sqrt{x}$. It appears that a radical function can be used to model the data.

7.1 Exercise Set

Find the square roots of each number.

1. 16

2. 225

3. 144

4. 9

5. 400

6. 81

7. 49

8. 900

Simplify.

9. $-\sqrt{\dfrac{49}{36}}$

10. $-\sqrt{\dfrac{361}{9}}$

11. $\sqrt{196}$

12. $\sqrt{441}$

13. $-\sqrt{\dfrac{16}{81}}$

14. $-\sqrt{\dfrac{81}{144}}$

15. $\sqrt{0.09}$

16. $\sqrt{0.36}$

17. $-\sqrt{0.0049}$

18. $\sqrt{0.0144}$

Identify the radicand and the index of each expression.

19. $5\sqrt{p^2 + 4}$

20. $-7\sqrt{y^2 - 8}$

21. $x^2y^3 \sqrt[3]{\dfrac{x}{y + 4}}$

22. $a^2b^3 \sqrt[3]{\dfrac{a}{a^2 - b}}$

For each function, find the specified function value, if it exists.

23. $f(y) = \sqrt{5y - 10}$; $f(6), f(2), f(1), f(-1)$

24. $g(x) = \sqrt{x^2 - 25}$; $g(-6), g(3), g(6), g(13)$

25. $p(z) = \sqrt{2z^2 - 20}$; $p(4), p(3), p(-5), p(0)$

26. $g(x) = -\sqrt{(x + 1)^2}$; $g(-3), g(4), g(-5)$

27. $g(x) = \sqrt{x^3 + 9}$; $g(-2), g(-3), g(3)$

28. $f(t) = \sqrt{t^3 - 10}$; $f(2), f(3), f(4)$

Simplify. Assume that variables can represent any real number.

29. $\sqrt{25t^2}$

30. $\sqrt{16x^2}$

31. $\sqrt{(-6b)^2}$

32. $\sqrt{(-7c)^2}$

33. $\sqrt{(5-b)^2}$

34. $\sqrt{(a+1)^2}$

35. $\sqrt{y^2 + 16y + 64}$

36. $\sqrt{x^2 - 4x + 4}$

37. $\sqrt{9x^2 - 30x + 25}$

38. $\sqrt{4x^2 + 28x + 49}$

39. $-\sqrt[4]{256}$

40. $\sqrt[4]{625}$

41. $-\sqrt[5]{7^5}$

42. $\sqrt[5]{-1}$

43. $\sqrt[5]{-\dfrac{1}{32}}$

44. $\sqrt[5]{-\dfrac{32}{243}}$

45. $\sqrt[8]{y^8}$

46. $\sqrt[6]{x^6}$

47. $\sqrt[4]{(7b)^4}$

48. $\sqrt[4]{(5a)^4}$

49. $\sqrt[12]{(-10)^{12}}$

50. $\sqrt[10]{(-6)^{10}}$

51. $\sqrt[1976]{(2a+b)^{1976}}$

52. $\sqrt[414]{(a+b)^{414}}$

53. $\sqrt{x^{12}}$

54. $\sqrt{a^{22}}$

55. $\sqrt{a^{14}}$

56. $\sqrt{x^{16}}$

Simplify. Assume that no radicands were formed by raising negative quantities to even powers.

57. $\sqrt{25t^2}$

58. $\sqrt{16x^2}$

59. $\sqrt{(7c)^2}$

60. $\sqrt{(6b)^2}$

61. $\sqrt{(5+b)^2}$

62. $\sqrt{(a+1)^2}$

63. $\sqrt{9x^2 + 36x + 36}$

64. $\sqrt{4x^2 + 8x + 4}$

65. $-\sqrt[3]{64}$

66. $\sqrt[3]{27}$

67. $\sqrt[4]{81x^4}$

68. $\sqrt[4]{16x^4}$

69. $-\sqrt[5]{-100,000}$

70. $\sqrt[3]{-216}$

71. $-\sqrt[3]{-64x^3}$

72. $-\sqrt[3]{-125y^3}$

73. $\sqrt{a^{14}}$

74. $\sqrt{a^{22}}$

75. $\sqrt{(x+3)^{10}}$

76. $\sqrt{(x-2)^8}$

For each function, find the specified function value, if it exists.

77. $f(x) = \sqrt[3]{x+1}$; $f(7), f(26), f(-9), f(-65)$

78. $g(x) = -\sqrt[3]{2x-1}$; $g(0), g(-62), g(-13), g(63)$

79. $g(t) = \sqrt[4]{t-3}$; $g(19), g(-13), g(1), g(84)$

80. $f(t) = \sqrt[4]{t+1}$; $f(0), f(15), f(-82), f(80)$

Determine the domain of each function described.

81. $f(x) = \sqrt{x-5}$

82. $g(x) = \sqrt{x+8}$

83. $g(x) = \sqrt[4]{5-x}$

84. $g(t) = \sqrt[3]{2t-5}$

85. $f(t) = \sqrt[5]{2t+9}$

86. $f(t) = \sqrt[6]{2t+5}$

87. $h(z) = -\sqrt[6]{5z+3}$

88. $d(x) = -\sqrt[4]{7x-5}$

Determine algebraically the domain of each function described. Then use a grapher to confirm your answer

and to estimate the range.

89. $f(x) = \sqrt{5-x}$

90. $g(x) = \sqrt{2x+1}$

91. $f(t) = 1 - \sqrt{x+1}$

92. $g(t) = 2 + \sqrt{3x-5}$

93. $g(x) = 3 + \sqrt{x^2+4}$

94. $f(x) = 5 - \sqrt{3x^2+1}$

In Exercises 95–98, match each function with one of the following graphs.

a)

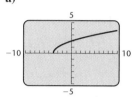

b)

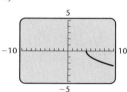

c)

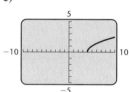

d)

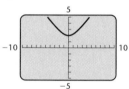

95. $f(x) = \sqrt{x-4}$

96. $g(x) = \sqrt{x+4}$

97. $h(x) = \sqrt{x^2+4}$

98. $f(x) = -\sqrt{x-4}$

In Exercises 99–102, determine whether a radical function would be a good model of the given situation.

99. *Farm Size.* The average size of United States' farms increased during the last half of the 20th century.

YEAR	AVERAGE SIZE OF FARM (IN ACRES)
1950	215
1960	302
1970	390
1980	445
1991	467
1992	468
1993	473

Source: The Macmillan Visual Almanac, 1996

100. *Wind Chill.* When the wind is blowing, the air temperature feels lower than the actual temperature. This is referred to as the wind-chill temperature. The following table lists wind-chill temperatures for various wind speeds at a thermometer reading of 15°F.

WIND SPEED (IN MILES PER HOUR)	WIND-CHILL TEMPERATURE (IN DEGREES FAHRENHEIT)
5	10
10	0
15	−10
20	−15
25	−20
30	−25
35	−30
40	−30

Source: The Handy Science Answer Book. Visible Ink Press, 1994.

101. *Telecommunications.* The total revenue of the telecommunications software market was projected in 1997 to increase to $19.6 billion by 2001.

YEAR	ANNUAL REVENUE (IN BILLIONS)
1997	$10.5
1998	13.2
1999	15.1
2000	17.5
2001	19.6

Source: USA Today, 1/23/98

102. *Electric Cars.* The number of cars powered by electricity increased between 1992 and 1996.

YEAR	NUMBER OF ELECTRIC CARS IN THE UNITED STATES
1992	1607
1993	1690
1994	2224
1995	2860
1996	3306

Source: Information Please Almanac, 1998

Skill Maintenance

Simplify. [1.4]

103. $(a^3 b^2 c^5)^3$

104. $(5a^7 b^8)(2a^3 b)$

Multiply. [5.2]

105. $(x - 3)(x + 3)$

106. $(a + bx)(a - bx)$

107. $(2x + 1)(x^2 - 3x + 1)$ **108.** $(5 - 2x + x^2)(x - 1)$

Synthesis

109. ◈ If the domain of $f = [1, \infty)$ and the range of $f = [2, \infty)$, find a possible expression for $f(x)$ and explain how such an expression is formulated.

110. ◈ Kelly obtains the following graph of
$$f(x) = \sqrt{x^2 - 4x - 12}$$
and concludes that the domain of f is $(-\infty, -2]$. Is she correct? If not, what mistake is she making?

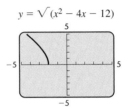

$y = \sqrt{(x^2 - 4x - 12)}$

111. ◈ Can the following situation be modeled using a radical function? Why or why not? "For each year, the yield increases. The amount of increase is smaller each year."

112. ◈ Can the following situation be modeled using a radical function? Why or why not? "For each year, the costs increase. The amount of increase is the same each year."

113. *Spaces in a Parking Lot.* A parking lot has attendants to park the cars. The number N of stalls needed for waiting cars before attendants can get to them is given by the formula $N = 2.5\sqrt{A}$, where A is the number of arrivals in peak hours. Find the number of spaces needed for the given number of arrivals in peak hours: **(a)** 25; **(b)** 36; **(c)** 49; **(d)** 64.

114. Find the domain of f if
$$f(x) = \frac{\sqrt{x + 3}}{\sqrt[4]{2 - x}}.$$

115. Find the domain of g if
$$g(x) = \frac{\sqrt[4]{5 - x}}{\sqrt[6]{x + 4}}.$$

7.2

Rational Numbers as Exponents

- *Rational Exponents*
- *Negative Rational Exponents*
- *Laws of Exponents*
- *Simplifying Radical Expressions*

In Section 1.1, we considered the natural numbers as exponents. Our discussion of exponents was expanded to include all integers in Section 1.4. In this section, we expand the study still further—to include all rational numbers. This will give meaning to expressions like $a^{1/3}$, $7^{-1/2}$, and $(3x)^{4/5}$. Such notation will help us simplify certain radical expressions.

Rational Exponents

Consider $a^{1/2} \cdot a^{1/2}$. If we still want to add exponents when multiplying, it must follow that $a^{1/2} \cdot a^{1/2} = a^{1/2 + 1/2}$, or a^1. This suggests that $a^{1/2}$ is a square root of a. Similarly, $a^{1/3} \cdot a^{1/3} \cdot a^{1/3} = a^{1/3+1/3+1/3}$, or a^1, so $a^{1/3}$ should mean $\sqrt[3]{a}$.

$$a^{1/n} = \sqrt[n]{a}$$

$a^{1/n}$ means $\sqrt[n]{a}$. When a is nonnegative, n can be any index. When a is negative, n must be odd.

Example 1   Rewrite without rational exponents: **(a)** $x^{1/2}$; **(b)** $(-8)^{1/3}$; **(c)** $(abc)^{1/5}$.

SOLUTION

a) $x^{1/2} = \sqrt{x}$

b) $(-8)^{1/3} = \sqrt[3]{-8} = -2$

c) $(abc)^{1/5} = \sqrt[5]{abc}$ ▬

Example 2   Rewrite with rational exponents: **(a)** $\sqrt[5]{7xy}$; **(b)** $\sqrt[7]{x^3y/9}$.

SOLUTION   Parentheses are required to indicate the base.

a) $\sqrt[5]{7xy} = (7xy)^{1/5}$

b) $\sqrt[7]{\dfrac{x^3y}{9}} = \left(\dfrac{x^3y}{9}\right)^{1/7}$ ▬

We can graph a radical function like the one described by $f(x) = \sqrt[4]{2x - 7}$ using rational exponents. When entering the radical expression, we use the ☐ key before the exponent. Both the exponent and the radicand should be placed in parentheses.

Example 3   Graph: $f(x) = \sqrt[4]{2x - 7}$.

SOLUTION   We rewrite the radical expression using a rational exponent:
$$f(x) = \sqrt[4]{2x - 7} = (2x - 7)^{1/4}.$$

Then we let $y = (2x - 7)^\wedge(1/4)$. Since $1/4 = 0.25$, this could also be written $y = (2x - 7)^\wedge(0.25)$. Be sure to enclose both the radicand and the exponent in parentheses.

Knowing the domain of the function can help us determine an appropriate viewing window. Since the index is even, the domain is the set of all x for which the radicand is nonnegative, or $\left[\frac{7}{2}, \infty\right)$. We choose a viewing window of $[-1, 10, -1, 5]$. This will show the axes and the first quadrant.

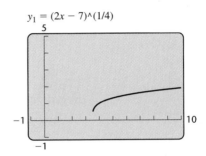

$y_1 = (2x - 7)^\wedge(1/4)$

How should we define $a^{2/3}$? If the property for multiplying exponents is to hold, we must have $a^{2/3} = (a^{1/3})^2$ and $a^{2/3} = (a^2)^{1/3}$. This would suggest that $a^{2/3} = (\sqrt[3]{a})^2$ and $a^{2/3} = \sqrt[3]{a^2}$. We make our definition accordingly.

Positive Rational Exponents

For any natural numbers m and n ($n \neq 1$) and any real number a for which $\sqrt[n]{a}$ exists,

$$a^{m/n} \quad \text{means} \quad (\sqrt[n]{a})^m, \quad \text{or} \quad \sqrt[n]{a^m}.$$

Example 4   Rewrite without rational exponents and simplify: **(a)** $27^{2/3}$; **(b)** $25^{3/2}$.

SOLUTION

a) $27^{2/3} = \sqrt[3]{27^2}$, or $(\sqrt[3]{27})^2$ It is easier to simplify using $(\sqrt[3]{27})^2$.
     $= 3^2$, or 9

b) $25^{3/2} = \sqrt[2]{25^3}$, or $(\sqrt{25})^3$ We generally omit the index 2.
     $= 5^3$, or 125   Taking the square root and cubing

Example 5   Rewrite with rational exponents: **(a)** $\sqrt[3]{9^4}$; **(b)** $(\sqrt[4]{7xy})^5$.

SOLUTION

a) $\sqrt[3]{9^4} = 9^{4/3}$

b) $(\sqrt[4]{7xy})^5 = (7xy)^{5/4}$ The index becomes the denominator of the rational exponent.

Rational roots of numbers can be approximated on a calculator.

Example 6 Approximate $\sqrt[5]{(-23)^3}$. Round to the nearest thousandth.

SOLUTION We first rewrite the expression using a rational exponent:

$$\sqrt[5]{(-23)^3} = (-23)^{3/5}.$$

Using a calculator, we have

$$(-23)^\wedge(3/5) \approx -6.562.$$

Negative Rational Exponents

Recall that $x^{-2} = \dfrac{1}{x^2}$. Negative rational exponents behave similarly.

Negative Rational Exponents

For any rational number m/n and any nonzero real number a for which $a^{m/n}$ exists,

$$a^{-m/n} \quad \text{means} \quad \frac{1}{a^{m/n}}.$$

Caution! A negative exponent does not indicate that the expression in which it appears is negative.

Example 7 Rewrite with positive exponents and, if possible, simplify.

a) $9^{-1/2}$ b) $(5xy)^{-4/5}$ c) $64^{-2/3}$

d) $4x^{-2/3}y^{1/5}$ e) $\left(\dfrac{3r}{7s}\right)^{-5/2}$

SOLUTION

a) $9^{-1/2} = \dfrac{1}{9^{1/2}}$ $9^{-1/2}$ is the reciprocal of $9^{1/2}$.

Since $9^{1/2} = \sqrt{9} = 3$, the answer simplifies to $\dfrac{1}{3}$.

b) $(5xy)^{-4/5} = \dfrac{1}{(5xy)^{4/5}}$ $(5xy)^{-4/5}$ is the reciprocal of $(5xy)^{4/5}$.

c) $64^{-2/3} = \dfrac{1}{64^{2/3}}$ $64^{-2/3}$ is the reciprocal of $64^{2/3}$.

Since $64^{2/3} = (\sqrt[3]{64})^2 = 4^2 = 16$, the answer simplifies to $\dfrac{1}{16}$.

d) $4x^{-2/3}y^{1/5} = 4 \cdot \dfrac{1}{x^{2/3}} \cdot y^{1/5} = \dfrac{4y^{1/5}}{x^{2/3}}$

e) In Section 1.4, we found that $(a/b)^{-n} = (b/a)^n$. This property holds for *any* negative exponent:

$$\left(\frac{3r}{7s}\right)^{-5/2} = \left(\frac{7s}{3r}\right)^{5/2}.$$

Finding the reciprocal of the base and changing the sign of the exponent

Laws of Exponents

The same laws hold for rational exponents as for integer exponents.

Laws of Exponents

For any real numbers a and b and any rational exponents m and n for which a^m, a^n, and b^m are defined:

1. $a^m \cdot a^n = a^{m+n}$ In multiplying, add exponents if the bases are the same.

2. $\dfrac{a^m}{a^n} = a^{m-n}$ In dividing, subtract exponents if the bases are the same. (Assume $a \neq 0$.)

3. $(a^m)^n = a^{m \cdot n}$ To raise a power to a power, multiply the exponents.

4. $(ab)^m = a^m b^m$ To raise a product to a power, raise each factor to the power and multiply.

Example 8   Use the laws of exponents to simplify.

a) $3^{1/5} \cdot 3^{3/5}$ **b)** $a^{1/4}/a^{1/2}$

c) $(7.2^{2/3})^{3/4}$ **d)** $(a^{-1/3}b^{2/5})^{1/2}$

SOLUTION

a) $3^{1/5} \cdot 3^{3/5} = 3^{1/5+3/5} = 3^{4/5}$ **Adding exponents**

b) $\dfrac{a^{1/4}}{a^{1/2}} = a^{1/4-1/2} = a^{1/4-2/4}$ **Subtracting exponents after finding a common denominator**

$$= a^{-1/4}, \text{ or } \frac{1}{a^{1/4}}$$

c) $(7.2^{2/3})^{3/4} = 7.2^{2/3 \cdot 3/4} = 7.2^{6/12}$ **Multiplying exponents**

$$= 7.2^{1/2}$$ **Using arithmetic to simplify the exponent**

d) $(a^{-1/3}b^{2/5})^{1/2} = a^{-1/3 \cdot 1/2} \cdot b^{2/5 \cdot 1/2}$ **Raising a product to a power and multiplying exponents**

$$= a^{-1/6}b^{1/5}, \text{ or } \frac{b^{1/5}}{a^{1/6}}$$

Simplifying Radical Expressions

Many radical expressions can be simplified using rational exponents.

To Simplify Radical Expressions:

1. Convert radical expressions to exponential expressions.

2. Use arithmetic and the laws of exponents to simplify.

3. Convert back to radical notation when appropriate.

Example 9 Use rational exponents to simplify.

a) $\sqrt[6]{(5x)^3}$ **b)** $\sqrt[5]{t^{20}}$

c) $(\sqrt[3]{ab^2c})^{12}$ **d)** $\sqrt{\sqrt[3]{x}}$

$y_1 = (5x)^\wedge(3/6),\ \ y_2 = \sqrt{(5x)}$

X	Y1	Y2
0	0	0
1	2.2361	2.2361
2	3.1623	3.1623
3	3.873	3.873
4	4.4721	4.4721
5	5	5
6	5.4772	5.4772

X = 0

SOLUTION

a) $\sqrt[6]{(5x)^3} = (5x)^{3/6}$ **Converting to exponential notation**

$\qquad\qquad = (5x)^{1/2}$ **Simplifying the exponent**

$\qquad\qquad = \sqrt{5x}$ **Returning to radical notation**

To check on a grapher, we let $y_1 = (5x)^\wedge(3/6)$ and $y_2 = \sqrt{(5x)}$, and compare values in a table. If we scroll through the table (see the figure at left), we see that $y_1 = y_2$, so our simplification is probably correct.

b) $\sqrt[5]{t^{20}} = t^{20/5}$ **Converting to exponential notation**

$\qquad = t^4$ **Simplifying the exponent**

c) $(\sqrt[3]{ab^2c})^{12} = (ab^2c)^{12/3}$ **Converting to exponential notation**

$\qquad\qquad = (ab^2c)^4$ **Simplifying the exponent**

$\qquad\qquad = a^4b^8c^4$ **Using the laws of exponents**

$y_1 = \sqrt{(\sqrt[3]{(x)})},\ \ y_2 = x^\wedge(1/6)$

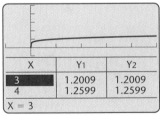

X	Y1	Y2
3	1.2009	1.2009
4	1.2599	1.2599

X = 3

d) $\sqrt{\sqrt[3]{x}} = \sqrt{x^{1/3}}$ **Converting the radicand to exponential notation**

$\qquad = (x^{1/3})^{1/2}$ **Try to go directly to this step.**

$\qquad = x^{1/6}$ **Using the laws of exponents**

$\qquad = \sqrt[6]{x}$ **Returning to radical notation**

We can check by graphing $y_1 = \sqrt{(\sqrt[3]{(x)})}$ and $y_2 = x^\wedge(1/6)$. The graphs coincide, as we see by scrolling through the table of values shown at left.   —

7.2 Exercise Set

Note: Assume for all exercises that even roots are of nonnegative quantities and that all denominators are nonzero.

Rewrite without rational exponents and, if possible, simplify.

1. $x^{1/4}$ **2.** $y^{1/5}$ **3.** $16^{1/2}$

4. $8^{1/3}$ **5.** $81^{1/4}$ **6.** $64^{1/6}$

7. $(xyz)^{1/3}$ **8.** $(ab)^{1/4}$ **9.** $(a^2b^2)^{1/5}$

10. $(x^3y^3)^{1/4}$ **11.** $a^{2/3}$ **12.** $b^{3/2}$

13. $16^{3/4}$ **14.** $4^{7/2}$ **15.** $49^{3/2}$

16. $9^{5/2}$ **17.** $(81x)^{3/4}$ **18.** $(125a)^{2/3}$

19. $(25x^4)^{3/2}$ **20.** $(9y^6)^{3/2}$

Rewrite with rational exponents.

21. $\sqrt[3]{20}$ **22.** $\sqrt[3]{19}$ **23.** $\sqrt{17}$

24. $\sqrt{6}$ **25.** $\sqrt{x^3}$ **26.** $\sqrt{a^5}$

27. $\sqrt[5]{m^2}$ **28.** $\sqrt[5]{n^4}$ **29.** $\sqrt[4]{cd}$ **91.** $\sqrt{(ab)^6}$ **92.** $\sqrt[4]{(xy)^{12}}$

30. $\sqrt[5]{xy}$ **31.** $\sqrt[5]{xy^2z}$ **32.** $\sqrt[3]{x^3y^2z^2}$ **93.** $(\sqrt[3]{x^2y^5})^{12}$ **94.** $(\sqrt[5]{a^2b^4})^{15}$

33. $(\sqrt{3mn})^3$ **34.** $(\sqrt[3]{7xy})^4$ **35.** $(\sqrt{8x^2y})^5$ **95.** $\sqrt[3]{\sqrt[3]{xy}}$ **96.** $\sqrt[5]{\sqrt{2a}}$

36. $(\sqrt[6]{2a^5b})^7$ **37.** $\dfrac{2x}{\sqrt[3]{z^2}}$ **38.** $\dfrac{3a}{\sqrt[5]{c^2}}$

Rewrite with positive rational exponents.

39. $x^{-1/3}$ **40.** $y^{-1/4}$ **41.** $(2rs)^{-3/4}$

42. $(5xy)^{-5/6}$ **43.** $\left(\dfrac{1}{10}\right)^{-2/3}$ **44.** $\left(\dfrac{1}{8}\right)^{-3/4}$

45. $\dfrac{1}{a^{-5/7}}$ **46.** $\dfrac{1}{a^{-3/5}}$

47. $2a^{3/4}b^{-1/2}c^{2/3}$ **48.** $5x^{-2/3}y^{4/5}z$

49. $\left(\dfrac{7x}{8yz}\right)^{-3/5}$ **50.** $\left(\dfrac{2ab}{3c}\right)^{-5/6}$

51. $\dfrac{7x}{\sqrt[3]{z}}$ **52.** $\dfrac{6a}{\sqrt[4]{b}}$

53. $\dfrac{5a}{3c^{-1/2}}$ **54.** $\dfrac{2z}{5x^{-1/3}}$

Graph.

55. $f(x) = \sqrt[4]{x+7}$ **56.** $g(x) = \sqrt[5]{4-x}$

57. $r(x) = \sqrt[3]{3x-2}$ **58.** $q(x) = \sqrt[6]{2x+3}$

59. $f(x) = \sqrt[6]{x^3}$ **60.** $g(x) = \sqrt[8]{x^2}$

Approximate. Round to the nearest thousandth.

61. $\sqrt[5]{9}$ **62.** $\sqrt[6]{13}$

63. $\sqrt[4]{10}$ **64.** $\sqrt[7]{-127}$

65. $\sqrt[3]{(-3)^5}$ **66.** $\sqrt[10]{(1.5)^6}$

Use the laws of exponents to simplify. Do not use negative exponents in any answers.

67. $5^{3/4} \cdot 5^{1/8}$ **68.** $11^{2/3} \cdot 11^{1/2}$ **69.** $\dfrac{3^{5/8}}{3^{-1/8}}$

70. $\dfrac{8^{7/11}}{8^{-2/11}}$ **71.** $(10^{3/5})^{2/5}$ **72.** $(5^{5/4})^{3/7}$

73. $a^{2/3} \cdot a^{5/4}$ **74.** $x^{3/4} \cdot x^{2/3}$ **75.** $(x^{2/3})^{-3/7}$

76. $(a^{-3/2})^{2/9}$ **77.** $(m^{2/3}n^{-1/4})^{1/2}$

78. $(x^{-1/3}y^{2/5})^{1/4}$

Use rational exponents to simplify. Write the answer in radical notation when appropriate.

79. $\sqrt[6]{a^2}$ **80.** $\sqrt[6]{t^4}$ **81.** $\sqrt[3]{x^{15}}$

82. $\sqrt[4]{a^{12}}$ **83.** $(\sqrt[3]{ab})^{15}$ **84.** $(\sqrt{xy})^{14}$

85. $\sqrt[8]{(3x)^2}$ **86.** $\sqrt[4]{(7a)^2}$ **87.** $(\sqrt[10]{3a})^5$

88. $(\sqrt[8]{2x})^6$ **89.** $\sqrt[4]{\sqrt{x}}$ **90.** $\sqrt[3]{\sqrt[6]{m}}$

Skill Maintenance

Solve.

97. $x^2 - 1 = 8$ [5.6]

98. $3x - 4 = 5x + 7$ [2.2]

99. $\dfrac{1}{x} + 2 = 5$ [6.4]

100. $x^2 = 49$ [5.6]

101. *Real-Estate Taxes.* For homes under $100,000, the real-estate transfer tax in Vermont is 0.5% of the selling price. Find the selling price of a home that had a transfer tax of $467.50. [2.3]

102. What numbers are their own squares? [5.7]

Synthesis

103. ◆ If $f(x) = (x+5)^{1/2}(x+7)^{-1/2}$, find the domain of f. Explain how you found your answer.

104. ◆ Explain why $\sqrt[3]{x^6} = x^2$ for any value of x, whereas $\sqrt[2]{x^6} = x^3$ only when $x \geq 0$.

105. ◆ Let $f(x) = 5x^{-1/3}$. Under what condition will we have $f(x) > 0$? Why?

106. ◆ If $g(x) = x^{1/n}$, in what way does the domain of g depend on whether n is odd or even?

Use rational exponents to simplify.

107. $\sqrt[5]{x^2y\sqrt{xy}}$ **108.** $\sqrt{x\sqrt[3]{x^2}}$

109. $\sqrt[4]{\sqrt[3]{8x^3y^6}}$

110. $\sqrt[12]{p^2 + 2pq + q^2}$

111. *Road-Pavement Messages.* In a psychological study, it was determined that the proper length L of the letters of a word printed on pavement is given by

$$L = \frac{0.000169d^{2.27}}{h},$$

where d is the distance of a car from the lettering and h is the height of the eye above the surface of the road. All units are in meters. This formula says that if a person is h meters above the surface of the road and is to be able to recognize a message d meters away, that message will be the most recognizable if the length of the letters is L. Find L to the nearest tenth of a meter, given d and h.

a) $h = 1$ m, $d = 60$ m
b) $h = 0.9906$ m, $d = 75$ m
c) $h = 2.4$ m, $d = 80$ m
d) $h = 1.1$ m, $d = 100$ m

112. *Dating Fossils.* The function $r(t) = 10^{-12}2^{-t/5700}$ expresses the ratio of carbon isotopes to carbon atoms in a fossil that is t years old. What ratio of carbon isotopes to carbon atoms would be present in a 1900-year-old bone?

113. *Physics.* The equation $m = m_0(1 - v^2c^{-2})^{-1/2}$, developed by Albert Einstein, is used to determine the mass m of an object that is moving v meters per second and has mass m_0 before the motion begins. The constant c is the speed of light, approximately 3×10^8 m/sec. Suppose that a particle with mass 8 mg is accelerated to a speed of $\frac{9}{5} \times 10^8$ m/sec. Without using a calculator, find the new mass of the particle.

114. Use a grapher in the SIMULTANEOUS mode to graph

$$y_1 = x^{1/2}, \qquad y_2 = 3x^{2/5},$$
$$y_3 = x^{4/7}, \quad \text{and} \quad y_4 = \tfrac{1}{5}x^{3/4}.$$

Then, looking only at coordinates, match each graph with its equation.

COLLABORATIVE CORNER

Focus: Functions and rational exponents

Time: 10–20 minutes

Group size: 3

Materials: Grapher and graph paper

In arithmetic, we have seen that $\frac{3}{5}$, $\frac{1}{10} \cdot 6$, and $6 \cdot \frac{1}{10}$ all represent the same number. Interestingly,

$$f(x) = x^{3/5},$$
$$g(x) = (x^{1/10})^6,$$
and $$h(x) = (x^6)^{1/10}$$

represent three *different* functions.

Activity

1. One group member should determine the domain of f and graph the function, a second group member should determine the domain of g and graph the function, and a third group member should determine the domain of h and graph the function. All graphs should be done first by hand, and then using a grapher.

2. Compare the three domains and graphs and check each other's work. How and why do the graphs differ?

3. Decide as a group which graph, if any, would best represent the graph of $k(x) = x^{6/10}$. (*Hint:* Study the definition of $a^{m/n}$ on p. 431 carefully.) Graph $k(x) = x^{6/10}$. Is the graph drawn by the grapher correct? Be prepared to explain your reasoning to the entire class.

7.3
Multiplying, Adding, and Subtracting Radical Expressions

- *Multiplying Radical Expressions*
- *Simplifying by Factoring*
- *Adding and Subtracting Radical Expressions*

Multiplying Radical Expressions

Note that $\sqrt{4}\sqrt{25} = 2 \cdot 5 = 10$. Also $\sqrt{4 \cdot 25} = \sqrt{100} = 10$. Likewise,

$$\sqrt[3]{27}\sqrt[3]{8} = 3 \cdot 2 = 6 \quad \text{and} \quad \sqrt[3]{27 \cdot 8} = \sqrt[3]{216} = 6.$$

These examples suggest the following.

The Product Rule for Radicals

For any real numbers $\sqrt[n]{a}$ and $\sqrt[n]{b}$,

$$\sqrt[n]{a} \cdot \sqrt[n]{b} = \sqrt[n]{a \cdot b}.$$

(To multiply, when the indices match, multiply the radicands.)

Example 1 Multiply.

a) $\sqrt{3} \cdot \sqrt{5}$

b) $\sqrt{x+3}\sqrt{x-3}$

c) $\sqrt[3]{4} \cdot \sqrt[3]{5}$

d) $\sqrt[4]{\dfrac{y}{5}} \cdot \sqrt[4]{\dfrac{7}{x}}$

SOLUTION

a) $\sqrt{3} \cdot \sqrt{5} = \sqrt{3 \cdot 5} = \sqrt{15}$

b) $\sqrt{x+3}\sqrt{x-3} = \sqrt{(x+3)(x-3)}$
$$= \sqrt{x^2 - 9}$$

Caution!
$$\sqrt{x^2 - 9} \neq \sqrt{x^2} - \sqrt{9}.$$

c) $\sqrt[3]{4}\sqrt[3]{5} = \sqrt[3]{4 \cdot 5} = \sqrt[3]{20}$

d) $\sqrt[4]{\dfrac{y}{5}} \cdot \sqrt[4]{\dfrac{7}{x}} = \sqrt[4]{\dfrac{y}{5} \cdot \dfrac{7}{x}} = \sqrt[4]{\dfrac{7y}{5x}}$

Although, as we saw in Example 1(b),
$$\sqrt{x+3}\sqrt{x-3} = \sqrt{x^2 - 9},$$

it is not true that
$$f(x) = \sqrt{x+3}\sqrt{x-3} \quad \text{and} \quad g(x) = \sqrt{x^2 - 9}$$

represent the same function. This is because the domains of the functions are not the same. As you can see from the graphs of the functions shown below, the domain of f is $[3, \infty)$, whereas the domain of g is $(-\infty, -3] \cup [3, \infty)$.

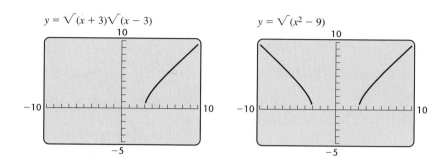

Note that the graphs do coincide where the domains coincide. We say that the expressions are equivalent because they represent the same number for all possible replacements for x in *both* expressions.

It is important to remember that the product rule for radicals applies only when radicals have the same index. When indices differ, rational exponents can be useful. Study the steps in the following example carefully.

Example 2   Multiply: $\sqrt{5x} \cdot \sqrt[4]{3y}$.

SOLUTION

$$\sqrt{5x} \cdot \sqrt[4]{3y} = (5x)^{1/2}(3y)^{1/4} \qquad \text{Converting to exponential notation}$$

$$= (5x)^{2/4}(3y)^{1/4} \qquad \text{Writing exponents with a common denominator. This is important.}$$

$$= [(5x)^2(3y)]^{1/4} \qquad \text{Using the laws of exponents}$$

$$= \sqrt[4]{(25x^2)(3y)} \qquad \text{Squaring } 5x \text{ and returning to radical notation}$$

$$= \sqrt[4]{75x^2y} \qquad \text{Multiplying under the radical}$$

Simplifying by Factoring

An integer p is a *perfect square* if there exists an integer q for which $q^2 = p$. We say that p is a *perfect cube* if $q^3 = p$ for some integer q. In general, p is a *perfect nth power* if $q^n = p$ for some integer q. The product rule allows us to simplify $\sqrt[n]{ab}$ when a or b is a perfect nth power.

Using the Product Rule to Simplify

$\sqrt[n]{ab} = \sqrt[n]{a} \cdot \sqrt[n]{b}$.

Thus to simplify $\sqrt{20}$, we note that 20 has 4, which is a perfect square, as a factor. Therefore,

$$\sqrt{20} = \sqrt{4 \cdot 5} \qquad \text{Factoring the radicand (4 is a perfect square)}$$

$$= \sqrt{4} \cdot \sqrt{5} \qquad \text{Factoring into two radicals}$$

$$= 2\sqrt{5}. \qquad \text{Taking the square root of 4}$$

To Simplify a Radical Expression by Factoring:

1. Express the radicand as the product of two factors, one of which is a perfect nth power (where n is the index).
2. Take the nth root of each factor.
3. Simplification is complete when no radicand has a factor that is a perfect nth power (where n is the index).

Example 3 Simplify by factoring: **(a)** $\sqrt{200}$; **(b)** $\sqrt[3]{32}$; **(c)** $\sqrt[4]{48}$; **(d)** $\sqrt{18x^2 y}$.

SOLUTION

a) $\sqrt{200} = \sqrt{100 \cdot 2} = \sqrt{100} \cdot \sqrt{2} = 10\sqrt{2}$ This is the largest perfect-square factor of 200.

Note: Had we not seen that 100 is the largest perfect-square factor of 200, we could have used the prime factorization:

$$2 \cdot 2 \cdot 2 \cdot 5 \cdot 5.$$

Each pair of identical factors makes a square, so

$$\sqrt{2 \cdot 2 \cdot 2 \cdot 5 \cdot 5} = \sqrt{2^2 \cdot 5^2 \cdot 2}$$

$$= 2 \cdot 5 \cdot \sqrt{2} = 10\sqrt{2}$$

b) $\sqrt[3]{32} = \sqrt[3]{8 \cdot 4} = \sqrt[3]{8} \cdot \sqrt[3]{4} = 2\sqrt[3]{4}$ This is the largest perfect-cube (third power) factor of 32.

c) $\sqrt[4]{48} = \sqrt[4]{16 \cdot 3} = \sqrt[4]{16} \cdot \sqrt[4]{3} = 2\sqrt[4]{3}$ This is the largest fourth-power factor of 48.

d) $\sqrt{18x^2 y} = \sqrt{9x^2 \cdot 2y}$ $9x^2$ is a perfect square.

$$= \sqrt{9x^2} \cdot \sqrt{2y} \qquad \text{Factoring into two radicals}$$

$$= |3x|\sqrt{2y}, \text{ or } 3|x|\sqrt{2y} \qquad \text{Taking the square root of } 9x^2$$

Example 4 If $f(x) = \sqrt{3x^2 - 6x + 3}$, find a simplified form for $f(x)$.

SOLUTION

$$f(x) = \sqrt{3x^2 - 6x + 3}$$
$$= \sqrt{3(x^2 - 2x + 1)}$$
$$= \sqrt{3(x - 1)^2}$$
$$= \sqrt{(x - 1)^2} \cdot \sqrt{3}$$
$$= |x - 1|\sqrt{3}$$

Factoring the radicand

Factoring into two radicals

Taking the square root of $(x - 1)^2$

We can check Example 4 by graphing $y_1 = \sqrt{3x^2 - 6x + 3}$ and $y_2 = |x - 1|\sqrt{3}$.

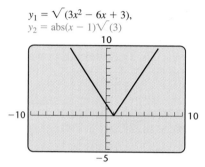

$y_1 = \sqrt{(3x^2 - 6x + 3)}$,
$y_2 = \text{abs}(x - 1)\sqrt{(3)}$

X	Y₁	Y₂
−2	5.1962	5.1962
−1	3.4641	3.4641
0	1.7321	1.7321
1	0	0
2	1.7321	1.7321
3	3.4641	3.4641
4	5.1962	5.1962

X = −2

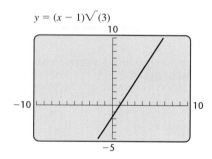

$y = (x - 1)\sqrt{(3)}$

It appears that the graphs coincide and scrolling through the table of values shows us that this is indeed the case. Note from the graph at left that the absolute-value sign is important since the graph of $y = (x - 1)\sqrt{3}$ is different from the graph of $y = \sqrt{3x^2 - 6x + 3}$.

In many situations that do not involve functions, it is safe to assume that no radicands were formed by raising negative quantities to even powers. We now make this assumption and thus discontinue the use of absolute-value notation when taking even roots unless functions are involved.

Example 5 Simplify: **(a)** $\sqrt{x^7y^{11}z^9}$; **(b)** $\sqrt[3]{16a^7b^{14}}$.

SOLUTION

a) There are many ways in which to factor $x^7y^{11}z^9$. Because of the second (square) root, we want to use the largest exponents that are divisible by 2:

$$\sqrt{x^7y^{11}z^9} = \sqrt{x^6 \cdot x \cdot y^{10} \cdot y \cdot z^8 \cdot z}$$

Identifying the largest even powers of x, y, and z

$$= \sqrt{x^6} \sqrt{y^{10}} \sqrt{z^8} \sqrt{xyz}$$

Factoring into several radicals

$$= x^3y^5z^4\sqrt{xyz}.$$

Note that $x^6 = (x^3)^2$, $y^{10} = (y^5)^2$, and $z^8 = (z^4)^2$. Assume $x, y, z \geq 0$.

CHECK: $(x^3y^5z^4\sqrt{xyz})^2 = (x^3)^2(y^5)^2(z^4)^2(\sqrt{xyz})^2$

$\qquad\qquad\qquad = x^6 \cdot y^{10} \cdot z^8 \cdot xyz$

$\qquad\qquad\qquad = x^7y^{11}z^9$

Our check shows that $x^3y^5z^4\sqrt{xyz}$ is the square root of $x^7y^{11}z^9$.

b) There are many ways in which to factor $16a^7b^{14}$. Because of the third (cube) root, we want to use the largest exponents that are divisible by 3:

$\sqrt[3]{16a^7b^{14}} = \sqrt[3]{8 \cdot 2 \cdot a^6 \cdot a \cdot b^{12} \cdot b^2}$ **Identifying the largest perfect-cube factors**

$\qquad\qquad = \sqrt[3]{8} \cdot \sqrt[3]{a^6} \cdot \sqrt[3]{b^{12}} \cdot \sqrt[3]{2ab^2}$ **Factoring into several radicals**

$\qquad\qquad = 2a^2b^4\sqrt[3]{2ab^2}$ **Taking cube roots. Note that $a^6 = (a^2)^3$ and $b^{12} = (b^4)^3$.**

CHECK: $(2a^2b^4\sqrt[3]{2ab^2})^3 = 2^3(a^2)^3(b^4)^3(\sqrt[3]{2ab^2})^3$

$\qquad\qquad\qquad\qquad = 8 \cdot a^6 \cdot b^{12} \cdot 2ab^2 = 16a^7b^{14}$

We see that $2a^2b^4\sqrt[3]{2ab^2}$ is the cube root of $16a^7b^{14}$. ▬

Example 5 shows that to simplify an *n*th root, we search for factors in the radicand that have exponents that are divisible by *n*.

Simplification can also be performed using rational exponents.

Example 6   Simplify: $\sqrt[5]{32r^8t^{12}}$.

SOLUTION   We have

$\sqrt[5]{32r^8t^{12}} = (2^5r^8t^{12})^{1/5}$ **Converting to exponential notation**

$\qquad\qquad = 2^1r^{8/5}t^{12/5}$ **Multiplying exponents**

$\qquad\qquad = 2 \cdot r^{1+3/5} \cdot t^{2+2/5}$ **Writing $\frac{8}{5}$ and $\frac{12}{5}$ as mixed numbers**

$\qquad\qquad = 2 \cdot r^1 \cdot r^{3/5} \cdot t^2 \cdot t^{2/5}$ **Factoring. Study this carefully.**

$\qquad\qquad = 2rt^2(r^3t^2)^{1/5}$ **Using the laws of exponents**

$\qquad\qquad = 2rt^2\sqrt[5]{r^3t^2}$.   **Returning to radical notation**

The check is left to the student. ▬

Adding and Subtracting Radical Expressions

When two radical expressions have the same indices and radicands, they are said to be **like radicals**. Like radicals can be combined (added or subtracted) in much the same way that we combined like terms earlier in this text.

Example 7   Simplify by combining like radical terms.

a) $6\sqrt{7} + 4\sqrt{7}$

b) $\sqrt[3]{2} - 7x\sqrt[3]{2} + 5\sqrt[3]{2}$

c) $6\sqrt[5]{4x} + 3\sqrt[5]{4x} - \sqrt[3]{4x}$

SOLUTION

a) $6\sqrt{7} + 4\sqrt{7} = (6 + 4)\sqrt{7}$ Using the distributive law
(factoring out $\sqrt{7}$)

$= 10\sqrt{7}$ You can think: 6 square roots of
7 plus 4 square roots of 7 results
in 10 square roots of 7.

b) $\sqrt[3]{2} - 7x\sqrt[3]{2} + 5\sqrt[3]{2} = (1 - 7x + 5)\sqrt[3]{2}$ Factoring out $\sqrt[3]{2}$

$= (6 - 7x)\sqrt[3]{2}$ These parentheses are
important!

c) $6\sqrt[5]{4x} + 3\sqrt[5]{4x} - \sqrt[3]{4x} = (6 + 3)\sqrt[5]{4x} - \sqrt[3]{4x}$ Try to do this step
mentally.

$= 9\sqrt[5]{4x} - \sqrt[3]{4x}$ Because the indices
differ, we are done.

Our ability to simplify radical expressions can help us to find like radicals where, at first, it may appear that none exists.

Example 8   Simplify by combining like radical terms, if possible.

a) $3\sqrt{8} - 5\sqrt{2}$

b) $9\sqrt{5} - 4\sqrt{3}$

c) $\sqrt[3]{2x^6y^4} + 7\sqrt[3]{2y}$

SOLUTION

a) $3\sqrt{8} - 5\sqrt{2} = 3\sqrt{4 \cdot 2} - 5\sqrt{2}$

$\left.\begin{array}{l} = 3\sqrt{4} \cdot \sqrt{2} - 5\sqrt{2} \\ = 3 \cdot 2 \cdot \sqrt{2} - 5\sqrt{2} \end{array}\right\}$ Simplifying $\sqrt{8}$

$= 6\sqrt{2} - 5\sqrt{2}$

$= \sqrt{2}$ Combining like radicals

b) $9\sqrt{5} - 4\sqrt{3}$ cannot be simplified.

c) $\sqrt[3]{2x^6y^4} + 7\sqrt[3]{2y} = \sqrt[3]{x^6y^3 \cdot 2y} + 7\sqrt[3]{2y}$

$\left.\begin{array}{l} = \sqrt[3]{x^6y^3} \cdot \sqrt[3]{2y} + 7\sqrt[3]{2y} \\ = x^2y \cdot \sqrt[3]{2y} + 7\sqrt[3]{2y} \end{array}\right\}$ Simplifying $\sqrt[3]{2x^6y^4}$

$= (x^2y + 7)\sqrt[3]{2y}$ Factoring to combine like
radical terms

7.3 | Exercise Set

Multiply.

1. $\sqrt{6} \ \sqrt{7}$

2. $\sqrt{5} \ \sqrt{7}$

3. $\sqrt[3]{2} \ \sqrt[3]{5}$

4. $\sqrt[4]{6} \ \sqrt[4]{3}$

5. $\sqrt{5a} \ \sqrt{3b}$

6. $\sqrt{2x} \ \sqrt{13y}$

7. $\sqrt[5]{9t^2} \ \sqrt[5]{2t}$

8. $\sqrt[5]{8y^3} \ \sqrt[5]{10y}$

9. $\sqrt{x - a} \ \sqrt{x + a}$

10. $\sqrt{y - b} \ \sqrt{y + b}$

11. $\sqrt[3]{0.5x} \ \sqrt[3]{0.2x}$

12. $\sqrt[3]{0.7y} \ \sqrt[3]{0.3y}$

13. $\sqrt{\dfrac{x}{5}} \ \sqrt{\dfrac{3}{y}}$

14. $\sqrt{\dfrac{7}{t}} \ \sqrt{\dfrac{s}{11}}$

15. $\sqrt[7]{\dfrac{x - 3}{4}} \ \sqrt[7]{\dfrac{5}{x + 2}}$

16. $\sqrt[6]{\dfrac{a}{b - 2}} \ \sqrt[6]{\dfrac{3}{b + 2}}$

Use rational exponents to write a single radical expression.

17. $\sqrt[3]{5}\,\sqrt{6}$

18. $\sqrt[3]{7}\cdot\sqrt[4]{5}$

19. $\sqrt{x}\,\sqrt[3]{7y}$ Vote

20. $\sqrt[3]{y}\,\sqrt[5]{3z}$

21. $\sqrt{x}\,\sqrt[3]{x-2}$

22. $\sqrt[4]{3x}\,\sqrt{y+4}$

23. $\sqrt[5]{yx^2}\,\sqrt{xy}$

24. $\sqrt{ab}\,\sqrt[3]{2a^2b^2}$

25. $\sqrt[4]{xy^2}\,\sqrt[3]{x^2y}$

26. $\sqrt[5]{a^2b^3}\,\sqrt[4]{a^2b}$

Simplify by factoring.

27. $\sqrt{27}$

28. $\sqrt{28}$

29. $\sqrt{12}$

30. $\sqrt{45}$

31. $\sqrt{8}$

32. $\sqrt{18}$

33. $\sqrt{44}$

34. $\sqrt{24}$

35. $\sqrt{36a^4b}$

36. $\sqrt{175y^8}$

37. $\sqrt[3]{8x^3y^2}$

38. $\sqrt[3]{27ab^6}$

39. $\sqrt[3]{-16x^6}$

40. $\sqrt[3]{-32a^6}$

Find a simplified form of f(x). Assume that x can be any real number.

41. $f(x)=\sqrt[3]{125x^5}$

42. $f(x)=\sqrt[3]{16x^6}$

43. $f(x)=\sqrt{49(x+5)^2}$

44. $f(x)=\sqrt{81(x-1)^2}$

45. $f(x)=\sqrt{5x^2-10x+5}$

46. $f(x)=\sqrt{2x^2+8x+8}$

Simplify. Assume that no radicands were formed by raising negative numbers to even powers.

47. $\sqrt{a^3b^4}$

48. $\sqrt{x^6y^9}$

49. $\sqrt[3]{x^5y^6z^{10}}$

50. $\sqrt[3]{a^6b^7c^{13}}$

51. $\sqrt[5]{-32a^7b^{11}}$

52. $\sqrt[4]{16x^5y^{11}}$

53. $\sqrt[5]{a^6b^{12}c^7}$

54. $\sqrt[5]{x^{13}y^8z^{17}}$

55. $\sqrt[4]{810x^9}$

56. $\sqrt[3]{-80a^{14}}$

Add or subtract. Simplify by combining like radical terms, if possible. Assume that all variables and radicands represent nonnegative numbers.

57. $3\sqrt{7}+2\sqrt{7}$

58. $8\sqrt{5}+9\sqrt{5}$

59. $4\sqrt[3]{y}+9\sqrt[3]{y}$

60. $9\sqrt[4]{t}-3\sqrt[4]{t}$

61. $8\sqrt{2}-6\sqrt{2}+5\sqrt{2}$

62. $2\sqrt{6}+8\sqrt{6}-3\sqrt{6}$

63. $9\sqrt[3]{7}-\sqrt{3}+4\sqrt[3]{7}+2\sqrt{3}$

64. $5\sqrt{7}-8\sqrt[4]{11}+\sqrt{7}+9\sqrt[4]{11}$

65. $8\sqrt{27}-3\sqrt{3}$

66. $9\sqrt{50}-4\sqrt{2}$

67. $3\sqrt{45}+7\sqrt{20}$

68. $5\sqrt{12}+16\sqrt{27}$

69. $3\sqrt[3]{16}+\sqrt[3]{54}$

70. $\sqrt[3]{27}-5\sqrt[3]{8}$

71. $\sqrt{5a}+2\sqrt{45a^3}$

72. $4\sqrt{3x^3}-\sqrt{12x}$

73. $\sqrt[3]{6x^4}+\sqrt[3]{48x}$

74. $\sqrt[3]{54x}-\sqrt[3]{2x^4}$

75. $\sqrt{4a-4}+\sqrt{a-1}$

76. $\sqrt{9y+27}+\sqrt{y+3}$

77. $\sqrt{x^3-x^2}+\sqrt{9x-9}$

78. $\sqrt{4x-4}=\sqrt{x^3-x^2}$

Find a simplified form for f(x). Assume x ≥ 0.

79. $f(x)=\sqrt{20x^2+4x^3}-3x\sqrt{45+9x}+\sqrt{5x^2+x^3}$

80. $f(x)=\sqrt{x^3-x^2}+\sqrt{9x^3-9x^2}-\sqrt{4x^3-4x^2}$

81. $f(x)=\sqrt[4]{x^5-x^4}+3\sqrt[4]{x^9-x^8}$

82. $f(x)=\sqrt[4]{16x^4+16x^5}-2\sqrt[4]{x^8+x^9}$

Not on the Test

Skill Maintenance

83. During a one-hour television show, there were 12 commercials. Some of the commercials were 30 sec long and the others were 60 sec long. If the number of 30-sec commercials was 6 less than the total number of minutes of commercial time during the show, how many 60-sec commercials were used? [3.3]

Add. [5.1]

84. $(5x^3-4x^2+x-7)+(9x^2-4x-9)$

85. $(7a^3b^2+5a^2b^2-a^2b)+(2a^3b^2-7a^2b+3ab)$

86. Multiply: $(2x-3)(2x+3)$. [5.2]

Factor.

87. $4x^2-49$ [5.6]

88. $2x^2-26x+72$ [5.4]

Synthesis

89. ◈ Why do we need to know how to multiply radical expressions before learning how to add them?

90. ◈ In what way(s) is combining like radical terms the same as combining like terms that are monomials?

91. ◈ Is the equation $\sqrt{(2x+3)^8}=(2x+3)^4$ always, sometimes, or never true? Why?

92. ◈ Rony is puzzled. When he uses a grapher to graph $y = \sqrt{x} \cdot \sqrt{x}$, he gets the following screen. Explain why Rony did not get the complete line $y = x$.

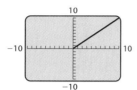

93. *Speed of a Skidding Car.* Police can estimate the speed at which a car was traveling by measuring its skid marks. The function

$$r(L) = 2\sqrt{5L}$$

can be used, where L is the length of a skid mark, in feet, and $r(L)$ is the speed, in miles per hour. Find the exact speed and an estimate (to the nearest tenth mile per hour) for the speed of a car that left skid marks **(a)** 20 ft long; **(b)** 70 ft long; **(c)** 90 ft long.

94. *Wind-Chill Temperature.* When the temperature is T degrees Celsius and the wind speed is v meters per second, the *wind-chill temperature, T_w,* is the temperature that it feels like if there were no wind. A formula for finding wind-chill temperature is

$$T_w = 33 - \frac{(10.45 + 10\sqrt{v} - v)(33 - T)}{22}.$$

Estimate the wind-chill temperature (to the nearest tenth of a degree) for the given actual temperatures and wind speeds.

a) $T = 7°C$, $v = 8$ m/sec
b) $T = 0°C$, $v = 12$ m/sec
c) $T = -5°C$, $v = 14$ m/sec
d) $T = -23°C$, $v = 15$ m/sec

Simplify. Assume that all radicands are nonnegative.

95. $(\sqrt{r^3 t})^7$

96. $(\sqrt[3]{25x^4})^4$

97. $\frac{1}{2}\sqrt{36a^5bc^4} - \frac{1}{2}\sqrt[3]{64a^4bc^6} + \frac{1}{6}\sqrt{144a^3bc^6}$

98. $7x\sqrt{(x+y)^3} - 5xy\sqrt{x+y} - 2y\sqrt{(x+y)^3}$

99. Graph the function given by

$$f(x) = \sqrt{(x-2)^2}.$$

What is the domain of f?

100. Graph the function given by

$$g(x) = \sqrt{(x+3)^2}.$$

What is the domain of g?

COLLABORATIVE CORNER

Focus: Radical equations and problem solving

Time: 15–25 minutes

Group size: 2–3

Materials: Calculators

The faster a car is traveling, the more distance it needs in order to come to a complete stop. Thus it is important for drivers to allow sufficient space between their vehicle and the vehicle in front of them. Police recommend that for each 10 mph of speed, a driver allow 1 car length. Thus a driver going 30 mph should have at least 3 car lengths between his or her vehicle and the one in front.

In Exercise 93, the function $r(L) = 2\sqrt{5L}$ was used to find the speed, in miles per hour, that a car was traveling when it left skid marks L feet long.

Activity

1. Each group member should estimate the length of a car in which he or she frequently travels. (Each should use a different length, if possible.)

2. Using a calculator as needed, each group member should complete the following table. Column 1 gives a car's speed s, column 2 lists the minimum amount of space between cars traveling s miles per hour, as recommended by police. Column 3 is the speed that a vehicle *could* travel were it forced to stop in the distance listed in column 2, using the above function.

3. Determine whether there are any speeds at which the "1 car length per 10 mph" guideline might not suffice. On what reasoning do you base your answer? Compare tables to determine how car length affects the results. What recommendations would your group make to a new driver?

COLUMN 1 s (IN MILES) PER HOUR)	COLUMN 2 $L(s)$ (IN FEET)	COLUMN 3 $r(L)$ (IN MILES PER HOUR)
20		
30		
40		
50		
60		
70		

7.4

Multiplying, Dividing, and Simplifying Radical Expressions

- *Multiplying and Simplifying*
- *Dividing and Simplifying*

Multiplying and Simplifying

We have already used the product rule for radicals, $\sqrt[n]{a} \cdot \sqrt[n]{b} = \sqrt[n]{ab}$, to find certain products. The product rule was also used to simplify certain radical expressions. For some radical expressions, it is possible to do both: First find a product and then simplify.

Example 1 Multiply and simplify.

a) $\sqrt{15}\,\sqrt{6}$ **b)** $3\sqrt[3]{25} \cdot 2\sqrt[3]{5}$ **c)** $\sqrt[4]{8x^3y^5}\,\sqrt[4]{4x^2y^3}$

SOLUTION

a) $\sqrt{15}\,\sqrt{6} = \sqrt{15 \cdot 6}$
$= \sqrt{90} = \sqrt{9 \cdot 10}$
$= 3\sqrt{10}$

b) $3\sqrt[3]{25} \cdot 2\sqrt[3]{5} = 6 \cdot \sqrt[3]{25 \cdot 5}$ **Multiplying radicands**
$= 6 \cdot \sqrt[3]{125}$ 125 is a perfect cube.
$= 6 \cdot 5$, or 30

c) $\sqrt[4]{8x^3y^5}\ \sqrt[4]{4x^2y^3} = \sqrt[4]{32x^5y^8}$ Multiplying radicands

$\phantom{\sqrt[4]{8x^3y^5}\ \sqrt[4]{4x^2y^3}} = \sqrt[4]{16x^4y^8 \cdot 2x}$ Factoring the radicand

$\phantom{\sqrt[4]{8x^3y^5}\ \sqrt[4]{4x^2y^3}} = \sqrt[4]{16}\ \sqrt[4]{x^4}\ \sqrt[4]{y^8}\ \sqrt[4]{2x}$ Factoring into radicals

$\phantom{\sqrt[4]{8x^3y^5}\ \sqrt[4]{4x^2y^3}} = 2xy^2\sqrt[4]{2x}$ Finding the fourth roots.
Assume $x \geq 0$.

The checks are left to the student.

As we saw earlier, when indices differ, rational exponents are helpful.

Example 2   Multiply and simplify: $\sqrt{x^3}\ \sqrt[3]{x}$.

SOLUTION   We have

$\sqrt{x^3}\ \sqrt[3]{x} = x^{3/2} \cdot x^{1/3}$ Converting to exponential notation

$\phantom{\sqrt{x^3}\ \sqrt[3]{x}} = x^{11/6}$ Adding exponents: $\frac{3}{2} + \frac{1}{3} = \frac{9}{6} + \frac{2}{6}$

$\phantom{\sqrt{x^3}\ \sqrt[3]{x}} = x^{1+5/6}$ Writing 11/6 as a mixed number

$\phantom{\sqrt{x^3}\ \sqrt[3]{x}} = x \cdot x^{5/6}$ Factoring

$\phantom{\sqrt{x^3}\ \sqrt[3]{x}} = x\sqrt[6]{x^5}.$ Converting back to radical notation

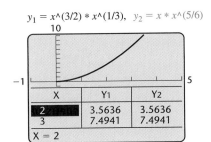

$y_1 = x^\wedge(3/2) * x^\wedge(1/3),\quad y_2 = x * x^\wedge(5/6)$

We can check by graphing $f(x) = \sqrt{x^3}\ \sqrt[3]{x}$ and $g(x) = x\sqrt[6]{x^5}$ as shown at left. The domain of both functions is $[0, \infty)$, and the graphs coincide for all x-values in the domain.

Dividing and Simplifying

Just as the root of a product can be expressed as the product of two roots, the root of a quotient can be expressed as the quotient of two roots. For example,

$$\sqrt[3]{\frac{27}{8}} = \frac{3}{2} \quad \text{and} \quad \frac{\sqrt[3]{27}}{\sqrt[3]{8}} = \frac{3}{2}.$$

This example suggests the following.

The Quotient Rule for Radicals
For any real numbers $\sqrt[n]{a}$ and $\sqrt[n]{b}$, $b \neq 0$,

$$\sqrt[n]{\frac{a}{b}} = \frac{\sqrt[n]{a}}{\sqrt[n]{b}}.$$

Remember that an nth root is simplified when its radicand has no factors that are perfect nth powers.

Example 3   Simplify by taking the roots of the numerator and the denominator.

a) $\sqrt[3]{\dfrac{27}{125}}$ **b)** $\sqrt{\dfrac{25}{y^2}}$ **c)** $\sqrt{\dfrac{16x^3}{y^8}}$ **d)** $\sqrt[3]{\dfrac{27y^{14}}{343x^3}}$

SOLUTION

a) $\sqrt[3]{\dfrac{27}{125}} = \dfrac{\sqrt[3]{27}}{\sqrt[3]{125}} = \dfrac{3}{5}$ — Taking the cube roots of the numerator and the denominator

b) $\sqrt{\dfrac{25}{y^2}} = \dfrac{\sqrt{25}}{\sqrt{y^2}} = \dfrac{5}{y}$ — Taking the square roots of the numerator and the denominator. Assume $y > 0$.

c) $\sqrt{\dfrac{16x^3}{y^8}} = \dfrac{\sqrt{16x^3}}{\sqrt{y^8}} = \dfrac{\sqrt{16x^2 \cdot x}}{\sqrt{y^8}} = \dfrac{4x\sqrt{x}}{y^4}$ Assume $x \geq 0, y \neq 0$.

d) $\sqrt[3]{\dfrac{27y^{14}}{343x^3}} = \dfrac{\sqrt[3]{27y^{14}}}{\sqrt[3]{343x^3}} = \dfrac{\sqrt[3]{27y^{12}y^2}}{\sqrt[3]{343x^3}} = \dfrac{\sqrt[3]{27y^{12}}\,\sqrt[3]{y^2}}{\sqrt[3]{343x^3}} = \dfrac{3y^4\sqrt[3]{y^2}}{7x}$

Assume $x \neq 0$.

If we read from right to left, the quotient rule tells us that to divide two radical expressions that have the same index, we can divide the radicands.

Example 4   Divide and, if possible, simplify.

a) $\dfrac{\sqrt{80}}{\sqrt{5}}$ **b)** $\dfrac{7\sqrt{2}}{5\sqrt{3}}$ **c)** $\dfrac{5\sqrt[3]{32}}{\sqrt[3]{2}}$

d) $\dfrac{\sqrt{72xy}}{2\sqrt{2}}$ **e)** $\dfrac{\sqrt[4]{3a^9b^3}}{\sqrt[4]{2b^{-1}}}$

SOLUTION

a) $\dfrac{\sqrt{80}}{\sqrt{5}} = \sqrt{\dfrac{80}{5}} = \sqrt{16} = 4$

> **Because the indices match, we can divide the radicands.**

b) $\dfrac{7\sqrt{2}}{5\sqrt{3}} = \dfrac{7}{5}\dfrac{\sqrt{2}}{\sqrt{3}} = \dfrac{7}{5} \cdot \sqrt{\dfrac{2}{3}}$

c) $\dfrac{5\sqrt[3]{32}}{\sqrt[3]{2}} = 5\sqrt[3]{\dfrac{32}{2}} = 5\sqrt[3]{16}$

$= 5\sqrt[3]{8 \cdot 2}$

$= 5\sqrt[3]{8}\,\sqrt[3]{2} = 5 \cdot 2\sqrt[3]{2}$

$= 10\sqrt[3]{2}$

d) $\dfrac{\sqrt{72xy}}{2\sqrt{2}} = \dfrac{1}{2}\dfrac{\sqrt{72xy}}{\sqrt{2}} = \dfrac{1}{2}\sqrt{\dfrac{72xy}{2}} = \dfrac{1}{2}\sqrt{36xy}$

$= \dfrac{1}{2}\sqrt{36}\,\sqrt{xy} = \dfrac{1}{2} \cdot 6\sqrt{xy}$

$= 3\sqrt{xy}$

e) $\dfrac{\sqrt[4]{3a^9b^3}}{\sqrt[4]{2b^{-1}}} = \sqrt[4]{\dfrac{3a^9b^3}{2b^{-1}}}$

$= \sqrt[4]{\dfrac{3a^9b^4}{2}} = \sqrt[4]{a^8b^4}\ \sqrt[4]{\dfrac{3a}{2}}$ Note that 8 is the largest power less than 9 that is divisible by the index 4.

$= a^2b\ \sqrt[4]{\dfrac{3a}{2}}$ Assume $a \geq 0$, $b > 0$.

Caution! Before multiplying or dividing radicands, make sure that both radicals have the same index.

When indices differ, we use rational exponents.

Example 5 Divide and, if possible, simplify.

a) $\dfrac{\sqrt[4]{(x+y)^3}}{\sqrt{x+y}}$ **b)** $\dfrac{\sqrt[3]{a^2b^4}}{\sqrt{ab}}$ **c)** $\dfrac{\sqrt[4]{x^3y^2}}{\sqrt[3]{x^2y}}$

SOLUTION

a) $\dfrac{\sqrt[4]{(x+y)^3}}{\sqrt{x+y}} = \dfrac{(x+y)^{3/4}}{(x+y)^{1/2}}$ Converting to exponential notation

$= (x+y)^{3/4-1/2}$ Since the bases match, we can subtract exponents.

$\left.\begin{array}{l} = (x+y)^{1/4} \\ = \sqrt[4]{x+y} \end{array}\right\}$ Converting back to radical notation

b) $\dfrac{\sqrt[3]{a^2b^4}}{\sqrt{ab}} = \dfrac{(a^2b^4)^{1/3}}{(ab)^{1/2}}$ Converting to exponential notation

$= \dfrac{a^{2/3}b^{4/3}}{a^{1/2}b^{1/2}}$ Using the product and power rules

$= a^{2/3-1/2}b^{4/3-1/2}$ Subtracting exponents

$\left.\begin{array}{l} = a^{1/6}b^{5/6} \\ = (ab^5)^{1/6} \\ = \sqrt[6]{ab^5} \end{array}\right\}$ Converting back to radical notation

c) $\dfrac{\sqrt[4]{x^3y^2}}{\sqrt[3]{x^2y}} = \dfrac{(x^3y^2)^{1/4}}{(x^2y)^{1/3}}$ Converting to exponential notation

$= \dfrac{x^{3/4}y^{2/4}}{x^{2/3}y^{1/3}}$ Using the product and power rules

$= x^{3/4-2/3}y^{2/4-1/3}$ Subtracting exponents

$\left.\begin{array}{l} = x^{1/12}y^{2/12} \\ = (xy^2)^{1/12} \\ = \sqrt[12]{xy^2} \end{array}\right\}$ Converting back to radical notation

7.4 Exercise Set

Multiply and simplify. Write all answers in radical notation.

1. $\sqrt{10}\,\sqrt{5}$

2. $\sqrt{6}\,\sqrt{3}$

3. $\sqrt{6}\,\sqrt{14}$

4. $\sqrt{15}\,\sqrt{21}$

5. $\sqrt[3]{2}\,\sqrt[3]{4}$

6. $\sqrt[3]{9}\,\sqrt[3]{3}$

7. $\sqrt[3]{5a^2}\,\sqrt[3]{2a}$

8. $\sqrt[3]{7x}\,\sqrt[3]{3x^2}$

9. $\sqrt{3x^3}\,\sqrt{6x^5}$

10. $\sqrt{5a^7}\,\sqrt{15a^3}$

11. $\sqrt[3]{s^2t^4}\,\sqrt[3]{s^4t^6}$

12. $\sqrt[3]{x^2y^4}\,\sqrt[3]{x^2y^6}$

13. $\sqrt[3]{(x+5)^2}\,\sqrt[3]{(x+5)^4}$

14. $\sqrt[3]{(a-b)^5}\,\sqrt[3]{(a-b)^7}$

15. $\sqrt[4]{12a^3b^7}\,\sqrt[4]{4a^2b^5}$

16. $\sqrt[4]{9x^7y^2}\,\sqrt[4]{9x^2y^9}$

17. $\sqrt[5]{x^3(y+z)^4}\,\sqrt[5]{x^3(y+z)^6}$

18. $\sqrt[5]{a^3(b-c)^4}\,\sqrt[5]{a^7(b-c)^4}$

19. $\sqrt{a}\,\sqrt[4]{a^3}$

20. $\sqrt[3]{x^2}\,\sqrt[6]{x^5}$

21. $\sqrt[5]{b^2}\,\sqrt{b^3}$

22. $\sqrt[4]{a^3}\,\sqrt[3]{a^2}$

23. $\sqrt{xy^3}\,\sqrt[3]{x^2y}$

24. $\sqrt[5]{a^3b}\,\sqrt{ab}$

25. $\sqrt[4]{9ab^3}\,\sqrt{3a^4b}$

26. $\sqrt{2x^3y^3}\,\sqrt[3]{4xy^2}$

27. $\sqrt[3]{xy^2z}\,\sqrt{x^3yz^2}$

28. $\sqrt{a^4b^3c^4}\,\sqrt[3]{ab^2c}$

29. $\sqrt{27a^5(b+1)}\,\sqrt[3]{81a(b+1)^4}$

30. $\sqrt{8x(y+z)^5}\,\sqrt[3]{4x^2(y+z)^2}$

Simplify by taking the roots of the numerator and the denominator.

31. $\sqrt{\dfrac{25}{36}}$

32. $\sqrt{\dfrac{100}{81}}$

33. $\sqrt[3]{\dfrac{64}{27}}$

34. $\sqrt[3]{\dfrac{343}{1000}}$

35. $\sqrt{\dfrac{49}{y^2}}$

36. $\sqrt{\dfrac{121}{x^2}}$

37. $\sqrt{\dfrac{25y^3}{x^4}}$

38. $\sqrt{\dfrac{36a^5}{b^6}}$

39. $\sqrt[3]{\dfrac{27a^4}{8b^3}}$

40. $\sqrt[3]{\dfrac{64x^7}{216y^6}}$

41. $\sqrt[4]{\dfrac{16a^4}{b^4c^8}}$

42. $\sqrt[4]{\dfrac{81x^4}{y^8z^4}}$

43. $\sqrt[4]{\dfrac{a^5b^8}{c^{10}}}$

44. $\sqrt[4]{\dfrac{x^9y^{12}}{z^6}}$

45. $\sqrt[5]{\dfrac{32x^6}{y^{11}}}$

46. $\sqrt[5]{\dfrac{243a^9}{b^{13}}}$

47. $\sqrt[6]{\dfrac{x^6y^8}{z^{15}}}$

48. $\sqrt[6]{\dfrac{a^9b^{12}}{c^{13}}}$

Divide and, if possible, simplify.

49. $\dfrac{\sqrt{35x}}{\sqrt{7x}}$

50. $\dfrac{\sqrt{28y}}{\sqrt{4y}}$

51. $\dfrac{\sqrt[3]{270}}{\sqrt[3]{10}}$

52. $\dfrac{\sqrt[3]{40}}{\sqrt[3]{5}}$

53. $\dfrac{\sqrt{40xy^3}}{\sqrt{8x}}$

54. $\dfrac{\sqrt{56ab^3}}{\sqrt{7a}}$

55. $\dfrac{\sqrt[3]{96a^4b^2}}{\sqrt[3]{12a^2b}}$

56. $\dfrac{\sqrt[3]{189x^5y^7}}{\sqrt[3]{7x^2y^2}}$

57. $\dfrac{\sqrt{100ab}}{5\sqrt{2}}$

58. $\dfrac{\sqrt{75ab}}{3\sqrt{3}}$

59. $\dfrac{\sqrt[4]{48x^9y^{13}}}{\sqrt[4]{3xy^{-2}}}$

60. $\dfrac{\sqrt[5]{64a^{11}b^{28}}}{\sqrt[5]{2ab^{-2}}}$

61. $\dfrac{\sqrt[3]{a^2}}{\sqrt[4]{a}}$

62. $\dfrac{\sqrt[3]{x^2}}{\sqrt[5]{x}}$

63. $\dfrac{\sqrt[4]{x^2y^3}}{\sqrt[3]{xy}}$

64. $\dfrac{\sqrt[5]{a^4b^2}}{\sqrt[3]{ab^2}}$

65. $\dfrac{\sqrt{ab^3c}}{\sqrt[5]{a^2b^3c^{-1}}}$

66. $\dfrac{\sqrt[5]{x^3y^4z^9}}{\sqrt{xy^{-2}z}}$

67. $\dfrac{\sqrt[4]{(3x-1)^3}}{\sqrt[5]{(3x-1)^3}}$

68. $\dfrac{\sqrt[3]{(2+5x)^2}}{\sqrt[4]{2+5x}}$

Skill Maintenance

Solve. [6.4]

69. $\dfrac{12x}{x-4} - \dfrac{3x^2}{x+4} = \dfrac{384}{x^2-16}$

70. $\dfrac{2}{3} + \dfrac{1}{t} = \dfrac{4}{5}$

71. The width of a rectangle is one-fourth the length. The area is twice the perimeter. Find the dimensions of the rectangle. [5.7]

72. The sum of a number and its square is 20. Find the number. [5.7]

73. Solve for a_1: $A = \dfrac{m}{a_2 - a_1}$. [6.7]

74. Solve for n: $3n + p = 4(m - n)$. [2.3]

Synthesis

75. ◈ Explain how common denominators can be used when dividing one radical expression by another. (*Hint:* See Example 5.)

76. ◈ Is the quotient of two irrational numbers always an irrational number? Why or why not?

77. ◈ Ramon *incorrectly* writes

$$x^{2/5} \cdot x^{3/2} = \sqrt[5]{x^3}.$$

What mistake do you suspect he is making?

78. ◈ After examining Exercise 15, Dyan (correctly) concludes that a and b are both nonnegative. Explain how she could reach this conclusion.

79. *Pendulums.* The *period* of a pendulum is the time it takes to complete one cycle, swinging to and fro. For a pendulum that is L centimeters long, the period T is given by the formula

$$T = 2\pi\sqrt{\frac{L}{980}},$$

where T is in seconds. Find, to the nearest hundredth of a second, the period of a pendulum of length **(a)** 65 cm; **(b)** 98 cm; **(c)** 120 cm. Use a calculator's $\boxed{\pi}$ key.

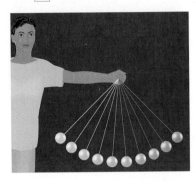

Perform the indicated operations.

80. $\dfrac{7\sqrt{a^2b}\ \sqrt{25xy}}{5\sqrt{a^{-4}b^{-1}}\ \sqrt{49x^{-1}y^{-3}}}$

81. $\dfrac{(\sqrt[3]{81mn^2})^2}{(\sqrt[3]{mn})^2}$

82. $\dfrac{\sqrt{44x^2y^9z}\ \sqrt{22y^9z^6}}{(\sqrt{11xy^8z^2})^2}$

83. $\dfrac{\sqrt{x^5 - 2x^4y} - \sqrt{xy^4 - 2y^5}}{\sqrt{xy^2 - 2y^3} + \sqrt{x^3 - 2x^2y}}$

84. Solve $\sqrt[3]{5x^{k+1}}\ \sqrt[3]{25x^k} = 5x^7$ for k.

85. Solve $\sqrt[5]{4a^{3k+2}}\ \sqrt[5]{8a^{6-k}} = 2a^4$ for k.

7.5

More with Multiplication and Division

- *Multiplication with Two or More Radical Terms*
- *Rationalizing Denominators and Numerators*

Multiplication with Two or More Radical Terms

Radical expressions often contain factors that have more than one term. The procedure for multiplying out such expressions is similar to finding products of polynomials.

Example 1   Multiply.

a) $\sqrt{3}(x - \sqrt{5})$ **b)** $\sqrt[3]{y}(\sqrt[3]{y^2} + \sqrt[3]{2})$

c) $(4\sqrt{3} + \sqrt{2})(\sqrt{3} - 5\sqrt{2})$ **d)** $(\sqrt{a} + \sqrt{b})(\sqrt{a} - \sqrt{b})$

SOLUTION

a) $\sqrt{3}(x - \sqrt{5}) = \sqrt{3} \cdot x - \sqrt{3} \cdot \sqrt{5}$ **Using the distributive law**

$\qquad\qquad\qquad = x\sqrt{3} - \sqrt{15}$ **Multiplying radicals**

b) $\sqrt[3]{y}(\sqrt[3]{y^2} + \sqrt[3]{2}) = \sqrt[3]{y} \cdot \sqrt[3]{y^2} + \sqrt[3]{y} \cdot \sqrt[3]{2}$ **Using the distributive law**

$\qquad\qquad\qquad\quad = \sqrt[3]{y^3} + \sqrt[3]{2y}$ **Multiplying radicals**

$\qquad\qquad\qquad\quad = y + \sqrt[3]{2y}$ **Simplifying $\sqrt[3]{y^3}$**

$$\underset{F}{} \quad \underset{O}{} \quad \underset{I}{} \quad \underset{L}{}$$

c) $(4\sqrt{3} + \sqrt{2})(\sqrt{3} - 5\sqrt{2}) = 4(\sqrt{3})^2 - 20\sqrt{3}\cdot\sqrt{2} + \sqrt{2}\cdot\sqrt{3} - 5(\sqrt{2})^2$

$$= 4\cdot3 - 20\sqrt{6} + \sqrt{6} - 5\cdot2 \qquad \textbf{Multiplying}$$
$$\textbf{radicals}$$

$$= 12 - 20\sqrt{6} + \sqrt{6} - 10$$

$$= 2 - 19\sqrt{6} \qquad \textbf{Combining like terms}$$

d) $(\sqrt{a} + \sqrt{b})(\sqrt{a} - \sqrt{b}) = (\sqrt{a})^2 - (\sqrt{b})^2$ **This is in the same form as a difference of two squares.**

$$= a - b \qquad \textbf{We assume } a, b \geq 0.$$

Note in Example 1(d) that the product of two radical expressions need not be a radical expression itself. Pairs of radical expressions like $\sqrt{a} + \sqrt{b}$ and $\sqrt{a} - \sqrt{b}$ or $5 - \sqrt{x}$ and $5 + \sqrt{x}$ are called **conjugates**. Conjugates will prove useful later in this section.

Example 2   If $f(x) = x^2$, find $f(a + \sqrt{5})$.

SOLUTION

$$f(a + \sqrt{5}) = (a + \sqrt{5})^2$$
$$= a^2 + 2a\sqrt{5} + (\sqrt{5})^2 \qquad \textbf{Squaring a binomial}$$
$$= a^2 + 2a\sqrt{5} + 5$$

As in our earlier work, when different indices appear, rational exponents are helpful.

Example 3   If $f(x) = \sqrt[3]{x^2}$ and $g(x) = \sqrt{x} + \sqrt[4]{5x}$, find $(f \cdot g)(x)$.

SOLUTION   Recall from Section 2.7 that $(f \cdot g)(x) = f(x) \cdot g(x)$. Thus,

$$(f \cdot g)(x) = \sqrt[3]{x^2}(\sqrt{x} + \sqrt[4]{5x}) \qquad \textbf{We assume } x \geq 0.$$

$$= x^{2/3}(x^{1/2} + (5x)^{1/4}) \qquad \textbf{Converting to exponential notation}$$

$$= x^{2/3}(x^{1/2} + 5^{1/4}x^{1/4}) \qquad \textbf{Using the laws of exponents}$$

$$= x^{2/3}\cdot x^{1/2} + x^{2/3}\cdot 5^{1/4}x^{1/4} \qquad \textbf{Using the distributive law}$$

$$= x^{2/3+1/2} + 5^{1/4}x^{2/3+1/4} \qquad \textbf{Adding exponents}$$

$$= x^{7/6} + 5^{3/12}x^{11/12} \qquad \textbf{Finding a common denominator}$$

$$= x^{1+1/6} + 5^{3/12}x^{11/12} \qquad \textbf{Writing a mixed number}$$

$$= x^1x^{1/6} + (5^3x^{11})^{1/12} \qquad \textbf{Using the laws of exponents}$$

$$= x\sqrt[6]{x} + \sqrt[12]{125x^{11}} \qquad \textbf{Converting back to radical notation}$$

Rationalizing Denominators and Numerators

When a radical expression appears in a denominator, it can be useful to find an equivalent expression in which the denominator no longer contains a radical.* The procedure for finding such an expression is called **rationalizing the denominator**. We carry this out by multiplying by 1.

Example 4 Rationalize each denominator.

a) $\dfrac{\sqrt{7}}{\sqrt{3}}$
  b) $\sqrt{\dfrac{4}{45b}}$

SOLUTION

a) The radicand in the denominator is 3. If the radicand were 9, the denominator would be a perfect square. We multiply by 1, using $\sqrt{3}/\sqrt{3}$:

$$\frac{\sqrt{7}}{\sqrt{3}} = \frac{\sqrt{7}}{\sqrt{3}} \cdot \frac{\sqrt{3}}{\sqrt{3}}$$

$$= \frac{\sqrt{21}}{\sqrt{9}} \qquad \text{Multiplying radicands. This radicand is now a perfect square.}$$

$$= \frac{\sqrt{21}}{3} \qquad \text{Simplifying}$$

b) We rewrite the expression as a quotient of two radicals. Then we simplify and multiply by 1:

$$\sqrt{\frac{4}{45b}} = \frac{\sqrt{4}}{\sqrt{45b}} = \frac{2}{3\sqrt{5b}} \qquad \text{Simplifying. We assume } b > 0.$$

$$= \frac{2}{3\sqrt{5b}} \cdot \frac{\sqrt{5b}}{\sqrt{5b}} \qquad \text{Multiplying by 1}$$

$$= \frac{2\sqrt{5b}}{3(\sqrt{5b})^2} \qquad \text{Try to do this step mentally.}$$

$$= \frac{2\sqrt{5b}}{3 \cdot 5b}$$

$$= \frac{2\sqrt{5b}}{15b}.$$

Sometimes in calculus it is necessary to rationalize a numerator. To do so, we multiply by 1 to make the radicand in the *numerator* a perfect power.

Example 5 Rationalize the numerator: $\sqrt{\dfrac{7}{5}}$.

---

*See Exercise 69 on p. 455.

SOLUTION

$$\sqrt{\frac{7}{5}} = \frac{\sqrt{7}}{\sqrt{5}} \qquad \text{Using the quotient rule for radicals}$$

$$= \frac{\sqrt{7}}{\sqrt{5}} \cdot \frac{\sqrt{7}}{\sqrt{7}} \qquad \text{Multiplying by 1.}$$

$$= \frac{\sqrt{49}}{\sqrt{35}} \longleftarrow \text{This radicand is now a perfect square.}$$

$$= \frac{7}{\sqrt{35}}$$

Recall from Example 1(d) that when two radical expressions are conjugates, their product contains no radicals. When rationalizing a denominator or a numerator that has two terms, we proceed much as we did in Examples 4 and 5. The difference is that now we use conjugates to construct the symbol for 1.

Example 6 Rationalize each denominator: **(a)** $\dfrac{4}{\sqrt{3} + x}$;

(b) $\dfrac{4 + \sqrt{2}}{\sqrt{5} - \sqrt{2}}$.

SOLUTION

a) $\dfrac{4}{\sqrt{3} + x} = \dfrac{4}{\sqrt{3} + x} \cdot \dfrac{\sqrt{3} - x}{\sqrt{3} - x}$ Multiplying by 1, using the conjugate of $\sqrt{3} + x$, which is $\sqrt{3} - x$

$$= \frac{4(\sqrt{3} - x)}{(\sqrt{3} + x)(\sqrt{3} - x)} \qquad \begin{array}{l}\text{Multiplying numerators and} \\ \text{denominators}\end{array}$$

$$= \frac{4(\sqrt{3} - x)}{(\sqrt{3})^2 - x^2} \qquad \text{Using FOIL in the denominator}$$

$$= \frac{4\sqrt{3} - 4x}{3 - x^2} \qquad \text{Simplifying}$$

b) $\dfrac{4 + \sqrt{2}}{\sqrt{5} - \sqrt{2}} = \dfrac{4 + \sqrt{2}}{\sqrt{5} - \sqrt{2}} \cdot \dfrac{\sqrt{5} + \sqrt{2}}{\sqrt{5} + \sqrt{2}}$ Multiplying by 1, using the conjugate of $\sqrt{5} - \sqrt{2}$, which is $\sqrt{5} + \sqrt{2}$

$$= \frac{(4 + \sqrt{2})(\sqrt{5} + \sqrt{2})}{(\sqrt{5} - \sqrt{2})(\sqrt{5} + \sqrt{2})} \qquad \begin{array}{l}\text{Multiplying nu-} \\ \text{merators and} \\ \text{denominators}\end{array}$$

$$= \frac{4\sqrt{5} + 4\sqrt{2} + \sqrt{2}\,\sqrt{5} + (\sqrt{2})^2}{(\sqrt{5})^2 - (\sqrt{2})^2} \qquad \text{Using FOIL}$$

$$= \frac{4\sqrt{5} + 4\sqrt{2} + \sqrt{10} + 2}{5 - 2} \qquad \begin{array}{l}\text{Squaring in the} \\ \text{denominator and} \\ \text{the numerator}\end{array}$$

$$= \frac{4\sqrt{5} + 4\sqrt{2} + \sqrt{10} + 2}{3}$$

```
(4+√ (2))/(√ (5)−√
(2))
              6.587801273
(4√ (5)+4√ (2)+√ (1
0)+2)/3
              6.587801273
▮
```

We can check by approximating the value of the original expression and the value of the rationalized expression as shown at left. Care must be taken to place the parentheses properly. The approximate values are the same, so our work is correct. ▬

To rationalize a numerator with more than one term, we use the conjugate of the numerator.

Example 7   Rationalize the numerator: $\dfrac{4 + \sqrt{2}}{\sqrt{5} - \sqrt{2}}$.

SOLUTION

$$\frac{4 + \sqrt{2}}{\sqrt{5} - \sqrt{2}} = \frac{4 + \sqrt{2}}{\sqrt{5} - \sqrt{2}} \cdot \frac{4 - \sqrt{2}}{4 - \sqrt{2}} \qquad \begin{array}{l}\textbf{Multiplying by 1, using}\\ \textbf{the conjugate of } 4 + \sqrt{2},\\ \textbf{which is } 4 - \sqrt{2}\end{array}$$

$$= \frac{16 - (\sqrt{2})^2}{4\sqrt{5} - \sqrt{5}\,\sqrt{2} - 4\sqrt{2} + (\sqrt{2})^2}$$

$$= \frac{14}{4\sqrt{5} - \sqrt{10} - 4\sqrt{2} + 2}$$

```
(4+√ (2))/(√ (5)−√
(2))
              6.587801273
14/(4√ (5)−√ (10)−
4√ (2)+2)
              6.587801273
▮
```

We check by comparing the values of the original expression and the expression with a rationalized numerator, as shown at left. ▬

7.5 Exercise Set

Multiply. Assume that all variables represent nonnegative real numbers. Note that Exercises 11–14 use conjugates.

1. $\sqrt{7}(3 - \sqrt{7})$

2. $\sqrt{3}(4 + \sqrt{3})$

3. $\sqrt{2}(\sqrt{3} - \sqrt{5})$

4. $\sqrt{5}(\sqrt{5} - \sqrt{2})$

5. $\sqrt{3}(2\sqrt{5} - 3\sqrt{4})$

6. $\sqrt{2}(3\sqrt{10} - 2\sqrt{2})$

7. $\sqrt[3]{2}(\sqrt[3]{4} - 2\sqrt[3]{32})$

8. $\sqrt[3]{3}(\sqrt[3]{9} - 4\sqrt[3]{21})$

9. $\sqrt[3]{a}(\sqrt[3]{a^2} + \sqrt[3]{24a^2})$

10. $\sqrt[3]{x}(\sqrt[3]{3x^2} - \sqrt[3]{81x^2})$

11. $(5 + \sqrt{6})(5 - \sqrt{6})$

12. $(2 - \sqrt{5})(2 + \sqrt{5})$

13. $(3 - 2\sqrt{7})(3 + 2\sqrt{7})$

14. $(4 + 3\sqrt{2})(4 - 3\sqrt{2})$

15. $(5 + \sqrt[3]{10})(3 - \sqrt[3]{10})$

16. $(\sqrt[3]{7} - 4)(\sqrt[3]{7} + 5)$

17. $(2\sqrt{7} - 4\sqrt{2})(3\sqrt{7} + 6\sqrt{2})$

18. $(4\sqrt{5} + 3\sqrt{3})(3\sqrt{5} - 4\sqrt{3})$

19. $(2\sqrt[3]{3} - \sqrt[3]{2})(\sqrt[3]{3} + 2\sqrt[3]{2})$

20. $(3\sqrt[4]{7} + \sqrt[4]{6})(2\sqrt[4]{9} - 3\sqrt[4]{6})$

21. $(\sqrt{3x} + \sqrt{y})^2$

22. $(\sqrt{t} - \sqrt{2r})^2$

23. $\sqrt[3]{x^2y}(\sqrt{xy} - \sqrt[5]{xy^3})$

24. $\sqrt[4]{a^2b}(\sqrt[3]{a^2b} - \sqrt[4]{a^2b^2})$

25. $(m + \sqrt[3]{n^2})(2m + \sqrt[4]{n})$

26. $(r - \sqrt[4]{s^3})(3r - \sqrt[5]{s})$

In Exercises 27–32, $f(x)$ and $g(x)$ are as given. Find $(f \cdot g)(x)$.

27. $f(x) = \sqrt[4]{x}$; $g(x) = \sqrt[4]{2x} - \sqrt[4]{x^{11}}$

28. $f(x) = \sqrt[4]{x^7} + \sqrt[4]{3x^2}$; $g(x) = \sqrt[4]{x}$

29. $f(x) = x + \sqrt{7}$; $g(x) = x - \sqrt{7}$

30. $f(x) = x - \sqrt{2}$; $g(x) = x + \sqrt{6}$

31. $f(x) = 2 - \sqrt{x}$; $g(x) = 1 - \sqrt{x}$

32. $f(x) = \sqrt{x} + \sqrt{3}$; $g(x) = \sqrt{x} + \sqrt{2}$

Let $f(x) = x^2$. Find each of the following.

33. $f(5 - \sqrt{2})$

34. $f(7 + \sqrt{3})$

35. $f(\sqrt{3} + \sqrt{5})$

36. $f(\sqrt{6} - \sqrt{3})$

37. $f(\sqrt{10} - \sqrt{5})$

38. $f(\sqrt{7} + \sqrt{8})$

Rationalize each denominator.

39. $\sqrt{\dfrac{5}{7}}$

40. $\sqrt{\dfrac{11}{6}}$

41. $\sqrt{\dfrac{7a}{18}}$

42. $\sqrt{\dfrac{3x}{10}}$ 　　**43.** $\sqrt{\dfrac{9}{20x^2y}}$ 　　**44.** $\sqrt{\dfrac{5}{32ab^2}}$

45. $\dfrac{5}{7 - \sqrt{2}}$ 　　　　**46.** $\dfrac{3}{5 + \sqrt{6}}$

47. $\dfrac{\sqrt{x}}{\sqrt{x} + \sqrt{y}}$ 　　　**48.** $\dfrac{\sqrt{b}}{\sqrt{a} - \sqrt{b}}$

49. $\dfrac{\sqrt{3} + 4\sqrt{5}}{\sqrt{3} - 2\sqrt{6}}$ 　　**50.** $\dfrac{10 - 2\sqrt{3}}{\sqrt{5} + 3\sqrt{3}}$

51. $\dfrac{5\sqrt{3} - 3\sqrt{2}}{3\sqrt{2} - 2\sqrt{3}}$ 　　**52.** $\dfrac{7\sqrt{2} + 4\sqrt{3}}{4\sqrt{3} - 3\sqrt{2}}$

Rationalize each numerator.

53. $\dfrac{\sqrt{5}}{\sqrt{7x}}$ 　　**54.** $\dfrac{\sqrt{10}}{\sqrt{3x}}$ 　　**55.** $\sqrt{\dfrac{14}{21}}$

56. $\sqrt{\dfrac{12}{15}}$ 　　**57.** $\sqrt{\dfrac{x^3y}{2}}$ 　　**58.** $\sqrt{\dfrac{ab^5}{3}}$

59. $\dfrac{\sqrt{5} + 2}{6}$ 　　　　**60.** $\dfrac{7 - \sqrt{3}}{4}$

61. $\dfrac{\sqrt{3} - 5}{\sqrt{2} + 5}$ 　　　**62.** $\dfrac{\sqrt{6} - 3}{\sqrt{3} + 7}$

63. $\dfrac{\sqrt{x} + \sqrt{y}}{\sqrt{x} - \sqrt{y}}$ 　　**64.** $\dfrac{\sqrt{x} - \sqrt{y}}{\sqrt{x} + \sqrt{y}}$

Skill Maintenance

Solve. [6.4]

65. $\dfrac{1}{2} - \dfrac{1}{3} = \dfrac{1}{t}$

66. $\dfrac{5}{x - 1} + \dfrac{9}{x^2 + x + 1} = \dfrac{15}{x^3 - 1}$

Divide and simplify. [6.1]

67. $\dfrac{2x^2 - x - 6}{x^2 + 4x + 3} \div \dfrac{2x^2 + x - 3}{x^2 - 1}$

68. $\dfrac{1}{x^3 - y^3} \div \dfrac{1}{(x - y)(x^2 + xy + y^2)}$

Synthesis

69. ◆ Explain why it is easier to approximate
$$\dfrac{\sqrt{2}}{2} \quad \text{than} \quad \dfrac{1}{\sqrt{2}}$$
if no calculator is available and $\sqrt{2} \approx 1.414213562$.

70. ◆ A student *incorrectly* claims that
$$\dfrac{5 + \sqrt{2}}{\sqrt{18}} = \dfrac{5 + \sqrt{1}}{\sqrt{9}} = \dfrac{5 + 1}{3}.$$
How could you convince the student that a mistake has been made? How would you explain the correct way of rationalizing the denominator?

Express each of the following as the product of two radical expressions.

71. $x - 5$ 　　　**72.** $y - 7$ 　　　**73.** $x - a$

Multiply.

74. $\sqrt{9 + 3\sqrt{5}} \; \sqrt{9 - 3\sqrt{5}}$

75. $(\sqrt{x + 2} - \sqrt{x - 2})^2$

For Exercises 76–81, assume that all radicands are positive and that no denominators are 0.

Rationalize each denominator.

76. $\dfrac{a - \sqrt{a + b}}{\sqrt{a + b} - b}$ 　　**77.** $\dfrac{\sqrt[3]{3a}}{\sqrt[3]{5c}}$

78. $\dfrac{\sqrt[3]{5y^4}}{\sqrt[3]{6x^4}}$

Rationalize each numerator.

79. $\dfrac{\sqrt{y + 18} - \sqrt{y}}{18}$

80. $\dfrac{\sqrt[3]{7}}{\sqrt[3]{2}}$

81. $\sqrt[3]{\dfrac{2a^5}{5b}}$

Simplify.

82. $\sqrt{a^2 - 3} - \dfrac{a^2}{\sqrt{a^2 - 3}}$

83. $5\sqrt{\dfrac{x}{y}} + 4\sqrt{\dfrac{y}{x}} - \dfrac{3}{\sqrt{xy}}$

84. $\dfrac{\dfrac{1}{\sqrt{w}} - \sqrt{w}}{\dfrac{\sqrt{w} + 1}{\sqrt{w}}}$

85. $\dfrac{1}{4 + \sqrt{3}} + \dfrac{1}{\sqrt{3}} + \dfrac{1}{\sqrt{3} - 4}$

86. ◆ Use what we know about factoring a difference of two cubes to present a method for rationalizing any denominator of the form $\sqrt[3]{a} - \sqrt[3]{b}$.

7.6
Solving Radical Equations

- *The Principle of Powers*
- *Equations with Two Radical Terms*

The Principle of Powers

A **radical equation** has variables in one or more radicands. Examples are

$$\sqrt[3]{2x} + 1 = 5,$$
$$\sqrt{a} + \sqrt{a - 2} = 7,$$

and

$$4 - \sqrt{3x + 1} = \sqrt{6 - x}.$$

To solve such equations, we need a new principle. Suppose an equation $a = b$ is true. If we square both sides, we get another true equation: $a^2 = b^2$. This can be generalized.

> ### The Principle of Powers
> If $a = b$, then $a^n = b^n$ for any exponent n.

The principle of powers is an "if–then" statement. The statement obtained by interchanging the two parts of the sentence—"if $a^n = b^n$ for some exponent n, then $a = b$"—is not always true.

Interactive Discovery

Solve each pair of equations graphically, and compare the solution sets.

1. $x = 3$; $x^2 = 9$
2. $x = -2$; $x^2 = 4$
3. $\sqrt{x} = 5$; $x = 25$
4. $\sqrt{x} = -3$; $x = 9$

We see that every solution of $x = a$ is a solution of $x^2 = a^2$, but not every solution of $x^2 = a^2$ is necessarily a solution of $x = a$. A similar statement can be made for any even exponent n; for example, the solution sets of $x^4 = 3^4$ and $x = 3$ are not the same.

When using the principle of powers, we will find every solution of the original equation, but we may also find possible answers that are *not* solutions of the original equation. For this reason, when we are raising both sides of an equation to an even power, we must check our potential solutions in the original equation.

Example 1 Solve:

$$\sqrt{x} - 3 = 4.$$

We have

$$\sqrt{x} - 3 = 4$$

$$\sqrt{x} = 7 \qquad \text{Adding 3 on both sides to isolate the radical}$$

$$(\sqrt{x})^2 = 7^2 \qquad \text{Using the principle of powers}$$

$$x = 49.$$

CHECK: $\dfrac{\sqrt{x} - 3 = 4}{\sqrt{49} - 3 \ ? \ 4}$
$7 - 3 \ \big|$
$4 \ \big| \ 4 \quad$ TRUE

The number 49 checks.

We let $f(x) = \sqrt{x} - 3$ and $g(x) = 4$. Since the domain of f is $[0, \infty)$, we try a viewing window of $[-1, 10, -10, 10]$.

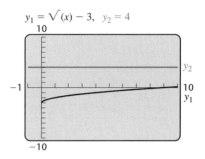

It appears as though the graphs may intersect at an x-value greater than 10. We adjust the viewing window to $[-10, 100, -5, 5]$ and find the point of intersection.

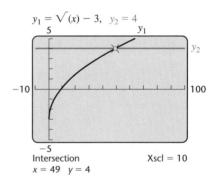

We have a solution of 49, which we can check by substituting into the original equation, as shown in the algebraic solution.
The solution is 49.

Caution! Raising both sides of an equation to an even power may not produce an equivalent equation. In this case, a check is essential.

Example 2 Solve: $\sqrt{x} - 5 = -7$.

We have

$\sqrt{x} - 5 = -7$

$\sqrt{x} = -2$ Adding 5 on both sides to isolate the radical

> The equation $\sqrt{x} = -2$ has no solution because the principal square root of a number is never negative. We continue as in Example 1 for comparison.

$(\sqrt{x})^2 = (-2)^2$ Using the principle of powers (squaring)

$x = 4$.

CHECK: $\dfrac{\sqrt{x} - 5 = -7}{\sqrt{4} - 5 \;?\; -7}$

$2 - 5$

$-3 \;\bigg|\; -7$ FALSE

The number 4 does not check.

We let $f(x) = \sqrt{x} - 5$ and $g(x) = -7$ and graph.

The graphs do not intersect. There is no solution.

$y_1 = \sqrt{(x)} - 5, \;\; y_2 = -7$

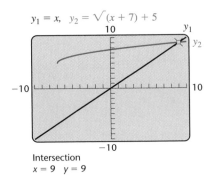

The equation $\sqrt{x} - 5 = -7$ has no real-number solution.   ▬

Example 3 Solve: $x = \sqrt{x + 7} + 5$.

We have

$x = \sqrt{x + 7} + 5$

$x - 5 = \sqrt{x + 7}$ Adding -5 on both sides to isolate the radical

$\left.\begin{array}{l} (x - 5)^2 = (\sqrt{x + 7})^2 \\ x^2 - 10x + 25 = x + 7 \end{array}\right\}$ Using the principle of powers; squaring both sides

$x^2 - 11x + 18 = 0$ Adding $-x - 7$ on both sides to write the quadratic equation in standard form

$(x - 9)(x - 2) = 0$ Factoring

$x = 9 \quad or \quad x = 2$.   Using the principle of zero products

We graph $y_1 = x$ and $y_2 = \sqrt{(x + 7)} + 5$ and determine any points of intersection.

$y_1 = x, \;\; y_2 = \sqrt{(x + 7)} + 5$

Intersection
$x = 9 \quad y = 9$

We obtain a solution of 9.   ▬

CHECK:

For 9:

$$x = \sqrt{x + 7} + 5$$

$$\begin{array}{c|c} 9 \ ? \ \sqrt{9 + 7} + 5 \\ \hline 9 & 9 \end{array} \qquad \text{TRUE}$$

For 2:

$$x = \sqrt{x + 7} + 5$$

$$\begin{array}{c|c} 2 \ ? \ \sqrt{2 + 7} + 5 \\ \hline 2 & 8 \end{array} \qquad \text{FALSE}$$

The number 9 checks but the number 2 does not.

It is important to isolate a radical term before using the principle of powers. Suppose in Example 3 that both sides of the equation were squared *before* isolating the radical. We then would have had the expression $(\sqrt{x + 7} + 5)^2$ or $x + 7 + 10\sqrt{x + 7} + 25$ in the third step of the algebraic solution, and the radical would have remained in the problem.

Example 4

Solve: $(2x + 1)^{1/3} + 5 = 0$.

ALGEBRAIC SOLUTION

We need not use radical notation to solve:

$$(2x + 1)^{1/3} + 5 = 0$$

$$(2x + 1)^{1/3} = -5 \qquad \text{Adding } -5 \text{ on both sides}$$

$$[(2x + 1)^{1/3}]^3 = (-5)^3 \qquad \text{Cubing both sides}$$

$$(2x + 1)^1 = (-5)^3 \qquad \text{Multiplying exponents. Try to do this mentally.}$$

$$2x + 1 = -125$$

$$2x = -126 \qquad \text{Adding } -1 \text{ on both sides}$$

$$x = -63.$$

Because both sides were raised to an *odd* power, we need check only the accuracy of our work.

GRAPHICAL SOLUTION

We let $f(x) = (2x + 1)^{1/3} + 5$ and determine any values of a for which $f(a) = 0$. We might start by graphing f using a standard viewing window.

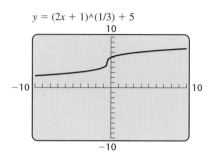

$y = (2x + 1)\wedge(1/3) + 5$

It appears that if the graph of f does have an x-intercept, it will be at a value of x less than -10. Let's try a viewing window of $[-100, 100, -10, 10]$.

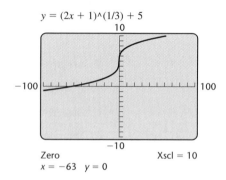

$y = (2x + 1)\wedge(1/3) + 5$

Zero Xscl = 10
$x = -63$ $y = 0$

We find that $f(x) = 0$ when $x = -63$.

Since both approaches yield the same solution, our work is probably correct. As a final check, the student can show that -63 checks in the original equation and is the solution. ▬

Equations with Two Radical Terms

A strategy for solving equations with two or more radical terms is as follows.

> **To Solve an Equation with Two or More Radical Terms:**
>
> **1.** Isolate one of the radical terms.
> **2.** Use the principle of powers.
> **3.** If a radical remains, perform steps (1) and (2) again.
> **4.** Check possible solutions in the original equation.

Example 5   Solve: $\sqrt{2x - 5} = 1 + \sqrt{x - 3}$.

ALGEBRAIC SOLUTION

We have

$$\sqrt{2x - 5} = 1 + \sqrt{x - 3}$$
$$(\sqrt{2x - 5})^2 = (1 + \sqrt{x - 3})^2 \qquad \text{One radical is already isolated. We square both sides.}$$

> **This is like squaring a binomial. We square 1 and then find twice the product of 1 and $\sqrt{x - 3}$ and then the square of $\sqrt{x - 3}$.**

$$2x - 5 = 1 + 2\sqrt{x - 3} + (\sqrt{x - 3})^2$$
$$2x - 5 = 1 + 2\sqrt{x - 3} + (x - 3)$$
$$x - 3 = 2\sqrt{x - 3} \qquad \text{Isolating the remaining radical term}$$
$$(x - 3)^2 = (2\sqrt{x - 3})^2 \qquad \text{Squaring both sides}$$
$$x^2 - 6x + 9 = 4(x - 3) \qquad \text{Remember to square both the 2 and the } \sqrt{x - 3} \text{ on the right side.}$$
$$x^2 - 6x + 9 = 4x - 12$$
$$x^2 - 10x + 21 = 0$$
$$(x - 7)(x - 3) = 0 \qquad \text{Factoring}$$
$$x = 7 \quad or \quad x = 3. \qquad \text{Using the principle of zero products}$$

The possible solutions are 7 and 3.

GRAPHICAL SOLUTION

We let $f(x) = \sqrt{2x - 5}$ and $g(x) = 1 + \sqrt{x - 3}$. The domain of f is $\left[\frac{5}{2}, \infty\right)$ (see Example 4 on p. 421) and the domain of g is $[3, \infty)$. Since the intersection of their domains is $[3, \infty)$, any point of intersection will occur in that interval. We choose a viewing window of $[-1, 10, -1, 5]$.

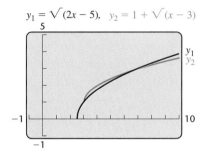

$y_1 = \sqrt{(2x - 5)}, \quad y_2 = 1 + \sqrt{(x - 3)}$

The graphs appear to intersect when x is approximately 3 and again when x is approximately 7. Using INTERSECT, we find that the points of intersection are indeed (3, 1) and (7, 3). To confirm that there are no other points of intersection, we could examine the graphs using a larger viewing window.

We leave it to the student to show that 7 and 3 both check in the original equation and are the solutions.

Caution!   A common error in solving equations like
$$\sqrt{2x - 5} = 1 + \sqrt{x - 3}$$
is to obtain $1 + (x - 3)$ as the square of the right side. This is wrong because $(A + B)^2 \neq A^2 + B^2$.

Example 6 Let $f(x) = \sqrt{x+5} - \sqrt{x-3}$. Find all x-values for which $f(x) = 2$.

┌── *ALGEBRAIC SOLUTION*

We must have $f(x) = 2$, or

$\sqrt{x+5} - \sqrt{x-3} = 2$.   Substituting for $f(x)$

To solve, we isolate one radical term and square both sides:

$\sqrt{x+5} = 2 + \sqrt{x-3}$ Adding $\sqrt{x-3}$; this isolates one of the radical terms.

$(\sqrt{x+5})^2 = (2 + \sqrt{x-3})^2$ Using the principle of powers (squaring both sides)

$x + 5 = 4 + 4\sqrt{x-3} + (x-3)$ Using $(A + B)^2 = A^2 + 2AB + B^2$

$5 = 1 + 4\sqrt{x-3}$ Adding $-x$ and combining like terms

$\left.\begin{array}{l} 4 = 4\sqrt{x-3} \\ 1 = \sqrt{x-3} \end{array}\right\}$ Isolating the remaining radical term

$1^2 = (\sqrt{x-3})^2$ Squaring both sides

$1 = x - 3$

$4 = x$.

The solution is 4.

┌── *GRAPHICAL SOLUTION*

We graph $y_1 = \sqrt{(x+5)} - \sqrt{(x-3)}$ and $y_2 = 2$ and determine the coordinates of any points of intersection.

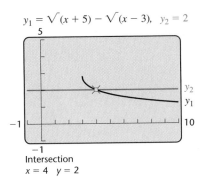

$y_1 = \sqrt{(x+5)} - \sqrt{(x-3)}, \quad y_2 = 2$

Intersection
$x = 4 \quad y = 2$

The solution is the x-coordinate of the point of intersection, which is 4.

CHECK: $f(4) = \sqrt{4+5} - \sqrt{4-3} = \sqrt{9} - \sqrt{1} = 3 - 1 = 2$.

We will have $f(x) = 2$ when $x = 4$.   ▬

7.6 | *Exercise Set*

Solve.

1. $\sqrt{5x+1} = 6$
2. $\sqrt{x+3} = 6$
3. $\sqrt{3x+1} = 7$
4. $\sqrt{2x-1} = 7$
5. $\sqrt{y+1} - 5 = 8$
6. $\sqrt{x-2} - 7 = -4$
7. $\sqrt[3]{x+5} = 2$
8. $\sqrt[3]{x-2} = 3$
9. $\sqrt[4]{y-3} = 2$
10. $\sqrt[4]{x+3} = 3$
11. $3\sqrt{x} = x$
12. $8\sqrt{y} = y$
13. $2y^{1/2} - 7 = 9$
14. $3x^{1/2} + 12 = 9$
15. $\sqrt[3]{x} = -3$
16. $\sqrt[3]{y} = -4$
17. $t^{1/3} - 2 = 3$
18. $x^{1/4} - 2 = 1$
19. $(x+2)^{1/2} = -4$
20. $(y-3)^{1/2} = -2$

21. $\sqrt[4]{2x+3} - 5 = -2$
22. $\sqrt[4]{3x+1} - 4 = -1$
23. $(y-7)^{1/4} = 3$
24. $(x+5)^{1/3} = 4$
25. $\sqrt{2t-7} = \sqrt{3t-12}$
26. $\sqrt{3t+4} = \sqrt{4t+3}$
27. $2(1-x)^{1/3} = 4^{1/3}$
28. $3(4-t)^{1/4} = 6^{1/4}$
29. $x = \sqrt{x-1} + 3$
30. $3 + \sqrt{5-x} = x$
31. $3 + \sqrt{z-6} = \sqrt{z+9}$
32. $\sqrt{4x-3} = 2 + \sqrt{2x-5}$
33. $\sqrt{20-x} + 8 = \sqrt{9-x} + 11$
34. $4 + \sqrt{10-x} = 6 + \sqrt{4-x}$
35. $\sqrt{x+2} + \sqrt{3x+4} = 2$
36. $\sqrt{6x+7} - \sqrt{3x+3} = 1$

37. If $f(t) = \sqrt{t-7}$, find a such that $f(a) = 7$.
38. If $g(t) = \sqrt{t+4}$, find a such that $g(a) = 1$.
39. If $f(x) = \sqrt[3]{6x+9} + 8$, find a such that $f(a) = 5$.
40. If $g(x) = \sqrt[3]{3x+6} + 2$, find a such that $g(a) = 3$.

41. If $f(x) = \sqrt{x} + \sqrt{x - 9}$, find x such that $f(x) = 1$.

42. If $g(x) = \sqrt{x} + \sqrt{x - 5}$, find x such that $g(x) = 5$.

43. If $g(x) = \sqrt{2x + 7} - \sqrt{x + 15}$, find a such that $g(a) = -1$.

44. If $f(x) = \sqrt{x - 2} - \sqrt{4x + 1}$, find a such that $f(a) = -3$.

45. If $f(x) = \sqrt{2x - 3}$ and $g(x) = \sqrt{x + 7} - 2$, find x such that $f(x) = g(x)$.

46. If $f(x) = 2\sqrt{3x + 6}$ and $g(x) = 5 + \sqrt{4x + 9}$, find x such that $f(x) = g(x)$.

47. If $f(t) = 4 - \sqrt{t - 3}$ and $g(t) = (t + 5)^{1/2}$, find a such that $f(a) = g(a)$.

48. If $f(t) = 7 + \sqrt{2t - 5}$ and $g(t) = 3(t + 1)^{1/2}$, find a such that $f(a) = g(a)$.

Skill Maintenance

49. Solve: [6.4]
$$\frac{3}{2x} + \frac{1}{x} = \frac{2x + 3.5}{3x}.$$

50. The base of a triangle is 2 in. longer than the height. The area is $31\frac{1}{2}$ in². Find the height and the base. [5.7]

Graph.

51. $f(x) = \frac{2}{5}x - 7$ [2.1]

52. $y > -3x + 5$ [4.4]

Synthesis

53. ◈ The principle of powers is an "if–then" statement that becomes false when the sentence parts are interchanged. Give an example of another such if–then statement.

54. ◈ Explain a method that could be used to solve $\sqrt{x} + \sqrt{2x - 1} - \sqrt{3x + 2} + \sqrt{2 + x} = 0$.

55. ◈ Is checking essential when the principle of powers is used with an odd power n? Why or why not?

56. ◈ Saul is trying to solve
$$\sqrt{2z - \frac{1}{4}} = \frac{1}{25}$$
with a grapher. Without resorting to trial and error, how can he determine a suitable viewing window for finding the solution?

Escape Velocity. *A formula for the escape velocity v of a satellite is*
$$v = \sqrt{2gr}\,\sqrt{\frac{h}{r + h}},$$
where g is the force of gravity, r is the planet or star's radius, and h is the height of the satellite above the planet or star's surface.

57. Solve for h.

58. Solve for r.

Sighting to the Horizon. *The function $D(h) = 1.2\sqrt{h}$ can be used to approximate the distance D, in miles, that a person can see to the horizon from a height h, in feet.*

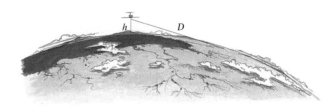

59. How far above sea level must a pilot fly in order to see a horizon that is 180 mi away?

60. How high above sea level must a sailor climb in order to see 10.2 mi out to sea?

Solve.

61. $\dfrac{x + \sqrt{x + 1}}{x - \sqrt{x + 1}} = \dfrac{5}{11}$

62. $\left(\dfrac{z}{4} - 5\right)^{2/3} = \dfrac{1}{25}$

63. $(z^2 + 17)^{3/4} = 27$

64. $\sqrt{\sqrt{y} + 49} = 7$

65. $x^2 - 5x - \sqrt{x^2 - 5x - 2} = 4$
(*Hint:* Let $u = x^2 - 5x - 2$.)

66. $\sqrt{8 - b} = b\sqrt{8 - b}$

Without graphing, determine the x-intercepts of the graphs given by each of the following.

67. $f(x) = \sqrt{x - 2} - \sqrt{x + 2} + 2$

68. $g(x) = 6x^{1/2} + 6x^{-1/2} - 37$

69. $f(x) = (x^2 + 30x)^{1/2} - x - (5x)^{1/2}$

Focus: Rational exponents, estimation, and solving
equations

Time: 20–30 minutes

Group size: 2–3

Materials: Calculators are required

Calistoga, California, has been a popular resort
town for over 100 years, famous for its naturally oc-
curring bubbling water. Thus residents were under-
standably unhappy with a proposal to raise the city
water rates over a period of three years: 30% in the
first year, 27% in the second, and 12% in the third.

Activity

1. Each group member should estimate the total
   amount of the percent increase in water rates at
   the end of the three-year period. Remember
   that the increase of 27% in the second year is
   applied to the new, higher rates that would exist
   after the 30% rate hike. Similarly, the 12% increase would apply to the rates in effect after
   the first two increases.

2. Determine, as a group, what the *exact* percent
   increase would be for the three-year period.
   Then find a multiplier that could be used to
   predict a household's new bill after the proposed increases.

3. Each group member should estimate what three
   equal rate increases would produce the same
   total increase at the end of three years. Check
   how well each estimate works.

4. Using a calculator and the multiplier found in
   part (2) above, determine, as a group, a precise
   answer to part (3).

5. What would benefit consumers most: three
   equal rate hikes, as determined in part (4), or
   the three increases originally proposed? Share
   the reasoning behind your group's answer to
   this question with the entire class.

7.7
Geometric Applications

- *Using the Pythagorean Theorem*
- *Two Special Triangles*
- *The Distance Formula*
- *Circles*

Using the Pythagorean Theorem

There are many kinds of problems that involve powers and roots. Many
also involve right triangles and the Pythagorean theorem, which we studied in Section 5.7.

Example 1 *Baseball.* A baseball diamond is actually a square 90 ft on
a side. Suppose that a catcher fields a ball along the third-base line 10 ft
from home plate. How far would the catcher's throw to first base be? Give
an exact answer and an approximation to three decimal places.

SOLUTION   We first make a drawing and let d = the distance, in feet, to
first base. Note that a right triangle is formed in which the length of the
leg from home to first base is 90 ft. The length of the leg from home to

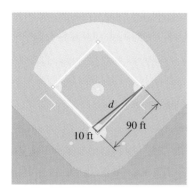

where the catcher fields the ball is 10 ft. We substitute these values into the Pythagorean equation to find d:

$$d^2 = 90^2 + 10^2$$
$$d^2 = 8100 + 100$$
$$d^2 = 8200$$
$$d = \sqrt{8200}.*$$

Exact answer: $d = \sqrt{8200}$ ft

Approximation: $d \approx 90.554$ ft   **Using a calculator**   ▬

Example 2 *Guy Wires.* The base of a 40-ft-long guy wire is located 15 ft from the telephone pole that it is anchoring. How high up the pole does the guy wire reach? Give an exact answer and an approximation to three decimal places.

SOLUTION   We make a drawing and let $h =$ the height on the pole that the guy wire reaches. A right triangle is formed in which the length of one leg is 15 ft and the length of the hypotenuse is 40 ft. Using the Pythagorean equation, we have

$$h^2 + 15^2 = 40^2$$
$$h^2 + 225 = 1600$$
$$h^2 = 1375$$
$$h = \sqrt{1375}.$$

Exact answer:
$$h = \sqrt{1375} \text{ ft}$$

Approximation:
$$h \approx 37.081 \text{ ft}$$
Using a calculator

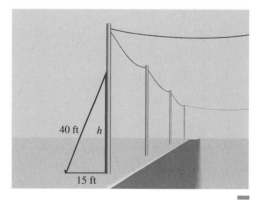

Two Special Triangles

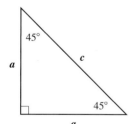

When both legs of a right triangle are the same size, we call the triangle an *isosceles right triangle*. If one leg of an isosceles right triangle has length a, we can find a formula for the length of the hypotenuse as follows:

$$c^2 = a^2 + b^2$$
$$c^2 = a^2 + a^2 \qquad \text{Because the triangle is isosceles, both legs are the same size: } a = b.$$
$$c^2 = 2a^2 \qquad \text{Combining like terms}$$
$$c = \sqrt{2a^2}$$
$$c = \sqrt{a^2 \cdot 2} = a\sqrt{2}. \qquad \text{Try to remember this formula.}$$

---

*Actually $d^2 = 8200$ also has $-\sqrt{8200}$ as a solution, but since the problems in this section all involve length, we concern ourselves only with positive answers.

Example 3   One leg of an isosceles right triangle measures 7 cm. Find the length of the hypotenuse. Give an exact answer and an approximation to three decimal places.

SOLUTION   We substitute:

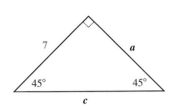

$$c = a\sqrt{2} \qquad \text{This equation should be memorized.}$$
$$= 7\sqrt{2}.$$

Exact answer: $c = 7\sqrt{2}$ cm

Approximation: $c \approx 9.899$ cm   **Using a calculator**   ▬

When the hypotenuse of an isosceles right triangle is known, the lengths of the legs can be found.

Example 4   The hypotenuse of an isosceles right triangle is 5 ft long. Find the length of a leg. Give an exact answer and an approximation to three decimal places.

SOLUTION   We replace c with 5 and solve for a:

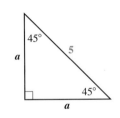

$$5 = a\sqrt{2} \qquad \text{Substituting 5 for } c \text{ in } c = a\sqrt{2}$$

$$\frac{5}{\sqrt{2}} = a \qquad \text{Dividing by } \sqrt{2} \text{ on both sides}$$

$$\frac{5\sqrt{2}}{2} = a. \qquad \text{Rationalize the denominator if necessary.}$$

Exact answer: $a = \dfrac{5}{\sqrt{2}}$ ft, or $\dfrac{5\sqrt{2}}{2}$ ft

Approximation: $a \approx 3.536$ ft   **Using a calculator**   ▬

A second special triangle is known as a 30°–60°–90° right triangle, so named because of the measures of its angles. Note that in an equilateral triangle, all sides have the same length and all angles are 60°. An altitude, drawn dashed in the figure, bisects, or splits, one angle and one side. Two 30°–60°–90° right triangles are thus formed.

Because of the way in which the altitude is drawn, if a represents the length of the shorter leg in a 30°–60°–90° right triangle, then $2a$ represents the length of the hypotenuse. We have

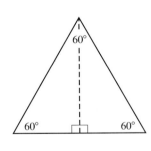

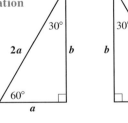

$$a^2 + b^2 = (2a)^2$$
$$\qquad \text{Using the Pythagorean equation}$$
$$a^2 + b^2 = 4a^2$$
$$b^2 = 3a^2$$
$$\qquad \text{Adding } -a^2$$
$$\qquad \text{on both sides}$$
$$b = \sqrt{3a^2}$$
$$= \sqrt{a^2 \cdot 3}$$
$$= a\sqrt{3}.$$

Example 5   The shorter leg of a 30°–60°–90° right triangle measures 8 in. Find the lengths of the other sides. Give exact answers and, where appropriate, an approximation to three decimal places.

SOLUTION   The hypotenuse is twice as long as the shorter leg, so we have

$$c = 2a$$ **This relationship should be memorized.**
$$= 2 \cdot 8$$
$$= 16 \text{ in.}$$

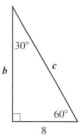

The length of the longer leg is the length of the shorter leg times $\sqrt{3}$. This gives us

$$b = a\sqrt{3}$$ **This should also be memorized.**
$$= 8\sqrt{3} \text{ in.}$$

Exact answer: $c = 16$ in., $b = 8\sqrt{3}$ in.
Approximation: $b \approx 13.856$ in.

Example 6   The length of the longer leg of a 30°–60°–90° right triangle is 14 cm. Find the length of the hypotenuse. Give an exact answer and an approximation to three decimal places.

SOLUTION   The length of the hypotenuse is twice the length of the shorter leg. We first find a, the length of the shorter leg, by using the length of the longer leg:

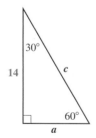

$$14 = a\sqrt{3}$$ **Substituting 14 for b in $b = a\sqrt{3}$**
$$\frac{14}{\sqrt{3}} = a.$$ **Dividing by $\sqrt{3}$**

Since the hypotenuse is twice as long as the shorter leg, we have

$$c = 2a$$
$$= 2 \cdot \frac{14}{\sqrt{3}}$$ **Substituting**
$$= \frac{28}{\sqrt{3}} \text{ cm.}$$

Exact answer: $c = \dfrac{28}{\sqrt{3}}$ cm, or $\dfrac{28\sqrt{3}}{3}$ cm if the denominator is rationalized.
Approximation: $c \approx 16.166$ cm

Lengths Within Isosceles and 30°–60°–90° Right Triangles

The length of the hypotenuse in an isosceles right triangle is the length of a leg times $\sqrt{2}$.

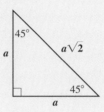

The length of the longer leg in a 30°–60°–90° right triangle is the length of the shorter leg times $\sqrt{3}$. The hypotenuse is twice as long as the shorter leg.

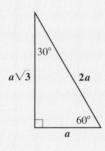

The Distance Formula

The Pythagorean theorem is also used to find the distance between two points.

The distance between (x_1, y_1) and (x_2, y_1) is $|x_2 - x_1|$.

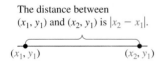

If two points are on a horizontal line, they have the same second coordinate. We can find the distance between them by subtracting their first coordinates. This difference may be negative, depending on the order in which we subtract. Thus, to make sure that we get a positive number, we take the absolute value of this difference. For example, the distance between $(3, 1)$ and $(-5, 1)$ is

$$|3 - (-5)| = |8| = 8$$

or $$|-5 - 3| = |-8| = 8.$$

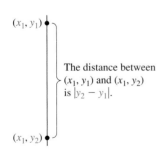

The distance between (x_1, y_1) and (x_1, y_2) is $|y_2 - y_1|$.

The distance between two general points on a horizontal line (x_1, y_1) and (x_2, y_1) is thus $|x_2 - x_1|$ or $|x_1 - x_2|$. Similarly, the distance between two points on a vertical line (x_1, y_1) and (x_1, y_2) is $|y_2 - y_1|$.

If two points are not on a horizontal line or a vertical line, then the distance between them can be viewed as the hypotenuse of a right triangle. If the two points are (x_1, y_1) and (x_2, y_2), then the point (x_2, y_1) is the other vertex of a right triangle. The lengths of the legs are $|x_2 - x_1|$ and $|y_2 - y_1|$.

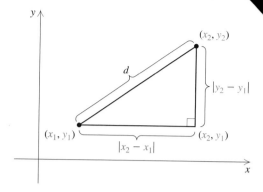

We can find d, the distance between (x_1, y_1) and (x_2, y_2), by using the Pythagorean theorem:

$$d^2 = |x_2 - x_1|^2 + |y_2 - y_1|^2.$$

Since the square of a number is the same as the square of its opposite, we need not include the absolute-value signs. Thus,

$$d^2 = (x_2 - x_1)^2 + (y_2 - y_1)^2.$$

Taking the principal square root, we obtain the distance between two points.

The Distance Formula

The distance d between any two points (x_1, y_1) and (x_2, y_2) is given by

$$d = \sqrt{(x_2 - x_1)^2 + (y_2 - y_1)^2}.$$

Example 7 Find the distance between $(5, -1)$ and $(-4, 6)$. Find an exact answer and an approximation to three decimal places.

SOLUTION We substitute into the distance formula:

$$d = \sqrt{(-4 - 5)^2 + [6 - (-1)]^2} \quad \textbf{Substituting}$$
$$= \sqrt{(-9)^2 + 7^2}$$
$$= \sqrt{130} \approx 11.402.$$

The distance is $\sqrt{130}$, or approximately 11.402.

Circles

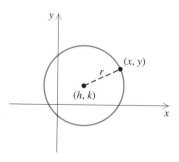

The distance formula is the basis for an equation for a circle. A **circle** is a set of points in a plane that are a fixed distance r, called the **radius**, from a fixed point (h, k), called the **center**.

Any point (x, y) that is on the circle is r units from the center (h, k).

We can write this distance using the distance formula as

$$r = \sqrt{(x - h)^2 + (y - k)^2}.$$

Squaring both sides gives the equation of a circle in standard form.

Equation of a Circle

The equation of a circle, centered at (h, k), with radius r, is given by

$$(x - h)^2 + (y - k)^2 = r^2.$$

Note that when $h = 0$ and $k = 0$, the circle is centered at the origin, and the equation of the circle is in the form

$$x^2 + y^2 = r^2.$$

Example 8 Find the center and the radius and then graph each circle.

a) $(x - 2)^2 + (y + 3)^2 = 4^2$

b) $x^2 + y^2 = 5$

SOLUTION

a) We write standard form:

$$(x - 2)^2 + [y - (-3)]^2 = 4^2.$$

The center is $(2, -3)$ and the radius is 4. The graph is easily drawn using a compass. Note that the center is not part of the graph of the circle.

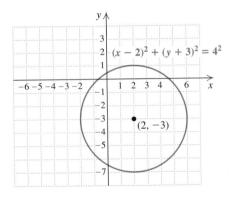

b) We first write the equation in standard form:

$$x^2 + y^2 = (\sqrt{5})^2.$$

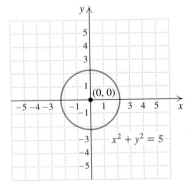

The center is $(0, 0)$ and the radius is $\sqrt{5}$. The graph is shown at the left.

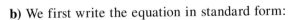

Example 9 Find an equation of the circle having center (4, 5) and radius 6.

SOLUTION Using the standard form, we obtain

$$(x - 4)^2 + (y - 5)^2 = 6^2, \qquad \textbf{Using } (x - h)^2 + (y - k)^2 = r^2$$

or

$$(x - 4)^2 + (y - 5)^2 = 36.$$

Many graphers will draw a circle given the center and the radius. In order for the graph to appear circular, the units on each axis should be the same length. The figure below shows the graph of the circle of Example 9.

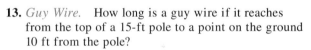

Circle (4, 5, 6)

7.7 Exercise Set

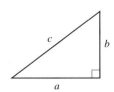

In a right triangle, find the length of the side not given. Give an exact answer and, where appropriate, an approximation to three decimal places.

1. $a = 5, b = 3$ **2.** $a = 8, b = 10$

3. $a = 7, b = 7$ **4.** $a = 10, b = 10$

5. $b = 12, c = 13$ **6.** $a = 5, c = 12$

7. $c = 6, a = \sqrt{5}$ **8.** $c = 8, a = 4\sqrt{3}$

9. $b = 1, c = \sqrt{13}$ **10.** $a = 1, c = \sqrt{20}$

11. $a = 1, c = \sqrt{n}$ **12.** $c = 2, a = \sqrt{n}$

In Exercises 13–20, give an exact answer and, where appropriate, an approximation to three decimal places.

13. *Guy Wire.* How long is a guy wire if it reaches from the top of a 15-ft pole to a point on the ground 10 ft from the pole?

14. *Softball.* A slow-pitch softball diamond is actually a square 65 ft on a side. How far is it from home to second base?

15. *Baseball.* Suppose the catcher in Example 1 makes a throw to second base. How far is that throw?

16. *Television Sets.* What does it mean to refer to a 20-in. TV set or a 25-in. TV set? Such units refer to the diagonal of the screen. A 20-in. TV set has a width of 16 in. What is its height?

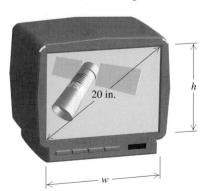

20 in.

17. *Television Sets.* A 25-in. TV set has a screen with a height of 15 in. What is its width? (See Exercise 16.)

18. *Speaker Placement.* A stereo receiver is in a corner of a 12-ft by 14-ft room. Speaker wire will run under a rug, diagonally, to a speaker in the far corner. If 4 ft of slack is required on each end, how long a piece of wire should be purchased?

19. *Distance over Water.* To determine the width of a pond, a surveyor locates two stakes at either end of the pond and uses instrumentation to place a third stake so that the distance across the pond is the length of a hypotenuse. If the third stake is 90 m from one stake and 70 m from the other, how wide is the pond?

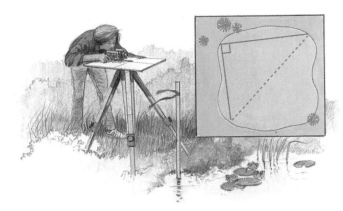

20. *Vegetable Garden.* Benito and Dominique are planting a 30-ft by 40-ft vegetable garden and are laying it out using string. They would like to know the length of a diagonal to make sure that right angles are formed. Find the length of a diagonal.

For each triangle, find the missing length(s). Give an exact answer and, where appropriate, an approximation to three decimal places.

21.

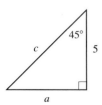

22.

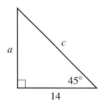

23.

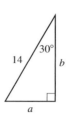

24.

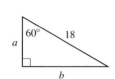

25.

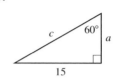

26.

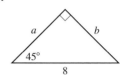

27.

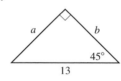

28.

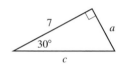

29.

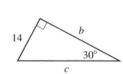

30.

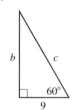

31.

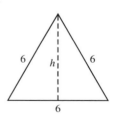

32.

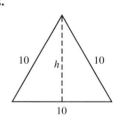

33.

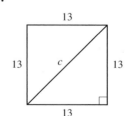

34.

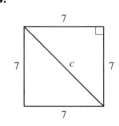

35.

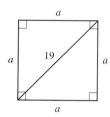

36.

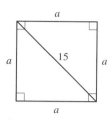

In Exercises 37–44, give an exact answer and, where appropriate, an approximation to three decimal places.

37. *Bridge Expansion.*   During the summer heat, a 2-mi bridge expands 2 ft in length. If we assume that the bulge occurs straight up the middle, how high is the bulge? (The answer may surprise you. Most bridges have expansion spaces to avoid such buckling.)

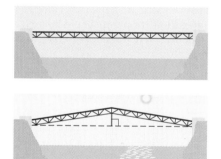

38. Triangle *ABC* has sides of lengths 25 ft, 25 ft, and 30 ft. Triangle *PQR* has sides of lengths 25 ft, 25 ft, and 40 ft. Which triangle has the greater area and by how much?

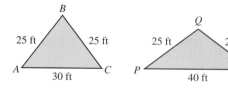

39. *Camping Tent.*   The entrance to a pup tent is the shape of an equilateral triangle. If the base of the tent is 4 ft wide, how tall is the tent?

40. Each side of a regular octagon has length *s*. Find a formula for the distance *d* between the parallel sides of the octagon.

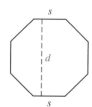

41. The diagonal of a square has length $8\sqrt{2}$ ft. Find the length of a side of the square.

42. The length and the width of a rectangle are given by consecutive integers. The area of the rectangle is 90 cm^2. Find the length of a diagonal of the rectangle.

43. Find all points on the *y*-axis of a Cartesian coordinate system that are 5 units from the point (3, 0).

44. Find all points on the *x*-axis of a Cartesian coordinate system that are 5 units from the point (0, 4).

Find the distance between each pair of points. Where appropriate, find an approximation to three decimal places.

45. (2, 7) and (6, 10)   **46.** (1, 10) and (7, 2)

47. (0, −7) and (3, −4)   **48.** (6, 2) and (6, −8)

49. (−5, 5) and (5, −5)   **50.** (5, 21) and (−3, 1)

51. (8.6, −3.4) and (−9.2, −3.4)

52. (5.9, 2) and (3.7, −7.7)

53. (0, $\sqrt{7}$) and ($\sqrt{6}$, 0)

54. (−$\sqrt{6}$, $\sqrt{2}$) and (0, 0)

55. (0, 0) and (*s*, *t*)

56. (*p*, *q*) and (0, 0)

Find the center and the radius of each circle.

57. $(x - 2)^2 + (y - 5)^2 = 10^2$

58. $x^2 + (y - 1)^2 = 5^2$

59. $(x + 2)^2 + y^2 = 64$

60. $(x - 7)^2 + (y + 1)^2 = 1$

61. $(x - 3)^2 + (y + 4)^2 = 7$

62. $(x + 1)^2 + (y - 1)^2 = 13$

Find an equation of the circle having the given center and radius.

63. (0, 3); 6   **64.** (2, 1); 4

65. (−5, −7); 1   **66.** (−4, 0); 3

67. (5, 7); $\sqrt{3}$ **68.** (−10, −10); $\sqrt{5}$

Skill Maintenance

Solve.

69. $x^2 - 11x + 24 = 0$ [5.4]

70. $2x^2 + 11x - 21 = 0$ [5.5]

71. $|3x - 5| = 7$ [4.3]

72. $|2x - 3| = |x + 7|$ [4.3]

Synthesis

73. ◆ Write a problem for a classmate to solve in which the solution is: "The height of the tepee is $5\sqrt{3}$ yd."

74. ◆ Write a problem for a classmate to solve in which the solution is: "The height of the window is $15\sqrt{3}$ ft."

75. A cube measures 5 cm on each side. How long is the diagonal that connects two opposite corners of the cube? Give an exact answer.

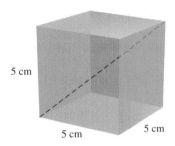

5 cm

5 cm   5 cm

76. *Roofing.* Kit's cottage, which is 24 ft wide and 32 ft long, needs a new roof. By counting clapboards that are 4 in. apart, Kit determines that the peak of the roof is 6 ft higher than the sides. If one packet of shingles covers 100 ft^2, how many packets will the job require?

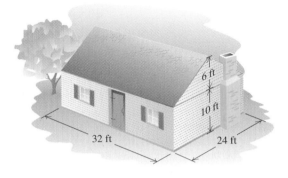

6 ft

10 ft

32 ft   24 ft

77. *Painting.* (Refer to Exercise 76.) A gallon of paint covers about 275 ft^2. If Kit's first floor is 10 ft high, how many gallons of paint should be bought to paint the house? What assumption(s) is made in your answer?

7.8
The Complex Numbers

- *Imaginary and Complex Numbers*
- *Addition and Subtraction*
- *Multiplication*
- *Conjugates and Division*
- *Powers of i*

Imaginary and Complex Numbers

Negative numbers do not have square roots in the real-number system. However, a larger number system that contains the real-number system is designed so that negative numbers *do* have square roots. That system is called the **complex-number system**, and it makes use of a number that is a square root of -1. We call this new number i.

> **The Number i**
> We define the number i such that $i = \sqrt{-1}$ and $i^2 = -1$.

To express roots of negative numbers in terms of i, we can use the fact that in the complex numbers, $\sqrt{-p} = \sqrt{-1}\sqrt{p}$ when p is a positive real number.

Example 1 Express in terms of i: **(a)** $\sqrt{-7}$; **(b)** $\sqrt{-16}$; **(c)** $-\sqrt{-13}$; **(d)** $-\sqrt{-50}$.

SOLUTION

> *i is **not** under the radical.*

a) $\sqrt{-7} = \sqrt{-1 \cdot 7} = \sqrt{-1} \cdot \sqrt{7} = i\sqrt{7}$, or $\sqrt{7}i$

b) $\sqrt{-16} = \sqrt{-1 \cdot 16} = \sqrt{-1} \cdot \sqrt{16} = i \cdot 4 = 4i$

c) $-\sqrt{-13} = -\sqrt{-1 \cdot 13} = -\sqrt{-1} \cdot \sqrt{13} = -i\sqrt{13}$, or $-\sqrt{13}i$

d) $-\sqrt{-50} = -\sqrt{-1} \cdot \sqrt{25} \cdot \sqrt{2} = -i \cdot 5 \cdot \sqrt{2} = -5i\sqrt{2}$, or $-5\sqrt{2}i$

Imaginary Numbers

An *imaginary number* is a number that can be written $a + bi$, where a and b are real numbers and $b \neq 0$.

Don't let the name "imaginary" fool you. Imaginary numbers appear in such fields as engineering and the physical sciences. The following are examples of imaginary numbers:

$5 + 4i$,   Here $a = 5$, $b = 4$.

$\sqrt{5} - \pi i$,   Here $a = \sqrt{5}$, $b = -\pi$.

$17i$.   Here $a = 0$, $b = 17$.

When a and b are real numbers and b is allowed to be 0, the number $a + bi$ is said to be **complex**.

Complex Numbers

A *complex number* is any number that can be written $a + bi$, where a and b are real numbers. (Note that a and b both can be 0.)

The following are examples of complex numbers:

$7 + 3i$ (here $a \neq 0$, $b \neq 0$); $4i$ (here $a = 0$, $b \neq 0$);

8 (here $a \neq 0$, $b = 0$); 0 (here $a = 0$, $b = 0$).

Imaginary numbers like $17i$ or $4i$, in which $a = 0$ and $b \neq 0$, have no real part and are called *pure imaginary numbers.*

Note that when $b = 0$, $a + 0i = a$, so every real number is a complex number. The relationships among various real and complex numbers are shown below.

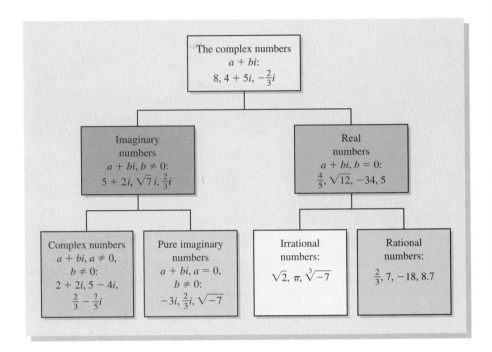

Observe that although $\sqrt{-7}$ and $\sqrt[3]{-7}$ are both complex numbers, $\sqrt{-7}$ is imaginary whereas $\sqrt[3]{-7}$ is real.

Addition and Subtraction

The complex numbers obey the commutative, associative, and distributive laws. Thus we can add and subtract them as we do binomials.

Example 2   Add or subtract and simplify.

a) $(8 + 6i) + (3 + 2i)$ **b)** $(4 + 5i) - (6 - 3i)$

SOLUTION

a) $(8 + 6i) + (3 + 2i) = (8 + 3) + (6i + 2i)$ **Combining the real parts and the imaginary parts**

$$= 11 + (6 + 2)i = 11 + 8i$$

b) $(4 + 5i) - (6 - 3i) = (4 - 6) + [5i - (-3i)]$ **Note that the 6 and the $-3i$ are both being subtracted.**

$$= -2 + 8i$$

Multiplication

For complex numbers, the property $\sqrt{a}\ \sqrt{b} = \sqrt{ab}$ does *not* hold in general, but it does hold when $a = -1$ and b is nonnegative. To multiply square roots of negative real numbers, we first express them in terms of i. For example,

$$\sqrt{-2} \cdot \sqrt{-5} = \sqrt{-1} \cdot \sqrt{2} \cdot \sqrt{-1} \cdot \sqrt{5}$$
$$= \quad i \quad \cdot \sqrt{2} \cdot \quad i \quad \cdot \sqrt{5}$$
$$= i^2\sqrt{10}$$
$$= -1\sqrt{10} = -\sqrt{10} \text{ is correct!}$$

Caution! With complex numbers, simply multiplying radicands is *incorrect*: $\sqrt{-2} \cdot \sqrt{-5} \neq \sqrt{10}$.

With this in mind, we can now multiply complex numbers.

Example 3 Multiply and simplify.

a) $\sqrt{-16} \cdot \sqrt{-25}$ 　　　　　　**b)** $\sqrt{-5} \cdot \sqrt{-7}$

c) $-3i \cdot 8i$ 　　　　　　　　　　**d)** $(1 + 2i)(1 + 3i)$

SOLUTION

a) $\sqrt{-16} \cdot \sqrt{-25} = \sqrt{-1} \cdot \sqrt{16} \cdot \sqrt{-1} \cdot \sqrt{25}$
$$= \quad i \quad \cdot \quad 4 \quad \cdot \quad i \quad \cdot \quad 5$$
$$= i^2 \cdot 20$$
$$= -1 \cdot 20 \qquad\qquad i^2 = -1$$
$$= -20$$

b) $\sqrt{-5} \cdot \sqrt{-7} = \sqrt{-1} \cdot \sqrt{5} \cdot \sqrt{-1} \cdot \sqrt{7}$ 　　**Try to do this step mentally.**
$$= \quad i \quad \cdot \sqrt{5} \cdot \quad i \quad \cdot \sqrt{7}$$
$$= i^2 \cdot \sqrt{35}$$
$$= -1 \cdot \sqrt{35} \qquad\qquad i^2 = -1$$
$$= -\sqrt{35}$$

c) $-3i \cdot 8i = -24 \cdot i^2$
$$= -24 \cdot (-1) \qquad\qquad i^2 = -1$$
$$= 24$$

d) $(1 + 2i)(1 + 3i) = 1 + 3i + 2i + 6i^2$ 　　**Multiplying each term of one number by every term of the other (FOIL)**
$$= 1 + 3i + 2i - 6 \qquad i^2 = -1$$
$$= -5 + 5i \qquad\qquad \textbf{Combining like terms}$$

Conjugates and Division

Conjugates of complex numbers are defined as follows.

Conjugate of a Complex Number
The *conjugate* of a complex number $a + bi$ is $a - bi$, and the *conjugate* of $a - bi$ is $a + bi$.

Example 4 Find the conjugate: **(a)** $-3 + 7i$; **(b)** $14 - 5i$; **(c)** $4i$.

SOLUTION

a) $-3 + 7i$ The conjugate is $-3 - 7i$.
b) $14 - 5i$ The conjugate is $14 + 5i$.
c) $4i$ The conjugate is $-4i$. Note that $4i = 0 + 4i$.   ▬

The product of a complex number and its conjugate is a real number.

Example 5 Multiply: $(5 + 7i)(5 - 7i)$.

SOLUTION

$$(5 + 7i)(5 - 7i) = 5^2 - (7i)^2 \quad \text{Using}$$
$$(A + B)(A - B) = A^2 - B^2$$
$$= 25 - 49i^2$$
$$= 25 - 49(-1) \quad i^2 = -1$$
$$= 25 + 49 = 74$$ ▬

Conjugates are used when dividing complex numbers. The procedure is much like that used to rationalize denominators in Section 7.5.

Example 6 Divide and simplify to the form $a + bi$: **(a)** $\dfrac{-5 + 9i}{1 - 2i}$; **(b)** $\dfrac{7 + 3i}{5i}$.

SOLUTION

a) To divide and simplify $(-5 + 9i)/(1 - 2i)$, we multiply by 1, using the conjugate of the denominator to form 1:

$$\frac{-5 + 9i}{1 - 2i} = \frac{-5 + 9i}{1 - 2i} \cdot \frac{1 + 2i}{1 + 2i} \qquad \text{Multiplying by 1 using the conjugate of the denominator in the symbol for 1}$$

$$= \frac{(-5 + 9i)(1 + 2i)}{(1 - 2i)(1 + 2i)}$$

$$= \frac{-5 - 10i + 9i + 18i^2}{1^2 - 4i^2} \qquad \text{Using FOIL}$$

Then,

$$= \frac{-5 - i - 18}{1 - 4(-1)} \qquad i^2 = -1$$

$$= \frac{-23 - i}{5}$$

$$= -\frac{23}{5} - \frac{1}{5}i \qquad \text{Writing in the form } a + bi$$

b) $\dfrac{7 + 3i}{5i} = \dfrac{7 + 3i}{5i} \cdot \dfrac{-5i}{-5i}$ **Multiplying by 1 using the conjugate of $0 + 5i$ in the symbol for 1**

$$= \frac{-35i - 15i^2}{-25i^2} \qquad \text{Multiplying}$$

$$= \frac{-35i - 15(-1)}{-25(-1)} \qquad i^2 = -1$$

$$= \frac{15 - 35i}{25}$$

$$= \frac{15}{25} - \frac{35}{25}i$$

$$= \frac{3}{5} - \frac{7}{5}i$$

Powers of *i*

Recall that -1 raised to an *even* power is 1, and -1 raised to an *odd* power is -1. Simplifying powers of *i* can then be done by using the fact that $i^2 = -1$ and expressing the given power of *i* in terms of i^2. Consider the following:

$$i, \text{ or } \sqrt{-1},$$
$$i^2 = -1,$$
$$i^3 = i^2 \cdot i = (-1)i = -i,$$
$$i^4 = (i^2)^2 = (-1)^2 = 1,$$
$$i^5 = i^4 \cdot i = (i^2)^2 \cdot i = (-1)^2 \cdot i = i,$$
$$i^6 = (i^2)^3 = (-1)^3 = -1.$$

Note that the powers of *i* cycle themselves through the values $i, -1, -i,$ and 1.

Example 7   Simplify: **(a)** i^{18}; **(b)** i^{24}; **(c)** i^{29}; **(d)** i^{75}.

SOLUTION

a) $i^{18} = (i^2)^9$ **b)** $i^{24} = (i^2)^{12}$

      $= (-1)^9 = -1$ $= (-1)^{12} = 1$

c) $i^{29} = i^{28}i^1$

$\quad = (i^2)^{14}i$

$\quad = (-1)^{14}i$

$\quad = 1 \cdot i = i$

d) $i^{75} = i^{74}i^1$

$\quad = (i^2)^{37}i$

$\quad = (-1)^{37}i$

$\quad = -1 \cdot i = -i$

7.8 Exercise Set

Express in terms of i.

1. $\sqrt{-25}$ **2.** $\sqrt{-36}$ **3.** $\sqrt{-13}$

4. $\sqrt{-19}$ **5.** $\sqrt{-18}$ **6.** $\sqrt{-98}$

7. $\sqrt{-3}$ **8.** $\sqrt{-4}$ **9.** $\sqrt{-81}$

10. $\sqrt{-27}$ **11.** $\sqrt{-300}$ **12.** $-\sqrt{-75}$

13. $-\sqrt{-49}$ **14.** $-\sqrt{-125}$

15. $4 - \sqrt{-60}$ **16.** $6 - \sqrt{-84}$

17. $\sqrt{-4} + \sqrt{-12}$ **18.** $-\sqrt{-76} + \sqrt{-125}$

Perform the indicated operation and simplify. Write each answer in the form a + bi.

19. $(4 + 7i) + (5 - 2i)$ **20.** $(5 + 2i) + (7 - i)$

21. $(-2 + 8i) + (5 + 3i)$ **22.** $(4 - 5i) + (3 + 9i)$

23. $(9 + 8i) - (5 + 3i)$ **24.** $(9 + 7i) - (2 + 4i)$

25. $(8 - 3i) - (9 + 2i)$ **26.** $(7 - 4i) - (5 - 3i)$

27. $(-2 + 6i) - (-7 + i)$ **28.** $(-5 - i) - (7 + 4i)$

29. $6i \cdot 5i$ **30.** $7i \cdot 6i$

31. $7i \cdot (-9i)$ **32.** $(-4i)(-6i)$

33. $\sqrt{-49}\,\sqrt{-25}$ **34.** $\sqrt{-36}\,\sqrt{-9}$

35. $\sqrt{-6}\,\sqrt{-7}$ **36.** $\sqrt{-5}\,\sqrt{-2}$

37. $\sqrt{-15}\,\sqrt{-10}$ **38.** $\sqrt{-6}\,\sqrt{-21}$

39. $2i(7 + 3i)$ **40.** $5i(2 + 6i)$

41. $-4i(6 - 5i)$ **42.** $-7i(3 - 4i)$

43. $(2 + 5i)(4 + 3i)$ **44.** $(1 + i)(3 + 2i)$

45. $(5 - 6i)(2 + 5i)$ **46.** $(6 - 5i)(3 + 4i)$

47. $(-4 + 5i)(3 - 4i)$ **48.** $(7 - 2i)(2 - 6i)$

49. $(7 - 3i)(4 - 7i)$ **50.** $(5 - 3i)(4 - 5i)$

51. $(-3 + 6i)(-3 + 4i)$ **52.** $(-2 + 3i)(-2 + 5i)$

53. $(2 + 9i)(-3 - 5i)$ **54.** $(-5 - 4i)(3 + 7i)$

55. $(5 - 2i)^2$ **56.** $(3 - 2i)^2$

57. $(4 + 2i)^2$ **58.** $(2 + 3i)^2$

59. $(-5 - 2i)^2$ **60.** $(-2 + 3i)^2$

61. $\dfrac{7}{2 - i}$ **62.** $\dfrac{4}{3 + i}$

63. $\dfrac{3i}{5 + 2i}$ **64.** $\dfrac{4i}{5 - 3i}$

65. $\dfrac{8}{9i}$ **66.** $\dfrac{5}{8i}$

67. $\dfrac{7 - 2i}{6i}$ **68.** $\dfrac{3 + 8i}{9i}$

69. $\dfrac{4 + 5i}{3 - 7i}$ **70.** $\dfrac{5 + 3i}{7 - 4i}$

71. $\dfrac{3 - 2i}{4 + 3i}$ **72.** $\dfrac{5 - 2i}{3 + 6i}$

Simplify.

73. i^7 **74.** i^{11} **75.** i^{24}

76. i^{35} **77.** i^{42} **78.** i^{64}

79. i^9 **80.** $(-i)^{71}$ **81.** $(-i)^6$

82. $(-i)^4$ **83.** $(5i)^3$ **84.** $(-3i)^5$

85. $i^2 + i^4$ **86.** $5i^5 + 4i^3$ **87.** $i^5 + i^7$

88. $i^{84} - i^{100}$

Skill Maintenance

Solve.

89. $\dfrac{x + 2}{x} + \dfrac{1}{x + 2} = \dfrac{4}{x^2 + 2x}$ [6.4]

90. $\dfrac{5}{t} - \dfrac{3}{2} = \dfrac{4}{7}$ [6.4]

91. $28 = 3x^2 - 17x$ [5.5]

92. $|3x + 7| < 22$ [4.3]

Synthesis

93. ◈ Is the product of two imaginary numbers always an imaginary number? Why or why not?

94. ◈ In what way(s) are conjugates of complex numbers similar to the conjugates used in Section 7.5?

95. ◈ Is the set of real numbers a subset of the complex numbers? Why or why not?

96. ◈ Is the union of the set of imaginary numbers and the set of real numbers the set of complex numbers? Why or why not?

97. A function g is given by

$$g(z) = \frac{z^4 - z^2}{z - 1}.$$

Find $g(2i)$; $g(i + 1)$; $g(2i - 1)$.

98. Evaluate

$$\frac{1}{w - w^2} \quad \text{for} \quad w = \frac{1 - i}{10}.$$

Simplify.

99. $\dfrac{i^5 + i^6 + i^7 + i^8}{(1 - i)^4}$

100. $(1 - i)^3(1 + i)^3$

101. $\dfrac{5 - \sqrt{5}i}{\sqrt{5}i}$

102. $\dfrac{6}{1 + \dfrac{3}{i}}$

103. $\left(\dfrac{1}{2} - \dfrac{1}{3}i\right)^2 - \left(\dfrac{1}{2} + \dfrac{1}{3}i\right)^2$

104. $\dfrac{i - i^{38}}{1 + i}$

CHAPTER

7 Summary and Review

Key Terms

Square root, p. 418
Principal square root, p. 418
Radical sign, p. 419
Radicand, p. 419
Radical expression, p. 419
Cube root, p. 423
nth root, p. 425
Index (pl., indices), p. 425
Odd root, p. 425
Even root, p. 425
Radical function, p. 426

Rational exponent, p. 431
Perfect square, p. 438
Perfect cube, p. 438
Perfect nth power, p. 438
Like radicals, p. 441
Conjugates, p. 451
Rationalizing, p. 452
Radical equation, p. 456
Isosceles right triangle, p. 465
30°–60°–90° right triangle,
  p. 466

Circle, p. 469
Radius, p. 469
Center, p. 469
Complex-number system,
  p. 475
Imaginary number, p. 475
Complex number, p. 475
Pure imaginary number,
  p. 475
Conjugate of a complex
  number, p. 478

Important Properties and Formulas

The number c is a square root of a if $c^2 = a$.
The number c is the cube root of a if $c^3 = a$.

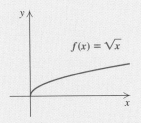

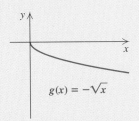

Domain of f: $[0, \infty)$ Domain of g: $[0, \infty)$
Range of f: $[0, \infty)$ Range of g: $(-\infty, 0]$

For any real number a:

a) $\sqrt[n]{a^n} = |a|$ when n is even. Unless a is known to be nonnegative, absolute-value notation is needed when n is even.

b) $\sqrt[n]{a^n} = a$ when n is odd. Absolute-value notation is not used when n is odd.

$a^{1/n}$ means $\sqrt[n]{a}$. When a is nonnegative, n can be any index. When a is negative, n must be odd.

For any natural numbers m and n ($n \neq 1$), and any real number a for which $\sqrt[n]{a}$ exists,

$$a^{m/n} \quad \text{means} \quad (\sqrt[n]{a})^m \quad \text{or} \quad \sqrt[n]{a^m}.$$

For any rational number m/n and any nonzero real number a for which $a^{m/n}$ exists,

$$a^{-m/n} \quad \text{means} \quad \frac{1}{a^{m/n}}.$$

For any real numbers a and b and any rational exponents m and n for which a^m, a^n, and b^m are defined:

1. $a^m \cdot a^n = a^{m+n}$ In multiplying, add exponents if the bases are the same.

2. $\dfrac{a^m}{a^n} = a^{m-n}$ In dividing, subtract exponents if the bases are the same. (Assume $a \neq 0$.)

3. $(a^m)^n = a^{m \cdot n}$ To raise a power to a power, multiply the exponents.

4. $(ab)^m = a^m b^m$ To raise a product to a power, raise each factor to the power and multiply.

The Product Rule for Radicals
For any real numbers $\sqrt[n]{a}$ and $\sqrt[n]{b}$,

$$\sqrt[n]{a}\,\sqrt[n]{b} = \sqrt[n]{a \cdot b}.$$

The Quotient Rule for Radicals
For any real numbers $\sqrt[n]{a}$ and $\sqrt[n]{b}$, $b \neq 0$,

$$\sqrt[n]{\frac{a}{b}} = \frac{\sqrt[n]{a}}{\sqrt[n]{b}}.$$

Some Ways to Simplify Radical Expressions
1. *Simplifying by factoring.* Factor the radicand and look for factors raised to powers that are divisible by the index.

Example: $\sqrt[3]{a^6 b} = \sqrt[3]{a^6}\,\sqrt[3]{b} = a^2\sqrt[3]{b}$

2. *Using rational exponents to simplify.* Convert to exponential notation and then use arithmetic and the laws of exponents to simplify the exponents. Then convert back to radical notation.

Example: $\sqrt[3]{p} \cdot \sqrt[4]{q^3} = p^{1/3} \cdot q^{3/4}$
$= p^{4/12} \cdot q^{9/12}$
$= \sqrt[12]{p^4 q^9}$

3. *Combining like radical terms.*

Example: $\sqrt{8} + 3\sqrt{2} = \sqrt{4} \cdot \sqrt{2} + 3\sqrt{2}$
$= 2\sqrt{2} + 3\sqrt{2}$
$= 5\sqrt{2}$

The Principle of Powers

If $a = b$, then $a^n = b^n$ for any exponent n.

A general strategy for solving equations with two or more radical terms is as follows.

1. Isolate one of the radical terms.
2. Use the principle of powers.
3. If a radical remains, repeat steps (1) and (2).
4. Check possible solutions in the original equation.

Special Triangles

The length of the hypotenuse in an isosceles right triangle is the length of a leg times $\sqrt{2}$.

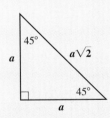

The length of the longer leg in a 30°–60°–90° right triangle is the length of the shorter leg times $\sqrt{3}$. The hypotenuse is twice as long as the shorter leg.

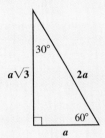

The Distance Formula

The distance d between any two points (x_1, y_1) and (x_2, y_2) is given by

$$d = \sqrt{(x_2 - x_1)^2 + (y_2 - y_1)^2}.$$

Equation of a Circle

The equation of a circle, centered at (h, k), with radius r, is given by

$$(x - h)^2 + (y - k)^2 = r^2.$$

A complex number is any number that can be written $a + bi$, where a and b are real numbers and $i = \sqrt{-1}$.

REVIEW EXERCISES

Simplify.

1. $\sqrt{\dfrac{49}{36}}$

2. $\sqrt{0.25}$

Let $f(x) = \sqrt{2x - 7}$. *Find the following.*

3. $f(16)$

4. The domain of f

5. Estimate the range of f.

Simplify. Assume that each variable can represent any real number.

6. $\sqrt{49a^2}$

7. $\sqrt{(c + 8)^2}$

8. $\sqrt{x^2 - 6x + 9}$

9. $\sqrt{4x^2 + 4x + 1}$

10. $\sqrt[5]{-32}$

11. $\sqrt[3]{-\dfrac{64x^6}{27}}$

12. $\sqrt[6]{64x^{12}}$

13. Rewrite with rational exponents: $(\sqrt[3]{5ab})^4$.

14. Rewrite without rational exponents: $(16a^8)^{3/4}$.

Use rational exponents to simplify. Assume $x, y \geq 0$.

15. $\sqrt{x^6y^{10}}$ **16.** $(\sqrt[3]{a^2b})^5$

Simplify. Do not use negative exponents in the answers.

17. $(x^{-2/3})^{3/5}$ **18.** $\dfrac{7^{-1/3}}{7^{-1/2}}$

19. If $f(x) = \sqrt{25(x-3)^2}$, find a simplified form for $f(x)$.

Perform the indicated operation and, if possible, simplify. Write all answers using radical notation.

20. $\sqrt{5x}\,\sqrt{3y}$ **21.** $\sqrt[3]{a^5b}\,\sqrt[3]{27b}$

22. $\sqrt[3]{ab}\,\sqrt[5]{a^3b^2}$ **23.** $\dfrac{\sqrt[3]{60xy^3}}{\sqrt[3]{10x}}$

24. $\dfrac{\sqrt{75x}}{2\sqrt{3}}$ **25.** $\dfrac{\sqrt[3]{x^2}}{\sqrt[4]{x}}$

26. $5\sqrt[3]{x} + 2\sqrt[3]{x}$ **27.** $2\sqrt{75} - 7\sqrt{3}$

28. $\sqrt[3]{8x^4} + \sqrt[3]{xy^6}$

29. $\sqrt{50} + 2\sqrt{18} + \sqrt{32}$

30. $(\sqrt{5} - 3\sqrt{8})(\sqrt{5} + 2\sqrt{8})$

31. If $f(x) = x^2$, find $f(a - \sqrt{2})$.

32. Rationalize the denominator:
$$\frac{5\sqrt{a}}{\sqrt{a} + \sqrt{b}}.$$

33. Rationalize the numerator of the expression in Exercise 32.

34. Solve: $1 + \sqrt{x} = \sqrt{3x - 3}$.

35. If $f(x) = \sqrt[4]{x + 2}$, find a such that $f(a) = 2$.

Solve. Give an exact answer and, where appropriate, an approximation to three decimal places.

36. The diagonal of a square has length $9\sqrt{2}$ cm. Find the length of a side of the square.

37. A bookcase is 5 ft tall and has a 7-ft diagonal brace, as shown. How wide is the bookcase?

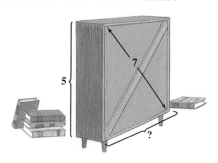

38. Find the missing lengths. Give exact answers and approximations to three decimal places.

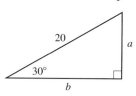

39. Find the distance between $(2, -5)$ and $(-1, -3)$. Give an exact answer and an approximation to three decimal places.

40. Find the center and the radius of the circle given by
$$(x - 12)^2 + (y + 4)^2 = 25.$$

41. Express in terms of i and simplify: $-\sqrt{-8}$.

42. Add: $(-4 + 3i) + (2 - 12i)$.

43. Subtract: $(4 - 7i) - (3 - 8i)$.

Multiply.

44. $(2 + 5i)(2 - 5i)$ **45.** i^{13}

46. $(6 - 3i)(2 - i)$

47. Divide and simplify to the form $a + bi$:
$$\frac{7 - 2i}{3 + 4i}.$$

Skill Maintenance

48. Find three consecutive positive integers such that the product of the first and second integers is 26 less than the product of the second and third integers.

49. Solve:
$$\frac{x}{x - 1} + \frac{3}{x + 1} = \frac{2}{x^2 - 1}.$$

50. Solve: $2x^2 + 3x - 27 = 0$.

51. Multiply and simplify:
$$\frac{x^2 + 3x}{x^2 - y^2} \cdot \frac{x^2 - xy + 2x - 2y}{x^2 - 9}.$$

Synthesis

52. ◈ Explain why $\sqrt[n]{x^n} = |x|$ when n is even, but $\sqrt[n]{x^n} = x$ when n is odd.

53. ◈ What is the difference between real numbers and complex numbers?

54. Solve: $\sqrt{11x + \sqrt{6 + x}} = 6$.

55. Simplify:
$$\frac{2}{1 - 3i} - \frac{3}{4 + 2i}.$$

CHAPTER 7 TEST

Simplify. Assume that variables can represent any real number.

1. $\sqrt{75}$

2. $\sqrt[3]{-\dfrac{8}{x^6}}$

3. $\sqrt{100a^2}$

4. $\sqrt{x^2 - 8x + 16}$

5. $\sqrt[5]{x^{12}y^8}$

6. $\sqrt{\dfrac{25x^2}{36y^4}}$

7. $\sqrt[3]{2x}\ \sqrt[3]{5y^2}$

8. $\dfrac{\sqrt[5]{x^3y^4}}{\sqrt[5]{xy^2}}$

9. $\sqrt[4]{x^3y^2}\ \sqrt{xy}$

10. $\dfrac{\sqrt[5]{a^2}}{\sqrt[4]{a}}$

11. $7\sqrt{2} - 2\sqrt{2}$

12. $\sqrt{x^4y} + \sqrt{9y^3}$

13. $(7 + \sqrt{x})(2 - 3\sqrt{x})$

14. If $f(x) = \sqrt{8 - 4x}$, determine the domain and estimate the range of f.

15. If $f(x) = x^2$, find $f(5 + \sqrt{2})$.

16. Rationalize the denominator:
$$\frac{\sqrt{2}}{3 - 5\sqrt{2}}.$$

17. Solve: $x = \sqrt{2x - 5} + 4$.

18. The shorter leg of a 30°–60°–90° right triangle measures 10 cm. Find the lengths of the other sides. Give exact answers and, where appropriate, approximations to three decimal places.

19. A referee jogs diagonally from one corner of a 50-ft by 90-ft basketball court to the far corner. How far does she jog? Give an exact answer and an approximation to three decimal places.

20. Find the distance between $(1.3, 2.7)$ and $(-1.3, -2.7)$. Give an exact answer and an approximation to three decimal places.

21. Find an equation of the circle having center $(5, 0)$ and radius $\sqrt{11}$.

22. Express in terms of i and simplify: $\sqrt{-50}$.

23. Subtract: $(7 + 8i) - (-3 + 6i)$.

24. Multiply: $\sqrt{-16}\ \sqrt{-36}$.

25. Multiply. Write the answer in the form $a + bi$.
$$(4 - i)^2$$

26. Divide and simplify to the form $a + bi$:
$$\frac{-3 + i}{2 - 7i}.$$

27. Simplify: i^{37}.

Skill Maintenance

28. Solve: $6x^2 = 13x + 5$.

29. Divide and simplify:
$$x^3 - 9x \div \frac{x - 4}{x^2 + 3x}.$$

30. Solve:
$$\frac{11x}{x + 3} + \frac{33}{x} + 12 = \frac{99}{x^2 + 3x}.$$

31. Find two consecutive even integers whose product is 288.

Synthesis

32. Solve:
$$\sqrt{2x - 2} + \sqrt{7x + 4} = \sqrt{13x + 10}.$$

33. Simplify:
$$\frac{1 - 4i}{4i(1 + 4i)^{-1}}.$$

Quadratic Functions and Equations 8

I n translating problem situations to mathematical language, we often obtain a function or equation containing a second-degree polynomial in one variable. Such functions or equations are said to be *quadratic*. In this chapter, we will study a variety of equations, inequalities, and applications for which we will need to solve quadratic equations or graph quadratic functions.

AIR QUALITY. Trees improve air quality in part by retaining airborne particles, called particulates, from the air. The following table shows the number of metric tons of particulates retained by trees in Chicago during the spring, summer, and autumn months. Find a quadratic function that fits the data.

MONTH		PARTICULATES RETAINED (IN METRIC TONS)
April	(0)	35
May	(1)	235
July	(3)	325
August	(4)	310
October	(6)	200
November	(7)	35

Source: The National Arbor Day Foundation

This problem appears as Example 4 in Section 8.8.

$$y = -23.54166667x^2 + 162.2583333x + 57.61666667$$

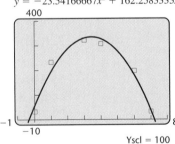

Yscl = 100

487

8.1
Quadratic Equations

- *The Principle of Square Roots*
- *Completing the Square*
- *Problem Solving*

In Chapter 5, we solved *quadratic equations* like $x^2 = 10 + 3x$ by graphing and by factoring. One way to solve by graphing is to rewrite the equation above, for example, as $x^2 - 3x - 10 = 0$ and then graph the *quadratic function* given by $f(x) = x^2 - 3x - 10$. The solutions of the equation, -2 and 5, are the first coordinates of the x-intercepts of the graph of f.

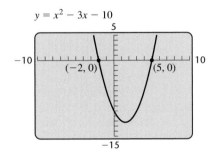

A quadratic equation will have no, one, or two real-number solutions. We can see this by examining the graphs of quadratic functions.

Interactive Discovery

Graph each function and determine the number of x-intercepts. Describe the shape of the graph of a quadratic function.

1. $f(x) = x^2$ **2.** $g(x) = -x^2$

3. $h(x) = (x - 2)^2$ **4.** $p(x) = 2x^2 + 1$

5. $f(x) = -1.5x^2 + x - 3$ **6.** $g(x) = 4x^2 - 2x - 7$

Although we will study graphs of quadratic functions in detail in Section 8.6, we can make some general observations here. The graph of a quadratic function is a cup-shaped curve called a *parabola*. It can open upward, like the graph of $f(x) = x^2$, or downward, like the graph of $g(x) = -x^2$.

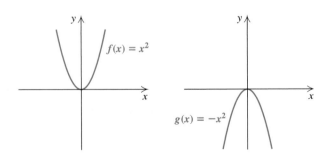

The graph of a quadratic function can have no, one, or two x-intercepts, as illustrated below. Thus a quadratic equation can have no, one, or two real-number roots.

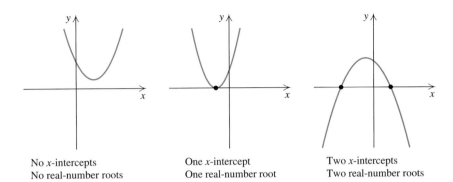

No x-intercepts
No real-number roots

One x-intercept
One real-number root

Two x-intercepts
Two real-number roots

To solve a quadratic equation by factoring, we write the equation in the *standard form* $ax^2 + bx + c = 0$, factor, and use the principle of zero products.

Example 1 Solve: $3x^2 = 2 - x$.

SOLUTION We have

$$3x^2 = 2 - x$$
$$3x^2 + x - 2 = 0 \qquad \text{Adding } -2 + x \text{ on both sides}$$
$$\text{to obtain standard form}$$
$$(3x - 2)(x + 1) = 0 \qquad \text{Factoring}$$
$$3x - 2 = 0 \quad or \quad x + 1 = 0 \qquad \text{Using the principle}$$
$$\text{of zero products}$$
$$3x = 2 \quad or \qquad x = -1$$
$$x = \tfrac{2}{3} \quad or \qquad x = -1.$$

We check by substituting -1 and $\tfrac{2}{3}$ into the original equation. A graphical solution (see the figure at left) provides another check.

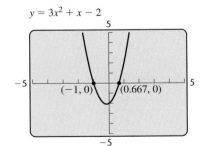

$y = 3x^2 + x - 2$

CHECK: For -1:

$$\frac{3x^2 = 2 - x}{\begin{array}{c|c} 3(-1)^2 \ ? \ 2 - (-1) \\ 3 \cdot 1 & 2 + 1 \\ 3 & 3 \end{array}} \quad \text{TRUE}$$

For $\tfrac{2}{3}$:

$$\frac{3x^2 = 2 - x}{\begin{array}{c|c} 3\left(\tfrac{2}{3}\right)^2 \ ? \ 2 - \tfrac{2}{3} \\ 3 \cdot \tfrac{4}{9} & \tfrac{6}{3} - \tfrac{2}{3} \\ \tfrac{4}{3} & \tfrac{4}{3} \end{array}} \quad \text{TRUE}$$

The solutions are -1 and $\tfrac{2}{3}$.

Example 2 Let $f(x) = x^2$. Find all x-values for which $f(x) = 25$.

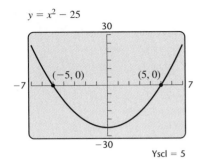

$y = x^2 - 25$

SOLUTION We are asked to find all x-values for which $f(x) = 25$. Thus we solve the equation $x^2 = 25$:

$$x^2 = 25$$
$$x^2 - 25 = 0 \qquad \text{Writing in standard form}$$
$$(x - 5)(x + 5) = 0 \qquad \text{Factoring}$$
$$x - 5 = 0 \quad or \quad x + 5 = 0 \qquad \text{Using the principle of zero products}$$
$$x = 5 \quad or \qquad x = -5.$$

Both -5 and 5 check in the original equation. The graph at left also confirms the solutions. The solutions are 5 and -5. ▬

The Principle of Square Roots

Consider the equation $x^2 = 25$ again. We know from Chapter 7 that the number 25 has two real-number square roots, namely, 5 and -5. Note that these are the solutions of the equation in Example 2. Thus square roots can provide a quick method for solving equations of the type $x^2 = k$.

The Principle of Square Roots

For any real number k, if $x^2 = k$, then $x = \sqrt{k}$ or $x = -\sqrt{k}$.

Example 3 Solve: $3x^2 = 6$.

SOLUTION We have

$$3x^2 = 6$$
$$x^2 = 2 \qquad \text{Multiplying by } \tfrac{1}{3}$$
$$x = \sqrt{2} \quad or \quad x = -\sqrt{2}. \qquad \text{Using the principle of square roots}$$

We often use the symbol $\pm\sqrt{2}$ to represent the two numbers $\sqrt{2}$ and $-\sqrt{2}$. We check as follows.

CHECK: For $\sqrt{2}$:

$$\begin{array}{c|c} 3x^2 = 6 \\ \hline 3(\sqrt{2})^2 \ ? \ 6 \\ 3 \cdot 2 \ & \\ 6 \ & \ 6 \qquad \text{TRUE} \end{array}$$

For $-\sqrt{2}$:

$$\begin{array}{c|c} 3x^2 = 6 \\ \hline 3(-\sqrt{2})^2 \ ? \ 6 \\ 3 \cdot 2 \ & \\ 6 \ & \ 6 \qquad \text{TRUE} \end{array}$$

The solutions are $\sqrt{2}$ and $-\sqrt{2}$, or $\pm\sqrt{2}$. ▬

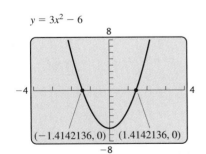

$y = 3x^2 - 6$

$(-1.4142136, 0)$ $(1.4142136, 0)$

Note that a graphical solution of Example 3, shown at left, yields approximate solutions, since $\sqrt{2}$ is irrational.

Quadratic equations that have no real-number solutions will have two imaginary-number solutions.

Example 4   Solve: $4x^2 + 9 = 0$.

SOLUTION   We have

$$4x^2 + 9 = 0$$

$$x^2 = -\frac{9}{4} \qquad \text{Isolating } x^2$$

$$x = \sqrt{-\frac{9}{4}} \quad \text{or} \quad x = -\sqrt{-\frac{9}{4}} \qquad \begin{array}{l}\textbf{Using the principle} \\ \textbf{of square roots}\end{array}$$

$$x = \sqrt{\frac{9}{4}} \sqrt{-1} \quad \text{or} \quad x = -\sqrt{\frac{9}{4}} \sqrt{-1}$$

$$x = \frac{3}{2}i \qquad \text{or} \quad x = -\frac{3}{2}i.$$

CHECK: Since the solutions are opposites and the equation has an x^2-term and no x-term, we can check both solutions at once.

$$
\begin{array}{c|c}
\multicolumn{2}{c}{4x^2 + 9 = 0} \\
\hline
4\left(\pm\frac{3}{2}i\right)^2 + 9 \ ? \ 0 & \\
4 \cdot \frac{9}{4} \cdot i^2 + 9 & \\
-9 + 9 & \\
0 & 0 \qquad \text{TRUE}
\end{array}
$$

The solutions are $\frac{3}{2}i$ and $-\frac{3}{2}i$, or $\pm\frac{3}{2}i$. The graph shown below indicates that there are no real-number solutions.

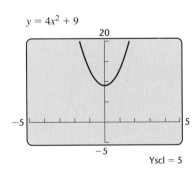

$y = 4x^2 + 9$

Yscl = 5

Equations like $(x - 2)^2 = 7$ can also be solved using the principle of square roots.

Example 5   Let $f(x) = (x - 2)^2$. Find all x-values for which $f(x) = 7$.

SOLUTION   We are asked to find all x-values for which

$$f(x) = 7,$$

or

$$(x - 2)^2 = 7. \qquad \textbf{Substituting } (x - 2)^2 \textbf{ for } f(x)$$

We can do so both algebraically and graphically.

The principle of square roots gives us

$$x - 2 = \sqrt{7} \qquad \text{or} \quad x - 2 = -\sqrt{7}$$

Using the principle of square roots

$$x = 2 + \sqrt{7} \quad \text{or} \qquad x = 2 - \sqrt{7}.$$

Thus the possible solutions are $2 + \sqrt{7}$ and $2 - \sqrt{7}$.

We graph the equation $y = (x - 2)^2 - 7$. If there are any real-number solutions of the equation, they will be the x-coordinates of the x-intercepts.

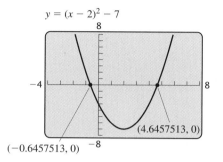

Rounded to the nearest thousandth, the solutions are -0.646 and 4.646.

As one check, note that $2 + \sqrt{7} \approx 4.646$ and $2 - \sqrt{7} \approx -0.646$, so both methods give us the same solution. As a complete check, we can evaluate the function for both possible solutions. This can be done by hand or using a grapher.

We have

$$f(2 + \sqrt{7}) = (2 + \sqrt{7} - 2)^2 = (\sqrt{7})^2 = 7.$$

Similarly,

$$f(2 - \sqrt{7}) = (2 - \sqrt{7} - 2)^2 = (-\sqrt{7})^2 = 7.$$

The numbers check.

One way to check the solutions on a grapher is to enter $y_1 = (x - 2)^2 - 7$ and then evaluate $y_1(2 + \sqrt{7})$ and $y_1(2 - \sqrt{7})$. The second calculation can be entered quickly by copying the first entry and editing it.

```
Y₁(2+√‾(7))
                              0
Y₁(2−√‾(7))
                              0
```

The solutions are $2 + \sqrt{7}$ and $2 - \sqrt{7}$, or $2 \pm \sqrt{7}$.   ▬

In Example 5, one side of the equation is the square of a binomial and the other side is a constant. Once an equation has been written in this form, we can proceed as we did in Example 5.

Example 6   Solve: $x^2 + 6x + 9 = 2$.

SOLUTION   We have

$$x^2 + 6x + 9 = 2$$ The left side is the square of a binomial.

$$(x + 3)^2 = 2$$
$$x + 3 = \sqrt{2} \qquad or \quad x + 3 = -\sqrt{2}$$ Using the principle of square roots
$$x = -3 + \sqrt{2} \quad or \qquad x = -3 - \sqrt{2}.$$

The solutions are $-3 + \sqrt{2}$ and $-3 - \sqrt{2}$, or $-3 \pm \sqrt{2}$.   ▬

Completing the Square

By using a method called *completing the square*, we can use the principle of square roots to solve *any* quadratic equation.

Example 7   Solve: $x^2 + 6x + 4 = 0$.

SOLUTION   We have

$$x^2 + 6x + 4 = 0$$
$$x^2 + 6x \qquad = -4$$ Adding -4 on both sides
$$x^2 + 6x + 9 = -4 + 9$$ Adding 9 on both sides. We explain this shortly.
$$(x + 3)^2 = 5$$ Factoring the perfect-square trinomial
$$x + 3 = \pm\sqrt{5}$$ Using the principle of square roots. Remember that $\pm\sqrt{5}$ represents two numbers.
$$x = -3 \pm \sqrt{5}.$$ Adding -3 on both sides

The solutions are $-3 + \sqrt{5}$ and $-3 - \sqrt{5}$, or $-3 \pm \sqrt{5}$.   ▬

Let's examine how the above solutions were found. The decision to add 9 on both sides in Example 7 was not made arbitrarily. We chose 9 because it made the left side a perfect-square trinomial. The 9 was obtained by taking half of the coefficient of x and squaring it—that is,

$$\left(\tfrac{1}{2} \cdot 6\right)^2 = 3^2, \quad or \quad 9.$$

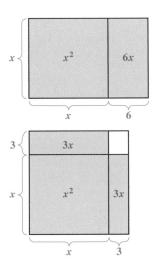

To help see why this procedure works, examine the drawings at left. Note that both figures represent the area $x^2 + 6x$. However, only the second figure, in which the $6x$ is halved, can be converted into a square with the addition of a constant term. The constant term, 9, can be interpreted as the area of the "missing" piece of the second diagram. It *completes* the square.

Example 8 Complete the square: **(a)** $x^2 + 14x$; **(b)** $x^2 - 5x$; **(c)** $x^2 + \frac{3}{4}x$.

SOLUTION

a) We take half of the coefficient of x and square it.

$$x^2 + 14x$$

→ Half of 14 is 7, and $7^2 = 49$. We add 49.

Thus, $x^2 + 14x + 49$ is a perfect-square trinomial. It is equivalent to $(x + 7)^2$. We must add 49 in order for $x^2 + 14x$ to become a perfect-square trinomial.

b) We take half of the coefficient of x and square it:

$$x^2 - 5x$$

→ $\frac{1}{2} \cdot (-5) = -\frac{5}{2}$, and $\left(-\frac{5}{2}\right)^2 = \frac{25}{4}$.

Thus, $x^2 - 5x + \frac{25}{4}$ is a perfect-square trinomial. It is equivalent to $\left(x - \frac{5}{2}\right)^2$. Note that for purposes of factoring, it is best to leave $\frac{25}{4}$ as a fraction.

c) We take half of the coefficient of x and square it:

$$x^2 + \frac{3}{4}x$$

→ $\frac{1}{2} \cdot \frac{3}{4} = \frac{3}{8}$, and $\left(\frac{3}{8}\right)^2 = \frac{9}{64}$.

Thus, $x^2 + \frac{3}{4}x + \frac{9}{64}$ is a perfect-square trinomial. It is equivalent to $\left(x + \frac{3}{8}\right)^2$.

We can now use the method of completing the square to solve equations similar to Example 7.

Example 9 Solve: $x^2 - 8x - 7 = 0$.

SOLUTION We have

$$x^2 - 8x - 7 = 0$$
$$x^2 - 8x = 7 \quad \text{Adding 7 on both sides. We can now complete the square on the left side.}$$
$$x^2 - 8x + 16 = 7 + 16 \quad \text{Adding 16 on both sides to complete the square: } \frac{1}{2}(-8) = -4, \text{ and } (-4)^2 = 16$$
$$(x - 4)^2 = 23 \quad \text{Factoring}$$
$$x - 4 = \pm\sqrt{23} \quad \text{Using the principle of square roots}$$
$$x = 4 \pm \sqrt{23}. \quad \text{Adding 4 on both sides}$$

The solutions are $4 - \sqrt{23}$ and $4 + \sqrt{23}$, or $4 \pm \sqrt{23}$. The checks are left to the student. ▬

Before we can complete the square, the coefficient of x^2 must be 1. When it is not 1, we divide by the coefficient of x^2 on both sides of the equation.

Example 10 Solve: $3x^2 + 7x - 2 = 0$.

SOLUTION We have

$$3x^2 + 7x - 2 = 0$$

$$3x^2 + 7x = 2 \qquad \text{Adding 2 on both sides}$$

$$x^2 + \frac{7}{3}x = \frac{2}{3} \qquad \text{Dividing by 3 on both sides}$$

$$x^2 + \frac{7}{3}x + \frac{49}{36} = \frac{2}{3} + \frac{49}{36} \qquad \text{Completing the square: } \left(\frac{1}{2} \cdot \frac{7}{3}\right)^2 = \frac{49}{36}$$

$$\left(x + \frac{7}{6}\right)^2 = \frac{73}{36} \qquad \text{Factoring and simplifying}$$

$$x + \frac{7}{6} = \pm \frac{\sqrt{73}}{6} \qquad \begin{array}{l}\text{Using the principle of square} \\ \text{roots and the quotient rule} \\ \text{for radicals}\end{array}$$

$$x = \frac{-7 \pm \sqrt{73}}{6}. \qquad \text{Adding } -\frac{7}{6} \text{ on both sides}$$

The solutions are

$$\frac{-7 - \sqrt{73}}{6}$$

and

$$\frac{-7 + \sqrt{73}}{6},$$

or

$$\frac{-7 \pm \sqrt{73}}{6}.$$

▬

The procedure used in Example 10 can be used to solve *any* quadratic equation.

> **To Solve a Quadratic Equation in x by Completing the Square:**
>
> **1.** Isolate the terms with variables on one side of the equation, and arrange them in descending order.
> **2.** The coefficient of x^2 must be 1. If it is not, divide by the coefficient of x^2 on both sides.
> **3.** Complete the square by taking half of the coefficient of x and adding its square on both sides.
> **4.** Express one side as the square of a binomial and simplify the other side.
> **5.** Use the principle of square roots.
> **6.** Solve for x by adding appropriately on both sides.

Problem Solving

If you put money in a savings account, the bank will pay you interest. As interest is paid into your account, the bank will start paying you interest on both the original amount and the interest already earned. This is called **compounding interest**. If interest is paid yearly, we say that it is **compounded annually**.

> **The Compound-Interest Formula**
>
> If an amount of money P is invested at interest rate r, compounded annually, then in t years, it will grow to the amount A given by
>
> $$A = P(1 + r)^t.$$

We can use quadratic equations to solve certain interest problems.

Example 11 *Investment Growth.* Rosa invested $4000 at interest rate r, compounded annually. In 2 yr, it grew to $4410. What was the interest rate?

SOLUTION

1. Familiarize. We are already familiar with the compound-interest formula. If we were not, we would need to consult an outside source.

2. Translate. The translation consists of substituting into the formula:

$$A = P(1 + r)^t$$
$$4410 = 4000(1 + r)^2.$$

3. Carry out. We solve for r both algebraically and graphically.

ALGEBRAIC SOLUTION

We have

$$4410 = 4000(1 + r)^2$$

$$\frac{4410}{4000} = (1 + r)^2 \qquad \text{Multiplying by } \tfrac{1}{4000}$$
$$\qquad\qquad\qquad\qquad \text{on both sides}$$

$$\frac{441}{400} = (1 + r)^2 \qquad \text{Simplifying}$$

$$\pm\sqrt{\frac{441}{400}} = 1 + r \qquad \text{Using the principle}$$
$$\qquad\qquad\qquad\qquad \text{of square roots}$$

$$\pm\frac{21}{20} = 1 + r \qquad \text{Simplifying}$$

$$-\frac{20}{20} \pm \frac{21}{20} = r$$

$$\frac{1}{20} = r \quad or \quad -\frac{41}{20} = r.$$

Since r represents an interest rate, we convert to decimal notation. We now have

$$r = 0.05 \quad or \quad r = -2.05.$$

GRAPHICAL SOLUTION

We let $y_1 = 4410$ and $y_2 = 4000(1 + x)^2$. It may take several tries to find an appropriate viewing window. Since $y_1 = 4410$, the vertical axis should extend above 4500. Using the window $[-3, 1, 0, 5000]$, we see that the graphs intersect at two points.

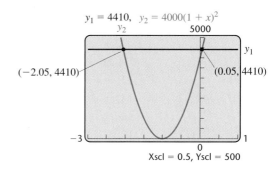

The first coordinates of the points of intersection are -2.05 and 0.05.

4. **Check.** Since the interest rate cannot be negative, we need only check 0.05, or 5%. If \$4000 were invested at 5% interest, compounded annually, then in 2 yr it would grow to $4000(1.05)^2$, or \$4410. The number 0.05 checks.

5. **State.** The interest rate was 5%.

8.1 Exercise Set

🔍 *Determine the number of real-number solutions of each equation from the given graph.*

1. $x^2 + x - 12 = 0$

$y = x^2 + x - 12$

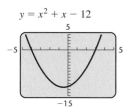

2. $-3x^2 - x - 7 = 0$

$y = -3x^2 - x - 7$

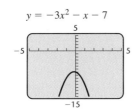

3. $4x^2 + 9 = 12x$

$y = 12x - 4x^2 - 9$

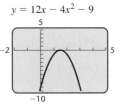

4. $2x^2 + 3 = 6x$

$y = 2x^2 + 3 - 6x$

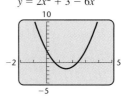

5. $5x^2 + 2x + 7 = 0$

$y = 5x^2 + 2x + 7$

6. $0.25x^2 + 0.05x + 0.01 = 0$

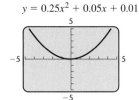

$$y = 0.25x^2 + 0.05x + 0.01$$

Solve.

7. $5x^2 = 15$ **8.** $7x^2 = 35$

9. $25x^2 + 4 = 0$ **10.** $9x^2 + 16 = 0$

11. $2x^2 - 3 = 0$ **12.** $3x^2 - 7 = 0$

13. $(x + 2)^2 = 49$ **14.** $(x - 1)^2 = 6$

15. $(a + 5)^2 = 8$ **16.** $(x - 13)^2 = 64$

17. $(x - 7)^2 = -4$ **18.** $(x + 1)^2 = -9$

19. $\left(x + \frac{3}{2}\right)^2 = \frac{7}{2}$ **20.** $\left(y + \frac{3}{4}\right)^2 = \frac{17}{16}$

21. $x^2 - 6x + 9 = 100$ **22.** $x^2 - 10x + 25 = 64$

23. Let $f(x) = (x - 7)^2$. Find x such that $f(x) = 16$.

24. Let $g(x) = (x - 2)^2$. Find x such that $g(x) = 25$.

25. Let $F(x) = (x - 3)^2$. Find x such that $F(x) = 13$.

26. Let $f(x) = (x + 3)^2$. Find x such that $f(x) = 17$.

27. Let $g(x) = x^2 + 14x + 49$. Find x such that $g(x) = 36$.

28. Let $F(x) = x^2 + 8x + 16$. Find x such that $F(x) = 9$.

Complete the square. Then write the perfect-square trinomial in factored form.

29. $x^2 + 10x$ **30.** $x^2 + 16x$ **31.** $x^2 - 6x$

32. $x^2 - 8x$ **33.** $x^2 + 9x$ **34.** $x^2 + 3x$

35. $x^2 - 3x$ **36.** $x^2 - 7x$ **37.** $x^2 + \frac{2}{3}x$

38. $x^2 + \frac{2}{5}x$ **39.** $x^2 - \frac{5}{6}x$ **40.** $x^2 - \frac{5}{3}x$

Solve by completing the square. Show your work.

41. $x^2 + 6x = 7$ **42.** $x^2 + 5x = -6$

43. $x^2 + 6x + 5 = 0$ **44.** $x^2 + 10x + 9 = 0$

45. $x^2 - 10x + 21 = 0$ **46.** $x^2 - 10x + 24 = 0$

47. $x^2 + 4x + 1 = 0$ **48.** $x^2 + 6x + 7 = 0$

49. $x^2 + 6x + 13 = 0$ **50.** $x^2 + 8x + 25 = 0$

51. $2x^2 - 5x - 3 = 0$ **52.** $3x^2 + 5x - 2 = 0$

53. $4x^2 + 8x + 3 = 0$ **54.** $9x^2 + 18x + 8 = 0$

55. $6x^2 - x = 15$ **56.** $6x^2 - x = 2$

57. $2x^2 + 4x + 1 = 0$ **58.** $2x^2 + 5x + 2 = 0$

Interest. Use $A = P(1 + r)^t$ to find the interest rate in Exercises 59–64. Refer to Example 11.

59. $2000 grows to $2420 in 2 yr

60. $2560 grows to $2890 in 2 yr

61. $1280 grows to $1805 in 2 yr

62. $1000 grows to $1440 in 2 yr

63. $6250 grows to $6760 in 2 yr

64. $6250 grows to $7290 in 2 yr

Free-Falling Objects. The formula $s = 16t^2$ is used to approximate the distance s, in feet, that an object falls freely from rest in t seconds. Use this formula for Exercises 65–68.

65. The CN Tower in Toronto, at 1815 ft, is the world's tallest self-supporting tower (no guy wires) (*Source: The Guinness Book of Records*). How long would it take an object to fall freely from the top?

66. Reaching 745 ft above the water, the towers of California's Golden Gate Bridge are the world's tallest bridge towers (*Source: The Guinness Book of Records*). How long would it take an object to fall freely from the top?

67. The Gateway Arch in St. Louis is 640 ft high. How long would it take an object to fall freely from the top?

68. The Sears Tower in Chicago is 1454 ft tall. How long would it take an object to fall freely from the top?

Skill Maintenance

Graph.

69. $f(x) = 5 - 2x$ [2.1]

70. $y = 7$ [2.5]

Simplify. [7.3]

71. $\sqrt[3]{270}$

72. $\sqrt{80}$

Let $f(x) = \sqrt{3x - 5}$. [7.1]

73. Find $f(10)$.

74. Find $f(18)$.

Synthesis

75. ◈ Explain in your own words a sequence of steps that can be used to solve any quadratic equation in the quickest way.

76. ◈ Write an interest-rate problem for a classmate to solve. Devise the problem so that the solution is "The loan was made at 7% interest."

77. ◈ Describe how to write a quadratic equation that can be solved algebraically but not graphically.

78. ◈ What would be better: to receive 3% interest every 6 months, or to receive 6% interest every 12 months? Why?

Find b such that each of the following is a perfect-square trinomial.

79. $x^2 + bx + 81$

80. $x^2 + bx + 49$

Solve.

81. $x(2x^2 - 9x - 56)(x^2 - 5) = 0$

82. $\left(x - \frac{1}{3}\right)\left(x^2 + \frac{1}{6}\right) + \left(x - \frac{1}{3}\right)\left(x^2 - \frac{2}{3}\right) = 0$

83. *Boating.* A barge and a fishing boat leave a dock at the same time, traveling at right angles to each other. The barge travels 7 km/h slower than the fishing boat. After 4 hr, the boats are 68 km apart. Find the speed of each vessel.

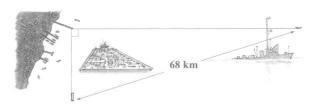

84. Find three consecutive integers such that the square of the first plus the product of the other two is 67.

8.2

The Quadratic Formula

• *Solving Using the Quadratic Formula*
• *The Discriminant*

There are at least two reasons for learning to complete the square. One is to enhance your ability to graph certain equations that appear later in this chapter. The other is to prove a general formula for solving quadratic equations.

Solving Using the Quadratic Formula

Each time you solve by completing the square, the procedure is the same. In mathematics, when a procedure must be repeated many times, a formula is often developed to speed up our work.

Consider any quadratic equation in standard form:

$$ax^2 + bx + c = 0, \quad a > 0.$$

Let's solve by completing the square. As the steps are performed, compare them with Example 10 in Section 8.1.

$$ax^2 + bx \quad = -c \qquad \text{Adding } -c \text{ on both sides}$$

$$x^2 + \frac{b}{a}x \quad = -\frac{c}{a} \qquad \text{Dividing by } a \text{ on both sides}$$

Half of $\dfrac{b}{a}$ is $\dfrac{b}{2a}$ and $\left(\dfrac{b}{2a}\right)^2$ is $\dfrac{b^2}{4a^2}$. We add $\dfrac{b^2}{4a^2}$ on both sides:

$$x^2 + \frac{b}{a}x + \frac{b^2}{4a^2} = -\frac{c}{a} + \frac{b^2}{4a^2}$$

Adding $\dfrac{b^2}{4a^2}$ to complete the square

$$\left(x + \frac{b}{2a}\right)^2 = -\frac{4ac}{4a^2} + \frac{b^2}{4a^2}$$

$$\left(x + \frac{b}{2a}\right)^2 = \frac{b^2 - 4ac}{4a^2}$$

Factoring on the left side; finding a common denominator on the right side

$$x + \frac{b}{2a} = \pm\frac{\sqrt{b^2 - 4ac}}{2a}$$

Using the principle of square roots and the quotient rule for radicals; since $a > 0$, $\sqrt{4a^2} = 2a$

$$x = \frac{-b \pm \sqrt{b^2 - 4ac}}{2a}.$$

Adding $-\dfrac{b}{2a}$ on both sides

A similar derivation yields the same result when a is negative. It is important that you remember the quadratic formula and know how to use it.

The Quadratic Formula

The solutions of $ax^2 + bx + c = 0$, $a \neq 0$, are given by

$$x = \frac{-b \pm \sqrt{b^2 - 4ac}}{2a}.$$

Example 1　Solve $5x^2 + 8x = -3$ using the quadratic formula.

SOLUTION　We first find standard form and determine a, b, and c:

$$5x^2 + 8x + 3 = 0;$$　Adding 3 on both sides

$$a = 5, \quad b = 8, \quad c = 3.$$

Next, we use the quadratic formula:

$$x = \frac{-b \pm \sqrt{b^2 - 4ac}}{2a}$$

$$= \frac{-8 \pm \sqrt{8^2 - 4 \cdot 5 \cdot 3}}{2 \cdot 5}$$　Substituting

> Be sure to write the fraction bar all the way across.

$$= \frac{-8 \pm \sqrt{64 - 60}}{10}$$

$$= \frac{-8 \pm \sqrt{4}}{10}$$

$$= \frac{-8 \pm 2}{10}.$$

Thus,

$$x = \frac{-8 + 2}{10} \quad or \quad x = \frac{-8 - 2}{10}$$

$$x = \frac{-6}{10} \quad or \quad x = \frac{-10}{10}$$

$$x = -\frac{3}{5} \quad or \quad x = -1.$$

The solutions are $-\frac{3}{5}$ and -1.

Because $5x^2 + 8x + 3$ can be factored as $(5x + 3)(x + 1)$, the quadratic formula may not have been the fastest way of solving Example 1. However, because the quadratic formula works for *any* quadratic equation, we need not spend much time struggling to solve a quadratic equation by factoring.

To Solve a Quadratic Equation:

1. Check for the form $ax^2 = p$ or $(x + k)^2 = d$. If it is in either of these forms, use the principle of square roots as in Section 8.1.
2. If it is not in the form of step (1), write it in standard form $ax^2 + bx + c = 0$.
3. Try factoring and using the principle of zero products.
4. If it is not possible to factor or factoring seems difficult, use the quadratic formula.

The solutions of a quadratic equation can always be found using the quadratic formula. They cannot always be found by factoring. Any real-number solutions can be found with a grapher, although the solutions may be approximations.

Recall from Section 5.1 that a **quadratic function** is a second-degree polynomial function in one variable.

Example 2 Given the quadratic function described by $f(x) = 5x^2 - 8x - 3$, find x such that $f(x) = 0$.

SOLUTION We substitute and solve for x:

$$f(x) = 0$$

$$5x^2 - 8x - 3 = 0 \qquad \text{Substituting; this cannot be solved by factoring.}$$

$$a = 5, \quad b = -8, \quad c = -3.$$

We then substitute into the quadratic formula:

$$x = \frac{-(-8) \pm \sqrt{(-8)^2 - 4 \cdot 5 \cdot (-3)}}{2 \cdot 5}$$

$$= \frac{8 \pm \sqrt{64 + 60}}{10}.$$

Thus,

$$x = \frac{8 \pm \sqrt{124}}{10}$$ **Note that 4 is a perfect-square factor of 124.**

$$= \frac{8 \pm \sqrt{4 \cdot 31}}{10}$$

$$= \frac{8 \pm 2\sqrt{31}}{10}$$

$$= \frac{2(4 \pm \sqrt{31})}{2 \cdot 5} = \frac{4 \pm \sqrt{31}}{5}.$$ **Removing a factor equal to 1: $\frac{2}{2} = 1$**

Caution! To avoid a common error, *factor the numerator and the denominator* when removing a factor equal to 1.

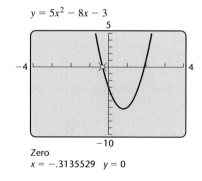

To check using a calculator, we enter $y_1 = 5x^2 - 8x - 3$ and evaluate y_1 for both possible solutions. Care must be taken to place parentheses properly around the radicand as well as around the entire numerator.

This equation can also be solved graphically by finding the zeros of the function $f(x) = 5x^2 - 8x - 3$. To compare the exact solutions found using the quadratic formula with the approximations found using a grapher, we must find approximations of the exact solutions.

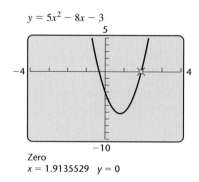

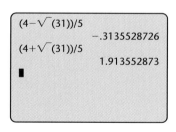

All real solutions can be found with both methods. The solutions are

$$\frac{4 + \sqrt{31}}{5} \quad \text{and} \quad \frac{4 - \sqrt{31}}{5}.$$

Some quadratic equations have solutions that are imaginary numbers.

Example 3 Solve: $x^2 + 2 = -x$.

SOLUTION We first find standard form:

$$x^2 + x + 2 = 0. \qquad \text{Adding } x \text{ on both sides}$$

Since we cannot solve by factoring, we use the quadratic formula with $a = 1$, $b = 1$, and $c = 2$:

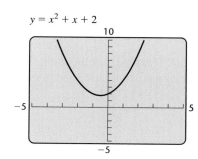

$y = x^2 + x + 2$

$$x = \frac{-1 \pm \sqrt{1^2 - 4 \cdot 1 \cdot 2}}{2 \cdot 1} \qquad \text{Substituting}$$

$$= \frac{-1 \pm \sqrt{1 - 8}}{2}$$

$$= \frac{-1 \pm \sqrt{-7}}{2}$$

$$= \frac{-1 \pm i\sqrt{7}}{2}.$$

Note from the graph at left that there are no x-intercepts of the graph of $f(x) = x^2 + x + 2$, and thus no real-number solutions of the equation. The solutions are

$$\frac{-1 + i\sqrt{7}}{2} \quad \text{and} \quad \frac{-1 - i\sqrt{7}}{2}.$$

Example 4 Solve: $2 + \dfrac{7}{x} = \dfrac{4}{x^2}$.

SOLUTION This is a rational equation similar to those in Section 6.4. Note that $x \neq 0$. Since the LCD is x^2, we multiply by x^2 on both sides:

$$x^2\left(2 + \frac{7}{x}\right) = x^2 \cdot \frac{4}{x^2}$$

$$2x^2 + 7x = 4 \qquad \text{Simplifying}$$

$$2x^2 + 7x - 4 = 0. \qquad \text{Subtracting 4 on both sides}$$

We have

$$a = 2, \quad b = 7, \quad c = -4.$$

Substituting then gives us

$$x = \frac{-7 \pm \sqrt{7^2 - 4 \cdot 2 \cdot (-4)}}{2 \cdot 2}$$

$$= \frac{-7 \pm \sqrt{49 + 32}}{4} = \frac{-7 \pm \sqrt{81}}{4}$$

$$= \frac{-7 \pm 9}{4}$$

$$x = \frac{-7 + 9}{4} = \frac{1}{2} \quad \text{or} \quad x = \frac{-7 - 9}{4} = -4. \qquad \begin{array}{l}\textbf{Both answers}\\\textbf{should check}\\\textbf{since } x \neq 0.\end{array}$$

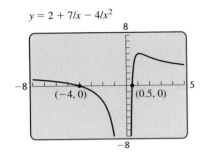

$y = 2 + 7/x - 4/x^2$

We leave it to the student to confirm that both $\frac{1}{2}$ and -4 check in the original equation. As another check, we see that the same solutions are found using a grapher, as shown in the figure at left. Note that the original equation is a *rational* equation, not a *quadratic* equation, so its graph is not a parabola.

The solutions are $\frac{1}{2}$ and -4.

Checking the solutions of Examples 2 and 3 can be cumbersome. Fortunately, when the quadratic formula has been used, checking for computational errors is usually sufficient.

The Discriminant

Sometimes in mathematics it is enough to know what *type* of number an equation will have for its solution(s), without actually solving the equation. For example, the radicand in the quadratic formula, $b^2 - 4ac$, determines what type of number the solutions of a quadratic equation will be. This expression is called the **discriminant**.

When $b^2 - 4ac$ simplifies to 0, the quadratic formula becomes

$$x = \frac{-b \pm \sqrt{0}}{2a}, \quad \text{or simply} \quad \frac{-b}{2a}.$$

Thus there is only one rational solution, $-b/2a$. In this case, the quadratic polynomial $ax^2 + bx + c$ is a perfect square, and the solution is sometimes called a *double root*.

Whenever the discriminant is positive, there will be two real-number solutions. When $b^2 - 4ac$ is a perfect square, as in Example 1, the solutions will be rational numbers. As we saw in Example 2, when $b^2 - 4ac$ is positive but not a perfect square, there will be two irrational solutions and they will be conjugates of each other (see p. 451).

Whenever the discriminant is negative, as in Example 3, there will be two imaginary-number solutions and they will be complex conjugates of each other.

The following table lists the various types of solutions.

DISCRIMINANT $b^2 - 4ac$	NATURE OF SOLUTIONS
0	Only one solution (double root); it is a rational number.

Positive
  Perfect square
  Not a perfect square
	Two different real-number solutions
	  Solutions are rational.
	  Solutions are irrational conjugates.

Negative	Two different imaginary-number solutions (complex conjugates)

Example 5 For each equation, determine what type of number the solutions will be and how many solutions exist.

a) $9x^2 - 12x + 4 = 0$ **b)** $x^2 + 5x + 8 = 0$ **c)** $2x^2 + 7x - 3 = 0$

SOLUTION

a) For $9x^2 - 12x + 4 = 0$, we have

$$a = 9, \quad b = -12, \quad c = 4.$$

We substitute and compute the discriminant:

$$b^2 - 4ac = (-12)^2 - 4 \cdot 9 \cdot 4$$
$$= 144 - 144 = 0.$$

There is just one solution, and it is rational. This tells us that $9x^2 - 12x + 4 = 0$ can be solved by factoring. The graph at left confirms that $9x^2 - 12x + 4 = 0$ has just one solution.

b) For $x^2 + 5x + 8 = 0$, we have

$$a = 1, \quad b = 5, \quad c = 8.$$

We substitute and compute the discriminant:

$$b^2 - 4ac = 5^2 - 4 \cdot 1 \cdot 8$$
$$= 25 - 32 = -7.$$

Since the discriminant is negative, there are two imaginary-number solutions that are complex conjugates of each other. As the graph at left shows, there are no x-intercepts of the graph of $y = x^2 + 5x + 8$, and thus no real solutions of the equation $x^2 + 5x + 8 = 0$.

c) For $2x^2 + 7x - 3 = 0$, we have

$$a = 2, \quad b = 7, \quad c = -3;$$
$$b^2 - 4ac = 7^2 - 4 \cdot 2(-3)$$
$$= 49 - (-24) = 73.$$

The discriminant is a positive number that is not a perfect square. Thus there are two irrational solutions that are conjugates of each other. From the graph at left, we see that there are two solutions of the equation. We cannot tell from simply observing the graph whether the solutions are rational or irrational.

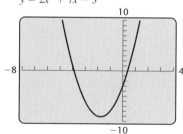

$y = 9x^2 - 12x + 4$

$y = x^2 + 5x + 8$

$y = 2x^2 + 7x - 3$

8.2 Exercise Set

Solve.

1. $x^2 + 7x + 4 = 0$

2. $x^2 - 7x - 3 = 0$

3. $3p^2 = -8p - 5$

4. $3u^2 = 18u - 6$

5. $x^2 - x + 2 = 0$

6. $x^2 - x + 1 = 0$

7. $x^2 + 13 = 6x$

8. $x^2 + 13 = 4x$

9. $h^2 + 4 = 6h$

10. $r^2 + 3r = 8$

11. $3 + \dfrac{8}{x} = \dfrac{1}{x^2}$

12. $2 + \dfrac{5}{x^2} = \dfrac{9}{x}$

13. $3x + x(x - 2) = 0$

14. $4x + x(x - 3) = 0$

15. $14x^2 + 9x = 0$

16. $19x^2 + 8x = 0$

17. $25x^2 - 20x + 4 = 0$

18. $36x^2 + 84x + 49 = 0$

19. $7x(x + 2) + 6 = 3x(x + 1)$

20. $5x(x - 1) + 2 = 4x(x - 2)$

21. $14(x - 4) - (x + 2) = (x + 2)(x - 4)$

22. $11(x - 2) + (x - 5) = (x + 2)(x - 6)$

23. $5x^2 = 13x + 17$ **24.** $25x = 3x^2 + 28$

25. $x^2 + 9 = 4x$ **26.** $x^2 + 7 = 3x$

27. $x + \dfrac{1}{x} = \dfrac{13}{6}$ **28.** $\dfrac{3}{x} + \dfrac{x}{3} = \dfrac{5}{2}$

29. Let $f(x) = 3x^2 - 5x - 1$. Find x such that $f(x) = 0$.

30. Let $g(x) = 4x^2 - 2x - 3$. Find x such that $g(x) = 0$.

31. Let
$$f(x) = \frac{7}{x} + \frac{7}{x + 4}.$$
Find x such that $f(x) = 1$.

32. Let
$$g(x) = \frac{2}{x} + \frac{2}{x + 3}.$$
Find x such that $g(x) = 1$.

33. Let
$$F(x) = \frac{x + 3}{x} \quad \text{and} \quad G(x) = \frac{x - 4}{3}.$$
Find x such that $F(x) = G(x)$.

34. Let
$$f(x) = x + 5 \quad \text{and} \quad g(x) = \frac{3}{x - 5}.$$
Find x such that $f(x) = g(x)$.

🔍 *Determine the sign of the discriminant for each equation from the given graph.*

35. $x^2 + x + 6 = 0$

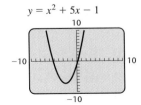

36. $x^2 + 5x - 1 = 0$

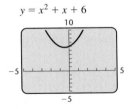

37. $9x^2 - 6x + 1 = 0$ **38.** $2x^2 - 3x + 5 = 0$

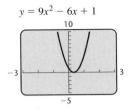
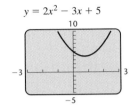

For each equation, determine, without solving the equation, what type of number the solutions are and how many solutions exist.

39. $x^2 - 4x + 3 = 0$ **40.** $x^2 + 6x + 5 = 0$

41. $x^2 + 5 = 0$ **42.** $x^2 + 3 = 0$

43. $x^2 - 2 = 0$

44. $x^2 - 5 = 0$

45. $4x^2 - 12x + 9 = 0$

46. $4x^2 + 8x - 5 = 0$

47. $x^2 - 2x + 4 = 0$

48. $x^2 + 4x + 6 = 0$

49. $a^2 + 11a + 28 = 0$

50. $t^2 - 8t + 16 = 0$

51. $6x^2 + 5x - 4 = 0$

52. $10x^2 - x - 2 = 0$

53. $9t^2 - 3t = 0$

54. $4m^2 + 7m = 0$

55. $x^2 + 5x = 7$

56. $x^2 + 4x = 7$

57. $2a^2 - 3a = -5$

58. $3a^2 + 5 = 7a$

59. $y^2 + \frac{9}{4} = 4y$

60. $x^2 = \frac{1}{2}x - \frac{3}{5}$

Skill Maintenance

61. *Coffee Beans.* Twin Cities Roasters has Kenyan coffee worth $6.75 a pound and Peruvian coffee worth $11.25 a pound. How much of each kind should be mixed to obtain a 50-lb mixture that is worth $8.55 a pound? [3.3]

Graph.

62. $y = -\frac{3}{7}x + 4$ [2.4]

63. $f(x) = -x - 3$ [2.1]

64. $5x - 2y = 8$ [2.5]

Synthesis

65. ◈ While solving a quadratic equation of the form $ax^2 + bx + c = 0$ using a grapher, Shawn-Marie gets the following screen.

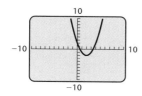

How could the discriminant help her check the graph?

66. ◈ Are there any equations that can be solved by the quadratic formula but not by completing the square? Why or why not?

67. ◈ Given the solutions of a quadratic equation, is it possible to reconstruct the original equation? Why or why not?

68. ◈ The list on p. 501 does not mention completing the square as a method of solving quadratic equations. Why not?

69. Show that the product of the solutions of $ax^2 + bx + c = 0$ is c/a.

For each equation under the given condition, (a) find k and (b) find the other solution.

70. $kx^2 - 2x + k = 0$; one solution is -3

71. $x^2 - kx + 2 = 0$; one solution is $1 + i$

72. Show that the sum of the solutions of $ax^2 + bx + c = 0$ is $-b/a$.

73. Find k, where
$$kx^2 - 4x + (2k - 1) = 0$$
and the product of the solutions is 3. (*Hint:* See Exercise 69.)

74. Find h and k, where $3x^2 - hx + 4k = 0$, the sum of the solutions is -12, and the product of the solutions is 20. (*Hint:* See Exercises 69 and 72.)

75. ◈ Explain how the results in Exercises 69 and 72 could be used.

Let $f(x) = \dfrac{x^2}{x - 2} + 1$ and $g(x) = \dfrac{4x - 2}{x - 2} + \dfrac{x + 4}{2}$.

76. Find the x-intercepts of the graph of g.

77. Find the x-intercepts of the graph of f.

78. Find x such that $f(x) = g(x)$.

Solve.

79. $x^2 - 0.75x - 0.5 = 0$

80. $z^2 + 0.84z - 0.4 = 0$

81. $\sqrt{2}x^2 + 5x + \sqrt{2} = 0$

82. $(1 + \sqrt{3})x^2 - (3 + 2\sqrt{3})x + 3 = 0$

83. $ix^2 - 2x + 1 = 0$

84. ◈ A discriminant that is a perfect square indicates that factoring can be used to solve the quadratic equation. Why?

8.3

Applications Involving Quadratic Equations

- *Solving Problems*
- *Solving Formulas*

Solving Problems

As we found in Section 6.5, some problems translate to rational equations. The solution of such rational equations can involve quadratic equations.

Example 1 *Motorcycle Travel.* Makita's motorcycle traveled 300 mi at a certain speed. Had she gone 10 mph faster, the trip would have taken 1 hr less. Find the speed of the motorcycle.

SOLUTION

1. Familiarize. We make a drawing, labeling it with the known and unknown information. As in Section 6.5, we can organize the data in a table. We let r and $t =$ the rate, in miles per hour, and time, in hours, respectively, for Makita's trip.

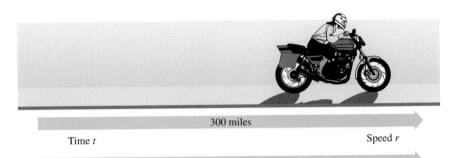

Time t 300 miles   Speed r

Time $t - 1$ 300 miles   Speed $r + 10$

DISTANCE	SPEED	TIME
300	r	t
300	$r + 10$	$t - 1$

Recall that the definition of speed, $r = d/t$, relates the three quantities.

2. Translate.   From the first line of the table, we obtain

$$r = \frac{300}{t}.$$

From the second line, we get

$$r + 10 = \frac{300}{t - 1}.$$

3. Carry out.   A system of equations has been formed. We substitute for r from the first equation into the second and solve the resulting equation:

$$\frac{300}{t} + 10 = \frac{300}{t - 1} \qquad \text{Substituting 300/}t \text{ for } r$$

$$t(t - 1) \cdot \left[\frac{300}{t} + 10 \right] = t(t - 1) \cdot \frac{300}{t - 1} \qquad \text{Multiplying by the LCD}$$

$$t(t - 1) \cdot \frac{300}{t} + t(t - 1) \cdot 10 = t(t - 1) \cdot \frac{300}{t - 1} \qquad \begin{array}{l} \text{Using the distributive law and removing factors that equal 1:} \\ \frac{t}{t} = 1; \frac{t - 1}{t - 1} = 1 \end{array}$$

then,

$$300(t - 1) + 10(t^2 - t) = 300t$$
$$300t - 300 + 10t^2 - 10t = 300t$$
$$10t^2 - 10t - 300 = 0 \qquad \text{Standard form}$$
$$t^2 - t - 30 = 0 \qquad \text{Multiplying by } \tfrac{1}{10}$$
$$(t - 6)(t + 5) = 0 \qquad \text{Factoring}$$
$$t = 6 \quad or \quad t = -5. \qquad \text{Principle of zero products}$$

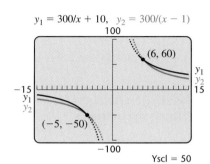

$y_1 = 300/x + 10, \quad y_2 = 300/(x - 1)$

(6, 60)

y_1
y_2

y_1
y_2

$(-5, -50)$

Yscl = 50

4. Check. As a partial check, we graph $y_1 = 300/x + 10$ and $y_2 = 300/(x - 1)$. The graph at left shows that the solutions are -5 and 6.

Note that we have solved for t, not r as required. Since negative time has no meaning here, we disregard the -5 and use 6 hr to find r:

$$r = \frac{300}{6} = 50 \text{ mph.}$$

Caution! Always make sure that you find the quantity asked for in the problem.

To see if 50 mph checks, we increase the speed 10 mph to 60 mph and see how long the trip would have taken at that speed:

$$t = \frac{d}{r} = \frac{300}{60} = 5 \text{ hr.}$$

This is 1 hr less than the trip actually took, so the answer checks.

5. State. Makita's motorcycle traveled at a speed of 50 mph.

Solving Formulas

Recall that to solve a formula for a certain letter, we use the principles for solving equations to get that letter alone on one side.

Example 2 *Period of a Pendulum.* The time T required for a pendulum of length l to swing back and forth (complete one period) is given by the formula $T = 2\pi\sqrt{l/g}$, where g is the gravitational constant. Solve for l.

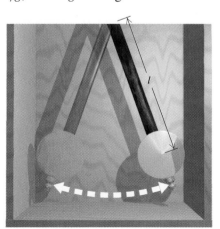

SOLUTION We have

$$T = 2\pi\sqrt{\frac{l}{g}} \qquad \text{This is a radical equation (see Section 7.6).}$$

$$T^2 = \left(2\pi\sqrt{\frac{l}{g}}\right)^2 \qquad \text{Principle of powers (squaring)}$$

$$T^2 = 2^2\pi^2\frac{l}{g}$$

$$gT^2 = 4\pi^2 l \qquad \text{Clearing fractions}$$

$$\frac{gT^2}{4\pi^2} = l. \qquad \text{Multiplying by } \frac{1}{4\pi^2}$$

We now have l alone on one side and l does not appear on the other side, so the formula is solved for l. ▬

In most formulas, variables represent nonnegative numbers, so we do not need to use absolute-value signs when taking square roots.

Example 3 *Hang Time.** An athlete's *hang time* is the amount of time that the athlete can remain airborne when jumping. A formula relating an athlete's vertical leap V, in inches, to hang time T, in seconds, is $V = 48T^2$. Solve for T.

SOLUTION

$$48T^2 = V$$

$$T^2 = \frac{V}{48} \qquad \text{Multiplying by } \frac{1}{48} \text{ to get } T^2 \text{ alone}$$

$$T = \sqrt{\frac{V}{48}} \qquad \begin{array}{l}\text{Using the principle of square roots.}\\ \text{Note that } T \geq 0.\end{array} \qquad ▬$$

---

*This formula is taken from an article by Peter Brancazio, "The Mechanics of a Slam Dunk," *Popular Mechanics*, November 1991. Courtesy of Professor Peter Brancazio, Brooklyn College.

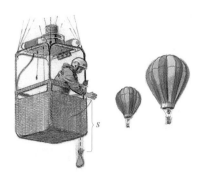

Example 4 *Falling Distance.* An object tossed downward with an initial speed (velocity) of v_0 will travel a distance of s meters, where $s = 4.9t^2 + v_0t$ and t is measured in seconds. Solve for t.

SOLUTION Since t is squared in one term and raised to the first power in the other term, the equation is quadratic in t.

$$4.9t^2 + v_0t = s$$

$$4.9t^2 + v_0t - s = 0 \qquad \text{Writing standard form}$$

$$a = 4.9, \quad b = v_0, \quad c = -s$$

$$t = \frac{-v_0 \pm \sqrt{v_0^2 - 4(4.9)(-s)}}{2(4.9)} \qquad \text{Using the quadratic formula}$$

Since the negative square root would yield a negative value for t, we use only the positive root:

$$t = \frac{-v_0 + \sqrt{v_0^2 + 19.6s}}{9.8}.$$

— ▬

The following list of steps should help you when solving formulas for a given letter. Try to remember that when solving a formula, you do the same things you would do to solve any equation.

To Solve a Formula for a Letter—Say, *b*:

1. Clear the fractions and use the principle of powers, as needed, until b does not appear in any radicand or denominator. (In some cases, you may clear the fractions first, and in some cases you may use the principle of powers first. Perform these steps until radicals containing b are gone and b is not in any denominator.)

2. Combine all terms with b^2 in them. Also combine all terms with b in them.

3. If b^2 does not appear, you can finish by using just the addition and multiplication principles as in Sections 2.3 and 6.7.

4. If b^2 appears but b does not, solve the equation for b^2. Then take square roots on both sides.

5. If there are terms containing both b and b^2, put the equation in standard form and use the quadratic formula.

8.3 | Exercise Set

Solve.

1. *Car Trips.*   During the first part of a trip, Meira's Honda traveled 120 mi at a certain speed. Meira then drove another 100 mi at a speed that was 10 mph slower. If Meira's total trip time was 4 hr, what was her speed on each part of the trip?

2. *Canoeing.*   During the first part of a canoe trip, Tim covered 60 km at a certain speed. He then traveled 24 km at a speed that was 4 km/h slower. If the total time for the trip was 8 hr, what was the speed on each part of the trip?

3. *Car Trips.*   Petra's Plymouth travels 200 mi at a certain speed. If the car had gone 10 mph faster, the trip would have taken 1 hr less. Find Petra's speed.

4. *Car Trips.*   Sandi's Subaru travels 280 mi at a certain speed. If the car had gone 5 mph faster, the trip would have taken 1 hr less. Find Sandi's speed.

5. *Air Travel.*   A Cessna flies 600 mi at a certain speed. A Beechcraft flies 1000 mi at a speed that is 50 mph faster, but takes 1 hr longer. Find the speed of each plane.

6. *Air Travel.*   A turbo-jet flies 50 mph faster than a super-prop plane. If a turbo-jet goes 2000 mi in 3 hr less time than it takes the super-prop to go 2800 mi, find the speed of each plane.

7. *Bicycling.*   Naoki bikes the 40 mi to Hillsboro at a certain speed. The return trip is made at a speed that is 6 mph slower. Total time for the round trip is 14 hr. Find Naoki's speed on each part of the trip.

8. *Car Speed.*   On a sales trip, Gail drives the 600 mi to Richmond at a certain speed. The return trip is made at a speed that is 10 mph slower. Total time for the round trip was 22 hr. How fast did Gail travel on each part of the trip?

9. *Navigation.*   The current in a typical Mississippi River shipping route flows at a rate of 4 mph. In order for a barge to travel 24 mi upriver and then return in a total of 5 hr, approximately how fast must the barge be able to travel in still water?

10. *Navigation.*   The Hudson River flows at a rate of 3 mph. A patrol boat travels 60 mi upriver and returns in a total time of 9 hr. What is the speed of the boat in still water?

11. *Filling a Pool.*   Two hoses are connected to a swimming pool. Working together, they can fill the pool in 4 hr. The larger hose, working alone, can fill the pool in 6 hr less time than the smaller one. How long would the smaller one take, working alone, to fill the pool?

12. *Filling a Tank.*   Two pipes are connected to the same tank. Working together, they can fill the tank in 2 hr. The larger pipe, working alone, can fill the tank in 3 hr less time than the smaller one. How long would the smaller one take, working alone, to fill the tank?

13. *Rowing.*   Dan rows 10 km upstream and 10 km back in a total time of 3 hr. The speed of the river is 5 km/h. Find Dan's speed in still water.

14. *Paddleboats.*   Ellen paddles 1 mi upstream and 1 mi back in a total time of 1 hr. The speed of the river is 2 mph. Find the speed of Ellen's paddleboat in still water.

Solve each formula for the indicated letter. Assume that all variables represent nonnegative numbers.

15. $A = 4\pi r^2$, for r
(Surface area of a sphere)

16. $A = 6s^2$, for s
(Surface area of a cube)

17. $A = 2\pi r^2 + 2\pi rh$, for r
(Surface area of a right cylindrical solid)

18. $F = \dfrac{Gm_1m_2}{r^2}$, for r
(Law of gravity)

19. $N = \dfrac{kQ_1Q_2}{s^2}$, for s
(Number of phone calls between two cities)

20. $A = \pi r^2$, for r
(Area of a circle)

21. $T = 2\pi\sqrt{\dfrac{l}{g}}$, for g
(A pendulum formula)

22. $a^2 + b^2 = c^2$, for b
(Pythagorean formula in two dimensions)

23. $a^2 + b^2 + c^2 = d^2$, for c
(Pythagorean formula in three dimensions)

24. $N = \dfrac{k^2 - 3k}{2}$, for k
(Number of diagonals of a polygon)

25. $s = v_0 t + \dfrac{gt^2}{2}$, for t
(A motion formula)

26. $A = \pi r^2 + \pi rs$, for r
(Surface area of a cone)

27. $N = \frac{1}{2}(n^2 - n)$, for n
(Number of games if n teams play each other once)

28. $A = A_0(1 - r)^2$, for r
(A business formula)

29. $V = 3.5\sqrt{h}$, for h
(Distance to horizon from a height)

30. $W = \sqrt{\dfrac{1}{LC}}$, for L
(An electricity formula)

31. $A = P_1(1 + r)^2 + P_2(1 + r)$, for r
(An investment formula)

32. $A = P_1\left(1 + \dfrac{r}{2}\right)^2 + P_2\left(1 + \dfrac{r}{2}\right)$, for r
(An investment formula)

Solve. Refer to Exercises 15–32 and Examples 2–4 for the appropriate formula.

33. *Falling Distance.*

a) An object is dropped 500 m from an airplane. How long does it take the object to reach the ground?

b) An object is thrown downward 500 m from the plane at an initial velocity of 30 m/sec. How long does it take the object to reach the ground?

c) How far will an object fall in 5 sec, when thrown downward at an initial velocity of 30 m/sec?

34. *Falling Distance.*

a) An object is dropped 75 m from an airplane. How long does it take the object to reach the ground?

b) An object is thrown downward with an initial velocity of 30 m/sec from a plane 75 m above the ground. How long does it take the object to reach the ground?

c) How far will an object fall in 2 sec, if thrown downward at an initial velocity of 30 m/sec?

35. *Bungee Jumping.* Jesse is tied to one end of a 40-m elasticized (bungee) cord. The other end of the cord is tied to the middle of a train trestle. If Jesse jumps off the bridge, for how long will he fall before the cord begins to stretch? (See Example 4 and let $v_0 = 0$.)

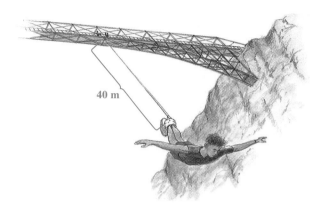

40 m

36. *Bungee Jumping.* Sheila is tied to a bungee cord (see Exercise 35) and falls for 2.5 sec before her cord begins to stretch. How long is the bungee cord?

37. *Hang Time.* Anfernee Hardaway of the Orlando Magic has a vertical leap of about 36 in.* What is his hang time?

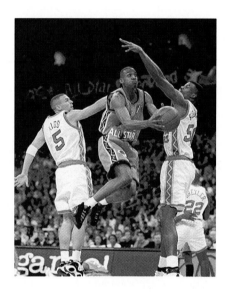

38. *League Schedules.* In a volleyball league, each team plays each of the other teams once. If a total of 66 games is played, how many teams are in the league?

39. *Downward Speed.* An object thrown downward from a 100-m cliff travels 51.6 m in 3 sec. What was the initial velocity of the object?

*Information provided by Orlando Magic media relations department.

40. *Downward Speed.* An object thrown downward from a 200-m cliff travels 91.2 m in 4 sec. What was the initial velocity of the object?

41. *Compound Interest.* A firm invests $3000 in a savings account for 2 yr. At the beginning of the second year, an additional $1700 is invested. If a total of $5253.70 is in the account at the end of the second year, what is the annual interest rate? (*Hint:* See Exercise 31.)

42. *Compound Interest.* A business invests $10,000 in a savings account for 2 yr. At the beginning of the second year, an additional $3500 is invested. If a total of $15,569.75 is in the account at the end of the second year, what is the annual interest rate? (*Hint:* See Exercise 31.)

Skill Maintenance

43. Solve:
$$\sqrt{3x + 1} = \sqrt{2x - 1} + 1. \ [7.6]$$

44. Add:
$$\frac{1}{x - 1} + \frac{1}{x^2 - 3x + 2}. \ [6.2]$$

45. Multiply and simplify:
$$\sqrt[3]{18y^3} \ \sqrt[3]{4x^2}. \ [7.3]$$

46. Divide and simplify:
$$\frac{x - 3}{x^2 - 5x + 6} \div \frac{7x + 14}{x^2 - 4}. \ [6.1]$$

Synthesis

47. ◆ Write a problem for a classmate to solve. Devise the problem so that **(a)** the solution is found after solving a rational equation and **(b)** the solution is "The express train travels 90 mph."

48. ◆ Under what circumstances would a negative value for *t*, time, have meaning?

49. ◆ Marti is tied to a bungee cord that is twice as long as the cord tied to Pedro. Will Marti's fall take twice as long as Pedro's before their cords begin to stretch? Why or why not? (See Exercises 35 and 36.)

50. Find *a* such that the reciprocal of $a - 1$ is $a + 1$.

51. *Purchasing.* A discount store bought a quantity of beach towels for $250 and sold all but 15 at a profit of $3.50 per towel. With the total amount received,

the manager could buy 4 more than twice as many as were bought before. Find the cost per towel.

52. Solve for *n*:
$$mn^4 - r^2pm^3 - r^2n^2 + p = 0.$$

53. *The Golden Rectangle.* For over 2000 yr, the proportions of a "golden" rectangle have been considered visually appealing. A rectangle of width *w* and length *l* is considered "golden" if
$$\frac{w}{l} = \frac{l}{w + l}.$$

Solve for *l*.

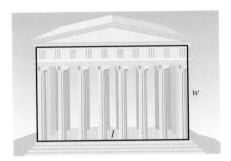

54. *Diagonal of a Cube.* Find a formula that expresses the length of the three-dimensional diagonal of a cube as a function of the cube's surface area.

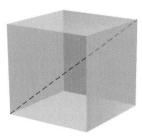

55. *Special Relativity.* Einstein found that if an object of mass m_0 is brought to a velocity *v*, its mass becomes
$$m = \frac{m_0}{\sqrt{1 - \dfrac{v^2}{c^2}}},$$
where *c* is the speed of light. Solve the formula for *c*.

56. *Surface Area.*   Find a formula that expresses the diameter of a right cylindrical solid as a function of its surface area and its height.

57. A sphere is inscribed in a cube as shown in the figure to the right. Express the surface area of the sphere as a function of the surface area S of the cube.

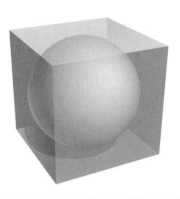

$$\frac{8.4}{}$$

More on Quadratic Equations

- *Equations Quadratic in Form*
- *Writing Equations from Solutions*

In Sections 8.1 and 8.2, we used the principle of square roots, the principle of zero products, and the quadratic formula to solve quadratic equations. In this section, we will use these approaches to solve equations that are *quadratic in form*. We will also use the principle of zero products, in reverse, to write equations with predetermined solutions.

Equations Quadratic in Form

Certain equations that are not actually quadratic can be thought of in such a way that they can be solved as though they are. For example, if x^4 is regarded as $(x^2)^2$, the equation $x^4 - 9x^2 + 8 = 0$ is "quadratic in x^2":

$$x^4 - 9x^2 + 8 = 0$$
$$(x^2)^2 - 9(x^2) + 8 = 0. \qquad \text{Thinking of } x^4 \text{ as } (x^2)^2$$

To make this clearer, we can substitute another variable for x^2. For example, if we let

$$u = x^2,$$

then

$$u^2 = x^4.$$

When we substitute u for x^2 and u^2 for x^4, we have

$$x^4 - 9x^2 + 8 = 0$$
$$u^2 - 9u + 8 = 0.$$

The equation $u^2 - 9u + 8 = 0$ can be solved by factoring or by the quadratic formula. Then, remembering that we have let $u = x^2$, we can solve for x. Equations that can be solved like this are said to be *in quadratic form*, or *reducible to quadratic*.

Example 1   Solve: $x^4 - 9x^2 + 8 = 0$.
We can solve both algebraically and graphically.

ALGEBRAIC SOLUTION

We first let $u = x^2$. Then we solve $x^4 - 9x^2 + 8 = 0$ by substituting u for x^2 and u^2 for x^4:

$$u^2 - 9u + 8 = 0$$
$$(u - 8)(u - 1) = 0 \qquad \text{Factoring}$$
$$u - 8 = 0 \quad or \quad u - 1 = 0 \qquad \text{Principle of zero products}$$
$$u = 8 \quad or \qquad u = 1.$$

Caution!   A common error is to solve for u and then forget to solve for x. Remember that you must find values for the *original* variable!

We replace u with x^2 and solve these equations:

$$x^2 = 8 \qquad or \quad x^2 = 1$$
$$x = \pm\sqrt{8} \quad or \qquad x = \pm 1$$
$$x = \pm 2\sqrt{2} \quad or \qquad x = \pm 1.$$

To check, note that for $x = 2\sqrt{2}$, we have $x^2 = 8$ and $x^4 = 64$. Similarly, for $x = -2\sqrt{2}$, we have $x^2 = 8$ and $x^4 = 64$. When $x = 1$, we have $x^2 = 1$ and $x^4 = 1$, and when $x = -1$, we have $x^2 = 1$ and $x^4 = 1$. Thus instead of making four checks, we need make only two.

CHECK:
For $\pm 2\sqrt{2}$:

$$\frac{x^4 - 9x^2 + 8 = 0}{(\pm 2\sqrt{2})^4 - 9(\pm 2\sqrt{2})^2 + 8 \ ? \ 0}$$
$$64 - 9 \cdot 8 + 8 \ \Big| $$
$$0 \ \Big| \ 0 \quad \text{TRUE}$$

For ± 1:

$$\frac{x^4 - 9x^2 + 8 = 0}{(\pm 1)^4 - 9(\pm 1)^2 + 8 \ ? \ 0}$$
$$1 - 9 + 8 \ \Big| $$
$$0 \ \Big| \ 0 \quad \text{TRUE}$$

The solutions are 1, -1, $2\sqrt{2}$, and $-2\sqrt{2}$.

GRAPHICAL SOLUTION

We let $f(x) = x^4 - 9x^2 + 8$ and determine the zeros of the function.

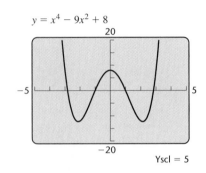

$y = x^4 - 9x^2 + 8$
Yscl = 5

It appears as though the graph has 4 x-intercepts. Recall from Chapter 5 that a fourth-degree function has at most 4 zeros, so we know that we have not missed any zeros. Using ZERO four times, we obtain the solutions -2.8284271, -1, 1, and 2.8284271. The solutions are ± 1 and, approximately, ± 2.8284271.

Note that since $2\sqrt{2} \approx 2.8284271$, the solutions found using both methods agree.

Example 1 can be solved directly by factoring:

$$x^4 - 9x^2 + 8 = 0$$
$$(x^2 - 1)(x^2 - 8) = 0$$
$$x^2 - 1 = 0 \quad or \quad x^2 - 8 = 0$$
$$x^2 = 1 \quad or \quad x^2 = 8$$
$$x = \pm 1 \quad or \quad x = \pm 2\sqrt{2}.$$

There is nothing wrong with this approach. However, in the examples that follow, you will note that it becomes increasingly difficult to solve the equation without first making a substitution.

Sometimes rational equations, radical equations, or equations containing fractional exponents are reducible to quadratic. It is especially important that answers to these equations be checked in the original equation.

Example 2   Solve: $x - 3\sqrt{x} - 4 = 0$.

SOLUTION   This radical equation could be solved using the method discussed in Section 7.6. However, if we think of x as $(\sqrt{x})^2$, we can regard the equation as "quadratic in $\sqrt{x}$."

We let $u = \sqrt{x}$ and $u^2 = x$:

$$x - 3\sqrt{x} - 4 = 0$$
$$u^2 - 3u - 4 = 0 \qquad \text{Substituting}$$
$$(u - 4)(u + 1) = 0$$
$$u = 4 \quad or \quad u = -1.$$

Now we replace u with $\sqrt{x}$ and solve these equations:

$$\sqrt{x} = 4 \quad or \quad \sqrt{x} = -1.$$

Squaring gives us $x = 16$ or $x = 1$ and also makes checking essential.

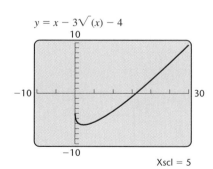

$y = x - 3\sqrt{(x)} - 4$

Xscl = 5

CHECK: For 16:

$$\begin{array}{c} x - 3\sqrt{x} - 4 = 0 \\ \hline 16 - 3\sqrt{16} - 4 \; ? \; 0 \\ 16 - 3 \cdot 4 - 4 \quad \big| \\ 0 \quad \big| \; 0 \quad \text{TRUE} \end{array}$$

For 1:

$$\begin{array}{c} x - 3\sqrt{x} - 4 = 0 \\ \hline 1 - 3\sqrt{1} - 4 \; ? \; 0 \\ 1 - 3 \cdot 1 - 4 \quad \big| \\ -6 \quad \big| \; 0 \quad \text{FALSE} \end{array}$$

The number 16 checks, but 1 does not. Had we noticed that $\sqrt{x} = -1$ has no solution (since principal roots are never negative), we could have solved only the equation $\sqrt{x} = 4$. The graph at left provides further evidence that although 16 is a solution, -1 is not. The solution is 16.   ▬

Example 3   Find the x-intercepts of the graph of $f(x) = (x^2 - 1)^2 - (x^2 - 1) - 2$.

SOLUTION   The x-intercepts occur where $f(x) = 0$ so we must have

$$(x^2 - 1)^2 - (x^2 - 1) - 2 = 0.$$

Since this equation is quadratic in $x^2 - 1$, we let $u = x^2 - 1$:

$$u^2 - u - 2 = 0 \qquad \qquad \text{Substituting in}$$
$$(x^2 - 1)^2 - (x^2 - 1) - 2 = 0$$
$$(u - 2)(u + 1) = 0$$
$$u = 2 \qquad or \qquad u = -1.$$

Now we replace u with $x^2 - 1$ and solve these equations:

$$x^2 - 1 = 2 \qquad or \quad x^2 - 1 = -1$$
$$x^2 = 3 \qquad or \qquad x^2 = 0$$
$$x = \pm\sqrt{3} \quad or \qquad x = 0.$$

The x-intercepts occur at $(-\sqrt{3}, 0)$, $(0, 0)$, and $(\sqrt{3}, 0)$. These solutions are confirmed by the graph below.

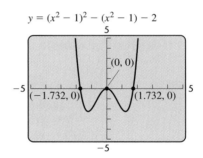

Sometimes great care must be taken in deciding what substitution to make.

Example 4   Solve: $m^{-2} - 6m^{-1} + 4 = 0$.

SOLUTION   We rewrite the equation using positive exponents:

$$\frac{1}{m^2} - \frac{6}{m} + 4 = 0.$$

If we let $u = 1/m$ and $u^2 = 1/m^2$, the equation is quadratic in $1/m$:

$$u^2 - 6u + 4 = 0 \qquad\qquad \text{Substituting}$$
$$u = \frac{-(-6) \pm \sqrt{(-6)^2 - 4 \cdot 1 \cdot 4}}{2 \cdot 1} \qquad \text{Using the quadratic formula}$$
$$= \frac{6 \pm \sqrt{20}}{2} = \frac{2 \cdot 3 \pm 2\sqrt{5}}{2} \Bigg\} \quad \text{Simplifying}$$
$$= 3 \pm \sqrt{5}.$$

Now we replace u with $1/m$ and solve:

$$\frac{1}{m} = 3 \pm \sqrt{5}$$
$$1 = m(3 \pm \sqrt{5}) \qquad \text{Multiplying by } m \text{ on both sides}$$
$$\frac{1}{3 \pm \sqrt{5}} = m. \qquad \text{Dividing by } 3 \pm \sqrt{5} \text{ on both sides}$$

CHECK: For $1/(3 - \sqrt{5})$:

$$m^{-2} - 6m^{-1} + 4 = 0$$

$$\left(\frac{1}{3 - \sqrt{5}}\right)^{-2} - 6\left(\frac{1}{3 - \sqrt{5}}\right)^{-1} + 4 \; ? \; 0$$

$$(3 - \sqrt{5})^2 - 6(3 - \sqrt{5}) + 4$$

$$9 - 6\sqrt{5} + 5 - 18 + 6\sqrt{5} + 4$$

$$\qquad\qquad\qquad\qquad\qquad 0 \;\bigg|\; 0 \quad \text{TRUE}$$

For $1/(3 + \sqrt{5})$:

$$m^{-2} - 6m^{-1} + 4 = 0$$

$$\left(\frac{1}{3 + \sqrt{5}}\right)^{-2} - 6\left(\frac{1}{3 + \sqrt{5}}\right)^{-1} + 4 \; ? \; 0$$

$$(3 + \sqrt{5})^2 - 6(3 + \sqrt{5}) + 4$$

$$9 + 6\sqrt{5} + 5 - 18 - 6\sqrt{5} + 4$$

$$\qquad\qquad\qquad\qquad\qquad 0 \;\bigg|\; 0 \quad \text{TRUE}$$

We could also check the solutions using a calculator by entering $y_1 = x^{-2} - 6x^{-1} + 4$ and evaluating y_1 for both numbers, as shown below.

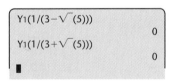

As another check, we solve the equation graphically. The first coordinates of the x-intercepts of the graph at left are approximate solutions.

Both numbers check. The solutions are $1/(3 - \sqrt{5})$ and $1/(3 + \sqrt{5})$, or approximately 1.309 and 0.191.   ▬

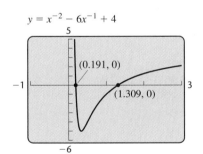

$y = x^{-2} - 6x^{-1} + 4$

(0.191, 0)

(1.309, 0)

Example 5   Solve: $t^{2/5} - t^{1/5} - 2 = 0$.

SOLUTION   Note that $t^{2/5}$ can be rewritten as $(t^{1/5})^2$. The equation can thus be written as $(t^{1/5})^2 - t^{1/5} - 2 = 0$. We let $u = t^{1/5}$ and solve the resulting equation:

$$u^2 - u - 2 = 0 \qquad \textbf{Substituting}$$

$$(u - 2)(u + 1) = 0$$

$$u = 2 \quad or \quad u = -1.$$

Now we replace u with $t^{1/5}$ and solve:

$$t^{1/5} = 2 \quad or \quad t^{1/5} = -1$$

$$t = 32 \quad or \quad t = -1. \qquad \textbf{Principle of powers; raising to the 5th power}$$

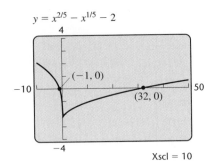

$y = x^{2/5} - x^{1/5} - 2$

$Xscl = 10$

CHECK:

For 32:

$$t^{2/5} - t^{1/5} - 2 = 0$$

$$\overline{32^{2/5} - 32^{1/5} - 2 \overset{?}{\,} 0}$$
$$(32^{1/5})^2 - 32^{1/5} - 2 \,\bigg|$$
$$2^2 - 2 - 2 \,\bigg|$$
$$0 \,\bigg|\, 0 \quad \text{TRUE}$$

For -1:

$$t^{2/5} - t^{1/5} - 2 = 0$$

$$\overline{(-1)^{2/5} - (-1)^{1/5} - 2 \overset{?}{\,} 0}$$
$$[(-1)^{1/5}]^2 - (-1)^{1/5} - 2 \,\bigg|$$
$$(-1)^2 - (-1) - 2 \,\bigg|$$
$$0 \,\bigg|\, 0 \quad \text{TRUE}$$

Both numbers check and are confirmed by the graph at left. The solutions are 32 and -1.

The following tips may prove useful.

When Solving an Equation That Is Quadratic in Form:

  1. Determine whether the equation is quadratic in form by identifying a term with a factor that is the square of a factor appearing in another term.

  2. Write down any substitutions that you are making.

  3. Whenever you make a substitution, be sure to solve for the variable that is used in the original equation.

  4. Check possible answers in the original equation.

Writing Equations from Solutions

We know by the principle of zero products that $(x - 2)(x + 3) = 0$ has solutions 2 and -3. If we know the solutions of an equation, we can write an equation, using the principle in reverse.

Example 6   Find an equation for the given solutions.

a) 3 and $-\frac{2}{5}$

b) $2i$ and $-2i$

c) $5\sqrt{7}$ and $-5\sqrt{7}$

d) -4, 0, and 1

SOLUTION

a)
$$x = 3 \quad or \quad x = -\tfrac{2}{5}$$
$$x - 3 = 0 \quad or \quad x + \tfrac{2}{5} = 0 \qquad \text{Getting 0's on one side}$$
$$(x - 3)\left(x + \tfrac{2}{5}\right) = 0 \qquad \text{Using the principle of zero products (multiplying)}$$
$$x^2 + \tfrac{2}{5}x - 3x - 3 \cdot \tfrac{2}{5} = 0 \qquad \text{Multiplying}$$
$$x^2 - \tfrac{13}{5}x - \tfrac{6}{5} = 0 \qquad \text{Combining like terms}$$
$$5\left(x^2 - \tfrac{13}{5}x - \tfrac{6}{5}\right) = 5 \cdot 0 \qquad \text{Multiplying by 5 on both sides to clear fractions}$$
$$5x^2 - 13x - 6 = 0 \qquad \text{Simplifying}$$

b)

$$x = 2i \quad or \quad x = -2i$$

$$x - 2i = 0 \quad or \quad x + 2i = 0 \qquad \text{Getting 0's on one side}$$

$$(x - 2i)(x + 2i) = 0 \qquad \begin{array}{l}\text{Using the principle of zero} \\ \text{products (multiplying)}\end{array}$$

$$x^2 - (2i)^2 = 0 \qquad \begin{array}{l}\text{Finding the product of} \\ \text{a sum and a difference}\end{array}$$

$$x^2 - 4i^2 = 0$$

$$x^2 + 4 = 0 \qquad\qquad i^2 = -1$$

c)

$$x = 5\sqrt{7} \quad or \quad x = -5\sqrt{7}$$

$$x - 5\sqrt{7} = 0 \quad or \quad x + 5\sqrt{7} = 0 \qquad \begin{array}{l}\text{Getting 0's on} \\ \text{one side}\end{array}$$

$$(x - 5\sqrt{7})(x + 5\sqrt{7}) = 0 \qquad \begin{array}{l}\text{Using the principle} \\ \text{of zero products}\end{array}$$

$$x^2 - (5\sqrt{7})^2 = 0 \qquad \begin{array}{l}\text{Finding the product of} \\ \text{a sum and a difference}\end{array}$$

$$x^2 - 25 \cdot 7 = 0$$

$$x^2 - 175 = 0$$

d)

$$x = -4 \quad or \quad x = 0 \quad or \quad x = 1$$

$$x + 4 = 0 \quad or \quad x = 0 \quad or \quad x - 1 = 0 \qquad \text{Getting 0's on one side}$$

$$(x + 4)x(x - 1) = 0 \qquad \begin{array}{l}\text{Using the principle} \\ \text{of zero products}\end{array}$$

$$x(x^2 + 3x - 4) = 0 \qquad \text{Multiplying}$$

$$x^3 + 3x^2 - 4x = 0$$

To check any of these equations, we can simply substitute one or more of the given solutions.

8.4 Exercise Set

Solve.

1. $x^4 - 5x^2 + 4 = 0$

2. $x^4 - 10x^2 + 9 = 0$

3. $x^4 - 12x^2 + 27 = 0$

4. $x^4 - 9x^2 + 20 = 0$

5. $4x^4 - 19x^2 + 12 = 0$

6. $9x^4 - 14x^2 + 5 = 0$

7. $x - 4\sqrt{x} - 1 = 0$

8. $x - 2\sqrt{x} - 6 = 0$

9. $(x^2 - 7)^2 - 3(x^2 - 7) + 2 = 0$

10. $(x^2 - 1)^2 - 5(x^2 - 1) + 6 = 0$

11. $(3 + \sqrt{x})^2 + 3(3 + \sqrt{x}) - 10 = 0$

12. $(1 + \sqrt{x})^2 + 5(1 + \sqrt{x}) + 6 = 0$

13. $x^{-2} - x^{-1} - 6 = 0$

14. $2x^{-2} - x^{-1} - 1 = 0$

15. $4x^{-2} + x^{-1} - 5 = 0$

16. $m^{-2} + 9m^{-1} - 10 = 0$

17. $t^{2/3} + t^{1/3} - 6 = 0$ (*Hint:* Let $u = t^{1/3}$.)

18. $w^{2/3} - 2w^{1/3} - 8 = 0$

19. $y^{1/3} - y^{1/6} - 6 = 0$

20. $t^{1/2} + 3t^{1/4} + 2 = 0$

21. $t^{1/3} + 2t^{1/6} = 3$

22. $m^{1/2} + 6 = 5m^{1/4}$

23. $(3 - \sqrt{x})^2 - 10(3 - \sqrt{x}) + 23 = 0$

24. $(5 + \sqrt{x})^2 - 12(5 + \sqrt{x}) + 33 = 0$

25. $16\left(\dfrac{x-1}{x-8}\right)^2 + 8\left(\dfrac{x-1}{x-8}\right) + 1 = 0$

26. $9\left(\dfrac{x+2}{x+3}\right)^2 - 6\left(\dfrac{x+2}{x+3}\right) + 1 = 0$

Find all x-intercepts of the given function f.

27. $f(x) = 5x + 13\sqrt{x} - 6$

28. $f(x) = 3x + 10\sqrt{x} - 8$

29. $f(x) = (x^2 - 3x)^2 - 10(x^2 - 3x) + 24$

30. $f(x) = (x^2 - 6x)^2 - 2(x^2 - 6x) - 35$

31. $f(x) = x^{2/5} + x^{1/5} - 6$

32. $f(x) = x^{1/2} - x^{1/4} - 6$

33. $f(x) = \left(\dfrac{x^2-2}{x}\right)^2 - 7\left(\dfrac{x^2-2}{x}\right) - 18$

34. $f(x) = \left(\dfrac{y^2-1}{y}\right)^2 - 4\left(\dfrac{y^2-1}{y}\right) - 12$

Write a quadratic equation having the given numbers as solutions.

35. $-7, 3$

36. $-6, 4$

37. 3, only solution
  (*Hint:* Use 3 as a
  solution twice.)

38. -5, only solution

39. $4, \frac{2}{3}$

40. $5, \frac{3}{4}$

41. $\frac{1}{2}, \frac{1}{3}$

42. $-\frac{1}{4}, -\frac{1}{2}$

43. $-0.6, 1.4$

44. $2.5, -0.4$

45. $-\sqrt{7}, \sqrt{7}$

46. $-\sqrt{3}, \sqrt{3}$

47. $3\sqrt{2}, -3\sqrt{2}$

48. $2\sqrt{5}, -2\sqrt{5}$

49. $3i, -3i$

50. $4i, -4i$

51. $5 - 2i, 5 + 2i$

52. $2 - 7i, 2 + 7i$

53. $2 - \sqrt{10}, 2 + \sqrt{10}$

54. $3 - \sqrt{14}, 3 + \sqrt{14}$

Write a third-degree equation having the given numbers as solutions.

55. $-3, 0, 4$

56. $-5, 0, 2$

57. $-1, 1, 2$

58. $-2, 2, 3$

Skill Maintenance

59. Multiply and simplify: $\sqrt{3x^2}\,\sqrt{3x^3}$. [7.4]

60. Solution A is 18% alcohol and solution B is 45% alcohol. How much of each should be mixed together to get 12 L of a solution that is 36% alcohol? [3.3]

61. *Insect Resistance to Pesticides.*   From 1948 to 1989, the use of pesticides in the United States increased. At the same time, the number of insect species resistant to these poisons has also increased.

  a) Use regression to find a linear function $n(t)$ that can be used to predict the number of insect species n that are resistant to pesticide t years after 1948.

  b) Use the function found in part (a) to estimate the year in which there were 100 insect species resistant to pesticide. [2.6]

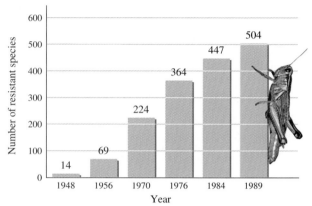

Source: U.S. Environmental Protection Agency

62. If $g(x) = x^2 - x$, find $g(a + 1)$. [5.2]

Synthesis

63. ◆ Describe a procedure that could be used to solve any equation of the form $ax^4 + bx^2 + c = 0$.

64. ◆ Describe a procedure that could be used to write an equation that is quadratic in $3x^2 + 1$. Then explain how the procedure could be adjusted to write equations that are quadratic in $3x^2 + 1$ and have no real-number solution.

65. ◆ While trying to solve $0.05x^4 - 0.8 = 0$ using a grapher, Murray gets the following screen.

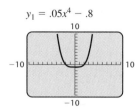

Can Murray find all solutions of this equation using a grapher? Why or why not?

Solve.

66. $3x^4 + 5x^2 - 1 = 0$

67. $5x^4 - 7x^2 + 1 = 0$

68. $(x^2 - 5x - 1)^2 - 18(x^2 - 5x - 1) + 65 = 0$

69. $(x^2 - 4x - 2)^2 - 13(x^2 - 4x - 2) + 30 = 0$

70. $\dfrac{x}{x-1} - 6\sqrt{\dfrac{x}{x-1}} - 40 = 0$

71. $\left(\sqrt{\dfrac{x}{x-3}}\right)^2 - 24 = 10\sqrt{\dfrac{x}{x-3}}$

72. $a^5(a^2 - 25) + 13a^3(25 - a^2) + 36a(a^2 - 25) = 0$

73. $a^3 - 26a^{3/2} - 27 = 0$

74. $x^6 - 28x^3 + 27 = 0$

75. $x^6 + 7x^3 - 8 = 0$

76. $x^4 - x^3 - 13x^2 + x + 12 = 0$

77. Suppose that $f(x) = ax^2 + bx + c$, with $f(-3) = 0$, $f\left(\tfrac{1}{2}\right) = 0$, and $f(0) = -12$. Find a, b, and c.

78. Find an equation for which $2 - \sqrt{3}$, $2 + \sqrt{3}$, $5 - 2i$, and $5 + 2i$ are solutions.

79. Find an equation for which $1 - \sqrt{5}$, $1 + \sqrt{5}$, $3 - 2i$, and $3 + 2i$ are solutions.

80. Find an equation quadratic in x^2 for which $\sqrt{2}$, $-\sqrt{2}$, $\sqrt{3}$, and $-\sqrt{3}$ are solutions.

81. The graph of an equation

$$y = ax^2 + bx + c$$

is shown below. Determine a, b, and c from the information given.

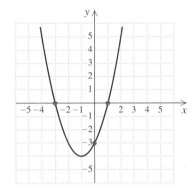

8.5
Variation and Problem Solving

- *Direct Variation*
- *Inverse Variation*
- *Combined Variation*

To extend our study of formulas and functions, we now examine three situations that frequently arise in problem solving: direct variation, inverse variation, and combined variation. These situations sometimes require the solution of a quadratic equation.

Direct Variation

A hair stylist earns \$18 per hour. In 1 hr, \$18 is earned. In 2 hr, \$36 is earned. In 3 hr, \$54 is earned, and so on. This gives rise to a set of ordered pairs of numbers:

$$(1, 18), (2, 36), (3, 54), (4, 72), \quad \text{and so on.}$$

The ratio of dollars to hours is $\frac{18}{1}$ in every case.

Whenever a situation gives rise to pairs of numbers in which the ratio is constant, we say that there is **direct variation**. Here earnings *vary directly* as the time:

$$E = 18t \quad \text{or, using function notation,} \quad E(t) = 18t.$$

> *Direct Variation*
>
> When a situation gives rise to a linear function of the form $f(x) = kx$, or $y = kx$, where k is a nonzero constant, we say that there is *direct variation*, that *y varies directly as x*, or that *y is proportional to x*. The number k is called the *variation constant*, or *constant of proportionality*.

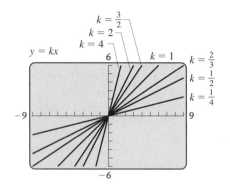

The graph of $y = kx$ always goes through the origin and, for $k > 0$, rises from left to right. (See the graph at left.) Note that as x increases, y increases.

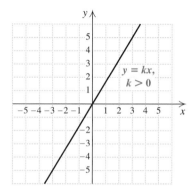

The graph of a function describing direct variation is always a straight line.

Example 1 Find the variation constant and an equation of variation if y varies directly as x, and $y = 32$ when $x = 2$.

SOLUTION We know that $(2, 32)$ is a solution of $y = kx$. Therefore,

$$32 = k \cdot 2 \qquad \textbf{Substituting}$$

$$\frac{32}{2} = k, \quad \text{or} \quad k = 16. \qquad \textbf{Solving for } k$$

The variation constant is 16. The equation of variation is $y = 16x$. The notation $y(x) = 16x$ or $f(x) = 16x$ can also be used. ▬

Example 2 *Water from Melting Snow.* The number of centimeters W of water produced from melting snow varies directly as the number of centimeters S of snow. Meteorologists know that under certain conditions, 150 cm of snow will melt to 16.8 cm of water. How many centimeters of water will replace 200 cm of snow under these conditions?

SOLUTION

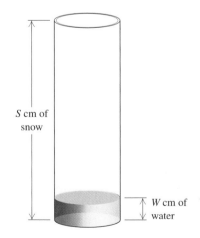

1. **Familiarize.** Because of the phrase "$W \ldots$ varies directly as $\ldots S$," we decide to express the amount of water as a function of the amount of snow. Thus, $W(S) = kS$, where k is the variation constant. Knowing that 150 cm of snow becomes 16.8 cm of water, we have $W(150) = 16.8$.

2. **Translate.** We find the variation constant using the data and then find the equation of variation:

$$W(S) = kS$$
$$W(150) = k \cdot 150 \qquad \text{Replacing } S \text{ with 150}$$
$$16.8 = k \cdot 150 \qquad \text{Replacing } W(150) \text{ with 16.8}$$
$$\frac{16.8}{150} = k \qquad \text{Solving for } k$$
$$0.112 = k. \qquad \text{This is the variation constant.}$$

The equation of variation is $W(S) = 0.112S$. This is the translation.

3. **Carry out.** To find how much water 200 cm of snow will become, we compute $W(200)$:

$$W(S) = 0.112S$$
$$W(200) = 0.112(200) \qquad \text{Substituting 200 for } S$$
$$W = 22.4.$$

4. **Check.** To check, we could reexamine all our calculations. Note that our answer seems reasonable since 200/22.4 and 150/16.8 are equal.

5. **State.** 200 cm of snow will melt into 22.4 cm of water. ▬

Inverse Variation

To see what we mean by inverse variation, consider the following situation.

A bus is traveling a distance of 20 mi. At a speed of 20 mph, the trip will take 1 hr. At 40 mph, it will take $\frac{1}{2}$ hr. At 60 mph, it will take $\frac{1}{3}$ hr, and so on. This gives rise to a set of pairs of numbers, all having the same product:

$$(20, 1), \left(40, \tfrac{1}{2}\right), \left(60, \tfrac{1}{3}\right), \left(80, \tfrac{1}{4}\right), \quad \text{and so on.}$$

Whenever a situation gives rise to pairs of numbers for which the product is constant, we say that there is **inverse variation**. The time t required for the bus to travel 20 mi at rate r is given by

$$t = \frac{20}{r} \quad \text{or, using function notation,} \quad t(r) = \frac{20}{r}.$$

Inverse Variation

When a situation gives rise to a function of the form $f(x) = k/x$, or $y = k/x$, where k is a nonzero constant, we say that there is *inverse variation*, that *y varies inversely as x*, or that *y is inversely proportional to x*. The number k is called the *variation constant*, or *constant of proportionality*.

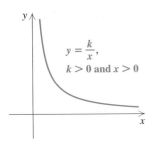

$y = \dfrac{k}{x}$,
$k > 0$ and $x > 0$

The graph of $y = k/x$, for $k > 0$ and $x > 0$, is like the one shown to the left. Note that as x increases, y decreases.

Example 3   Find the variation constant and an equation of variation if y varies inversely as x, and $y = 32$ when $x = 0.2$.

SOLUTION   We know that $(0.2, 32)$ is a solution of

$$y = \frac{k}{x}.$$

Therefore,

$$32 = \frac{k}{0.2} \qquad \textbf{Substituting}$$

$$(0.2)32 = k$$

$$6.4 = k. \qquad \textbf{Solving for } k$$

The variation constant is 6.4. The equation of variation is

$$y = \frac{6.4}{x}.$$

There are many problems that translate to an equation of inverse variation.

Example 4   *Building a Shed.* The time t required to do a job varies inversely as the number of people P who work on the job (assuming that all work at the same rate). It takes 4 hr for 12 people to build a woodshed. How long would it take 3 people to complete the same job?

SOLUTION

1. **Familiarize.**   Because of the phrase "$t \ldots$ varies inversely as $\ldots P$," we express the amount of time required, in hours, as a function of the number of people working. Therefore, we have $t(P) = k/P$. From the information given, we know that $t(12) = 4$. That is, 12 people take 4 hr to build the shed.

2. **Translate.** We find the variation constant using the data and then find the equation of variation:

$$t(P) = \frac{k}{P} \qquad \text{Using function notation}$$

$$t(12) = \frac{k}{12} \qquad \text{Replacing } P \text{ with 12}$$

$$4 = \frac{k}{12} \qquad \text{Replacing } t(12) \text{ with 4}$$

$$48 = k. \qquad \text{Solving for } k, \text{ the variation constant}$$

The equation of variation is $t(P) = 48/P$. This is the translation.

3. **Carry out.** To determine how long it would take 3 people to complete the job, we compute $t(3)$:

$$t(P) = \frac{48}{P}$$

$$t(3) = \frac{48}{3} \qquad \text{Substituting 3 for } P$$

$$t = 16. \qquad t = 16 \text{ when } P = 3$$

4. **Check.** We could now recheck each step. Note that, as expected, as the number of people working goes *down*, the time required for the job goes *up*.

5. **State.** It will take 3 people 16 hr to build a woodshed. ▬

We can tell from the graph of a group of data whether two quantities are directly or inversely proportional.

Example 5 Determine whether each graph seems to represent direct variation, inverse variation, or neither.

a) Let $x =$ the number of calories per minute that Terry burns in a particular exercise and $y =$ the number of minutes Terry must exercise in order to burn 250 calories.

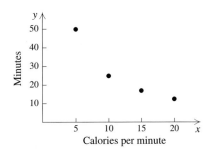

Calories per minute

b) Let x = the number of shopping centers in the United States and y = the total square footage of retail space.

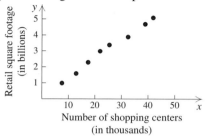

Source: International Council of Shopping Centers

c) Let x = age and y = the percentage of people of age x who participate in vigorous physical activity.

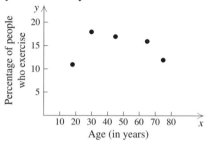

Source: Centers for Disease Control and Prevention,
National Center for Chronic Disease Prevention and Health Promotion

SOLUTION

a) As the number of calories per minute increases, the number of minutes Terry must exercise decreases. A curve drawn through the data points would look like the graph of $y = k/x$ for some choice of k. Thus the graph appears to represent inverse variation.

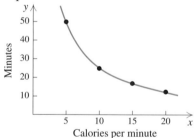

b) As the number of shopping centers increases, the total square footage increases. Since this is true, and the data points fall approximately on a straight line, the graph represents direct variation.

c) As age increases, the percentage of people that age who exercise vigorously first increases, then decreases. No equation of the form $y = kx$ or $y = k/x$ can model these data. Thus neither direct nor inverse variation is represented.

Combined Variation

Often one variable varies directly or inversely with more than one other variable. For example, in the formula for the volume of a right circular cylinder, $V = \pi r^2 h$, we say that V varies *jointly* as h and the square of r.

Joint Variation

y varies *jointly* as x and z if, for some nonzero constant k, $y = kxz$.

Example 6   Find an equation of variation if y varies jointly as x and z, and $y = 30$ when $x = 2$ and $z = 3$.

SOLUTION   We have

$$y = kxz,$$

so

$$30 = k \cdot 2 \cdot 3$$
$$k = 5. \qquad \text{The variation constant is 5.}$$

The equation of variation is $y = 5xz$.

Example 7   Find an equation of variation if y varies jointly as x and z and inversely as the square of w, and $y = 105$ when $x = 3$, $z = 20$, and $w = 2$.

SOLUTION   The equation of variation is of the form

$$y = k \cdot \frac{xz}{w^2},$$

so, substituting, we have

$$105 = k \cdot \frac{3 \cdot 20}{2^2}$$
$$105 = k \cdot 15$$
$$k = 7.$$

Thus,

$$y = 7 \cdot \frac{xz}{w^2}.$$

8.5 Exercise Set

Find the variation constant and an equation of variation if y varies directly as x and the following conditions exist.

1. $y = 28$ when $x = 7$

2. $y = 5$ when $x = 12$

3. $y = 3.4$ when $x = 2$

4. $y = 2$ when $x = 5$

5. $y = 30$ when $x = 8$

6. $y = 1$ when $x = \frac{1}{3}$

7. $y = 0.8$ when $x = 0.5$

8. $y = 0.6$ when $x = 0.4$

Solve.

9. *Ohm's Law.* The electric current I, in amperes, in a circuit varies directly as the voltage V. When 15 volts are applied, the current is 5 amperes. What is the current when 18 volts are applied?

10. *Hooke's Law.* Hooke's law states that the distance d that a spring is stretched by a hanging object varies directly as the mass m of the object. If the distance is 20 cm when the mass is 3 kg, what is the distance when the mass is 5 kg?

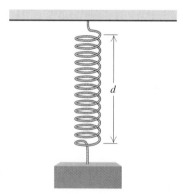

11. *Use of Aluminum Cans.* The number N of aluminum cans used each year varies directly as the number of people using the cans. If 250 people use 60,000 cans in one year, how many cans are used each year in Dallas, which has a population of 1,008,000?

12. *Weekly Allowance.* According to Fidelity Investments *Investment Vision Magazine*, the average weekly allowance A of children varies directly as their grade level, G. In a recent year, the average allowance of a 9th-grade student was $9.66 per week. What was the average weekly allowance of a 4th-grade student?

13. *Lead Pollution.* The average U.S. community of population 12,500 released about 385 tons of lead into the environment in a recent year. (*Source: Conservation Matters,* Autumn 1995 issue. Boston: Conservation Law Foundation, p. 30.) How many tons were released nationally? Use 250,000,000 as the U.S. population.

14. *Relative Aperture.* The relative aperture, or f-stop, of a 23.5-mm lens is directly proportional to the focal length F of the lens. If a 150-mm focal length has an f-stop of 6.3, find the f-stop of a 23.5-mm lens with a focal length of 80 mm.

Find the variation constant and an equation of variation in which y varies inversely as x, and the following conditions exist.

15. $y = 6$ when $x = 10$

16. $y = 16$ when $x = 4$

17. $y = 4$ when $x = 3$

18. $y = 4$ when $x = 9$

19. $y = 12$ when $x = 3$

20. $y = 9$ when $x = 5$

21. $y = 27$ when $x = \frac{1}{3}$

22. $y = 81$ when $x = \frac{1}{9}$

Solve.

23. *Pumping Rate.* The time t required to empty a tank varies inversely as the rate r of pumping. If a pump can empty a tank in 45 min at the rate of 600 kL/min, how long will it take the pump to empty the same tank at the rate of 1000 kL/min?

24. *Current and Resistance.* The current I in an electrical conductor varies inversely as the resistance R of the conductor. If the current is $\frac{1}{2}$ ampere when the resistance is 240 ohms, what is the current when the resistance is 540 ohms?

25. *Volume and Pressure.* The volume V of a gas varies inversely as the pressure P upon it. The volume of a gas is 200 cm^3 under a pressure of 32 kg/cm^2. What will be its volume under a pressure of 40 kg/cm^2?

26. *Rate of Travel.* The time t required to drive a fixed distance varies inversely as the speed r. It takes 5 hr at a speed of 80 km/h to drive a fixed distance. How long will it take to drive the same distance at a speed of 70 km/h?

🔍 *In Exercises 27–32, determine whether each graph appears to represent direct variation, inverse variation, or neither.*

27.

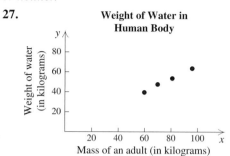

Weight of Water in Human Body

28.

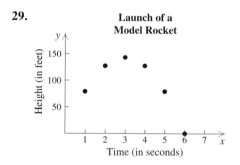

Building a Brick Wall

29.

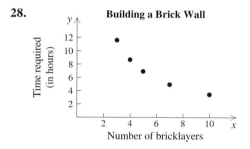

Launch of a Model Rocket

30.

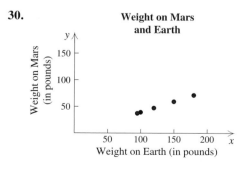

Weight on Mars and Earth

31.

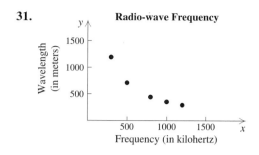

Radio-wave Frequency

32.

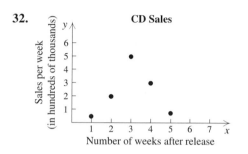

CD Sales

Find an equation of variation in which:

33. y varies directly as the square of x, and $y = 6$ when $x = 3$.

34. y varies directly as the square of x, and $y = 0.15$ when $x = 0.1$.

35. y varies inversely as the square of x, and $y = 6$ when $x = 3$.

36. y varies inversely as the square of x, and $y = 0.15$ when $x = 0.1$.

37. y varies jointly as x and z, and $y = 56$ when $x = 14$ and $z = 8$.

38. y varies directly as x and inversely as z, and $y = 4$ when $x = 12$ and $z = 15$.

39. y varies jointly as x and the square of z, and $y = 105$ when $x = 14$ and $z = 5$.

40. y varies jointly as x and z and inversely as w, and $y = \frac{3}{2}$ when $x = 2$, $z = 3$, and $w = 4$.

41. y varies jointly as w and the square of x and inversely as z, and $y = 49$ when $w = 3$, $x = 7$, and $z = 12$.

42. y varies directly as x and inversely as w and the square of z, and $y = 4.5$ when $x = 15$, $w = 5$, and $z = 2$.

Solve.

43. *Stopping Distance of a Car.* The stopping distance d of a car after the brakes have been applied varies directly as the square of the speed r. If a car traveling 60 mph can stop in 200 ft, how fast can a car travel and still stop in 72 ft?

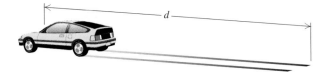

44. *Volume of a Gas.*   The volume V of a given mass of a gas varies directly as the temperature T and inversely as the pressure P. If $V = 231$ cm^3 when $T = 42°$ and $P = 20$ kg/cm^2, what is the volume when $T = 30°$ and $P = 15$ kg/cm^2?

45. *Intensity of Light.*   The intensity I of light from a light bulb varies inversely as the square of the distance d from the bulb. Suppose I is 90 W/m^2 (watts per square meter) when the distance is 5 m. How much *further* would it be to a point the intensity is 40 W/m^2?

46. *Intensity of a Signal.*   The intensity I of a television signal varies inversely as the square of the distance d from the transmitter. If the intensity is 25 W/m^2 at a distance of 2 km, how far from the transmitter are you when the intensity is 2.56 W/m^2?

47. *Weight of an Astronaut.*   The weight W of an object varies inversely as the square of the distance d from the center of Earth. At sea level (6400 km from the center of Earth), an astronaut weighs 100 lb. How far *above Earth* must the astronaut be in order to weigh 64 lb?

48. *Electrical Resistance.*   At a fixed temperature, the resistance R of a wire varies directly as the length l and inversely as the square of its diameter d. If the resistance is 0.1 ohm when the diameter is 1 mm and the length is 50 cm, what is the diameter when the resistance is 1 ohm and the length is 2000 cm?

49. *Atmospheric Drag.*   Wind resistance, or atmospheric drag, tends to slow down moving objects. Atmospheric drag varies jointly as an object's surface area A and velocity v. If a car traveling at a speed of 40 mph with a surface area of 37.8 ft^2 experiences a drag of 222 N (Newtons), how fast must a car with 51 ft^2 of surface area travel in order to experience a drag force of 430 N?

50. *Drag Force.*   The drag force F on a boat varies jointly as the wetted surface area A and the square of the velocity of the boat. If a boat going 6.5 mph experiences a drag force of 86 N when the wetted

surface area is 41.2 ft^2, how fast must a boat with 28.5 ft^2 of wetted surface area go in order to experience a drag force of 94 N?

Skill Maintenance

51. Give an equation for a line with slope $-\frac{2}{3}$ and y-intercept $(0, -5)$. [2.4]

52. Give an equation in point–slope form for the line passing through the point $(4, 7)$ with slope $-\frac{2}{7}$. [2.6]

Simplify. [6.3]

53. $\dfrac{\dfrac{1}{ab} - \dfrac{2}{bc}}{\dfrac{3}{ab} + \dfrac{4}{bc}}$

54. $\dfrac{\dfrac{3}{x^2 - 4}}{\dfrac{1}{3x + 6} - \dfrac{2}{4x - 8}}$

55. If $f(x) = x^3 - 2x^2$, find $f(3)$. [2.1]

56. Multiply: $(3x - 2y)^2$. [5.2]

Synthesis

57. ◆ If y varies directly as x^2, explain why doubling x would not cause y to be doubled as well.

58. ◆ If y varies directly as x and x varies inversely as z, how does y vary with regard to z? Why?

59. ◆ Write a variation problem for a classmate to solve. Design the problem so that the answer is "When Simone studies for 8 hr a week, her quiz score is 92."

60. ◆ Suppose that the number of customer complaints is inversely proportional to the number of employees hired. Will a firm reduce the number of complaints more by expanding from 5 to 10 employees, or from 20 to 25? Explain. Consider using a graph to help justify your answer.

61. If y varies inversely as the cube of x and x is multiplied by 0.5, what is the effect on y?

Describe, in words, the variation given by the equation.

62. $Q = \dfrac{kp^2}{q^3}$

63. $W = \dfrac{km_1 M_1}{d^2}$

64. *Tension of a Musical String.*   The tension T on a string in a musical instrument varies jointly as the string's mass per unit length m, the square of its length l, and the square of its fundamental frequency f. A 2-m–long string of mass 5 gm/m with a fundamental frequency of 80 has a tension of

100 N. How long should the same string be if its tension is going to be changed to 72 N?

65. *The Gravity Model.* It has been determined that the average number of telephone calls in a day N, between two cities, is directly proportional to the populations P_1 and P_2 of the cities and inversely proportional to the square of the distance between the cities. This model is called the *gravity model* because the equation of variation resembles the equation that applies to Newton's law of gravitation.

   a) In 1994, the population of Indianapolis was 752,279 and the population of Cincinnati was 358,170. The average number of daily phone calls between the two cities was 11,153. Find the value k and write the equation of variation given that the cities are 174 km apart.

   b) In 1994, the average number of daily phone calls between Indianapolis and New York was 4270, and the population of New York was 7,333,153. Estimate the distance between Indianapolis and New York.

66. *Volume and Cost.* A peanut butter jar in the shape of a right circular cylinder is 4 in. high and 3 in. in diameter and sells for $1.20. If we assume that cost is proportional to volume, how much should a jar 6 in. high and 6 in. in diameter cost?

67. *Golf Distance Finder.* A device used in golf to estimate the distance d to a hole measures the size s that the 7-ft pin *appears* to be in a viewfinder. The viewfinder uses the principle, diagrammed here, that s gets bigger when d gets smaller. If $s = 0.56$ in. when $d = 50$ yd, find an equation of variation that expresses d as a function of s. What is d when $s = 0.40$ in.?

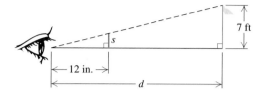

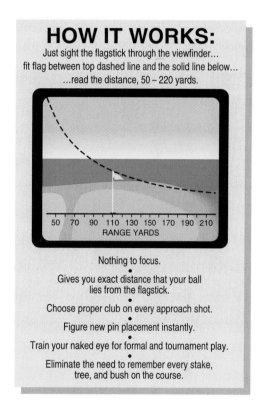

HOW IT WORKS:
Just sight the flagstick through the viewfinder...
fit flag between top dashed line and the solid line below...
...read the distance, 50 – 220 yards.

RANGE YARDS

Nothing to focus.

Gives you exact distance that your ball lies from the flagstick.

Choose proper club on every approach shot.

Figure new pin placement instantly.

Train your naked eye for formal and tournament play.

Eliminate the need to remember every stake, tree, and bush on the course.

COLLABORATIVE CORNER

Focus: Direct variation and estimation

Time: 15 minutes

Group size: 2 or 3 and entire class

The National Park Service's estimates of crowd sizes for static (stationary) mass demonstrations vary directly as the area covered by the crowd. Park Service officials have found that at basic "shoulder-to-shoulder" demonstrations, 1 acre of land (about 45,000 ft^2) holds about 9000 people. Using aerial photographs, officials impose a grid to estimate the total area covered by the demonstrators. Once this has been accomplished, estimates of crowd size can be prepared.

(continued)

Activity

1. In the grid imposed on the photograph below, each square represents 10,000 ft². Estimate the size of the crowd photographed. Then compare your group's estimate with those of other groups. What might explain discrepancies between estimates? List ways in which your group's estimate could be made more accurate.

2. Park Service officials use an "acceptable margin of error" of no more than 20%. Using all estimates from part (1) above and allowing for error, find a range of values within which you feel certain that the actual crowd size lies.

3. The Million-Man March of 1995 was not a static demonstration because of a periodic turnover of people in attendance (many people stayed for only part of the day's festivities). How might you change your methodology to compensate for this complication?

8.6
Quadratic Functions and Their Graphs

- *Graphs of $f(x) = ax^2$*
- *Graphs of $f(x) = a(x - h)^2$*
- *Graphs of $f(x) = a(x - h)^2 + k$*

We saw in Section 8.1 that the graph of a quadratic function is a cup-shaped curve called a parabola. In this section, we develop techniques for graphing such functions.

Graphs of $f(x) = ax^2$

The most basic quadratic function is $f(x) = x^2$.

Example 1   Graph: $f(x) = x^2$.

SOLUTION   We choose some values for x and compute $f(x)$ for each. Then we plot the ordered pairs and connect them with a smooth curve.

x	$f(x) = x^2$	$(x, f(x))$
-3	9	$(-3, 9)$
-2	4	$(-2, 4)$
-1	1	$(-1, 1)$
0	0	$(0, 0)$
1	1	$(1, 1)$
2	4	$(2, 4)$
3	9	$(3, 9)$

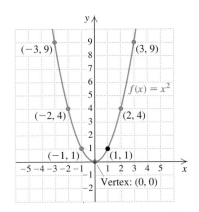

All quadratic functions have graphs similar to the one in Example 1. Each parabola is symmetric with respect to a vertical line known as the parabola's *axis of symmetry*. In the graph of $f(x) = x^2$, the y-axis (or the line $x = 0$) is the axis of symmetry. Were the paper folded on this line, the two halves of the curve would match. The point $(0, 0)$ is known as the *vertex* of this parabola.

Since x^2 is a polynomial, we know that the domain of $f(x) = x^2$ is the set of all real numbers, or $(-\infty, \infty)$. The degree of x^2, 2, is even. We see from the graph and the table in Example 1 that the range of f is $\{x \mid x \geq 0\}$, or $[0, \infty)$. The function has a *minimum value* of 0. This minimum value occurs when $x = 0$.

Now let's consider a quadratic function of the form $g(x) = ax^2$. How does the constant a affect the graph of the function?

Interactive Discovery

Graph each of the following functions, along with the function given by $f(x) = x^2$, in a $[-5, 5, -10, 10]$ window. For each function, answer the following questions.

a) What is the vertex of the graph of g?

b) What is the axis of symmetry of the graph of g?

c) Does the parabola open upward or downward?

d) What is the range of g?

e) Is there a minimum or a maximum value of g?

f) Is the parabola narrower or wider than the parabola given by $y = x^2$?

1. $g(x) = 3x^2$ **2.** $g(x) = \frac{1}{3}x^2$

3. $g(x) = 0.2x^2$ **4.** $g(x) = -x^2$

5. $g(x) = -4x^2$ **6.** $g(x) = -\frac{2}{3}x^2$

Find a rule that describes the effect of multiplying x^2 by a when $a > 1$ and when $0 < a < 1$.

Find a rule that describes the effect of multiplying x^2 by a when $a < -1$ and when $-1 < a < 0$.

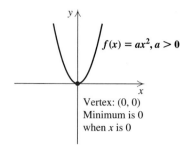

Vertex: (0, 0)
Minimum is 0
when x is 0

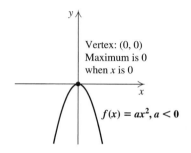

Vertex: (0, 0)
Maximum is 0
when x is 0

We can generalize the results as follows.

Graphing f(x) = ax²

The graph of $f(x) = ax^2$ is a parabola with $x = 0$ as its axis of symmetry. Its vertex is the origin.

For $a > 0$, the parabola opens upward. The range of the function is $[0, \infty)$. A minimum value of 0 occurs when $x = 0$.

For $a < 0$, the parabola opens downward. The range of the function is $(-\infty, 0]$. A maximum value of 0 occurs when $x = 0$.

If $|a|$ is greater than 1, the parabola is narrower than $y = x^2$.

If $|a|$ is between 0 and 1, the parabola is wider than $y = x^2$.

Example 2 Graph: $g(x) = \frac{1}{2}x^2$.

SOLUTION The function is in the form $g(x) = ax^2$, where $a = \frac{1}{2}$. The vertex is (0, 0) and the axis of symmetry is $x = 0$. Since $\frac{1}{2} > 0$, the parabola opens upward. The parabola is wider than $y = x^2$ since $\left|\frac{1}{2}\right|$ is between 0 and 1.

To graph the function, we calculate and plot points. Because we know the general shape of the graph, we can sketch the graph using only a few points.

x	$g(x) = \frac{1}{2}x^2$
-3	$\frac{9}{2}$
-2	2
-1	$\frac{1}{2}$
0	0
1	$\frac{1}{2}$
2	2
3	$\frac{9}{2}$

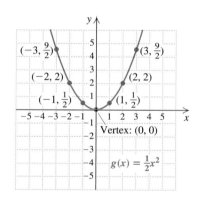

Note in Example 2 that since a parabola is symmetric, we can calculate function values for inputs to one side of the vertex and then plot the mirror images of those points on the other "half" of the graph.

Graphs of f(x) = a(x − h)²

Why not now consider graphs of

$$f(x) = ax^2 + bx + c,$$

where b and c are not both 0? In effect, we will do that, but in a disguised form. It turns out to be convenient to first graph $f(x) = a(x - h)^2$.* This allows us to observe similarities to the graphs drawn above.

Interactive Discovery

Graph each of the following pairs of functions in a $[-5, 5, -10, 10]$ window. For each function, answer the following questions.

a) What is the vertex of the graph of g?

b) What is the axis of symmetry of the graph of g?

c) Is the graph of g narrower, wider, or the same shape as the graph of f?

d) What is the range of g?

e) What is the minimum or maximum value of $g(x)$, and where does it occur?

1. $f(x) = x^2$; $g(x) = (x - 3)^2$ **2.** $f(x) = x^2$; $g(x) = (x + 4)^2$

3. $f(x) = 2x^2$; $g(x) = 2(x + 1)^2$ **4.** $f(x) = -x^2$; $g(x) = -(x - 2)^2$

5. $f(x) = \frac{1}{2}x^2$; $g(x) = \frac{1}{2}\left(x - \frac{3}{2}\right)^2$ **6.** $f(x) = -3x^2$; $g(x) = -3(x + 2)^2$

Describe the effect of h on the graph of $g(x) = a(x - h)^2$.

The graph of $g(x) = a(x - h)^2$ looks just like the graph of $f(x) = ax^2$, except that it is moved, or *translated*, $|h|$ units.

Graphing $f(x) = a(x - h)^2$

The graph of $f(x) = a(x - h)^2$ has the same shape as the graph of $y = ax^2$.

If h is positive, the graph of $y = ax^2$ is shifted h units to the right.

If h is negative, the graph of $y = ax^2$ is shifted $|h|$ units to the left.

The vertex is $(h, 0)$, and the axis of symmetry is $x = h$.

For $a > 0$, the range of f is $[0, \infty)$. A minimum value of 0 occurs when $x = h$.

For $a < 0$, the range of f is $(-\infty, 0]$. A maximum value of 0 occurs when $x = h$.

Example 3 Graph $g(x) = -2(x + 3)^2$, and determine the maximum value of $g(x)$ and where it occurs.

---

*The letters h and k are often used to name functions, in which case the notation $h(x)$ and $k(x)$ is used. When h and k appear in expressions like $f(x) = a(x - h)^2 + k$, assume that they represent constants.

SOLUTION We rewrite the equation as $g(x) = -2[x - (-3)]^2$. In this case, $a = -2$ and $h = -3$, so the graph looks like that of $y = 2x^2$ translated 3 units to the left and, since $-2 < 0$, flipped upside down. The vertex is $(-3, 0)$, and the axis of symmetry is $x = -3$. Plotting points as needed, we obtain the graph shown below.

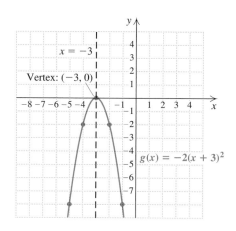

The maximum value of $g(x)$ is 0 when $x = -3$.   ▬

Graphs of $f(x) = a(x - h)^2 + k$

Given a graph of $f(x) = a(x - h)^2$, what happens if we add a constant k?

Interactive Discovery

For each pair of functions given below:

a) Create a table of values for both functions. How do the values of $f(x)$ and $g(x)$ compare?

b) Graph both functions using the same viewing window. Do both graphs have the same axis of symmetry? Do both graphs have the same vertex? Do both functions have the same range?

1. $f(x) = 3(x - 2)^2$; $g(x) = 3(x - 2)^2 + 10$
2. $f(x) = 3(x - 2)^2$; $g(x) = 3(x - 2)^2 - 5$
3. $f(x) = -\frac{1}{2}(x + 1)^2$; $g(x) = -\frac{1}{2}(x + 1)^2 + 4$

Describe the effect of k on the graph of $g(x) = a(x - h)^2 + k$.

If we add a positive constant k to the graph of $f(x) = a(x - h)^2$, the graph is moved up. If we add a negative constant k, the graph is moved down. The axis of symmetry for the parabola remains at $x = h$, but the vertex will be at (h, k).

Graphing $f(x) = a(x - h)^2 + k$

The graph of $f(x) = a(x - h)^2 + k$ has the same shape as the graph of $y = a(x - h)^2$.

If k is positive, the graph of $y = a(x - h)^2$ is shifted k units up.

If k is negative, the graph of $y = a(x - h)^2$ is shifted $|k|$ units down.

The vertex is (h, k), and the axis of symmetry is $x = h$.

The domain of f is $(-\infty, \infty)$.

For $a > 0$, the range of f is $[k, \infty)$. The minimum value of $f(x)$ is k, which occurs when $x = h$.

For $a < 0$, the range of f is $(-\infty, k]$. The maximum value of $f(x)$ is k, which occurs when $x = h$.

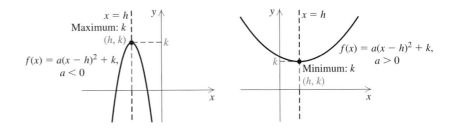

Example 4 Let $g(x) = (x - 3)^2 - 5$.

a) Find the vertex and the axis of symmetry.

b) Find the minimum or maximum function value and the x-value at which it occurs.

c) Find the range of g.

d) Sketch the graph of g.

SOLUTION

a) We write the function in the form $g(x) = a(x - h)^2 + k$:

$$g(x) = (x - 3)^2 + (-5).$$

We have $a = 1$, $h = 3$, and $k = -5$. The vertex is thus $(3, -5)$, and the axis of symmetry is $x = 3$.

b) Since $a > 0$, the parabola opens upward and there is a minimum function value. That minimum value is -5, and it occurs when $x = 3$.

c) The range of g is $[-5, \infty)$.

d) The graph will look like that of $f(x) = (x - 3)^2$, shifted 5 units down.

You can confirm this by plotting some points. For instance, $g(4) = (4 - 3)^2 - 5 = -4$, whereas $f(4) = (4 - 3)^2 = 1$.

x	$g(x) = (x - 3)^2 - 5$
0	4
1	-1
2	-4
3	-5	← Vertex
4	-4
5	-1
6	4

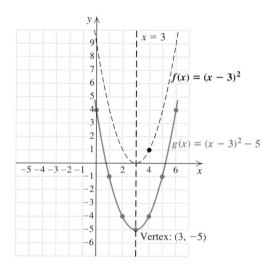

Example 5 Let $h(x) = -\frac{1}{2}(x + 4)^2 + 3$.

a) Find the vertex and the axis of symmetry.

b) Find the minimum or maximum function value and the x-value at which it occurs.

c) Find the range of h.

d) Sketch the graph of h.

SOLUTION

a) We first express the function in the equivalent form

$$h(x) = -\tfrac{1}{2}[x - (-4)]^2 + 3.$$

The vertex is $(-4, 3)$, and the axis of symmetry is $x = -4$.

b) Since $-\frac{1}{2} < 0$, we know that 3, the second coordinate of the vertex, is the maximum function value. It occurs when $x = -4$.

c) The range of h is $(-\infty, 3]$.

d) The graph looks like that of $f(x) = -\frac{1}{2}x^2$ translated 4 units to the left and 3 units up. We compute a few points as needed and graph the function.

x	$h(x) = -\frac{1}{2}(x + 4)^2 + 3$	
-6	1	
-4	3	← **Vertex**
-2	1	
0	-5	

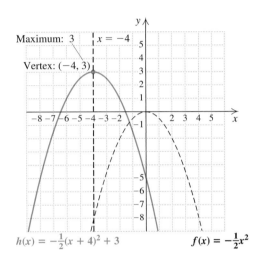

$h(x) = -\frac{1}{2}(x + 4)^2 + 3$ $\qquad$ $f(x) = -\frac{1}{2}x^2$

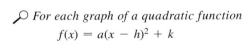

8.6 | *Exercise Set*

🔎 *For each graph of a quadratic function*

$$f(x) = a(x - h)^2 + k$$

in Exercises 1–6:

a) *Tell whether a is positive or negative.*
b) *Determine the vertex.*
c) *Determine the axis of symmetry.*
d) *Determine the range.*

1.

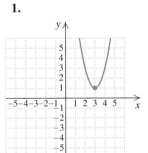

2.

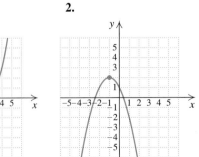

3.

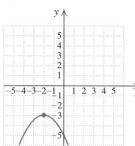

4.

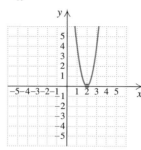

5.

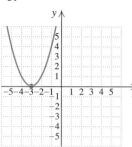

6.

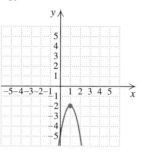

○ In Exercises 7–12, match each function with one of the following graphs.

a)

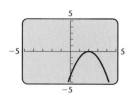

b)

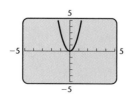

c)

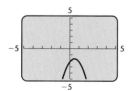

d)

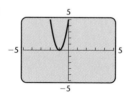

e)

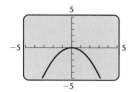

f)

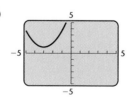

7. $f(x) = 3x^2$

8. $g(x) = -\frac{1}{2}x^2$

9. $h(x) = -(x - 2)^2$

10. $f(x) = 5(x + 1)^2$

11. $g(x) = \frac{2}{3}(x + 3)^2 + 1$

12. $h(x) = -2\left(x - \frac{1}{2}\right)^2 - \frac{5}{3}$

Graph by hand.

13. $f(x) = x^2$

14. $f(x) = -x^2$

15. $f(x) = -2x^2$

16. $f(x) = -3x^2$

17. $g(x) = \frac{1}{4}x^2$

18. $g(x) = \frac{1}{3}x^2$

19. $h(x) = -\frac{1}{3}x^2$

20. $h(x) = -\frac{1}{4}x^2$

21. $f(x) = \frac{3}{2}x^2$

22. $f(x) = \frac{5}{2}x^2$

For each of the following, graph the function, label the vertex, and draw the axis of symmetry.

23. $g(x) = (x + 1)^2$

24. $g(x) = (x + 4)^2$

25. $f(x) = (x - 2)^2$

26. $f(x) = (x - 1)^2$

27. $f(x) = -(x + 4)^2$

28. $f(x) = -(x - 2)^2$

29. $f(x) = 2(x + 1)^2$

30. $f(x) = 2(x + 4)^2$

31. $h(x) = -\frac{1}{2}(x - 3)^2$

32. $h(x) = -\frac{3}{2}(x - 2)^2$

33. $f(x) = \frac{1}{2}(x - 1)^2$

34. $f(x) = \frac{1}{3}(x + 2)^2$

For each of the following, graph the function by hand and find the vertex, the axis of symmetry, and the maximum value or the minimum value.

35. $f(x) = (x - 5)^2 + 1$

36. $f(x) = (x + 3)^2 - 2$

37. $f(x) = (x + 1)^2 - 2$

38. $g(x) = -(x - 2)^2 - 4$

39. $h(x) = -2(x - 1)^2 - 3$

40. $h(x) = -2(x + 1)^2 + 4$

41. $f(x) = 2(x + 4)^2 + 1$

42. $f(x) = 2(x - 5)^2 - 3$

43. $g(x) = -\frac{3}{2}(x - 1)^2 + 2$

44. $g(x) = \frac{3}{2}(x + 2)^2 - 1$

○ *Without graphing, find the vertex, the axis of symmetry, the maximum value or the minimum value, and the range.*

45. $f(x) = 8(x - 9)^2 + 5$

46. $f(x) = 10(x + 5)^2 - 8$

47. $h(x) = -\frac{2}{7}(x + 6)^2 + 11$

48. $h(x) = -\frac{3}{11}(x - 7)^2 - 9$

49. $f(x) = 5\left(x + \frac{1}{4}\right)^2 - 13$

50. $f(x) = 6\left(x - \frac{1}{4}\right)^2 + 19$

51. $f(x) = \sqrt{2}(x + 4.58)^2 + 65\pi$

52. $f(x) = 4\pi(x - 38.2)^2 - \sqrt{34}$

Skill Maintenance

Solve each system. [3.2]

53. $3x + 4y = -19,$
$\quad 7x - 6y = -29$

54. $5x + 7y = 9,$
$\quad 3x - 4y = -11$

Complete the square. [8.1]

55. $x^2 + 5x +$ _____

56. $x^2 - 9x +$ _____

Synthesis

57. ◈ While trying to graph $y = -\frac{1}{2}x^2 + 3x + 1$, Omar gets the following screen.

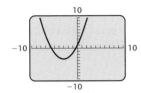

How can Omar tell at a glance that a mistake has been made?

58. ◈ Explain, without plotting points, why the graph of $y = (x + 2)^2$ looks like the graph of $y = x^2$ translated 2 units to the left.

59. ◈ Explain, without plotting points, why the graph of $y = x^2 - 4$ looks like the graph of $y = x^2$ translated 4 units down.

60. ◈ If the graphs of $f(x) = a_1(x - h_1)^2 + k_1$ and $g(x) = a_2(x - h_2)^2 + k_2$ have the same shape, what, if anything, can you conclude about the a's, the h's, and the k's? Why?

For each of the following, write the equation of the parabola that has the shape of $f(x) = 2x^2$ or $g(x) = -2x^2$ and has a maximum or minimum value at the specified point.

61. Minimum: $(5, 0)$

62. Minimum: $(2, 0)$

63. Maximum: $(-4, 0)$

64. Maximum: $(0, 3)$

65. Maximum: $(3, 8)$

66. Minimum: $(-2, 3)$

Find an equation for a quadratic function F that satisfies the following conditions.

67. The graph of F is the same shape as the graph of f, where $f(x) = 3(x + 2)^2 + 7$, and $F(x)$ is a minimum at the same point that $g(x) = -2(x - 5)^2 + 1$ is a maximum.

68. The graph of F is the same shape as the graph of f, where $f(x) = -\frac{1}{3}(x - 2)^2 + 7$, and $F(x)$ is a maximum at the same point that $g(x) = 2(x + 4)^2 - 6$ is a minimum.

69. The graph of F is the same shape as the graph of f, where $f(x) = -\frac{1}{2}x^2$, and $F(x)$ is a maximum at the same point that $g(x) = \frac{3}{4}(x + 3)^2 - 4$ is a minimum.

70. The graph of F is the same shape as the graph of f, where $f(x) = 5x^2$, and $F(x)$ is a minimum at the same point that $g(x) = -3\left(x + \frac{1}{2}\right)^2 + \frac{2}{3}$ is a maximum.

Functions other than parabolas can be translated. For a function $f(x)$, if we replace x with $x - h$, where h is a constant, the graph will be moved horizontally. If we add a constant k to a function $f(x)$, the graph will be moved vertically.

Use the graph of the function $y = f(x)$ below for Exercises 71–76.

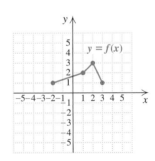

Draw a graph of each of the following.

71. $y = f(x - 1)$

72. $y = f(x + 2)$

73. $y = f(x) + 2$

74. $y = f(x) - 3$

75. $y = f(x + 3) - 2$

76. $y = f(x - 3) + 1$

COLLABORATIVE CORNER

Focus: Graphing quadratic functions

Time: 15–20 minutes

Group size: 6

Activity

1. On each of six scraps of paper, write one of the following equations:

$$y = \tfrac{1}{2}(x - 3)^2 + 1; \qquad y = \tfrac{1}{2}(x - 1)^2 + 3;$$
$$y = \tfrac{1}{2}(x + 1)^2 - 3; \qquad y = \tfrac{1}{2}(x + 3)^2 + 1;$$
$$y = \tfrac{1}{2}(x + 3)^2 - 1; \qquad y = \tfrac{1}{2}(x + 1)^2 + 3.$$

2. Fold each scrap of paper and mix up the six scraps in a hat or bag. Then, one by one, each group member should select one of the equations.

3. Each group member should carefully graph by hand the equation selected. Make the graph large enough so that when it is finished, it can be easily viewed by the rest of the group. Be sure to scale the axes, but **do not label the graph with the equation used.**

4. When all group members have drawn a graph, place the graphs in a pile. The group should then agree on the correct equation for each graph *with no help from the person who drew the graph.*

5. Compare your group's labeled graphs with those of other groups to reach consensus within the class on the correct label for each graph.

8.7
More About Graphing Quadratic Functions

- *Completing the Square*
- *Finding the Vertex Using a Grapher*
- *Finding Intercepts*

Completing the Square

By *completing the square* (see Section 8.1), we can rewrite any polynomial $ax^2 + bx + c$ in the form $a(x - h)^2 + k$. This ability, combined with the procedures discussed in Section 8.6, enables us to graph any quadratic function.

Example 1 Graph: $g(x) = x^2 - 6x + 4$.

SOLUTION We have

$$g(x) = x^2 - 6x + 4$$
$$= (x^2 - 6x) + 4.$$

Since we are using function notation, we do not complete the square by adding a number on both sides of the equals sign. Instead, we add and subtract the same number within the parentheses—in effect, adding 0 to the expression.

To complete the square inside the parentheses, we take half the

x-coefficient, $\frac{1}{2} \cdot (-6) = -3$, and square it to get $(-3)^2 = 9$. Then we add $9 - 9$ inside the parentheses:

$$g(x) = (x^2 - 6x + 9 - 9) + 4 \qquad \text{The effect is of adding 0.}$$
$$= (x^2 - 6x + 9) + (-9 + 4) \qquad \begin{array}{l}\text{Using the associative law} \\ \text{of addition to regroup}\end{array}$$
$$= (x - 3)^2 - 5. \qquad \begin{array}{l}\text{Factoring and} \\ \text{simplifying}\end{array}$$

This equation was graphed in Example 4 on p. 539–540. The vertex is $(3, -5)$, and the axis of symmetry is $x = 3$.

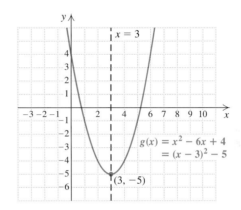

When the leading coefficient is not 1, we factor out that number from the first two terms. Then we complete the square.

Example 2 Graph $f(x) = 3x^2 + 12x + 13$. Find the maximum or minimum function value.

SOLUTION Since the coefficient of x^2 is not 1, we need to factor out that number—in this case, 3—from the first two terms. Remember that we want the form $f(x) = a(x - h)^2 + k$:

$$f(x) = 3x^2 + 12x + 13$$
$$= 3(x^2 + 4x) + 13.$$

Now we complete the square as before. We take half of the x-coefficient, $\frac{1}{2} \cdot 4 = 2$, and square it: $2^2 = 4$. Then we add $4 - 4$ inside the parentheses:

$$f(x) = 3(x^2 + 4x + 4 - 4) + 13. \qquad \begin{array}{l}\textbf{Adding 4 − 4, or 0, inside} \\ \textbf{the parentheses}\end{array}$$

The distributive law allows us to separate the -4 from the perfect-square trinomial so long as it is multiplied by 3:

$$f(x) = 3(x^2 + 4x + 4) + 3(-4) + 13 \qquad \begin{array}{l}\textbf{This leaves a perfect-} \\ \textbf{square trinomial in} \\ \textbf{the parentheses.}\end{array}$$

$$= 3(x + 2)^2 + 1. \qquad \begin{array}{l}\textbf{Factoring and} \\ \textbf{simplifying}\end{array}$$

The vertex is $(-2, 1)$, and the axis of symmetry is $x = -2$. The coefficient of x^2 is 3, so the graph is narrow and opens upward. The minimum function value is 1. We choose a few x-values on either side of the vertex, compute y-values, and then graph the parabola.

x	$f(x) = 3(x + 2)^2 + 1$
-2	1
-3	4
-1	4

←————— **Vertex**

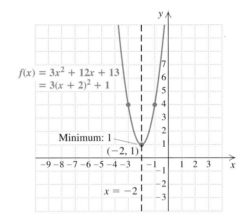

$f(x) = 3x^2 + 12x + 13$
$= 3(x + 2)^2 + 1$

Minimum: 1
$(-2, 1)$

$x = -2$

Example 3 Graph $f(x) = -2x^2 + 10x - 7$. Find the maximum or minimum function value.

SOLUTION We first find the vertex by completing the square. To do so, we factor out -2 from the first two terms of the expression. This makes the coefficient of x^2 inside the parentheses 1:

$$f(x) = -2x^2 + 10x - 7$$
$$= -2(x^2 - 5x) - 7.$$

Now we complete the square as before. We take half of the x-coefficient and square it to get $\frac{25}{4}$. Then we add $\frac{25}{4} - \frac{25}{4}$ inside the parentheses:

$$f(x) = -2\left(x^2 - 5x + \tfrac{25}{4} - \tfrac{25}{4}\right) - 7$$
$$= -2\left(x^2 - 5x + \tfrac{25}{4}\right) + (-2)\left(-\tfrac{25}{4}\right) - 7 \qquad \text{Multiplying by } -2 \text{, using the distributive law, and regrouping}$$
$$= -2\left(x - \tfrac{5}{2}\right)^2 + \tfrac{11}{2}. \qquad \text{Factoring and simplifying}$$

The vertex is $\left(\frac{5}{2}, \frac{11}{2}\right)$, and the axis of symmetry is $x = \frac{5}{2}$. The coefficient of x^2, -2, is negative, so the graph opens downward. The maximum function value is $\frac{11}{2}$ when $x = \frac{5}{2}$. We plot a few points on either side of the vertex, including the y-intercept, $f(0)$, and graph the parabola.

x	$f(x)$	
$\frac{5}{2}$	$\frac{11}{2}$	← Vertex
0	-7	← *y*-intercept
1	1	
4	1	

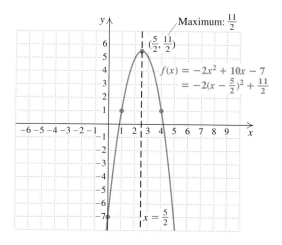

The method used in Examples 1–3 can be generalized to find a formula for locating the vertex. We complete the square as follows:

$$f(x) = ax^2 + bx + c$$
$$= a\left(x^2 + \frac{b}{a}x\right) + c. \qquad \text{Factoring } a \text{ out of the first two terms. Check by multiplying.}$$

Half of the *x*-coefficient, $\dfrac{b}{a}$, is $\dfrac{b}{2a}$. We square it to get $\dfrac{b^2}{4a^2}$ and add $\dfrac{b^2}{4a^2} - \dfrac{b^2}{4a^2}$ inside the parentheses. Then we distribute the *a* and regroup terms:

$$f(x) = a\left(x^2 + \frac{b}{a}x + \frac{b^2}{4a^2} - \frac{b^2}{4a^2}\right) + c$$
$$= a\left(x^2 + \frac{b}{a}x + \frac{b^2}{4a^2}\right) + a\left(-\frac{b^2}{4a^2}\right) + c \qquad \text{Using the distributive law}$$
$$= a\left(x + \frac{b}{2a}\right)^2 + \frac{-b^2}{4a} + \frac{4ac}{4a} \qquad \text{Factoring and finding a common denominator}$$
$$= a\left[x - \left(-\frac{b}{2a}\right)\right]^2 + \frac{4ac - b^2}{4a}.$$

Thus we have the following.

The Vertex of a Parabola

The vertex of the parabola given by $f(x) = ax^2 + bx + c$ is

$$\left(-\frac{b}{2a}, \frac{4ac - b^2}{4a}\right), \quad \text{or} \quad \left(-\frac{b}{2a}, f\left(-\frac{b}{2a}\right)\right).$$

The x-coordinate of the vertex is $-b/(2a)$. The axis of symmetry is $x = -b/(2a)$. The second coordinate of the vertex is most commonly found by computing $f\left(-\dfrac{b}{2a}\right)$.

Let's reexamine Example 3 to see how we could have found the vertex directly. From the formula above,

$$\text{the } x\text{-coordinate of the vertex is } -\frac{b}{2a} = -\frac{10}{2(-2)} = \frac{5}{2}.$$

Substituting $\frac{5}{2}$ into $f(x) = -2x^2 + 10x - 7$, we find the second coordinate of the vertex:

$$\begin{aligned}
f\left(\tfrac{5}{2}\right) &= -2\left(\tfrac{5}{2}\right)^2 + 10\left(\tfrac{5}{2}\right) - 7 \\
&= -2\left(\tfrac{25}{4}\right) + 25 - 7 \\
&= -\tfrac{25}{2} + 18 \\
&= -\tfrac{25}{2} + \tfrac{36}{2} = \tfrac{11}{2}.
\end{aligned}$$

The vertex is $\left(\frac{5}{2}, \frac{11}{2}\right)$. The axis of symmetry is $x = \frac{5}{2}$.

We have actually developed two methods for finding the vertex. One is by completing the square and the other is by using a formula. We can also find the vertex using a grapher.

Finding the Vertex Using a Grapher

We have seen that we can find the maximum or minimum value of a quadratic function by determining the coordinates of the vertex of the graph of the function. Using a grapher, we can find the vertex of the graph by determining the maximum or minimum value of the function.

Example 4   Use a grapher to determine the vertex of the graph of the function given by $f(x) = -5x^2 - 14x + 3$.

SOLUTION   The coefficient of x^2 is negative, so the parabola opens downward and the function has a maximum value. We enter and graph the function, choosing a viewing window that will show the vertex.

We find the coordinates of the vertex using a MAXIMUM feature, often found in the CALC menu. The grapher calculates the maximum function

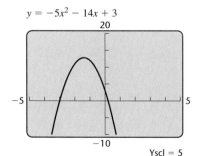

$y = -5x^2 - 14x + 3$

Yscl = 5

value over a specified interval, so we must enter the left and right end-points, or bounds, of the interval, as well as a guess close to the vertex.

$y = -5x^2 - 14x + 3$

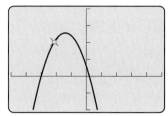

Left Bound?
$x = -2.12766$ $y = 10.152558$

$y = -5x^2 - 14x + 3$

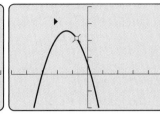

Right Bound?
$x = -.7446809$ $y = 10.652784$

$y = -5x^2 - 14x + 3$

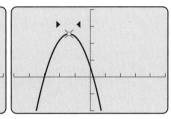

Guess?
$x = -1.382979$ $y = 12.798551$

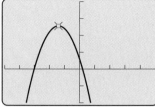

Maximum
$x = -1.400002$ $y = 12.8$

We see that the maximum function value is 12.8 when $x = -1.4$. The coordinates of the vertex are $(-1.4, 12.8)$. ▬

Finding Intercepts

The points at which a graph crosses an axis are called intercepts. We saw in Chapter 2 and again in Example 3 that the y-intercept occurs at $f(0)$. For $f(x) = ax^2 + bx + c$, the y-intercept is simply $(0, c)$. To find x-intercepts, we look for points where $y = 0$ or $f(x) = 0$. To find the x-intercepts of a quadratic function $f(x) = ax^2 + bx + c$, we solve the equation

$$0 = ax^2 + bx + c.$$

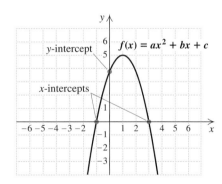

Example 5 Find the x- and y-intercepts of the graph of $f(x) = x^2 - 2x - 2$.

SOLUTION The y-intercept is simply $(0, f(0))$, or $(0, -2)$. To find the x-intercepts, we solve the equation

$$0 = x^2 - 2x - 2.$$

We are unable to factor $x^2 - 2x - 2$, so we use the quadratic formula and get $x = 1 \pm \sqrt{3}$. Thus the x-intercepts are $(1 - \sqrt{3}, 0)$ and $(1 + \sqrt{3}, 0)$.

If graphing, we would approximate, to get $(-0.7, 0)$ and $(2.7, 0)$. ▬

We summarize what we know about the graph of a quadratic function.

The Graph of a Quadratic Function Given by
$$f(x) = ax^2 + bx + c \text{ or } f(x) = a(x - h)^2 + k$$

The graph is a parabola.

The vertex is (h, k) or $\left(-\dfrac{b}{2a}, f\left(-\dfrac{b}{2a} \right) \right)$.

The axis of symmetry is $x = h$.

The y-intercept of the graph is $(0, c)$.

The x-intercepts can be found by solving $ax^2 + bx + c = 0$.
    If $b^2 - 4ac > 0$, there are two x-intercepts.
    If $b^2 - 4ac = 0$, there is one x-intercept.
    If $b^2 - 4ac < 0$, there are no x-intercepts.

The domain of the function is $(-\infty, \infty)$.

If a is positive: The graph opens upward.
                 The function has a minimum value, given by k.
                     This occurs when $x = h$.
                 The range of the function is $[k, \infty)$.

If a is negative: The graph opens downward.
                 The function has a maximum value, given by k.
                     This occurs when $x = h$.
                 The range of the function is $(-\infty, k]$.

8.7 Exercise Set

For each quadratic function, (a) write the function in the form $f(x) = a(x - h)^2 + k$ and (b) find the vertex and the axis of symmetry.

1. $f(x) = x^2 - 4x + 5$ **2.** $f(x) = x^2 + 6x + 13$

3. $f(x) = -x^2 + 3x - 10$ **4.** $f(x) = x^2 + 5x + 4$

5. $f(x) = 2x^2 - 7x + 1$

6. $f(x) = -2x^2 + 5x - 1$

For each quadratic function, (a) find the vertex and the axis of symmetry and (b) graph the function.

7. $f(x) = x^2 + 2x - 5$ **8.** $g(x) = x^2 + 4x + 5$

9. $f(x) = x^2 + 8x + 20$

10. $f(x) = x^2 - 10x + 21$

11. $h(x) = 2x^2 + 16x + 25$

12. $h(x) = 2x^2 - 16x + 23$

13. $f(x) = -x^2 + 2x + 5$

14. $f(x) = -x^2 - 2x + 7$

15. $g(x) = x^2 + 7x - 1$

16. $g(x) = x^2 + 3x + 5$

17. $h(x) = x^2 - 9x$

18. $h(x) = x^2 + x$

19. $f(x) = -2x^2 - 6$

20. $f(x) = -3x^2 + 2$

21. $f(x) = -3x^2 + 5x - 2$

22. $f(x) = -3x^2 - 7x + 2$

23. $h(x) = \frac{1}{2}x^2 + 4x + \frac{19}{3}$

24. $h(x) = \frac{1}{2}x^2 - 3x + 2$

Use a grapher to find the vertex of the graph of each function.

25. $f(x) = x^2 + x - 6$ **26.** $f(x) = x^2 + 2x - 5$

27. $f(x) = 5x^2 - x + 1$

28. $f(x) = -4x^2 - 3x + 7$

29. $f(x) = -0.2x^2 + 1.4x - 6.7$

30. $f(x) = 0.5x^2 + 2.4x + 3.2$

Find the x- and y-intercepts. If no x-intercepts exist, state this.

31. $f(x) = x^2 - 6x + 3$

32. $f(x) = x^2 + 5x + 2$

33. $g(x) = -x^2 + 2x + 3$

34. $g(x) = x^2 - 6x + 9$

35. $f(x) = x^2 - 3x + 4$

36. $f(x) = x^2 - 7x - 2$

37. $h(x) = -x^2 + 4x - 4$

38. $h(x) = 2x^2 - 4x + 6$

39. $f(x) = 4x^2 - 12x + 3$

40. $f(x) = x^2 - x + 2$

For each quadratic function, find **(a)** *the maximum or minimum value and* **(b)** *the x- and y-intercepts. Round to the nearest hundredth.*

41. $f(x) = 2.31x^2 - 3.135x - 5.89$

42. $f(x) = -18.8x^2 + 7.92x + 6.18$

43. $g(x) = -1.25x^2 + 3.42x - 2.79$

44. $g(x) = 0.45x^2 - 1.72x + 12.92$

Skill Maintenance

Solve. [7.6]

45. $\sqrt{4x - 4} = \sqrt{x + 4} + 1$

46. $\sqrt{5x - 4} + \sqrt{13 - x} = 7$

47. Solve for n: $A = \dfrac{3Q + nT}{n}$. [6.7]

48. Simplify: [6.3]

$$\dfrac{\dfrac{3}{x} - \dfrac{2}{x - 1}}{\dfrac{5}{x^3 - x}}.$$

Synthesis

49. ◈ Is it possible for the graph of a quadratic function to have only one x-intercept if the vertex is off the x-axis? Why or why not?

50. ◈ Does the graph of every quadratic function have a y-intercept? Why or why not?

51. ◈ Suppose that the graph of $f(x) = ax^2 + bx + c$ has $(x_1, 0)$ and $(x_2, 0)$ as x-intercepts. Explain why the graph of $g(x) = -ax^2 - bx - c$ will also have $(x_1, 0)$ and $(x_2, 0)$ as x-intercepts.

52. Graph the function

$$f(x) = x^2 - x - 6.$$

Then use the graph to approximate solutions to each of the following equations.

a) $x^2 - x - 6 = 2$

b) $x^2 - x - 6 = -3$

53. Graph the function

$$f(x) = \frac{x^2}{8} + \frac{x}{4} - \frac{3}{8}.$$

Then use the graph to approximate solutions to each of the following equations.

a) $\dfrac{x^2}{8} + \dfrac{x}{4} - \dfrac{3}{8} = 0$

b) $\dfrac{x^2}{8} + \dfrac{x}{4} - \dfrac{3}{8} = 1$

c) $\dfrac{x^2}{8} + \dfrac{x}{4} - \dfrac{3}{8} = 2$

Find an equivalent equation of the type

$$f(x) = a(x - h)^2 + k.$$

54. $f(x) = mx^2 - nx + p$

55. $f(x) = 3x^2 + mx + m^2$

56. A quadratic function has $(-1, 0)$ as one of its intercepts and $(3, -5)$ as its vertex. Find an equation for the function.

57. A quadratic function has $(4, 0)$ as one of its intercepts and $(-1, 7)$ as its vertex. Find an equation for the function.

Graph.

58. $f(x) = |x^2 - 1|$

59. $f(x) = |x^2 - 3x - 4|$

60. $f(x) = |2(x - 3)^2 - 5|$

8.8
Applications and Models Involving Quadratic Functions

- *Maximum and Minimum Problems*
- *Fitting Quadratic Functions to Data*

Let's look now at some of the many situations in which quadratic functions are used for problem solving.

Maximum and Minimum Problems

We have seen that for any quadratic function f, the value of $f(x)$ at the vertex is either a maximum or a minimum. Thus problems in which a quantity must be maximized or minimized can often be solved by finding the coordinates of a vertex. This assumes that the problem can be modeled with a quadratic function.

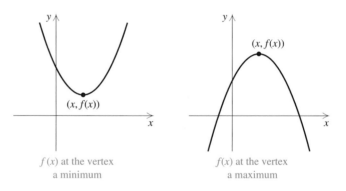

$f(x)$ at the vertex
a minimum

$f(x)$ at the vertex
a maximum

Example 1 *Fenced-in Land.* What are the dimensions of the largest rectangular pen that a farmer can enclose with 64 m of fence?

SOLUTION

1. **Familiarize.** We make a drawing and label it. Recall these important formulas:

Perimeter: $2w + 2l$;
Area: $l \cdot w$.

To get a better understanding of the problem, we can look at some possible dimensions for a rectangular pen that can be enclosed with 64 m of fence. We choose a length and a width that give a perimeter of 64, and calculate the corresponding area. We can express the width in terms of the length to help us make appropriate choices:

$$2l + 2w = 64 \qquad \text{The perimenter is 64 m.}$$
$$l + w = 32 \qquad \text{Dividing by 2}$$
$$w = 32 - l. \qquad \text{Subtracting } l$$

If we chose a length l, the width w will be $32 - l$.

We can make a table by hand or using a grapher. To use a grapher, we let x represent the length of the pen. Then the width is given by $y_1 = 32 - x$, and the area is $y_2 = x * y_1$.

l	w	A
22	10	220
20	12	240
18	14	252
.	.	.
.	.	.
.	.	.

} What choice of l and w will maximize A?

X	Y₁	Y₂
22	10	220
21.5	10.5	225.75
21	11	231
20.5	11.5	235.75
20	12	240
19.5	12.5	243.75
19	13	247

X = 22

} What choice of X will maximize Y2?

2. Translate. We have two equations: One guarantees that the perimeter is 64 m; the other expresses area in terms of length and width.

$$2w + 2l = 64$$
$$A = l \cdot w$$

3. Carry out. We solve the system of equations both algebraically and graphically.

┌-- *ALGEBRAIC SOLUTION*

We need to express A as a function of l or w but not both. To do so, we solve $2w + 2l = 64$ for w, as we did in the *Familiarize* step, to obtain $w = 32 - l$. Substituting for w in the second equation, we get a quadratic function, $A(l)$, or just A:

$$A = (32 - l)l \qquad \text{Substituting for } w$$
$$= -l^2 + 32l. \qquad \textbf{This is a parabola opening downward,}$$
$$\textbf{so a maximum value exists.}$$

Completing the square, we get

$$A = -(l^2 - 32l + 256 - 256)$$
$$= -(l - 16)^2 + 256.$$

The maximum function value, 256, occurs when $l = 16$ and $w = 32 - 16$, or 16.

┌-- *GRAPHICAL SOLUTION*

As in the *Familiarize* step, we let x represent the length of the pen. Then the width is given by $y_1 = 32 - x$, and the area is $y_2 = x * y_1$. The maximum area is the second coordinate of the vertex of the graph $y_2 = x * y_1$, or $y_2 = x(32 - x)$.

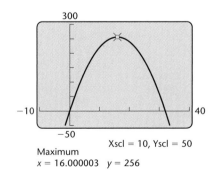

Xscl = 10, Yscl = 50
Maximum
$x = 16.000003$ $y = 256$

The maximum is 256 when the length $x = 16$ and the width $y_1 = 32 - 16$, or 16.

4. **Check.** Note that 256 is greater than any of the values for A found in the *Familiarize* step. To be more certain, we could check values other than those used in that step. For example, if $l = 15$, then $w = 32 - 15 = 17$, and $A = 15 \cdot 17 = 255$. The same value for the area results if $l = 17$ and $w = 15$. Since 256 is greater than 255, it looks as though 256 is the maximum.

5. **State.** The largest rectangular pen that can be enclosed is 16 m by 16 m.

Example 2 What is the minimum product of two numbers that differ by 5? What are the numbers?

SOLUTION

1. **Familiarize.** We try some pairs of numbers that differ by 5 and compute their products:

$$1 \cdot 6 = 6,$$
$$0 \cdot 5 = 0,$$
$$(-1) \cdot 4 = -4.$$

We suspect that one of the two numbers will be negative and the other positive. Let $x =$ the larger number and $x - 5$ the other number.

2. **Translate.** We represent the product of the two numbers by

$$p = x(x - 5), \quad \text{or} \quad p(x) = x^2 - 5x.$$

3. **Carry out.** The function $p(x) = x^2 - 5x$ represents a parabola opening upward.

ALGEBRAIC SOLUTION

Completing the square, we get

$$p(x) = x^2 - 5x + \frac{25}{4} - \frac{25}{4}$$
$$= \left(x - \frac{5}{2}\right)^2 - \frac{25}{4}.$$

The minimum function value of $-\frac{25}{4}$ occurs when one number is $\frac{5}{2}$ and the other number is $\frac{5}{2} - 5$, or $-\frac{5}{2}$.

GRAPHICAL SOLUTION

We graph the function and find the minimum value of the function.

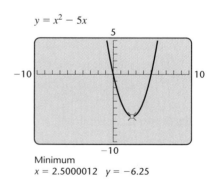

Because of the method the grapher uses and the choices of a left and right bound, the x-coordinate of the vertex is shown as 2.5000012. The exact value is 2.5, as can be seen algebraically. The vertex is (2.5, −6.25). The minimum value of −6.25 occurs when one number is 2.5 and the other is 2.5 − 5 = −2.5.

4. Check. First, note that $\frac{5}{2} = 2.5$ and $-\frac{5}{2} = -2$
same solution using both methods. To check, we
near $\frac{5}{2}$, yield smaller values of $x(x-5)$. Note
$x - 5 = -3$ and $2(-3) = -6$. Also note t
$x - 5 = -2$ and $3(-2) = -6$. Thus, since $-$
that $x(x-5)$ is minimized when x is $\frac{5}{2}$.

5. State. The minimum product of two numbers that differ by 5 is
$-\frac{25}{4}$, or -6.25. This minimum occurs when $\frac{5}{2}$ and $-\frac{5}{2}$ are multiplied.

Fitting Quadratic Functions to Data

We can now model some real-world situations using quadratic functions.
To determine what kind of function might fit a set of data, we graph the
data and compare the result with the basic graphs in our library of functions, shown below.

Linear function:
$f(x) = mx + b$

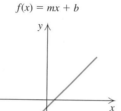

Absolute-value function:
$f(x) = |x|$

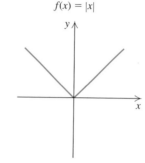

Rational function:
$f(x) = \frac{1}{x}$

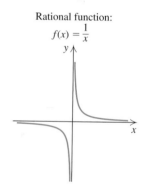

Radical function:
$f(x) = \sqrt{x}$

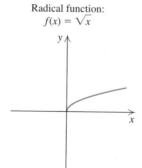

Quadratic function:
$f(x) = ax^2 + bx + c, \, a > 0$

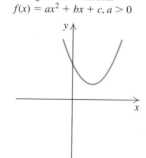

Quadratic function:
$f(x) = ax^2 + bx + c, \, a < 0$

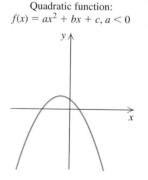

We see that in order for a quadratic function to fit a set of data, the
data must rise and then fall, or fall and then rise, in a manner resembling a
parabola. Data that resemble one half of a parabola might also be modeled

using a quadratic function with a restricted domain, as illustrated in the figure below.

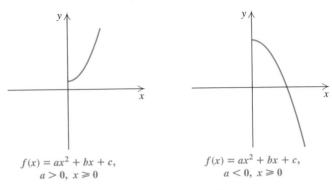

$$f(x) = ax^2 + bx + c,$$
$$a > 0,\ x \geq 0$$

$$f(x) = ax^2 + bx + c,$$
$$a < 0,\ x \geq 0$$

Example 3 Determine whether a linear or a quadratic function can be used to model each of the following situations.

a) *Air quality.* The number of metric tons of particulates retained in Chicago by trees

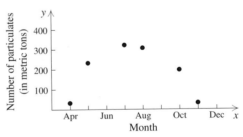

Source: The National Arbor Day Foundation

b) *Population.*

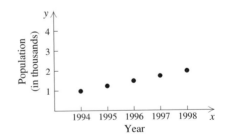

c) *Sales.*

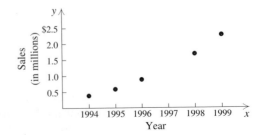

d) *Number of driver fatalities by age.*

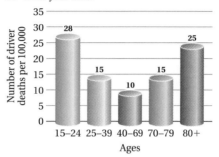

Driver Fatalities by Age

Number of licensed drivers per 100,000 who died in motor vehicle accidents in 1990. The fatality rates for both the 70–79 group and the 80+ age group were lower than for the 15- to 24-year-olds.

Source: National Highway Traffic Administration

e) *Work stoppages.*

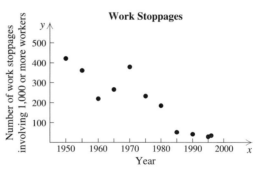

Sources: U.S. Bureau of Labor Statistics; *The Wall Street Journal Almanac,* 1998

SOLUTION

a) The data rise and then fall, resembling a parabola that opens downward. The situation could be modeled by a quadratic function $f(x) = ax^2 + bx + c, a < 0$.

b) The data rise steadily. The situation could be modeled by a linear function $f(x) = mx + b$.

c) Since the data appear nearly linear, we might use a linear function to model the situation. The data also resemble the right half of a parabola that opens upward. We could use a quadratic function $f(x) = ax^2 + bx + c, a > 0$, as a model, for $x > 0$.

d) The data fall and then rise, resembling a parabola that opens upward. A quadratic function $f(x) = ax^2 + bx + c, a > 0$, might be used as a model for this situation.

e) The data fall, then rise, then fall, then rise again. None of the functions listed would be a good model for this situation. Another polynomial function might be a good model (see Section 5.7). ▬

Whenever a certain quadratic function models a situation, a function that fits *exactly* can be determined if exactly three data points are given. If more than three data points are given, regression can be used to determine the "best-fit" quadratic function.

Example 4 *Air Quality.* Trees improve air quality in part by retaining airborne particles, called particulates, from the air. The following table shows the number of metric tons of particulates retained by trees in Chicago during the spring, summer, and autumn months. As we saw in Example 3(a), we can model this situation using a quadratic function. To do so, we let $x =$ the number of months after April.

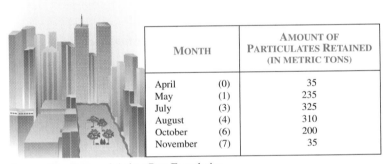

MONTH		AMOUNT OF PARTICULATES RETAINED (IN METRIC TONS)
April	(0)	35
May	(1)	235
July	(3)	325
August	(4)	310
October	(6)	200
November	(7)	35

Source: The National Arbor Day Foundation

a) Use the data points (0, 35), (4, 310), and (6, 200) to fit a quadratic function to the data.

b) Use the function from part (a) to estimate the number of metric tons of particulates retained in September.

c) Use regression to fit a quadratic function to all the given data.

d) Use the function from part (c) to estimate the number of metric tons of particulates retained in September.

e) The actual amount of particulates retained in September is 280 metric tons. What function provides the better estimation?

SOLUTION

a) The statement of the problem leads us to look for a function of the form

$$Y1(x) = ax^2 + bx + c,$$

where $Y1(x)$ is the number of metric tons of particulates retained by trees x months after April. We need values for a, b, and c.

We substitute the given values of x and Y1:

$$35 = a \cdot 0^2 + b \cdot 0 + c, \quad \text{Using the data point (0, 35)}$$
$$310 = a \cdot 4^2 + b \cdot 4 + c, \quad \text{Using the data point (4, 310)}$$
$$200 = a \cdot 6^2 + b \cdot 6 + c. \quad \text{Using the data point (6, 200)}$$

After simplifying, we see that we need to solve the system

$$35 = c, \qquad (1)$$
$$310 = 16a + 4b + c, \qquad (2)$$
$$200 = 36a + 6b + c. \qquad (3)$$

We know from equation (1) that $c = 35$. Substituting that value into equations (2) and (3), we have

$$310 = 16a + 4b + 35,$$
$$200 = 36a + 6b + 35.$$

Simplifying, we obtain the system of equations

$$275 = 16a + 4b, \qquad (4)$$
$$165 = 36a + 6b. \qquad (5)$$

To solve, we multiply equation (4) by -3 and equation (5) by 2 and add:

$$-825 = -48a - 12b,$$
$$\underline{330 = \quad 72a + 12b}$$
$$-495 = \quad 24a$$

Solving for a, we obtain

$$-20.625 = a.$$

Next, we solve for b, using equation (4) above:

$$275 = 16(-20.625) + 4b \quad \text{Substituting in equation (4)}$$
$$275 = -330 + 4b$$
$$605 = 4b$$
$$151.25 = b.$$

Thus the function $Y1(x) = -20.625x^2 + 151.25x + 35$ fits the three points given. As a partial check, we graph the function along with the data, as shown in the figure at the left. The function goes through the three given points.

b) Since September is 5 months after April, we evaluate Y1(5):

$$Y1(5) = -20.625(5)^2 + 151.25(5) + 35$$
$$= 275.625.$$

In September, the trees retained approximately 275.6 metric tons of particulates.

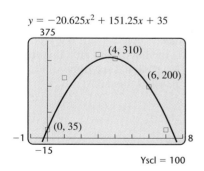

$y = -20.625x^2 + 151.25x + 35$

375

(4, 310)

(6, 200)

(0, 35)

-1 8

-15

Yscl = 100

c) To fit a quadratic function Y2(x) to the data using regression, we enter the data into the grapher and choose the QUADREG option in the STAT CALC menu.

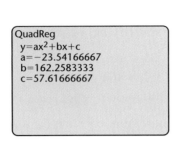

```
QuadReg
y=ax²+bx+c
a=-23.54166667
b=162.2583333
c=57.61666667
```

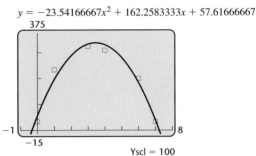

$y = -23.54166667x^2 + 162.2583333x + 57.61666667$

Yscl = 100

We obtain the quadratic function Y2(x) = $-23.54166667x^2 + 162.2583333x + 57.61666667$. We graph Y2($x$), along with the data points.

d) To estimate the number of metric tons of particulates retained in September, we evaluate Y2(5) and obtain Y2(5) ≈ 280.4.

e) The quadratic function Y1(x) found by considering only three of the data points gives us an estimate of 275.6 metric tons of particulates retained in September. Using the function Y2(x), found with regression, we estimate that 280.4 metric tons were retained in September. Although both models yielded a reasonable estimate, the function Y2(x) gives us a solution closer to the actual value, 280 tons. ▬

8.8 Exercise Set

Solve.

1. *Architecture.* An architect is designing a rectangular family room with a perimeter of 56 ft. What dimensions will yield the maximum area? What is the maximum area?

2. *Stained-Glass Window Design.* An artist is designing a rectangular stained-glass window with a perimeter of 84 in. What dimensions will yield the maximum area?

3. What is the minimum product of two numbers that differ by 6? What are the numbers?

4. What is the maximum product of two numbers that add to 18? What numbers yield this product?

5. What is the maximum product of two numbers that add to -12? What numbers yield this product?

6. What is the minimum product of two numbers that differ by 9? What are the numbers?

7. *Patio Design.* A stone mason has enough stones to enclose a rectangular patio with 60 ft of perimeter, assuming that the attached house forms one side of the rectangle. What is the maximum area that the mason can enclose? What should the dimensions of the patio be in order to yield this area?

8. *Garden Design.* A farmer decides to enclose a rectangular garden, using the side of a barn as one side of the rectangle. What is the maximum area that the farmer can enclose with 40 ft of fence? What should the dimensions of the garden be in order to yield this area?

9. *Molding Plastics.* Economite Plastics plans to produce a one-compartment vertical file by bending the long side of an 8-in. by 14-in. sheet of plastic along two lines to form a U shape. How tall should the file be in order to maximize the volume that the file can hold?

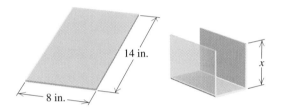

10. *Composting.* A rectangular compost container is to be formed in a corner of a fenced yard, with 8 ft of chicken wire completing the other two sides of the rectangle. If the chicken wire is 3 ft high, what dimensions of the base will maximize the container's volume?

11. *Minimizing Cost.* Aki's Bicycle Designs has determined that when x hundred bicycles are built, the average cost per bicycle is given by

$$C(x) = 0.1x^2 - 0.7x + 2.425,$$

where $C(x)$ is in hundreds of dollars. How many bicycles should the shop build in order to minimize the average cost per bicycle?

12. *Corral Design.* A rancher needs to enclose two adjacent rectangular corrals, one for sheep and one for cattle. If a river forms one side of the corrals and 180 yd of fencing is available, what is the largest total area that can be enclosed?

Maximizing Profit. Recall that total profit P is the difference between total revenue R and total cost C. Given the following total-revenue and total-cost functions, find the total profit, the maximum value of the total profit, and the value of x at which it occurs.

13. $R(x) = 1000x - x^2$,
 $C(x) = 3000 + 20x$

14. $R(x) = 200x - x^2$,
 $C(x) = 5000 + 8x$

In Exercises 15–20, determine whether a linear or a quadratic model might be appropriate. If neither seems appropriate, state this.

15.

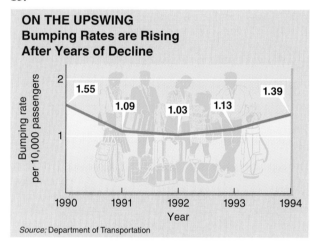

ON THE UPSWING
Bumping Rates are Rising
After Years of Decline

Source: Department of Transportation

16.

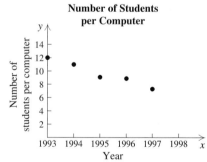

Number of Students per Computer

Source: Market Data Retrieval, Inc.

17.

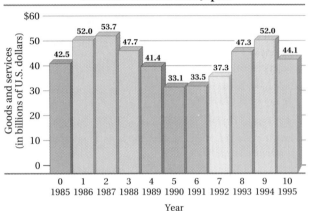

U.S. Trade Deficit with Japan

18.

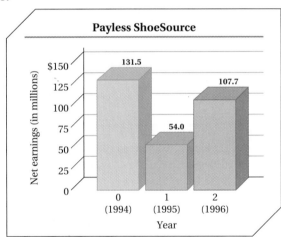

Source: Payless 1996 Annual Report

19.

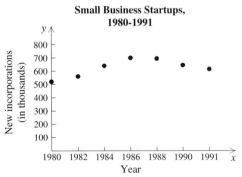

Small Business Startups, 1980-1991

Sources: U.S. Small Business Administration; *The MacMillan Visual Almanac*

20.

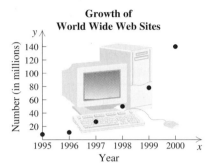

Growth of World Wide Web Sites

Find the quadratic function that fits each group of data points.

21. $(1, 4), (-1, -2), (2, 13)$

22. $(1, 4), (-1, 6), (-2, 16)$

23. $(2, 0), (4, 3), (12, -5)$

24. $(-3, -30), (3, 0), (6, 6)$

25. a) Find a quadratic function that fits the following data.

TRAVEL SPEED (IN KILOMETERS PER HOUR)	NUMBER OF NIGHTTIME ACCIDENTS (FOR EVERY 200 MILLION KILOMETERS DRIVEN)
60	400
80	250
100	250

b) Use the function to estimate the number of nighttime accidents that occur at 50 km/h.

26. a) Find a quadratic function that fits the following data.

TRAVEL SPEED (IN KILOMETERS PER HOUR)	NUMBER OF DAYTIME ACCIDENTS (FOR EVERY 200 MILLION KILOMETERS DRIVEN)
60	100
80	130
100	200

b) Use the function to estimate the number of daytime accidents that occur at 50 km/h.

27. *Archery.* The Olympic flame tower at the 1992 Summer Olympics was lit at a height of about 27 m by a flaming arrow that was launched about 63 m from the base of the tower. Given that the arrow landed about 63 m beyond the tower, find a quadratic function that expresses the height h of the arrow as a function of the distance d that it traveled horizontally.

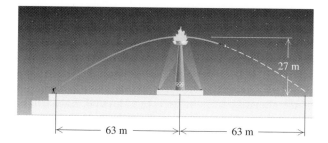

28. *Earnings.* The graph in Exercise 18 shows the net earnings of Payless Shoe Source for three years.

  a) Let $x =$ the number of years after 1994. Find a quadratic function $f(x)$ that fits the data.
  b) Use the function to estimate the net earnings in 1998.

29. *Air Quality.* Using the data for April, July, and November from Example 4:

  a) Find a quadratic function that could be used to predict the number of metric tons of particulates retained by trees in Chicago x months after April.
  b) Use the function to estimate the amount of particulates retained in September.
  c) Compare the estimate in part (b) with the estimate found using regression in Example 4.

30. *Air Quality.* Using the data for May, October, and November from Example 4:

  a) Find a quadratic function that could be used to predict the number of metric tons of particulates retained by trees in Chicago x months after April.
  b) Use the function to estimate the amount of particulates retained in September.
  c) Compare the estimate in part (b) with the estimate found using regression in Example 4.

31. *Airline Bumping Rates.* The graph in Exercise 15 shows the bumping rate per 10,000 passengers from 1990 through 1994.

  a) Let $x =$ the number of years after 1990. Use regression to find a quadratic function $b(x)$ that fits the data.
  b) Use the function to estimate the bumping rate in 1996.

32. *World Wide Web Sites.* The following table shows the growth of world wide web sites from 1995 through 2000. Use regression to find a quadratic function $w(x)$ that can be used to predict the number of world wide web sites x years after 1995.

YEAR	NUMBER OF WORLD WIDE WEB SITES (IN MILLIONS)
1995	5
1996	8
1997	30
1998	50
1999	80
2000	140

33. *Small Business.* The following table shows the number of new small-business incorporations for various years. Use regression to find a quadratic function $f(x)$ that can be used to predict the number of new incorporations x years after 1980.

YEAR	NUMBER OF NEW INCORPORATIONS (IN THOUSANDS)
1980	520
1982	560
1984	640
1986	700
1988	695
1990	645
1991	615

Source: U.S. Small Business Administration

34. *Household Income.* The following table shows the median U.S. household income for people of various ages. Use regression to find a quadratic function $f(x)$ that can be used to predict the median household income of people of age x.

AGE	MEDIAN INCOME IN 1996
19.5	$21,438
29.5	35,888
39.5	44,420
49.5	50,472
59.5	39,815
65	19,448

Source: U.S. Bureau of the Census; The Conference Board: Simmons Bureau of Labor Statistics

Skill Maintenance

Simplify.

35. $\dfrac{x}{x^2 + 17x + 72} - \dfrac{8}{x^2 + 15x + 56}$ [6.2]

36. $\dfrac{x^2 - 9}{x^2 - 8x + 7} \div \dfrac{x^2 + 6x + 9}{x^2 - 1}$ [6.1]

Synthesis

37. ◆ Explain how the leading coefficient of a quadratic function can be used to determine whether a maximum or a minimum function value exists.

38. ◆ Explain what restrictions should be placed on the quadratic functions developed in Exercises 13 and 25 and why such restrictions are needed.

39. *Norman Window.* A *Norman window* is a rectangle with a semicircle on top. Big Sky Windows is designing a Norman window that will require 24 ft of trim. What dimensions will allow the maximum amount of light to enter a house?

40. *Minimizing Area.* A 36-in. piece of string is cut into two pieces. One piece is used to form a circle while the other is used to form a square. How should the string be cut so that the sum of the areas is a minimum?

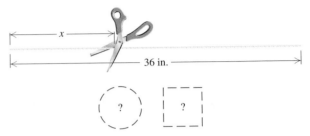

41. *Crop Yield.* An orange grower finds that she gets an average yield of 40 bushels (bu) per tree when she plants 20 trees on an acre of ground. Each time she adds a tree to an acre, the yield per tree decreases by 1 bu, due to congestion. How many trees per acre should she plant for maximum yield?

42. *Cover Charges.* When the owner of Sweet Sounds charges a $5 cover charge, an average of 100 people will attend a show. For each 25¢ increase in admission price, the average number attending decreases by 1. What should the owner charge in order to make the most money?

43. *Trajectory of a Launched Object.* The height above the ground of a launched object is a quadratic function of the time that it is in the air. Suppose that a flare is launched from a cliff 64 ft above sea level. If 3 sec after being launched the flare is again level with the cliff, and if 2 sec after that it lands in the sea, what is the maximum height that the flare will reach?

44. *Bridge Design.* The cables supporting a straight-line suspension bridge are nearly parabolic in shape. Suppose that a suspension bridge is being designed to cross a river that is 160 ft wide and that the vertical cables are 30 ft above road level at the midpoint of the bridge and 80 ft above road level at a point 50 ft from the midpoint of the bridge. How long are the longest vertical cables?

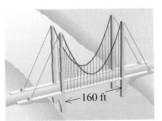

Focus: Modeling quadratic functions

Time: 20–30 minutes

Group size: 3 or 4

Materials: Graphers

The Panasonic Portable Stereo System RX-DT680® has a counter for finding locations on an audiocassette. When a fully wound cassette with 45 min of music on a side begins to play, the counter is at 0. After 15 min of music has played, the counter reads 250 and after 35 min, it reads 487. When the 45-min side is finished playing, the counter reads 590.

Activity

1. The paragraph above describes four ordered pairs of the form (counter number, minutes played). Three pairs are enough to find a function of the form

$$T(n) = an^2 + bn + c,$$

where $T(n)$ is the time, in minutes, that the tape has run at counter reading n hundred. Each group member should select a different set of three points from the four given and then fit a quadratic function to the data.

2. Of the 3 or 4 functions found in part (1) above, which fits the data "best"? One way to answer this is to see how well each function predicts other pairs. The same counter used above reads 432 after a 45-min tape has played for 30 min. Which function comes closest to predicting this?

3. Use regression to fit a quadratic function to the four pairs originally listed. How well does this function predict the reading after 30 min.?

4. If a class member has access to a Panasonic System RX-DT680, see how well the functions developed above predict the counter readings for a tape that has played for 5 or 10 min.

8.9
Polynomial and Rational Inequalities

• *Quadratic and Other Polynomial Inequalities*
• *Rational Inequalities*

Quadratic and Other Polynomial Inequalities

Inequalities like the following are called **polynomial inequalities**:

$$x^3 - 5x > x^2 + 7, \qquad 3x - 1 \le 5, \qquad 5x^2 - 3x + 2 \ge 0.$$

Second-degree polynomial inequalities in one variable are called **quadratic inequalities**. To solve polynomial inequalities, we often focus attention on where the outputs of a polynomial function are positive and where they are negative.

Example 1 Solve: $x^2 + 3x - 10 > 0$.

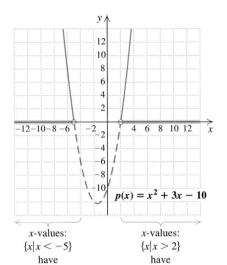

x-values:
$\{x \mid x < -5\}$
have
positive
y-values.

x-values:
$\{x \mid x > 2\}$
have
positive
y-values.

SOLUTION   Consider the "related" function $p(x) = x^2 + 3x - 10$ and its graph. Its graph opens upward since the leading coefficient is positive.

Values of y will be positive to the left and right of the x-intercepts, as shown. To find the intercepts, we set the polynomial equal to 0 and solve:

$$x^2 + 3x - 10 = 0$$
$$(x + 5)(x - 2) = 0$$
$$x + 5 = 0 \quad or \quad x - 2 = 0$$
$$x = -5 \quad or \quad \quad x = 2.$$

Thus the solution set of the inequality is $(-\infty, -5) \cup (2, \infty)$.   ▬

Any inequality with 0 on one side can be solved by considering a graph of the related function and finding intercepts as in Example 1. Sometimes the quadratic formula is needed to find the intercepts.

Example 2   Solve: $x^2 - 2x \leq 2$.

SOLUTION   We first find standard form with 0 on one side:

$$x^2 - 2x - 2 \leq 0. \qquad \text{This is equivalent to the original inequality.}$$

The graph of $p(x) = x^2 - 2x - 2$ is a parabola opening upward. Values of $p(x)$ are negative for x-values between the x-intercepts. We find the x-intercepts by solving $p(x) = 0$:

$$x = \frac{-b \pm \sqrt{b^2 - 4ac}}{2a}$$
$$= \frac{-(-2) \pm \sqrt{(-2)^2 - 4 \cdot 1(-2)}}{2 \cdot 1}$$
$$= \frac{2 \pm \sqrt{12}}{2} = \frac{2 \pm 2\sqrt{3}}{2}$$
$$= 1 \pm \sqrt{3}.$$

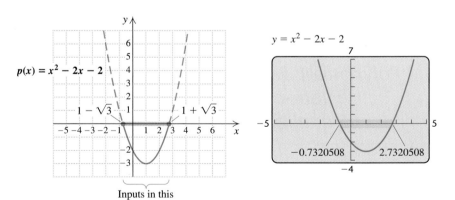

$p(x) = x^2 - 2x - 2$

$1 - \sqrt{3}$ $1 + \sqrt{3}$

Inputs in this interval have negative or 0 outputs.

$y = x^2 - 2x - 2$

-0.7320508 2.7320508

We can also find the x-intercepts using a grapher. Because the numbers are irrational, this procedure will result in an approximation.

At the x-intercepts, the value of $p(x)$ is 0. Thus the solution set of the inequality is

$$[1 - \sqrt{3}, 1 + \sqrt{3}], \quad \text{or approximately} \quad [-0.7320508, 2.7320508].$$

We can solve a polynomial inequality without graphing. Note in Example 2 that the important information came from the location of the x-intercepts and the sign of $p(x)$ on each side of those intercepts. In the next example, we solve a third-degree polynomial inequality, without graphing, by locating the x-intercepts, or zeros, of p and then using *test points* to determine the sign of $p(x)$ over each interval of the x-axis.

Example 3   Solve: $5x^3 + 10x^2 - 15x > 0$.

SOLUTION   We first solve the related equation:

$$5x^3 + 10x^2 - 15x = 0$$
$$5x(x^2 + 2x - 3) = 0$$
$$5x(x + 3)(x - 1) = 0$$
$$5x = 0 \quad or \quad x + 3 = 0 \quad or \quad x - 1 = 0$$
$$x = 0 \quad or \quad\quad x = -3 \quad or \quad\quad x = 1.$$

We see that if $p(x) = 5x^3 + 10x^2 - 15x$, then the zeros of p are -3, 0, and 1. These zeros divide the number line, or x-axis, into four intervals: A, B, C, and D.

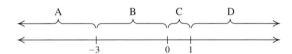

Next, using one convenient test value from each interval, we determine the sign of $p(x)$ over that interval. We choose -4 for a test value from interval A, -1 from interval B, 0.5 from interval C, and 2 from interval D. We enter $y_1 = 5x^3 + 10x^2 - 15x$ and use a table to evaluate the polynomial for each test value.

X	Y1	
−4	−100	
−1	20	
.5	−4.375	
2	50	

X = −4

We are interested only in the signs of each function value. From the table, we see that $p(-4)$ and $p(0.5)$ are negative and that $p(-1)$ and $p(2)$ are positive. We indicate on the number line the sign of $p(x)$ in each interval.

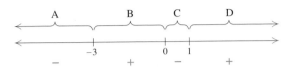

Recall that we are looking for all x for which $5x^3 + 10x^2 - 15x > 0$. The calculations above indicate that $p(x)$ is positive for any number in intervals B and D. The solution set of the original inequality is $(-3, 0) \cup (1, \infty)$.

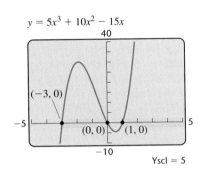

The method of Example 3 works because polynomial function values can change signs only when the graph of the function crosses the x-axis. The graph of $p(x) = 5x^3 + 10x^2 - 15x$ illustrates the solution of Example 3. We can see from the graph at left that the function values are positive in the intervals $(-3, 0)$ and $(1, \infty)$.

Example 4   Solve: $x^4 + x^3 - 2x^2 \le 0$.

SOLUTION   We first solve the related equation:

$$x^4 + x^3 - 2x^2 = 0$$
$$x^2(x^2 + x - 2) = 0$$
$$x^2(x + 2)(x - 1) = 0$$
$$x^2 = 0 \quad or \quad x + 2 = 0 \quad or \quad x - 1 = 0$$
$$x = 0 \quad or \quad x = -2 \quad or \quad x = 1.$$

The function $p(x) = x^4 + x^3 - 2x^2$ has zeros at -2, 0, and 1. We graph the function and determine the sign of $p(x)$ over each interval of the number line.

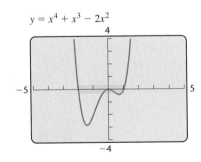

We see from the graph that $p(x)$ is negative in the intervals $(-2, 0)$ and $(0, 1)$. Since the inequality is also true when the function is 0, we include the endpoints of the intervals. The solution is

$$[-2, 0] \cup [0, 1], \quad \text{or simply} \quad [-2, 1].$$

To Solve a Polynomial Inequality by Factoring:

1. Get 0 on one side and solve the related polynomial equation $p(x) = 0$ by factoring.
2. Use the numbers found in step (1) to divide the number line into intervals.
3. Using a test point from each interval or the graph of the related function, determine the sign of $p(x)$ over each interval.
4. Select the interval(s) for which the inequality is satisfied and write interval notation for the solution set. Include the endpoints when $\leq$ or $\geq$ is used.

Note that if the polynomial cannot be factored, the grapher can be used to at least approximate the x-intercepts of the graph of the function, as in Example 2.

Rational Inequalities

Inequalities involving rational expressions are called **rational inequalities**. Like polynomial inequalities, rational inequalities can be solved using graphs or test values.

Example 5 Solve: $\dfrac{x - 3}{x + 4} \geq 2.$

SOLUTION We first find an equivalent inequality with 0 on one side. This can be done by subtracting 2 on both sides:

$$\frac{x - 3}{x + 4} - 2 \geq 0.$$

If $r(x) = \dfrac{x - 3}{x + 4} - 2$, the solution set of the inequality is all values of x for which $r(x) \geq 0$.

We first find all values of x for which $r(x) = 0$:

$$\frac{x - 3}{x + 4} - 2 = 0$$

$$(x + 4)\left(\frac{x - 3}{x + 4} - 2\right) = (x + 4)(0)$$

Multiplying by the LCD, $x + 4$, on both sides

$$(x + 4)\left(\frac{x - 3}{x + 4}\right) - (x + 4)(2) = 0$$

Using the distributive law

$$x - 3 - (2x + 8) = 0$$

$$x - 3 - 2x - 8 = 0$$

$$-11 = x.$$

Solving for x

In the case of a rational inequality, we also need to determine those values that make any denominator 0. We set the denominator equal to 0 and solve:

$$\left.\begin{array}{l} x + 4 = 0 \\ x = -4. \end{array}\right\}$$

This tells us that -4 is not in the domain of r.

We use -11 and -4 to divide the number line into intervals:

Next, we evaluate $r(x)$ for a test number in each interval to see where the function is positive. We test -15, -8, and 1:

X	Y1	
-15	-.3636	
-8	.75	
1	-2.4	
X = -15		

$r(-15)$ is negative.
$r(-8)$ is positive.
$r(1)$ is negative.

We indicate on the number line the sign of $r(x)$ in each interval.

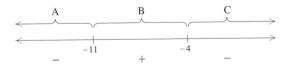

The solution set includes the interval B. The endpoint -11 is included since the inequality symbol is $\geq$ and $r(-11) = 0$. The number -4 is *not* included since $(x - 3)/(x + 4)$ is undefined for $x = -4$. Thus the solution set of the original inequality is $[-11, -4)$.

We graph the function

$$r(x) = \frac{x - 3}{x + 4} - 2.$$

The inequality is true for those values of x such that $r(x) = 0$ or $r(x) > 0$.

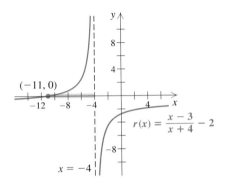

The x-intercept of the graph is $(-11, 0)$. Thus, $r(-11) = 0$, and -11 is a solution of the inequality. The function is positive over the interval $(-11, -4)$. Thus the solution set is $[-11, -4)$.

570

Note that the solution of the equation $r(x) = 0$ corresponds to the x-intercept of the graph of $r(x)$, and the value of x that makes the denominator 0 corresponds to the vertical asymptote of the graph, $x = -4$.

To Solve a Rational Inequality:

1. Get 0 on one side and solve the related rational equation $r(x) = 0$.

2. Find any x-values that make a denominator 0.

3. Use the numbers found in steps (1) and (2) to divide the number line into intervals.

4. Using a test point from each interval or the graph of the related function, determine the sign of $r(x)$ over each interval.

5. Select the interval(s) for which the inequality is satisfied and write interval notation for the solution set. If the inequality symbol is $\leq$ or $\geq$, then the solutions to step (1) should be included in the solution set.

8.9 | *Exercise Set*

🔍 *Determine the solution set of each inequality from the given graph.*

1. $2x^2 + 5x - 12 \leq 0$

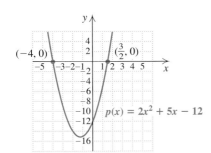

2. $3x^2 + 14x + 8 < 0$

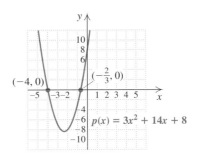

3. $x^4 + 12x > 3x^3 + 4x^2$

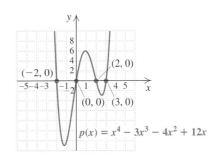

4. $x^4 + x^3 \geq 6x^2$

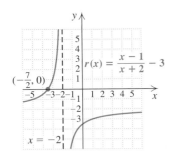

5. $\dfrac{x-1}{x+2} < 3$

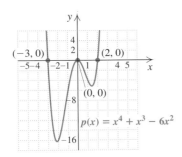

6. $\dfrac{2x-1}{x-5} \geq 1$

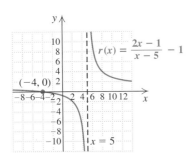

Solve. Round any approximations to the nearest hundredth.

7. $(x+4)(x-3) > 0$ **8.** $(x-5)(x+2) > 0$

9. $(x+7)(x-2) \leq 0$ **10.** $(x-1)(x+4) \leq 0$

11. $x^2 - x - 2 < 0$ **12.** $x^2 + x - 2 < 0$

13. $9 - x^2 \leq 0$ **14.** $4 - x^2 \geq 0$

15. $x^2 - 2x + 1 \geq 0$ **16.** $x^2 + 6x + 9 < 0$

17. $x^2 - 4x < 12$ **18.** $x^2 + 6x > -8$

19. $3x(x+2)(x-2) < 0$ **20.** $5x(x+1)(x-1) > 0$

21. $(x+3)(x-2)(x+1) > 0$

22. $(x-1)(x+2)(x-4) < 0$

23. $(x+3)(x+2)(x-1) < 0$

24. $(x-2)(x-3)(x+1) < 0$

25. $4.32x^2 - 3.54x - 5.34 \leq 0$

26. $7.34x^2 - 16.55x - 3.89 \geq 0$

27. $x^3 - 2x^2 - 5x + 6 < 0$

28. $\frac{1}{3}x^3 - x + \frac{2}{3} > 0$

29. $\dfrac{1}{x+7} < 0$ **30.** $\dfrac{1}{x+4} > 0$

31. $\dfrac{x+1}{x-3} \geq 0$ **32.** $\dfrac{x-2}{x+5} \leq 0$

33. $\dfrac{3x+2}{x-3} \leq 0$ **34.** $\dfrac{5-2x}{4x+3} \leq 0$

35. $\dfrac{x+1}{2x-3} > 1$ **36.** $\dfrac{x-1}{x-2} < 1$

37. $\dfrac{(x-2)(x+1)}{x-5} \leq 0$ **38.** $\dfrac{(x+4)(x-1)}{x+3} \geq 0$

39. $\dfrac{x}{x+3} \geq 0$ **40.** $\dfrac{x-2}{x} \leq 0$

41. $\dfrac{x-5}{x} < 1$ **42.** $\dfrac{x}{x-1} > 2$

43. $\dfrac{x-1}{(x-3)(x+4)} \leq 0$ **44.** $\dfrac{x+2}{(x-2)(x+7)} \geq 0$

45. $4 < \dfrac{1}{x}$ **46.** $\dfrac{1}{x} \leq 5$

Skill Maintenance

47. Multiply and simplify: $\sqrt[5]{a^2 b}\,\sqrt[3]{ab^2}$. [7.4]

48. The perimeter of an equilateral triangle is the same as that of a square. If the sides of the triangle are 2 units longer than the sides of the square, how long is each side of the square? [3.3]

Synthesis

49. ◆ Explain how any quadratic inequality can be solved by examining a parabola that opens upward.

50. ◆ Describe a method that could be used to create quadratic inequalities that have no solution.

Solve.

51. $x^2 + 2x > 4$ **52.** $x^4 + 2x^2 \geq 0$

53. $x^4 + 3x^2 \leq 0$ **54.** $\left|\dfrac{x+2}{x-1}\right| \leq 3$

55. *Total Profit.* Derex, Inc., determines that its total-profit function is given by
$$P(x) = -3x^2 + 630x - 6000.$$

a) Find all values of x for which Derex makes a profit.
b) Find all values of x for which Derex loses money.

56. *Height of a Thrown Object.* The function

$$S(t) = -16t^2 + 32t + 1920$$

gives the height S, in feet, of an object thrown from a cliff that is 1920 ft high. Here t is the time, in seconds, that the object is in the air.

a) For what times does the height exceed 1920 ft?
b) For what times is the height less than 640 ft?

57. *Number of Handshakes.* There are n people in a room. The number N of possible handshakes by the people is given by the function

$$N(n) = \frac{n(n-1)}{2}.$$

For what number of people n is $66 \le N \le 300$?

58. *Number of Diagonals.* A polygon with n sides has D diagonals, where D is given by the function

$$D(n) = \frac{n(n-3)}{2}.$$

Find the number of sides n if $27 \le D \le 230$.

Use a grapher to graph each function and find solutions of $f(x) = 0$. Then solve the inequalities $f(x) < 0$ and $f(x) \ge 0$.

59. $f(x) = x + \dfrac{1}{x}$

60. $f(x) = x - \sqrt{x}, x \ge 0$

61. $f(x) = x^4 - 4x^3 - x^2 + 16x - 12$

62. $f(x) = \dfrac{x^3 + x^2 - 2x}{x^2 + x - 6}$

63. *Population.* The population of downtown Indianapolis began to rise in the 1990s after having been on the decline for two decades. This growth was due at least in part to renewed development in the downtown area. The following chart lists the population for several years.

YEAR	DOWNTOWN POPULATION
1970	22,834
1980	13,070
1990	12,179
1996	13,000

Source: Indianapolis Star

a) Use regression to find a quadratic function $p(x)$ that can be used to predict the population of downtown Indianapolis x years after 1970.
b) Use the function to estimate the years for which the population of downtown Indianapolis is greater than 20,000. In other words, solve $p(x) > 20,000$.

CHAPTER 8 Summary and Review

Key Terms

Quadratic equation, p. 488
Quadratic function, p. 488
Parabola, p. 488
Principle of square roots, p. 490
Completing the square, p. 493
Compounding interest annually, p. 496
The quadratic formula, p. 500

Discriminant, p. 504
Quadratic in form, p. 515
Direct variation, p. 523
Variation constant, p. 524
Constant of proportionality, p. 525
Inverse variation, p. 525
Joint variation, p. 529

Axis of symmetry, p. 535
Vertex, p. 535
Minimum value, p. 535
Maximum value, p. 536
Translated, p. 537
Polynomial inequality, p. 565
Quadratic inequality, p. 565
Rational inequality, p. 569

Important Properties and Formulas

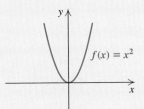

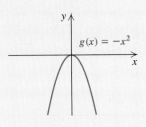

The Principle of Square Roots

For any real number k, if $x^2 = k$, then $x = \sqrt{k}$ or $x = -\sqrt{k}$.

The Quadratic Formula

The solutions of $ax^2 + bx + c = 0$, $a \neq 0$, are given by

$$x = \frac{-b \pm \sqrt{b^2 - 4ac}}{2a}.$$

To solve a quadratic equation:

1. Check for the form $ax^2 = p$ or $(x + k)^2 = d$. If it is in either of these forms, use the principle of square roots.
2. If it is not in the form of step (1), write it in standard form $ax^2 + bx + c = 0$.
3. Try factoring and using the principle of zero products.
4. If it is not possible to factor or factoring seems difficult, use the quadratic formula.

The solutions of a quadratic equation can always be found using the quadratic formula. They cannot always be found by factoring. Any real-number solutions can be found with a grapher, although the solutions may be approximations.

DISCRIMINANT $b^2 - 4ac$	NATURE OF SOLUTIONS
0	Only one solution (double root); it is a rational number (One x-intercept)
Positive	Two different real-number solutions (Two x-intercepts)
Perfect square	Solutions are rational.
Not a perfect square	Solutions are irrational conjugates.
Negative	Two different imaginary-number solutions (complex conjugates) (No x-intercepts)

Variation

y varies directly as x if there is some nonzero constant k such that $y = kx$.

y varies inversely as x if there is some nonzero constant k such that $y = k/x$.

y varies jointly as x and z if there is some nonzero constant k such that $y = kxz$.

The graph of $f(x) = ax^2$ is a parabola with $x = 0$ as its axis of symmetry. Its vertex is the origin.

For $a > 0$, the parabola opens upward; for $a < 0$, the parabola opens downward.

If $|a|$ is greater than 1, the parabola is narrower than $y = x^2$.

If $|a|$ is between 0 and 1, the parabola is wider than $y = x^2$.

The graph of $f(x) = a(x - h)^2$ has the same shape as the graph of $y = ax^2$.

If h is positive, the graph of $y = ax^2$ is shifted h units to the right.

If h is negative, the graph of $y = ax^2$ is shifted $|h|$ units to the left.

The vertex is $(h, 0)$, and the axis of symmetry is $x = h$.

The graph of $f(x) = a(x - h)^2 + k$ has the same shape as the graph of $y = a(x - h)^2$.

If k is positive, the graph of $y = a(x - h)^2$ is shifted k units up.

If k is negative, the graph of $y = a(x - h)^2$ is shifted $|k|$ units down.

The vertex is (h, k), and the axis of symmetry is $x = h$.

The domain of f is $(-\infty, \infty)$.

For $a > 0$, the range of f is $[k, \infty)$. The minimum value of $f(x)$ is k, which occurs when $x = h$.

For $a < 0$, the range of f is $(-\infty, k]$. The maximum value of $f(x)$ is k, which occurs when $x = h$.

The vertex of the parabola given by $f(x) = ax^2 + bx + c$ is

$$\left(-\frac{b}{2a}, \frac{4ac - b^2}{4a} \right), \quad \text{or} \quad \left(-\frac{b}{2a}, f\left(-\frac{b}{2a} \right) \right).$$

The x-coordinate of the vertex is $-b/(2a)$. The axis of symmetry is $x = -b/(2a)$.

REVIEW EXERCISES

1. Given the following graph of $f(x) = ax^2 + bx + c$.

  a) State the number of real-number solutions of $ax^2 + bx + c = 0$.

  b) State whether a is positive or negative.

  c) Determine the minimum value of f.

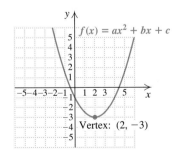

Vertex: $(2, -3)$

Solve.

2. $2x^2 - 7 = 0$

3. $14x^2 + 5x = 0$

4. $x^2 - 12x + 36 = 9$

5. $x^2 - 5x + 9 = 0$

6. $x(3x + 4) = 4x(x - 1) + 15$

7. $x^2 - 5x - 2 = 0$
  Give exact solutions and approximations to two decimal places.

8. Let $f(x) = 4x^2 - 3x - 1$. Find x such that $f(x) = 0$.

Complete the square. Then write the perfect-square trinomial in factored form.

9. $x^2 - 12x$ **10.** $x^2 + \frac{3}{5}x$

Solve by completing the square. Show your work.

11. $x^2 + 2x - 8 = 0$ **12.** $x^2 - 6x + 1 = 0$

13. 2500 grows to 3025 in 2 yr. Use the formula $A = P(1 + r)^t$ to find the interest rate.

14. The Peachtree Center Plaza in Atlanta, Georgia, is 723 ft tall. Use the formula $s = 16t^2$ to approximate how long it would take an object to fall from the top.

For each equation, determine what type of number the solutions will be.

15. $x^2 + 3x - 6 = 0$ **16.** $x^2 + 2x + 5 = 0$

Solve.

17. The Columbia River flows at a rate of 2 mph for the length of a popular boating route. In order for a motorized dinghy to travel 3 mi upriver and then return in a total of 4 hr, how fast must the boat be able to travel in still water?

18. Working together, Jean and Stacy can cut and split a cord of wood in 4 hr. Working alone, Jean takes 6 hr longer than Stacy. How long would it take Stacy to complete this job alone?

19. Find all x-intercepts of the graph of
$$f(x) = x^4 - 13x^2 + 36.$$

Solve.

20. $15x^{-2} - 2x^{-1} - 1 = 0$

21. $(x^2 - 4)^2 - (x^2 - 4) - 6 = 0$

22. Write a quadratic equation having the solutions $5i$ and $-5i$.

23. Write a quadratic equation having -4 as its only solution.

24. The power P expended by heat in an electric circuit of fixed resistance varies directly as the square of the current C in the circuit. A circuit expends 180 watts when a current of 6 amperes is flowing. What is the current in the circuit when it is expending 125 watts in heat?

25. A warning dye is used by people in lifeboats to aid searching airplanes. The radius r of the circle formed by the dye varies directly as the square root of the volume V. It is found that 4 L of dye will spread to a circle of radius 5 m. How much dye is needed to form a circle with a 20-m radius?

26. Find an equation of variation in which y varies jointly as x and z and inversely as w, and $y = 9$ when $x = 3$, $z = 2$, and $w = 5$.

27. a) Graph: $f(x) = -3(x + 2)^2 + 4$.
   b) Label the vertex.
   c) Draw the axis of symmetry.
   d) Find the maximum or the minimum value.

28. Given the function $f(x) = 2x^2 - 12x + 23$.
   a) Find the vertex and the axis of symmetry.
   b) Graph the function.

29. Find the x- and y-intercepts of
$$f(x) = x^2 - 9x + 14.$$

30. Solve $N = 3\pi\sqrt{1/p}$ for p.

31. Solve $2A + T = 3T^2$ for T.

32. What is the minimum product of two numbers whose difference is 22? What numbers yield this product?

33. Find the quadratic function that fits the data points $(0, -2)$, $(1, 3)$, and $(3, 7)$.

Determine which, if any, of the following sets of graphed data could be modeled by a quadratic function.

34.

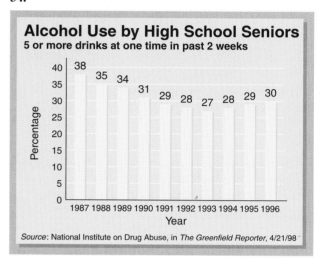

Alcohol Use by High School Seniors
5 or more drinks at one time in past 2 weeks

Source: National Institute on Drug Abuse, in *The Greenfield Reporter*, 4/21/98

35.

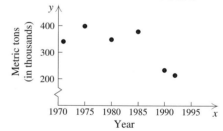

Chlorofluorocarbon (CFC) Production in the United States

Source: U.S. Environmental Projection Agency; *The MacMillan Visual Almanac*

36. *Alcohol Use.* The graph in Exercise 34 shows the percentages, from 1987 through 1996, of high school seniors who had 5 or more drinks at one time during a given two-week period. Use regression to find a quadratic function $f(x)$ that can be used to predict the percentage of such seniors x years after 1987.

Solve.

37. $x^3 - 3x > 2x^2$

38. $\dfrac{x - 5}{x + 3} \le 0$

Skill Maintenance

39. Metal alloy A is 75% silver. Metal alloy B is 25% silver. How much of each should be mixed in order to produce 300 kg of an alloy that is 60% silver?

40. Solve: $\sqrt{5x - 1} + \sqrt{2x} = 5$.

41. Add: $\dfrac{x}{x^2 - 3x + 2} + \dfrac{2}{x^2 - 5x + 6}$.

42. Multiply and simplify: $\sqrt[3]{9t^6}\sqrt[3]{3s^4t^9}$.

Synthesis

43. ◈ Explain how the x-intercepts of a quadratic function can be used to help find the maximum or minimum value of the function.

44. ◈ Suppose that the quadratic formula is used to solve a quadratic equation. If the discriminant is a perfect square, could factoring have been used to solve the equation? Why or why not?

45. ◈ What is the greatest number of solutions that an equation of the form $ax^4 + bx^2 + c = 0$ can have? Why?

46. ◈ Discuss two ways in which completing the square was used in this chapter.

47. A quadratic function has x-intercepts at -3 and 5. If the y-intercept is at -7, find an equation for the function.

48. Find h and k if, for $3x^2 - hx + 4k = 0$, the sum of the solutions is 20 and the product is 80.

49. The average of two positive integers is 171. One of the numbers is the square root of the other. Find the integers.

Chapter 8 Test

1. Given the following graph of $f(x) = ax^2 + bx + c$.

  a) State the number of real-number solutions of $ax^2 + bx + c = 0$.

  b) State whether a is positive or negative.

  c) Determine the maximum value of f.

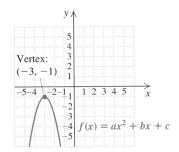

Vertex: $(-3, -1)$

$f(x) = ax^2 + bx + c$

Solve.

2. $3x^2 - 16 = 0$

3. $4x(x - 2) - 3x(x + 1) = -18$

4. $x^2 + x + 1 = 0$

5. $x^{-2} - x^{-1} = \dfrac{3}{4}$

6. $x^2 + 3x = 5$

Give exact solutions and approximations to two decimal places.

7. Let $f(x) = 12x^2 - 19x - 21$. Find x such that $f(x) = 0$.

Complete the square. Then write the perfect-square trinomial in factored form.

8. $x^2 + 14x$

9. $x^2 - \dfrac{2}{7}x$

Solve by completing the square. Show your work.

10. $x^2 + 3x - 18 = 0$

11. $x^2 + 10x + 15 = 0$

12. Determine the type of number that the solutions of $x^2 + 5x + 17 = 0$ will be.

Solve.

13. The Connecticut River flows at a rate of 4 km/h for the length of a popular scenic route. In order for a cruiser to travel 60 km upriver and then return in a total of 8 hr, how fast must the boat be able to travel in still water?

14. Two pipes can fill a tank in $1\frac{1}{2}$ hr. One pipe requires 4 hr longer running alone to fill the tank than the other. How long would it take for the faster pipe, working alone, to fill the tank?

15. Find all x-intercepts of the graph of
$$f(x) = (x^2 + 4x)^2 + 2(x^2 + 4x) - 3.$$

16. Write a quadratic equation having solutions $\sqrt{13}$ and $-\sqrt{13}$.

17. The surface area of a balloon varies directly as the square of its radius. The area is 3.4 in² when the radius is 5 in. What is the area when the radius is 7 in.?

18. a) Graph:
$$f(x) = 4(x - 3)^2 + 5.$$

b) Label the vertex.
c) Draw the axis of symmetry.
d) Find the maximum or the minimum function value.

19. Given the function $f(x) = 2x^2 + 4x - 6$.

a) Find the vertex and the axis of symmetry.
b) Graph the function.

20. Find the x- and y-intercepts of
$$f(x) = x^2 - x - 6.$$

21. Solve $V = \frac{1}{3}\pi(R^2 + r^2)$ for r.

22. What is the minimum product of two numbers having a difference of 8?

23. Find the quadratic function that fits the data points $(0, 0)$, $(3, 0)$, and $(5, 2)$.

Determine which, if any, of the following sets of graphed data could be modeled by a quadratic function.

24.

25.

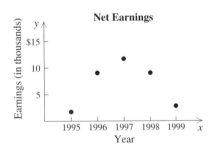

26. *River Depth.* Typically rivers are deeper in the middle, with the depth decreasing to 0 at the edges. A hydrologist measures the depths D, in feet, of a river at distances x, in feet, from one bank. The depths are listed in the following table.

DISTANCE, x, FROM THE RIVERBANK (IN FEET)	DEPTH, D, OF THE RIVER (IN FEET)
0	0
15	10.2
25	17
50	20
90	7.2

a) Graph the data and determine whether a quadratic function might fit the situation.
b) If the data appear to be quadratic, use regression to find a quadratic function $D(x)$ that could be used to estimate the depth of the river x feet from the bank.

27. Solve: $x^2 + 5x \leq 6$.

Skill Maintenance

28. Solve: $\sqrt{x + 3} = x - 3$.

29. Simplify: $\sqrt[4]{2a^2b^3}\,\sqrt[4]{a^4b}$. (Assume $a, b \geq 0$.)

30. Subtract and simplify:
$$\frac{x}{x^2 + 15x + 56} - \frac{7}{x^2 + 13x + 42}.$$

31. The perimeter of a hexagon with all six sides the same length is the same as the perimeter of a square. One side of the hexagon is 3 less than a side of the square. Find the perimeter of each polygon.

Synthesis

32. One solution of $kx^2 + 3x - k = 0$ is -2. Find the other solution.

33. Find a fourth-degree polynomial equation, with integer coefficients, for which $2 - \sqrt{3}$ and $5 - i$ are solutions.

34. Find a polynomial equation, with integer coefficients, for which 5 is a repeated root and $\sqrt{2}$ and $\sqrt{3}$ are solutions.

Exponential and Logarithmic Functions 9

The functions that we consider in this chapter are interesting not only from a purely intellectual point of view, but also for their rich applications to many fields. We will look at applications such as compound interest and population growth, to name just two.

The basis of the theory centers on functions having variable exponents (*exponential functions*). Results follow from those functions and their properties.

APPLICATION

GROWTH OF ZEBRA MUSSEL POPULATIONS. Zebra mussels, inadvertently imported from Europe, began fouling North American waters in 1988. These mussels are so prolific that lake and river bottoms, as well as water intake pipes, can become blanketed with them, altering an entire ecosystem. In 1996, a portion of the Mississippi River contained an average of 1 zebra mussel per square mile. There were 10 per square mile by 1997, 300 by 1998, and 9000 by 1999. Let z = the number of zebra mussels per square mile and x the number of years since 1996. Use regression to fit an exponential function $z(x)$ to the data and graph the function. (*Source:* Based on an interview with Dr. Gerald Mackie of the Department of Zoology at the University of Geulph in Ontario)

This problem appears as Example 8 in Section 9.7.

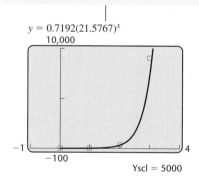

$y = 0.7192(21.5767)^x$

Yscl = 5000

9.1
Exponential Functions

- *Graphing Exponential Functions*
- *Equations with x and y Interchanged*
- *Applications of Exponential Functions*

Consider the graph below. The rapidly rising curve approximates the graph of an *exponential function*. We now consider such functions and some of their applications.

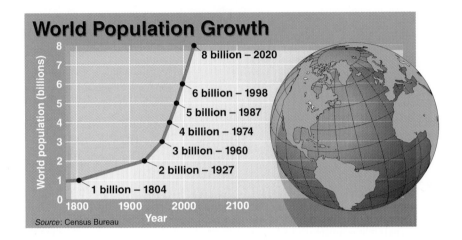

Graphing Exponential Functions

In Chapter 7, we studied exponential expressions with rational-number exponents, such as

$$5^{1/4}, \qquad 3^{-3/4}, \qquad 7^{2.34}, \qquad 5^{1.73}.$$

For example, $5^{1.73}$, or $5^{173/100}$, represents the 100th root of 5 raised to the 173rd power. What about expressions with irrational exponents, such as $5^{\sqrt{3}}$ or $7^{-\pi}$? To attach meaning to $5^{\sqrt{3}}$, consider a rational approximation, r, of $\sqrt{3}$. As r gets closer to $\sqrt{3}$, the value of 5^r gets closer to some real number p.

r closes in on $\sqrt{3}$.	5^r closes in on some real number p.
$1.7 < r < 1.8$	$15.426 \approx 5^{1.7} < p < 5^{1.8} \approx 18.119$
$1.73 < r < 1.74$	$16.189 \approx 5^{1.73} < p < 5^{1.74} \approx 16.452$
$1.732 < r < 1.733$	$16.241 \approx 5^{1.732} < p < 5^{1.733} \approx 16.267$

We define $5^{\sqrt{3}}$ to be the number p. To eight decimal places,

$$5^{\sqrt{3}} \approx 16.24245082.$$

Any positive irrational exponent can be defined in a similar way. Negative irrational exponents are then defined using reciprocals. Thus, so long as a is positive, a^x has meaning for *any* real number x. The general laws of exponents still hold, but we will not prove that here. We now define an *exponential function*.

Exponential Function

The function $f(x) = a^x$, where a is a positive constant, $a \neq 1$, is called the *exponential function*, base a.

We require the base a to be positive to avoid the imaginary numbers that would result from taking even roots of negative numbers. The restriction $a \neq 1$ is made to exclude the constant function $f(x) = 1^x$, or $f(x) = 1$.

The following are examples of exponential functions:

$$f(x) = 2^x, \qquad f(x) = \left(\tfrac{1}{3}\right)^x, \qquad f(x) = 5^{-3x}.$$

> Note that
> $5^{-3x} = (5^{-3})^x$.

In contrast to polynomial functions like $f(x) = x^2$ or $g(x) = 5x^3 - 7$, exponential functions have a variable exponent. Because of this, graphs of exponential functions rise or fall dramatically.

Example 1   Graph the exponential function $y = f(x) = 2^x$.

SOLUTION   We compute some function values, thinking of y as $f(x)$, and list the results in a table. It is a good idea to start by letting $x = 0$.

$$f(0) = 2^0 = 1; \qquad f(-1) = 2^{-1} = \frac{1}{2^1} = \frac{1}{2};$$
$$f(1) = 2^1 = 2;$$
$$f(2) = 2^2 = 4; \qquad f(-2) = 2^{-2} = \frac{1}{2^2} = \frac{1}{4};$$
$$f(3) = 2^3 = 8;$$
$$f(-3) = 2^{-3} = \frac{1}{2^3} = \frac{1}{8}.$$

Next, we plot these points and connect them with a smooth curve.

x	y, or $f(x)$
0	1
1	2
2	4
3	8
-1	$\frac{1}{2}$
-2	$\frac{1}{4}$
-3	$\frac{1}{8}$

The curve comes very close to the x-axis, but does not touch or cross it.

Be sure to plot enough points to determine how steeply the curve rises.

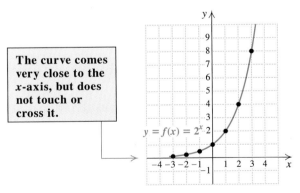

Note that as x increases, the function values increase without bound. As x decreases, the function values decrease, getting very close to 0. The *x*-axis, or the line $y = 0$, is a horizontal *asymptote*, meaning that the curve gets closer and closer to this line the further we move to the left. ▬

Example 2 Graph the exponential function $y = f(x) = \left(\frac{1}{2}\right)^x$.

SOLUTION We compute some function values, thinking of y as $f(x)$, and list the results in a table. Before we do this, note that

$$y = f(x) = \left(\tfrac{1}{2}\right)^x = (2^{-1})^x = 2^{-x}.$$

Then we have

$$f(0) = 2^{-0} = 1;$$

$$f(1) = 2^{-1} = \frac{1}{2^1} = \frac{1}{2};$$

$$f(2) = 2^{-2} = \frac{1}{2^2} = \frac{1}{4};$$

$$f(3) = 2^{-3} = \frac{1}{2^3} = \frac{1}{8};$$

$$f(-1) = 2^{-(-1)} = 2^1 = 2;$$

$$f(-2) = 2^{-(-2)} = 2^2 = 4;$$

$$f(-3) = 2^{-(-3)} = 2^3 = 8.$$

x	y, or $f(x)$
0	1
1	$\frac{1}{2}$
2	$\frac{1}{4}$
3	$\frac{1}{8}$
-1	2
-2	4
-3	8

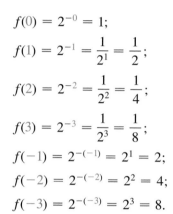

$y = f(x) = \left(\frac{1}{2}\right)^x$

Next, we plot these points and connect them with a smooth curve. Note that this curve is a mirror image, or *reflection*, of the above graph of $y = 2^x$ across the *y*-axis. The line $y = 0$ is again an asymptote. ▬

Interactive Discovery

1. Graph each of the following functions, and determine whether the graph increases or decreases from left to right. How can you tell from the number a in $f(x) = a^x$ whether the graph of f increases or decreases?
   a) $f(x) = 3^x$
   b) $g(x) = 4^x$
   c) $h(x) = 1.5^x$
   d) $r(x) = \left(\frac{1}{3}\right)^x$
   e) $t(x) = 0.75^x$

2. Compare the graphs of f, g, and h. Which curve is steepest?

3. Compare the graphs of r and t. Which curve is steeper?

We can make the following observations.

A. When $a > 1$, the graph of $f(x) = a^x$ increases from left to right. The greater the value of a, the steeper the curve. (See the figure on the left at the top of the next page.)

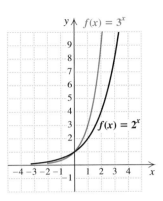
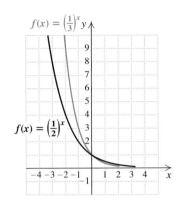

B. When $0 < a < 1$, the graph of $f(x) = a^x$ decreases from left to right. For smaller values of a, the curve becomes steeper. (See the figure on the right, above.)

C. All graphs of $f(x) = a^x$ go through the y-intercept $(0, 1)$.

D. The domain of $f(x) = a^x$ is $(-\infty, \infty)$. The range is $(0, \infty)$.

Example 3 Graph: $y = f(x) = 2^{x-2}$.

SOLUTION We construct a table of values. Then we plot the points and connect them with a smooth curve. Here $x - 2$ is the *exponent*.

$$f(0) = 2^{0-2} = 2^{-2} = \frac{1}{2^2} = \frac{1}{4} \qquad f(-1) = 2^{-1-2} = 2^{-3} = \frac{1}{2^3} = \frac{1}{8}$$

$$f(1) = 2^{1-2} = 2^{-1} = \frac{1}{2^1} = \frac{1}{2} \qquad f(-2) = 2^{-2-2} = 2^{-4} = \frac{1}{2^4} = \frac{1}{16}$$

$$f(2) = 2^{2-2} = 2^0 = 1$$
$$f(3) = 2^{3-2} = 2^1 = 2$$
$$f(4) = 2^{4-2} = 2^2 = 4$$

x	y, or $f(x)$
0	$\frac{1}{4}$
1	$\frac{1}{2}$
2	1
3	2
4	4
-1	$\frac{1}{8}$
-2	$\frac{1}{16}$

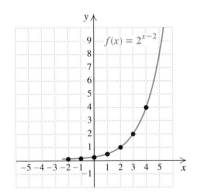

The graph looks just like the graph of $y = 2^x$, but it is translated 2 units to the right. The y-intercept of $y = 2^x$ is $(0, 1)$. The y-intercept of $y = 2^{x-2}$ is $\left(0, \frac{1}{4}\right)$. The line $y = 0$ is again the asymptote. ▬

Equations with x and y Interchanged

It will be helpful in later work to be able to graph an equation in which the x and the y in $y = a^x$ are interchanged.

Example 4   Graph: $x = 2^y$.

SOLUTION   Note that x is alone on one side of the equation. To find ordered pairs that are solutions, we choose values for y and then compute values for x:

For $y = 0$, $x = 2^0 = 1$.

For $y = 1$, $x = 2^1 = 2$.

For $y = 2$, $x = 2^2 = 4$.

For $y = 3$, $x = 2^3 = 8$.

For $y = -1$, $x = 2^{-1} = \dfrac{1}{2^1} = \dfrac{1}{2}$.

For $y = -2$, $x = 2^{-2} = \dfrac{1}{2^2} = \dfrac{1}{4}$.

For $y = -3$, $x = 2^{-3} = \dfrac{1}{2^3} = \dfrac{1}{8}$.

x	y
1	0
2	1
4	2
8	3
$\frac{1}{2}$	-1
$\frac{1}{4}$	-2
$\frac{1}{8}$	-3

(1) Choose values for y.
(2) Compute values for x.

We plot the points and connect them with a smooth curve.

This curve does not touch or cross the y-axis.

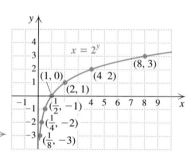

Note too that this curve looks just like the graph of $y = 2^x$, except that it is reflected across the line $y = x$, as shown here.

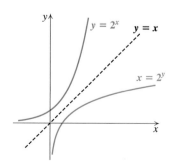

Applications of Exponential Functions

Example 5 *Interest Compounded Annually.* The amount of money A that a principal P will be worth after t years at interest rate i, compounded annually, is given by the formula

$$A = P(1 + i)^t.$$ **You might review Example 11 in Section 8.1.**

Suppose that \$100,000 is invested at 8% interest, compounded annually.

a) Find a function for the amount in the account after t years.

b) Find the amount of money in the account at $t = 0$, $t = 4$, $t = 8$, and $t = 10$.

c) Graph the function.

SOLUTION

a) If $P = \$100,000$ and $i = 8\% = 0.08$, we can substitute these values and form the following function:

$$A(t) = \$100,000(1 + 0.08)^t$$
$$= \$100,000(1.08)^t.$$

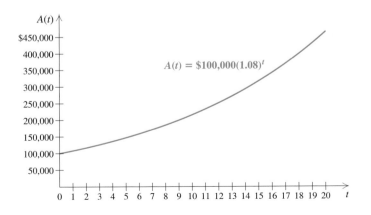

X	Y₁
0	100000
4	**136049**
8	185093
10	215892

Y₁ = 136048.896

b) An efficient way to find the function values using a grapher is to let $y_1 = 100{,}000(1.08)^x$ and use the TABLE feature with Indpnt set to Ask, as shown at left. We highlight a table entry if we want to view the function value to more decimal places than is shown in the table.

Alternatively, we can use $y_1(\ )$ notation to evaluate the function for each value. Using either procedure gives us

$$A(0) = \$100{,}000,$$
$$A(4) \approx \$136{,}048.90,$$
$$A(8) \approx \$185{,}093.02,$$
and $A(10) \approx \$215{,}892.50.$

c) We can use the function values computed in part (b), and others if we wish, to draw the graph by hand. Whether we are graphing by hand or using a grapher, the axes will be scaled differently because of the large numbers.

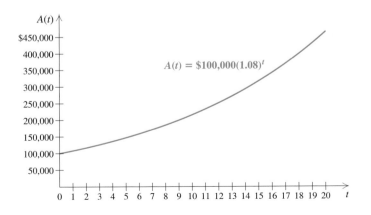

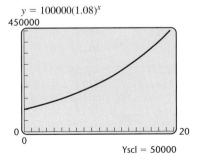

9.1 | Exercise Set

In each of Exercises 1–4 is the graph of a function $f(x) = a^x$. Determine from the graph whether $a > 1$ or $0 < a < 1$.

1.

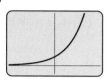

2.

3.

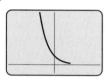

4.

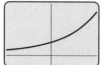

Graph.

5. $y = f(x) = 2^x$ **6.** $y = f(x) = 3^x$

7. $y = 5^x$ **8.** $y = 6^x$

9. $y = 2^{x-1}$ **10.** $y = 2^{x+1}$

11. $y = 3^{x+2}$ **12.** $y = 3^{x-2}$

13. $y = 2^x - 1$ **14.** $y = 2^x + 3$

15. $y = 1.7^x$ **16.** $y = 4.8^x$

17. $y = \left(\frac{1}{2}\right)^x$ **18.** $y = \left(\frac{1}{3}\right)^x$

19. $y = \left(\frac{1}{5}\right)^x$ **20.** $y = \left(\frac{1}{4}\right)^x$

21. $y = 0.15^x$ **22.** $y = 0.98^x$

23. $y = 2^{2x-1}$ **24.** $y = 3^{4-x}$

25. $x = 3^y$ **26.** $x = 6^y$

27. $x = \left(\frac{1}{2}\right)^y$ **28.** $x = \left(\frac{1}{3}\right)^y$

29. $x = 5^y$ **30.** $x = 4^y$

Graph both equations using the same set of axes.

31. $y = 3^x$, $x = 3^y$

32. $y = 2^x$, $x = 2^y$

33. $y = \left(\frac{1}{2}\right)^x$, $x = \left(\frac{1}{2}\right)^y$

34. $y = \left(\frac{1}{4}\right)^x$, $x = \left(\frac{1}{4}\right)^y$

In Exercises 35–40, match each equation with one of the following graphs.

a)

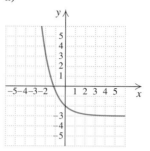

b)

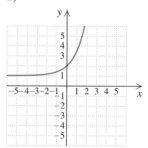

c)

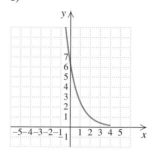

d)

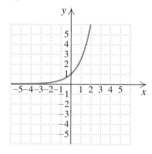

e)

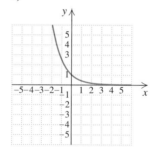

f)

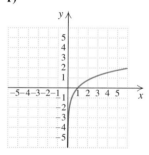

35. $y = \left(\frac{5}{2}\right)^x$ **36.** $y = \left(\frac{2}{5}\right)^x$

37. $x = \left(\frac{2}{5}\right)^y$ **38.** $y = \left(\frac{2}{5}\right)^x - 3$

39. $y = \left(\frac{2}{5}\right)^{x-2}$ **40.** $y = \left(\frac{5}{2}\right)^x + 1$

Solve.

41. *Cases of AIDS.* The total number of Americans who have contracted AIDS, in thousands, can be approximated by the exponential function

$$N(t) = 100(1.4)^t,$$

where $t = 0$ corresponds to 1989.

a) According to the function, how many Americans had been infected as of 1993?

b) Estimate the total number of Americans who had been infected as of 1998.

c) Graph the function.

42. *Growth of Bacteria.* The bacteria *Escherichi coli* are commonly found in the human bladder. Suppose that 3000 of the bacteria are present at time $t = 0$. Then t minutes later, the number of bacteria present will be

$$N(t) = 3000(2)^{t/20}.$$

a) How many bacteria will be present after 10 min? 20 min? 30 min? 40 min? 60 min?

b) Graph the function.

43. *Recycling Aluminum Cans.* It is estimated that $\frac{2}{3}$ of all aluminum cans distributed will be recycled each year. A beverage company distributes 250,000 cans. The number still in use after time t, in years, is given by the exponential function

$$N(t) = 250,000\left(\tfrac{2}{3}\right)^t.$$

a) How many cans are still in use after 0 yr? 1 yr? 4 yr? 10 yr?

b) Graph the function.

44. *Salvage Value.* A photocopier is purchased for $5200. Its value each year is about 80% of the value of the preceding year. Its value, in dollars, after t years is given by the exponential function

$$V(t) = 5200(0.8)^t.$$

a) Find the value of the machine after 0 yr, 1 yr, 2 yr, 5 yr, and 10 yr.

b) Graph the function.

45. *World Demand for Lumber.* The world demand for lumber is increasing exponentially. The amount of timber N, in billions of cubic feet, consumed t years after 1997, can be approximated by the exponential function

$$N(t) = 62(1.018)^t,$$

where $t = 0$ corresponds to 1997.

a) How much timber was consumed in 1997? in 1998?

b) Estimate the amount of timber to be consumed in 2000 and 2010.

c) Graph the function.

46. *Typing Speed.* Ali is studying typing. After he has studied for t hours, Ali's speed, in number of words per minute, is given by the exponential function

$$S(t) = 200[1 - (0.99)^t].$$

a) Predict Ali's speed after he has studied for 10 hr, 40 hr, and 80 hr.

b) Graph the function.

Skill Maintenance

47. Multiply and simplify: $x^{-5} \cdot x^3$. [1.4]

48. Simplify: $(x^{-3})^4$. [1.4]

49. Divide and simplify: $\dfrac{x^{-3}}{x^4}$. [1.4]

50. Simplify: 5^0. [1.4]

Synthesis

51. ◈ Suppose that $1000 is invested for 5 yr at 7% interest, compounded annually. In what year will the most interest be earned? Why?

52. ◈ Without using a calculator, explain why 2^π must be greater than 8 but less than 16.

53. ◈ Consider any exponential function of the form $f(x) = a^x$ with $a > 1$. Will it always follow that $f(3) - f(2) > f(2) - f(1)$, and, in general, $f(n + 2) - f(n + 1) > f(n + 1) - f(n)$? Why or why not? (*Hint:* Think graphically.)

54. ◈ Why was it necessary to discuss irrational exponents before graphing exponential functions?

Determine which of the two numbers is larger.

55. $\pi^{1.3}$ or $\pi^{2.4}$ 　　**56.** $\sqrt{8^3}$ or $8^{\sqrt{3}}$

Graph.

57. $y = 2^x + 2^{-x}$ 　　**58.** $y = \left|\left(\tfrac{1}{2}\right)^x - 1\right|$

59. $y = |2^x - 2|$ 　　**60.** $y = 2^{-(x-1)^2}$

61. $y = |2^{x^2} - 1|$ 　　**62.** $y = 3^x + 3^{-x}$

Graph both equations using the same set of axes.

63. $y = 3^{-(x-1)}, \ x = 3^{-(y-1)}$

64. $y = 1^x, \ x = 1^y$

Graph each function and estimate its range.

65. $f(x) = 2^x - 5$

66. $f(x) = 3^{x-4}$

67. $g(x) = 5^{2-x}$

68. $g(x) = 5^{x+1} + 3$

COLLABORATIVE CORNER

Focus: Car loans and exponential functions

Time: 30 minutes

Group size: 2

Materials: Calculators with exponentiation keys

The formula

$$M = \frac{Pr}{1 - (1 + r)^{-n}}$$

is used to determine the payment size, M, when a loan of P dollars is to be repaid in n equally sized monthly payments. Here r represents the monthly interest rate. Loans repaid in this fashion are said to be *amortized* (spread out equally) over a period of n months.

Activity

1. Suppose that one group member is selling the other a car for $2600, financed at 1% interest per month for 24 months. What should be the size of each monthly payment?

2. Suppose both group members are shopping for the same model new car. To save time, each group member visits a different dealer. One dealer offers the car for $13,000 at 10.5% interest (0.00875 monthly interest) for 60 months (no down payment). The other dealer offers the same car for $12,000, but at 12% interest (0.01 monthly interest) for 48 months (no down payment).

   a) Determine the monthly payment size for each offer (remember to use the *monthly* interest rates). Then determine the total amount paid for the car under each offer. How much of each total is interest?

   b) Work together to find the annual interest rate for which the total cost of 60 monthly payments for the $13,000 car would equal the total amount paid for the $12,000 car (as found in part (a) above).

9.2
Composite and Inverse Functions

- *Composite Functions*
- *Inverses and One-to-One Functions*
- *Finding Formulas for Inverses*
- *Graphing Functions and Their Inverses*
- *Inverse Functions and Composition*

Composite Functions

In the real world, functions frequently occur in which some quantity depends on a variable that, in turn, depends on another variable. For instance, the number of employees hired by a firm may depend on the firm's profits, which may in turn depend on the number of items the firm produces. Functions like this are called **composite functions**.

For example, the function *g* that gives a correspondence between women's shoe sizes in the United States and those in Italy is given by $g(x) = 2x + 24$, where *x* is the U.S. size and $g(x)$ is the Italian size. Thus a U.S. size 4 corresponds to a shoe size of $g(4) = 2 \cdot 4 + 24$, or 32, in Italy.

There is also a function that gives a correspondence between women's shoe sizes in Italy and those in Britain. The function is given by $f(x) = \frac{1}{2}x - 14$, where *x* is the Italian size and $f(x)$ is the corresponding British size. Thus an Italian size 32 corresponds to a British size $f(32) = \frac{1}{2} \cdot 32 - 14$, or 2.

It seems reasonable to conclude that a shoe size of 4 in the United States corresponds to a size of 2 in Britain and that some function *h* describes this correspondence. Can we find a formula for *h*? If we look at the following tables, we might guess that such a formula is $h(x) = x - 2$, and that is indeed correct. But, for more complicated formulas, we would need to use algebra.

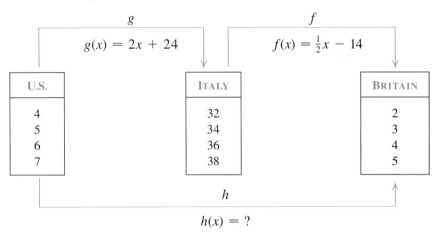

$$h(x) = ?$$

Size *x* shoes in the United States correspond to size $g(x)$ shoes in Italy, where

$$g(x) = 2x + 24.$$

Size *n* shoes in Italy correspond to size $f(n)$ shoes in Britain. Thus size $g(x)$ shoes in Italy correspond to size $f(g(x))$ shoes in Britain. Since the *x* in the expression $f(g(x))$ represents a U.S. shoe size, we can find the British shoe size that corresponds to a U.S. size *x* as follows:

$$f(g(x)) = f(2x + 24) = \frac{1}{2} \cdot (2x + 24) - 14 \qquad \text{Using } g(x) \text{ as an input}$$

$$= x + 12 - 14 = x - 2.$$

This gives a formula for *h*: $h(x) = x - 2$. Thus a shoe size of 4 in the United States corresponds to a shoe size of $h(4) = 4 - 2$, or 2, in Britain. The function *h* is called the *composition* of *f* and *g* and is denoted $f \circ g$ (read "*f* composed with *g*" or "*f* circle *g*").

> **Composition of Functions**
>
> The *composite function f ∘ g*, the *composition* of *f* and *g*, is defined as
>
> $$f \circ g(x) = f(g(x)).$$

Example 1 Given $f(x) = 3x$ and $g(x) = 1 + x^2$.

a) Find $f \circ g(5)$ and $g \circ f(5)$.

b) Find $f \circ g(x)$ and $g \circ f(x)$.

SOLUTION   Consider each function separately:

$$f(x) = 3x \qquad \text{This function multiplies each input by 3.}$$

and

$$g(x) = 1 + x^2. \qquad \text{This function squares an input and then adds 1.}$$

a) To find $f \circ g(5)$, we first find $g(5)$ by substituting in the formula for g: Square 5 and add 1, to get 26. We then use 26 as an input for f:

$$\begin{aligned} f \circ g(5) = f(g(5)) &= f(1 + 5^2) \\ &= f(26) = 3 \cdot 26 = 78. \end{aligned}$$

To find $g \circ f(5)$, we first find $f(5)$ by substituting into the formula for f: Multiply 5 by 3, to get 15. We then use 15 as an input for g:

$$\begin{aligned} g \circ f(5) = g(f(5)) &= g(3 \cdot 5) \qquad \text{Note that } f(5) = 3 \cdot 5 = 15. \\ &= g(15) = 1 + 15^2 = 1 + 225 = 226. \end{aligned}$$

b) We find $f \circ g(x)$ by substituting $g(x)$ for x in the equation for $f(x)$:

$$\begin{aligned} f \circ g(x) = f(g(x)) &= f(1 + x^2) \qquad \text{Substituting } 1 + x^2 \text{ for } g(x) \\ &= 3(1 + x^2) = 3 + 3x^2. \qquad \textit{These} \text{ parentheses indicate multiplication.} \end{aligned}$$

To find $g \circ f(x)$, we substitute $f(x)$ for x in the equation for $g(x)$:

$$\begin{aligned} g \circ f(x) = g(f(x)) &= g(3x) \qquad \text{Substituting } 3x \text{ for } f(x) \\ &= 1 + (3x)^2 = 1 + 9x^2. \end{aligned}$$

As a check, note that $g \circ f(5) = 1 + 9 \cdot 5^2 = 1 + 9 \cdot 25 = 226$, as we expected from part (a) above.   ▬

Example 1 shows that, in general, $f \circ g(5) \neq g \circ f(5)$ and $f \circ g(x) \neq g \circ f(x)$.

When we enter $y_2(y_1)$ on a grapher, we are finding the composition $y_2 \circ y_1$. We can then use graphs and tables to check a formula for a composition of functions.

Example 2 Given $f(x) = \sqrt{x}$ and $g(x) = x - 1$, find $f \circ g(x)$ and $g \circ f(x)$.

X	Y3	Y4
1	0	0
1.5	.70711	.70711
2	1	1
2.5	1.2247	1.2247
3	1.4142	1.4142
3.5	1.5811	1.5811
4	1.7321	1.7321
X = 1		

X	Y5	Y6
1	0	0
1.5	.22474	.22474
2	.41421	.41421
2.5	.58114	.58114
3	.73205	.73205
3.5	.87083	.87083
4	1	1
X = 1		

SOLUTION We have

$$f \circ g(x) = f(g(x)) = f(x - 1) = \sqrt{x - 1};$$
$$g \circ f(x) = g(f(x)) = g(\sqrt{x}) = \sqrt{x} - 1.$$

To check using a grapher, we let $y_1 = \sqrt{}(x)$ and $y_2 = x - 1$. Then $f \circ g = y_1(y_2)$ and $g \circ f = y_2(y_1)$. Next, we let $y_3 = \sqrt{}(x - 1)$, the expression we obtained for $f \circ g$, and $y_4 = y_1(y_2)$. The first table at left indicates that $y_3 = y_4$ and that our work is correct.

Similarly, to check that $g \circ f(x) = \sqrt{x} - 1$, we let $y_5 = \sqrt{}(x) - 1$ and $y_6 = y_2(y_1)$. The second table indicates that $y_5 = y_6$. ▬

In some applications, one needs to recognize how a function can be regarded as the composition of two "simpler" functions.

Example 3 If $h(x) = (7x + 3)^2$, find $f(x)$ and $g(x)$ such that $h(x) = f \circ g(x)$.

SOLUTION To find $h(x)$, we can think of two steps: forming $7x + 3$ and then squaring. This suggests that $g(x) = 7x + 3$ and $f(x) = x^2$. We check by forming the composition:

$$h(x) = f \circ g(x) = f(g(x))$$
$$= f(7x + 3) = (7x + 3)^2.$$

This is probably the most "obvious" answer to the question. There can be other less obvious answers. For example, if

$$f(x) = (x - 1)^2$$

and $\quad g(x) = 7x + 4,$

then $\quad h(x) = f \circ g(x) = f(g(x)) = f(7x + 4)$
$$= (7x + 4 - 1)^2 = (7x + 3)^2.$$ ▬

Inverses and One-to-One Functions

Let's consider the following two functions. We think of them as relations, or correspondences.

Cost of a 60-Second Super Bowl Commercial, by Year

DOMAIN (SET OF INPUTS)	RANGE (SET OF OUTPUTS)
1981 ⟶	$550,000
1983 ⟶	$800,000
1988 ⟶	$1,350,000
1991 ⟶	$1,600,000
1995 ⟶	$2,200,000
1997 ⟶	$2,400,000

U.S. Senators and Their States

DOMAIN (SET OF INPUTS)	RANGE (SET OF OUTPUTS)
Wellstone ⟶	Minnesota
Grams	
Mack ⟶	Florida
Graham	
Lautenberg ⟶	New Jersey
Torricelli	

Suppose we reverse the arrows. We obtain what is called the **inverse relation**. Are these inverse relations functions?

Cost of a 60-Second Super Bowl Commercial, by Year

RANGE (SET OF OUTPUTS)	DOMAIN (SET OF INPUTS)
1981 ⟵	$550,000
1983 ⟵	$800,000
1988 ⟵	$1,350,000
1991 ⟵	$1,600,000
1995 ⟵	$2,200,000
1997 ⟵	$2,400,000

U.S. Senators and Their States

RANGE (SET OF OUTPUTS)	DOMAIN (SET OF INPUTS)
Wellstone ⟵	Minnesota
Grams ⟵	
Mack ⟵	Florida
Graham ⟵	
Lautenberg ⟵	New Jersey
Torricelli ⟵	

We see that the inverse of the first correspondence is a function, but that the inverse of the second correspondence is not a function.

Recall that for each input, a function provides exactly one output. However, a function can have the same output for two or more different inputs. Thus it is possible for different inputs to correspond to the same output. Only when this possibility is *excluded* will the inverse be a function.

In the Super Bowl function, different inputs have different outputs. It is an example of a **one-to-one function**. In the U.S. Senator function, *Wellstone* and *Grams* are both paired with *Minnesota*. Thus the U.S. Senator function is not one-to-one.

One-to-One Function

A function f is *one-to-one* if different inputs have different outputs. That is, if for any $a \neq b$, we have $f(a) \neq f(b)$, the function f is one-to-one. If a function is one-to-one, then its inverse correspondence is also a function.

Recall that a graph represents a function if it passes the vertical-line test. Thus if the graph of the inverse of a function passes the vertical-line test, the inverse is also a function. How can we tell from the graph of a function, without drawing its inverse, whether it is one-to-one and the inverse correspondence is a function?

Example 4   Shown here is the graph of an exponential function. Determine whether the function is one-to-one and thus has an inverse that is a function.

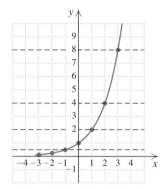

SOLUTION   A function is one-to-one if different inputs have different outputs. In other words, no two x-values will have the same y-value. For this function, we cannot find two x-values that have the same y-value.

Equivalently, no horizontal line can be drawn so that it crosses the graph more than once. The function is one-to-one so its inverse is a function.

The Horizontal-Line Test

A function is one-to-one, and thus has an inverse that is a function, if no horizontal line can cross its graph more than once.

Example 5 Determine whether the function $f(x) = x^2$ is one-to-one and thus has an inverse that is a function.

SOLUTION The graph of $f(x) = x^2$ is shown here. Many horizontal lines cross the graph more than once—in particular, the line $y = 4$. Note that where the line crosses, the first coordinates are -2 and 2. Although these are different inputs, they have the same output. That is, $-2 \neq 2$, but

$$f(-2) = (-2)^2 = 4 = 2^2 = f(2).$$

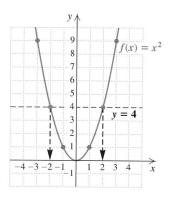

$y_1 = x^2$

X	Y1	
-2	4	
2	4	

X = -2

The table at left also shows that $f(-2) = f(2)$. Thus the function is not one-to-one and no inverse function exists.

Finding Formulas for Inverses

When the inverse of f is also a function, it is denoted f^{-1} (read "f-inverse").

Caution! The -1 in f^{-1} is *not* an exponent!

Suppose that a function is described by a formula. If it has an inverse that is a function, how do we find a formula for the inverse? For any equation in two variables, if we interchange the variables, we obtain an equation of the inverse correspondence. If it is a function, we proceed as follows to find a formula for f^{-1}.

To Find a Formula for f^{-1}:

First make sure that f is one-to-one. Then:

1. Replace $f(x)$ with y.
2. Interchange x and y. (This gives the inverse function.)
3. Solve for y.
4. Replace y with $f^{-1}(x)$. (This is inverse function notation.)

Example 6   Determine whether each function is one-to-one and if it is, find a formula for $f^{-1}(x)$: **(a)** $f(x) = x + 2$; **(b)** $f(x) = 2x - 3$.

SOLUTION

a) The graph of $f(x) = x + 2$ is shown below. It passes the horizontal-line test, so it is one-to-one. Thus its inverse is a function.

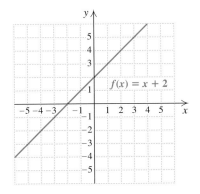

1. Replace $f(x)$ with y: $y = x + 2$.
2. Interchange x and y: $x = y + 2$.   **This gives the inverse function.**

3. Solve for y: $x - 2 = y$.
4. Replace y with $f^{-1}(x)$: $f^{-1}(x) = x - 2$.   **We also "reversed" the sides of the equation.**

In this case, the function f added 2 to all inputs. Thus, to "undo" f, the function f^{-1} must subtract 2 from its inputs.

b) The function $f(x) = 2x - 3$ is also linear. Any linear function that is not constant will pass the horizontal-line test. Thus, f is one-to-one.

1. Replace $f(x)$ with y: $y = 2x - 3$.
2. Interchange x and y: $x = 2y - 3$.
3. Solve for y: $x + 3 = 2y$

$$\frac{x + 3}{2} = y.$$

4. Replace y with $f^{-1}(x)$: $f^{-1}(x) = \dfrac{x+3}{2}$.   ▬

Let's consider inverses of functions in terms of a function machine. Suppose that a one-to-one function f is programmed into a machine. If the machine has a reverse switch, when the switch is thrown, the machine performs the inverse function f^{-1}. Inputs then enter at the opposite end, and the entire process is reversed.

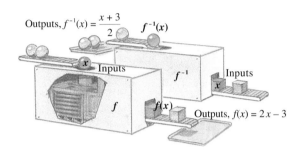

Consider $f(x) = 2x - 3$ and $f^{-1}(x) = \dfrac{x+3}{2}$ from Example 6(b). For the input 5,

$$f(5) = 2 \cdot 5 - 3$$
$$= 10 - 3 = 7.$$

The output is 7. Now we use 7 for the input in the inverse:

$$f^{-1}(7) = \frac{7+3}{2} = \frac{10}{2} = 5.$$

The function f takes 5 to 7. The inverse function f^{-1} takes the number 7 back to 5.

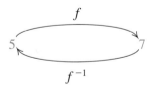

Graphing Functions and Their Inverses

How do the graphs of a function and its inverse compare?

Example 7   Graph $f(x) = 2x - 3$ and $f^{-1}(x) = (x + 3)/2$ on the same set of axes. Then compare.

SOLUTION   The graph of each function follows. Note that the graph of f^{-1} can be drawn by reflecting the graph of f across the line $y = x$. That is, if we graph $f(x) = 2x - 3$ in wet ink and fold the paper along the line

$y = x$, the graph of $f^{-1}(x) = (x + 3)/2$ will appear as the impression made by f.

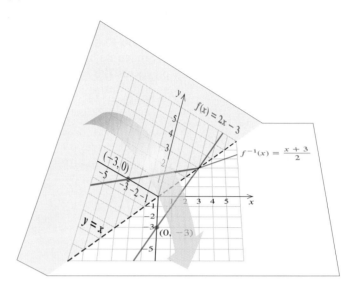

When x and y are interchanged to find a formula for the inverse, we are, in effect, flipping the graph of $f(x) = 2x - 3$ over the line $y = x$. For example, when the coordinates of the y-intercept of the graph of f, $(0, -3)$, are reversed, we get the x-intercept of the graph of f^{-1}, $(-3, 0)$.

Visualizing Inverses

The graph of f^{-1} is a reflection of the graph of f across the line $y = x$.

Example 8 Consider $g(x) = x^3 + 2$.

a) Determine whether the function is one-to-one.

b) If it is one-to-one, find a formula for its inverse.

c) Graph the inverse, if it exists.

SOLUTION

a) The graph of $g(x) = x^3 + 2$ is shown to the right. It passes the horizontal-line test and thus has an inverse.

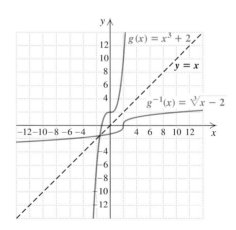

b) **1.** Replace $g(x)$ with y: $\qquad y = x^3 + 2.$

2. Interchange x and y: $\qquad x = y^3 + 2.$

3. Solve for y: $\qquad x - 2 = y^3$

$\qquad\qquad\qquad\qquad\qquad \sqrt[3]{x - 2} = y.$ **Since a number has only one cube root, we can solve for y.**

4. Replace y with $g^{-1}(x)$: $\qquad g^{-1}(x) = \sqrt[3]{x - 2}.$

c) To find the graph, we reflect the graph of $g(x) = x^3 + 2$ across the line $y = x$, as we did in Example 7. It can also be found by substituting into $g^{-1}(x) = \sqrt[3]{x - 2}$ and plotting points. The graphs of g and g^{-1} are shown together in the figure at the bottom of page 596.

We can check the graph of $g^{-1}(x)$ using a grapher by graphing $g(x)$ and drawing its inverse. In order to compare the graphs, we may need to adjust the viewing window so that the units shown on both axes are the same length. This is called *squaring* a viewing window. Many graphers have a Zsquare option in the ZOOM menu that will automatically square the viewing window.

To graph g^{-1}, we use the DrawInv option of the DRAW menu. The graph of $g(x) = x^3 + 2$ and its inverse are shown at right in a viewing window that was squared from a standard viewing window. The line $y = x$ is also shown.

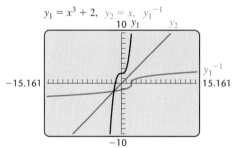

We could also verify that $g^{-1}(x) = \sqrt[3]{x - 2}$ by graphing $y = \sqrt[3]{x - 2}$ and comparing its graph with the inverse of g drawn by the grapher.

Inverse Functions and Composition

Suppose that we use some input x for the function f and find its output, $f(x)$. The function f^{-1} will then take that output back to x. Similarly, if we begin with an input x for the function f^{-1} and find its output, $f^{-1}(x)$, the original function f will then take that output back to x. This is summarized as follows.

Composition and Inverses

If a function f is one-to-one, then f^{-1} is the unique function for which

$$f^{-1} \circ f(x) = x \quad \text{and} \quad f \circ f^{-1}(x) = x.$$

Example 9 Let $f(x) = 2x + 1$. Show that

$$f^{-1}(x) = \frac{x-1}{2}.$$

SOLUTION   We find $f^{-1} \circ f(x)$ and $f \circ f^{-1}(x)$ and check to see that each is x.

$$f^{-1} \circ f(x) = f^{-1}(f(x)) = f^{-1}(2x+1)$$

$$= \frac{(2x+1)-1}{2}$$

$$= \frac{2x}{2} = x$$

$$f \circ f^{-1}(x) = f(f^{-1}(x)) = f\left(\frac{x-1}{2}\right)$$

$$= 2 \cdot \frac{x-1}{2} + 1$$

$$= x - 1 + 1 = x$$

To check using a grapher, we let $y_1 = 2x + 1$, $y_2 = (x-1)/2$, $y_3 = y_1(y_2)$, and $y_4 = y_2(y_1)$. If y_2 is the inverse of y_1, then $y_3 = x$ and $y_4 = x$. The table at left shows that this is true.   ▬

X	Y3	Y4
1	1	1
1.5	1.5	1.5
2	2	2
2.5	2.5	2.5
3	3	3
3.5	3.5	3.5
4	4	4

X = 1

9.2 Exercise Set

Find $f \circ g(x)$ and $g \circ f(x)$.

1. $f(x) = 3x^2 - 1$; $g(x) = 2x + 3$
2. $f(x) = 4x + 3$; $g(x) = 2x^2 - 5$
3. $f(x) = 4x^2 - 1$; $g(x) = 2/x$
4. $f(x) = 3/x$; $g(x) = 2x^2 + 3$
5. $f(x) = x^2 - 3$; $g(x) = x^2 + 1$
6. $f(x) = 1/x^2$; $g(x) = x + 2$

Find $f(x)$ and $g(x)$ such that $h(x) = f \circ g(x)$. Answers may vary.

7. $h(x) = (7 - 5x)^2$
8. $h(x) = 4(3x - 1)^2 + 9$
9. $h(x) = (3x^2 - 7)^5$
10. $h(x) = \sqrt{5x + 2}$
11. $h(x) = \dfrac{2}{x - 3}$
12. $h(x) = \dfrac{3}{x} + 4$
13. $h(x) = \dfrac{1}{\sqrt{7x + 2}}$
14. $h(x) = \sqrt{x - 7} - 3$

15. $h(x) = \dfrac{x^3 + 1}{x^3 - 1}$
16. $h(x) = (\sqrt{x} + 5)^4$

Determine whether each function is one-to-one.

17. $f(x) = x - 5$
18. $f(x) = 5 - 2x$
19. $f(x) = x^2 + 1$
20. $f(x) = 1 - x^2$
21. $g(x) = 3^x$
22. $g(x) = \left(\frac{1}{2}\right)^x$
23. $g(x) = |x|$
24. $h(x) = |x| - 1$

*For each function, **(a)** determine whether it is one-to-one and **(b)** if it is one-to-one, find a formula for the inverse.*

25. $f(x) = x + 6$
26. $f(x) = x + 7$
27. $f(x) = 3 - x$
28. $f(x) = 9 - x$
29. $g(x) = x - 5$
30. $g(x) = x - 8$
31. $f(x) = 4x$
32. $f(x) = 7x$
33. $g(x) = 4x + 3$
34. $g(x) = 4x + 7$
35. $h(x) = 5$
36. $h(x) = -2$
37. $f(x) = \dfrac{1}{x}$
38. $f(x) = \dfrac{3}{x}$
39. $f(x) = \dfrac{2x + 1}{3}$
40. $f(x) = \dfrac{3x + 2}{5}$

41. $f(x) = x^3 - 5$ **42.** $f(x) = x^3 + 2$

43. $g(x) = (x - 2)^3$ **44.** $g(x) = (x + 7)^3$

45. $f(x) = \sqrt{x}$ **46.** $f(x) = \sqrt{x - 1}$

47. $f(x) = 2x^2 + 1, \ x \geq 0$

48. $f(x) = 3x^2 - 2, \ x \geq 0$

Graph each function and its inverse using the same set of axes.

49. $f(x) = \frac{1}{3}x - 2$ **50.** $g(x) = x + 4$

51. $f(x) = x^3$ **52.** $f(x) = x^3 - 1$

53. $y = 2^x$ **54.** $y = 3^x$

55. $y = \left(\frac{2}{3}\right)^x$ **56.** $y = \left(\frac{1}{2}\right)^x$

57. $f(x) = 3 - x^2, \ x \geq 0$

58. $f(x) = x^2 - 1, \ x \leq 0$

In Exercises 59–62, use a grapher to help determine whether or not the given functions are inverses of each other.

59. $f(x) = 0.75x^2 + 2; \ g(x) = \sqrt{\dfrac{4(x - 2)}{3}}$

60. $f(x) = 1.4x^3 + 3.2; \ g(x) = \sqrt[3]{\dfrac{x - 3.2}{1.4}}$

61. $f(x) = \sqrt{2.5x + 9.25};$
$g(x) = 0.4x^2 - 3.7, \ x \geq 0$

62. $f(x) = 0.8x^{1/2} + 5.23;$
$g(x) = 1.25(x^2 - 5.23), \ x \geq 0$

○ *In Exercises 63 and 64, match the graph of each function in Column A with the graph of its inverse in Column B.*

63.

Column A   Column B

(1)   A.

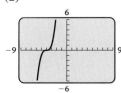   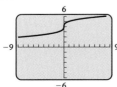

(2)   B.

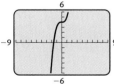   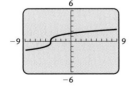

Column A   Column B

(3)   C.

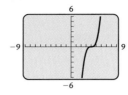   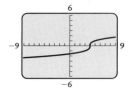

(4)   D.

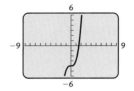   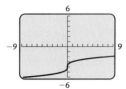

64.

Column A   Column B

(1)   A.

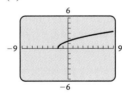   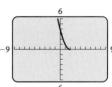

(2)   B.

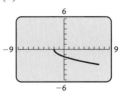   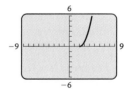

(3)   C.

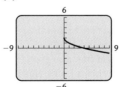   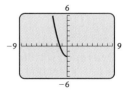

(4)   D.

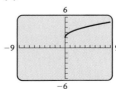   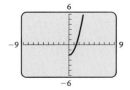

65. Let $f(x) = \frac{4}{5}x$. Show that
$$f^{-1}(x) = \frac{5}{4}x.$$

66. Let $f(x) = (x + 7)/3$. Show that
$$f^{-1}(x) = 3x - 7.$$

67. Let $f(x) = (1 - x)/x$. Show that
$$f^{-1}(x) = \frac{1}{x + 1}.$$

68. Let $f(x) = x^3 - 5$. Show that
$$f^{-1}(x) = \sqrt[3]{x + 5}.$$

69. *Dress Sizes in the United States and France.* A size-6 dress in the United States is size 38 in France. A function that converts dress sizes in the United States to those in France is
$$f(x) = x + 32.$$

a) Find the dress sizes in France that correspond to sizes 8, 10, 14, and 18 in the United States.

b) Determine whether this function has an inverse that is a function. If so, find a formula for the inverse.

c) Use the inverse function to find dress sizes in the United States that correspond to sizes 40, 42, 46, and 50 in France.

70. *Dress Sizes in the United States and Italy.* A size-6 dress in the United States is size 36 in Italy. A function that converts dress sizes in the United States to those in Italy is
$$f(x) = 2(x + 12).$$

a) Find the dress sizes in Italy that correspond to sizes 8, 10, 14, and 18 in the United States.

b) Determine whether this function has an inverse that is a function. If so, find a formula for the inverse.

c) Use the inverse function to find dress sizes in the United States that correspond to sizes 40, 44, 52, and 60 in Italy.

Skill Maintenance

71. Find an equation of variation if y varies directly as x, and $y = 7.2$ when $x = 0.8$. [8.5]

72. Find an equation of variation if y varies inversely as x, and $y = 3.5$ when $x = 6.1$. [8.5]

Simplify. [1.4]

73. $(a^3b^2)^5(a^2b^7)$

74. $(x^5y^3z^2)(x^2yz^2)^3$

Synthesis

75. ◈ Mathematicians usually try to select "logical" words when forming definitions. Does the term "one-to-one" seem logical? Why or why not?

76. ◈ Does the constant function $f(x) = 4$ have an inverse that is a function? If so, find a formula. If not, explain why.

77. ◈ An organization determines that the cost per person of chartering a bus is given by the function
$$C(x) = \frac{100 + 5x}{x},$$
where x is the number of people in the group and $C(x)$ is in dollars. Determine $C^{-1}(x)$ and explain how this inverse function could be used.

78. ◈ The function $V(t) = 750(1.2)^t$ is used to predict the value, $V(t)$, of a certain rare stamp t years from 1997. Do not calculate $V^{-1}(t)$, but explain how V^{-1} could be used.

79. *Dress Sizes in France and Italy.* Use the information in Exercises 69 and 70 to find a function for the French dress size that corresponds to a size x dress in Italy.

80. ◈ Is function composition associative? That is, is it true that for any choices of f, g, and h,
$$f \circ (g \circ h) = (f \circ g) \circ h?$$
Why or why not?

81. Graph each function and draw its inverse. Estimate the domains and the ranges of the function and its inverse. How do they compare?

a) $f(x) = 2^x$

b) $g(x) = \sqrt{x + 1}$

c) $h(x) = \dfrac{1}{x - 2}$

9.3
Logarithmic Functions

- *Graphs of Logarithmic Functions*
- *Common Logarithms*
- *Converting Exponential and Logarithmic Equations*
- *Solving Certain Logarithmic Equations*

We are now ready to study inverses of exponential functions. These functions have many applications and are referred to as *logarithm*, or *logarithmic*, *functions*.

Graphs of Logarithmic Functions

Consider the exponential function $f(x) = 2^x$. Like all exponential functions, f is one-to-one. Can a formula for f^{-1} be found?

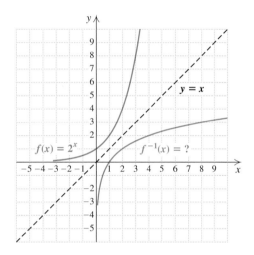

To answer this, we use the method of Section 9.2:

1. Replace $f(x)$ with y: $\quad\quad y = 2^x$.

2. Interchange x and y: $\quad\quad x = 2^y$.

3. Solve for y: $\quad\quad\quad\quad\quad y =$ the power to which we raise 2 to get x.

4. Replace y with $f^{-1}(x)$: $\quad f^{-1}(x) =$ the power to which we raise 2 to get x.

Note that we cannot use any of our current algebraic techniques to solve for y. Thus we now define a new symbol to replace the words "the power to which we raise 2 to get x":

> $\log_2 (x)$, **read "the logarithm, base 2, of x," or "log, base 2, of x," means "the power to which we raise 2 to get x."**

Thus if $f(x) = 2^x$, then $f^{-1}(x) = \log_2 (x)$. Note that $f^{-1}(8) = \log_2 (8) = 3$, because 3 is *the power to which we raise 2 to get* 8. In practice, we often omit the parentheses in a logarithmic expression if the meaning is clear.

For example, we generally write $\log_2 x$ for $\log_2 (x)$. Many graphers, however, require the insertion of the parentheses.

Although expressions like $\log_2 13$ can be only approximated, we must remember that $\log_2 13$ represents *the power to which we raise 2 to get* 13. That is, $2^{\log_2 13} = 13$.

For any exponential function $f(x) = a^x$, the inverse is called a **logarithmic function, base a.** The graph of the inverse can, of course, be drawn by reflecting the graph of $f(x) = a^x$ across the line $y = x$. It will be helpful to remember that the inverse of $f(x) = a^x$ is given by $f^{-1}(x) = \log_a x$. Normally, we use a number a that is greater than 1 for the logarithm base.

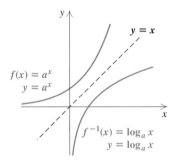

The Meaning of $\log_a x$

For $x > 0$ and a a positive constant other than 1, $\log_a x$ is the number to which a is raised to get x. Thus,

$$a^{\log_a x} = x \qquad \text{or equivalently,} \qquad \text{if } y = \log_a x, \text{ then } a^y = x.$$

It is important to remember that *the logarithm of a number is an exponent*. It might help to repeat to yourself several times: "The logarithm, base a, of a number x is the power to which a must be raised in order to get x."

Example 1   Simplify: **(a)** $\log_{10} 1000$; **(b)** $7^{\log_7 13}$; **(c)** $\log_4 1$.

Solution

a) Think of the meaning of $\log_{10} 1000$. It is the exponent to which we raise 10 to get 1000. That exponent is 3. Therefore, $\log_{10} 1000 = 3$.

b) It is important to remember what $\log_7 13$ is:

$\log_7 13$ is the power to which 7 is raised to get 13.

Thus, since $\log_7 13$ is the power to which 7 is raised to get 13,

$7^{\log_7 13} = 13$.

c) We ask ourselves: "To what power do we raise 4 in order to get 1?" That power is 0 (recall that $4^0 = 1$). Thus, $\log_4 1 = 0$.   ▬

The following is a comparison of exponential and logarithmic functions.

EXPONENTIAL FUNCTION	LOGARITHMIC FUNCTION
$y = a^x$ $f(x) = a^x$ $a > 0, a \neq 1$ The y-intercept is $(0, 1)$. The input x can be any real number. The domain is $(-\infty, \infty)$. The range is $(0, \infty)$. $y > 0$ (Outputs are positive.)	$x = a^y$ $f(x) = \log_a x$ $a > 0, a \neq 1$ The x-intercept is $(1, 0)$. The output y can be any real number. The range is $(-\infty, \infty)$. The domain is $(0, \infty)$. $x > 0$ (Inputs are positive.)

Example 2 Graph: $y = f(x) = \log_5 x$.

SOLUTION   The equation $y = \log_5 x$ is equivalent to $5^y = x$. We can find ordered pairs that are solutions by choosing values for y and computing the x-values.

For $y = 0$, $x = 5^0 = 1$.
For $y = 1$, $x = 5^1 = 5$.
For $y = 2$, $x = 5^2 = 25$.
For $y = -1$, $x = 5^{-1} = \frac{1}{5}$.
For $y = -2$, $x = 5^{-2} = \frac{1}{25}$.

(1) Select y.
(2) Compute x.

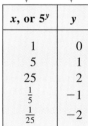

x, or 5^y	y
1	0
5	1
25	2
$\frac{1}{5}$	-1
$\frac{1}{25}$	-2

This table shows the following:

$\log_5 1 = 0;$
$\log_5 5 = 1;$
$\log_5 25 = 2;$
$\log_5 \frac{1}{5} = -1;$
$\log_5 \frac{1}{25} = -2.$

These can all be checked using the equations above.

We plot the set of ordered pairs and connect the points with a smooth curve. The graph of $y = 5^x$ is shown only for reference.

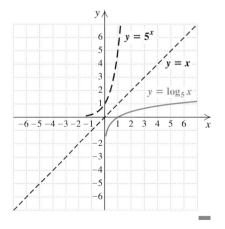

Common Logarithms

Some logarithm bases are easier to use than others. Base-10 logarithms, called **common logarithms**, are useful because they have the same base as our "commonly" used decimal system. Before calculators became so widely available, common logarithms were helpful when performing tedious calculations. In fact, that is why logarithms were invented.

The abbreviation **log**, with no base written, is understood to mean logarithm base 10, or a common logarithm. Thus,

$$\log 17 \quad \text{means} \quad \log_{10} 17.$$

$y_1 = \log(x)$

X	Y1
11.25	1.0512
11.26	1.0515
11.27	1.0519
11.28	1.0523
11.29	1.0527
11.3	1.0531
11.31	1.0535

X = 11.25

Before the advent of calculators, tables were developed to list common logarithms. A portion of such a table is shown at left. Today we approximate common logarithms using calculators. Logarithms with other bases are discussed in Section 9.5.

Example 3 Find log 828.

SOLUTION   We press log , then enter 828 enclosed in parentheses, and then press ENTER , as shown at left. We find that

$$\log 828 \approx 2.9180. \qquad \text{Rounded to four decimal places}$$

log(828)
    2.918030337

We can partially check this by noting that since

$$\log 100 = \log_{10} 100 = 2$$

and $\log 1000 = \log_{10} 1000 = 3,$

we expect log 828 to be between 2 and 3. Since $2 < 2.9180 < 3$, our answer is reasonable.

The inverse of a logarithmic function is an exponential function. Because of this, on many calculators the log key doubles as the 10ˣ key after a 2nd key has been pressed.

Example 4 We find log 0.372 and check by raising 10 to that power.

SOLUTION   We press log , then enter 0.372 in parentheses, and then press ENTER , as shown at left. We find that

$$\log 0.372 \approx -0.4295. \qquad \text{Rounded to four decimal places}$$

log(.372)
    −.4294570601
10^(−.4295)
    .3719632212

To check, we press 10ˣ , followed by −0.4295 and ENTER , which gives us

$$10^{-0.4295} \approx 0.372. \qquad \begin{array}{l}\text{The answer is approximate because}\\ \text{we rounded the logarithm.}\end{array}$$

Example 5 Find $10^{3.417}$ and check by finding the common logarithm of the answer.

SOLUTION   We press 10ˣ , followed by 3.417 in parentheses and ENTER , which gives us

$$10^{3.417} \approx 2612.1614. \qquad \text{Rounded to four decimal places}$$

```
10^(3.417)
              2612.161354
log(2612.1614)
              3.417000008
```

Recall that $\log_{10} x$ is the exponent to which we raise 10 to get x. Thus, if our answer is correct, we should have $\log_{10} 2612.1614 \approx 3.417$. Using a calculator as shown at left, we find that

$$\log 2612.1614 \approx 3.417.$$

Our answer checks.

We can use a grapher to show that $f(x) = \log_{10} x$ and $g(x) = 10^x$ are inverses of each other. If g is the inverse of f, then

$$g[f(x)] = g[\log x] = 10^{\log x} = x.$$

We let $y_1 = \log_{10} x = \log x$ and $y_2 = 10^{\wedge} y_1$ and examine a table of values. The table on the left below shows that $y_2 = x$, as expected.

X	Y1	Y2
1	0	1
1.1	.04139	1.1
1.2	.07918	1.2
1.3	.11394	1.3
1.4	.14613	1.4
1.5	.17609	1.5
1.6	.20412	1.6
X = 1		

X	Y3	Y4
1	10	1
1.1	12.589	1.1
1.2	15.849	1.2
1.3	19.953	1.3
1.4	25.119	1.4
1.5	31.623	1.5
1.6	39.811	1.6
X = 1		

Similarly, we can show that $f[g(x)] = x$ by letting $y_3 = 10^{\wedge} x$ and $y_4 = \log y_3$. The table on the right above shows that $y_4 = x$.

A grapher can quickly draw graphs of logarithmic functions, base 10. It is important, however, to remember to place parentheses properly when entering logarithmic expressions.

Example 6 Graph: $f(x) = \log \dfrac{x}{5} + 1$.

SOLUTION We enter $y = \log (x/5) + 1$. (Note that on some graphers, the left parenthesis is supplied automatically when $\boxed{\log}$ is pressed.) When choosing a viewing window, recall that logarithms of negative numbers are not defined. The grapher provides the graph at left.

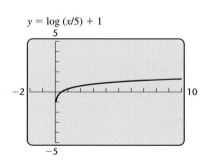

$y = \log (x/5) + 1$

Converting Exponential and Logarithmic Equations

We use the definition of logarithm to convert from *exponential equations* to *logarithmic equations*:

$$y = \log_a x \quad \textbf{is equivalent to} \quad a^y = x.$$

Caution! **Be sure to memorize this relationship!** It is probably the most important definition in the chapter. Many times this definition will serve as a justification for a property we are considering.

Example 7 Convert each to a logarithmic equation: **(a)** $8 = 2^x$; **(b)** $y^{-1} = 4$; **(c)** $a^b = c$.

SOLUTION

a) $8 = 2^x$ is equivalent to $x = \log_2 8$ The exponent is the logarithm.

     The base remains the same.

b) $y^{-1} = 4$ is equivalent to $-1 = \log_y 4$

c) $a^b = c$ is equivalent to $b = \log_a c$ ▬

We also use the definition of logarithm to convert from logarithmic equations to exponential equations.

Example 8 Convert each to an exponential equation: **(a)** $y = \log_3 5$; **(b)** $-2 = \log_a 7$; **(c)** $a = \log_b d$.

SOLUTION

a) $y = \log_3 5$ is equivalent to $3^y = 5$ The logarithm is the exponent.

     The base remains the same.

b) $-2 = \log_a 7$ is equivalent to $a^{-2} = 7$

c) $a = \log_b d$ is equivalent to $b^a = d$ ▬

Solving Certain Logarithmic Equations

Some logarithmic equations can be solved by converting to exponential equations.

Example 9   Solve: **(a)** $\log_2 x = -3$; **(b)** $\log_x 16 = 2$.

SOLUTION

a) $\log_2 x = -3$

$\quad\quad 2^{-3} = x$ Converting to an exponential equation

$\quad\quad \frac{1}{8} = x$ Computing 2^{-3}

CHECK: $\log_2 \frac{1}{8} = -3$ since $2^{-3} = \frac{1}{8}$.

The solution is $\frac{1}{8}$.

b) $\log_x 16 = 2$

$\quad\quad x^2 = 16$ Converting to an exponential equation

$x = 4$ *or* $x = -4$ Principle of square roots

CHECK: $\log_4 16 = 2$ because $4^2 = 16$. Thus, 4 is a solution. Because all logarithm bases must be positive, -4 cannot be a solution. Logarithm bases must be positive because logarithms are defined using exponential functions that require positive bases. The solution is 4.   ▬

One method for solving certain logarithmic and exponential equations relies on the following property, which results from the fact that exponential functions are one-to-one.

> ### *The Principle of Exponential Equality*
> For any real number b, where $b \neq -1, 0$ or 1,
>
> $\qquad b^x = b^y \quad$ is equivalent to $\quad x = y.$

Example 10   Solve: **(a)** $\log_{10} 1000 = x$; **(b)** $\log_4 1 = t$.

SOLUTION

a) We convert $\log_{10} 1000 = x$ to exponential form and solve:

$\qquad 10^x = 1000 \qquad$ Converting to an exponential equation

$\qquad 10^x = 10^3 \qquad$ Writing 1000 as a power of 10

$\qquad\quad x = 3. \qquad$ Equating exponents

CHECK: This equation can also be solved using the definition of logarithm, exactly as we did in Example 1(a). Since in both cases we find that $\log_{10} 1000 = 3$, we have a check. The solution is 3.

b) We convert $\log_4 1 = t$ to exponential form and solve:

$\qquad 4^t = 1 \qquad$ Converting to an exponential equation

$\qquad 4^t = 4^0 \qquad$ Writing 1 as a power of 4

$\qquad\quad t = 0. \qquad$ Equating exponents

CHECK: As in part (a), this equation can be solved using the definition of logarithm. This is precisely what we did in Example 1(c). Since in both cases we find that $\log_4 1 = 0$, we have a check. The solution is 0.

Example 10(b) illustrates an important property of logarithms.

> ### $\log_a 1$
> The logarithm, base a, of 1 is always 0: $\log_a 1 = 0$.

This follows from the fact that $a^0 = 1$ is equivalent to the logarithmic equation $\log_a 1 = 0$. Thus, $\log_{10} 1 = 0$, $\log_7 1 = 0$, and so on.

Another property results from the fact that $a^1 = a$. This is equivalent to the equation $\log_a a = 1$.

> ### $\log_a a$
> The logarithm, base a, of a is always 1: $\log_a a = 1$.

Thus, $\log_{10} 10 = 1$, $\log_8 8 = 1$, and so on.

9.3 Exercise Set

Graph.

1. $y = \log_2 x$
2. $y = \log_{10} x$
3. $y = \log_7 x$
4. $y = \log_3 x$
5. $f(x) = \log_4 x$
6. $f(x) = \log_6 x$
7. $f(x) = \log_{1/2} x$
8. $f(x) = \log_{2.5} x$

Graph both functions using the same set of axes.

9. $f(x) = 3^x, \; f^{-1}(x) = \log_3 x$
10. $f(x) = 4^x, \; f^{-1}(x) = \log_4 x$

Use a calculator to find each of the following rounded to four decimal places. Then check by raising 10 to that power.

11. $\log 4$
12. $\log 5$
13. $\log 13,400$
14. $\log 93,100$
15. $\log 0.527$
16. $\log 0.493$

Use a calculator to find each of the following rounded to four decimal places. Then check by finding the common logarithm of your answer.

17. $10^{2.3}$
18. $10^{0.173}$
19. $10^{-2.9523}$
20. $10^{4.8982}$
21. $10^{0.0012}$
22. $10^{-3.89}$

Graph.

23. $\log (x + 2)$
24. $\log (x - 5)$
25. $\log (1 - 2x)$
26. $\log (3x + 2.7)$
27. $\log (x^2)$
28. $\log (x^2 + 1)$

Convert to logarithmic equations.

29. $10^4 = 10,000$
30. $10^2 = 100$
31. $5^{-3} = \frac{1}{125}$
32. $4^{-5} = \frac{1}{1024}$
33. $8^{1/3} = 2$
34. $16^{3/4} = 8$
35. $10^{0.3010} = 2$
36. $10^{0.4771} = 3$
37. $m^n = r$
38. $p^k = 3$
39. $e^2 = 7.3891$
40. $e^{-4} = 0.0183$

Convert to exponential equations.

41. $t = \log_3 8$
42. $h = \log_7 10$
43. $\log_5 25 = 2$
44. $\log_6 6 = 1$
45. $\log_{10} 0.1 = -1$
46. $\log_{10} 0.01 = -2$
47. $\log_{10} 7 = 0.845$
48. $\log_{10} 3 = 0.4771$
49. $\log_c m = 17$
50. $\log_b n = 23$
51. $\log_e 0.25 = -1.3863$
52. $\log_e 0.989 = -0.0111$
53. $\log_r T = -x$
54. $\log_c M = -w$

Solve.

55. $\log_3 x = 4$
56. $\log_4 x = 2$
57. $\log_x 125 = 3$
58. $\log_x 64 = 3$
59. $\log_2 16 = x$
60. $\log_5 25 = x$
61. $\log_3 27 = x$
62. $\log_4 16 = x$
63. $\log_x 8 = 1$
64. $\log_x 7 = 1$
65. $\log_6 x = 0$
66. $\log_9 x = 1$
67. $\log_2 x = -1$
68. $\log_3 x = -2$
69. $\log_8 x = \frac{2}{3}$
70. $\log_{32} x = \frac{2}{5}$

Find each of the following.

71. $\log_{10} 10,000$
72. $\log_{10} 100,000$
73. $\log_{10} 1$
74. $\log_{10} 10$
75. $\log_5 625$
76. $\log_6 1$
77. $\log_5 \frac{1}{25}$
78. $\log_4 64$
79. $\log_3 3$
80. $\log_2 \frac{1}{16}$
81. $\log_7 1$
82. $\log_2 2$
83. $6^{\log_6 15}$
84. $7^{\log_7 23}$
85. $\log_{27} 9$
86. $\log_8 2$
87. $\log_b b^7$
88. $\log_n n^8$

Skill Maintenance

Simplify. [6.3]

89. $\dfrac{\dfrac{3}{x} - \dfrac{2}{xy}}{\dfrac{2}{x^2} + \dfrac{1}{xy}}$

90. $\dfrac{\dfrac{4 + x}{x^2 + 2x + 1}}{\dfrac{3}{x + 1} - \dfrac{2}{x + 2}}$

Rename without using exponents.

91. 8^{-4} [1.4]
92. $x^{1/5}$ [7.2]
93. $t^{-1/3}$ [7.2]
94. 5^1 [1.4]

Synthesis

95. ◆ Express in words what number is represented by $\log_b c$.
96. ◆ Is it true that $2 = b^{\log_b 2}$? Why or why not?
97. ◆ Would a manufacturer be pleased or unhappy if sales of a product grew logarithmically? Why?
98. ◆ Explain why the number $\log_2 13$ must be between 3 and 4.

99. Graph both equations using the same set of axes:
$$y = \left(\tfrac{3}{2}\right)^x, \qquad y = \log_{3/2} x.$$

Graph.

100. $y = \log_2 (x - 1)$

101. $y = \log_3 |x + 1|$

Solve.

102. $|\log_3 x| = 2$

103. $\log_{125} x = \tfrac{2}{3}$

104. $\log_4 (3x - 2) = 2$

105. $\log_8 (2x + 1) = -1$

106. $\log_{10} (x^2 + 21x) = 2$

Simplify.

107. $\log_{1/4} \tfrac{1}{64}$

108. $\log_{1/5} 25$

109. $\log_{81} 3 \cdot \log_3 81$

110. $\log_{10} (\log_4 (\log_3 81))$

111. $\log_2 (\log_2 (\log_4 256))$

112. Show that $b^x = b^y$ is *not* equivalent to $x = y$ for $b = 0$, $b = 1$, or $b = -1$.

9.4

Properties of Logarithmic Functions

- *Logarithms of Products*
- *Logarithms of Powers*
- *Logarithms of Quotients*
- *Using the Properties Together*

Logarithmic functions are important in many applications and in more advanced mathematics. We now establish some basic properties that are useful in manipulating expressions involving logarithms.

Interactive Discovery

For each of the following, use a calculator to determine which is the equivalent logarithmic expression.

1. $\log (20 \cdot 5)$
  a) $(\log 20)(\log 5)$
  b) $(\log 20)5$
  c) $\log 20 + \log 5$

2. $\log 20^5$
  a) $(\log 20)(\log 5)$
  b) $(\log 20)5$
  c) $\log 20 - \log 5$

3. $\log \left(\dfrac{20}{5}\right)$
  a) $(\log 20)(\log 5)$
  b) $(\log 20)/(\log 5)$
  c) $\log 20 - \log 5$

In this section, we will state and prove the patterns you may have observed. As their proofs reveal, the properties of logarithms are related to the properties of exponents.

Logarithms of Products

The first property we discuss is reminiscent of the property $a^m \cdot a^n = a^{m+n}$.

> **The Product Rule for Logarithms**
>
> For any positive numbers M, N, and a ($a \neq 1$),
>
> $$\log_a MN = \log_a M + \log_a N.$$
>
> (The logarithm of a product is the sum of the logarithms of the factors.)

Example 1   Express as a sum of logarithms: $\log_2 (4 \cdot 16)$.

SOLUTION   We have

$$\log_2 (4 \cdot 16) = \log_2 4 + \log_2 16. \qquad \textbf{Using the product rule}$$

As a check, note that

$$\log_2 (4 \cdot 16) = \log_2 64$$
$$= 6$$

and that

$$\log_2 4 + \log_2 16 = 2 + 4$$
$$= 6.$$

Example 2   Express as a single logarithm: $\log_{10} 0.01 + \log_{10} 1000$.

SOLUTION   We have

$$\log_{10} 0.01 + \log_{10} 1000 = \log_{10} (0.01 \times 1000) \qquad \textbf{Using the product rule}$$
$$= \log_{10} 10.$$

The check is left to the student.

A Proof of the Product Rule: Let $\log_a M = x$ and $\log_a N = y$. Converting to exponential equations, we have $a^x = M$ and $a^y = N$.

Now we multiply the latter two equations, to obtain

$$MN = a^x \cdot a^y, \quad \text{or} \quad MN = a^{x+y}.$$

Converting back to a logarithmic equation, we get

$$\log_a MN = x + y.$$

Recalling what x and y represent, we get

$$\log_a MN = \log_a M + \log_a N.$$

Logarithms of Powers

The second basic property is related to the property $(a^m)^n = a^{mn}$.

The Power Rule for Logarithms

For any positive numbers M and a ($a \neq 1$), and any real number p,

$$\log_a M^p = p \cdot \log_a M.$$

(The logarithm of a power of M is the exponent times the logarithm of M.)

Example 3   Express as a product: **(a)** $\log_a 9^{-5}$; **(b)** $\log_7 \sqrt[3]{x}$.

SOLUTION

a) $\log_a 9^{-5} = -5 \log_a 9$ **Using the power rule**

b) $\log_7 \sqrt[3]{x} = \log_7 x^{1/3}$ **Writing exponential notation**

$\phantom{\log_7 \sqrt[3]{x}} = \frac{1}{3} \log_7 x$ **Using the power rule**   ▬

A Proof of the Power Rule: Let $x = \log_a M$. We then convert to an exponential equation, to get $a^x = M$. Raising both sides to the pth power, we obtain

$$(a^x)^p = M^p, \quad \text{or} \quad a^{xp} = M^p.$$

Converting back to a logarithmic equation gives us

$$\log_a M^p = xp.$$

But $x = \log_a M$, so substituting, we have

$$\log_a M^p = (\log_a M)p = p \cdot \log_a M.$$ ▬

Logarithms of Quotients

The third property that we study is similar to the property $\dfrac{a^m}{a^n} = a^{m-n}$.

The Quotient Rule for Logarithms

For any positive numbers M, N, and a ($a \neq 1$),

$$\log_a \frac{M}{N} = \log_a M - \log_a N.$$

(The logarithm of a quotient is the logarithm of the dividend minus the logarithm of the divisor.)

Example 4   Express as a difference of logarithms: $\log_t (6/U)$.

SOLUTION

$$\log_t \frac{6}{U} = \log_t 6 - \log_t U \qquad \text{Using the quotient rule}$$

▬

Example 5 Express as a single logarithm: $\log_b 17 - \log_b 27$.

SOLUTION

$$\log_b 17 - \log_b 27 = \log_b \frac{17}{27} \qquad \text{Using the quotient rule "in reverse"}$$

A Proof of the Quotient Rule: Our proof uses both the product and power rules:

$$\log_a \frac{M}{N} = \log_a MN^{-1} \qquad \text{Rewriting } \frac{M}{N} \text{ with a negative exponent}$$
$$= \log_a M + \log_a N^{-1} \qquad \text{Using the product rule}$$
$$= \log_a M + (-1)\log_a N \qquad \text{Using the power rule}$$
$$= \log_a M - \log_a N.$$

Using the Properties Together

Example 6 Express in terms of logarithms of x, y, and z.

a) $\log_b \dfrac{x^3}{yz}$

b) $\log_a \sqrt[4]{\dfrac{xy}{z^3}}$

SOLUTION

a) $\log_b \dfrac{x^3}{yz} = \log_b x^3 - \log_b yz \qquad$ Using the quotient rule

$$= 3 \log_b x - \log_b yz \qquad \text{Using the power rule}$$
$$= 3 \log_b x - (\log_b y + \log_b z) \qquad \text{Using the product rule. Because of the subtraction, parentheses are essential!}$$
$$= 3 \log_b x - \log_b y - \log_b z \qquad \text{Using the distributive law}$$

b) $\log_a \sqrt[4]{\dfrac{xy}{z^3}} = \log_a \left(\dfrac{xy}{z^3}\right)^{1/4} \qquad$ Writing exponential notation

$$= \frac{1}{4} \cdot \log_a \frac{xy}{z^3} \qquad \text{Using the power rule}$$
$$= \frac{1}{4}(\log_a xy - \log_a z^3) \qquad \text{Using the quotient rule. Parentheses are important.}$$
$$= \frac{1}{4}(\log_a x + \log_a y - 3 \log_a z) \qquad \text{Using the product and power rules}$$

Caution! When subtraction or multiplication precedes use of the product or quotient rule, parentheses are needed, as in Example 6.

Example 7 Express as a single logarithm.

a) $\dfrac{1}{2} \log_a x - 7 \log_a y + \log_a z$

b) $\log_a \dfrac{b}{\sqrt{x}} + \log_a \sqrt{bx}$

SOLUTION

a) $\dfrac{1}{2} \log_a x - 7 \log_a y + \log_a z$

$\qquad = \log_a x^{1/2} - \log_a y^7 + \log_a z \qquad$ Using the power rule

$\qquad = (\log_a \sqrt{x} - \log_a y^7) + \log_a z \qquad$ Using parentheses to emphasize the order of operations; $x^{1/2} = \sqrt{x}$

$\qquad = \log_a \dfrac{\sqrt{x}}{y^7} + \log_a z \qquad$ Using the quotient rule

$\qquad = \log_a \dfrac{z\sqrt{x}}{y^7} \qquad$ Using the product rule

b) $\log_a \dfrac{b}{\sqrt{x}} + \log_a \sqrt{bx} = \log_a \dfrac{b \cdot \sqrt{bx}}{\sqrt{x}} \qquad$ Using the product rule

$\qquad\qquad = \log_a b\sqrt{b} \qquad$ Removing a factor equal to 1: $\dfrac{\sqrt{x}}{\sqrt{x}} = 1$

$\qquad\qquad = \log_a b^{3/2}, \text{ or } \dfrac{3}{2} \log_a b \qquad$ Since $b\sqrt{b} = b^1 \cdot b^{1/2}$

If we know the logarithms of two different numbers (to the same base), the properties allow us to calculate other logarithms.

Example 8 Given $\log_a 2 = 0.301$ and $\log_a 3 = 0.477$, find each of the following.

a) $\log_a 6$ **b)** $\log_a \frac{2}{3}$ **c)** $\log_a 81$

d) $\log_a \frac{1}{3}$ **e)** $\log_a 2a$ **f)** $\log_a 5$

SOLUTION

a) $\log_a 6 = \log_a (2 \cdot 3) = \log_a 2 + \log_a 3 \qquad$ Using the product rule

$\qquad\qquad = 0.301 + 0.477 = 0.778$

b) $\log_a \frac{2}{3} = \log_a 2 - \log_a 3 \qquad$ Using the quotient rule

$\qquad\quad = 0.301 - 0.477 = -0.176$

c) $\log_a 81 = \log_a 3^4 = 4 \log_a 3 \qquad$ Using the power rule

$\qquad\qquad = 4(0.477) = 1.908$

d) $\log_a \frac{1}{3} = \log_a 1 - \log_a 3 \qquad$ Using the quotient rule

$\qquad\quad = 0 - 0.477 = -0.477$

e) $\log_a 2a = \log_a 2 + \log_a a \qquad$ Using the product rule

$\qquad\qquad = 0.301 + 1 = 1.301$

f) $\log_a 5$ *cannot be found using these properties.*
$\quad (\log_a 5 \neq \log_a 2 + \log_a 3)$

A final property follows from the product rule: Since $\log_a a^k = k \log_a a$, and $\log_a a = 1$, we have $\log_a a^k = k$.

The Logarithm of the Base to a Power

For any base a,

$$\log_a a^k = k.$$

(The logarithm, base a, of a to a power is the power.)

This property also follows from the definition of logarithm: k is the power to which you raise a in order to get a^k.

Example 9   Simplify: **(a)** $\log_3 3^7$; **(b)** $\log_{10} 10^{-5.2}$.

SOLUTION

a) $\log_3 3^7 = 7$ **7 is the power to which you raise 3 in order to get 3^7.**

b) $\log_{10} 10^{-5.2} = -5.2$

We summarize the properties covered in this section as follows.

For any positive numbers M, N, and a ($a \neq 1$):

$$\log_a MN = \log_a M + \log_a N; \qquad \log_a M^p = p \cdot \log_a M;$$

$$\log_a \frac{M}{N} = \log_a M - \log_a N; \qquad \log_a a^k = k$$

Caution!   Keep in mind that, in general,

$$\log_a (M + N) \neq \log_a M + \log_a N, \qquad \log_a MN \neq (\log_a M)(\log_a N),$$
$$\log_a (M - N) \neq \log_a M - \log_a N,$$
$$\log_a (M/N) \neq (\log_a M) \div (\log_a N)$$

9.4 | Exercise Set

Express as a sum of logarithms.

1. $\log_3 (81 \cdot 27)$

2. $\log_2 (16 \cdot 32)$

3. $\log_4 (64 \cdot 16)$

4. $\log_5 (25 \cdot 125)$

5. $\log_c xyz$

6. $\log_t 3ab$

Express as a single logarithm.

7. $\log_a 5 + \log_a 14$

8. $\log_b 65 + \log_b 2$

9. $\log_c t + \log_c y$

10. $\log_t H + \log_t M$

Express as a product.

11. $\log_a t^7$

12. $\log_{10} y^7$

13. $\log_b C^{-3}$

14. $\log_c M^{-5}$

Express as a difference of logarithms.

15. $\log_2 \dfrac{64}{16}$ **16.** $\log_3 \dfrac{27}{9}$

17. $\log_b \dfrac{m}{n}$ **18.** $\log_a \dfrac{y}{x}$

Express as a single logarithm.

19. $\log_a 15 - \log_a 7$ **20.** $\log_b 42 - \log_b 7$

Express in terms of logarithms of w, x, y, and z.

21. $\log_a x^2 y^3 z$ **22.** $\log_a xy^4 z^3$

23. $\log_b \dfrac{xy^2}{z^3}$ **24.** $\log_b \dfrac{x^2 y^5}{w^4 z^7}$

25. $\log_b \dfrac{xy^2}{wz^3}$ **26.** $\log_b \dfrac{w^2 x}{y^3 z}$

27. $\log_a \sqrt{\dfrac{x^6}{y^5 z^8}}$ **28.** $\log_c \sqrt[3]{\dfrac{x^4}{y^3 z^2}}$

29. $\log_a \sqrt[3]{\dfrac{x^6 y^3}{a^2 z^7}}$ **30.** $\log_a \sqrt[4]{\dfrac{x^8 y^{12}}{a^3 z^5}}$

Express as a single logarithm and, if possible, simplify.

31. $4 \log_a x + 3 \log_a y$ **32.** $2 \log_b m + \frac{1}{2} \log_b n$

33. $\log_a x^2 - 2 \log_a \sqrt{x}$ **34.** $\log_a \dfrac{a}{\sqrt{x}} - \log_a \sqrt{ax}$

35. $\frac{1}{2} \log_a x + 3 \log_a y - 2 \log_a x$

36. $\log_a 2x + 3(\log_a x - \log_a y)$

37. $\log_a (x^2 - 4) - \log_a (x - 2)$

38. $\log_a (2x + 10) - \log_a (x^2 - 25)$

Given $\log_b 3 = 1.099$ and $\log_b 5 = 1.609$. If possible, find each of the following.

39. $\log_b 15$ **40.** $\log_b \frac{5}{3}$ **41.** $\log_b \frac{3}{5}$

42. $\log_b \frac{1}{3}$ **43.** $\log_b \frac{1}{5}$ **44.** $\log_b \sqrt{b}$

45. $\log_b \sqrt{b^3}$ **46.** $\log_b 3b$ **47.** $\log_b 6$

48. $\log_b 45$ **49.** $\log_b 75$ **50.** $\log_b 20$

Simplify.

51. $\log_4 4^9$ **52.** $\log_5 5^4$

53. $\log_{11} 11^m$ **54.** $\log_p p^{-2}$

Find each of the following.

55. $\log_5 (125 \cdot 625)$ **56.** $\log_3 (9 \cdot 81)$

57. $\log_2 \left(\dfrac{128}{16}\right)$ **58.** $\log_3 \left(\dfrac{243}{27}\right)$

Skill Maintenance

Compute and simplify. Express answers in the form $a + bi$ where $i^2 = -1$. [7.8]

59. i^{29} **60.** $(2 + i)^2$ **61.** $5i(2 - i)$

62. i^{34} **63.** $(5 - 3i)^2$ **64.** $7i(i - 4)$

Synthesis

65. ◈ Is it possible to express $\log_b \dfrac{x}{5}$ as a difference of two logarithms without using the quotient rule? Why or why not?

66. ◈ Is it true that $\log_a x + \log_b x = \log_{ab} x$? Why or why not?

67. ◈ A student *incorrectly* reasons that

$$\log_b \frac{1}{x} = \log_b \frac{x}{xx}$$
$$= \log_b x - \log_b x + \log_b x = \log_b x.$$

What mistake has the student made?

Express as a single logarithm and, if possible, simplify.

68. $\log_a (x^8 - y^8) - \log_a (x^2 + y^2)$

69. $\log_a (x + y) + \log_a (x^2 - xy + y^2)$

Express as a sum or difference of logarithms.

70. $\log_a \sqrt{1 - s^2}$ **71.** $\log_a \dfrac{c - d}{\sqrt{c^2 - d^2}}$

72. If $\log_a x = 2$, $\log_a y = 3$, and $\log_a z = 4$, what is

$$\log_a \dfrac{\sqrt[3]{x^2 z}}{\sqrt[3]{y^2 z^{-2}}}?$$

73. If $\log_a x = 2$, what is $\log_a (1/x)$?

74. If $\log_a x = 2$, what is $\log_{1/a} x$?

Classify each of the following as true or false. Assume a, x, P, and $Q > 0$.

75. $\log_a \left(\dfrac{P}{Q}\right)^x = x \log_a P - \log_a Q$

76. $\log_a (Q + Q^2) = \log_a Q + \log_a (Q + 1)$

77. Use graphs to show that
$$\log x^2 \neq \log x \cdot \log x.$$

9.5

Natural Logarithms and Changing Bases

- *The Base e and Natural Logarithms*
- *Changing Logarithm Bases*
- *Graphs of Exponential and Logarithmic Functions*

Any positive number other than 1 can serve as the base of a logarithmic function. However, there are logarithm bases that fit into certain applications more naturally than others. We have already looked at logarithms with one such base, base-10 logarithms, or common logarithms. Another logarithm base widely used today is an irrational number named *e*.

The Base e and Natural Logarithms

When interest is computed *n* times a year, the compound interest formula is

$$A = P\left(1 + \frac{r}{n}\right)^{nt},$$

where *A* is the amount that an initial investment *P* will be worth after *t* years at interest rate *r*. Suppose that $1 is invested at 100% interest for 1 year (no bank would pay this). The preceding formula becomes a function *A* defined in terms of the number of compounding periods *n*:

$$A(n) = \left(1 + \frac{1}{n}\right)^{n}.$$

Interactive Discovery

What happens to the function values of $A(n) = \left(1 + \frac{1}{n}\right)^{n}$ as *n* gets larger? To find out, fill in the following table. Round each entry to six decimal places.

n	$A(n) = \left(1 + \dfrac{1}{n}\right)^{n}$
1 (compounded annually)	$\left(1 + \dfrac{1}{1}\right)^{1}$, or $2.00
2 (compounded semiannually)	$\left(1 + \dfrac{1}{2}\right)^{2}$, or $?
3	
4 (compounded quarterly)	
12 (compounded monthly)	
100	
365 (compounded daily)	
8760 (compounded hourly)	

Which of the following statements appears to be true?

- $A(n)$ gets very large as n gets very large.
- $A(n)$ gets very small as n gets very large.
- $A(n)$ approaches a certain number as n gets very large.

The numbers in the table approach a very important number in mathematics, called e. Because e is irrational, its decimal representation does not terminate or repeat.

The Number e
$$e \approx 2.7182818284 \ldots$$

Logarithms base e are called **natural logarithms**, or **Napierian logarithms**, in honor of John Napier (1550–1617), who first "discovered" logarithms.

The abbreviation "ln" is generally used to denote natural logarithms. Thus,

$$\ln 53 \quad \text{means} \quad \log_e 53.$$

Example 1   Find $\ln 4568$.

SOLUTION   We press $\boxed{\text{ln}}$, then the number, and then $\boxed{\text{ENTER}}$ as shown at left. We find that

$$\ln 4568 \approx 8.4268. \qquad \textbf{Rounded to four decimal places}$$

On many calculators, the $\boxed{\text{ln}}$ key doubles as the $\boxed{e^x}$ key after a $\boxed{\text{2nd}}$ key has been pressed.

Example 2   Find $e^{-1.524}$, and check by finding the natural logarithm of the answer.

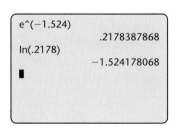

SOLUTION   We press $\boxed{e^x}$, then -1.524, and then $\boxed{\text{ENTER}}$ as shown at left. Since $e^{-1.524}$ is irrational, our answer is approximate:

$$e^{-1.524} \approx 0.2178. \qquad \textbf{Rounded to four decimal places}$$

To check, we find $\ln 0.2178$:

$$\ln 0.2178 \approx -1.524. \qquad \textbf{−1.524 is the power to which we raise e to get 0.2178.}$$

Changing Logarithm Bases

Most calculators can find both common logarithms and natural logarithms. To find a logarithm with some other base, a conversion formula is needed.

The Change-of-Base Formula

For any logarithm bases a and b, and any positive number M,

$$\log_b M = \frac{\log_a M}{\log_a b}.$$

Proof: Let $x = \log_b M$. Then,

$$b^x = M \qquad \text{Rewriting } x = \log_b M \text{ in exponential form}$$

$$\log_a b^x = \log_a M \qquad \text{Taking the logarithm, base } a, \text{ on both sides}$$

$$x \log_a b = \log_a M \qquad \text{Using the power rule for logarithms}$$

$$x = \frac{\log_a M}{\log_a b}. \qquad \text{Solving for } x$$

But at the outset we stated that $x = \log_b M$. Thus, by substitution, we have

$$\log_b M = \frac{\log_a M}{\log_a b},$$

which is the change-of-base formula. ▬

Example 3 Find $\log_5 8$ using common logarithms.

SOLUTION We use the change-of-base formula with $a = 10$, $b = 5$, and $M = 8$:

$$\log_5 8 = \frac{\log_{10} 8}{\log_{10} 5} \qquad \text{Substituting into } \log_b M = \frac{\log_a M}{\log_a b}$$

$$\approx \frac{0.9031}{0.6990} \qquad \text{Using the } \boxed{\text{log}} \text{ key twice}$$

$$\approx 1.2920. \qquad \text{When using a calculator, you need not round before dividing.}$$

The figure below shows the computation using a grapher.

log(8)/log(5)
1.292029674

To check, we use the $\boxed{\wedge}$ key to verify that $5^{1.2920} \approx 8$. ▬

We can also use base e for a conversion.

Example 4 Find $\log_4 31$ using natural logarithms.

SOLUTION   Substituting e for a, 4 for b, and 31 for M, we have

$$\log_4 31 = \frac{\log_e 31}{\log_e 4} \qquad \text{Substituting into } \log_b M = \frac{\log_a M}{\log_a b}$$

$$= \frac{\ln 31}{\ln 4} \qquad \text{Using the } \boxed{\text{ln}} \text{ key twice}$$

$$\approx 2.4771.$$

The use of a grapher is illustrated in the figure at left.

ln(31)/ln(4)

2.477098155

Graphs of Exponential and Logarithmic Functions

Example 5   Graph $f(x) = e^x$ and $g(x) = e^{-x}$.

SOLUTION   We use a calculator with an $\boxed{e^x}$ key to find approximate values of e^x and e^{-x}. Using these values, we can graph the functions.

x	e^x	e^{-x}
0	1	1
1	2.7	0.4
2	7.4	0.1
−1	0.4	2.7
−2	0.1	7.4

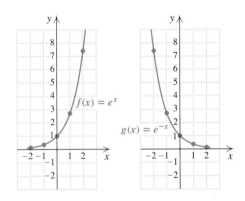

Example 6   Graph: $f(x) = e^{-0.5x}$.

SOLUTION   To graph this by hand, we find some solutions with a calculator, plot them, and then draw the graph. To graph using a grapher, we enter $y = e^{\wedge}(-0.5x)$. When choosing a viewing window, you may wish to show more of the y-axis since the function is exponential.

x	$e^{-0.5x}$
0	1
1	0.6
2	0.4
3	0.2
−1	1.6
−2	2.7
−3	4.5

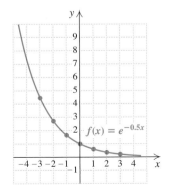

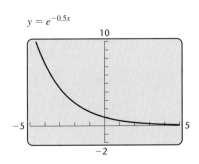

Example 7 Graph: **(a)** $g(x) = \ln x$; **(b)** $f(x) = \ln (x + 3)$.

SOLUTION

a) We find some solutions with a calculator and then draw the graph. As expected, the graph is a reflection across the line $y = x$ of the graph of $y = e^x$.

x	$\ln x$
1	0
4	1.4
7	1.9
0.5	−0.7

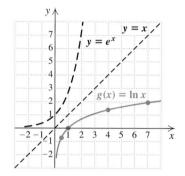

b) To graph by hand, we find some solutions with a calculator, plot them, and draw the graph. To graph using a grapher, we enter $y = \ln (x + 3)$. Since a logarithmic function is the inverse of an exponential function, we choose a viewing window that shows more of the x-axis than the y-axis.

x	$\ln (x + 3)$
0	1.1
1	1.4
2	1.6
3	1.8
4	1.9
−1	0.7
−2	0
−2.5	−0.7

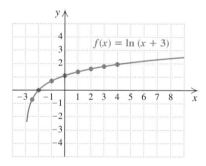

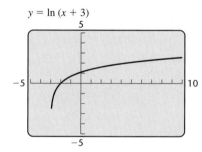

The graph of $y = \ln (x + 3)$ is the graph of $y = \ln x$ translated 3 units to the left. ▬

Logarithmic functions with bases other than 10 or e can be drawn on a grapher using the change-of-base formula.

Example 8 Graph: $f(x) = \log_7 x + 2$.

SOLUTION   We use the change-of-base formula with natural logarithms. (We would get the same graph if we used common logarithms.) Note that to find a function value, we find the logarithm, base 7, of x and add 2 to the result. This is different from finding the logarithm, base 7, of the quantity $(x + 2)$.

$$f(x) = \log_7 x + 2$$
$$= \frac{\ln x}{\ln 7} + 2$$

We graph $y = \ln(x)/\ln(7) + 2$.

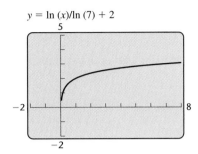

$y = \ln(x)/\ln(7) + 2$

9.5 | Exercise Set

Find each of the following rounded to four decimal places.

1. $\ln 5$ **2.** $\ln 2$ **3.** $\ln 62$

4. $\ln 30$ **5.** $\ln 4365$ **6.** $\ln 901.2$

7. $\ln 0.0062$ **8.** $\ln 0.00073$ **9.** $e^{2.71}$

10. $e^{3.06}$ **11.** $e^{-3.49}$ **12.** $e^{-2.64}$

Find each of the following.

13. e^0 **14.** $\ln 1$ **15.** $\ln e^2$

16. $\ln e^6$ **17.** $\ln e^{-3.5}$ **18.** $\ln \dfrac{e^2}{e^6}$

Find each of the following using the change-of-base formula. Round to four decimal places.

19. $\log_6 100$ **20.** $\log_3 100$ **21.** $\log_2 100$

22. $\log_7 100$ **23.** $\log_7 65$ **24.** $\log_5 42$

25. $\log_{0.5} 5$ **26.** $\log_{0.1} 3$ **27.** $\log_2 0.2$

28. $\log_2 0.08$ **29.** $\log_\pi 58$ **30.** $\log_\pi 200$

Graph.

31. $f(x) = e^x$ **32.** $f(x) = e^{-x}$

33. $f(x) = e^{-5x}$ **34.** $f(x) = e^{2x}$

35. $f(x) = e^{x-1}$ **36.** $f(x) = e^{-x} - 3$

37. $f(x) = e^x + 3$ **38.** $f(x) = e^{x-2}$

39. $f(x) = 2e^{-0.5x}$ **40.** $f(x) = 2e^{0.5x}$

41. $f(x) = \ln(x + 1)$ **42.** $f(x) = \ln(x - 2)$

43. $f(x) = 2 \ln x$ **44.** $f(x) = 3 \ln x$

45. $f(x) = \ln x + 2$ **46.** $f(x) = \ln x - 3$

Write an equivalent expression for the function that could be graphed using a grapher. Then graph the function using a grapher.

47. $f(x) = \log_5 x$ **48.** $f(x) = \log_3 x$

49. $f(x) = \log_2(x - 5)$ **50.** $f(x) = \log_5(2x + 1)$

51. $f(x) = \log_3 x + x$ **52.** $f(x) = \log_2 x - x + 1$

Skill Maintenance

Solve.

53. $4x^2 - 25 = 0$ [5.6]   **54.** $5x^2 - 7x = 0$ [5.3]

55. $17x - 15 = 0$ [2.2]   **56.** $9 - 13x = 0$ [2.2]

57. $x^{1/2} - 6x^{1/4} + 8 = 0$ [8.4]

58. $2y - 7\sqrt{y} + 3 = 0$ [8.4]

Synthesis

59. ◈ In an attempt to solve $\ln x = 1.5$, Emma gets the following graph.

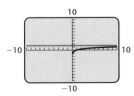

How can Emma tell at a glance that she has made a mistake?

60. ◈ Without referring to a calculator, explain why $\log 87 < \ln 10$ is a true statement.

61. ◈ Explain how the graph of $f(x) = e^x$ could be used to graph the function given by $g(x) = 1 + \ln x$.

62. ◆ Without drawing a graph or calculating any pairs, explain how the graphs of $f(x) = \ln |x|$ and $g(x) = |\ln x|$ differ.

63. Find a formula for converting common logarithms to natural logarithms.

64. Find a formula for converting natural logarithms to common logarithms.

Solve for x.

65. $\log (275x^2) = 38$

66. $\log (492x) = 5.728$

67. $\dfrac{3.01}{\ln x} = \dfrac{28}{4.31}$

68. $\log 692 + \log x = \log 3450$

For each function given below, (a) determine the domain, (b) set an appropriate window, and (c) draw the graph.

69. $f(x) = 7.4e^x \ln x$

70. $f(x) = 3.4 \ln x - 0.25e^x$

71. $f(x) = 5.3 \ln (x - 2.1)$

72. $f(x) = 2x^3 \ln x$

9.6
Solving Exponential and Logarithmic Equations

- *Solving Exponential Equations*
- *Solving Logarithmic Equations*

Solving Exponential Equations

Equations with variables in exponents, such as $5^x = 12$ and $2^{7x} = 64$, are called **exponential equations.** In Section 9.3, we solved certain exponential equations by using the principle of exponential equality. We restate that principle below.

The Principle of Exponential Equality

For any real number b, where $b \neq -1$, 0, or 1,

$$b^x = b^y \quad \text{is equivalent to} \quad x = y.$$

Example 1 Solve: $4^{3x-5} = 16$.

SOLUTION Note that $16 = 4^2$. Thus we can write each side as a power of the same number:

$$4^{3x-5} = 4^2.$$

Since the base is the same, 4, the exponents must be the same. Thus,

$$3x - 5 = 2 \qquad \textbf{Equating exponents}$$
$$3x = 7$$
$$x = \tfrac{7}{3}.$$

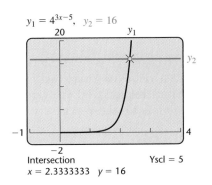

$y_1 = 4^{3x-5}, \ y_2 = 16$

Intersection
$x = 2.3333333 \quad y = 16$
Yscl = 5

CHECK:

$$
\begin{array}{c|c}
4^{3x-5} = 16 \\
\hline
4^{3 \cdot 7/3 - 5} \ \overset{?}{} \ 16 \\
4^{7-5} \\
4^2 \quad \big| \quad 16 \quad \text{TRUE}
\end{array}
$$

As another check, we solve the equation graphically. We graph $y_1 = 4\char94(3x - 5)$ and $y_2 = 16$ and determine the x-coordinates of any points of intersection. As the figure at left indicates, the graphs intersect at the point $(2.3333333, 16)$. Since $\frac{7}{3} \approx 2.3333333$, the answer checks. The solution is $\frac{7}{3}$.

When it does not seem possible to write both sides of an equation as powers of the same base, we can use the following principle along with the properties developed in Section 9.4.

The Principle of Logarithmic Equality

For any logarithm base a, and for $x, y > 0$,

$$x = y \quad \text{is equivalent to} \quad \log_a x = \log_a y.$$

Because calculators can generally find only common or natural logarithms (without resorting to the change-of-base formula), we usually take the common or natural logarithm on both sides of the equation.

Example 2   Solve: $5^x = 12$.

ALGEBRAIC SOLUTION

We have

$$5^x = 12$$

$\log 5^x = \log 12$ Using the principle of logarithmic equality to take the common logarithm on both sides. Natural logarithms also would work.

$x \log 5 = \log 12$ Using the power rule for logarithms

$x = \dfrac{\log 12}{\log 5}$

Caution! This is not $\log 12 - \log 5$!

$\approx 1.544.$ Using a calculator and rounding to three decimal places

The answer is approximately 1.544.

GRAPHICAL SOLUTION

We graph $y_1 = 5\char94 x$ and $y_2 = 12$.

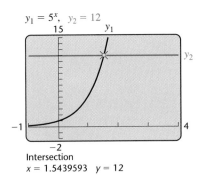

$y_1 = 5^x, \ y_2 = 12$

Intersection
$x = 1.5439593 \quad y = 12$

Rounded to three decimal places, the x-coordinate of the point of intersection is 1.544.

Since $5^{1.544} \approx 12$, we have a check. The solution is $\log 12 / \log 5$, or approximately 1.544.

Example 3 Solve: $e^{0.06t} = 1500$.

┌── *ALGEBRAIC SOLUTION*

We take the natural logarithm on both sides:

$\ln e^{0.06t} = \ln 1500$ **Taking the natural logarithm on both sides**

$0.06t = \ln 1500$ **Finding the logarithm of the base to a power:** $\log_a a^k = k$

$t = \dfrac{\ln 1500}{0.06}$ **Solving for** t

$\approx 121.887.$ **Using a calculator and rounding to three decimal places**

The answer is approximately 121.887.

┌── *GRAPHICAL SOLUTION*

We graph $y_1 = e\wedge(0.06x)$ and $y_2 = 1500$. Since $y_2 = 1500$, we choose a value for Ymax that is greater than 1500. It may require trial and error to choose appropriate units for the x-axis.

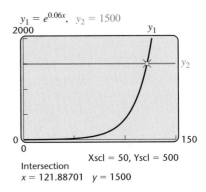

Rounded to three decimal places, the x-coordinate of the point of intersection is 121.887.

We check by substituting 121.887 for t in the original equation. The solution is approximately 121.887. ▬

Some equations, like the one in Example 3, are more readily solved algebraically. There are other exponential equations for which we do not have the tools to solve algebraically, but we can nevertheless solve graphically.

Example 4 Solve: $xe^{3x-1} = 5$.

SOLUTION   We graph $y_1 = xe\wedge(3x - 1)$ and $y_2 = 5$ and determine the coordinates of any points of intersection.

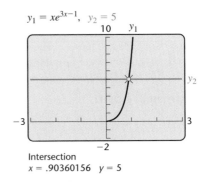

The x-coordinate of the point of intersection is approximately 0.90360156. Thus the solution is approximately 0.904. ▬

Solving Logarithmic Equations

Equations containing logarithmic expressions are called **logarithmic equations**. We saw in Section 9.3 that certain logarithmic equations can be solved by converting them into exponential equations.

Example 5   Solve: $\log_4 (8x - 6) = 3$.

SOLUTION   We write an equivalent exponential equation:

$$4^3 = 8x - 6$$
$$64 = 8x - 6$$
$$70 = 8x \qquad \text{Adding 6 on both sides}$$
$$x = \frac{70}{8}, \text{ or } \frac{35}{4}.$$

The check is left to the student. The solution is $\frac{35}{4}$.

Often the properties for logarithms are needed.

Example 6   Solve.

a) $\log x + \log (x - 3) = 1$

b) $\log_2 (x + 7) - \log_2 (x - 7) = 3$

c) $\log_7 (x + 1) + \log_7 (x - 1) = \log_7 8$

a)

ALGEBRAIC SOLUTION

As an aid in solving, we write in the base, 10.

$$\log_{10} x + \log_{10} (x - 3) = 1$$
$$\log_{10} [x(x - 3)] = 1 \qquad \text{Using the product rule for logarithms to obtain a single logarithm}$$
$$x(x - 3) = 10^1 \qquad \text{Writing an equivalent exponential equation}$$
$$x^2 - 3x = 10$$
$$x^2 - 3x - 10 = 0$$
$$(x + 2)(x - 5) = 0 \qquad \text{Factoring}$$
$$x + 2 = 0 \quad or \quad x - 5 = 0 \qquad \text{Using the principle of zero products}$$
$$x = -2 \quad or \quad x = 5$$

The possible solutions are -2 and 5.

GRAPHICAL SOLUTION

We graph $y_1 = \log (x) + \log (x + 3)$ and $y_2 = 1$ and determine the coordinates of any points of intersection.

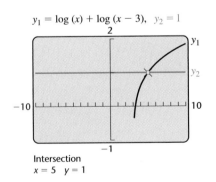

$y_1 = \log (x) + \log (x - 3), \ y_2 = 1$

Intersection
$x = 5 \quad y = 1$

There is one point of intersection, $(5, 1)$. The solution is the first coordinate of that point, or 5.

Note that the algebraic approach resulted in 2 possible solutions and the graphical approach in only 1. We suspect that -2 is not a solution of the equation. To check, we substitute both -2 and 5 in the original equation.

For -2:

$$\frac{\log x + \log (x - 3) = 1}{\log (-2) + \log (-2 - 3) \overset{?}{=} 1}$$

The number -2 *does not check* because negative numbers do not have logarithms.

For 5:

$$\frac{\log x + \log (x - 3) = 1}{\log 5 + \log (5 - 3) \overset{?}{=} 1}$$
$$\log 5 + \log 2 \ \big|$$
$$\log 10 \ \big|$$
$$1 \ \big| \ 1 \quad \text{TRUE}$$

The solution is 5.

b)

ALGEBRAIC SOLUTION

We have

$\log_2 (x + 7) - \log_2 (x - 7) = 3$

$$\log_2 \frac{x + 7}{x - 7} = 3 \qquad \text{Using the quotient rule for logarithms to obtain a single logarithm}$$

$$\frac{x + 7}{x - 7} = 2^3 \qquad \text{Writing an equivalent exponential equation}$$

$$\frac{x + 7}{x - 7} = 8$$

$$x + 7 = 8(x - 7) \qquad \text{Multiplying by the LCD, } x - 7$$

$$x + 7 = 8x - 56 \qquad \text{Using the distributive law}$$

$$63 = 7x$$

$$9 = x. \qquad \text{Dividing by 7}$$

GRAPHICAL SOLUTION

We first use the change-of-base formula to write the base-2 logarithms using common logarithms. Then we graph and determine the coordinates of any points of intersection. (We could use natural logarithms if we wish.)

$y_1 = \log (x + 7)/\log (2) - \log (x - 7)/\log (2),$
$y_2 = 3$

Intersection
$x = 9 \quad y = 3$

The graphs intersect at (9, 3). We have a solution of 9.

CHECK:
$$\frac{\log_2 (x + 7) - \log_2 (x - 7) = 3}{\log_2 (9 + 7) - \log_2 (9 - 7) \overset{?}{=} 3}$$
$$\log_2 16 - \log_2 2 \ \big|$$
$$4 - 1 \ \big|$$
$$3 \ \big| \ 3 \quad \text{TRUE}$$

The solution is 9.

c)

r-- *ALGEBRAIC SOLUTION*

We have

$\log_7 (x + 1) + \log_7 (x - 1) = \log_7 8$

$\log_7 (x^2 - 1) = \log_7 8$ **Using the product rule for logarithms: $(x + 1)(x - 1) = x^2 - 1$**

$x^2 - 1 = 8$ **Using the principle of logarithmic equality. Study this step carefully.**

$x^2 - 9 = 0$

$(x - 3)(x + 3) = 0$ **Solving the quadratic equation**

$x = 3 \quad or \quad x = -3.$

Thus we see that there are two possible solutions, 3 and −3.

r-- *GRAPHICAL SOLUTION*

Using the change-of-base formula, we graph

$$y_1 = \log (x + 1)/\log (7) + \log (x - 1)/\log (7)$$

and

$$y_2 = \log (8)/\log (7).$$

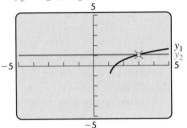

$y_1 = \log (x + 1)/\log (7) + \log (x - 1)/\log (7)$,
$y_2 = \log (8)/\log (7)$

Intersection
$x = 3 \quad y = 1.0686216$

The graphs intersect at (3, 1.0686216). We have a solution of 3.

The graphical approach indicates that 3 is a solution but −3 is not. The student should confirm that 3 checks in the equation, but −3 does not. The solution is 3. ▬

9.6 Exercise Set

Solve. Where appropriate, include approximations to the nearest thousandth as well as exact answers.

1. $3^x = 81$

2. $2^x = 8$

3. $4^x = 256$

4. $5^x = 125$

5. $2^{x+3} = 32$

6. $4^{3x} = 64$

7. $5^{3x} = 625$

8. $3^{5-x} = 27$

9. $4^{2x-1} = 64$

10. $5^{2x-3} = 25$

11. $3^{2x^2} \cdot 3^{5x} = 27$

12. $3^{4x} \cdot 3^{x^2} = \frac{1}{27}$

13. $2^x = 13$

14. $2^x = 19$

15. $4^x = 7$

16. $8^x = 10$

17. $e^t = 100$

18. $e^t = 1000$

19. $e^{-0.07t} = 0.08$

20. $e^{0.03t} = 5$

21. $2^x = 3^{x-1}$

22. $5^x = 3^{x+1}$

23. $e^{0.5x} - 7 = 2x + 6$

24. $e^{-x} - 3 = x^2$

25. $20 - (1.7)^x = 0$

26. $125 - (4.5)^y = 0$

27. $\log_5 x = 4$

28. $\log_3 x = 3$

29. $\log x = 3$

30. $\log x = 1$

31. $2 \log x = -6$

32. $4 \log x = -8$

33. $\ln x = 1$

34. $\ln x = 2$

35. $5 \ln x = -15$

36. $3 \ln x = -3$

37. $\log_2 (8 - 6x) = 5$

38. $\log_5 (2x - 7) = 3$

39. $\log (x + 9) + \log x = 1$

40. $\log (x - 9) + \log x = 1$

41. $\log x - \log (x + 7) = 1$

42. $\log x - \log (x + 3) = -1$

43. $\log_4 (x + 3) - \log_4 (x - 5) = 2$

44. $\log_2 (x + 3) + \log_2 (x - 3) = 4$

45. $\log_7 (x + 2) + \log_7 (x + 1) = \log_7 6$

46. $\log_6 (x + 3) + \log_6 (x + 2) = \log_6 20$

47. $\log_5 (x + 4) + \log_5 (x - 4) = 2$

48. $\log_{14} (x + 3) + \log_{14} (x - 2) = 1$

49. $\log_2 (x - 2) + \log_2 x = 3$

50. $\log_4 (x + 6) - \log_4 x = 2$

51. $\ln 3x = 3x - 8$

52. $\ln (x^2) = -x^2$

53. Find the value of x for which the natural logarithm is the same as the common logarithm.

54. Find all values of x for which the common logarithm of the square of x is the same as the square of the common logarithm of x.

55. ⌕ If $x = (\log_{125} 5)^{\log_5 125}$, what is the value of $\log_3 x$?

56. ⌕ If $2^x = 16$ and $3^y = 27$, what is the value of $x + y$?

⌕ *Solve.*

57. $\log x^{\log x} = 25$ **58.** $\log_5 |x| = 4$

Skill Maintenance

Simplify.

59. $(125x^7y^{-2}z^6)^{-2/3}$ [7.2]

60. i^{79} [7.8]

61. $(3 + 5i)^2$ [7.8]

Solve.

62. $E = mc^2$, for c (Assume $E, m, c > 0$.) [8.3]

63. $x^4 + 400 = 104x^2$ [8.4]

64. $x(x - 3) = 5$ [8.2]

Synthesis

65. ◈ Can the principle of logarithmic equality be expanded to include all functions? That is, will the statement "$m = n$ is equivalent to $f(m) = f(n)$" be true for any function f? Why or why not?

66. ◈ Explain how examining the domain of the logarithmic functions in Example 6 could help in checking possible solutions.

67. ◈ In Example 3, we took the natural logarithm on both sides. What would have happened had we used common logarithms instead?

68. ◈ Christina finds that the solution of $\log_3 (x + 4) = 1$ is -1, but rejects -1 as an answer. What mistake is she making?

Solve.

69. $27^x = 81^{2x-3}$ **70.** $8^x = 16^{3x+9}$

71. $\log_x (\log_3 27) = 3$ **72.** $\log_6 (\log_2 x) = 0$

73. $x \log \frac{1}{8} = \log 8$ **74.** $\log_5 \sqrt{x^2 - 9} = 1$

75. $2^{x^2+4x} = \frac{1}{8}$ **76.** $\log (\log x) = 5$

77. $\log \sqrt{2x} = \sqrt{\log 2x}$

78. $3^{2x} - 8 \cdot 3^x + 15 = 0$

79. $(81^{x-2})(27^{x+1}) = 9^{2x-3}$

80. $3^{2x} - 3^{2x-1} = 18$

9.7

Applications and Models Using Exponential and Logarithmic Functions

- *Applications of Logarithmic Functions*
- *Applications of Exponential Functions*

We now consider applications of exponential and logarithmic functions.

Applications of Logarithmic Functions

Example 1 *Sound Levels.* To measure the "loudness" of any particular sound, the decibel scale is used. The loudness L, in decibels (dB), of a sound is given by

$$L = 10 \cdot \log \frac{I}{I_0},$$

where I is the intensity of the sound, in watts per square meter (W/m²), and $I_0 = 10^{-12}$ W/m². (I_0 is approximately the intensity of the softest sound that can be heard.)

a) It is common for the intensity of sound at live performances of rock music to reach 10^{-1} W/m² (even higher close to the stage). How loud, in decibels, is this sound level?

b) Audiologists and physicians recommend that earplugs be worn when one is exposed to sounds in excess of 90 dB. What is the intensity of such sounds?

SOLUTION

a) To find the loudness, in decibels, we use the above formula:

$$L = 10 \cdot \log \frac{I}{I_0}$$

$$= 10 \cdot \log \frac{10^{-1}}{10^{-12}} \qquad \text{Substituting}$$

$$= 10 \cdot \log 10^{11} \qquad \text{Subtracting exponents}$$

$$= 10 \cdot 11 \qquad \qquad \log 10^a = a$$

$$= 110.$$

The volume of the music is 110 decibels.

b) We substitute and solve for I:

$$L = 10 \cdot \log \frac{I}{I_0}$$

$$90 = 10 \cdot \log \frac{I}{10^{-12}} \qquad \text{Substituting}$$

$$9 = \log \frac{I}{10^{-12}} \qquad \text{Dividing by 10 on both sides}$$

$$9 = \log I - \log 10^{-12} \qquad \text{Using the quotient rule for logarithms}$$

$$9 = \log I - (-12) \qquad \log 10^a = a$$

$$-3 = \log I \qquad \text{Adding } -12 \text{ on both sides}$$

$$10^{-3} = I. \qquad \text{Converting to an exponential equation}$$

Earplugs would be recommended for sounds with intensities exceeding 10^{-3} W/m².

Example 2   *Chemistry: pH of Liquids.* **In chemistry the pH of a liquid is defined as follows:**

$$pH = -\log [H^+],$$

where $[H^+]$ is the hydrogen ion concentration in moles per liter.

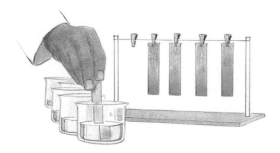

a) The hydrogen ion concentration of human blood is normally about 3.98×10^{-8} moles per liter. Find the pH.

b) The pH of seawater is about 8.3. Find the hydrogen ion concentration.

SOLUTION

a) To find the pH of blood, we use the above formula:

$$
\begin{aligned}
pH &= -\log [H^+] \\
&= -\log [3.98 \times 10^{-8}] \\
&\approx -(-7.400117) \qquad \textbf{Using a calculator} \\
&\approx 7.4.
\end{aligned}
$$

The pH of human blood is normally about 7.4.

b) We substitute and solve for $[H^+]$:

$$8.3 = -\log [H^+] \qquad \textbf{Using pH} = -\log[H^+]$$
$$-8.3 = \log [H^+] \qquad \textbf{Dividing by −1 on both sides}$$
$$10^{-8.3} = [H^+] \qquad \textbf{Converting to an exponential equation}$$
$$5.01 \times 10^{-9} \approx [H^+]. \qquad \textbf{Using a calculator; writing scientific notation}$$

The hydrogen ion concentration of seawater is about 5.01×10^{-9} moles per liter.   ▬

Applications of Exponential Functions

Example 3   *Interest Compounded Annually.* **Suppose that $30,000 is invested at 8% interest, compounded annually. In t years, it will grow to the amount A given by the function**

$$A(t) = 30{,}000(1.08)^t.$$

(See Example 5 in Section 9.1.)

a) How long will it take to accumulate $150,000 in the account?

b) Find the amount of time it takes for the $30,000 to double itself.

SOLUTION

a) We set $A(t) = 150,000$ and solve for t:

$$150,000 = 30,000(1.08)^t$$

$$\frac{150,000}{30,000} = 1.08^t \qquad \text{Dividing by 30,000 on both sides}$$

$$5 = 1.08^t$$

$$\log 5 = \log 1.08^t \qquad \text{Taking the common logarithm on both sides}$$

$$\log 5 = t \log 1.08 \qquad \text{Using the power rule for logarithms}$$

$$\frac{\log 5}{\log 1.08} = t \qquad \text{Dividing by log 1.08 on both sides}$$

$$20.9 \approx t. \qquad \text{Using a calculator}$$

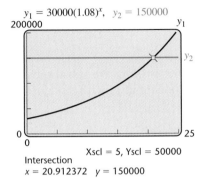

$y_1 = 30000(1.08)^x, \quad y_2 = 150000$

Xscl = 5, Yscl = 50000

Intersection
$x = 20.912372 \quad y = 150000$

We check by solving graphically, as shown in the figure at left. It will take about 20.9 yr for the $30,000 to grow to $150,000.

b) To find the *doubling time* T, we replace $A(t)$ with 60,000, t with T, and solve for T:

$$60,000 = 30,000(1.08)^T$$

$$2 = (1.08)^T \qquad \text{Dividing by 30,000 on both sides}$$

$$\log 2 = \log (1.08)^T \qquad \text{Taking the common logarithm on both sides}$$

$$\log 2 = T \log 1.08 \qquad \text{Using the power rule for logarithms}$$

$$T = \frac{\log 2}{\log 1.08} \approx 9.0. \qquad \text{Using a calculator}$$

The doubling time is about 9 yr.   ▬

Like investments, populations often grow exponentially.

Exponential Growth

An **exponential growth model** is a function of the form

$$P(t) = P_0 e^{kt}, \quad k > 0,$$

where P_0 is the population at time 0, $P(t)$ is the population at time t, and k is the **exponential growth rate** for the situation. The **doubling time** is the amount of time necessary for the population to double in size.

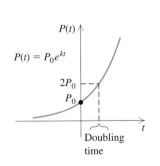

The exponential growth rate is the rate of growth of a population at any *instant* in time. Since the population is continually growing, the percent of total growth after one year will exceed the exponential growth rate.

Example 4   *U.S. Population Growth.* The population of the United States was 267 million in 1998 and the exponential growth rate was 0.8% per year (*Source:* U.S. Bureau of the Census).

a) Find the exponential growth model.

b) Estimate the population in 2010.

SOLUTION

a) In 1998, at $t = 0$, the population was 267 million. We substitute 267 for P_0 and 0.8%, or 0.008, for k in the exponential growth model. This gives the exponential growth function

$$P(t) = 267e^{0.008t}.$$

The graph of the function is shown below.

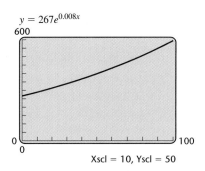

b) In 2010, we have $t = 12$ (since 12 yr have passed since 1998). To find the population 2010, we evaluate $P(12)$:

$$P(12) = 267e^{0.008(12)} \qquad \text{Using } P(t) = 267e^{0.008t} \text{ from part (a)}$$
$$\approx 294 \text{ million.}$$

The population of the United States will reach approximately 294 million in 2010.   ▬

Example 5   *Spread of AIDS.* The number of people $N(t)$ infected with a contagious disease at time t usually increases exponentially. Through 1985, 8249 cases of AIDS had been reported in the United States. By the end of 1994, the cumulative number had grown to 441,528. (*Source:* Centers for Disease Control)

a) Find the exponential growth rate and the exponential growth function.

b) Assuming exponential growth, predict the year in which the 1,000,000th case will occur.

SOLUTION

a) We use the equation $N(t) = N_0 e^{kt}$, where t is the number of years since 1985. In 1985, at $t = 0$, a total of 8249 cases had been reported. We substitute 8249 for N_0:

$$N(t) = 8249 e^{kt}.$$

To find the exponential growth rate k, note that 9 yr later, at the end of 1994, the total number of cases had grown to 441,528:

$$
\left.
\begin{aligned}
N(9) &= 8249 e^{k \cdot 9} \\
441{,}528 &= 8249 e^{9k}
\end{aligned}
\right\} \qquad \text{Substituting}
$$

$$53.525 \approx e^{9k} \qquad\qquad \text{Dividing by 8249 on both sides}$$

$$\ln 53.525 \approx \ln e^{9k} \qquad\qquad \text{Taking the natural logarithm on both sides}$$

$$3.9801 \approx 9k \qquad\qquad \ln e^a = a$$

$$0.442 \approx k.$$

The exponential growth function is $N(t) = 8249 e^{0.442t}$, where t is measured in years since the end of 1985.

b) To predict when the 1,000,000th case will occur, we replace $N(t)$ with 1,000,000 and solve for t:

$$1{,}000{,}000 = 8249 e^{0.442t}$$

$$121.227 \approx e^{0.442t} \qquad\qquad \text{Dividing by 8249 on both sides}$$

$$\ln 121.227 \approx \ln e^{0.442t} \qquad\qquad \text{Taking the natural logarithm on both sides}$$

$$4.798 \approx 0.442t \qquad\qquad \ln e^a = a$$

$$10.86 \approx t.$$

Rounding up to 11 yr, we see that, according to this model, by the end of 1985 + 11, or 1996, the 1,000,000th case occurred. (In reality, the spread of AIDS slowed in the mid-90s.)

In some real-life situations, a quantity or population is *decreasing* or *decaying* exponentially.

Exponential Decay

An **exponential decay model** is a function of the form

$$P(t) = P_0 e^{-kt}, \quad k > 0,$$

where P_0 is the quantity present at time 0, $P(t)$ is the amount present at time t, and k is the **decay rate**. The **half-life** is the amount of time necessary for half of the quantity to decay.

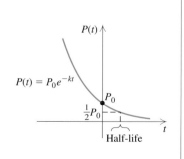

Example 6   *Carbon Dating.* The radioactive element carbon-14 has a half-life of 5750 yr. The percentage of carbon-14 present in the remains of animal bones can be used to determine age. How old is an animal bone that has lost 40% of its carbon-14?

SOLUTION   We first find k. To do so, we use the concept of half-life. When $t = 5750$ (the half-life), $P(t)$ will be half of P_0. Then

$$0.5P_0 = P_0 e^{-k(5750)}$$

$$0.5 = e^{-5750k} \qquad \text{Dividing by } P_0 \text{ on both sides}$$

$$\ln 0.5 = \ln e^{-5750k} \qquad \text{Taking the natural logarithm on both sides}$$

$$\ln 0.5 = -5750k$$

$$\frac{\ln 0.5}{-5750} = k$$

$$0.00012 \approx k.$$

Now we have a function for the decay of carbon-14:

$$P(t) = P_0 e^{-0.00012t}. \qquad \text{This completes the first part of our solution.}$$

(*Note:* This equation can be used for any subsequent carbon-dating problem.) If an animal bone has lost 40% of its carbon-14 from an initial amount P_0, then 60% of P_0 is still present. To find the age t of the bone, we solve this equation for t:

$$0.6P_0 = P_0 e^{-0.00012t} \qquad \text{We want to find } t \text{ for which } P(t) = 0.6P_0.$$

$$0.6 = e^{-0.00012t} \qquad \text{Dividing by } P_0 \text{ on both sides}$$

$$\ln 0.6 = \ln e^{-0.00012t} \qquad \text{Taking the natural logarithm on both sides}$$

$$-0.5108 \approx -0.00012t$$

$$t \approx \frac{-0.5108}{-0.00012}$$

$$\approx 4257.$$

The animal bone is about 4257 yr old.   ▬

We can now add the exponential functions $f(t) = P_0 e^{kt}$ and $f(t) = P_0 e^{-kt}$, $k > 1$, and the logarithmic function $f(x) = \log_b x$, $b > 1$, to our library of functions.

By looking at the graph of a set of data, we can tell whether a population or other quantity is growing or decaying exponentially.

Linear function:
$f(x) = mx + b$

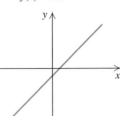

Absolute-value function:
$f(x) = |x|$

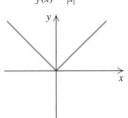

Rational function:
$f(x) = \dfrac{1}{x}$

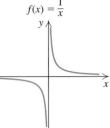

Radical function:
$f(x) = \sqrt{x}$

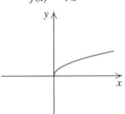

Quadratic function:
$f(x) = ax^2 + bx + c, a > 0$

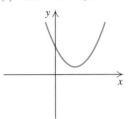

Quadratic function:
$f(x) = ax^2 + bx + c, a < 0$

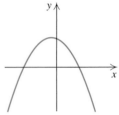

Exponential growth function:
$f(t) = P_0 e^{kt}, k > 0$

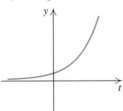

Exponential decay function:
$f(t) = P_0 e^{-kt}, k > 0$

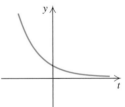

Logarithmic function:
$f(x) = \log_b x, b > 1$

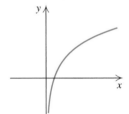

Example 7 For each of the following graphs, determine whether an exponential function might fit the data.

a)

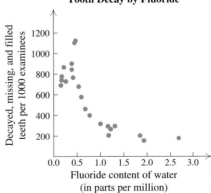

**Reduction of
Tooth Decay by Fluoride**

Source: Dudley, Brian A.C., Mathematical
and Biological Interrelations

b)

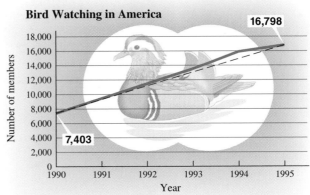

Bird Watching in America

Number of members

16,798

7,403

1990 1991 1992 1993 1994 1995

Year

Source: American Birding Association

c)

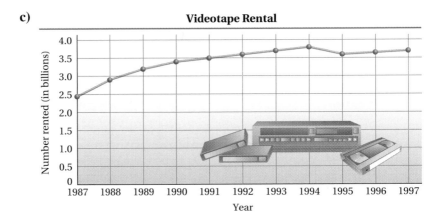

Videotape Rental

Number rented (in billions)

1987 1988 1989 1990 1991 1992 1993 1994 1995 1996 1997

Year

Source: Adams Media Research, Carmel Valley, California

d)

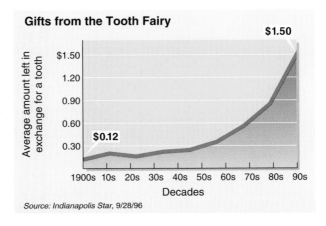

Gifts from the Tooth Fairy

Average amount left in exchange for a tooth

$1.50

$0.12

1900s 10s 20s 30s 40s 50s 60s 70s 80s 90s

Decades

Source: Indianapolis Star, 9/28/96

SOLUTION

a) As the fluoride content in the water increases, tooth decay decreases. The amount of decrease gets smaller as the amount of fluoride increases. It appears that an exponential decay function might fit the data.

b) The number of members of the American Birding Association increased between 1990 and 1995. The amount of increase was steady, however, and it does not appear that an exponential function could be used to model the data.

c) The number of videotape rentals increased between 1987 and 1997. The amount of increase was smaller during the later years. An exponential function cannot be used to model this situation.

d) The average amount left in exchange for a tooth increased from the 1910s to the 1990s. The amount of increase was larger for later decades. An exponential growth function could be used to model this situation. ▬

Many graphers can use regression to fit an exponential function to a set of data. Often the feature will be an option labeled ExpReg in the STAT CALC menu. The graph may return a function of the form $f(x) = ab^x$ instead of $f(x) = ae^{kx}$.

To convert this function to an exponential function with base e, note that if

$$ab^x = ae^{kx},$$

then

$$b^x = e^{kx}$$

and thus

$$b = e^k.$$

Solving for k, we have

$$\ln b = \ln e^k$$
$$\ln b = k.$$

Example 8 *Growth of Zebra Mussel Populations.* Zebra mussels, inadvertently imported from Europe, began fouling North American waters in 1988. These mussels are so prolific that lake and river bottoms, as well as water intake pipes, can become blanketed with them, altering an entire ecosystem. In 1996, a portion of the Mississippi River contained an average of 1 zebra mussel per square mile. There were 10 per square mile by 1997, 300 by 1998, and 9000 by 1999.*

*Many thanks to Dr. Gerald Mackie of the Department of Zoology at the University of Guelph in Ontario for the background information for this example.

a) Determine whether the zebra mussel population can be modeled by an exponential function.

b) Use regression to fit an exponential function to the data, and graph the function.

c) Determine the exponential growth rate.

d) Use the exponential function to predict the number of mussels per square mile in 2002.

SOLUTION

a) We let t = the number of years since 1996, and enter the data. From the graph, it appears that an exponential growth function could be used to model the zebra mussel population.

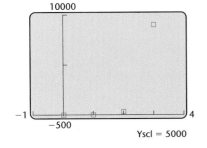

b) Using regression and rounding the coefficients to four decimal places, we find the exponential function

$$f(x) = 0.7192(21.5767)^x.$$

The graph of the function, along with the data points, is shown below.

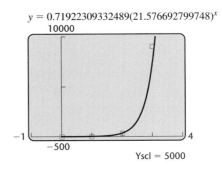

$y = 0.71922309332489(21.576692799748)^x$

c) Since ln (21.5767) ≈ 3.07, we can write the function as

$$P(t) = 0.7192e^{3.07t}.$$

The exponential growth rate is 307% per year.

d) To predict the number of zebra mussels per square mile in 2002, we evaluate $P(6)$. If $y_1 = P(t)$, we can use the grapher to evaluate the function.

The number of zebra mussels per square mile in 2002 will be about 75,572,635.

<table>
</table>

9.7 Exercise Set

Solve.

1. *Credit-Card Spending from Thanksgiving to Christmas.* The amount S, in billions of dollars, charged on credit cards between Thanksgiving and Christmas each year has been increasing exponentially according to the function

$$S(t) = 52.4(1.16)^t,$$

where t is the number of years since 1990 (*Source:* RAM Research Group, National Credit Counseling Services).

a) Estimate the amount of credit-card spending between Thanksgiving and Christmas in 2000.
b) In what year will spending be $1 trillion?
c) What is the doubling time?

2. *College Tuition Costs.* The cost of tuition, books, room, and board at a state university is projected to follow the exponential function

$$C(t) = 11,054(1.06)^t,$$

where C is the cost, in dollars, and t is the number

of years after 2000 (*Source:* College Board, Senate Labor Committee).

a) Estimate college costs in 2005.
b) In what year will costs be $21,000?
c) What is the doubling time of college costs?

3. *Student Loan Repayment.* A college loan of $29,000 is made at 8% interest, compounded annually. After t years, the amount due, A, is given by the function

$$A(t) = 29,000(1.08)^t.$$

a) After what amount of time will the amount due reach $40,000?
b) Find the doubling time.

4. *Spread of a Rumor.* The number of people who have heard a rumor increases exponentially. If all who hear a rumor repeat it to two people a day, and if 20 people start the rumor, the number of people N who have heard the rumor after t days is given by

$$N(t) = 20(3)^t.$$

a) After what amount of time will 1000 people have heard the rumor?
b) What is the doubling time for the number of people who have heard the rumor?

5. *Recycling Aluminum Cans.* Approximately two-thirds of all aluminum cans distributed will be recycled each year. A beverage company distributes 250,000 cans. The number still in use after t years is given by the function

$$N(t) = 250{,}000\left(\tfrac{2}{3}\right)^t.$$

a) After how many years will 60,000 cans still be in use?

b) After what amount of time will only 1000 cans still be in use?

6. *Salvage Value.* A color photocopier is purchased for $5200. Its value each year is about 80% of its value in the preceding year. Its value in dollars after t years is given by the exponential function

$$V(t) = 5200(0.8)^t.$$

a) After what amount of time will the salvage value be $1200?

b) After what amount of time will the salvage value be half the original value?

Use the pH formula given in Example 2 for Exercises 7–10.

7. *Chemistry.* The hydrogen ion concentration of fresh-brewed coffee is about 1.3×10^{-5} moles per liter. Find the pH.

8. *Chemistry.* The hydrogen ion concentration of milk is about 1.6×10^{-7} moles per liter. Find the pH.

9. *Medicine.* When the pH of a patient's blood drops below 7.4, a condition called *acidosis* sets in. Acidosis can be deadly when the patient's pH reaches 7.0. What would the hydrogen ion concentration of the patient's blood be at that point?

10. *Medicine.* When the pH of a patient's blood rises above 7.4, a condition called *alkalosis* sets in. Alkalosis can be deadly when the patient's pH reaches 7.8. What would the hydrogen ion concentration of the patient's blood be at that point?

Use the decibel formula given in Example 1 for Exercises 11–14.

11. *Acoustics.* The intensity of sound in normal conversation is about 3.2×10^{-6} W/m^2. How loud in decibels is this sound level?

12. *Acoustics.* The intensity of sound of a riveter at work is about 3.2×10^{-3} W/m^2. How loud in decibels is this sound level?

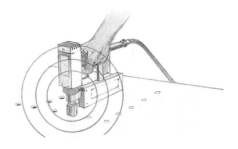

13. *Music.* The rock group Phish recently performed and sound measurements of 105 dB were recorded. What is the intensity of such sounds?

14. *Audiology.* Overexposure to excessive sound levels can diminish one's hearing to the point where the softest sound that is audible is 28 dB. What is the intensity of such a sound?

When interest is compounded continuously, the balance in an account after t years is given by $P_0(t) = P_0 e^{kt}$, where P_0 is the initial investment and k is the interest rate. Use this model for Exercises 15 and 16.

15. *Interest Compounded Continuously.* Suppose that P_0 is invested in a savings account where interest is compounded continuously at 6% per year.

a) Express $P(t)$ in terms of P_0 and 0.06.

b) Suppose that $5000 is invested. What is the balance after 1 yr? after 2 yr?

c) When will an investment of $5000 double itself?

16. *Interest Compounded Continuously.* Suppose that P_0 is invested in a savings account where interest is compounded continuously at 5% per year.

a) Express $P(t)$ in terms of P_0 and 0.05.

b) Suppose that $1000 is invested. What is the balance after 1 yr? after 2 yr?

c) When will an investment of $1000 double itself?

17. *Population Growth.* In 1997, the population of

Canada was 29.1 million and the exponential growth rate was 0.8% per year.

a) Find the exponential growth function.
b) Predict the population of Canada in 2010.
c) When will the population of Canada reach 33 million?

18. *Population Growth.* In 1997, the population of Saudi Arabia was 20.1 million and the exponential growth rate was 3.4% per year.

a) Find the exponential growth function.
b) Predict the population of Saudi Arabia in 2010.
c) When will the population of Saudi Arabia reach 40 million?

19. *Advertising.* A model for advertising response is given by

$$N(a) = 2000 + 500 \log a, \quad a \geq 1,$$

where $N(a)$ is the number of units sold and a is the amount spent on advertising, in thousands of dollars.

a) How many units were sold after spending $1000 ($a = 1$) on advertising?
b) How many units were sold after spending $8000?
c) Graph the function.
d) How much would have to be spent in order to sell 5000 units?

20. *Forgetting.* Students in an English class took a final exam. They took equivalent forms of the exam at monthly intervals thereafter. The average score $S(t)$, in percent, after t months was found to be given by

$$S(t) = 68 - 20 \log (t + 1), \quad t \geq 0.$$

a) What was the average score when they initially took the test, $t = 0$?
b) What was the average score after 4 months? after 24 months?
c) Graph the function.
d) After what time t was the average score 50?

21. *Public Health.* In 1995, an outbreak of Herpes infected 17 people in a large community. By 1996, the number of those infected had grown to 29.

a) Find an exponential growth function that fits the data.
b) Predict the number of people who will be infected in 2001.

22. *Heart Transplants.* In 1967, Dr. Christiaan Barnard of South Africa stunned the world by performing the first heart transplant. There was 1 transplant in 1967. In 1987, there were 1418 such transplants.

a) Find an exponential growth function that fits the data.
b) Use the function to predict the number of heart transplants in 2002.

23. *Lumber Demand.* The exponential growth rate of the demand for lumber in the world is 1.8% per year. When will the demand be double that of 1999?

24. *Oil Demand.* The exponential growth rate of the demand for oil in the United States is 10% per year. When will the demand be double that of 1998?

25. *Decline of Long-Playing Records.* The sales S of long-playing records has declined considerably in recent years because of the advent of the compact disc. There were 205 million LP records sold in 1983, and 1.2 million sold in 1993 (*Source:* Recording Industry Association of America). Assume that the sales are decreasing according to the exponential decay model.

a) Find the value k, and write an exponential function that describes the number of long-playing records sold after time t, in years.
b) Estimate the sales of LP records in the year 2001.
c) In what year (theoretically) will only 1 long-playing record be sold?

26. *Decline in Beef Consumption.* In 1985, the annual consumption of beef was about 80 lb per person. In 1996, it was about 67 lb per person. Assume that consumption is decreasing according to the exponential decay model.

a) Find the value k, and write an equation that describes beef consumption after time t, in years.
b) Estimate the consumption of beef in the year 2001.
c) In what year (theoretically) will the annual consumption of beef be 20 lb per person?

27. *Archaeology.* When archaeologists found the Dead Sea scrolls, they determined that the linen wrapping had lost 22.3% of its carbon-14. How old is the linen wrapping? (See Example 6.)

28. *Archaeology.* In 1996, researchers found an ivory tusk that had lost 18% of its carbon-14. How old was the tusk? (See Example 6.)

29. *Chemistry.* The decay rate of iodine-131 is 9.6% per day. What is its half-life?

30. *Chemistry.* The decay rate of krypton-85 is 6.3% per year. What is its half-life?

31. *Home Construction.* The chemical urea formaldehyde was found in some insulation used in houses built during the mid- to late-60s. Unknown at the time was the fact that urea formaldehyde emitted toxic fumes as it decayed. The half-life of urea formaldehyde is 1 yr. What is its decay rate?

32. *Plumbing.* Lead pipes and solder are often found in older buildings. Unfortunately, as lead decays, toxic chemicals can get in the water resting in the pipes. The half-life of lead is 22 yr. What is its decay rate?

33. *Value of a Sports Card.* Because he objected to smoking, and because his first baseball card was issued in cigarette packs, the great shortstop Honus Wagner halted production of his card before many were produced. One of these cards was sold in 1991 for $451,000 and again in 1996 for $640,500. For the following questions, assume that the card's value increases exponentially.

a) Find an exponential function $V(t)$, if $V_0 = 451,000$.
b) Predict the card's value in 2002.
c) What is the doubling time for the value of the card?
d) In what year will the value of the card first exceed $1,000,000?

34. *Value of a Painting.* The Van Gogh painting *Irises,* shown below, sold for $84,000 in 1947 and was sold again for $53,900,000 in 1987. Assume that the growth in the value V of the painting is exponential.

Van Gogh's *Irises*, a 28-in. by 32-in. oil on canvas.

a) Find the exponential growth rate k, and determine the exponential growth function, assuming $t = 0$ in 1947.
b) Estimate the value of the painting in 2008.
c) What is the doubling time for the value of the painting?
d) How long after 1947 will the value of the painting be $1 billion?

In Exercises 35–38, determine whether an exponential function might fit the data.

35.

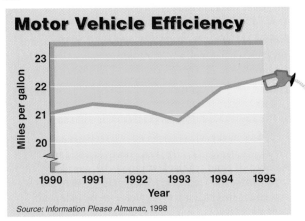

Source: Information Please Almanac, 1998

36.

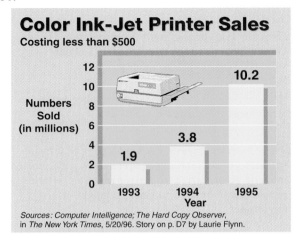

Color Ink-Jet Printer Sales
Costing less than $500

Numbers Sold (in millions)

1993: 1.9
1994: 3.8
1995: 10.2

Year

Sources: *Computer Intelligence; The Hard Copy Observer,* in *The New York Times,* 5/20/96. Story on p. D7 by Laurie Flynn.

37. *$110,000 Loan Repayment.*

TIME OF LOAN (IN YEARS)	MONTHLY LOAN PAYMENT
5	$2283.42
10	1393.43
20	989.70
30	885.08
40	848.50

38. *World Bicycle Production.*

YEAR	WORLD BICYCLE PRODUCTION (IN MILLIONS)
1989	95
1990	90
1991	96
1992	103
1993	108
1994	111
1995	114

Source: United Nations Interbike Directory

39. *Sales of Color Ink-Jet Printers.* The sales of color ink-jet printers has been growing exponentially, as shown in the graph of Exercise 36.

a) Use regression to fit an exponential function $f(x) = ab^x$ to the data, where x is the number of years since 1993.
b) Determine the exponential growth rate.
c) Predict the sales of color ink-jet printers in 2004.

40. *Loan Repayment.* The amount of a monthly mortgage payment for a $110,000 loan decreases exponentially as the time of the loan increases, as shown in the table in Example 37.

a) Use regression to fit an exponential function $f(x) = ab^x$ to the data.
b) Determine the exponential decay rate.
c) Estimate the amount of a monthly mortgage payment for a 25-yr loan.

41. *Cell Phones.* The number of people using cell phones is growing exponentially, as shown by the data in the following table.

YEAR	NUMBER OF SUBSCRIBERS
1985	340,123
1987	1,230,855
1989	3,508,944
1991	7,557,148
1994	24,134,421
1996	44,000,000

Source: Cellular Telecommunications Industry Association

a) Use regression to fit an exponential function $P(t) = P_0 e^{kt}$ to the data, where t is the number of years since 1985 and P is in millions.
b) Predict the number of cell-phone subscribers in 2005.

42. *World Population.* The population of the world is growing exponentially, as shown by the graph on p. 580.

a) Use regression to fit an exponential function $P(t) = P_0 e^{kt}$ to the data, where t is the number of years since 1804 and P is in billions.
b) Predict the world population in 2050.

43. *Fluoride Water Content.* The number of decayed, missing, and fitted teeth in dental patients decreases exponentially as the fluoride content of the water increases, as indicated by the following data from one study.

FLUORIDE CONTENT OF WATER (IN PARTS PER MILLION)	NUMBER OF DECAYED, MISSING, OR FILLED TEETH PER 100 PARTICIPANTS IN STUDY
0.2	805
0.4	410
0.7	360
1	350
1.7	270

Source: Dudley, Brian A.C. *Mathematical and Biological Interrelations,* Chichester: John Wiley & Sons, 1977

a) Use regression to fit an exponential function $f(x) = ab^x$ to the data.
b) Determine the exponential decay rate.
c) Estimate the number of decayed, missing, or filled teeth per 100 patients if the fluoride count of the water is 2.5 ppm.

44. *Spread of AIDS.* In 1985, a total of 8249 cases of AIDS was reported in the United States; in 1988, a total of 31,001 cases, in 1989, a total of 33,722 cases; and in 1990, a total of 41,595 cases.

a) Use regression to fit an exponential function $f(x) = ab^x$ to the data, where x is the number of years since 1985.
b) Determine the exponential growth rate.
c) Predict the number of AIDS cases reported in 2005.

Skill Maintenance

Simplify. [6.3]

45. $\dfrac{\dfrac{x-5}{x+3}}{\dfrac{x}{x-3}+\dfrac{2}{x+3}}$

46. $\dfrac{\dfrac{3}{a}+\dfrac{5}{b}}{\dfrac{2}{a^2}-\dfrac{4}{b^2}}$

47. $\dfrac{6a^3b^{-7}}{8a^{-5}b^{-10}}$

48. $\dfrac{a^{-1}+b^{-2}}{a^{-2}-b^{-1}}$

Synthesis

49. ◈ Use the models developed in Example 4 and Exercise 41 to predict both the population of the United States and the number of cell-phone subscribers in 2005. Do the predictions make sense? Why or why not?

50. ◈ *Atmospheric Pressure.* Atmospheric pressure P at altitude a is given by
$$P = P_0e^{-0.00005a},$$
where P_0 is the pressure at sea level ≈ 14.7 lb/in^2 (pounds per square inch). Explain how a barometer, or some other device for measuring atmospheric pressure, can be used to find the height of a skyscraper.

51. ◈ By the end of 1996, the cumulative number of cases of AIDS in the United States had grown to approximately 607,308 (*Source:* Based on interview with the Vermont AIDS hotline). Use the information in Example 5 to determine whether an exponential function or a linear function better fits the data. Then explain how you would predict the cumulative number of cases of AIDS in 2005.

52. ◈ Examine the restriction on t in Exercise 20.

a) What upper limit might be placed on t?
b) In practice, would this upper limit ever be enforced? Why or why not?

Logistic Curves. Realistically, most quantities that are growing exponentially eventually level off. The quantity may continue to increase, but at a decreasing rate. This pattern of growth can be modeled by a logistic function
$$f(x) = \frac{c}{(1+ae^{-bx})}.$$

The general shape of this family of functions is shown by the following graph.

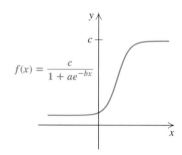

Many graphers can fit a logistic function to a set of data.

53. *Telephones.*   The percentage of telephones in U.S. households for various years is shown in the following table.

YEAR	PERCENTAGE OF HOUSEHOLDS WITH TELEPHONES
1920	35.0
1930	40.9
1950	61.8
1960	78.3
1970	90.5
1980	92.9
1990	94.8
1996	93.9

Source: U.S. Bureau of the Census; Federal Communications Commission

a) Graph the data from 1920 to 1970. Do the data appear to be growing exponentially?
b) Use regression and only the data from 1920 to 1970 to find an exponential function $f(x) = ab^x$ that could be used to estimate the percentage of U.S. households with telephones x years after 1920.
c) Use the function found in part (b) to predict the percentage of U.S. households with telephones in 2010. Does the estimate make sense?
d) Using all the data in the table, find a logistic function

$$f(x) = \frac{c}{1 + ae^{-bx}}$$

that could be used to estimate the percentage of U.S. households with telephones x years after 1920.
e) Use the function found in part (d) to predict the percentage of U.S. households with telephones in 2010. Does the estimate make sense?

54. *Home Videos.*   The following table shows the total revenue from sales and rentals of home videos for various years.

YEAR	TOTAL REVENUE (IN BILLIONS)
1985	$ 3.41
1990	9.81
1995	14.83
2000	18.18

Source: Paul Kagan Associates, Inc.

a) Find a logistic function

$$f(x) = \frac{c}{1 + ae^{-bx}}$$

that could be used to estimate the total revenue of the home video industry x years after 1985.
b) Use the function from part (a) to predict the total revenue from home videos in 2005.

55. *Supply and Demand.*   The supply and demand for the sale of stereos by Sound Ideas are given by

$$S(x) = e^x \quad \text{and} \quad D(x) = 162{,}755e^{-x},$$

where $S(x)$ is the price at which the company is willing to supply x stereos and $D(x)$ is the demand price for a quantity of x stereos. Find the equilibrium point. (For reference, see Section 3.7.)

COLLABORATIVE CORNER

Focus: Models

Time: 20 minutes

Group size: 3–6

Sometimes simply looking at the graph of a particular set of data will give a good indication of an appropriate model for the data. Often, more than one type of model is a possibility. One way to determine which is the best model is to calculate several possible functions, graph them using the data, and determine which one seems to best fit the data. Another indication of a good model is its ability to predict another, known, data point not included in the regression.

Activity

The following table shows the amount that television networks have either paid or agreed to pay for the television rights to the Summer Olympic Games.

Year	Amount Paid for Television Rights (in Millions)
1968	$ 4.5
1972	7.5
1976	25
1980	87
1984	225
1988	300
1992	401
1996	456
2000	705
2004	793

Source: NBC Sports

1. Each group member should enter and graph the data. The group should then agree on at least two possible models for the data from the following list: linear, quadratic, cubic, quartic, or exponential. (See the library of functions on the inside front cover.)

2. Use regression to calculate each of the models chosen in part (1), and graph each model along with the data. Determine, as a group, which model appears to fit the data best.

3. NBC has agreed to pay $894 million for the television rights to the 2008 Summer Olympics. Determine, as a group, which model best predicts this amount.

CHAPTER

9 *Summary and Review*

Key Terms

Exponential function, p. 581
Asymptote, p. 582
Composite function, p. 588
Inverse relation, p. 592
One-to-one function, p. 592
Horizontal-line test, p. 593
Logarithmic function, p. 602
Common logarithm, p. 604

Natural logarithm, p. 617
Exponential equation, p. 622
Logarithmic equation, p. 625
Exponential growth model,
  p. 631
Exponential growth rate,
  p. 631

Doubling time, p. 631
Exponential decay model,
  p. 633
Exponential decay rate,
  p. 633
Half-life, p. 633

Important Properties and Formulas

Exponential function: $f(x) = a^x$
Interest compounded annually: $A = P(1 + i)^t$
Composition of f and g: $f \circ g(x) = f(g(x))$

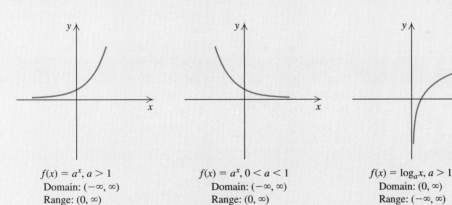

$f(x) = a^x, a > 1$
Domain: $(-\infty, \infty)$
Range: $(0, \infty)$

$f(x) = a^x, 0 < a < 1$
Domain: $(-\infty, \infty)$
Range: $(0, \infty)$

$f(x) = \log_a x, a > 1$
Domain: $(0, \infty)$
Range: $(-\infty, \infty)$

To Find a Formula for the Inverse of a Function

Make sure that the function f is one-to-one. Then:

1. Replace $f(x)$ with y.

2. Interchange x and y.

3. Solve for y.

4. Replace y with $f^{-1}(x)$.

For $x > 0$ and a a positive constant other than 1, $\log_a x$ is the number to which a is raised to get x. Thus, $a^{\log_a x} = x$, or equivalently, if $y = \log_a x$, then $a^y = x$.

The Principle of Exponential Equality

For any real number b, $b \neq -1, 0,$ or 1:

$$b^x = b^y \quad \text{is equivalent to} \quad x = y.$$

The Principle of Logarithmic Equality

For any logarithm base a, and for $x, y > 0$:

$$x = y \quad \text{is equivalent to} \quad \log_a x = \log_a y.$$

Properties of logarithms:

$$\log_a MN = \log_a M + \log_a N, \qquad \log_a \frac{M}{N} = \log_a M - \log_a N,$$

$$\log_a M^p = p \cdot \log_a M, \qquad \log_a 1 = 0,$$
$$\log_a a = 1, \qquad \log_a a^k = k,$$

$$\log M = \log_{10} M, \qquad \log_b M = \frac{\log_a M}{\log_a b}$$

$$\ln M = \log_e M,$$

$e \approx 2.7182818284\ldots$
Exponential growth: $\qquad\qquad P(t) = P_0 e^{kt}, k > 0$
Exponential decay: $\qquad\qquad\; P(t) = P_0 e^{-kt}, k > 0$
Interest compounded continuously: $\; P(t) = P_0 e^{kt}$, where P_0 is the principal invested for t years at interest rate k
Carbon dating: $\qquad\qquad\qquad P(t) = P_0 e^{-0.00012t}$

REVIEW EXERCISES

Graph.

1. $f(x) = 3^{x-1}$

2. $x = 3^{y-1}$

3. $y = \log_3 x$

4. $f(x) = e^{x+1}$

5. Find $f \circ g(x)$ and $g \circ f(x)$ if $f(x) = x^2$ and $g(x) = 3x - 5$.

6. If $h(x) = \sqrt{4 - 7x}$, find $f(x)$ and $g(x)$ such that $h(x) = f \circ g(x)$. Answers may vary.

7. Determine whether $f(x) = 4 - x^2$ is one-to-one.

Find a formula for the inverse of each function.

8. $f(x) = x + 2$

9. $g(x) = \dfrac{2x - 3}{7}$

10. $f(x) = 8x^3$

Convert to exponential equations.

11. $\log_4 16 = x$

12. $\log_{1/2} 8 = -3$

Convert to logarithmic equations.

13. $10^4 = 10,000$ **14.** $25^{1/2} = 5$

Find each of the following.

15. $\log_3 9$ **16.** $\log_{10} \frac{1}{10}$

Express in terms of logarithms of x, y, and z.

17. $\log_a x^4 y^2 z^3$ **18.** $\log_a \frac{xy}{z^2}$

19. $\log \sqrt[4]{\dfrac{z^2}{x^3 y}}$ **20.** $\log_q \left(\dfrac{x^2 y^{1/3}}{z^4} \right)$

Express as a single logarithm.

21. $\log_a 8 + \log_a 15$

22. $\log_a 72 - \log_a 12$

23. $\frac{1}{2} \log a - \log b - 2 \log c$

24. $\frac{1}{3}[\log_a x - 2 \log_a y]$

Simplify.

25. $\log_m m$ **26.** $\log_m 1$

27. $\log_m m^{17}$ **28.** $\log_m m^{-7}$

Given $\log_a 2 = 1.8301$ and $\log_a 7 = 5.0999$, find each of the following.

29. $\log_a 14$ **30.** $\log_a \frac{2}{7}$ **31.** $\log_a 28$

32. $\log_a 3.5$ **33.** $\log_a \sqrt{7}$ **34.** $\log_a \frac{1}{4}$

Find each of the following.

35. $\log 0.00627$ **36.** $\log 72,800,000$

37. $10^{1.789}$ **38.** $10^{-3.65}$

39. $\ln 23,912.2$ **40.** $\ln 0.06774$

41. $e^{-0.98}$ **42.** $e^{2.91}$

Find each of the following logarithms using the change-of-base formula.

43. $\log_5 2$ **44.** $\log_{12} 70$

Solve. Where appropriate, include approximations to the nearest ten-thousandth.

45. $\log_3 x = -2$ **46.** $\log_x 32 = 5$

47. $\log x = -4$ **48.** $3 \ln x = -6$

49. $4^{2x-5} = 16$ **50.** $2^{x^2} \cdot 2^{4x} = 32$

51. $4^x = 8.3$ **52.** $e^{-0.1t} = 0.03$

53. $\ln (x + 2) = x$

54. $\log_4 x + \log_4 (x - 6) = 2$

55. $\log x + \log (x - 15) = 2$

56. $\log_3 (x - 4) = 3 - \log_3 (x + 4)$

57. *Forgetting.* In a business class, students were tested at the end of the course on a final exam. They were tested again after 6 months. The forgetting formula was determined to be

$$S(t) = 62 - 18 \log (t + 1),$$

where t is the time, in months, after taking the first test.

a) Determine the average score when they first took the test (when $t = 0$).
b) What was the average score after 6 months?
c) After what time was the average score 34?

58. *Aeronautics.* There were 6 commercial space launches in 1994 and 12 in 1996.* Assume that the number is growing exponentially.

a) Find k and write the exponential growth function.
b) Predict the number of launches in 2003.
c) When will there be 192 annual scheduled launches?

59. The population of Riverton doubled in 16 yr. What was the exponential growth rate?

60. How long will it take $7600 to double itself if it is invested at 8.4%, compounded continuously?

61. How old is a skeleton that has lost 34% of its carbon-14? (Use $P(t) = P_0 e^{-0.00012t}$.)

62. The intensity of the sound of water at the foot of the Niagara Falls is about 10^{-3} W/m².† How loud in decibels is this sound level?

$$\left(\text{Use } L = 10 \cdot \log \frac{I}{I_0}, \text{ where } I_0 = 10^{-12} \text{ W/m}^2 \right)$$

63. *E. Coli Outbreaks.* The number of *E. coli* outbreaks reported in the United States is growing exponentially. In 1982, 47 cases were reported. This number increased to 70 in 1984, 246 in 1989, and 1000 in 1993. (*Source: Indianapolis Star*)

a) Use regression to fit an exponential function $f(x) = ab^x$ to the data, where x is the number of years since 1982.
b) Determine the exponential growth rate.
c) Predict the number of *E. coli* cases reported in 2005.

---

*Source: Telephone interview, U.S. Department of Transportation, Federal Aviation Administration, 1/6/97.
†*Sound and Hearing,* Life Science Library. (New York: Time Incorporated, 1965), p. 173.

Skill Maintenance

64. Solve $aT^2 + bT = Q$ for T.

65. Solve: $x^4 + 80 = 21x^2$.

66. Divide: $\dfrac{4 - 5i}{1 + 3i}$.

67. Simplify:
$$\frac{\dfrac{1}{ab} - \dfrac{2}{bc}}{\dfrac{2}{ac} + \dfrac{3}{ab}}.$$

Synthesis

68. ◆ Explain why negative numbers do not have logarithms.

69. ◆ Explain why taking the natural or common logarithm on each side of an equation produces an equivalent equation.

Solve.

70. $\ln (\ln x) = 3$

71. $2^{x^2 + 4x} = \frac{1}{8}$

72. $5^{x+y} = 25$,
$2^{2x-y} = 64$

CHAPTER 9 TEST

Graph.

1. $f(x) = 2^{x+3}$

2. $f(x) = \log_7 x$

3. Find $f \circ g(x)$ and $g \circ f(x)$ if $f(x) = x + x^2$ and $g(x) = 5x - 2$.

4. Determine whether $f(x) = 2 - |x|$ is one-to-one.

Find a formula for the inverse.

5. $f(x) = 4x - 3$

6. $f(x) = (x + 1)^3$

Convert to logarithmic equations.

7. $4^{-3} = x$

8. $256^{1/2} = 16$

Convert to exponential equations.

9. $\log_4 16 = 2$

10. $m = \log_7 49$

11. Express in terms of logarithms of a, b, and c:
$$\log \frac{a^3 b^{1/2}}{c^2}.$$

12. Express as a single logarithm:
$$\tfrac{1}{3} \log_a x + 2 \log_a z.$$

Simplify.

13. $\log_5 125$

14. $\log_t t^{23}$

15. $\log_p p$

16. $\log_c 1$

Given $\log_a 2 = 0.301$, $\log_a 6 = 0.778$, and $\log_a 7 = 0.845$, find each of the following.

17. $\log_a \frac{2}{7}$

18. $\log_a 12$

19. $\log_a 21$

Find each of the following.

20. $\log 0.0123$

21. $10^{3.8}$

22. $\ln 0.01234$

23. $e^{4.68}$

24. Find $\log_{18} 31$ using the change-of-base formula.

Solve. Where appropriate, include approximations to the nearest ten-thousandth.

25. $\log_x 25 = 2$

26. $\log_4 x = \frac{1}{2}$

27. $\log x = 4$

28. $5^{4-3x} = 125$

29. $7^x = 1.2$

30. $\ln x = \frac{1}{4}$

31. $\log (x - 3) + \log (x + 1) = \log 5$

32. The average walking speed R of people living in a city of population P, in thousands, is given by $R = 0.37 \ln P + 0.05$, where R is in feet per second.

a) The population of Akron, Ohio, is 225,000. Find the average walking speed.

b) A city has an average walking speed of 2.6 ft/sec. Find the population.

33. The population of Pakistan was 132 million in 1997, and the exponential growth rate was 1.7% per year.

a) Write an exponential function describing the population of Pakistan.

b) What will the population be in 2010?

c) When will the population be 200 million?

d) What is the doubling time?

34. An investment with interest compounded

continuously doubled itself in 15 yr. What is the interest rate?

35. How old is an animal bone that has lost 43% of its carbon-14? (Use $P(t) = P_0 e^{-0.00012t}$.)

36. The hydrogen ion concentration of water is 1.0×10^{-7} moles per liter. What is the pH? (Use pH $= -\log [\text{H}^+]$.)

37. *Social Security.* The number of workers per social security beneficiary is declining exponentially. In 1941, there were 41.9 workers for every beneficiary. That number declined to 16.5 in 1950, 5.1 in 1960, 3.7 in 1970, and 3.4 in 1990. (*Source: The Wall Street Journal Almanac 1998*)

a) Use regression to fit an exponential function $f(x) = ab^x$ to the data, where x is the number of years since 1941.
b) Determine the exponential decay rate.
c) Predict the number of workers per social security beneficiary in 2010.

Skill Maintenance

38. Solve: $y - 9\sqrt{y} + 8 = 0$.

39. Solve $S = at^2 - bt$ for t.

40. Multiply: $(2 + 5i)(2 - 5i)$.

41. Simplify:
$$\frac{\dfrac{1}{x^2 - 4}}{\dfrac{1}{x + 2} + \dfrac{1}{x - 2}}.$$

Synthesis

42. Solve: $\log_5 |2x - 7| = 4$.

43. If $\log_a x = 2$, $\log_a y = 3$, and $\log_a z = 4$, find
$$\log_a \frac{\sqrt[3]{x^2 z}}{\sqrt[3]{y^2 z^{-1}}}.$$

CUMULATIVE REVIEW 1–9

1. Evaluate $\dfrac{x^0 + y}{-z}$ for $x = 6$, $y = 9$, and $z = -5$.

Simplify.

2. $\left| -\frac{5}{2} + \left(-\frac{7}{2} \right) \right|$
3. $(-2x^2 y^{-3})^{-4}$
4. $(-5x^4 y^{-3} z^2)(-4x^2 y^2)$
5. $\dfrac{3x^4 y^6 z^{-2}}{-9x^4 y^2 z^3}$
6. $2x - 3 - 2[5 - 3(2 - x)]$
7. $3^3 + 2^2 - (32 \div 4 - 16 \div 8)$

Solve.

8. $8(2x - 3) = 6 - 4(2 - 3x)$
9. $4x - 3y = 15,$
$3x + 5y = 4$
10. $x + y - 3z = -1,$
$2x - y + z = 4,$
$-x - y + z = 1$
11. $x(x - 3) = 10$

12. $\dfrac{7}{x^2 - 5x} - \dfrac{2}{x - 5} = \dfrac{4}{x}$
13. $\sqrt{x - 1} = \sqrt{x + 4} - 1$
14. $\sqrt[3]{2x} = 1$
15. $3x^2 + 75 = 0$
16. $x^4 - 13x^2 + 36 = 0$
17. $\log_8 x = 1$ **18.** $\log_x 49 = 2$
19. $9^x = 27$ **20.** $3^{5x} = 7$
21. $x^2 + 4x > 5$
22. $\log x - \log (x - 8) = 1$
23. If $f(x) = x^2 + 6x$, find a such that $f(a) = 11$.
24. If $f(x) = |2x - 3|$, find all x for which $f(x) \geq 9$.

Solve.

25. $D = \dfrac{ab}{b + a}$, for a
26. $M = \dfrac{2}{3}(A + B)$, for B

In Exercises 27–30, match each function with one of the following graphs.

a)

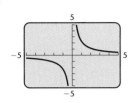

b)

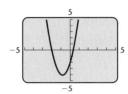

c)

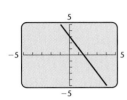

d)

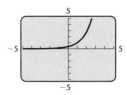

27. $f(x) = -2x + 3.1$

28. $p(x) = 3x^2 + 5x - 2$

29. $h(x) = \dfrac{2.3}{x}$

30. $g(x) = e^{x-1}$

31. Find the domain of the function f given by

$$f(x) = \frac{-4}{3x^2 - 5x - 2}.$$

32. Lowell King, a farmer in the greater Indianapolis metropolitan area, is losing farmland to urban development. He farmed 1400 acres in 1994 and 950 in 1996 (*Source: Indianapolis Star, 7/14/96*). Let A = the number of acres farmed and t the number of years since 1994.

a) Find a linear function $A(t)$ that fits the data.

b) Use the function of part (a) to predict the number of acres farmed in 2000.

For each of the following sets of data, (a) graph the data and determine whether a linear, quadratic, or exponential function would best model the situation and (b) use regression to fit a function to the data.

33.

YEAR	NATIONAL DEBT (IN TRILLIONS)
1988	$2.60
1990	3.20
1992	4.05
1994	4.75
1996	5.22

Source: U.S. Department of the Treasury

34.

YEAR	NUMBER OF TEENS TO BEGIN SMOKING DAILY (PER 1000 PREVIOUSLY NONSMOKING TEENS)
1977	67
1983	44
1988	51.2
1996	77.0

Source: Centers for Disease Control and Prevention

Solve.

35. The perimeter of a rectangular garden is 112 m. The length is 16 m more than the width. Find the length and the width.

36. In triangle ABC, the measure of angle B is three times the measure of angle A. The measure of angle C is 105° greater than the measure of angle A. Find the angle measures.

37. Jenny can build a shed from a lumber kit in 10 hr. Phil can build the same shed in 12 hr. How long would it take Phil and Jenny, working together, to build the shed?

38. Swim Clean is 30% muriatic acid. Pure Swim is 80% muriatic acid. How many liters of each should be mixed together in order to get 100 L of a solution that is 50% muriatic acid?

39. A fishing boat with a trolling motor can move at a speed of 5 km/h in still water. The boat travels 42 km downstream in the same time that it takes to travel 12 km upstream. What is the speed of the stream?

40. What is the minimum product of two numbers whose difference is 14? What are the numbers that yield this product?

Students in a biology class just took a final exam. A formula for determining what the average exam grade will be t months later is

$$S(t) = 78 - 15 \log (t + 1).$$

41. The average score when the students first took the test occurs when $t = 0$. Find the students' average score on the final exam.

42. What would the average score be on a retest after 4 months?

The population of Ukraine was 51 million in 1997, and the exponential decay rate was −0.1% per year.

43. Write an exponential function describing the growth of the population of Ukraine.

44. Predict what the population will be in 2005.

45. y varies directly as the square of x and inversely as z, and $y = 2$ when $x = 5$ and $z = 100$. What is y when $x = 3$ and $z = 4$?

Perform the indicated operations and simplify.

46. $(5p^2q^3 - 4p^3q + 6pq - p^2 + 3)$
$\qquad\qquad + (2p^2q^3 + 2p^3q + p^2 - 5pq - 9)$

47. $(11x^2 - 6x - 3) - (3x^2 + 5x - 2)$

48. $(3x^2 - 2y)^2$

49. $(5a + 3b)(2a - 3b)$

50. $\dfrac{x^2 + 8x + 16}{2x + 6} \div \dfrac{x^2 + 3x - 4}{x^2 - 9}$

51. $\dfrac{1 + \dfrac{3}{x}}{x - 1 - \dfrac{12}{x}}$

52. $\dfrac{3}{x + 6} - \dfrac{2}{x^2 - 36} + \dfrac{4}{x - 6}$

Factor.

53. $xy - 2xz + xw$ 　　　 **54.** $1 - 2x + x^2$

55. $6x^2 + 8xy - 8y^2$ 　　　 **56.** $x^4 - 4x^3 + 7x - 28$

57. $a^2 - 10a + 25 - 81b^2$

58. $2m^2 + 12mn + 18n^2$

59. $3x^2 - 3x - 18$

60. For the function described by
$$h(x) = -3x^2 + 4x + 8,$$
find $h(-2)$.

61. Divide: $(x^4 - 5x^3 + 2x^2 - 6) \div (x - 3)$.

62. Multiply $(5.2 \times 10^4)(3.5 \times 10^{-6})$. Write scientific notation for the answer.

63. Divide:
$$\dfrac{3.4 \times 10^5}{6.8 \times 10^{-9}}.$$
Write scientific notation for the answer.

For the radical expressions that follow, assume that all variables represent positive numbers.

64. Divide and simplify:
$$\dfrac{\sqrt[3]{40xy^8}}{\sqrt[3]{5xy}}.$$

65. Multiply and simplify:
$$\sqrt{7xy^3} \cdot \sqrt{28x^2y}.$$

66. Rewrite without rational exponents: $(27a^6b)^{4/3}$.

67. Rationalize the denominator:
$$\dfrac{3 - \sqrt{y}}{2 - \sqrt{y}}.$$

68. Find the center and the radius of the circle given by
$$(x - 5)^2 + (y + 1)^2 = 49.$$

69. Multiply these complex numbers:
$$(1 + i\sqrt{3})(6 - 2i\sqrt{3}).$$

70. Divide these complex numbers:
$$\dfrac{3 - 2i}{4 - 3i}.$$

71. Find the inverse of f if $f(x) = 7 - 2x$.

72. Find a linear function with a graph that contains the points $(0, -3)$ and $(-1, 2)$.

73. Find an equation of the line with a y-intercept of $(0, -1)$ that is perpendicular to the line whose equation is $2x + y = 6$.

Graph.

74. $5x = 15 + 3y$

75. $y = 2x^2 - 4x - 1$

76. $y = \log_3 x$

77. $y = 3^x$

78. $-2x - 3y \le 6$

79. Graph:
$$f(x) = 2(x + 3)^2 + 1.$$
a) Label the vertex.
b) Draw the axis of symmetry.
c) Find the maximum or minimum value.

80. Express in terms of logarithms of a, b, and c:
$$\log\left(\dfrac{a^2c^3}{b}\right).$$

81. Express as a single logarithm:
$$3 \log x - \tfrac{1}{2} \log y - 2 \log z.$$

82. Convert to an exponential equation: $\log_a 5 = x$.

83. Convert to a logarithmic equation: $x^3 = t$.

Find each of the following.

84. $\log 0.05566$ 　　　 **85.** $10^{2.89}$

86. $\ln 12.78$ 　　　 **87.** $e^{-1.4}$

Synthesis

Solve.

88. $\dfrac{5}{3x - 3} + \dfrac{10}{3x + 6} = \dfrac{5x}{x^2 + x - 2}$

89. $\log \sqrt{3x} = \sqrt{\log 3x}$

90. A train travels 280 mi at a certain speed. If the speed had been increased by 5 mph, the trip could have been made in 1 hr less time. Find the actual speed.

Sequences, Series, and the Binomial Theorem

10

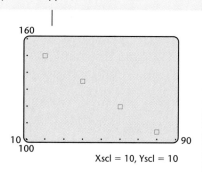

The first three sections of this chapter are devoted to a study of *sequences and series*. A sequence is simply an ordered list. For example, when a baseball coach writes a batting order, a sequence is being formed. When the members of a sequence are numbers, we can discuss their sum. Such a sum is called a *series*.

Section 11.4 presents the *binomial theorem*, which is used to expand expressions of the form $(a + b)^n$. Such an expression is itself a series.

APPLICATION

AEROBIC EXERCISE.   In exercising, your target heart rate decreases as age increases. Graph the data in the following table, and state whether or not the numbers are part of an arithmetic sequence. If so, find a formula for the general term of the sequence.

AGE	TARGET HEART RATE (IN BEATS PER MINUTE)
20	150
40	135
60	120
80	105

The points lie on a straight line, so the numbers are part of an arithmetic sequence.

This problem appears as Exercise 56 in Section 10.2.

Xscl = 10, Yscl = 10

10.1   Sequences and Series
10.2   Arithmetic Sequences and Series
10.3   Geometric Sequences and Series
10.4   The Binomial Theorem
       SUMMARY AND REVIEW
       TEST

10.1
Sequences and Series

- *Sequences*
- *Finding the General Term*
- *Sums and Series*
- *Sigma Notation*
- *Graphs of Sequences*

Sequences

Suppose that $1000 is invested at 8%, compounded annually. The amounts to which the account will grow after 1 year, 2 years, 3 years, and so on, form the following sequence of numbers:

$$\$1080.00, \quad \$1166.40, \quad \$1259.71, \quad \$1360.49, \ldots.$$

We can regard this as a function that pairs 1 with $1080.00, 2 with $1166.40, 3 with $1259.71, and so on. A **sequence** is thus a function, where the domain is a set of consecutive positive integers beginning with 1.

If we continue computing the amounts in the account forever, we obtain an **infinite sequence**, with function values

$$\$1080.00, \$1166.40, \$1259.71, \$1360.49, \$1469.33, \$1586.87, \ldots.$$

The three dots at the end indicate that the sequence goes on without stopping. If we stop after a certain number of years, we obtain a **finite sequence**:

$$\$1080.00, \$1166.40, \$1259.71, \$1360.49.$$

Sequences

An *infinite sequence* is a function having for its domain the set of positive integers: $\{1, 2, 3, 4, 5, \ldots\}$.

A *finite sequence* is a function having for its domain a set of positive integers $\{1, 2, 3, 4, 5, \ldots, n\}$, for some positive integer n.

As another example, consider the sequence given by

$$a(n) = 2^n, \quad \text{or} \quad a_n = 2^n.$$

The notation a_n means the same as $a(n)$.

Some function values (also called *terms* of the sequence) follow:

$$a_1 = 2^1 = 2,$$
$$a_2 = 2^2 = 4,$$
$$a_3 = 2^3 = 8,$$
$$a_6 = 2^6 = 64.$$

The first term of the sequence is a_1, the fifth term is a_5, and the nth term, or **general term**, is a_n. This sequence can also be denoted in the following ways:

$$2, 4, 8, \ldots; \quad \text{or}$$
$$2, 4, 8, \ldots, 2^n, \ldots.$$

The 2^n emphasizes that the nth term of this sequence is found by raising 2 to the nth power.

To enter a sequence in a grapher, we first select the SEQUENCE mode. The variable n is used instead of x, and the functions are named u and v instead of Y1 and Y2.

Example 1   Find the first 4 terms and the 13th term of the sequence for which the general term is given by $a_n = (-1)^n n^2$.

--- BY HAND

We have $a_n = (-1)^n n^2$, so

$$a_1 = (-1)^1 \cdot 1^2 = -1,$$
$$a_2 = (-1)^2 \cdot 2^2 = 4,$$
$$a_3 = (-1)^3 \cdot 3^2 = -9,$$
$$a_4 = (-1)^4 \cdot 4^2 = 16,$$
$$a_{13} = (-1)^{13} \cdot 13^2 = -169.$$

--- USING A GRAPHER

We let $u(n) = (-1)^\wedge n * n^2$.

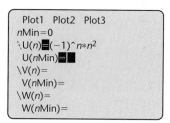

We set up a table with Indpnt set to Ask, and then supply 1, 2, 3, 4, and 13 as values for n.

n	$u(n)$	
1	-1	
2	4	
3	-9	
4	16	
13	-169	
$n =$		

Note in Example 1 that the expression $(-1)^n$ causes the signs of the terms to alternate between positive and negative, depending on whether n is even or odd. This kind of sequence is called an **alternating sequence**.

We may want to write the terms of a sequence using fractional notation. Many graphers have a seq(feature that will write the terms as a list. We can then convert the entire list to fractional notation.

Example 2   Use a grapher to find the first 5 terms of the sequence for which the general term is given by $a_n = n/(n + 1)^2$.

SOLUTION   To use the seq(feature, we must supply the formula for the general term, the variable, and the values of n for the first and last terms that we wish to calculate.

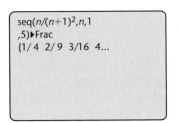

Here we used ▶Frac to write each term in the sequence in fractional notation.

The first three terms are listed on the screen; the remaining can be found by pressing $\boxed{\blacktriangleright}$. We have

$$a_1 = \tfrac{1}{4}, \qquad a_2 = \tfrac{2}{9}, \qquad a_3 = \tfrac{3}{16}, \qquad a_4 = \tfrac{4}{25}, \quad \text{and} \quad a_5 = \tfrac{5}{36}.$$

Finding the General Term

When only the first few terms of a sequence are known, it is impossible to be certain what the general term is, but a prediction can be made by looking for a pattern.

Example 3   For each sequence, predict the general term.

a) $1, 4, 9, 16, 25, \ldots$

b) $2, 4, 8, \ldots$

c) $-1, 2, -4, 8, -16, \ldots$

SOLUTION

a) $1, 4, 9, 16, 25, \ldots$

These are squares of consecutive positive integers, so the general term may be n^2.

b) $2, 4, 8, \ldots$

We regard the pattern as powers of 2, so 16 is the next term and 2^n is the general term. The sequence is then written with more terms as

$$2, 4, 8, 16, 32, 64, 128, \ldots.$$

c) $-1, 2, -4, 8, -16, \ldots$

These are powers of 2 with alternating signs, so the general term may be $(-1)^n[2^{n-1}]$. To check, note that 8 is the fourth term, and $(-1)^4[2^{4-1}] = 1 \cdot 2^3 = 8$.

In part (b) above, suppose that the second term is found by adding 2, the third term by adding 4, the next term by adding 6, and so on. In this case, 14 would be the next term and the sequence would be

$$2, 4, 8, 14, 22, 32, 44, 58, \ldots.$$

This illustrates that the fewer terms we are given, the greater the uncertainty about the nth term.

Sums and Series

Series

Given the infinite sequence

$$a_1, a_2, a_3, a_4, \ldots, a_n, \ldots,$$

the sum of the terms

$$a_1 + a_2 + a_3 + \cdots + a_n + \cdots$$

is called an *infinite series*. A *partial sum*, sometimes called an *nth partial sum*, is the sum of the first n terms:

$$a_1 + a_2 + a_3 + \cdots + a_n.$$

A partial sum is also called a *finite series* and is denoted S_n.

Example 4 For the sequence $-2, 4, -6, 8, -10, 12, -14$, find: **(a)** S_1; **(b)** S_3; **(c)** S_6.

SOLUTION

a) $S_1 = -2$

b) $S_3 = -2 + 4 + (-6) = -4$

c) $S_6 = -2 + 4 + (-6) + 8 + (-10) + 12 = 6$ ▬

We can use a grapher to find partial sums of a sequence for which the general term is given by a formula.

Example 5 Use a grapher to find S_1, S_2, S_3, and S_4 for the sequence in which the general term is given by $a_n = (-1)^n/(n + 1)$.

SOLUTION If the terms of the sequence are given as a list, the grapher can list the partial sums of the sequence. Some graphers use a cumsum(feature. This option lists the cumulative, or partial, sums for a sequence defined using the seq(feature.

```
cumSum(seq((−1)^
n/(n+1),n,1,4))▶
Frac
{−1/2 −1/6 −5/1...
█
```

Here we used ▶Frac to write each sum in fractional notation.

The first two sums, S_1 and S_2, are listed on the screen; S_3 and S_4 can be seen by pressing ▶. We have

$$S_1 = -\tfrac{1}{2}, \qquad S_2 = -\tfrac{1}{6}, \qquad S_3 = -\tfrac{5}{12}, \quad \text{and} \quad S_4 = -\tfrac{13}{60}. \qquad \blacksquare$$

Sigma Notation

When the general term of a sequence is known, the Greek letter Σ (sigma) can be used to write a series. For example, the sum of the first four terms of the sequence $3, 5, 7, 9, \ldots, 2k + 1, \ldots$ can be named as follows, using *sigma notation*, or *summation notation*:

$$\sum_{k=1}^{4} (2k + 1).$$

This is read "the sum, as k goes from 1 to 4, of $(2k + 1)$." The letter k is called the *index of summation*. The index of summation need not start at 1.

Example 6 Find and evaluate each sum.

a) $\displaystyle\sum_{k=1}^{5} k^2$ **b)** $\displaystyle\sum_{k=4}^{6} (-1)^k(2k)$ **c)** $\displaystyle\sum_{k=0}^{3} (2^k + 5)$

SOLUTION

a) $\displaystyle\sum_{k=1}^{5} k^2 = 1^2 + 2^2 + 3^2 + 4^2 + 5^2 = 1 + 4 + 9 + 16 + 25 = 55$

         Evaluate k^2 for all integers from 1 through 5. Then add.

b) $\displaystyle\sum_{k=4}^{6} (-1)^k(2k) = (-1)^4(2 \cdot 4) + (-1)^5(2 \cdot 5) + (-1)^6(2 \cdot 6)$

$$= 8 - 10 + 12 = 10$$

c) $\displaystyle\sum_{k=0}^{3} (2^k + 5) = (2^0 + 5) + (2^1 + 5) + (2^2 + 5) + (2^3 + 5)$

$$= 6 + 7 + 9 + 13 = 35 \qquad \blacksquare$$

Example 7 Write sigma notation for each sum.

a) $1 + 4 + 9 + 16 + 25$ **b)** $-1 + 3 - 5 + 7$

c) $3 + 9 + 27 + 81 + \cdots$

SOLUTION

a) $1 + 4 + 9 + 16 + 25$

This is a sum of squares, $1^2 + 2^2 + 3^2 + 4^2 + 5^2$, so the general term is k^2. Sigma notation is

$$\sum_{k=1}^{5} k^2. \qquad \text{The sum starts with } 1^2 \text{ and ends with } 5^2.$$

Answers may vary here. For example, another—perhaps less obvious—way of writing $1 + 4 + 9 + 16 + 25$ is

$$\sum_{k=2}^{6} (k - 1)^2.$$

b) $-1 + 3 - 5 + 7$

Except for the alternating signs, this is the sum of the first four positive odd numbers. Note that $2k - 1$ is a formula for the kth positive odd number, and $(-1)^k = 1$ when k is even and $(-1)^k = -1$ when k is odd. The general term is thus $(-1)^k(2k - 1)$, beginning with $k = 1$. Sigma notation is

$$\sum_{k=1}^{4} (-1)^k(2k - 1).$$

To check, we can evaluate $(-1)^k(2k - 1)$ using 1, 2, 3, and 4. Then we can write the sum of the four terms. We leave this to the student.

c) $3 + 9 + 27 + 81 + \cdots$

This is a sum of powers of 3, and it is also an infinite series. We use the symbol ∞ for infinity and write the series using sigma notation:

$$\sum_{k=1}^{\infty} 3^k.$$

Graphs of Sequences

Because the domain of a sequence is a set of integers, the graph of a sequence is a set of points that are not connected. We can use the DOT mode to graph a sequence when a formula for the general term is known.

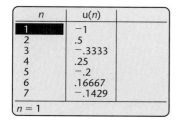

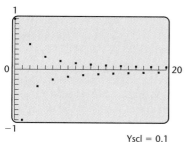

Yscl = 0.1

Example 8 Graph the sequence for which the general term is given by $a_n = (-1)^n/n$.

SOLUTION We let $u(n) = (-1)^\wedge n/n$ and set the graph mode to DOT. You may wish to examine a table of values to help set up the window dimensions for the graph. From the table at the top left, the terms appear to be between -1 and 1.

For the window settings, we must determine nMin and nMax, as well as Xmin, Xmax, Ymin, and Ymax. Here nMin is the smallest value of n for which we wish to evaluate the sequence, and nMax is the largest value. We let nMin = 1 and nMax = 15.

If the grapher also allows us to set PlotStart and PlotStep, we set each of those to 1.

We see from the graph at the bottom left that the absolute value of the terms gets smaller as n gets larger. The graph also illustrates that this is an alternating sequence.

10.1 Exercise Set

In each of the following, the nth term of a sequence is given. In each case, find the first 4 terms; the 10th term, a_{10}; and the 15th term, a_{15}. Then graph the sequence.

1. $a_n = 5n - 3$

2. $a_n = 2n + 5$

3. $a_n = \dfrac{n}{n + 2}$

4. $a_n = n^2 + 1$

5. $a_n = n^2 - 2n$

6. $a_n = \dfrac{n^2 - 1}{n^2 + 1}$

7. $a_n = n + \dfrac{1}{n}$

8. $a_n = \left(-\dfrac{1}{2}\right)^{n-1}$

9. $a_n = (-1)^n n^2$

10. $a_n = (-1)^n(n + 3)$

11. $a_n = (-1)^{n+1}(3n - 5)$

12. $a_n = (-1)^n(n^3 - 1)$

Find the indicated term of each sequence.

13. $a_n = (3n + 4)(2n - 5)$; a_9

14. $a_n = (3n + 2)^2$; a_6

15. $a_n = (-1)^{n-1}(3.4n - 17.3)$; a_{12}

16. $a_n = (-2)^{n-2}(45.68 - 1.2n)$; a_{23}

17. $a_n = \log 10^n$; a_{43}

18. $a_n = \ln e^n$; a_{67}

Predict the general term, or nth term, a_n, of each sequence. Answers may vary.

19. $1, 3, 5, 7, 9, \ldots$

20. $3, 9, 27, 81, 243, \ldots$

21. $-2, 6, -18, 54, \ldots$

22. $-2, 3, 8, 13, 18, \ldots$

23. $\frac{1}{2}, \frac{2}{3}, \frac{3}{4}, \frac{4}{5}, \frac{5}{6}, \ldots$

24. $1, \sqrt{3}, \sqrt{5}, \sqrt{7}, 3, \ldots$

25. $\sqrt{3}, 3, 3\sqrt{3}, 9, 9\sqrt{3}, \ldots$

26. $1 \cdot 2, 2 \cdot 3, 3 \cdot 4, 4 \cdot 5, \ldots$

27. $-1, -4, -7, -10, -13, \ldots$

28. $\log 1, \log 10, \log 100, \log 1000, \ldots$

Find the indicated partial sum for each sequence.

29. $1, -2, 3, -4, 5, -6, \ldots$; S_7

30. $1, -3, 5, -7, 9, -11, \ldots$; S_8

31. $2, 4, 6, 8, \ldots$; S_5

32. $1, \frac{1}{4}, \frac{1}{9}, \frac{1}{16}, \frac{1}{25}, \ldots$; S_5

Rename and evaluate each sum.

33. $\displaystyle\sum_{k=1}^{5} \dfrac{1}{2k}$

34. $\displaystyle\sum_{k=1}^{6} \dfrac{1}{2k - 1}$

35. $\displaystyle\sum_{k=0}^{4} 3^k$

36. $\displaystyle\sum_{k=4}^{7} \sqrt{2k + 1}$

37. $\displaystyle\sum_{k=1}^{8} \dfrac{k}{k + 1}$

38. $\displaystyle\sum_{k=1}^{4} \dfrac{k - 2}{k + 3}$

39. $\displaystyle\sum_{k=1}^{5} (-1)^k$

40. $\displaystyle\sum_{k=1}^{5} (-1)^{k+1}$

41. $\displaystyle\sum_{k=1}^{8} (-1)^{k+1} 2^k$

42. $\displaystyle\sum_{k=1}^{7} (-1)^k 4^{k+1}$

43. $\displaystyle\sum_{k=0}^{5} (k^2 - 2k + 3)$

44. $\displaystyle\sum_{k=0}^{5} (k^2 - 3k + 4)$

Rewrite each sum using sigma notation. Answers may vary.

45. $\dfrac{2}{3} + \dfrac{3}{4} + \dfrac{4}{5} + \dfrac{5}{6} + \dfrac{6}{7}$

46. $3 + 6 + 9 + 12 + 15$

47. $1 + 4 + 9 + 16 + 25 + 36$

48. $\dfrac{1}{1^2} + \dfrac{1}{2^2} + \dfrac{1}{3^2} + \dfrac{1}{4^2} + \dfrac{1}{5^2}$

49. $4 - 9 + 16 - 25 + \cdots + (-1)^n n^2$

50. $9 - 16 + 25 + \cdots + (-1)^{n+1} n^2$

51. $5 + 10 + 15 + 20 + 25 + \cdots$

52. $7 + 14 + 21 + 28 + 35 + \cdots$

53. $\dfrac{1}{1 \cdot 2} + \dfrac{1}{2 \cdot 3} + \dfrac{1}{3 \cdot 4} + \dfrac{1}{4 \cdot 5} + \cdots$

54. $\dfrac{1}{1 \cdot 2^2} + \dfrac{1}{2 \cdot 3^2} + \dfrac{1}{3 \cdot 4^2} + \dfrac{1}{4 \cdot 5^2} + \cdots$

Skill Maintenance

Simplify. [9.4]

55. $6^{\log_6 29}$

56. $9^{\log_9 43}$

57. $\log_3 3$

58. $\log_3 1$

59. $\log_3 3^7$

60. $\log_c c$

Synthesis

61. ◆ The sequence $1, 4, 9, 16, \ldots$ can be written as $f(x) = x^2$ with the domain of $f = \{x \mid x \text{ is an integer and } x > 0\}$. Explain how the graph of f would compare with the graph of $y = x^2$.

62. ◆ Explain why the equation

$$\sum_{k=1}^{n} (a_k + b_k) = \sum_{k=1}^{n} a_k + \sum_{k=1}^{n} b_k$$

is true for any positive integer n. What laws are used to justify this result?

63. ◆ Consider the sums

$$\sum_{k=1}^{5} 3k^2 \quad \text{and} \quad 3\sum_{k=1}^{5} k^2.$$

a) Which is easier to evaluate and why?
b) Is it true that

$$\sum_{k=1}^{n} ca_k = c\sum_{k=1}^{n} a_k?$$

Why or why not?

Some sequences are given by a recursive *definition. The value of the first term, a_1, is given, and then we are told how to find any subsequent term from the term preceding it. Find the first six terms of each of the following recursively defined sequences.*

64. $a_1 = 1$, $a_{n+1} = 5a_n - 2$
65. $a_1 = 0$, $a_{n+1} = a_n^2 + 3$

66. *Cell Biology.* A single cell of bacterium divides into two every 15 min. Suppose that the same rate of division is maintained for 4 hr. Give a sequence that lists the number of cells after successive 15-min periods.

67. *Value of a Copier.* The value of a color photocopier is $5200. Its scrap value each year is 75% of its value the year before. Give a sequence that lists the scrap value of the machine at the start of each year for a 10-yr period.

68. *Hourly Wages.* Katrina is paid $8.20 an hour for filling orders for River's Bend Publishing. Each year she receives a $0.40 hourly raise. Give a sequence that lists Katrina's hourly salary over a 10-yr period.

Find the first 5 terms of each sequence; then find S_5.

69. $a_n = \dfrac{1}{2^n} \log 1000^n$ **70.** $a_n = i^n$, $i = \sqrt{-1}$

71. Find all values for x that solve the following:

$$\sum_{k=1}^{x} i^k = -1.$$

72. The nth term of a sequence is given by

$$a_n = n^5 - 14n^4 + 6n^3 + 416n^2 - 655n - 1050.$$

Use a grapher with a TABLE feature to determine what term in the sequence is 6144.

73. To define a sequence recursively on a grapher (see Exercises 64 and 65), we specify a value for the first term and enter a pattern. Use recursion to determine how many handshakes will occur if a group of 50 people shake hands with one another. To develop the recursion formula, begin with a group of 2 and determine how many additional handshakes occur with the arrival of each new group member. (See the Collaborative Corner following Exercise Set 5.1 on p. 282.)

10.2
Arithmetic Sequences and Series

• *Arithmetic Sequences*
• *Sum of the First n Terms of an Arithmetic Sequence*
• *Problem Solving*

In this section, we concentrate on sequences and series that are said to be arithmetic (pronounced ar-ith-MET-ik).

Arithmetic Sequences

In an **arithmetic sequence**, any term (other than the first) can be found by adding the same number to its preceding term. For example, the sequence 2, 5, 8, 11, 14, 17,... is arithmetic because adding 3 to any term produces the next term. In other words, the difference between any term and the preceding one is 3. Arithmetic sequences are also called *arithmetic progressions.*

Arithmetic Sequence

A sequence is *arithmetic* if there exists a number d, called the *common difference*, such that $a_{n+1} = a_n + d$ for any integer $n \geq 1$.

Example 1 For each arithmetic sequence, identify the first term, a_1, and the common difference, d.

a) 4, 9, 14, 19, 24, . . .   **b)** 27, 20, 13, 6, -1, -8, . . .

SOLUTION   To find a_1, we simply use the first term listed. To find d, we choose any term beyond the first and subtract the preceding term from it.

SEQUENCE	FIRST TERM, a_1	COMMON DIFFERENCE, d
a) 4, 9, 14, 19, 24, . . .	4	$5 \longleftarrow 9 - 4 = 5$
b) 27, 20, 13, 6, -1, -8, . . .	27	$-7 \longleftarrow 20 - 27 = -7$

To find the common difference, we subtracted a_1 from a_2. Had we subtracted a_2 from a_3 or a_3 from a_4, we would have found the same values for d. As a check, we simply add d to any term and see if the next term results.

CHECK: **a)** $4 + 5 = 9$, $9 + 5 = 14$, $14 + 5 = 19$,
            $19 + 5 = 24$

        b) $27 + (-7) = 20$, $20 + (-7) = 13$, $13 + (-7) = 6$,
            $6 + (-7) = -1$, $-1 + (-7) = -8$

To find a formula for the general, or nth, term of any arithmetic sequence, we denote the common difference by d and write out the first few terms:

a_1,

$a_2 = a_1 + d$,

$a_3 = a_2 + d = (a_1 + d) + d = a_1 + 2d$,   Substituting for a_2

$a_4 = a_3 + d = (a_1 + 2d) + d = a_1 + 3d$.   Substituting for a_3

Note that the coefficient of d in each case is 1 less than the subscript.

Generalizing, we obtain the following formula.

> **To Find a_n for an Arithmetic Sequence:**
>
> The nth term of an arithmetic sequence with common difference d is
>
> $$a_n = a_1 + (n - 1)d, \quad \text{for any integer } n \geq 1.$$

Example 2 Find the 14th term of the arithmetic sequence 4, 7, 10, 13, ...

SOLUTION   First we note that $a_1 = 4$, $d = 3$, and $n = 14$. Using the formula for the nth term of an arithmetic sequence, we have

$$a_n = a_1 + (n - 1)d$$
$$a_{14} = 4 + (14 - 1) \cdot 3 = 4 + 13 \cdot 3 = 4 + 39 = 43.$$

The 14th term is 43. ▬

Example 3 For the sequence in Example 2, which term is 301? That is, find n if $a_n = 301$.

SOLUTION   We substitute into the formula for the nth term of an arithmetic sequence and solve for n:

$$a_n = a_1 + (n - 1)d$$
$$301 = 4 + (n - 1) \cdot 3$$
$$301 = 4 + 3n - 3$$
$$300 = 3n$$
$$100 = n.$$

The term 301 is the 100th term of the sequence. ▬

Given two terms and their positions in an arithmetic sequence, we can construct the sequence.

Example 4 The 3rd term of an arithmetic sequence is 8, and the 16th term is 47. Find a_1 and d and construct the sequence.

SOLUTION   We know that $a_3 = 8$ and $a_{16} = 47$. Thus we would have to add d thirteen times to get from 8 to 47. That is,

$$8 + 13d = 47. \qquad \text{a_3 and a_{16} are 13 terms apart.}$$

Solving $8 + 13d = 47$, we obtain

$$13d = 39$$
$$d = 3.$$

We subtract d twice from a_3 to get to a_1. Thus,

$$a_1 = 8 - 2 \cdot 3 = 2. \qquad \text{a_1 and a_3 are 2 terms apart.}$$

The sequence is 2, 5, 8, 11, ... Note that we could have subtracted d 15 times from a_{16} in order to find a_1. ▬

In general, d should be subtracted $(n - 1)$ times from a_n in order to find a_1.

What will the graph of an arithmetic sequence look like?

Interactive Discovery

Graph each of the following arithmetic sequences. What pattern do you observe?

1. $a_n = -5 + (n - 1)3$

2. $a_n = \frac{1}{2} + (n - 1)\frac{3}{2}$

3. $a_n = 60 + (n - 1)(-5)$

4. $a_n = -1.7 + (n - 1)(-0.2)$

The pattern you may have observed above is true in general:

The graph of an arithmetic sequence is a set of points that lie on a straight line.

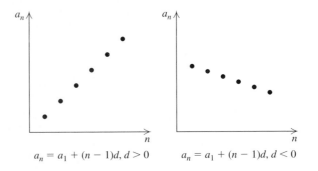

$a_n = a_1 + (n - 1)d, d > 0 \qquad a_n = a_1 + (n - 1)d, d < 0$

Sum of the First n Terms of an Arithmetic Sequence

When the terms of an arithmetic sequence are added, an **arithmetic series** is formed. To find a formula for computing S_n when the series is arithmetic, we denote the first n terms as follows:

This is the next-to-last term. If you add d to this term, the result is a_n.

$$a_1, (a_1 + d), (a_1 + 2d), \ldots, (a_n - 2d), (a_n - d), a_n$$

This term is two terms back from the last. If you add d to this term, you get the next-to-last term, $a_n - d$.

Thus, S_n is given by

$$S_n = a_1 + (a_1 + d) + (a_1 + 2d) + \cdots + (a_n - 2d) + (a_n - d) + a_n.$$

Reversing the order of addition, we have

$$S_n = a_n + (a_n - d) + (a_n - 2d) + \cdots + (a_1 + 2d) + (a_1 + d) + a_1.$$

Adding corresponding terms on each side of the above equations, we get

$$2S_n = [a_1 + a_n] + [(a_1 + d) + (a_n - d)] + [(a_1 + 2d) + (a_n - 2d)]$$
$$+ \cdots + [(a_n - 2d) + (a_1 + 2d)] + [(a_n - d) + (a_1 + d)]$$
$$+ [a_n + a_1].$$

This simplifies to

$$2S_n = [a_1 + a_n] + [a_1 + a_n] + [a_1 + a_n]$$
$$+ \cdots + [a_n + a_1] + [a_n + a_1] + [a_n + a_1].$$

Since $(a_1 + a_n)$ is being added n times, it follows that

$$2S_n = n(a_1 + a_n).$$

This leads to the following formula.

To Find S_n for an Arithmetic Sequence:

The sum of the first n terms of an arithmetic sequence is given by

$$S_n = \frac{n}{2}(a_1 + a_n).$$

Example 5   Find the sum of the first 100 positive even numbers.

SOLUTION   The sum is

$$2 + 4 + 6 + \cdots + 198 + 200.$$

This is the sum of the first 100 terms of the arithmetic sequence for which

$$a_1 = 2, \qquad n = 100, \quad \text{and} \quad a_n = 200.$$

Substituting in the formula

$$S_n = \frac{n}{2}(a_1 + a_n),$$

we get

$$S_{100} = \frac{100}{2}(2 + 200)$$
$$= 50(202) = 10{,}100.$$

The above formula is useful when we know the first and last terms, a_1

and a_n. To find S_n when a_n is unknown, but a_1, n, and d are known, we must first calculate a_n.

Example 6 Find the sum of the first 15 terms of the arithmetic sequence 4, 7, 10, 13, ...

SOLUTION Note that

$$a_1 = 4, \quad n = 15, \quad \text{and} \quad d = 3.$$

Before using the formula for S_n, we find a_{15}:

$$a_{15} = 4 + (15 - 1)3 \qquad \textbf{Substituting into the formula for } a_n$$
$$= 4 + 14 \cdot 3 = 46.$$

Thus, knowing that $a_{15} = 46$, we have

$$S_{15} = \tfrac{15}{2}(4 + 46) \qquad \textbf{Using the formula for } S_n$$
$$= \tfrac{15}{2}(50) = 375.$$

Problem Solving

For some problem-solving situations, the translation may involve sequences or series. We look at some examples.

Example 7 *Hourly Wages.* Chris takes a job, starting with an hourly wage of $14.25, and is promised a raise of 15¢ per hour every 2 months for 5 years. At the end of 5 years, what will be Chris's hourly wage?

SOLUTION

1. **Familiarize.** It helps to write down the hourly wage for several two-month time periods.

   Beginning: 14.25,
   After two months: 14.40,
   After four months: 14.55,

   and so on.

   What appears is a sequence of numbers: 14.25, 14.40, 14.55, ... Since the same amount is added each time, the sequence is arithmetic.
   We ask ourselves what we know about arithmetic sequences. The pertinent formulas are

   $$a_n = a_1 + (n - 1)d$$

   and

   $$S_n = \frac{n}{2}(a_1 + a_n).$$

In this case, we are not looking for a sum, so it is probably the first formula that will give us our answer. We want to determine the last term in a sequence. To do so, we need to know a_1, n, and d. From our list above, we see that

$$a_1 = 14.25 \quad \text{and} \quad d = 0.15.$$

What is n? That is, how many terms are in the sequence? Each year there are 6 raises, since Chris gets a raise every 2 months. There are 5 years, so the total number of raises will be $5 \cdot 6$, or 30. There will be 31 terms: the original wage and 30 increased rates.

2. **Translate.**   We want to find a_n for the arithmetic sequence in which $a_1 = 14.25$, $n = 31$, and $d = 0.15$.

3. **Carry out.**   Substituting in the formula for a_n gives us

$$a_{31} = 14.25 + (31 - 1) \cdot 0.15$$
$$= 18.75.$$

4. **Check.**   We can check by redoing the calculations or we can calculate in a slightly different way for another check. For example, at the end of a year, there will be 6 raises, for a total raise of $0.90. At the end of 5 years, the total raise will be $5 \times \$0.90$, or $4.50. If we add that to the original wage of $14.25, we obtain $18.75. The answer checks.

5. **State.**   At the end of 5 years, Chris's hourly wage will be $18.75.

Example 7 is one in which the calculations or the translation could be done in a number of ways. There is often a variety of ways in which a problem can be solved. You should use the one that is best or easiest for you. In this chapter, however, our approaches emphasize sequences and series and their related formulas.

Example 8 *Telephone Pole Storage.* A stack of telephone poles has 30 poles in the bottom row. There are 29 poles in the second row, 28 in the next row, and so on. How many poles are in the stack if there are 5 poles in the top row?

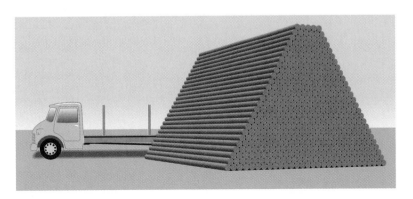

SOLUTION

1. Familiarize. A picture will help in this case. The following figure shows the ends of the poles and the way in which they stack. There are 30 poles on the bottom, and we see that there will be one fewer in each succeeding row. How many rows will there be?

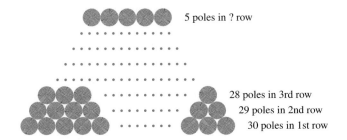

5 poles in ? row

28 poles in 3rd row
29 poles in 2nd row
30 poles in 1st row

Note that there are $30 - 1 = 29$ logs in the 2nd row, $30 - 2 = 28$ logs in the 3rd row, $30 - 3 = 27$ logs in the 4th row, and so on. The pattern leads to $30 - 25 = 5$ logs in the 26th row.

The situation is represented by the equation

$$30 + 29 + 28 + \cdots + 5.$$ **There are 26 terms in this series.**

Thus we have an arithmetic series. We recall the formula

$$S_n = \frac{n}{2}(a_1 + a_n).$$

2. Translate. We want to find the sum of the first 26 terms of an arithmetic sequence in which $a_1 = 30$ and $a_{26} = 5$.

3. Carry out. Substituting into the above formula gives us

$$S_{26} = \frac{26}{2}(30 + 5)$$
$$= 13 \cdot 35 = 455.$$

4. Check. In this case, we can check the calculations by doing them again. A longer, harder way would be to do the entire addition:

$$30 + 29 + 28 + \cdots + 5.$$

5. State. There are 455 poles in the stack.   —

10.2 | *Exercise Set*

Find the first term and the common difference.

1. 3, 8, 13, 18, . . .

2. 1.06, 1.12, 1.18, 1.24, . . .

3. 6, 2, −2, −6, . . .

4. −9, −6, −3, 0, . . .

5. $\frac{3}{2}$, $\frac{9}{4}$, 3, $\frac{15}{4}$, . . .

6. $\frac{3}{5}$, $\frac{1}{10}$, −$\frac{2}{5}$, . . .

7. \$2.12, \$2.24, \$2.36, \$2.48, . . .

8. \$214, \$211, \$208, \$205, . . .

9. Find the 12th term of the arithmetic sequence 3, 7, 11, ...

10. Find the 11th term of the arithmetic sequence
0.07, 0.12, 0.17, ...

11. Find the 17th term of the arithmetic sequence
7, 4, 1, ...

12. Find the 14th term of the arithmetic sequence
$3, \frac{7}{3}, \frac{5}{3}, \ldots$.

13. Find the 13th term of the arithmetic sequence
\$1200, \$964.32, \$728.64, ...

14. Find the 10th term of the arithmetic sequence
\$2345.78, \$2967.54, \$3589.30, ...

15. In the sequence of Exercise 9, what term is 107?

16. In the sequence of Exercise 10, what term is 1.67?

17. In the sequence of Exercise 11, what term is -296?

18. In the sequence of Exercise 12, what term is -27?

19. Find a_{17} when $a_1 = 2$ and $d = 5$.

20. Find a_{20} when $a_1 = 14$ and $d = -3$.

21. Find a_1 when $d = 4$ and $a_8 = 33$.

22. Find a_1 when $d = 8$ and $a_{11} = 26$.

23. Find n when $a_1 = 5$, $d = -3$, and $a_n = -76$.

24. Find n when $a_1 = 25$, $d = -14$, and $a_n = -507$.

25. For an arithmetic sequence in which $a_{17} = -40$ and
$a_{28} = -73$, find a_1 and d. Write the first five terms
of the sequence.

26. In an arithmetic sequence, $a_{17} = \frac{25}{3}$ and $a_{32} = \frac{95}{6}$.
Find a_1 and d. Write the first five terms of the
sequence.

27. Find the sum of the first 20 terms of the arithmetic
series $1 + 5 + 9 + 13 + \cdots$.

28. Find the sum of the first 14 terms of the arithmetic
series $11 + 7 + 3 + \cdots$.

29. Find the sum of the first 300 natural numbers.

30. Find the sum of the first 400 natural numbers.

31. Find the sum of the even numbers from 2 to 100,
inclusive.

32. Find the sum of the odd numbers from 1 to 99,
inclusive.

33. Find the sum of all multiples of 6 from 6 to 102,
inclusive.

34. Find the sum of all multiples of 4 that are between
14 and 523.

35. An arithmetic series has $a_1 = 2$ and $d = 5$. Find S_{20}.

36. An arithmetic series has $a_1 = 7$ and $d = -3$. Find S_{32}.

Solve.

37. *Band Formations.* The Duxbury marching band
has 14 marchers in the front row, 16 in the second

row, 18 in the third row, and so on, for 15 rows.
How many marchers are in the last row? How many
marchers are there altogether?

38. *Gardening.* A gardener is making a triangular
planting, with 39 plants in the front row, 35 in the
second row, 31 in the third row, and so on. If the
pattern is consistent, how many plants will be in the
last row? How many plants will there be altogether?

39. *Telephone Pole Piles.* How many poles will be in
a pile of telephone poles if there are 50 in the first
layer, 49 in the second, and so on, until there are 6
in the last layer?

40. *Accumulated Savings.* If 10¢ is saved on
October 1, another 20¢ on October 2, another 30¢
on October 3, and so on, how much is saved during
October? (October has 31 days.)

41. *Accumulated Savings.* Renata saves money in an
arithmetic sequence: \$600 the first year, another
\$700 the second, and so on, for 20 yr. How much
does she save in all (disregarding interest)?

42. *Spending.* Jacob spent \$30 on August 1, \$50 on
August 2, \$70 on August 3, and so on. How much
did Jacob spend in August? (August has 31 days.)

43. *Auditorium Design.* Theaters are often built with
more seats per row as the rows move toward the
back. The Sanders Amphitheater has 20 seats in the
first row, 22 in the second, 24 in the third, and so
on, for 19 rows. How many seats are in the
amphitheater?

44. *Accumulated Savings.* Shirley sets up an investment such that it will return $5000 the first year, $6125 the second year, $7250 the third year, and so on, for 25 yr. How much in all is received from the investment?

Skill Maintenance

Convert to an exponential equation. [9.3]

45. $\log_a P = k$ **46.** $\ln t = a$

Convert to a logarithmic equation. [9.3]

47. $e^t = 0.1579$ **48.** $2^6 = 64$

Synthesis

49. ◈ It is said that as a young child, the mathematician Karl F. Gauss (1777–1855) was able to compute the sum $1 + 2 + 3 + \cdots + 100$ very quickly in his head. Explain how Gauss might have done this and present a formula for the sum of the first n natural numbers. (*Hint:* $1 + 99 = 100$.)

50. ◈ The sum of the first n terms of an arithmetic sequence is given by

$$S_n = \frac{n}{2}[2a_1 + (n-1)d].$$

Use the formulas for a_n and S_n to explain how this equation was developed.

51. Find a formula for the sum of the first n consecutive odd numbers starting with 1:

$$1 + 3 + 5 + \cdots + (2n - 1).$$

52. Find three numbers in an arithmetic sequence for which the sum of the first and third is 10 and the product of the first and second is 15.

53. In an arithmetic sequence, $a_1 = \$8760$ and $d = -\$798.23$. Find the first 10 terms of the sequence.

54. Find the sum of the first 10 terms of the sequence given in Exercise 53.

55. Prove that if p, m, and q are consecutive terms in an arithmetic sequence, then

$$m = \frac{p + q}{2}.$$

In Exercises 56–59, graph the data in each table, and state whether or not the graph could be the graph of an arithmetic sequence. If it can, find a formula for the general term of the sequence.

56. *Aerobic Exercise.*

AGE	TARGET HEART RATE (IN BEATS PER MINUTE)
20	150
40	135
60	120
80	105

57. *Book Sales in the United States.*

YEAR	BOOK SALES (IN BILLIONS)
1993	23
1994	24
1995	25
1996	26

Source: Book Industry Trends

58. *Ozone Levels.*

YEAR	LEVEL OF OZONE (IN PARTS PER BILLION)
1991	2981
1992	3133
1993	3148
1994	3138
1995	3124

Source: National Oceanic and Atmospheric Administration

59. *Coca-Cola Revenue.*

YEAR	REVENUE (IN BILLIONS)
1991	$465
1992	656
1993	687
1994	724
1995	762

Source: Coca-Cola Bottling Consolidated Annual Report

60. *Straight-Line Depreciation.* A company buys a color photocopier for $5200 on January 1 of a given year. The machine is expected to last for 8 yr, at the end of which time its *trade-in*, or *salvage*, value will be $1100. If the company figures the decline in value to be the same each year, then the trade-in values, after t years, $0 \leq t \leq 8$, form an arithmetic sequence given by

$$a_t = C - t\left(\frac{C - S}{N}\right),$$

where C is the original cost of the item, N is the years of expected life, and S is the salvage value.

a) Find the formula for a_t for the straight-line depreciation of the copier.
b) Find the salvage value after 0 yr, 1 yr, 2 yr, 3 yr, 4 yr, 7 yr, and 8 yr.
c) Find a formula that expresses a_t recursively.

10.3
Geometric Sequences and Series

- *Geometric Sequences*
- *Sum of the First n Terms of a Geometric Sequence*
- *Infinite Geometric Series*
- *Problem Solving*

In an arithmetic sequence, a certain number is added to each term to get the next term. When each term in a sequence is *multiplied* by a certain number to get the next term, the sequence is **geometric**. In this section, we examine geometric sequences (or progressions) and *geometric series*.

Geometric Sequences

Consider the sequence

$$2, 6, 18, 54, 162, \ldots.$$

If we multiply each term by 3, we obtain the next term. The multiplier is called the *common ratio* because it is found by dividing any term by the preceding term.

Geometric Sequence

A sequence is *geometric* if there exists a number r, called the *common ratio*, for which

$$\frac{a_{n+1}}{a_n} = r, \quad \text{or} \quad a_{n+1} = a_n \cdot r \quad \text{for any integer } n \geq 1.$$

Example 1 For each geometric sequence, find the common ratio.

a) 3, 6, 12, 24, 48, ...
b) 3, −6, 12, −24, 48, −96, ...
c) $5200, $3900, $2925, $2193.75, ...

SOLUTION

	SEQUENCE	COMMON RATIO	
a)	3, 6, 12, 24, 48, ...	2	$\frac{6}{3} = 2$, $\frac{12}{6} = 2$, and so on
b)	3, −6, 12, −24, 48, −96, ...	−2	$\frac{-6}{3} = -2$, $\frac{12}{-6} = -2$, and so on
c)	$5200, $3900, $2925, $2193.75, ...	0.75	$\frac{\$3900}{\$5200} = 0.75$, $\frac{\$2925}{\$3900} = 0.75$

To develop a formula for the general, or *n*th, term of a geometric sequence, let a_1 be the first term and let r be the common ratio. We write out the first few terms as follows:

$$a_1,$$
$$a_2 = a_1 r,$$
$$a_3 = a_2 r = (a_1 r)r = a_1 r^2, \qquad \text{Substituting } a_1 r \text{ for } a_2$$
$$a_4 = a_3 r = (a_1 r^2)r = a_1 r^3. \qquad \text{Substituting } a_1 r^2 \text{ for } a_3$$

Note that the exponent is 1 less than the subscript.

Generalizing, we obtain the following.

To Find a_n for a Geometric Sequence:

The *n*th term of a geometric sequence is given by

$$a_n = a_1 r^{n-1}, \quad \text{for any integer } n \geq 1.$$

Example 2 Find the 7th term of the geometric sequence 4, 20, 100, ...

SOLUTION First we note that

$$a_1 = 4 \quad \text{and} \quad n = 7.$$

To find the common ratio, we can divide any term (other than the first) by the term preceding it. Since the second term is 20 and the first is 4,

$$r = \frac{20}{4}, \quad \text{or } 5.$$

The formula

$$a_n = a_1 r^{n-1}$$

gives us

$$a_7 = 4 \cdot 5^{7-1} = 4 \cdot 5^6 = 4 \cdot 15{,}625 = 62{,}500.$$

Example 3 Find the 10th term of the geometric sequence

$$64, \ -32, \ 16, \ -8, \ldots .$$

SOLUTION First we note that

$$a_1 = 64, \qquad n = 10, \quad \text{and} \quad r = \frac{-32}{64} = -\frac{1}{2}.$$

Then, using the formula for the nth term of a geometric series, we have

$$a_{10} = 64 \cdot \left(-\frac{1}{2}\right)^{10-1} = 64 \cdot \left(-\frac{1}{2}\right)^9 = 2^6 \cdot \left(-\frac{1}{2^9}\right) = -\frac{1}{2^3} = -\frac{1}{8}.$$

What does the graph of a geometric series look like?

Interactive Discovery

Graph each of the following geometric series. What patterns, if any, do you observe?

1. $a_n = 5 \cdot 2^{n-1}$
2. $a_n = \frac{1}{2} \cdot \left(\frac{5}{4}\right)^{n-1}$
3. $a_n = 2.3 \cdot (0.75)^{n-1}$
4. $a_n = 3\left(\frac{9}{10}\right)^{n-1}$

The pattern you may have observed is true in general for $r > 0$.

The graph of a geometric series is a set of points that lie on the graph of an exponential function.

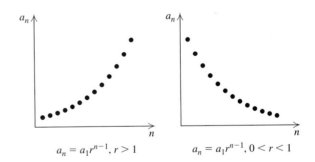

$$a_n = a_1 r^{n-1}, r > 1 \qquad\qquad a_n = a_1 r^{n-1}, 0 < r < 1$$

Sum of the First n Terms of a Geometric Sequence

We next develop a formula for S_n when a sequence is geometric:

$$a_1, \ a_1r, \ a_1r^2, \ a_1r^3, \ldots, \ a_1r^{n-1}, \ \ldots .$$

The **geometric series** S_n is given by

$$S_n = a_1 + a_1r + a_1r^2 + \cdots + a_1r^{n-2} + a_1r^{n-1}. \qquad \textbf{(1)}$$

Multiplying by r on both sides gives us

$$rS_n = a_1r + a_1r^2 + a_1r^3 + \cdots + a_1r^{n-1} + a_1r^n. \qquad \textbf{(2)}$$

When we subtract corresponding sides of equation (2) from equation (1), the terms shown in color drop out, leaving

$$S_n - rS_n = a_1 - a_1 r^n,$$

or

$$S_n(1 - r) = a_1(1 - r^n). \quad \text{Factoring}$$

Dividing by $1 - r$ on both sides gives us the following formula.

> **To Find S_n for a Geometric Sequence:**
> The sum of the first n terms of a geometric sequence is given by
> $$S_n = \frac{a_1(1 - r^n)}{1 - r}, \quad \text{for any } r \neq 1.$$

Example 4 Find the sum of the first 7 terms of the geometric sequence 3, 15, 75, 375, ...

SOLUTION First we note that

$$a_1 = 3, \quad n = 7, \quad \text{and} \quad r = \frac{15}{3} = 5.$$

Then, substituting in the formula $S_n = \dfrac{a_1(1 - r^n)}{1 - r}$, we have

$$\begin{aligned} S_7 &= \frac{3(1 - 5^7)}{1 - 5} \\ &= \frac{3(1 - 78{,}125)}{-4} \\ &= \frac{3(-78{,}124)}{-4} \\ &= 58{,}593. \end{aligned}$$

Infinite Geometric Series

Suppose we consider the sum of the terms of an infinite geometric sequence, such as 2, 4, 8, 16, 32, ... We get what is called an **infinite geometric series:**

$$2 + 4 + 8 + 16 + 32 + \cdots.$$

Here, as n grows larger and larger, the sum of the first n terms, S_n, becomes larger and larger without bound. There are also infinite series that get closer and closer to some specific number. Here is an example:

$$\frac{1}{2} + \frac{1}{4} + \frac{1}{8} + \frac{1}{16} + \cdots + \frac{1}{2^n} + \cdots.$$

Let's consider S_n for the first five values of n:

$$S_1 = \tfrac{1}{2} \qquad\qquad\qquad = \tfrac{1}{2} = 0.5,$$

$$S_2 = \tfrac{1}{2} + \tfrac{1}{4} \qquad\qquad = \tfrac{3}{4} = 0.75,$$

$$S_3 = \tfrac{1}{2} + \tfrac{1}{4} + \tfrac{1}{8} \qquad = \tfrac{7}{8} = 0.875,$$

$$S_4 = \tfrac{1}{2} + \tfrac{1}{4} + \tfrac{1}{8} + \tfrac{1}{16} = \tfrac{15}{16} = 0.9375.$$

> **The denominator of the sum is 2^n, where n is the subscript of S. The numerator is $2^n - 1$.**

Thus, for this particular series, we have

$$S_n = \frac{2^n - 1}{2^n} = \frac{2^n}{2^n} - \frac{1}{2^n} = 1 - \frac{1}{2^n}.$$

Note that the value of S_n is less than 1 for any value of n, but as n gets larger and larger, the values of S_n get closer and closer to 1. We say that 1 is the *limit* of S_n and that 1 is the sum of this infinite geometric sequence. An infinite geometric series is denoted S_∞.

How can we tell if an infinite geometric series has a sum?

Interactive Discovery

For each of the following geometric sequences, **(a)** determine the common ratio, **(b)** list the first 10 partial sums, S_1 through S_{10}, and **(c)** estimate, if possible, S_∞.

1. $a_n = \left(\tfrac{1}{3}\right)^{n-1}$
2. $a_n = \left(-\tfrac{7}{2}\right)^{n-1}$
3. $a_n = 4(2)^{n-1}$
4. $a_n = 1.3(-0.4)^{n-1}$

What relationship do you observe between the common ratio and the existence of S_∞?

It can be shown (but we will not do it here) that the pattern you may have observed is true in general.

> **The sum of the terms of an infinite geometric sequence exists if and only if $|r| < 1$.**

To find a formula for the sum of an infinite geometric sequence, we first consider the sum of the first n terms:

$$S_n = \frac{a_1(1 - r^n)}{1 - r} = \frac{a_1 - a_1 r^n}{1 - r}. \qquad \textbf{Using the distributive law}$$

For $|r| < 1$, it follows that values of r^n get closer and closer to 0 as n gets larger. (Check this by selecting a number between -1 and 1 and finding larger and larger powers on a calculator.) As r^n gets closer and closer to 0, so does $a_1 r^n$. Thus, S_n gets closer and closer to $a_1/(1 - r)$.

> **The Limit of an Infinite Geometric Series**
> When $|r| < 1$, the limit of an infinite geometric series is given by
> $$S_\infty = \frac{a_1}{1-r}. \quad \text{(For } |r| \geq 1, \text{ no limit exists.)}$$

Example 5 Determine whether each series has a limit. If one exists, find it.

a) $1 + 3 + 9 + 27 + \cdots$ 　　　　　　**b)** $-2 + 1 - \frac{1}{2} + \frac{1}{4} - \frac{1}{8} + \cdots$

SOLUTION

a) Here $r = 3$, so $|r| = |3| = 3$. Since $|r| \not< 1$, the series does *not* have a limit.

b) Here $r = -\frac{1}{2}$, so $|r| = |-\frac{1}{2}| = \frac{1}{2}$. Since $|r| < 1$, the series *does* have a limit. We find the limit by substituting into the formula for S_∞:

$$S_\infty = \frac{-2}{1 - \left(-\frac{1}{2}\right)} = \frac{-2}{\frac{3}{2}} = -\frac{4}{3}.$$

Example 6 Find fractional notation for $0.63636363\ldots$.

SOLUTION　We can express this as

$$0.63 + 0.0063 + 0.000063 + \cdots.$$

This is an infinite geometric series, where $a_1 = 0.63$ and $r = 0.01$. Since $|r| < 1$, this series has a limit:

$$S_\infty = \frac{a_1}{1-r} = \frac{0.63}{1 - 0.01} = \frac{0.63}{0.99} = \frac{63}{99}.$$

Thus fractional notation for $0.63636363\ldots$ is $\frac{63}{99}$, or $\frac{7}{11}$.

Problem Solving

For some problem-solving situations, the translation may involve geometric sequences or series.

Example 7 *Daily Wages.* Suppose someone offered you a job for the month of September (30 days) under the following conditions. You will be paid $0.01 for the first day, $0.02 for the second, $0.04 for the third, and so on, doubling your previous day's salary each day. How much would you earn? (Would you take the job? Make a guess before reading further.)

SOLUTION

1. **Familiarize.** You earn $0.01 the first day, $0.01(2)$ the second day, $0.01(2)(2)$ the third day, and so on. Since each day's wages are a constant multiple of the previous day's wage, a geometric sequence is formed.

2. Translate. The amount earned is the geometric series

$$\$0.01 + \$0.01(2) + \$0.01(2^2) + \$0.01(2^3) + \cdots + \$0.01(2^{29}),$$

where

$$a_1 = \$0.01, \qquad n = 30, \quad \text{and} \quad r = 2.$$

3. Carry out. Using the formula

$$S_n = \frac{a_1(1 - r^n)}{1 - r},$$

we have

$$S_{30} = \frac{\$0.01(1 - 2^{30})}{1 - 2}$$

$$= \frac{\$0.01(-1{,}073{,}741{,}823)}{-1} \qquad \textbf{Using a calculator}$$

$$= \$10{,}737{,}418.23.$$

4. Check. The calculations can be repeated as a check.

5. State. The pay exceeds \$10.7 million for the month. Most people would probably take the job! ▬

Example 8 *Loan Repayment.* A student loan is in the amount of \$6000. Interest is to be 9% compounded annually, and the entire amount is to be paid after 10 yr. How much is to be paid back?

SOLUTION

1. Familiarize. Suppose we let P represent any principal amount. At the end of one year, the amount owed will be $P + 0.09P$, or $1.09P$. That amount will be the principal for the second year. The amount owed at the end of the second year will be $1.09 \times$ New principal $= 1.09(1.09P)$, or 1.09^2P. Thus the amount owed at the beginning of successive years is as follows:

$$\overset{\textstyle ①}{\underset{P,}{\downarrow}} \quad \overset{\textstyle ②}{\underset{1.09P,}{\downarrow}} \quad \overset{\textstyle ③}{\underset{1.09^2P,}{\downarrow}} \quad \overset{\textstyle ④}{\underset{1.09^3P,}{\downarrow}} \quad \text{and so on.}$$

We have a geometric sequence. The amount owed at the beginning of the 11th year will be the amount owed at the end of the 10th year.

2. Translate. We have a geometric sequence with $a_1 = 6000$, $r = 1.09$, and $n = 11$. The appropriate formula is

$$a_n = a_1 r^{n-1}.$$

3. Carry out. We substitute and calculate:

$$a_{11} = \$6000(1.09)^{11-1} = \$6000(1.09)^{10}$$

$$\approx \$6000(2.3673637) \qquad \textbf{Using a calculator to}$$
$$\textbf{approximate } 1.09^{10}$$

$$\approx \$14{,}204.18. \qquad \textbf{Rounded to the nearest hundredth}$$

4. **Check.** A check, by repeating the calculations, is left to the student.

5. **State.** A total of $14,204.18 is to be paid back at the end of 10 yr.

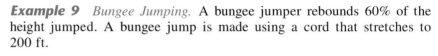

Example 9 *Bungee Jumping.* A bungee jumper rebounds 60% of the height jumped. A bungee jump is made using a cord that stretches to 200 ft.

a) After jumping and then rebounding 9 times, how far has a bungee jumper traveled upward (the total rebound distance)?

b) Approximately how far will a jumper have traveled upward (bounced) before coming to rest?

SOLUTION

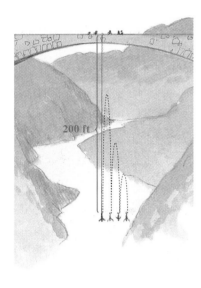

1. **Familiarize.** Let's do some calculations and look for a pattern.

First fall:	200 ft
First rebound:	0.6×200, or 120 ft
Second fall:	120 ft, or 0.6×200
Second rebound:	0.6×120, or $0.6(0.6 \times 200)$, which is 72 ft
Third fall:	72 ft, or $0.6(0.6 \times 200)$
Third rebound:	0.6×72, or $0.6(0.6(0.6 \times 200))$, which is 43.2 ft

The rebound distances form a geometric sequence:

①   ②   ③   ④
↓   ↓   ↓   ↓
$120, \quad 0.6 \times 120, \quad 0.6^2 \times 120, \quad 0.6^3 \times 120, \dots.$

2. **Translate.**

a) The total rebound distance after 9 bounces is the sum of a geometric sequence. The first term is 120 and the common ratio is 0.6. There will be 9 terms, so we can use the formula

$$S_n = \frac{a_1(1 - r^n)}{1 - r}.$$

b) Theoretically, the jumper will never stop bouncing. Realistically, the bouncing will eventually stop. To approximate the actual distance bounced, we consider an infinite number of bounces and use the formula

$$S_\infty = \frac{a_1}{1 - r}.$$

3. **Carry out.**

a) We substitute into the formula and calculate:

$$S_9 = \frac{120[1 - (0.6)^9]}{1 - 0.6} \approx 297.$$

b) We substitute and calculate:

$$S_\infty = \frac{120}{1 - 0.6}$$
$$= 300.$$

4. Check. We can do the calculations again.

5. State.

a) In 9 bounces, the bungee jumper will have traveled upward a total distance of about 297 ft.

b) The jumper will travel upward about 300 ft before coming to rest.

10.3 *Exercise Set*

Find the common ratio for each geometric sequence.

1. 5, 10, 20, 40, ...

2. 2, 6, 18, 54, ...

3. 5, −5, 5, −5, ...

4. −5, −0.5, −0.05, −0.005, ...

5. $\frac{1}{2}, -\frac{1}{4}, \frac{1}{8}, -\frac{1}{16}, \ldots$

6. $\frac{2}{3}, -\frac{4}{3}, \frac{8}{3}, -\frac{16}{3}, \ldots$

7. 75, 15, 3, $\frac{3}{5}$, ...

8. 12, −4, $\frac{4}{3}$, $-\frac{4}{9}$, ...

9. $\frac{1}{m}, \frac{3}{m^2}, \frac{9}{m^3}, \frac{27}{m^4}, \ldots$

10. $4, \frac{4m}{5}, \frac{4m^2}{25}, \frac{4m^3}{125}, \ldots$

Find the indicated term for each geometric sequence.

11. 5, 10, 20, ...; the 7th term

12. 2, 8, 32, ...; the 9th term

13. 3, $3\sqrt{2}$, 6, ...; the 9th term

14. 4, $4\sqrt{3}$, 12, ...; the 8th term

15. $-\frac{8}{243}, \frac{8}{81}, -\frac{8}{27}, \ldots$; the 10th term

16. $\frac{7}{625}, \frac{-7}{125}, \frac{7}{25}, \ldots$; the 13th term

17. $1000, $1080, $1166.40, ...; the 12th term

18. $1000, $1070, $1144.90, ...; the 11th term

Find the nth, or general, term for each geometric sequence.

19. 1, 3, 9, ...

20. 25, 5, 1, ...

21. 1, −1, 1, −1, ...

22. 2, 4, 8, ...

23. $\frac{1}{x}, \frac{1}{x^2}, \frac{1}{x^3}, \ldots$

24. $5, \frac{5m}{2}, \frac{5m^2}{4}, \ldots$

For Exercises 25–32, use the formula for S_n to find the indicated sum.

25. S_7 for the geometric series $7 + 14 + 28 + \cdots$

26. S_6 for the geometric series $16 - 8 + 4 - \cdots$

27. S_7 for the geometric series $\frac{1}{18} - \frac{1}{6} + \frac{1}{2} - \cdots$

28. S_5 for the geometric series $6 + 0.6 + 0.06 + \cdots$

29. S_8 for the series $1 + x + x^2 + x^3 + \cdots$

30. S_{10} for the series $1 + x^2 + x^4 + x^6 + \cdots$

31. S_{16} for the geometric sequence
$$\$200, \$200(1.06), \$200(1.06)^2, \ldots$$

32. S_{23} for the geometric sequence
$$\$1000, \$1000(1.08), \$1000(1.08)^2, \ldots$$

Determine whether each infinite geometric series has a limit. If a limit exists, find it.

33. $9 + 3 + 1 + \cdots$

34. $8 + 4 + 2 + \cdots$

35. $7 + 3 + \frac{9}{7} + \cdots$

36. $12 + 9 + \frac{27}{4} + \cdots$

37. $3 + 15 + 75 + \cdots$

38. $2 + 3 + \frac{9}{2} + \cdots$

39. $4 - 6 + 9 - \frac{27}{2} + \cdots$

40. $-6 + 3 - \frac{3}{2} + \frac{3}{4} - \cdots$

41. $0.43 + 0.0043 + 0.000043 + \cdots$

42. $0.37 + 0.0037 + 0.000037 + \cdots$

43. $\$500(1.02)^{-1} + \$500(1.02)^{-2} + \$500(1.02)^{-3} + \cdots$

44. $\$1000(1.08)^{-1} + \$1000(1.08)^{-2} + \$1000(1.08)^{-3} + \cdots$

Find fractional notation for each infinite sum. (These are geometric series.)

45. $0.7777\ldots$

46. $0.2222\ldots$

47. $8.3838\ldots$

48. $7.4747\ldots$

49. $0.15151515\ldots$

50. $0.12121212\ldots$

Solve. Use a calculator as needed for evaluating formulas.

51. *Rebound Distance.* A ping-pong ball is dropped from a height of 20 ft and always rebounds one-fourth of the distance fallen. How high does it rebound the 6th time?

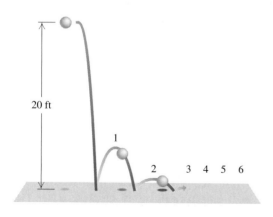

20 ft

52. *Rebound Distance.* Approximate the total of the rebound heights of the ball in Exercise 51.

53. *Population Growth.* Yorktown has a current population of 100,000, and the population is increasing by 3% each year. What will the population be in 15 yr?

54. *Doubling Time.* How long will it take for the population of Yorktown to double? (See Exercise 53.)

55. *Amount Owed.* Gilberto borrows $15,000. The loan is to be repaid in 13 yr at 8.5% interest, compounded annually. How much will be repaid at the end of 13 yr?

56. *Shrinking Population.* A population of 5000 fruit flies is dying off at a rate of 4% per minute. How many flies will be alive after 15 min?

57. *Shrinking Population.* For the population of fruit flies in Exercise 56, how long will it take for only 1800 fruit flies to remain alive? (See Exercise 56 and use logarithms.) Round to the nearest minute.

58. *Investing.* Leslie is saving money in a retirement account. At the beginning of each year, she invests $1000 at 7%, compounded annually. How much will be in the retirement fund at the end of 40 yr?

59. *Rebound Distance.* A superball dropped from the top of the Washington Monument (556 ft high) rebounds three-fourths of the distance fallen. How far (up and down) will the ball have traveled when it hits the ground for the 6th time?

60. *Rebound Distance.* Approximate the total distance that the ball of Exercise 59 will have traveled when it comes to rest.

61. *Stacking Paper.* Construction paper is about 0.02 in. thick. Beginning with just one piece, a stack is doubled again and again 10 times. Find the height of the final stack.

62. *Monthly Earnings.* Suppose you accepted a job for the month of February (28 days) under the following conditions. You will be paid $0.01 the first day, $0.02 the second, $0.04 the third, and so on, doubling your previous day's salary each day. How much would you earn?

In Exercises 63–68, state whether the graph is that of an arithmetic sequence or a geometric sequence.

63.

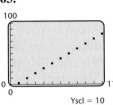

Yscl = 10

64.

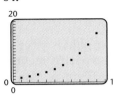

65.

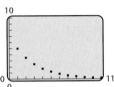

66.

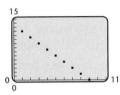

67.

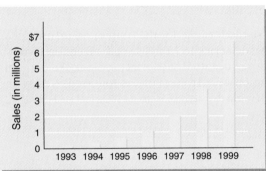

68.

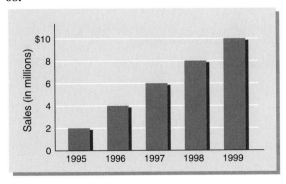

Skill Maintenance

Solve the system.

69. $5x - 2y = -3,$
$2x + 5y = -24$
[3.2]

70. $x - 2y + 3z = 4,$
$2x - y + z = -1,$
$4x + y + z = 1$
[3.4]

Synthesis

71. ◈ Write a problem for a classmate to solve. Devise the problem so that a geometric series is involved and the solution is "The total amount in the bank is
$$\$900(1.08)^{40},$$
or about \$19,550."

72. ◈ The infinite series
$$S_\infty = 2 + \frac{1}{2} + \frac{1}{2 \cdot 3} + \frac{1}{2 \cdot 3 \cdot 4}$$
$$+ \frac{1}{2 \cdot 3 \cdot 4 \cdot 5} + \frac{1}{2 \cdot 3 \cdot 4 \cdot 5 \cdot 6}$$
$$+ \cdots$$

is not geometric, but it does have a sum. Using S_1, S_2, S_3, S_4, S_5, and S_6, make a conjecture about the value of S_∞ and explain your reasoning.

73. ◈ Explain how the graph of a geometric sequence can be used to determine whether a geometric series has a limit.

74. Find the sum of the first n terms of
$$x^2 - x^3 + x^4 - x^5 + \cdots.$$

75. Find the sum of the first n terms of
$$1 + x + x^2 + x^3 + \cdots.$$

76. The sides of a square are each 16 cm long. A second square is inscribed by joining the midpoints of the sides, successively. In the second square we repeat the process, inscribing a third square. If this process is continued indefinitely, what is the sum of all of the areas of all the squares? (*Hint:* Use an infinite geometric series.)

COLLABORATIVE CORNER

Focus: Geometric Series

Time: 30 minutes

Group size: 2

Materials: Graphing calculators

Activity *

1. One group member ("the seller") has a car for sale and is asking $3500. The second ("the buyer") offers $1500. The seller splits the difference ($2000 ÷ 2 = $1000) and lowers the price to $2500. The buyer then splits the difference again ($1000 ÷ 2 = $500) and counters with $2000. Continue in this manner and stop when you are able to agree on the car's selling price to the nearest penny.

2. What should the buyer's initial offer be in order to achieve a purchase price of $2000? (Check several guesses to find the appropriate initial offer.)

---

*This activity is based on the article, "Bargaining Theory, or Zeno's Used Cars," by James C. Kirby, *The College Mathematics Journal,* **27**(4), September 1996.

3. The seller's price in the bargaining above can be modeled recursively (see Exercises 64, 65, and 73 in Section 10.1) by the sequence

$$a_1 = 3500, \qquad a_n = a_{n-1} - \frac{d}{2^{2n-3}},$$

where d is the difference between the initial price and the first offer. Use this recursively defined sequence to solve parts (1) and (2) with the SEQUENTIAL FUNCTION mode and the TABLE feature of a grapher.

4. The first four terms in the sequence in part (3) can be written as

$$a_0, \quad a_0 - \frac{d}{2}, \quad a_0 - \frac{d}{2} - \frac{d}{8},$$

$$a_0 - \frac{d}{2} - \frac{d}{8} - \frac{d}{32}.$$

Use the formula for the limit of an infinite geometric series to find a simple algebraic formula for the eventual sale price, P, when the bargaining process from above is followed. Verify the formula by using it to solve parts (1) and (2) above.

10.4 The Binomial Theorem

- *Binomial Expansion Using Pascal's Triangle*
- *Binomial Expansion Using Factorial Notation*

Binomial Expansion Using Pascal's Triangle

Consider the following expanded powers of $(a + b)^n$:

$$
\begin{aligned}
(a + b)^0 &= 1 \\
(a + b)^1 &= a + b \\
(a + b)^2 &= a^2 + 2a^1b^1 + b^2 \\
(a + b)^3 &= a^3 + 3a^2b^1 + 3a^1b^2 + b^3 \\
(a + b)^4 &= a^4 + 4a^3b^1 + 6a^2b^2 + 4a^1b^3 + b^4 \\
(a + b)^5 &= a^5 + 5a^4b^1 + 10a^3b^2 + 10a^2b^3 + 5a^1b^4 + b^5.
\end{aligned}
$$

Each expansion is a polynomial. There are some patterns to be noted:

1. There is one more term than the power of the binomial, n. That is, there are $n + 1$ terms in the expansion of $(a + b)^n$.

2. In each term, the sum of the exponents is the power to which the binomial is raised.

3. The exponents of a start with n, the power of the binomial, and decrease to 0. The last term has no factor of a. The first term has no factor of b, so powers of b start with 0 and increase to n.

4. The coefficients start at 1, increase through certain values, and then decrease through these same values back to 1. Let's study the coefficients further.

Suppose we wish to expand $(a + b)^8$. The patterns we noticed above indicate 9 terms in the expansion:

$$a^8 + c_1 a^7 b + c_2 a^6 b^2 + c_3 a^5 b^3 + c_4 a^4 b^4 + c_5 a^3 b^5 + c_6 a^2 b^6 + c_7 ab^7 + b^8.$$

How can we determine the values for the c's? One method seems to be the easiest, but is not always. It involves writing down the coefficients in a triangular array as follows. We form what is known as **Pascal's triangle**:

$$
\begin{array}{ll}
(a + b)^0: & \qquad\qquad\qquad 1 \\
(a + b)^1: & \qquad\qquad 1 \quad\quad 1 \\
(a + b)^2: & \qquad\quad 1 \quad 2 \quad 1 \\
(a + b)^3: & \quad\quad 1 \quad 3 \quad 3 \quad 1 \\
(a + b)^4: & \quad 1 \quad 4 \quad 6 \quad 4 \quad 1 \\
(a + b)^5: & 1 \quad 5 \quad 10 \quad 10 \quad 5 \quad 1
\end{array}
$$

There are many patterns in the triangle. Find as many as you can.

Perhaps you discovered a way to write the next row of numbers, given the numbers in the row above it. There are always 1's on the outside. Each remaining number is the sum of the two numbers above:

$$
\begin{array}{ccccccccccccc}
 & & & & & & 1 \\
 & & & & & 1 & & 1 \\
 & & & & 1 & & 2 & & 1 \\
 & & & 1 & & 3 & & 3 & & 1 \\
 & & 1 & & 4 & & 6 & & 4 & & 1 \\
 & 1 & & 5 & & 10 & & 10 & & 5 & & 1 \\
1 & & 6 & & 15 & & 20 & & 15 & & 6 & & 1
\end{array}
$$

We see that in the bottom (seventh) row

the 1st and last numbers are 1;

the 2nd number is $1 + 5$, or 6;

the 3rd number is $5 + 10$, or 15;

the 4th number is $10 + 10$, or 20;

the 5th number is $10 + 5$, or 15; and

the 6th number is $5 + 1$, or 6.

Thus the expansion of $(a + b)^6$ is

$$(a + b)^6 = 1a^6 + 6a^5b + 15a^4b^2 + 20a^3b^3 + 15a^2b^4 + 6ab^5 + 1b^6.$$

To expand $(a + b)^8$, we complete two more rows of Pascal's triangle:

$$
\begin{array}{ccccccccccccccccc}
 & & & & & & & & 1 & & & & & & & & \\
 & & & & & & & 1 & & 1 & & & & & & & \\
 & & & & & & 1 & & 2 & & 1 & & & & & & \\
 & & & & & 1 & & 3 & & 3 & & 1 & & & & & \\
 & & & & 1 & & 4 & & 6 & & 4 & & 1 & & & & \\
 & & & 1 & & 5 & & 10 & & 10 & & 5 & & 1 & & & \\
 & & 1 & & 6 & & 15 & & 20 & & 15 & & 6 & & 1 & & \\
 & 1 & & 7 & & 21 & & 35 & & 35 & & 21 & & 7 & & 1 & \\
1 & & 8 & & 28 & & 56 & & 70 & & 56 & & 28 & & 8 & & 1
\end{array}
$$

Thus the expansion of $(a + b)^8$ is

$$
\begin{aligned}
(a + b)^8 = {} & 1a^8 + 8a^7b + 28a^6b^2 + 56a^5b^3 + 70a^4b^4 \\
& + 56a^3b^5 + 28a^2b^6 + 8ab^7 + 1b^8.
\end{aligned}
$$

We can generalize our results as follows:

The Binomial Theorem (Form 1)

For any binomial $a + b$ and any natural number n,

$$
\begin{aligned}
(a + b)^n = {} & c_0a^nb^0 + c_1a^{n-1}b^1 + c_2a^{n-2}b^2 + \cdots \\
& + c_{n-1}a^1b^{n-1} + c_na^0b^n,
\end{aligned}
$$

where the numbers $c_0, c_1, c_2, \ldots, c_n$ are from the $(n + 1)$st row of Pascal's triangle.

Example 1   Expand: $(u - v)^5$.

SOLUTION   Using the binomial theorem, we have $a = u$, $b = -v$, and $n = 5$. We use the 6th row of Pascal's triangle: $1 \quad 5 \quad 10 \quad 10 \quad 5 \quad 1$. Thus,

$$
\begin{aligned}
(u - v)^5 &= [u + (-v)]^5 \qquad \text{Rewriting } u - v \text{ as a sum} \\
&= 1(u)^5 + 5(u)^4(-v)^1 + 10(u)^3(-v)^2 + 10(u)^2(-v)^3 \\
&\quad + 5(u)^1(-v)^4 + 1(-v)^5 \\
&= u^5 - 5u^4v + 10u^3v^2 - 10u^2v^3 + 5uv^4 - v^5.
\end{aligned}
$$

Note that the signs of the terms alternate between $+$ and $-$. When $-v$ is raised to an odd power, the sign is $-$.

Example 2 Expand: $\left(2t + \dfrac{3}{t}\right)^6$.

SOLUTION Note that $a = 2t$, $b = 3/t$, and $n = 6$. We use the 7th row of Pascal's triangle: 1   6   15   20   15   6   1. Thus,

$$\left(2t + \frac{3}{t}\right)^6 = 1(2t)^6 + 6(2t)^5\left(\frac{3}{t}\right)^1 + 15(2t)^4\left(\frac{3}{t}\right)^2 + 20(2t)^3\left(\frac{3}{t}\right)^3$$

$$+ 15(2t)^2\left(\frac{3}{t}\right)^4 + 6(2t)^1\left(\frac{3}{t}\right)^5 + 1\left(\frac{3}{t}\right)^6$$

$$= 64t^6 + 6(32t^5)\left(\frac{3}{t}\right) + 15(16t^4)\left(\frac{9}{t^2}\right) + 20(8t^3)\left(\frac{27}{t^3}\right)$$

$$+ 15(4t^2)\left(\frac{81}{t^4}\right) + 6(2t)\left(\frac{243}{t^5}\right) + \frac{729}{t^6}$$

$$= 64t^6 + 576t^4 + 2160t^2 + 4320 + 4860t^{-2} + 2916t^{-4}$$

$$+ 729t^{-6}.$$

Binomial Expansion Using Factorial Notation

The disadvantage in using Pascal's triangle is that we must compute all the preceding rows in the table to obtain the row needed for the expansion. The following method avoids this difficulty. It will also enable us to find a specific term—say, the 8th term—without computing all the other terms in the expansion. This method is useful in such courses as finite mathematics, calculus, and statistics.

To develop the method, we need some new notation. Products of successive natural numbers, such as $6 \cdot 5 \cdot 4 \cdot 3 \cdot 2 \cdot 1$ and $8 \cdot 7 \cdot 6 \cdot 5 \cdot 4 \cdot 3 \cdot 2 \cdot 1$, have a special notation. For the product $6 \cdot 5 \cdot 4 \cdot 3 \cdot 2 \cdot 1$, we write 6!, read "6 factorial."

Factorial Notation

For any natural number n,

$$n! = n(n - 1)(n - 2)\cdots(3)(2)(1).$$

Here are some examples:

$$6! = 6 \cdot 5 \cdot 4 \cdot 3 \cdot 2 \cdot 1 = 720,$$
$$5! = 5 \cdot 4 \cdot 3 \cdot 2 \cdot 1 = 120,$$
$$4! = 4 \cdot 3 \cdot 2 \cdot 1 = 24,$$
$$3! = 3 \cdot 2 \cdot 1 = 6,$$
$$2! = 2 \cdot 1 = 2,$$
$$1! = 1 = 1.$$

We also define 0! to be 1 for reasons explained shortly.

To simplify expressions like

$$\frac{8!}{5!\,3!},$$

note that

$$8! = 8 \cdot 7 \cdot 6 \cdot 5 \cdot 4 \cdot 3 \cdot 2 \cdot 1 = 8 \cdot 7! = 8 \cdot 7 \cdot 6! = 8 \cdot 7 \cdot 6 \cdot 5!,$$

and so on.

Example 3   Simplify: $\dfrac{8!}{5!\,3!}$.

SOLUTION

$$\frac{8!}{5!\,3!} = \frac{8 \cdot 7 \cdot 6 \cdot 5!}{5! \cdot 3 \cdot 2 \cdot 1} = 8 \cdot 7 \qquad \begin{array}{l} \textbf{Removing a factor equal to 1:} \\[4pt] \dfrac{6 \cdot 5!}{5! \cdot 3 \cdot 2} = 1 \end{array}$$

$$= 56.$$

```
8!/(5!3!)
                    56
```

Factorials can be evaluated using a grapher. To evaluate the expression in Example 3, we enter $8!/(5!\,3!)$. The parentheses in the denominator are necessary.

The following notation is used in our second formulation of the binomial theorem.

$\dbinom{n}{r}$ **Notation**

For n, r nonnegative integers with $n \geq r$,

$$\binom{n}{r}, \quad \text{read "}n \text{ choose } r\text{," } \quad \text{means} \quad \frac{n!}{(n-r)!\,r!}.$$

Example 4   Simplify: **(a)** $\dbinom{7}{2}$; **(b)** $\dbinom{9}{6}$; **(c)** $\dbinom{6}{6}$.

SOLUTION

a) $\dbinom{7}{2} = \dfrac{7!}{(7-2)!\,2!}$

$$= \frac{7!}{5!\,2!} = \frac{7 \cdot 6 \cdot 5!}{5! \cdot 2 \cdot 1}$$

$$= \frac{7 \cdot 6}{2}$$

$$= 7 \cdot 3$$

$$= 21$$

b) $\dbinom{9}{6} = \dfrac{9!}{3!\,6!}$

$= \dfrac{9 \cdot 8 \cdot 7 \cdot 6!}{3 \cdot 2 \cdot 1 \cdot 6!}$

$= \dfrac{9 \cdot 8 \cdot 7}{3 \cdot 2}$

$= 3 \cdot 4 \cdot 7$

$= 84$

c) $\dbinom{6}{6} = \dfrac{6!}{0!\,6!} = \dfrac{6!}{1 \cdot 6!}$ Since $0! = 1$

$= \dfrac{6!}{6!}$

$= 1$

Now we can restate the binomial theorem using our new notation.

The Binomial Theorem (Form 2)

For any binomial $a + b$ and any natural number n,

$$(a + b)^n = \binom{n}{0}a^n + \binom{n}{1}a^{n-1}b + \binom{n}{2}a^{n-2}b^2 + \cdots + \binom{n}{n}b^n.$$

Example 5 Expand: $(3x + y)^4$.

SOLUTION We use the binomial theorem (Form 2) with $a = 3x$, $b = y$, and $n = 4$:

$$(3x + y)^4 = \binom{4}{0}(3x)^4 + \binom{4}{1}(3x)^3 y + \binom{4}{2}(3x)^2 y^2 + \binom{4}{3}(3x)y^3$$

$$+ \binom{4}{4}y^4$$

$$= \dfrac{4!}{4!\,0!}3^4 x^4 + \dfrac{4!}{3!\,1!}3^3 x^3 y + \dfrac{4!}{2!\,2!}3^2 x^2 y^2 + \dfrac{4!}{1!\,3!}3xy^3 + \dfrac{4!}{0!\,4!}y^4$$

$$= 81x^4 + 108x^3 y + 54x^2 y^2 + 12xy^3 + y^4. \text{Simplifying}$$

Example 6 Expand:

$$(x^2 - 2y)^5.$$

SOLUTION   In this case, $a = x^2$, $b = -2y$, and $n = 5$:

$$(x^2 - 2y)^5 = \binom{5}{0}(x^2)^5 + \binom{5}{1}(x^2)^4(-2y) + \binom{5}{2}(x^2)^3(-2y)^2$$

$$+ \binom{5}{3}(x^2)^2(-2y)^3 + \binom{5}{4}(x^2)(-2y)^4 + \binom{5}{5}(-2y)^5$$

$$= \frac{5!}{5!\,0!}x^{10} + \frac{5!}{4!\,1!}x^8(-2y) + \frac{5!}{3!\,2!}x^6(-2y)^2 + \frac{5!}{2!\,3!}x^4(-2y)^3$$

$$+ \frac{5!}{1!\,4!}x^2(-2y)^4 + \frac{5!}{0!\,5!}(-2y)^5$$

$$= x^{10} - 10x^8y + 40x^6y^2 - 80x^4y^3 + 80x^2y^4 - 32y^5.$$

Note that in the binomial theorem (Form 2), $\binom{n}{0}a^nb^0$ gives us the first term, $\binom{n}{1}a^{n-1}b^1$ gives us the second term, $\binom{n}{2}a^{n-2}b^2$ gives us the third term, and so on. This can be generalized to give a method for finding a specific term without writing the entire expansion.

Finding a Specific Term
The $(r + 1)$st term of $(a + b)^n$ is

$$\binom{n}{r}a^{n-r}b^r.$$

Example 7   Find the 5th term in the expansion of $(2x - 3y)^7$.

SOLUTION   First, we note that $5 = 4 + 1$. Thus, $r = 4$, $a = 2x$, $b = -3y$, and $n = 7$. Then the 5th term of the expansion is

$$\binom{7}{4}(2x)^{7-4}(-3y)^4, \quad \text{or} \quad \frac{7!}{3!\,4!}(2x)^3(-3y)^4, \quad \text{or} \quad 22{,}680x^3y^4.$$

It is because of the binomial theorem that $\binom{n}{r}$ is called a *binomial coefficient*. We can now explain why 0! is defined to be 1. In the binomial expansion, we want $\binom{n}{0}$ to equal 1 and we also want the definition

$$\binom{n}{r} = \frac{n!}{(n-r)!\,r!}$$

to hold for all whole numbers n and r. Thus we must have

$$\binom{n}{0} = \frac{n!}{(n-0)!\,0!} = \frac{n!}{n!\,0!} = 1.$$

This is satisfied only if 0! is defined to be 1.

10.4 | Exercise Set

Simplify.

1. 8!

2. 9!

3. 10!

4. 11!

5. $\dfrac{7!}{4!}$

6. $\dfrac{8!}{6!}$

7. $\dfrac{10!}{7!}$

8. $\dfrac{9!}{5!}$

9. $\dbinom{8}{2}$

10. $\dbinom{7}{4}$

11. $\dbinom{10}{6}$

12. $\dbinom{9}{5}$

13. $\dbinom{20}{18}$

14. $\dbinom{30}{3}$

15. $\dbinom{35}{2}$

16. $\dbinom{40}{38}$

Expand. Use both of the methods shown in this section.

17. $(m + n)^5$

18. $(a - b)^4$

19. $(x - y)^6$

20. $(p + q)^7$

21. $(x^2 - 3y)^5$

22. $(3c - d)^7$

23. $(3c - d)^6$

24. $(t^{-2} + 2)^6$

25. $(x - y)^3$

26. $(x - y)^5$

27. $\left(x + \dfrac{2}{y}\right)^9$

28. $\left(3s + \dfrac{1}{t}\right)^9$

29. $(a^2 - b^3)^5$

30. $(x^3 - 2y)^5$

31. $(\sqrt{3} - t)^4$

32. $(\sqrt{5} + t)^6$

33. $(x^{-2} + x^2)^4$

34. $\left(\dfrac{1}{\sqrt{x}} - \sqrt{x}\right)^6$

Find the indicated term for each binomial expression.

35. 3rd, $(a + b)^6$

36. 6th, $(x + y)^7$

37. 12th, $(a - 3)^{14}$

38. 11th, $(x - 2)^{12}$

39. 5th, $(2x^3 + \sqrt{y})^8$

40. 4th, $\left(\dfrac{1}{b^2} + c\right)^7$

41. Middle, $(2u - 3v^2)^{10}$

42. Middle two, $(\sqrt{x} + \sqrt{3})^5$

Skill Maintenance

Solve. [9.6]

43. $\log_2 x + \log_2(x - 2) = 3$

44. $\log_3(x + 2) - \log_3(x - 2) = 2$

45. $e^t = 280$

46. $\log_5 x^2 = 2$

Synthesis

47. ◆ Devise two problems requiring the use of the binomial theorem. Design the problems so that one is solved more easily using Form 1 and the other is solved more easily using Form 2. Then explain what makes one form easier to use than the other in each case.

48. Show that there are exactly $\dbinom{5}{3}$ ways of forming a subset of size 3 from a set of 5 elements.

49. *Baseball.* At one point in a recent season, Mike Piazza of the Los Angeles Dodgers had a batting average of 0.313. Suppose that he came to bat 5 times in a game. The probability of his getting exactly 3 hits is the 3rd term of the binomial expansion of $(0.313 + 0.687)^5$. Find that term and use a calculator to estimate the probability.

50. *Widows or Divorcees.* The probability that a woman will be either widowed or divorced is 85%. Suppose that 8 women are interviewed. The probability that exactly 5 of them will be either widowed or divorced in their lifetime is the 6th term of the binomial expansion of $(0.15 + 0.85)^8$. Find that term and use a calculator to estimate the probability.

51. *Baseball.* In reference to Exercise 49, the probability that Piazza will get *at most* 3 hits is found by adding the last 4 terms of the binomial expansion of $(0.313 + 0.687)^5$. Find these terms and use a calculator to estimate the probability.

52. *Widows or Divorcees.* In reference to Exercise 50, the probability that *at least* 6 of the women will be widowed or divorced is found by adding the last three terms of the binomial expansion of $(0.15 + 0.85)^8$. Find these terms and use a calculator to estimate the probability.

53. Prove that
$$\binom{n}{r} = \binom{n}{n - r}$$
for any whole numbers n and r. Assume $r \le n$.

54. Find the term of
$$\left(\frac{3x^2}{2} - \frac{1}{3x}\right)^{12}$$
that does not contain x.

55. Find the middle term of $(x^2 - 6y^{3/2})^6$.

56. Find the ratio of the 4th term of
$$\left(p^2 - \frac{1}{2}p\sqrt[3]{q}\right)^5$$
to the 3rd term.

57. Find the term containing $\frac{1}{x^{1/6}}$ of
$$\left(\sqrt[3]{x} - \frac{1}{\sqrt{x}}\right)^7.$$

58. What is the degree of $(x^2 + 3)^4$?

CHAPTER

10 *Summary and Review*

Key Terms

Sequence, p. 656
Infinite sequence, p. 656
Finite sequence, p. 656
General term, p. 657
Alternating sequence, p. 657
Series, p. 659
Infinite series, p. 659
Partial sum, p. 659
nth partial sum, p. 659

Finite series, p. 659
Sigma notation, p. 660
Summation notation, p. 660
Index of summation, p. 660
Arithmetic sequence, p. 663
Arithmetic progression, p. 664
Common difference, p. 664
Arithmetic series, p. 666
Geometric sequence, p. 673

Geometric progression, p. 673
Common ratio, p. 673
Geometric series, p. 675
Infinite geometric series,
    p. 676
Pascal's triangle, p. 685
Binomial theorem, p. 686
Factorial, p. 687
Binomial coefficient, p. 690

Important Properties and Formulas

Arithmetic sequence: $a_{n+1} = a_n + d$

nth term of an arithmetic sequence: $a_n = a_1 + (n-1)d$

Sum of the first n terms of an arithmetic sequence: $S_n = \dfrac{n}{2}(a_1 + a_n)$

Geometric sequence: $a_{n+1} = a_n \cdot r$

nth term of a geometric sequence: $a_n = a_1 r^{n-1}$

Sum of the first n terms of a geometric sequence: $S_n = \dfrac{a_1(1 - r^n)}{1 - r}$

Limit of an infinite geometric series: $S_\infty = \dfrac{a_1}{1 - r}, \quad |r| < 1$

Factorial notation: $n! = n(n-1)(n-2) \cdots 3 \cdot 2 \cdot 1$

Binomial coefficient: $\dbinom{n}{r} = \dfrac{n!}{(n-r)!\,r!}$

Binomial theorem: $(a+b)^n = \dbinom{n}{0}a^n + \dbinom{n}{1}a^{n-1}b + \dbinom{n}{2}a^{n-2}b^2 + \cdots + \dbinom{n}{n}b^n$

$(r+1)$st term of $(a+b)^n$: $\dbinom{n}{r}a^{n-r}b^r$

REVIEW EXERCISES

Find the first four terms; the 8th term, a_8; and the 12th term, a_{12}.

1. $a_n = 4n - 3$

2. $a_n = \dfrac{n-1}{n^2+1}$

Predict the general term. Answers may vary.

3. $-2, -4, -6, -8, -10, \ldots$

4. $1, 4, 9, 16, 25, \ldots$

Rename and evaluate each sum.

5. $\displaystyle\sum_{k=1}^{5} (-2)^k$

6. $\displaystyle\sum_{k=2}^{7} (1 - 2k)$

Rewrite using sigma notation.

7. $4 + 8 + 12 + 16 + 20$

8. $\dfrac{-1}{2} + \dfrac{1}{4} + \dfrac{-1}{8} + \dfrac{1}{16} + \dfrac{-1}{32}$

9. Find the 14th term of the arithmetic sequence $-6, 1, 8, \ldots$.

10. Find d when $a_1 = 11$ and $a_{10} = 35$. Assume an arithmetic sequence.

11. Find a_1 and d when $a_{12} = 25$ and $a_{24} = 40$. Assume an arithmetic sequence.

12. Find the sum of the first 17 terms of the arithmetic series $-8 + (-11) + (-14) + \cdots$.

13. Find the sum of all the multiples of 6 from 12 to 318, inclusive.

14. Find the 20th term of the geometric sequence $2, 2\sqrt{2}, 4, \ldots$.

15. Find the common ratio of the geometric sequence $2, \frac{4}{3}, \frac{8}{9}, \ldots$.

16. Find the nth term of the geometric sequence $-2, 2, -2, \ldots$.

17. Find the nth term of the geometric sequence $3, \frac{3}{4}x, \frac{3}{16}x^2, \dots$.

18. Find S_6 for the geometric series
$$3 + 12 + 48 + \cdots.$$

19. Find S_{12} for the geometric series
$$3x - 6x + 12x - \cdots.$$

Determine whether each infinite geometric series has a limit. If a limit exists, find it.

20. $6 + 3 + 1.5 + 0.75 + \cdots$

21. $7 - 4 + \frac{16}{7} - \cdots$

22. $2 + (-2) + 2 + (-2) + \cdots$

23. $0.04 + 0.08 + 0.16 + 0.32 + \cdots$

24. $\$2000 + \$1900 + \$1805 + \$1714.75 + \cdots$

25. Find fractional notation for $0.555555\dots$.

26. Find fractional notation for $1.39393939\dots$.

Solve.

27. You take a job, starting with an hourly wage of $11.40. You are promised a raise of 20¢ per hour every 3 mos for 8 yr. At the end of 8 yr, what will be your hourly wage?

28. A stack of logs has 42 poles in the bottom row. There are 41 poles in the second row, 40 poles in the third row, and so on, ending with 1 pole in the top row. How many poles are in the stack?

29. A student loan is in the amount of $10,000. Interest is 7%, compounded annually, and the amount is to be paid off in 12 yr. How much is to be paid back?

30. Find the total rebound distance of a ball, given that it is dropped from a height of 12 m and each rebound is one third of the preceding one.

Simplify.

31. $8!$

32. $\binom{8}{3}$

33. Find the 3rd term of $(a + b)^{20}$.

34. Expand: $(x - 2y)^4$.

Skill Maintenance

Solve.

35. $3x - y = 7,$
    $2x + 3y = 5$

36. $\log (x + 5) - \log x = 1$

Simplify.

37. $7^{\log_7 13}$

38. $\log_4 \frac{1}{4}$

Synthesis

39. ◆ What happens to the terms of a geometric sequence with $|r| < 1$ as n gets larger? Why?

40. ◆ Compare the two forms of the binomial theorem given in the text. Under what circumstances would one be more useful than the other?

41. Find the sum of the first n terms of the geometric series $1 - x + x^2 - x^3 + \cdots$.

42. Expand: $(x^{-3} + x^3)^5$.

CHAPTER 10 TEST

1. Find the first 5 terms and the 16th term of a sequence with general term $a_n = 6n - 5$.

2. Predict the general term of the sequence
$$\frac{4}{3}, \frac{4}{9}, \frac{4}{27}, \dots.$$

3. Rename and evaluate:
$$\sum_{k=1}^{5} (3 - 2^k).$$

4. Rewrite using sigma notation:
$$1 + 8 + 27 + 64 + 125.$$

5. Find the 12th term, a_{12}, of the arithmetic sequence $9, 4, -1, \dots$.

Assume arithmetic sequences for Questions 6 and 7.

6. Find the common difference d when $a_1 = 9$ and $a_7 = 11\frac{1}{4}$.

7. Find a_1 and d when $a_5 = 16$ and $a_{10} = -3$.

8. Find the sum of all the multiples of 12 from 24 to 240, inclusive.

9. Find the 6th term of the geometric sequence $72, 18, 4\frac{1}{2}, \dots$.

10. Find the common ratio of the geometric sequence $22\frac{1}{2}, 15, 10, \dots$.

11. Find the nth term of the geometric sequence $3, -9, 27, \dots$.

12. Find the sum of the first nine terms of the geometric series

$$(1 + x) + (2 + 2x) + (4 + 4x) + \cdots.$$

Determine whether each infinite geometric series has a limit. If a limit exists, find it.

13. $0.5 + 0.25 + 0.125 + \cdots$

14. $0.5 + 1 + 2 + 4 + \cdots$

15. $\$1000 + \$80 + \$6.40 + \cdots$

16. Find fractional notation for $0.85858585\ldots$.

17. An auditorium has 31 seats in the first row, 33 seats in the second row, 35 seats in the third row, and so on, for 18 rows. How many seats are in the 17th row?

18. A stack of poles has 52 poles in the bottom row. There are 51 poles in the second row, 50 poles in the third row, and so on, ending with 1 pole in the top row. How many poles are in the stack?

19. Each week the price of a $15,000 boat will be reduced 5% of the previous week's price. If we assume that it is not sold, what will be the price after 10 weeks?

20. Find the total rebound distance of a ball that is dropped from a height of 18 m, with each rebound two thirds of the preceding one.

21. Simplify: $\begin{pmatrix} 13 \\ 11 \end{pmatrix}$.

22. Expand: $(x^2 - 3y)^5$.

23. Find the 4th term in the expansion of $(a + x)^{12}$.

Skill Maintenance

Solve.

24. $4^{2x-3} = 64$

25. $\log_x \frac{1}{9} = -2$

26. Solve:

$$y = 3x + 5,$$
$$2x + 5y = 8.$$

27. Convert to an exponential equation: $\log x = 1.5$.

Synthesis

28. Find a formula for the sum of the first n even natural numbers:

$$2 + 4 + 6 + \cdots + 2n.$$

29. Find the sum of the first n terms of

$$1 + \frac{1}{x} + \frac{1}{x^2} + \frac{1}{x^3} + \cdots.$$

CUMULATIVE REVIEW 1–10

Simplify.

1. $(-9x^2y^3)(5x^4y^{-7})$

2. $|-3.5 + 9.8|$

3. $2y - [3 - 4(5 - 2y) - 3y]$

4. $(10 \cdot 8 - 9 \cdot 7)^2 - 54 \div 9 - 3$

5. Evaluate

$$\frac{ab - ac}{bc}$$

for $a = -2$, $b = 3$, and $c = -4$.

Perform the indicated operations and simplify.

6. $(5a^2 - 3ab - 7b^2) - (2a^2 + 5ab + 8b^2)$

7. $(-3x^2 + 4x^3 - 5x - 1) + (9x^3 - 4x^2 + 7 - x)$

8. $(2a - 1)(3a + 5)$

9. $(3a^2 - 5y)^2$

10. $\dfrac{1}{x - 2} - \dfrac{4}{x^2 - 4} + \dfrac{3}{x + 2}$

11. $\dfrac{x^2 - 6x + 8}{3x + 9} \cdot \dfrac{x + 3}{x^2 - 4}$

12. $\dfrac{x - 1}{x^2 - 9} \div \dfrac{x^2 - 1}{3x - 9}$

13. $\dfrac{x - \dfrac{a^2}{x}}{1 + \dfrac{a}{x}}$

Factor.

14. $4x^2 - 12x + 9$

15. $-2t^2 + 2t + 12$

16. $a^3 + 3a^2 - ab - 3b$

17. $15y^4 + 33y^2 - 36$

18. For the function described by
$$f(x) = 3x^2 - 4x,$$
find $f(-2)$.

19. Divide:
$$(7x^4 - 5x^3 + x^2 - 4) \div (x - 2).$$

Solve.

20. $9(x - 1) - 3(x - 2) = 1$

21. $\dfrac{6}{x} + \dfrac{6}{x + 2} = \dfrac{5}{2}$

22. $2x + 1 > 5 \ or \ x - 7 \le 3$

23. $5x + 3y = 2,$
$3x + 5y = -2$

24. $x + y - z = 0,$
$3x + y + z = 6,$
$x - y + 2z = 5$

25. $3\sqrt{x - 1} = 5 - x$

26. $x^4 - 29x^2 + 100 = 0$

27. $e^{-t} = 1.703$

28. $3x^2 + 5x = 2$

29. $5^x = 8$

30. $\log (x^2 - 25) - \log (x + 5) = 3$

31. $\log_4 x = -2$ **32.** $7^{2x+3} = 49$

33. $|2x - 1| \le 5$ **34.** $7x^2 + 14 = 0$

35. $x^2 + 4x = 3$ **36.** $y^2 + 3y > 10$

37. Let $f(x) = x^2 - 2x$. Find a such that $f(a) = 48$.

Solve.

38. A music club offers two types of membership. Limited members pay a fee of $10 a year and can buy CDs for $10 each. Preferred members pay $20 a year and can buy CDs for $7.50 each. For what numbers of annual CD purchases would it be less expensive to be a preferred member?

39. Find three consecutive integers whose sum is 198.

40. A pentagon with all five sides the same size has a perimeter equal to that of an octagon in which all eight sides are the same size. One side of the pentagon is 2 less than three times one side of the octagon. What is the perimeter of each figure?

41. A chemist has two solutions of ammonia and water. Solution A is 6% ammonia and solution B is 2% ammonia. How many liters of each solution are needed in order to obtain 80 L of a solution that is 3.2% ammonia?

42. An airplane can fly 190 mi with the wind in the same time it takes to fly 160 mi against the wind. The speed of the wind is 30 mph. How fast can the plane fly in still air?

43. Bianca can do a certain job in 21 min. Dahlia can do the same job in 14 min. How long would it take to do the job if the two worked together?

44. The centripetal force F of an object moving in a circle varies directly as the square of the velocity v and inversely as the radius r of the circle. If $F = 8$ when $v = 1$ and $r = 10$, what is F when $v = 2$ and $r = 16$?

45. A farmer wants to fence in a rectangular area next to a river. (Note that no fence will be needed along the river.) What is the area of the largest region that can be fenced in with 100 ft of fencing?

Graph.

46. $3x - y = 6$

47. $y = \log_2 x$

48. $2x - 3y < -6$

49. Graph $f(x) = \sqrt{4 - x}$, and use the graph to estimate the domain and the range of f.

50. Graph: $f(x) = -2(x - 3)^2 + 1$.
   a) Label the vertex.
   b) Draw the axis of symmetry.
   c) Find the maximum or minimum value.
   d) Find the domain and the range of f.

Solve.

51. $V = P - Prt$, for r

52. $I = \dfrac{R}{R + r}$, for R

53. Find a linear equation whose graph contains the point $(-1, 4)$ and is perpendicular to the line whose equation is $3x - y = 6$.

Find the domain of each function.

54. $f(x) = \sqrt{5 - 3x}$

55. $g(x) = \dfrac{x - 4}{x^2 - 2x + 1}$

56. Multiply $(8.9 \times 10^{-17})(7.6 \times 10^4)$. Write scientific notation for the answer.

57. Multiply and simplify: $\sqrt{8x} \, \sqrt{8x^3y}$.

58. Simplify: $(25x^{4/3}y^{1/2})^{3/2}$.

59. Divide and simplify:
$$\frac{\sqrt[3]{15x}}{\sqrt[3]{3y^2}}.$$

60. Rationalize the denominator:
$$\frac{1 - \sqrt{x}}{1 + \sqrt{x}}.$$

61. Write a single radical expression:
$$\frac{\sqrt[3]{(x + 1)^5}}{\sqrt{(x + 1)^3}}.$$

62. Multiply these complex numbers:
$$(3 + 2i)(4 - 7i).$$

63. Write a quadratic equation whose solutions are $5\sqrt{2}$ and $-5\sqrt{2}$.

64. Express as a single logarithm:
$$\tfrac{2}{3} \log_a x - \tfrac{1}{2} \log_a y + 5 \log_a z.$$

65. Convert to an exponential equation: $\log_a c = 5$.

Calling Cards. *The sale of prepaid long-distance telephone cards brought in $75 million in 1993 and $800 million in 1995 (Source: Associated Press article in the* Indianapolis Star, *7/14/96). Assume that the business is growing exponentially.*

66. Write an exponential function describing the sales of the cards t years after 1993.

67. Predict the amount of sales in 2000.

68. Find the distance between the points $(-1, -5)$ and $(2, -1)$.

69. Find the 21st term of the arithmetic sequence $19, 12, 5, \ldots$.

70. Find the sum of the first 25 terms of the arithmetic series $-1 + 2 + 5 + \cdots$.

71. Find the general term of the geometric sequence $16, 4, 1, \ldots$.

72. Find the 7th term of $(a - 2b)^{10}$.

73. Find the sum of the first nine terms of the geometric series $x + 1.5x + 2.25x + \cdots$.

74. On Mark's 9th birthday, his grandmother opened a savings account for him with $100. The account draws 6% interest, compounded annually. If Mark neither adds to nor withdraws any money from the bank, how much will be in the account on his 18th birthday?

For each of the following graphs, **(a)** *determine whether a linear, quadratic, or exponential function would best fit the situation, and* **(b)** *use regression to fit a function to the data.*

75.

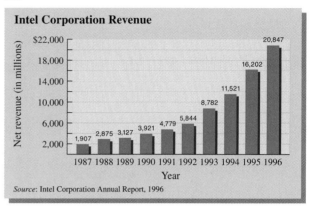

Source: Intel Corporation Annual Report, 1996

76.

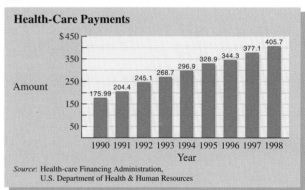

Source: Health-care Financing Administration, U.S. Department of Health & Human Resources

77.

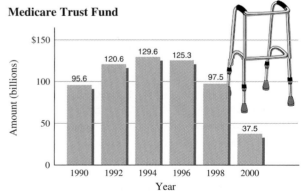

Source: Health-care Financing Administration, U.S. Department of Health & Human Resources

Synthesis

Solve.

78. $\dfrac{9}{x} - \dfrac{9}{x + 12} = \dfrac{108}{x^2 + 12x}$

79. $\log_2 (\log_3 x) = 2$

80. y varies directly as the cube of x and x is multiplied by 0.5. What is the effect on y?

81. Divide these complex numbers:

$$\frac{2\sqrt{6} + 4\sqrt{5}i}{2\sqrt{6} - 4\sqrt{5}i}.$$

82. Diaphantos, a famous mathematician, spent $\frac{1}{6}$ of his life as a child, $\frac{1}{12}$ as an adolescent, and $\frac{1}{7}$ as a bachelor. Five years after he was married, he had a son who died 4 years before his father at half his father's final age. How long did Diaphantos live?

Answers

Chapter 1

EXERCISE SET 1.1

1. 10   **3.** 16   **5.** 23   **7.** 387   **9.** 5   **11.** 57
13. 15 ft^2 **15.** 6.4 m^2 **17. (a)** Yes; **(b)** no; **(c)** no
19. (a) No; **(b)** yes; **(c)** yes   **21. (a)** Yes; **(b)** no; **(c)** no
23. {a, e, i, o, u}   **25.** {1, 3, 5, 7, . . .}
27. {7, 14, 21, 28, . . .}
29. $\{x \mid x$ is an odd number between 10 and 30$\}$
31. $\{x \mid x$ is a whole number less than 5$\}$
33. $\{x \mid x$ is a multiple of 5 between 7 and 79$\}$
35. False   **37.** True   **39.** False   **41.** True   **43.** True
45. True   **47.** 8   **49.** 9   **51.** 6.2   **53.** 0   **55.** $1\frac{7}{8}$
57. 4.21   **59.** True   **61.** False   **63.** True   **65.** True
67. True   **69.** False   **71.** 15   **73.** 41.4494
75. 25.125   **77.** 13,778   **79.** ◆   **81.** ◆
83. $x - 10 = 4$; answers may vary   **85. (a)** Yes; **(b)** yes;
(c) no   **87.** True   **89.** True   **91.** {0}
93. {5, 10, 15, . . .}
95. {. . . , −6, −4, −2, 0, 2, 4, 6, . . .}

97.

INTERACTIVE DISCOVERY P. 15

−6.72, +, −, −; −323, −, +, −; 266.2, −, −, +

EXERCISE SET 1.2

1. 17   **3.** −11   **5.** −3.2   **7.** $-\frac{11}{35}$ **9.** −8.5   **11.** $\frac{5}{9}$
13. −4.5   **15.** 0   **17.** −6.4   **19.** −7.29   **21.** $4\frac{1}{3}$
23. 0   **25.** −7   **27.** 2.7   **29.** −1.79   **31.** 0   **33.** 2
35. −5   **37.** 4   **39.** −17   **41.** −3.1   **43.** $-\frac{11}{10}$

45. 2.9   **47.** 7.9   **49.** −28   **51.** 24   **53.** −21
55. $-\frac{3}{7}$ **57.** 0   **59.** 5.44   **61.** 5   **63.** −5   **65.** −73
67. 0   **69.** $\frac{1}{5}$ **71.** $-\frac{1}{9}$ **73.** $\frac{3}{2}$ **75.** $-\frac{11}{3}$ **77.** $\frac{5}{6}$
79. $-\frac{6}{5}$ **81.** $\frac{1}{36}$ **83.** $-\frac{12}{7}$ **85. (a)**   **87. (d)**   **89.** 27
91. $-\frac{6}{11}$ **93.** $-\frac{65}{7}$ **95.** 117   **97.** 28
99. −1.025641026   **101.** 110.3420204   **103.** −12.86
105. 5   **107.** ◆   **109.** ◆   **111.** $(3 - 8)^2 + 9 = 34$
113. $5 \cdot 2^3 \div (3 - 4)^4 = 40$ **115.** −6.2

EXERCISE SET 1.3

1. $8y + 3x$; $x^3 + 8y$; $3x + y^8$ **3.** $y(7x)$; $(x7)y$ **5.** $3(xy)$
7. $(x + 2y) + 5$ **9.** $3a + 3$ **11.** $4x - 4y$
13. $-10a - 15b$ **15.** $2ab - 2ac + 2ad$
17. $2\pi rh + 2\pi r$ **19.** $5(x + y)$ **21.** $3(p - 3)$
23. $7(x - 3y)$ **25.** $2(x - y + z)$ **27.** $x(y + 1)$
29. $a(b + c - d)$ **31.** $4a, -5b, 6$ **33.** $2x^2, -6x, 7$
35. $9a$ **37.** $-2rt$ **39.** $9x^2$ **41.** $11a$ **43.** $-8t$
45. $10x$ **47.** $8x - 2x^2$ **49.** $9a^2 + 2a$ **51.** $22x + 18$
53. $4t^3 - 5t^2 + 2t$ **55.** $-a - 5$ **57.** $m + 1$
59. $5d - 12$ **61.** $-7x + 14$ **63.** $-9x + 21$
65. $44a - 22$ **67.** $-100a - 90$ **69.** $-12y - 145$
71. 3   **73.** 35   **75.** ◆   **77.** ◆
79. $5(a + bc)$
    $= 5a + 5(bc)$ Distributive
    $= 5(bc) + 5a$ Commutative, addition
    $= (bc)5 + a5$ Commutative, multiplication
    $= (cb)5 + a5$ Commutative, multiplication
    $= c(b5) + a5$ Associative, multiplication
81. $-42x - 360y - 276$ **83.** $4x - f$

INTERACTIVE DISCOVERY P. 26

1. (b)   **2. (c)**   **3. (a)**   **4. (d)**

INTERACTIVE DISCOVERY P. 27

1. (c)   **2. (a)**   **3. (b)**

EXERCISE SET 1.4

1. 7^7 **3.** a^3 **5.** $18x^7$ **7.** $21m^{13}$ **9.** $x^{10}y^{10}$ **11.** a^6
13. $2x^3$ **15.** m^5n^4 **17.** $-3x^4y^6z^6$ **19.** 81 **21.** -81
23. $\dfrac{1}{25}$ **25.** $-\dfrac{1}{25}$ **27.** -1 **29.** $\dfrac{1}{n^6}$ **31.** $\dfrac{1}{(4xy)^5}$
33. $\dfrac{2a^2}{b^5}$ **35.** $\dfrac{1}{3x^5z^4}$ **37.** 3^{-4} **39.** $\dfrac{1}{x^{-5}}$ **41.** $\dfrac{6}{x^{-2}}$
43. $(5y)^{-3}$ **45.** $\dfrac{y^{-4}}{3}$ **47.** 9^{-7}, or $\dfrac{1}{9^7}$ **49.** a **51.** 1
53. $6a^{-4}b^{-9}$, or $\dfrac{6}{a^4b^9}$ **55.** 12^{-12}, or $\dfrac{1}{12^{12}}$ **57.** 9^2, or 81
59. a^5 **61.** $\dfrac{1}{3}a^{10}b^{-9}c^{-2}$, or $\dfrac{a^{10}}{3b^9c^2}$ **63.** a^6 **65.** 8^{-12},
or $\dfrac{1}{8^{12}}$ **67.** 6^{12} **69.** $x^{15}y^5$ **71.** $7a^6b^8$ **73.** $\dfrac{x^{-4}y^{10}}{9}$, or
$\dfrac{y^{10}}{9x^4}$ **75.** $\dfrac{5a^4b}{2}$ **77.** $\dfrac{8x^9y^3}{27}$ **79.** 1 **81.** 2.6×10^{12}
83. 2.63×10^{-7} **85.** 3.09×10^{12} **87.** 8.02×10^{-9}
89. 0.00005 **91.** 92,400,000 **93.** 0.07034
95. 90,010,000,000 **97.** 3.38×10^{-4}
99. 2.6732×10^{-11} **101.** 4.646174×10^{-30}
103. 1.4663385×10^{-1} **105.** 2.5×10^3
107. 3.0×10^{11} **109.** 3.018297×10^{-13}
111. 1.8550947 **113.** 3.2129032×10
115. 7.2×10^{-34} **117.** 4.30098×10^{-6}
119. 30.96 **121.** $2(4x - 5)$ **123.** ◈ **125.** ◈
127. 12^{6b-2ab} **129.** $-5x^{2b}y^{-2a}$ **131.** $\frac{2}{27}$
133. 1.25×10^{22} **135.** 1

EXERCISE SET 1.5

1.

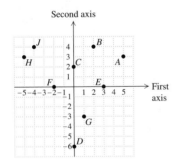

3.

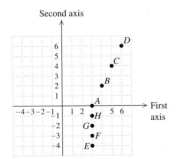

5.

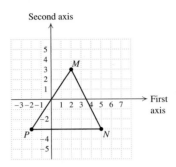

triangle; 21 sq. units
7. III **9.** II **11.** I **13.** IV **15.** Yes **17.** No
19. Yes **21.** Yes **23.** Yes **25.** No **27.** Yes
29. No

31.

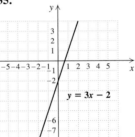

33.

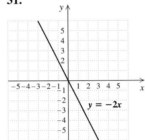

$y = x + 3$

35.

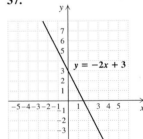

$y = 3x - 2$

37.

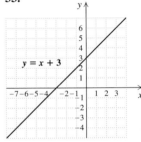

$y = -2x + 3$

39.

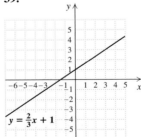

$y = \frac{2}{3}x + 1$

41.

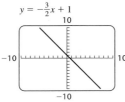

$$y = -\frac{3}{2}x + 1$$

43.

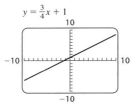

$$y = \frac{3}{4}x + 1$$

45.

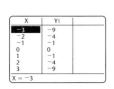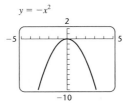

$$y = -x^2$$

47.

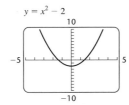

$$y = x^2 - 2$$

49.

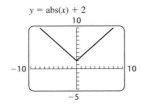

$$y = \text{abs}(x) + 2$$

51.

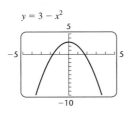

$$y = 3 - x^2$$

53.

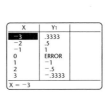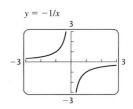

$$y = -1/x$$

55. (b)   **57.** (a)   **59.** (b)   **61.** Exercies **31**–**43** and **55** are linear.   **63.** Yes   **65.** No   **67.** ◈   **69.** ◈
71. ◈   **73.** (2, 4), (−5, −3); 49 square units   **75.** −2
77. [0, 30, −50, 10] with Xscl = 10 and Yscl = 10; answers may vary

EXERCISE SET 1.6

1. Let n represent the number; $7 + n$, or $n + 7$ **3.** Let t represent the number; $12t$ **5.** Let x represent the number; $0.65x$, or $\frac{65}{100}x$ **7.** Let y represent the number; $2y - 9$
9. Let s represent the number; $0.1s + 8$, or $8 + 0.1s$
11. Let m and n represent the numbers; $m - n - 1$
13. $90 \div 4$, or $\frac{90}{4}$ **15.** Let x and $x + 7$ represent the numbers; $x + (x + 7) = 65$ **17.** Let x represent the number; $128 = 0.4x$ **19.** Let a represent the number; $\frac{a}{4} = 12.3$ **21.** Let w represent the rectangle's width; $21 = 2(2w) + 2w$ **23.** Let p represent the original price; $377 = p - 0.35p$ **25.** Let n represent the number of hours Rhonda rode; $25 = 15n$ **27.** Let x represent the measure of the smallest angle; $x + (x + 1) + (x + 2) = 180$ **29.** Let t represent the number of minutes to reach cruising altitude; $29{,}000 - 8{,}000 = 3500t$ **31.** Let x be the measure of the second angle; $3x + x + (2x - 12) = 180$ **33.** Let n be the first odd number; $n + 2(n + 2) + 3(n + 4) = 70$
35. Let s be the length of a side of the smaller square; $4s + 4 \cdot 2s = 100$ **37.** Let x be the first number; $x + (3x - 6) + \left[\frac{2}{3}(3x - 6) + 2\right] = 172$ **39.** Let x be the score on the next test:
$$\frac{93 + 89 + 72 + 80 + 96 + x}{6} = 88$$
41. 75   **43.** 1,000,000   **45.** 1991, 1995

47. 3.5 drinks

49.

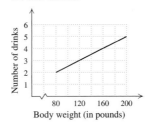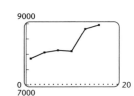

51. 2.3 million   **53.** 1991, 1995   **55.** 16   **57.** z^4
59. ◈   **61.** ◈   **63. (a)** IV; **(b)** III; **(c)** I; **(d)** II

REVIEW EXERCISES, CHAPTER 1

1. [1.1] 28   **2.** [1.1] 60.614425
3. [1.1] {2, 4, 6, 8, 10, 12}; {$x\,|\,x$ is an even integer
between 1 and 13}   **4.** [1.1] 3150 cm^2 **5.** [1.1] **(a)** No;
(b) yes   **6.** [1.1] **(a)** Yes; **(b)** yes   **7.** [1.1] 7.3
8. [1.1] 4.09   **9.** [1.1] 0   **10.** [1.2] -13.1
11. [1.2] $-\frac{23}{35}$ **12.** [1.2] $\frac{7}{15}$ **13.** [1.2] -11.5
14. [1.2] $-\frac{1}{6}$ **15.** [1.2] -5.4 **16.** [1.2] 6.3
17. [1.2] $-\frac{5}{12}$ **18.** [1.2] -9.1 **19.** [1.2] $-\frac{21}{4}$
20. [1.2] 4.01   **21.** [1.3] $a + 5$ **22.** [1.3] $y7$
23. [1.3] $y + 5x$, or $x5 + y$, or $y + x5$
24. [1.3] $4 + (a + b)$ **25.** [1.3] $x(y7)$
26. [1.3] $7m(n + 2)$ **27.** [1.3] $6x^3 - 8x^2 + 2$
28. [1.3] $47x - 60$ **29.** [1.4] $-10a^5b^8$ **30.** [1.4] $4xy^6$
31. [1.4] 1; 28.09; -28.09 **32.** [1.4] 3^3, or 27
33. [1.4] $125a^6$ **34.** [1.4] $-\dfrac{a^9}{8b^6}$ **35.** [1.4] $\dfrac{z^8}{x^4y^6}$
36. [1.4] $\dfrac{b^{16}}{16a^{20}}$ **37.** [1.2] $-\frac{16}{7}$ **38.** [1.2] 0.7128
39. [1.4] 1.03×10^{-7} **40.** [1.4] 3.086×10^{13}
41. [1.4] 3.741×10^7 **42.** [1.4] 1.9672131×10^{-6}
43. [1.5] Yes   **44.** [1.5] No   **45.** [1.5] Yes
46. [1.5] No
47. [1.5]   **48.** [1.5]

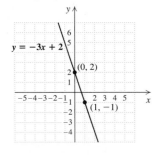

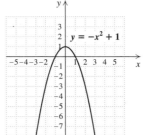

49. [1.5]   **50.** [1.5]

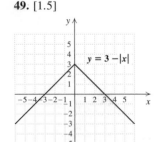

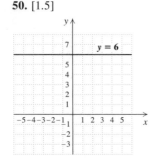

51. [1.5] 3, 1, -1, -3, -1, 1, 3

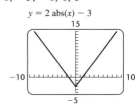

52. [1.6] Let n represent the number; $2n - 13 = 21$
53. [1.6] Let n represent the larger number;
$n + (n - 17) = 115$ **54.** [1.6] Let x be the measure of
the second angle; $3x + x + 2x = 180$
55. [1.6] **(a)** 60 million; **(b)** 1975; **(c)** increase;
(d) 20 million
56. [1.6]

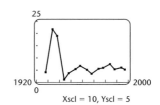

57. [1.1], [1.5] ◈ A solution of $y = 2x + 1$ is an
ordered pair; there are infinitely many such solutions.
There is only one solution of $3x + 5 = 2$; it is a
number.   **58.** [1.1] ◈ Without a standard set of rules,
an expression could be interpreted to have more than
one value.   **59.** [1.2], [1.4] $-\frac{23}{24}$ **60.** [1.4] $3^{-2a+2b-8ab}$
61. [1.3] $a2 + cb + cd + ad = ad + a2 + cb + cd =$
$a(d + 2) + c(b + d)$ **62.** [1.2] 0.56556555655556...;
answers may vary

TEST, CHAPTER 1

1. [1.1] -47 **2.** [1.1] 3.75 cm^2 **3.** [1.1] **(a)** No; **(b)** yes;
(c) no   **4.** [1.2] -41 **5.** [1.2] -3.7 **6.** [1.2] -2.11
7. [1.2] -14.2 **8.** [1.2] -43.2 **9.** [1.2] -33.92
10. [1.2] $-\frac{19}{12}$ **11.** [1.2] $\frac{5}{49}$ **12.** [1.2] 6   **13.** [1.2] $-\frac{4}{3}$
14. [1.2] $-\frac{3}{2}$ **15.** [1.3] $y + 7x$, or $x7 + y$, or $y + x7$
16. [1.3] $-3y - 29$ **17.** [1.3] $3x + 8$ **18.** [1.4] $-\dfrac{72}{x^{10}y^6}$
19. [1.4] $-\frac{1}{9}$ **20.** [1.4] $\dfrac{y^8}{36x^4}$ **21.** [1.4] $\dfrac{x^6}{4y^8}$ **22.** [1.4] 1
23. [1.4] 2.0091×10^{-7} **24.** [1.4] 2.0×10^{10}
25. [1.4] 3.0573066×10^4 **26.** [1.5] Yes   **27.** [1.5] No

28. [1.5]

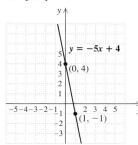

$y = -5x + 4$

(0, 4)

(1, −1)

29. [1.5]

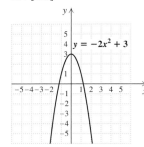

$y = -2x^2 + 3$

30. [1.5]

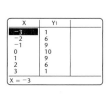

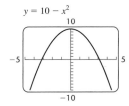

$y = 10 - x^2$

31. [1.6] Let m and n represent the numbers; $mn + 3$, or $3 + mn$ **32.** [1.6] Let t be the score on the sixth test;
$$\frac{94 + 80 + 76 + 91 + 75 + t}{6} = 85$$ **33.** [1.6] 45 mph

34. [1.6] 7 mpg   **35.** [1.4] $16^c x^{6ac} y^{2bc+2c}$

36. [1.4] $-9a^3$ **37.** [1.4] $\dfrac{4}{7y^2}$

Chapter 2

EXERCISE SET 2.1

1. No   **3.** Yes   **5.** Yes   **7.** Function
9. A relation but not a function   **11.** Function
13. **(a)** 3; **(b)** $\{-4, -3, -2, -1, 0, 1, 2\}$; **(c)** $-2, 0$;
(d) $\{1, 2, 3, 4\}$ **15.** **(a)** About $\frac{5}{2}$; **(b)** $\{x \mid -3 \le x \le 5\}$;
(c) about $\frac{7}{3}$; **(d)** $\{y \mid 1 \le y \le 4\}$ **17.** **(a)** About $\frac{9}{4}$;
(b) $\{x \mid -4 \le x \le 3\}$; **(c)** about 0; **(d)** $\{y \mid -5 \le y \le 4\}$
19. **(a)** About $\frac{3}{2}$; **(b)** $\{x \mid -5 \le x \le 2\}$;
(c) about $\frac{7}{6}$; **(d)** $\{y \mid -3 \le y \le 5\}$ **21.** **(a)** 2;
(b) $\{x \mid -4 \le x \le 4\}$; **(c)** $\{x \mid 0 < x \le 2\}$; **(d)** $\{1, 2, 3, 4\}$
23. Yes   **25.** Yes   **27.** No   **29.** No
31. **(a)** 1; **(b)** -3; **(c)** -6; **(d)** 9; **(e)** 0.185; **(f)** $a + 3$
33. **(a)** 0; **(b)** 1; **(c)** 57; **(d)** 9.858; **(e)** $5t^2 + 4t$;
(f) $20a^2 + 8a$ **35.** **(a)** $\frac{3}{5}$; **(b)** $\frac{1}{3}$; **(c)** $\frac{4}{7}$; **(d)** 0; **(e)** 0.49378;
(f) $\dfrac{x - 1}{2x - 1}$ **37.** **(a)** Yes; **(b)** yes; **(c)** yes; **(d)** no
39. **(a)** No; **(b)** yes; **(c)** yes; **(d)** yes

41. Domain: $\{x \mid x$ is a real number$\}$, range: $\{y \mid y$ is a real number$\}$ **43.** Domain: $\{x \mid x \ge 4\}$, range: $\{y \mid y \ge 0\}$
45. Domain: $\{x \mid x$ is a real number$\}$, range: $\{y \mid y \ge -6\}$
47. Domain: $\{x \mid x$ is a real number$\}$, range: $\{y \mid y \ge 1\}$
49. $4\sqrt{3}$ cm^2 **51.** 113.097 in^2 **53.** 14°F
55. 159.9 cm
57. About 55 feet

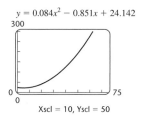

$y = 0.084x^2 - 0.851x + 24.142$

Xscl = 10, Yscl = 50

59. About 2.5 hr

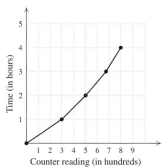

Time (in hours)

Counter reading (in hundreds)

61. About 64,000

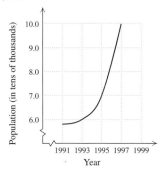

Population (in tens of thousands)

Year

63. About $300,000

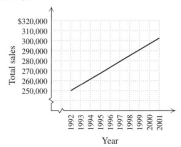

Total sales

Year

65. $-6x + 10$ **67.** $44x - 26$ **69.** $-10x + 71$
71. ◈   **73.** ◈   **75.** 26; 99   **77.** About 2.285 cm
79. About 22 mm   **81.** ◈
83.

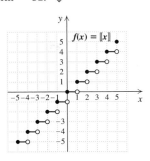

85. Bicycling 14 mph for 1 hr

EXERCISE SET 2.2

1. 7   **3.** 3   **5.** 9   **7.** 1   **9.** 6   **11.** 1.33333   **13.** Yes
15. Yes   **17.** No   **19.** No   **21.** 14.6   **23.** 8   **25.** 18
27. 5   **29.** 8   **31.** 21   **33.** 2   **35.** 2   **37.** 2   **39.** 7
41. 5   **43.** 2   **45.** 2   **47.** $\frac{49}{9}$ **49.** -4.17619 **51.** $\frac{4}{5}$
53. 5   **55.** $-\frac{1}{2}$ **57.** $\varnothing$; contradiction   **59.** $\mathbb{R}$; identity
61. $\{0\}$; conditional   **63.** $\mathbb{R}$; identity
65. $\{1, 2, 3, 4, 5, 6, 7, 8, 9\}$; $\{x \mid x$ is a positive integer less than 10$\}$ **67.** Let x and y represent the numbers; $xy - 3$
69. ◈   **71.** ◈   **73.** 8   **75.** $\frac{224}{29}$ **77.** 1   **79.** $-6, 2$
81. $-1, 2$

EXERCISE SET 2.3

1. 13, 19   **3.** $\frac{3}{2}$ **5.** Length: 5 ft; width: $\frac{5}{2}$ ft   **7.** $\frac{100}{3}$ sec
9. $\frac{29}{13}$ hr   **11.** 45°, 45°, 90°   **13.** 4900 megawatts
15. $\frac{1}{5}$ lb   **17.** $21\frac{1}{3}$ pt   **19.** 38   **21.** $450
23. 6.79×10^8 km   **25.** 2.2×10^{-3} lb
27. 8 light years   **29.** 7.90×10^7
31. 4.49×10^4 km/h   **33.** About 8.5 cm   **35.** 9 ft
37. About 7.7 hr   **39.** $t = \dfrac{d}{r}$ **41.** $a = \dfrac{F}{m}$
43. $h = \dfrac{V}{lw}$ **45.** $k = Ld^2$ **47.** $n = \dfrac{G - w}{150}$
49. $l = p - 2w - 2h$ **51.** $y = \dfrac{C - Ax}{B}$
53. $F = \dfrac{9}{5}C + 32$ **55.** $b_2 = \dfrac{2A}{h} - b_1$ **57.** $m = \dfrac{Fr}{v^2}$
59. $n = \dfrac{q_1 + q_2 + q_3}{A}$ **61.** $d_1 = d_2 - vt$
63. $m = \dfrac{r}{1 + np}$ **65.** $a = \dfrac{y}{b - c^2}$ **67.** $y = \dfrac{3 - x}{2}$
69. $y = x + 7$ **71.** $y = -3x - 10$
73. $y = \dfrac{x + 1 - x^2}{4}$

75.

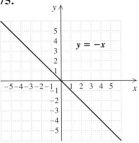

$y = -x$

77.

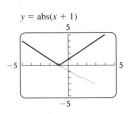

$y = \text{abs}(x + 1)$

79. ◈   **81.** ◈   **83.** 10   **85.** $110,000
87. $a = \dfrac{2s - 2v_i t}{t^2}$ **89.** $T_2 = \dfrac{P_2 V_2 T_1}{P_1 V_1}$ **91.** $b = \dfrac{ac}{1 + c}$

93.

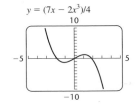

$y = (2x^3 + 7)/5$

95.

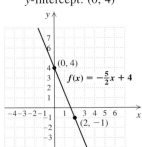

$y = (7x - 2x^3)/4$

EXERCISE SET 2.4

1. Slope: 4; y-intercept: $(0, 5)$
3. Slope: -2; y-intercept: $(0, -6)$
5. Slope: $-\frac{3}{8}$; y-intercept: $(0, -0.2)$
7. Slope: 0.5; y-intercept: $(0, -9)$
9. Slope: 43; y-intercept: $(0, 197)$ **11.** $f(x) = \frac{2}{3}x - 7$
13. $f(x) = -4x + 2$ **15.** $f(x) = -\frac{7}{9}x + 3$
17. $f(x) = 5x + \frac{1}{2}$ **19.** 2   **21.** -2 **23.** $-\frac{1}{3}$
25. The weight is increasing at a rate of 0.5 lb per bag.
27. The distance from home is increasing at a rate of 0.25 km/min.   **29.** The distance from the finish line is decreasing at a rate of $6\frac{2}{3}$ m/sec.   **31.** 12 km/h
33. 0.6 ton/hr   **35.** 300 ft/min
37. (a) II; **(b)** IV; **(c)** III; **(d)** I
39. Slope: $\frac{5}{2}$;   **41.** Slope: $-\frac{5}{2}$;
y-intercept: $(0, 1)$ y-intercept: $(0, 4)$

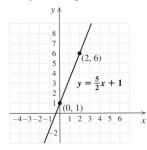

43. Slope: 2;
    y-intercept: $(0, -5)$

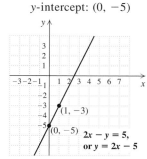

$2x - y = 5$,
or $y = 2x - 5$

45. Slope: $-\frac{2}{7}$;
    y-intercept: $(0, 1)$

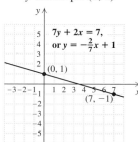

$7y + 2x = 7$,
or $y = -\frac{2}{7}x + 1$

47. Slope: -0.25;
    y-intercept: $(0, 2)$

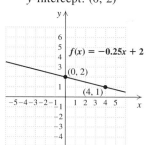

$f(x) = -0.25x + 2$

49. Slope: $\frac{4}{5}$;
    y-intercept: $(0, -2)$

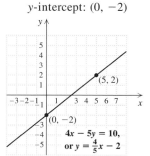

$4x - 5y = 10$,
or $y = \frac{4}{5}x - 2$

51. Slope: $\frac{5}{4}$;
    y-intercept: $(0, -2)$

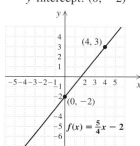

$f(x) = \frac{5}{4}x - 2$

53. Slope: $-\frac{3}{4}$;
    y-intercept: $(0, 3)$

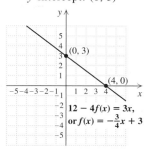

$12 - 4f(x) = 3x$,
or $f(x) = -\frac{3}{4}x + 3$

55. (a)   **(b)** 4 chirps per minute

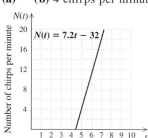

$N(t) = 7.2t - 32$

57.

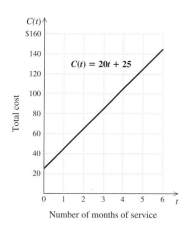

$C(t) = 20t + 25$

Number of months of service

$145

59.

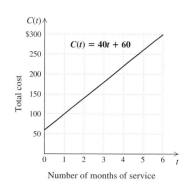

$C(t) = 40t + 60$

Number of months of service

$280; $\{t \mid t \geq 0\}$

61.

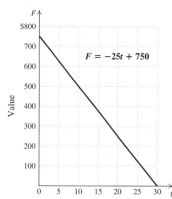

$F = -25t + 750$

Number of months after purchase

$\{t \mid 0 \leq t \leq 30\}$

63. 0.75 signifies that the cost per mile is $0.75; 2 signifies that the minimum cost of a taxi ride is $2.
65. 2.6 signifies that sales increase $2.6 billion per year after 1975; 17.8 signifies that sales in 1975 were

$17.8 billion.　**67.** $\frac{3}{20}$ signifies that the life expectancy of American women increases $\frac{3}{20}$ yr per year after 1950; 72 signifies that the life expectancy of American women was 72 yr in 1950.　**69. (a)** II; **(b)** IV; **(c)** I; **(d)** III　**71.** $45x + 54$　**73.** $-26m^7n^4$　**75.** ◈　**77.** ◈

79. Slope: $-\dfrac{r}{p}$; y-intercept: $\left(0, \dfrac{s}{p}\right)$　**81.** Since (x_1, y_1) and (x_2, y_2) are two points on the graph of $y = mx + b$, then $y_1 = mx_1 + b$ and $y_2 = mx_2 + b$. Using the definition of slope, we have

$$\begin{aligned}\text{Slope} &= \frac{y_2 - y_1}{x_2 - x_1}\\ &= \frac{(mx_2 + b) - (mx_1 + b)}{x_2 - x_1}\\ &= \frac{m(x_2 - x_1)}{x_2 - x_1}\\ &= m.\end{aligned}$$

83. False　**85.** False
87.

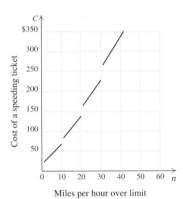

89. (a) III; **(b)** IV; **(c)** I; **(d)** II　**91.** Linear
93. Not linear

EXERCISE SET 2.5

1. Undefined　**3.** 3　**5.** 0　**7.** $\frac{5}{7}$　**9.** Undefined
11. $\frac{1}{2}$　**13.** Undefined　**15.** -2　**17.** 0　**19.** 0
21. 3
23.

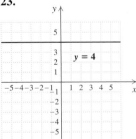

25.

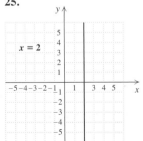

27.

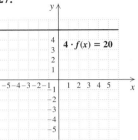

29.

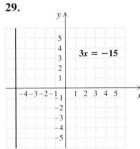

31.

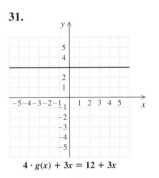

33.

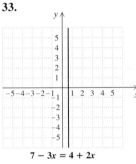

35.

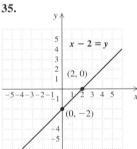

37.

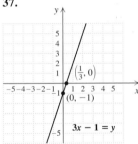

39.

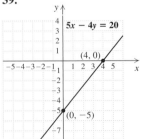

41.

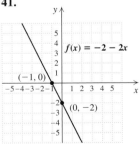

43.

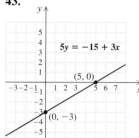

$5y = -15 + 3x$

(5, 0)

(0, −3)

45.

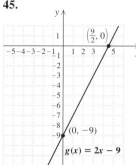

$\left(\frac{9}{2}, 0\right)$

(0, −9)

$g(x) = 2x - 9$

5.

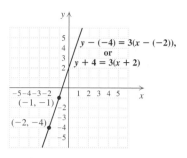

$y - (-4) = 3(x - (-2))$,
or
$y + 4 = 3(x + 2)$

(−1, −1)

(−2, −4)

7.

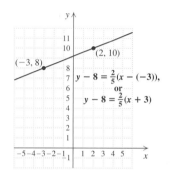

(−3, 8)   (2, 10)

$y - 8 = \frac{2}{5}(x - (-3))$,
or
$y - 8 = \frac{2}{5}(x + 3)$

47.

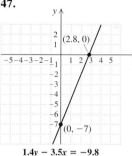

(2.8, 0)

(0, −7)

$1.4y - 3.5x = -9.8$

49.

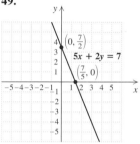

$\left(0, \frac{7}{2}\right)$

$5x + 2y = 7$

$\left(\frac{7}{5}, 0\right)$

9. $\frac{2}{7}$; (1, 3)   **11.** −5; (7, −2)   **13.** $-\frac{5}{3}$; (−2, 1)
15. $f(x) = 5x - 13$ **17.** $f(x) = -\frac{2}{3}x - \frac{13}{3}$
19. $f(x) = -0.6x - 1.8$ **21.** $f(x) = -6x + 2.4$
23. $f(x) = \frac{1}{2}x + \frac{7}{2}$ **25.** $f(x) = 1.5x - 6.75$
27. $f(x) = \frac{1}{2}x - 2$ **29.** $f(x) = \frac{3}{2}x$
31. (a) $R(t) = -0.075t + 46.8$; (b) 41.625 sec; 41.4 sec;
(c) 2021   **33.** (a) $A(t) = 5.8625t + 132.7$;
(b) $226.5 million   **35.** (a) $N(t) = 4.46t + 33.9$;
(b) 82.96 million tons   **37.** (a) $A(t) = -0.375t + 76.4$;
(b) 71.9 million acres   **39.** (a) $E(t) = 0.02t + 78.8$;
(b) 79.08   **41.** (a) $f(x) = 0.3326x + 81.2167$; (b) 85.87
43. (a) $f(x) = -6.055x + 99.5336$; (b) $14.76
45. (a) $f(x) = 77.9845x + 7728.5271$; (b) $8508   **47.** $\frac{1}{12}$
49. $-\frac{5}{36}$ **51.** ◈   **53.** ◈   **55.** $1940   **57.** $11,000
59. 21.1°C   **61.** $y = 3x + 7$ **63.** $y = -3x + 12$
65. (a) $g(x) = x - 8$; (b) −10; (c) 83   **67.** 7

51. (c)   **53.** (d)   **55.** Yes   **57.** No   **59.** Yes
61. Yes   **63.** No   **65.** $f(x) = 3x + 9$
67. $f(x) = -2x - 1$ **69.** $f(x) = -\frac{2}{5}x - \frac{1}{3}$
71. $f(x) = 12$ **73.** $f(x) = -x + 4$ **75.** $f(x) = \frac{3}{2}x - 9$
77. $f(x) = -\frac{1}{5}x + \frac{1}{5}$ **79.** Linear; slope is $\frac{5}{3}$
81. Linear; slope is 0   **83.** Nonlinear   **85.** Nonlinear
87. Nonlinear   **89.** Nonlinear   **91.** $6x - 3y + 21$
93. $3(3x - 5y)$ **95.** $\frac{1}{3}(x + 2y - 4)$ **97.** ◈   **99.** ◈
101. $4x - 5y = 20$ **103.** Linear   **105.** Linear
107. The slope of equation B is $\frac{1}{2}$ the slope of equation A.
109. $a = 7, b = -3$ **111.** (a) Yes; (b) no

EXERCISE SET 2.6

1.

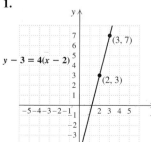

$y - 3 = 4(x - 2)$

(3, 7)

(2, 3)

3.

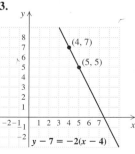

(4, 7)

(5, 5)

$y - 7 = -2(x - 4)$

EXERCISE SET 2.7

1. 1   **3.** −41   **5.** 12   **7.** $\frac{13}{18}$ **9.** 5   **11.** 2
13. $x^2 - x + 1$ **15.** 21   **17.** 5   **19.** 42
21. $-\frac{3}{4}$ **23.** $\frac{1}{6}$ **25.** 1550; in January 1993, there were
1550 new cases of AIDS diagnosed among people born in
1960 or later.   **27.** $T(6) \approx 4800$, $G(6) \approx 3550$,
$F(6) \approx 1250$; $(G + F)(6) = G(6) + F(6) \approx 3550 +$
$1250 \approx 4800 \approx T(6)$.   **29.** 44; a total of about
44 million passengers used Newark and LaGuardia
airports in 1992.   **31.** 8; about 8 million more

passengers used Kennedy airport than LaGuardia airport in 1992.   **33.** 72; a total of about 72 million passengers used Newark, LaGuardia, and Kennedy airports in 1993.
35. (a) $\{x \,|\, x$ is a real number *and* $x \neq 3\}$; **(b)** $\{x \,|\, x$ is a real number *and* $x \neq 5\}$; **(c)** $\mathbb{R}$; **(d)** $\mathbb{R}$; **(e)** $\{x \,|\, x$ is a real number *and* $x \neq \frac{5}{2}\}$; **(f)** $\mathbb{R}$ **37.** $\mathbb{R}$
39. $\{x \,|\, x$ is a real number *and* $x \neq 2\}$
41. $\{x \,|\, x$ is a real number *and* $x \neq 1\}$
43. $\{x \,|\, x$ is a real number *and* $x \neq 2$ *and* $x \neq -4\}$
45. $\{x \,|\, x$ is a real number *and* $x \neq 3\}$
47. $\{x \,|\, x$ is a real number *and* $x \neq 4\}$
49. $\{x \,|\, x$ is a real number *and* $x \neq 4$ *and* $x \neq 5\}$
51. $\{x \,|\, x$ is a real number *and* $x \neq -1$ *and* $x \neq -\frac{5}{2}\}$
53. 4; 3   **55.** $\{x \,|\, 0 \leq x \leq 9\}$; $\{x \,|\, 3 \leq x \leq 10\}$; $\{x \,|\, 3 \leq x \leq 9\}$; $\{x \,|\, 3 \leq x \leq 9\}$
57.

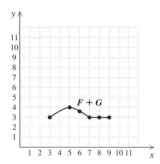

59. 270   **61.** 7.03×10^{-6} **63.**
65. $\{x \,|\, x$ is a real number *and* $-1 < x < 5$ *and* $x \neq \frac{3}{2}\}$
67. 5; 15; $\frac{2}{3}$ **69.** $\{x \,|\, x$ is a real number *and* $x \neq -\frac{5}{2}$ *and* $x \neq -3$ *and* $x \neq -1$ *and* $x \neq 1\}$
71. Answers may vary.

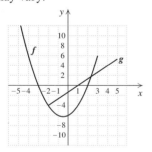

73.   **75.** The domain of $y_3 = \dfrac{2.5x + 1.5}{x - 3}$ is $\{x \,|\, x$ is a real number *and* $x \neq 3\}$. The CONNECTED mode graph contains the line $x = 3$, whereas the DOT mode graph contains no points having 3 as the first coordinate. Thus the DOT mode graph represents y_3 more accurately.
77. (a) IV; **(b)** I; **(c)** II; **(d)** III

REVIEW EXERCISES, CHAPTER 2

1. [2.1] **(a)** 3; **(b)** $\{x \,|\, -2 \leq x \leq 4\}$; **(c)** -1;
(d) $\{y \,|\, 1 \leq y \leq 5\}$ **2.** [2.1] Domain: $\mathbb{R}$;
range: $\{y \,|\, y \geq -1\}$ **3.** [2.1] \$18,185   **4.** [2.2] 12
5. [2.2] 9   **6.** [2.2] 6.6   **7.** [2.2] $\frac{27}{2}$ **8.** [2.2] $-\frac{4}{11}$
9. [2.2] $\mathbb{R}$; identity   **10.** [2.2] $\varnothing$; contradiction
11. [2.2] 0.46170   **12.** [2.3] 93   **13.** [2.3] 55 in.
14. [2.3] $20\frac{1}{6}°$, $50\frac{1}{3}°$, $109\frac{1}{2}°$ **15.** [2.3] 1.422×10^4 mm^3, or 1.422×10^{-5} m^3 **16.** [2.3] $m = PS$
17. [2.3] $x = \dfrac{c}{m - r}$ **18.** [2.3] $y = \dfrac{5x - 7}{6}$
19. [2.4] Slope: -4; y-intercept: $(0, -9)$
20. [2.4] Slope: $\frac{1}{3}$; y-intercept: $\left(0, -\frac{7}{6}\right)$ **21.** [2.4] $\frac{4}{7}$
22. [2.4] Undefined   **23.** [2.4] Personal income is increasing at a rate of \$2500 per year since high school.
24. [2.4] $f(x) = \frac{2}{7}x - 6$
25. [2.5]

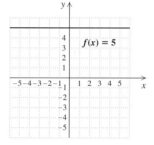

26. [2.5]

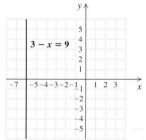

27. [2.5]

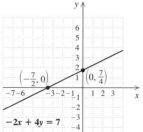

28. [2.5]

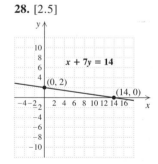

29. [2.5] The window should show both intercepts: (98, 0) and (0, 14). One possibility is $[-10, 120, -5, 15]$, with Xscl = 10.   **30.** [2.5] Perpendicular   **31.** [2.5] Parallel
32. [2.5] Linear   **33.** [2.5] Linear   **34.** [2.5] Nonlinear
35. [2.5] Nonlinear   **36.** [2.6] $y - 4 = -2(x - (-3))$, or $y - 4 = -2(x + 3)$ **37.** [2.6] $f(x) = \frac{4}{3}x + \frac{7}{3}$
38. [2.6] **(a)** $W(t) = 0.0875t + 0.75$; **(b)** about \$5.13
39. [2.6] **(a)** $f(x) = 52.35x + 372.7$; **(b)** \$1,210,300,000
40. [2.7] **(a)** $\{x \,|\, x$ is a real number *and* $x \neq 7\}$; **(b)** $\mathbb{R}$
41. [2.1] -6 **42.** [2.1] 26   **43.** [2.7] 102
44. [2.7] -17 **45.** [2.7] $-\frac{9}{2}$ **46.** [2.7] $\mathbb{R}$

47. [2.7] $\{x \mid x \text{ is a real number } and \text{ } x \neq 2\}$ **48.** [1.2] $\frac{2}{15}$
49. [1.4] $25a^6b^2$
50. [1.5]

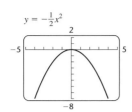

$y = -\frac{1}{2}x^2$

51. [1.3] $2(xy - 4x + 2)$ **52.** [2.1] ◆ For any function, each member of the domain corresponds to *exactly one* member of the range. Thus, for any function, each member of the domain corresponds to *at least one* member of the range. Therefore, a function is a relation. In a relation, every member of the domain corresponds to *at least one*, but not necessarily *exactly one*, member of the range. Therefore, a relation may or may not be a function.   **53.** [2.5] ◆ The slope of a line is the rise between two points on the line divided by the run between those points. For a vertical line, there is no run between any two points, and division by 0 is undefined; therefore, the slope is undefined. For a horizontal line, there is no rise between any two points, so the slope is 0/run, or 0.   **54.** [2.5] $(0, -9)$ **55.** [2.5] $-\frac{9}{2}$
56. [2.6] $f(x) = 3.09x + 3.75$ **57.** [2.4] **(a)** III; **(b)** IV; **(c)** I; **(d)** II

TEST, CHAPTER 2

1. [2.1] **(a)** 1; **(b)** $\{x \mid -3 \leq x \leq 4\}$; **(c)** 3; **(d)** $\{y \mid -1 \leq y \leq 2\}$ **2.** [2.1] Domain: $\{x \mid x \geq 0\}$; range: $\{y \mid y \geq -1\}$ **3.** [2.1] \$33.4 billion   **4.** [2.2] 43
5. [2.2] -2 **6.** [2.2] $\mathbb{R}$; identity   **7.** [2.3] 18, 20, 22
8. [2.3] 20.175% increase   **9.** [2.3] About 4.2×10^8 mi
10. [2.3] $l = \dfrac{S - 2wh}{2h + 2w}$ **11.** [2.3] $y = \dfrac{x - 7 - 3x^2}{2}$
12. [2.4] Slope: $-\frac{3}{5}$; y-intercept: $(0, 12)$
13. [2.4] Slope: $-\frac{2}{5}$; y-intercept: $\left(0, -\frac{7}{5}\right)$ **14.** [2.4] $\frac{5}{8}$
15. [2.4] 0   **16.** [2.4] $f(x) = -5x - 1$
17. [2.4]   **18.** [2.5]

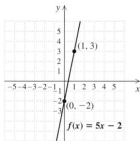

$f(x) = 5x - 2$

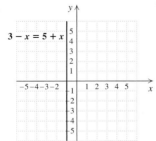

$3 - x = 5 + x$

19. [2.5]

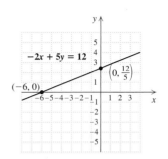

$-2x + 5y = 12$

$(-6, 0)$ $\left(0, \frac{12}{5}\right)$

20. [2.5] Yes   **21.** [2.5] Parallel
22. [2.5] Perpendicular   **23.** [2.5] (a), (c)
24. [2.6] $y - (-4) = 4(x - (-2))$, or $y + 4 = 4(x + 2)$
25. [2.6] $f(x) = -x + 2$
26. [2.6] **(a)** $C(m) = 0.3m + 25$; **(b)** \$175
27. [2.6] **(a)** $f(x) = 0.86x + 12.76$; **(b)** 26.52 million
28. [2.7] $\{x \mid x \text{ is a real number } and \text{ } x \neq \frac{5}{2}\}$
29. (a) [2.1] 5; **(b)** [2.7] -130; **(c)** [2.7] $\{x \mid x \text{ is a real number } and \text{ } x \neq -\frac{4}{3}\}$ **30.** [1.2] $-\frac{14}{15}$ **31.** [1.4] a^6b^{12}
32. [1.5]

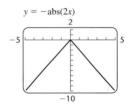

$y = -\text{abs}(2x)$

33. [1.3] $3x(1 - 2y)$ **34.** [2.1], [2.4] **(a)** 30 mi;
(b) 15 mph   **35.** [2.5], [2.6] $y = \frac{2}{5}x + \frac{16}{5}$
36. [2.5], [2.6] $y = -\frac{5}{2}x - \frac{11}{2}$ **37.** [2.5] $s = -\frac{3}{2}r + \frac{27}{2}$
38. [2.7] $h(x) = 7x - 2$

Chapter 3

EXERCISE SET 3.1

1. Let $x =$ the larger number and $y =$ the smaller number; $x - y = 11$, $3x + 2y = 123$ **3.** Let $x =$ the number of \$8.50 brushes sold and $y =$ the number of \$9.75 brushes sold; $x + y = 45$, $8.50x + 9.75y = 398.75$
5. Let x and y represent the angles; $x + y = 180$, $y = 2x - 3$ **7.** Let $x =$ the number of field goals and $y =$ the number of free throws; $x + y = 18$, $2x + y = 30$ **9.** Let $h =$ the number of vials of Humulin sold and $n =$ the number of vials of Novolin; $h + n = 65$, $15.75h + 12.95n = 959.35$ **11.** Let $l =$ the length and $w =$ the width; $2l + 2w = 228$, $w = l - 42$ **13.** Let $l =$ the number of units of lumber produced and $p =$ the number of units of plywood

produced; $p = 2l$, $25l + 40p = 10,920$ **15.** Let $w =$ the number of wins and $t =$ the number of ties; $2w + t = 60$, $w = t + 9$ **17.** Let $x =$ the number of ounces of lemon juice and $y =$ the number of ounces of linseed oil; $y = 2x$, $x + y = 32$ **19.** Yes   **21.** No   **23.** Yes   **25.** Yes   **27.** $(4, 1)$ **29.** $(2, -1)$ **31.** $(4, 3)$ **33.** $(-3, -2)$ **35.** $\left(\frac{5}{2}, -2\right)$ **37.** $(-3, 2)$ **39.** $(1.53, 2.58)$ **41.** $(7, 2)$ **43.** $(4, 0)$ **45.** $(2.23, 1.14)$ **47.** No solution   **49.** $(-0.73, 0)$ **51.** $\{(x, y)|\ y = 3 - x\}$ **53.** No solution   **55.** All except 47 and 53   **57.** 51   **59.** -3 **61.** $\frac{9}{20}$ **63.** $3(x - 7)$ **65.** ◈   **67.** ◈   **69.** From 1991 to 1995   **71.** Answers may vary. **(a)** $x + y = 6$, $x - y = 4$; **(b)** $x + y = 1$, $2x + 2y = 3$; **(c)** $x + y = 1$, $2x + 2y = 2$ **73.** $A = -\frac{17}{4}$, $B = -\frac{12}{5}$ **75.** Let x and $y =$ the number of years that Lou and Juanita have taught at the university, respectively; $x + y = 46$, $x - 2 = 2.5(y - 2)$ **77.** Let s and $v =$ the number of ounces of baking soda and vinegar needed, respectively; $s = 4v$, $s + v = 16$ **79.** $(0, 0)$, $(1, 1)$ **81.** (c)   **83.** (b)

EXERCISE SET 3.2

1. $(-4, 3)$ **3.** $(-3, -15)$ **5.** $(2, -7)$ **7.** $(4, -1)$ **9.** $(3, -2)$ **11.** $\left(\frac{25}{23}, -\frac{11}{23}\right)$ **13.** No solution   **15.** $(1, 2)$ **17.** $(3, 0)$ **19.** $\left(\frac{1}{2}, -5\right)$ **21.** $\left(\frac{10}{21}, \frac{11}{14}\right)$ **23.** $(1, 3)$ **25.** $(2, 3)$ **27.** $(-2, 3)$ **29.** $(12, 15)$ **31.** $\{(x, y)|\ 3x + 5y = 7\}$ **33.** $(-2, -9)$ **35.** $(30, 6)$ **37.** $\{(x, y)|\ x = 2 + 3y\}$ **39.** No solution   **41.** $(140, 60)$ **43.** $\{(x, y)|\ 2x + 3y = 5\}$ **45.** (d)   **47.** (b)   **49. (a)** $a(t) = 12.45t + 113.9$; **(b)** 238   **51.** $\mathbb{R}$ **53.** ◈   **55.** ◈   **57.** $m = -\frac{1}{2}$, $b = \frac{5}{2}$ **59.** $a = 5$, $b = 2$ **61.** $\left(-\frac{32}{17}, \frac{38}{17}\right)$ **63.** $\left(-\frac{1}{5}, \frac{1}{10}\right)$

EXERCISE SET 3.3

1. 29; 18   **3.** 32 at \$8.50; 13 at \$9.75   **5.** $119°$, $61°$ **7.** Field goals: 12; free throws: 6   **9.** Humulin: 42; Novolin: 23   **11.** Length: 78 ft; width: 36 ft   **13.** Lumber: 104; plywood: 208   **15.** 23 wins; 14 ties   **17.** Lemon juice: $10\frac{2}{3}$ oz; linseed oil: $21\frac{1}{3}$ oz   **19.** Scientific calculators: 18; graphing calculators: 27   **21.** Kenyan: 8 lb; Sumatran: 12 lb   **23.** Deep Thought: 12 lb; Oat Dream: 8 lb   **25.** 4 L of 25%; 6 L of 50%   **27.** \$7500 at 6%; \$4500 at 9%   **29.** Whole milk: $169\frac{3}{13}$ lb; cream: $30\frac{10}{13}$ lb   **31.** Length: 265 ft; width: 165 ft   **33.** 7 \$5 bills; 15 \$1 bills   **35.** 375 km   **37.** 14 km/h   **39.** About 1489 mi   **41. (a)** $n(x) = -3.6714x + 84.8571$, $u(x) = 3.6714x + 15.1429$; **(b)** 1989   **43.** 2012   **45.** $x + 29$ **47.** $y = -\frac{3}{4}x - \frac{7}{2}$ **49.** -1.1014 **51.** ◈   **53.** ◈   **55.** Burl: 40; son: 20

57. Width: $\frac{102}{5}$ in.; length: $\frac{288}{5}$ in.   **59.** $4\frac{4}{7}$ L   **61.** 180   **63.** First train: 36 km/h; second train: 54 km/h   **65.** Brown: 0.8 gal; neutral: 0.2 gal   **67.** $P(x) = \dfrac{0.1 + x}{1.5}$ (This expresses the percent as a decimal quantity.)

EXERCISE SET 3.4

1. No   **3.** $(4, 0, 2)$ **5.** $(2, -2, 2)$ **7.** $(3, -2, 1)$ **9.** $(7, -3, -4)$ **11.** $(2, 1, 3)$ **13.** $(2, -5, 6)$ **15.** The equations are dependent.   **17.** $(3, -5, 8)$ **19.** $\left(\frac{3}{5}, \frac{2}{3}, -3\right)$ **21.** $\left(4, \frac{1}{2}, -\frac{1}{2}\right)$ **23.** $(17, 9, 79)$ **25.** $\left(\frac{1}{4}, -\frac{1}{2}, -\frac{1}{4}\right)$ **27.** $\left(\frac{98}{5}, \frac{304}{5}, \frac{498}{5}\right)$ **29.** No solution   **31.** The equations are dependent.   **33.** $2a + 9$ **35.** $b = a - \dfrac{2K}{t}$, or $\dfrac{at - 2K}{t}$ **37.** ◈   **39.** ◈   **41.** $(1, -1, 2)$ **43.** $(-3, -1, 0, 4)$ **45.** $\left(-\frac{1}{2}, -1, -\frac{1}{3}\right)$ **47.** 14   **49.** $z = 8 - 2x - 4y$

EXERCISE SET 3.5

1. 16, 19, 22   **3.** 8, 21, -3 **5.** $32°$, $96°$, $52°$ **7.** Automatic transmission: \$865; power door locks: \$520; air conditioning: \$375   **9.** A: 1500; B: 1900; C: 2300   **11.** 10-oz cups: 8; 14-oz cups: 20; 20-oz cups: 6   **13.** First fund: \$45,000; second fund: \$10,000; third fund: \$25,000   **15.** Roast beef: 2; baked potato: 1; broccoli: 2   **17.** Asian-American: 385; African-American: 200; Caucasian: 154   **19.** 2-point field goals: 32; 3-point field goals: 5; foul shots: 13   **21.** 2   **23.** $\dfrac{a^3}{b}$ **25.** $\{x|\ x$ is a real number *and* $x \neq -7\}$ **27.** ◈   **29.** 464   **31.** adults: 5; student: 1; children: 94   **33.** $180°$

EXERCISE SET 3.6

1. $(2, -1)$ **3.** $(-4, 3)$ **5.** $\left(\frac{3}{2}, \frac{5}{2}\right)$ **7.** $\left(\frac{3}{2}, -4, 3\right)$ **9.** $(2, -2, 1)$ **11.** $\left(4, \frac{1}{2}, -\frac{1}{2}\right)$ **13.** $\left(\frac{93}{34}, -\frac{7}{34}, \frac{67}{34}\right)$ **15.** $(-0.5332, -1.2933, -0.6189)$ **17.** $(7, 4, 5, 6)$ **19.** 4 dimes; 30 nickels   **21.** 5 lb of \$4.05; 10 lb of \$2.70   **23.** \$400 at 7%; \$500 at 8%; \$1600 at 9%   **25.** 6.9   **27.** -32 **29.** ◈   **31.** 1324

EXERCISE SET 3.7

1. (a) $P(x) = 45x - 270,000$; **(b)** 6000 units   **3. (a)** $P(x) = 50x - 120,000$; **(b)** 2400 units   **5. (a)** $P(x) = 80x - 10,000$; **(b)** 125 units   **7. (a)** $P(x) = 18x - 16,000$; **(b)** 889 units   **9. (a)** $P(x) = 75x - 195,000$; **(b)** 2600 units

11. ($70, 300) **13.** ($22, 474) **15.** ($50, 6250)
17. ($10, 1070) **19. (a)** $C(x) = 22{,}500 + 40x$;
(b) $R(x) = 85x$; **(c)** $P(x) = 45x - 22{,}500$;
(d) $112,500 profit, $4500 loss;
(e) (500 lamps, $42,500) **21. (a)** $C(x) = 16{,}404 + 6x$;
(b) $R(x) = 18x$; **(c)** $P(x) = 12x - 16{,}404$; **(d)** $19,596
profit, $4404 loss; **(e)** (1367 dozen caps, $24,606)
23. (a) $8.74; **(b)** 24,509 units
25. (a) $S(p) = 15.97p - 1.05$;
(b) $D(p) = -11.26p + 41.16$; **(c)** ($1.55, 23.7 million jars)
27.

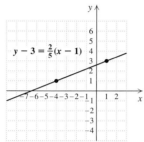

$$y - 3 = \tfrac{2}{5}(x - 1)$$

29. $\frac{5}{2}$ **31.** ◆ **33.** 308 pairs

REVIEW EXERCISES, CHAPTER 3

1. [3.1] $(-2, 1)$ **2.** [3.1] $(8.4, 4.8)$ **3.** [3.2] $\left(-\frac{4}{5}, \frac{2}{5}\right)$
4. [3.2] No solution **5.** [3.2] $\left(-\frac{11}{15}, -\frac{43}{30}\right)$
6. [3.2] $\left(\frac{6}{7}, \frac{11}{14}\right)$ **7.** [3.2] $\left(\frac{76}{17}, -\frac{2}{119}\right)$ **8.** [3.2] $(2, 2)$
9. [3.2] $\{(x, y) \mid 3x + 4y = 6\}$ **10.** [3.3] CD: $14;
cassette: $9 **11.** [3.3] 4 hr **12.** [3.3] 5 L of each
13. [3.4] $(4, -8, 10)$ **14.** [3.4] The equations
are dependent. **15.** [3.4] $(2, 0, 4)$
16. [3.2] No solution **17.** [3.4] $\left(\frac{8}{9}, -\frac{2}{3}, \frac{10}{9}\right)$
18. [3.5] A: 90°; B: 67.5°; C: 22.5° **19.** [3.5] 641
20. [3.5] 7 $20 bills; 3 $5 bills; 4 $1 bills
21. [3.6] $\left(55, -\frac{89}{2}\right)$ **22.** [3.6] $(-1, 1, 3)$
23. [3.6] $\left(-\frac{32}{15}, -\frac{439}{45}, -\frac{421}{45}\right)$ **24.** [3.7] ($3, 81)
25. [3.7] **(a)** $C(x) = 175x + 35{,}000$; **(b)** $R(x) = 300x$;
(c) $P(x) = 125x - 35{,}000$; **(d)** $115,000 profit, $10,000
loss; **(e)** 280 beds **26.** [2.2] $-\frac{4}{3}$ **27.** [2.3] $t = \dfrac{Q}{a - 4}$
28. [2.4] $f(x) = -\frac{1}{3}x - \frac{9}{10}$ **29.** [2.3] 69, 76, 83
30. [3.5] ◆ To solve a problem involving four variables,
go through the *Familiarize* and *Translate* steps as usual.
The resulting system of equations can be solved using the
elimination method just as for three variables but likely
with more steps. **31.** [3.4] ◆ A system of equations can
be both dependent and inconsistent if it is equivalent to a
system with fewer equations that has no solution. An
example is a system of three equations in three unknowns

in which two of the equations represent the same plane,
and the third represents a parallel plane.
32. [3.1] $(0, 2)$, $(1, 3)$ **33.** [3.5] $a = -\frac{2}{3}$, $b = -\frac{4}{3}$,
$c = 3$; $f(x) = -\frac{2}{3}x^2 - \frac{4}{3}x + 3$

TEST, CHAPTER 3

1. [3.1] $(-7, -11)$ **2.** [3.1] $(2, 1)$
3. [3.2] $\left(3, -\frac{11}{3}\right)$ **4.** [3.2] $\left(\frac{15}{7}, -\frac{18}{7}\right)$
5. [3.2] $\left(-\frac{3}{2}, -\frac{3}{2}\right)$ **6.** [3.2] No solution
7. [3.3] Length: 30; width: 18 **8.** [3.5] Mortgage:
$74,000; car loan: $600; credit card bill: $700
9. [3.4] The equations are dependent.
10. [3.4] $\left(2, -\frac{1}{2}, -1\right)$ **11.** [3.4] No solution
12. [3.4] $(0, 1, 0)$ **13.** [3.6] $\left(\frac{34}{107}, -\frac{104}{107}\right)$
14. [3.6] $(3, 1, -2)$ **15.** [3.6] $\left(\frac{202}{9}, \frac{281}{18}, \frac{490}{9}\right)$
16. [3.5] 3.5 hr **17.** [3.7] ($3, 55)
18. [3.7] **(a)** $C(x) = 40{,}000 + 30x$; **(b)** $R(x) = 80x$;
(c) $P(x) = 50x - 40{,}000$; **(d)** $35,000 profit;
(e) (800 rackets, $64,000) **19.** [2.3] $35
20. [2.3] $a = \dfrac{P + 3b}{4}$ **21.** [2.2] $\frac{9}{5}$
22. [2.6] $y + 1 = -\frac{5}{8}(x - 5)$, or $y = -\frac{5}{8}x + \frac{17}{8}$
23. [2.4], [3.3] $m = 7$, $b = 10$ **24.** [3.5] Adult: 1346;
senior citizen: 335; child: 1651

CUMULATIVE REVIEW, CHAPTERS 1–3

1. [2.2] 14.87 **2.** [2.2] -22 **3.** [2.2] -42.9
4. [2.2] 20 **5.** [2.2] $-\frac{21}{4}$ **6.** [2.2] 6 **7.** [2.2] -5
8. [2.2] $\frac{10}{9}$ **9.** [2.2] $-\frac{32}{5}$ **10.** [2.2] $\frac{18}{17}$ **11.** [1.4] x^{11}
12. [1.4] $-\dfrac{40x}{y^5}$ **13.** [1.4] $-288x^4y^{18}$ **14.** [1.4] y^{10}
15. [1.4] $-\dfrac{2a^{11}}{5b^{33}}$ **16.** [1.4] $\dfrac{81x^{36}}{256y^8}$
17. [1.4] 1.11735×10^6 **18.** [1.4] 4.0×10^6
19. [2.3] $b = \dfrac{2A}{h} - t$, or $\dfrac{2A - ht}{h}$ **20.** [1.5] Yes

21. [2.4]

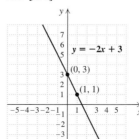

$y = -2x + 3$
$(0, 3)$
$(1, 1)$

22. [1.5]

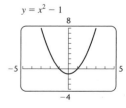

$y = x^2 - 1$

23. [2.5]

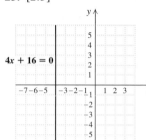

$4x + 16 = 0$

24. [2.5]

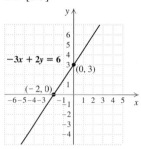

$-3x + 2y = 6$

$(0, 3)$

$(-2, 0)$

25. [2.4] Slope: $\frac{9}{4}$; y-intercept: $(0, -3)$ **26.** [2.4] $\frac{4}{3}$
27. [2.6] $y = -3x - 5$ **28.** [2.6] $y = -\frac{1}{10}x + \frac{12}{5}$
29. [2.5] Parallel   **30.** [2.5] $f(x) = -\frac{1}{3}x - 11$
31. [2.1] $\{-5, -3, -1, 1, 3\}$; $\{-3, -2, 1, 4, 5\}$; -2; 3
32. [2.1] $\left\{x \mid x \text{ is a real number } and \ x \neq \frac{1}{2}\right\}$
33. [2.1] -31 **34.** [2.1] 3 **35.** [2.7] 7
36. [2.7] $8a^2 + 4a - 4$ **37.** [3.2] $(1, 1)$
38. [3.2] $(-2, 3)$ **39.** [3.2] $\left(-3, \frac{2}{5}\right)$
40. [3.4] $(-3, 2, -4)$ **41.** [3.4] $(0, -1, 2)$
42. [3.2] $\left(-\frac{33}{56}, -\frac{169}{224}\right)$ **43.** [3.4] $\left(\frac{11}{15}, \frac{2}{15}, \frac{16}{15}\right)$
44. [3.3] $8, 18$ **45.** [3.3] Soakem: $48\frac{8}{9}$ oz; Rinsem:
$71\frac{1}{9}$ oz   **46.** [2.3] $3, 5, 7$ **47.** [2.3] 90
48. [3.3] $l = 10$ cm; $w = 6$ cm   **49.** [3.3] 19 nickels;
15 dimes   **50.** [3.5] $120   **51.** [3.5] 23 wins, 33 losses,
8 ties   **52.** [3.5] Cookie: 90; banana: 80; yogurt: 165
53. (a) [1.6]

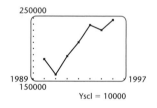

250000

1989

150000

1997

Yscl = 10000

(b) [2.6] $f(x) = 13917x + 166,205$;
(c) [2.6] $305,375 million
54. [1.4] $-12x^{2a}y^{b+y+3}$
55. [2.6] $151,000   **56.** [2.4], [3.3] $m = -\frac{5}{9}, b = -\frac{2}{9}$

Chapter 4

EXERCISE SET 4.1

1. No, no, no, yes   **3.** No, no, yes, no
5. $\{y \mid y < 5\}$, $(-\infty, 5)$
7. $\{x \mid x \geq -4\}$, $[-4, \infty)$

9. $\{t \mid t > -2\}$, $(-2, \infty)$ **11.** $\{x \mid x \leq -5\}$, $(-\infty, -5]$

13. $\{x \mid x > -5\}$, or $(-5, \infty)$

15. $\{a \mid a \leq -20\}$, or $(-\infty, -20]$

17. $\{y \mid y > -9\}$, or $(-9, \infty)$

19. $\{y \mid y \leq 14\}$, or $(-\infty, 14]$

21. $\{t \mid t < -9\}$, or $(-\infty, -9)$

23. $\{x \mid x < 50\}$, or $(-\infty, 50)$

25. $\{y \mid y \geq -0.4\}$, or $[-0.4, \infty)$

27. $\left\{y \mid y \geq \frac{9}{10}\right\}$, or $\left[\frac{9}{10}, \infty\right)$

29. $\{y \mid y > 3\}$, or $(3, \infty)$

31. $\{x \mid x \leq 4\}$, or $(-\infty, 4]$

33. $\left\{x \mid x < -\frac{2}{5}\right\}$, or $\left(-\infty, -\frac{2}{5}\right)$

35. $\{x \mid x \geq 11.25\}$, or $[11.25, \infty)$

37. $\{x \mid x < 4\}$, or $(-\infty, 4)$
39. $\{x \mid x \geq 3.2\}$, or $[3.2, \infty)$ **41.** $\{m \mid m \leq 3.3\}$, or
$(-\infty, 3.3]$ **43.** $\left\{y \mid y < \frac{13}{185}\right\}$, or $\left(-\infty, \frac{13}{185}\right)$
45. $\left\{x \mid x \leq -\frac{23}{2}\right\}$, or $\left(-\infty, -\frac{23}{2}\right]$ **47.** $\{y \mid y < 5\}$, or
$(-\infty, 5)$ **49.** $\left\{x \mid x \leq \frac{4}{7}\right\}$, or $\left(-\infty, \frac{4}{7}\right]$
51. Years after 2001   **53.** Years after 2001
55. $11,500 or more   **57.** Calls longer than 1.5 min
59. Gross sales greater than $7000

61. Values of n greater than $85\frac{5}{7}$

63. (a) $\left\{x\middle|\ x < 8181\frac{9}{11}\right\}$; (b) $\left\{x\middle|\ x > 8181\frac{9}{11}\right\}$

65.

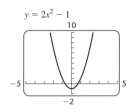

$y = 2x^2 - 1$

67. -2 **69.** 16   **71.** ◈   **73.** ◈

75. $\left\{y\middle|\ y \geq -\dfrac{10}{b + 4}\right\}$ **77.** $\left\{x\middle|\ x < \dfrac{4c + 3d}{6c - 2d}\right\}$

79. $\left\{x\middle|\ x < \dfrac{-3a + 2d}{c - (4a - 5d)}\right\}$ **81.** False; $-3 < -2$, but $9 > 4$.   **83.** ◈   **85.** ∅

INTERACTIVE DISCOVERY, P. 240

Domain of f: $(-\infty, 3]$; domain of g: $[-1, \infty)$; domain of $f + g = $ domain of $f - g = $ domain of $f \cdot g = [-1, 3]$

EXERCISE SET 4.2

1. $\{9, 11\}$ **3.** $\{1, 5, 10, 15, 20\}$ **5.** $\{b\}$
7. $\{r, s, t, u, v\}$ **9.** $\{3, 5, 7\}$
11. $(2, 7)$;

13. $[-6, -2]$;

15. $(-\infty, -2) \cup (1, \infty)$;

17. $(-\infty, -1] \cup (4, \infty)$;

19. $(-5, 3]$;

21. $(-2, 4)$;

23. $(-\infty, 5) \cup (7, \infty)$;

25. $(-\infty, -4] \cup [5, \infty)$;

27. $[-6, 5)$;

29. $[3, 7)$;

31. $(-\infty, 5)$;

33. $(-\infty, \infty)$;

35. $(7, \infty)$;

37. $\{t\middle|\ -5 < t < 5\}$, or $(-5, 5)$

39. $\{a\middle|\ -2 \leq a < 2\}$, or $[-2, 2)$

41. $\{x\middle|\ x \leq -9\ or\ x \geq -2\}$, or $(-\infty, -9] \cup [-2, \infty)$

43. $\{x\middle|\ 1 \leq x \leq 3\}$, or $[1, 3]$

45. $\left\{x\middle|\ -\frac{7}{2} < x \leq \frac{11}{2}\right\}$, or $\left(-\frac{7}{2}, \frac{11}{2}\right]$

47. $\{x\middle|\ x \leq 1\ or\ x \geq 3\}$, or $(-\infty, 1] \cup [3, \infty)$

49. $\{x\middle|\ x < 3\ or\ x > 4\}$, or $(-\infty, 3) \cup (4, \infty)$

51. $\{a\middle|\ a < -5\}$, or $(-\infty, -5)$

53. $\mathbb{R}$, or $(-\infty, \infty)$

55. $\{t\middle|\ t \leq 6\}$, or $(-\infty, 6]$

57. $\mathbb{R}$, or $(-\infty, \infty)$ **59.** ∅   **61.** $\{x\middle|\ -1 < x < 6\}$, or $(-1, 6)$ **63.** $(-\infty, 5) \cup (5, \infty)$ **65.** $[-4, \infty)$
67. $\left(-\infty, \frac{5}{2}\right) \cup \left(\frac{5}{2}, \infty\right)$ **69.** $(-\infty, 4]$ **71.** $[-3, 4]$
73. $\left[\frac{3}{4}, \infty\right)$ **75.** $\left[-\frac{2}{3}, 5\right) \cup (5, \infty)$ **77.** $\left(-\frac{16}{13}, -\frac{41}{13}\right)$
79. $-\frac{27}{7}$
81.

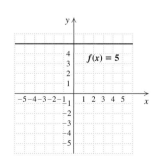

$f(x) = 5$

83. ◈   **85.** ◈   **87.** More than 1 and less than 5
89. $0\ \text{ft} \leq d \leq 198\ \text{ft}$ **91.** From 2005 to 2010

93. $\left\{a\middle| -\frac{3}{2} \le a \le 1\right\}$, or $\left[-\frac{3}{2}, 1\right]$;

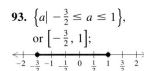

95. $\{x| -4 < x \le 1\}$, or $(-4, 1]$;

97. True **99.** False **101.** $\left[-\frac{5}{2}, 1\right) \cup (1, \infty)$
103. Left to the student.

INTERACTIVE DISCOVERY, P. 244

Solutions of $|x| = 4$ are -4 and 4; solution of $|x| = 0$ is 0; $|x| = -7$ has no solution

INTERACTIVE DISCOVERY, P. 248

1. $(-4, 4)$ **2.** $(-\infty, -4) \cup (4, \infty)$ **3.** $[-4, 4]$
4. $(-\infty, -4] \cup [4, \infty)$

EXERCISE SET 4.3

1. $\{-5, 1\}$ **3.** $\{x| -5 < x < 1\}$, or $(-5, 1)$
5. $\{x| x \le -5 \text{ or } x \ge 1\}$, or $(-\infty, -5] \cup [1, \infty)$
7. $\{-7, 7\}$ **9.** $\varnothing$ **11.** $\{-8.6, 8.6\}$ **13.** $\{0\}$
15. $\left\{-1, \frac{1}{5}\right\}$ **17.** $\varnothing$ **19.** $\{-9, 9\}$ **21.** $\{-2, 2\}$
23. $\{6, 8\}$ **25.** $\left\{\frac{3}{2}, \frac{7}{2}\right\}$ **27.** $\left\{-\frac{4}{3}, 4\right\}$ **29.** $\{-8.3, 8.3\}$
31. $\left\{-\frac{8}{3}, 4\right\}$ **33.** 11 **35.** 33 **37.** 34 **39.** $\{1, 11\}$
41. $\left\{\frac{3}{2}\right\}$ **43.** $\left\{-4, -\frac{10}{9}\right\}$ **45.** $\mathbb{R}$ **47.** $\{1\}$
49. $\{a| -6 \le a \le 6\}$, or $[-6, 6]$

51. $\{x| x < -7 \text{ or } x > 7\}$, or $(-\infty, -7) \cup (7, \infty)$

53. $\{t| t < 0 \text{ or } t > 0\}$, or $(-\infty, 0) \cup (0, \infty)$

55. $\{x| -2 < x < 8\}$, or $(-2, 8)$

57. $\{x| x \le -2 \text{ or } x \ge 8\}$, or $(-\infty, -2] \cup [8, \infty)$

59. $\{y| y \text{ is a real number}\}$, or $\mathbb{R}$, or $(-\infty, \infty)$

61. $\left\{a\middle| a \le -\frac{1}{3} \text{ or } a \ge 3\right\}$, or $\left(-\infty, -\frac{1}{3}\right] \cup [3, \infty)$

63. $\{y| -9 < y < 15\}$, or $(-9, 15)$

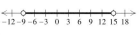

65. $\{x| x \le -8 \text{ or } x \ge 0\}$, or $(-\infty, -8] \cup [0, \infty)$

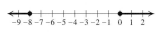

67. $\left\{y\middle| y < -\frac{4}{3} \text{ or } y > 4\right\}$, or $\left(-\infty, -\frac{4}{3}\right) \cup (4, \infty)$;

69. $\varnothing$

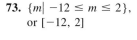

71. $\left\{x\middle| x \le -\frac{2}{15} \text{ or } x \ge \frac{14}{15}\right\}$, or $\left(-\infty, -\frac{2}{15}\right] \cup \left[\frac{14}{15}, \infty\right)$

73. $\{m| -12 \le m \le 2\}$, or $[-12, 2]$

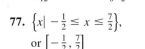

75. $\{a| -6 < a < 0\}$, or $(-6, 0)$

77. $\left\{x\middle| -\frac{1}{2} \le x \le \frac{7}{2}\right\}$, or $\left[-\frac{1}{2}, \frac{7}{2}\right]$

79. $\left\{x\middle| x \le -\frac{7}{3} \text{ or } x \ge 5\right\}$, or $\left(-\infty, -\frac{7}{3}\right] \cup [5, \infty)$

81. $\{x| -4 < x < 5\}$, or $(-4, 5)$

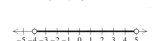

83. child's: 118; adult's: 132 **85.** 68 **87.** ◈ **89.** ◈
91. $\left\{x\middle| x \ge \frac{5}{2}\right\}$, or $\left[\frac{5}{2}, \infty\right)$ **93.** $\left\{-\frac{1}{4}, 1\right\}$
95. $\{x| -4 \le x \le -1 \text{ or } 3 \le x \le 6\}$, or $[-4, -1] \cup [3, 6]$ **97.** $\varnothing$ **99.** $\{0.45\}$
101. $\{x| -20 \le x \le -2.666667\}$, or $[-20, -2.666667]$
103. $\{x| x > 5.4\}$, or $(5.4, \infty)$ **105.** $|y| \le 5$
107. $|x| > 4$ **109.** $|x + 2| < 3$
111. $|x - 5| > 1$, or $|5 - x| > 1$ **113.** ◈

EXERCISE SET 4.4

1. Yes **3.** No
5.

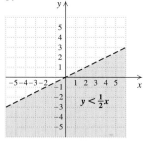

$y < \frac{1}{2}x$

7.

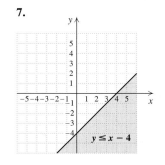

$y \le x - 4$

9.

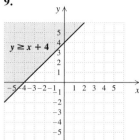

$y \geq x + 4$

11.

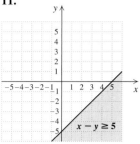

$x - y \geq 5$

25.

$y > x + 3.5$

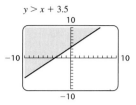

27.

$8x - 2y < 11$

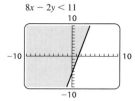

13.

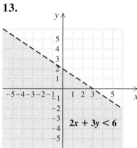

$2x + 3y < 6$

15.

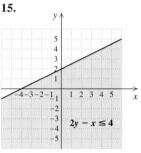

$2y - x \leq 4$

29.

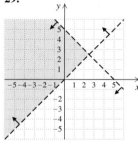

31.

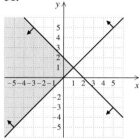

17.

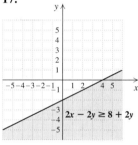

$2x - 2y \geq 8 + 2y$

19.

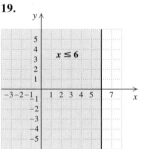

$x \leq 6$

33.

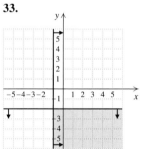

35.

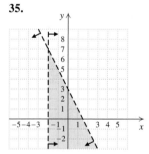

21.

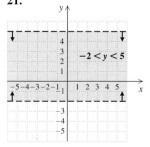

$-2 < y < 5$

23.

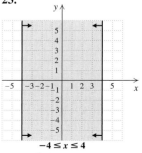

$-4 \leq x \leq 4$

37.

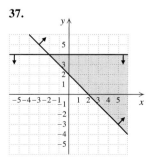

39.

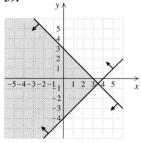

41.

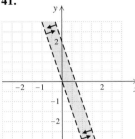

43.

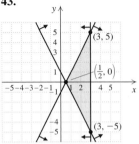

63. $0 < w \le 62$,
$0 < h \le 62$,
$62 + 2w + 2h \le 108$,
or $w + h \le 23$

65. $2w + t \ge 60$,
$w \ge 0$,
$t \ge 0$

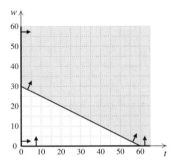

45.

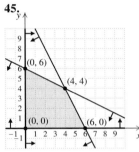

47.

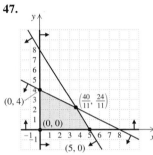

67. $y \le x$,
$y \le 2$
69. $y \le x + 2$,
$y \le -x + 4$,
$y \ge 0$

49.

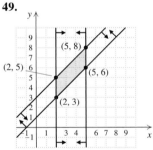

51. 15, 20   **53.** $\frac{10}{17}$

REVIEW EXERCISES, CHAPTER 4

1. [4.1] $\{x|\ x \le -4\}$,
or $(-\infty, -4]$;

2. [4.1] $\{a|\ a \le -21\}$,
or $(-\infty, -21]$;

3. [4.1] $\{y|\ y \ge -7\}$,
or $[-7, \infty)$;

4. [4.1] $\left\{y|\ y > -\frac{15}{4}\right\}$,
or $\left(-\frac{15}{4}, \infty\right)$;

55. ◈   **57.** ◈

5. [4.1] $\{y|\ y > -30\}$,
or $(-30, \infty)$;

6. [4.1] $\{x|\ x > -3\}$,
or $(-3, \infty)$;

59.

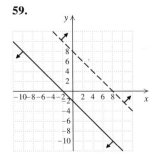

61.

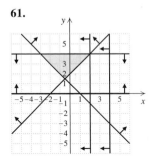

7. [4.1] $\{x|\ x < -3\}$,
or $(-\infty, -3)$;

8. [4.1] $\left\{y|\ y > -\frac{220}{23}\right\}$,
or $\left(-\frac{220}{23}, \infty\right)$;

9. [4.1] $\left\{x|\ x \le -\frac{5}{2}\right\}$,
or $\left(-\infty, -\frac{5}{2}\right]$;

10. [4.1] $\{x \mid x \le 4\}$, or $(-\infty, 4]$ **11.** [4.1] More than 125 hr   **12.** [4.1] $1500   **13.** [4.2] $\{1, 5, 9\}$
14. [4.2] $\{1, 2, 3, 5, 6, 9\}$
15. [4.2] ; $(-5, 3]$

16. [4.2] ; $(-\infty, \infty)$

17. [4.2] $\{x \mid -7 < x \le 2\}$,
   or $(-7, 2]$

18. [4.2] $\left\{x \mid -\frac{5}{4} < x < \frac{5}{2}\right\}$,
   or $\left(-\frac{5}{4}, \frac{5}{2}\right)$

19. [4.2] $\{x \mid x < -3 \ or \ x > 1\}$, or $(-\infty, -3) \cup (1, \infty)$

20. [4.2] $\{x \mid x < -11 \ or \ x \ge -6\}$, or
$(-\infty, -11) \cup [-6, \infty)$

21. [4.2] $\{x \mid x \le -6 \ or \ x \ge 8\}$, or $(-\infty, -6] \cup [8, \infty)$

22. [4.2] $\left\{x \mid x < -\frac{2}{5} \ or \ x > \frac{8}{5}\right\}$, or $\left(-\infty, -\frac{2}{5}\right) \cup \left(\frac{8}{5}, \infty\right)$

23. [4.2] $[3, \infty)$ **24.** [4.3] $\{-6, 6\}$
25. [4.3] $\{x \mid x \le -3.5 \ or \ x \ge 3.5\}$, or
$(-\infty, -3.5] \cup [3.5, \infty)$ **26.** [4.3] $\{-5, 9\}$
27. [4.3] $\left\{x \mid -\frac{17}{2} < x < \frac{7}{2}\right\}$, or $\left(-\frac{17}{2}, \frac{7}{2}\right)$
28. [4.3] $\left\{x \mid x \le -\frac{11}{3} \ or \ x \ge \frac{19}{3}\right\}$, or $\left(-\infty, -\frac{11}{3}\right] \cup \left[\frac{19}{3}, \infty\right)$
29. [4.3] $\left\{-14, \frac{4}{3}\right\}$ **30.** [4.3] $\varnothing$
31. [4.3] $\{x \mid -12 \le x \le 4\}$, or $[-12, 4]$
32. [4.3] $\{x \mid x < 0 \ or \ x > 10\}$, or $(-\infty, 0) \cup (10, \infty)$
33. [4.3] $\varnothing$

34. [4.4]

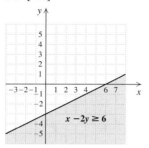

35. [4.4]

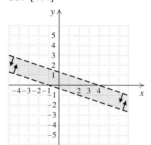

36. [4.4]

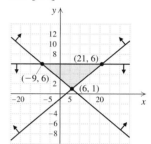

37. [3.2] $\left(0, \frac{5}{2}\right)$ **38.** [2.2] -22
39. [2.4]

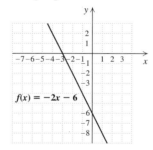

40. [3.3] 38,592 ft^2
41. [4.3] ◈ The equation $|x| = p$ has two solutions when p is positive because x can be either p or $-p$. The same equation has no solution when p is negative because no number has a negative absolute value.   **42.** [4.4] ◈ The solution set of a system of inequalities is all ordered pairs that make *all* the individual inequalities true. This consists of ordered pairs that are common to all the individual solution sets, or the intersection of the graphs.
43. [4.3] $\left\{x \mid -\frac{8}{3} \le x \le -2\right\}$, or $\left[-\frac{8}{3}, -2\right]$
44. [4.1] False; $-4 < 3$ is true, but $(-4)^2 < 9$ is false.
45. [4.3] $|d - 1.1| \le 0.03$

TEST, CHAPTER 4

1. [4.1] $\{x \mid x < 14\}$, or $(-\infty, 14)$

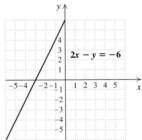

2. [4.1] $\{y \mid y > -50\}$, or $(-50, \infty)$

3. [4.1] $\{y \mid y \le -2\}$, or $(-\infty, -2]$

4. [4.1] $\left\{a \mid a \le \frac{11}{5}\right\}$, or $\left(-\infty, \frac{11}{5}\right]$

5. [4.1] $\left\{x \mid x > \frac{5}{2}\right\}$, or $\left(\frac{5}{2}, \infty\right)$

6. [4.1] $\left\{x \mid x \le \frac{7}{4}\right\}$, or $\left(-\infty, \frac{7}{4}\right]$

7. [4.1] $\{x \mid x > 1\}$, or $(1, \infty)$
8. [4.1] More than 166 mi   **9.** [4.1] Less than or equal to
2.5 hr   **10.** [4.2] $\{3, 5\}$ **11.** [4.2] $\{1, 3, 5, 7, 9, 11, 13\}$
12. [4.2] $(-\infty, 7]$
13. [4.2] $\{x \mid -1 < x < 6\}$,
   or $(-1, 6)$

14. [4.2] $\left\{x \mid -\frac{2}{5} < x \le \frac{9}{5}\right\}$,
   or $\left(-\frac{2}{5}, \frac{9}{5}\right]$

15. [4.2] $\{x \mid x < 3 \text{ or } x > 6\}$, or $(-\infty, 3) \cup (6, \infty)$

16. [4.2] $\left\{x \mid x < -4 \text{ or } x > -\frac{5}{2}\right\}$, or
$(-\infty, -4) \cup \left(-\frac{5}{2}, \infty\right)$

17. [4.2] $\left\{x \mid 4 \le x < \frac{15}{2}\right\}$,
   or $\left[4, \frac{15}{2}\right)$

18. [4.3] $\{-9, 9\}$

19. [4.3] $\{x \mid x < -3 \text{ or } x > 3\}$, or $(-\infty, -3) \cup (3, \infty)$

20. [4.3] $\left\{x \mid -\frac{7}{8} < x < \frac{11}{8}\right\}$, or $\left(-\frac{7}{8}, \frac{11}{8}\right)$

21. [4.3] $\left\{x \mid x \le -\frac{13}{5} \text{ or } x \ge \frac{7}{5}\right\}$, or $\left(-\infty, -\frac{13}{5}\right] \cup \left[\frac{7}{5}, \infty\right)$

22. [4.3] $\varnothing$
23. [4.2] $\left\{x \mid x < \frac{1}{2} \text{ or } x > \frac{7}{2}\right\}$, or $\left(-\infty, \frac{1}{2}\right) \cup \left(\frac{7}{2}, \infty\right)$

24. [4.3] $\{1\}$
25. [4.4]   **26.** [4.4]

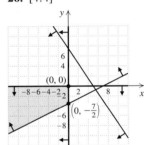

27. [2.2] $-\frac{33}{2}$ **28.** [3.2] $\left(-\frac{127}{8}, \frac{89}{8}\right)$
29. [2.5]

$2x - y = -6$

30. [3.3] 30-sec: 10; 60-sec: 5   **31.** [4.3] $[-1, 0] \cup [4, 6]$
32. [4.2] $\left(\frac{1}{5}, \frac{4}{5}\right)$

Chapter 5

INTERACTIVE DISCOVERY P. 274

1. (a) Odd; **(b)** $(-\infty, \infty)$ **2. (a)** Even; **(b)** $[-19, \infty)$
3. (a) Odd; **(b)** $(-\infty, \infty)$ **4. (a)** Odd; **(b)** $(-\infty, \infty)$

5. (a) Even; **(b)** $(-\infty, 7]$ **6. (a)** Odd; **(b)** $(-\infty, \infty)$

EXERCISE SET 5.1

1. Yes   **3.** No   **5.** Yes   **7.** 5, 3, 2, 1, 0; 5
9. 5, 6, 2, 1, 0; 6
11. $-4y^3 - 6y^2 + 7y + 15$; $-4y^3$; -4
13. $-a^7 + 8a^5 + 5a^3 - 19a^2 + a$; $-a^7$; -1
15. $-9 + 7x - 5x^2 + 3x^4$
17. $3xy^3 + x^2y^2 - 9x^3y + 2x^4$ **19.** 45; 5
21. -45; $-8\frac{19}{27}$ **23.** -23 **25.** -12 **27.** 9
29. 1320   **31.** 144 ft   **33.** \$18,750   **35.** \$8375
37. 399   **39.** About 340 mg   **41.** $M(5) \approx 65$
43. 56.5 in^2 **45.** $[0, \infty)$ **47.** $(-\infty, 5]$ **49.** $(-\infty, \infty)$
51. $[-6.7, \infty)$ **53.** Odd   **55.** Even
57. $-2a + 6 + 2a^3$ **59.** $-6a^2b - 2b^2$
61. $9x^2 + 2xy + 15y^2$ **63.** $9a + 4b - c$
65. $-2x^2 + x - xy - 1$ **67.** $9r^2 + 9r - 9$
69. $-\frac{5}{24}xy - \frac{27}{20}x^3y^2 + 1.4y^3$
71. $-(5x^3 - 7x^2 + 3x - 6)$; $-5x^3 + 7x^2 - 3x + 6$
73. $-(-12y^5 + 4ay^4 - 7by^2)$; $12y^5 - 4ay^4 + 7by^2$
75. $13x - 6$ **77.** $-4x^2 - 3x + 13$ **79.** $2a - 4b + 3c$
81. $-2x^2 + 6x$ **83.** $x^4 - x^2 - 1$ **85.** \$9700
87. (a), (c)   **89.** (b), (d)
91.

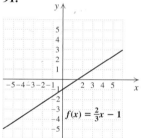

93. $3y - 6$ **95.** ◈   **97.** ◈
99. $68x^5 - 81x^4 - 22x^3 + 52x^2 + 2x + 250$
101. $45x^5 - 8x^4 + 208x^3 - 176x^2 + 116x - 25$
103. 494.55 cm^3
105. $200\pi rh + 20{,}000\pi r^2$ cm^2, or $0.02\pi rh + 2\pi r^2$ m^2
107. $x^{6a} - 5x^{5a} - 4x^{4a} + x^{3a} - 2x^{2a} + 8$

INTERACTIVE DISCOVERY, P. 286

2 and 5 are identities; 1, 3, and 4 are not.

INTERACTIVE DISCOVERY, P. 287

1 is an identity; 2, 3, and 4 are not.

EXERCISE SET 5.2

1. $21a^3$ **3.** $-20x^3y$ **5.** $-10x^5y^6$ **7.** $16x - 8x^2$
9. $15c^3d^2 - 25c^2d^3$ **11.** $6x^2 + 7x - 20$
13. $m^2 - mn - 6n^2$ **15.** $a^4 - 5a^2b^2 + 6b^4$
17. $x^3 - 64$ **19.** $a^4 + 5a^3 - 2a^2 - 9a + 5$

21. $4a^3b^2 + 4a^3b - 10a^2b^2 - 2a^2b + 3ab^3 + 7ab^2 - 6b^3$
23. $x^2 - \frac{3}{4}x + \frac{1}{8}$ **25.** $3x^2 - 1.5xy - 15y^2$
27. $12x^4 - 21x^3 - 14x^2 + 35x - 10$
29. $a^2 + 13a + 40$ **31.** $y^2 - y - 12$
33. $x^2 + 6x + 9$ **35.** $x^2 - 4xy + 4y^2$
37. $2x^2 + 13x + 18$ **39.** $100a^2 - 2.4ab + 0.0144b^2$
41. $4a^2 + \frac{4}{3}a + \frac{1}{9}$ **43.** $4x^6 - 12x^3y^2 + 9y^4$
45. $a^4b^4 + 2a^2b^2 + 1$ **47.** $16x^2 - 8x + 1$
49. $c^2 - 4$ **51.** $4a^2 - 1$ **53.** $x^6 - y^2z^2$
55. $-m^2n^2 + m^4$, or $m^4 - m^2n^2$ **57.** $x^4 - 1$
59. $a^4 - 2a^2b^2 + b^4$ **61.** $a^2 + 2ab + b^2 - 1$
63. $4x^2 + 12xy + 9y^2 - 16$ **65.** $A = P + 2Pi + Pi^2$
67. (a) $t^2 + 3t - 4$; **(b)** $5h + 2ah + h^2$
69. (a) $-4t^2 - 4t + 1$; **(b)** $-2ah - h^2$
71. (a) $2p^2 - 19p + 50$; **(b)** $4ah + 2h^2 + h$
73. 8   **75.** A: 75; B: 84; C: 63
77. ◈   **79.** ◈
81.

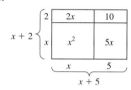

83. $a^{2n}b^{6n^2}$ **85.** z^{5n^5} **87.** $a^{xw+xz}b^{yw+yz}$
89. $-a^4 - 2a^3b + 25a^2 + 2ab^3 - 25b^2 + b^4$
91. $y^{12} - 6y^{10} + 15y^8 - 20y^6 + 15y^4 - 6y^2 + 1$
93. $\frac{4}{9}x^2 - \frac{1}{9}y^2 - \frac{2}{3}y - 1$ **95.** $16x^4 + 4x^2y^2 + y^4$
97. $x^{4a} - y^{4b}$ **99.** $x^{a^2-b^2}$

INTERACTIVE DISCOVERY, P. 294

1. 1   **2.** 0   **3.** 2   **4.** 3   **5.** 2   **6.** 1

INTERACTIVE DISCOVERY, P. 295

The zeros of $f(x)$ are 0 and -4. The zero of $g(x)$ is 0. The zero of $h(x)$ is -4.

EXERCISE SET 5.3

1. $-3, 5$ **3.** $-3, 1$ **5.** $-4, 2$ **7.** 0, 5   **9.** 1, 3
11. 10, 15   **13.** 0, 1, 2   **15.** $-15, 6, 12$
17. $-0.42857, 0.33333$ **19.** $-5, 9$ **21.** $-0.5, 7$
23. $-1, 0, 3$ **25.** III   **27.** I   **29.** Expression
31. Equation   **33.** Equation   **35.** $2t(4t + 1)$
37. $y^2(y + 9)$ **39.** $5x^2(1 - 3x^2)$ **41.** $4xy(x - 3y)$
43. $3(y^2 - y - 3)$ **45.** $2a(3b - 2d + 6c)$
47. $3x^2y^4z^2(3xy^2 - 4x^2z^2 + 5yz)$ **49.** $-5(x - 3)$
51. $-6(y + 12)$ **53.** $-2(x^2 - 2x + 6)$
55. $-3y(y^2 - 4y + 5)$ **57.** $-(x^2 - 5x + 9)$
59. (a) $h(t) = -16t(t - 6)$; **(b)** $h(2) = 128$
61. $N(x) = \frac{1}{6}x(x^2 + 3x + 2)$ **63.** $\pi r(2h + r)$
65. $R(x) = 0.4x(700 - x)$ **67.** $(b - 2)(a + c)$

69. $(x + 7)(2x - 3)$ **71.** $(x - y)(a^2 - 5)$
73. $(c + d)(a + b)$ **75.** $(b - 1)(b^2 + 2)$
77. $(a - 3)(a^2 - 2)$ **79.** $12x(2x^2 - 3x + 6)$
81. $x^3(x - 1)(x^2 + 1)$ **83.** $(y^2 + 3)(2y^2 + 5)$
85. $-1, 0$ **87.** $0, 3$ **89.** $-3, 0$ **91.** $-\frac{1}{3}, 0$
93.

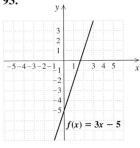

$f(x) = 3x - 5$

95. Countryside: 9.375 lb; Mystic: 15.625 lb
97. ◆   **99.** ◆   **101.** $(x - 5)(x + 3)$
103. $(x + 4)(x - 2)$
105. $x^5y^4 + x^4y^6 = x^3y(x^2y^3 + xy^5)$
107. $(x^2 - x + 5)(r + s)$ **109.** $x(5 - y)(x + 2 + 3z)$
111. $2x^a(x^{2a} + 4 + 2x^a)$ **113.** $x^a(4x^b + 7x^{-b})$

EXERCISE SET 5.4

1. $(x + 2)(x + 6)$ **3.** $(t - 3)(t - 5)$
5. $(x - 9)(x + 3)$ **7.** $2(n - 5)(n - 5)$, or $2(n - 5)^2$
9. $a(a + 9)(a - 8)$ **11.** $(x + 9)(x + 5)$
13. $(x + 5)(x - 2)$ **15.** $3(x + 2)(x + 3)$
17. $-(x - 8)(x + 7)$, or $(-x + 8)(x + 7)$, or
$(8 - x)(7 + x)$ **19.** $-y(y - 8)(y + 4)$, or
$y(-y + 8)(y + 4)$, or $y(8 - y)(4 + y)$
21. $(x^2 + 16)(x^2 - 5)$ **23.** Prime
25. $(p - 8q)(p + 3q)$
27. $(y + 4z)(y + 4z)$, or $(y + 4z)^2$
29. $(p^2 + 1)(p^2 + 79)$ **31.** $-6, -2$ **33.** 5
35. $-9, 0, 8$ **37.** $-5, 1$ **39.** $-3, 2$ **41.** $-5, 9$
43. $-1, 0, 3$ **45.** $-9, 5$ **47.** $0, 9$ **49.** $-5, 0, 8$
51. $-4, 5$ **53.** $-7, 5$ **55.** $(x + 22)(x - 12)$
57. $(x + 16)(x + 24)$ **59.** $(x - 38)(x + 64)$
61. $f(x) = x^2 - x - 2$ **63.** $f(x) = x^2 + 17x + 70$
65. $f(x) = x^3 - 3x^2 + 2x$
67.

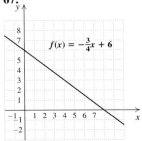

$f(x) = -\frac{3}{4}x + 6$

69. $\frac{80}{7}$ **71.** ◆   **73.** ◆
75. $-1, 3$; $\{x| -2 < x < 4\}$, or $(-2, 4)$
77. Answers may vary; $f(x) = 5x^3 - 20x^2 + 5x + 30$
79. 6.90   **81.** 3.48   **83.** $\left(y + \frac{4}{7}\right)\left(y - \frac{2}{7}\right)$
85. $(x^a + 8)(x^a - 3)$ **87.** $76, -76, 28, -28, 20,$
-20
89. $(x - 365)$

EXERCISE SET 5.5

1. $(3x + 5)(2x - 5)$ **3.** $y(2y - 3)(5y + 4)$
5. $2(4a - 1)(3a - 1)$ **7.** $(7y + 4)(5y + 2)$
9. $2(5t - 3)(t + 1)$ **11.** $4(x - 4)(2x + 1)$
13. $x^2(2x - 3)(7x + 1)$ **15.** $4(3a - 4)(a + 1)$
17. $(3x + 4)(3x + 1)$ **19.** $(4 + 3z)(2 - 3z)$
21. $xy(6y + 5)(3y - 2)$ **23.** $(x - 2)(24x + 1)$
25. $3x(7x + 3)(3x + 4)$ **27.** $(4x^2 - 3)(6x^2 + 5)$
29. $-\frac{5}{3}, \frac{5}{2}$ **31.** $-\frac{4}{3}, \frac{2}{3}$ **33.** $-\frac{4}{3}, -\frac{3}{7}, 0$ **35.** $\frac{2}{3}, 2$
37. $-2, -\frac{3}{4}, 0$ **39.** $-\frac{1}{3}, \frac{5}{2}$ **41.** $-\frac{7}{4}, \frac{4}{3}$ **43.** $-\frac{1}{2}, 7$
45. $-8, -4$ **47.** $-4, \frac{3}{2}$ **49.** $-9, -3$
51. $\{x| \ x$ is a real number $and \ x \neq 5 \ and \ x \neq -1\}$
53. $\left\{x| \ x$ is a real number $and \ x \neq 0 \ and \ x \neq \frac{1}{2}\right\}$
55. $\{x| \ x$ is a real number $and \ x \neq 0 \ and \ x \neq 2 \ and$
$x \neq 5\}$ **57.** $228 \ ft^2$ **59.** Slope: $\frac{4}{3}$; y-intercept: $\left(0, -\frac{8}{3}\right)$
61. $\left\{x| -\frac{33}{5} \leq x \leq 9\right\}$, or $\left[-\frac{33}{5}, 9\right]$ **63.** ◆   **65.** ◆
67. $(2x + 15)(2x + 45)$ **69.** $3x(x - 18)(x + 68)$
71. $-\frac{11}{8}, -\frac{1}{4}, \frac{2}{3}$ **73.** 1   **75.** $(2a^2b^3 + 5)(a^2b^3 - 4)$
77. $(2x^a + 1)(2x^a - 3)$

EXERCISE SET 5.6

1. $(x + 4)^2$ **3.** $(a - 8)^2$ **5.** $2(a + 2)^2$ **7.** $(y - 6)^2$
9. $a(a + 12)^2$ **11.** $2(4x + 3)^2$ **13.** $(5a - 8)^2$
15. $(0.5x + 0.3)^2$ **17.** $(p - q)^2$ **19.** $(a^2 - 5)^2$
21. $(5a - 3b)^2$ **23.** $(t^4 + s^4)^2$ **25.** $(x + 4)(x - 4)$
27. $(p + 7)(p - 7)$ **29.** $(ab + 9)(ab - 9)$
31. $6(x + y)(x - y)$ **33.** $7x(y^2 + z^2)(y + z)(y - z)$
35. $a(2a + 7)(2a - 7)$
37. $3(x^4 + y^4)(x^2 + y^2)(x + y)(x - y)$
39. $a^2(3a + 5b^2)(3a - 5b^2)$ **41.** $\left(\frac{1}{5} + x\right)\left(\frac{1}{5} - x\right)$
43. $(a + b + 3)(a + b - 3)$
45. $(6 + x + y)(6 - x - y)$
47. $(m - 7)(m + 2)(m - 2)$ **49.** $(a + b)(a - b)(a - 2)$
51. -4 **53.** $-4, 4$ **55.** 1   **57.** 6   **59.** $-3, 3$
61. $-\frac{1}{5}, \frac{1}{5}$ **63.** $-5, -1, 1, 5$ **65.** $-1, 1, 3$
67. $-1.541, 4.541$ **69.** $-3.871, -0.129$
71. $-2.414, -1, 0.414$ **73.** 6   **75.** $(2, -1, 3)$
77. $\left\{x| -\frac{4}{7} \leq x \leq 2\right\}$, or $\left[-\frac{4}{7}, 2\right]$ **79.** ◆   **81.** ◆
83. **(a)** $\pi h(R + r)(R - r)$; **(b)** $3,014,400 \ cm^3$
85. $(x^a + y)(x^a - y)$ **87.** $(3x^n - 1)^2$
89. $5(c^{50} + 4d^{50})(c^{25} + 2d^{25})(c^{25} - 2d^{25})$

91. $(x + 4)(x^2 - 4x + 16)$ **93.** $(z - 1)(z^2 + z + 1)$
95. $2(3x + 1)(9x^2 - 3x + 1)$
97. $a(b + 5)(b^2 - 5b + 25)$
99. $5(x - 2z)(x^2 + 2xz + 4z^2)$
101. $(x + 0.1)(x^2 - 0.1x + 0.01)$
103. $8(2x^2 - t^2)(4x^4 + 2x^2t^2 + t^4)$
105. $2y(y - 4)(y^2 + 4y + 16)$

EXERCISE SET 5.7

1. $-13, 12$ **3.** Length: 12 cm; width: 7 cm   **5.** 6 cm
7. 3 cm   **9.** 5 ft   **11.** Height: 6 ft; base: 4 ft
13. 16, 18, 20   **15.** 150 ft by 200 ft   **17.** 41 ft
19. Length: 100 m; width: 75 m   **21.** 11 sec   **23.** 6
25. No   **27.** No   **29.** Yes
31. (a) $f(x) = 0.00000018x^4 - 0.00005260x^3 + 0.00387914x^2 - 0.00794849x + 2.12144793$;
(b) 2.686 million; (c) 1879, 1966
33. (a) $w(x) = 6.125x^2 - 5.36785714x + 9.60714286$;
(b) 1307 million; (c) 2001
35. Faster car: 54 mph; slower car: 39 mph
37. $\left\{x \mid x > \frac{10}{3}\right\}$, or $\left(\frac{10}{3}, \infty\right)$ **39.** ◈   **41.** ◈
43. 14 cm by 28 cm   **45.** About 5.7 sec

REVIEW EXERCISES, CHAPTER 5

1. [5.1] 7, 11, 3, 0; 11
2. [5.1] $-5x^3 + 2x^2 + 4x - 7$; $-5x^3$; -5
3. [5.1] $-3x^2 + 2x^3 + 3x^6y - 7x^8y^3$
4. [5.1] 0; -6 **5.** [5.1] 4   **6.** [5.1] (a) $\mathbb{R}$;
(b) $[-6.25, \infty)$ **7.** [5.1] $-x^2y - 2xy^2$
8. [5.1] $ab + 4 + 12ab^2$
9. [5.1] $-x^3 + 2x^2 + 5x + 2$
10. [5.1] $-8xy^2 + 4xy + 13x^2y$ **11.** [5.1] $9x - 7$
12. [5.1] $-2a + 6b + 7c$ **13.** [5.1] $6x^2 - 7xy + 3y^2$
14. [5.2] $-18x^3y^4$ **15.** [5.2] $x^8 - x^6 + 5x^2 - 3$
16. [5.2] $8a^2b^2 + 2abc - 3c^2$ **17.** [5.2] $4x^2 - 25y^2$
18. [5.2] $4x^2 - 20xy + 25y^2$ **19.** [5.2] $2x^2 + 5x - 3$
20. [5.2] $x^3 - 125$ **21.** [5.2] $x^2 - \frac{1}{2}x + \frac{1}{18}$
22. [5.2] $t^2 + 4t + 4$ **23.** [5.4] $(x - 32)(x - 18)$
24. [5.3] $3y^2(3y^2 - 1)$
25. [5.3] $3x(5x^3 - 6x^2 + 7x - 3)$
26. [5.4] $(a - 9)(a - 3)$ **27.** [5.5] $(3m + 2)(m + 4)$
28. [5.6] $(5x + 2)^2$ **29.** [5.6] $4(y + 2)(y - 2)$
30. [5.4] $x(x - 2)(x + 7)$ **31.** [5.3] $(a + 2b)(x - y)$
32. [5.3] $(y + 2)(3y^2 - 5)$
33. [5.6] $(a^2 + 9)(a + 3)(a - 3)$
34. [5.4] $4(x^4 + x^2 + 5)$
35. [5.3] $y(y^4 + 1)$ **36.** [5.3] $2z^6(z^2 - 8)$
37. [5.5] $(3t + p)(2t + 5p)$ **38.** [5.6] $4(3x - 5)^2$
39. [5.6] $(x + 3)(x - 3)(x + 2)$
40. (a) [5.1] Odd; (b) [5.3] $-2, 1, 5$ **41.** [5.4] 4, 7
42. [5.6] 10   **43.** [5.5] $\frac{2}{3}, \frac{3}{2}$ **44.** [5.3] 0, $\frac{7}{4}$
45. [5.6] $-4, 4$ **46.** [5.6] $-4, 4, 5$
47. [5.4] 12, 15   **48.** [5.4] $-4, 11$

49. [5.5] $\left\{x \mid x \text{ is a real number } and\ x \neq -7\ and\ x \neq \frac{2}{3}\right\}$
50. [5.7] 5   **51.** [5.7] 3, 5, 7; $-7, -5, -3$
52. [5.7] 5 in. by 8 in.
53. [5.7] (a) $m(x) = -0.17897727x^4 + 3.51704546x^3 - 22.73768939x^2 + 51.35200216x + 5.83928571$;
(b) \$15.9 million (c) 9
54. [3.4] $(0, -2, 7)$ **55.** [4.1] $\left\{x \mid x > -\frac{7}{3}\right\}$, or $\left(-\frac{7}{3}, \infty\right)$
56. [4.3] $\left\{x \mid -\frac{4}{3} \leq x \leq 8\right\}$, or $\left[-\frac{4}{3}, 8\right]$
57. [2.4]

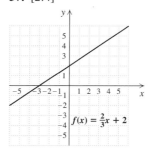

$f(x) = \frac{2}{3}x + 2$

58. [3.5] Multiple-choice: 31; true–false: 26; fill-in: 13
59. [5.3] ◈   The zeros of a polynomial function are the x-coordinates of the points where the graph of the function crosses the x-axis.
60. [5.3] ◈   The principle of zero products states that if a product is equal to 0, at least one of the factors must be 0. If a product is nonzero, we cannot conclude that any one of the factors is a particular value.
61. [5.6] $2(x - y)(x^2 + xy + y^2)$
62. [5.6] $(ab + cd)(a^2b^2 - abcd + c^2d^2)$
63. [5.2], [5.6] $a^3 - b^3 + 3b^2 - 3b + 1$ **64.** [5.2] z^{5n^5}

TEST, CHAPTER 5

1. [5.1] 4   **2.** [5.1] $5y^4 + 3y^3 - 4y - 2$
3. [5.1] $-4a^3$ **4.** [5.1] 4; 2   **5.** [5.1] $2ah + h^2 - 5h$
6. [5.1] $3xy + 3xy^2$ **7.** [5.1] $-3x^3 + 3x^2 - 6y - 7y^2$
8. [5.1] $7m^3 + 2m^2n + 3mn^2 - 7n^3$ **9.** [5.1] $6a - 8b$
10. [5.1] $2y^2 + 5y + y^3$ **11.** [5.2] $64x^3y^3$
12. [5.2] $12a^2 - 4ab - 5b^2$ **13.** [5.2] $x^3 - 2x^2y + y^3$
14. [5.2] $-3m^4 - 13m^3 + 5m^2 + 26m - 10$
15. [5.2] $16y^2 - 72y + 81$ **16.** [5.2] $x^2 - 4y^2$
17. [5.3] $5x^2(3 - x^2)$ **18.** [5.6] $(y + 5)(y + 2)(y - 2)$
19. [5.4] $(p - 14)(p + 2)$ **20.** [5.5] $(6m + 1)(2m + 3)$
21. [5.6] $(3y + 5)(3y - 5)$
22. [5.4] $(x - 22)(x + 15)$ **23.** [5.6] $(3x - 5)^2$
24. [5.6] $(x^4 + y^4)(x^2 + y^2)(x + y)(x - y)$
25. [5.5] $2(4x - 1)(3x - 5)$
26. [5.6] $5(2a - b)(2a + b)$
27. [5.3] $4xy(y^3 + 9x + 2x^2y - 4)$
28. (a) [5.3] $-3, -1, 2, 4$; (b) [5.1] even
29. [5.3] (a) $\mathbb{R}$; (b) $-2.5, 8$; (c) $[-55, \infty)$
30. [5.4] $-3, 6$ **31.** [5.6] $-5, 5$ **32.** [5.5] $-7, -\frac{3}{2}$

33. [5.3] $-\frac{1}{3}$, 0   **34.** [5.6] 9
35. [5.4] -3.372, 0, 2.372   **36.** [5.5] 0, 5
37. [5.5] $\{x|\ x$ is a real number $and\ x \neq -1\}$
38. [5.7] Length: 8 cm; width: 5 cm
39. [5.7] **(a)** $0.03135826x^3 - 3.21956840x^2 +$
$101.17516232x - 886.92962439$; **(b)** 100; **(c)** 20, 30
40. [4.3] $\left\{x|\ -6 < x < \frac{2}{3}\right\}$, or $\left(-6, \frac{2}{3}\right)$
41. [4.1] $\{x|\ x > 1\}$, or $(1, \infty)$ **42.** [3.4] $(5, -1, -2)$
43. [2.4] Slope: $-\frac{1}{3}$; y-intercept: $\left(0, \frac{8}{3}\right)$
44. [5.2], [5.3] **(a)** $x^5 + x + 1$;
(b) $(x^2 + x + 1)(x^3 - x^2 + 1)$
45. [5.5] $(3x^n + 4)(2x^n - 5)$

Chapter 6

EXERCISE SET 6.1

1. 0; -184; does not exist   **3.** $-\frac{9}{4}$; does not exist; $-\frac{11}{9}$
5. $\dfrac{5x(x - 3)}{5x(x + 2)}$ **7.** $\dfrac{(t - 2)(-1)}{(t + 3)(-1)}$ **9.** $\dfrac{3}{x}$ **11.** $\dfrac{2}{3t^4}$
13. $a - 5$ **15.** $\dfrac{3}{5a - 6}$ **17.** $\dfrac{5}{x}$ **19.** $-\dfrac{1}{2}$
21. $\dfrac{8}{t + 2}$ **23.** $-\dfrac{1}{1 + 2t}$ **25.** $\dfrac{a - 5}{a + 5}$ **27.** $\dfrac{x + 8}{x - 4}$
29. $\dfrac{4 + t}{4 - t}$ **31.** $x = -5$ **33.** None   **35.** $x = -3$
37. $x = 4, x = 2$ **39.** (b)   **41.** (f)   **43.** (a)
45. $\dfrac{7b^2}{6a^4}$ **47.** $\dfrac{3x^2}{25}$ **49.** $\dfrac{y + 4}{2}$ **51.** $\dfrac{(x + 4)(x - 4)}{x(x + 3)}$
53. $-\dfrac{a + 1}{2 + a}$ **55.** $\dfrac{-2}{t^4(t + 2)(t + 3)}$
57. $\dfrac{(x + 5)(2x + 3)}{7x}$ **59.** $6x^4y^7$ **61.** $\dfrac{6}{x^5}$
63. $\dfrac{(x + 2)(x + 4)}{x^7}$ **65.** $\dfrac{-5x - 2}{x - 3}$ **67.** $-\dfrac{1}{y^3}$
69. $\dfrac{(x + 4)(x + 2)}{3(x - 5)}$ **71.** $\dfrac{y(y^2 + 3)}{(y + 3)(y - 2)}$ **73.** $\left(\dfrac{7}{2}, \dfrac{5}{2}\right)$
75. $\dfrac{16}{3}$ **77.** ◆   **79.** ◆
81.

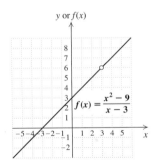

y or $f(x)$

$f(x) = \dfrac{x^2 - 9}{x - 3}$

83. $\dfrac{(d - 1)(d - 5)}{5d(d + 5)}$ **85.** $\dfrac{m - t}{m + t + 1}$
87. $\dfrac{x^2 + xy + y^2 + x + y}{x - y}$
89. Domain: $(-\infty, -2) \cup (-2, 1) \cup (1, \infty)$;
range: $(-\infty, 2) \cup (2, 3) \cup (3, \infty)$
91. Domain: $(-\infty, -1) \cup (-1, 1) \cup (1, \infty)$;
range: $(-\infty, -1] \cup (0, \infty)$

EXERCISE SET 6.2

1. $\dfrac{4}{a}$ **3.** $-\dfrac{1}{a^2b}$ **5.** 2   **7.** $\dfrac{3y + 5}{y - 2}$ **9.** $-\dfrac{1}{a + 5}$
11. $a + b$ **13.** $\dfrac{11}{x}$ **15.** $\dfrac{2}{x + 4}$ **17.** $\dfrac{1}{t^2 + 4}$
19. $\dfrac{2a^2 - a + 14}{(a - 4)(a + 3)}$ **21.** $\dfrac{3x - 1}{x + 1}$ **23.** $\dfrac{x + y}{x - y}$
25. $\dfrac{3x^2 + 7x + 14}{(2x - 5)(x - 1)(x + 2)}$ **27.** $\dfrac{8x + 1}{(x + 1)(x - 1)}$
29. $\dfrac{-x + 34}{20(x + 2)}$ **31.** $\dfrac{x - 5}{(x + 5)(x + 3)}$
33. $\dfrac{y}{(y - 2)(y - 3)}$ **35.** $\dfrac{7x + 1}{x - y}$
37. $\dfrac{-2a^2 - 7a - 10}{(a + 4)(a - 4)(a + 6)}$ **39.** $\dfrac{4t^2 - 2t - 14}{(t + 2)(t - 2)}$
41. $-\dfrac{y}{(y + 3)(y - 1)}$ **43.** $\dfrac{-14y^2 - 3y + 3}{(2y + 1)(2y - 1)}$
45. $\dfrac{-6x + 42}{(x - 3)(x - 2)(x + 1)(x + 3)}$ **47.** $\dfrac{3y^6z^7}{7x^5}$
49. Dimes: 2 rolls; nickels: 5 rolls; quarters: 5 rolls
51. ◆   **53.** ◆   **55.** 12
57. $x^4(x^2 + 1)(x + 1)(x - 1)(x^2 + x + 1)(x^2 - x + 1)$
59. $8a^4, 8a^4b, 8a^4b^2, 8a^4b^3, 8a^4b^4, 8a^4b^5, 8a^4b^6, 8a^4b^7$
61. $\dfrac{x^4 + 4x^3 - 2x^2}{(x + 2)(x - 2)(x + 5)}$
63. $\dfrac{x(x + 5)}{x + 2}, x \neq 0, x \neq -5, x \neq 2$
65. $\{x|\ x$ is a real number $and\ x \neq -2\ and\ x \neq 2\ and$
$x \neq -5\ and\ x \neq 0\}$
67. $\dfrac{5y + 23}{5 - 2y}$, or $\dfrac{-5y - 23}{2y - 5}$ **69.** $-4t^4$
71. Domain: $(-\infty, -1) \cup (-1, \infty)$; range: $(5, \infty)$

EXERCISE SET 6.3

1. $\dfrac{5a + 1}{1 - 2a}$ **3.** $\dfrac{x^2 - 1}{x^2 + 1}$ **5.** $\dfrac{3y + 4x}{4y - 3x}$ **7.** $\dfrac{x + y}{x}$
9. $\dfrac{3x^2 - x^2y}{2xy^2 - y^2}$ **11.** $\dfrac{1}{a - b}$ **13.** $\dfrac{-1}{x(x + h)}$
15. $\dfrac{(a - 2)(a - 7)}{(a + 1)(a - 6)}$ **17.** $\dfrac{3y - 1}{7y - 5}$ **19.** $\dfrac{a^2 - 3a - 6}{a^2 - 2a - 3}$

21. $\dfrac{x+2}{x+3}$ **23.** $\dfrac{2a-3}{a+1}$ **25.** $\dfrac{7-3x}{3-2x}$ **27.** $\dfrac{1}{y+3}$

29. $-y$ **31.** $\dfrac{2(5a^2+4a+12)}{5(a^2+6a+18)}$ **33.** $\dfrac{(2x+1)(x+2)}{2x(x-1)}$

35. $\dfrac{2x^2-11x-27}{2x^2+21x+13}$ **37.** 22

39.

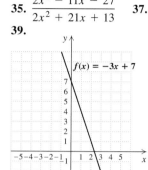
$f(x)=-3x+7$

41. \$34   **43.** ◆   **45.** ◆   **47.** $\dfrac{b-a}{ab}$ **49.** $\dfrac{2x+1}{3x+2}$

51. $\dfrac{1+a}{2+a}$ **53.** $\dfrac{-3(2x+h)}{x^2(x+h)^2}$ **55.** $\dfrac{1}{(1-x-h)(1-x)}$

57. $\{x\,|\,x$ is a real number *and* $x\neq0$ *and* $x\neq-2$ *and* $x\neq2\}$ **59.** $\dfrac{x^4}{81}$; $\{x\,|\,x$ is a real number *and* $x\neq3\}$

EXERCISE SET 6.4

1. $\frac{51}{5}$ **3.** 144   **5.** $\frac{40}{9}$ **7.** 5   **9.** No solution
11. $-4,-1$ **13.** No solution   **15.** 11   **17.** 2
19. 2, 3   **21.** -23 **23.** -1 **25.** $-\frac{5}{2},3$ **27.** 14
29. $\frac{3}{4}$ **31.** 4   **33.** 3   **35.** No solution   **37.** $-\frac{7}{3}$
39. $\frac{1}{7}$ **41.** $(9x^2+y^2)(3x+y)(3x-y)$
43. $\{x\,|\,x<-1\ or\ x>5\}$, or $(-\infty,-1)\cup(5,\infty)$
45. 16 and 18, -18 and -16 **47.** ◆   **49.** ◆
51. -8 **53.** $\{a\,|\,a$ is a real number *and* $a\neq-1$ *and* $a\neq1\}$ **55.** 0.0854697   **57.** Yes

EXERCISE SET 6.5

1. 2   **3.** $-3,-2$ **5.** -7 and -6, 6 and 7   **7.** $1\frac{5}{7}$ hr
9. $8\frac{4}{7}$ hr   **11.** 2.475 hr   **13.** 6 hr   **15.** Sara: 6 hr;
Kate: 3 hr   **17.** Zsuzanna: $\frac{4}{3}$ hr; Stan: 4 hr
19. 300 min, or 5 hr   **21.** New machine: $22\frac{1}{2}$ hr;
old machine: 45 hr   **23.** 12 mph   **25.** 5.2 ft/sec
27. Freight: 66 mph; passenger: 80 mph
29. Local: 35 mph; express: 42 mph   **31.** 9 km/h
33. Jaime: 23 km/h; Mara: 15 km/h
35. 5 m per minute   **37.** 20 mph
39. $\{x\,|\,x$ is a real number *and* $x\neq5$ *and* $x\neq-1\}$
41. (a) $c(x)=4.8533333x+46.26$; (b) 2003
43. ◆   **45.** ◆   **47.** 40 min   **49.** 2250   **51.** $14\frac{7}{8}$ mi
53. 30 mi   **55.** $8\frac{2}{11}$ min after 10:30   **57.** $51\frac{3}{7}$ mph

EXERCISE SET 6.6

1. $4x^4+3x^3-6$ **3.** $4a^2+a-\dfrac{3}{7}-\dfrac{2}{a}$
5. $13y-\dfrac{9}{2}-\dfrac{4}{y}$ **7.** $-6x^5+9x^2+1$
9. $-2pq+3p-4q$ **11.** $x+7$
13. $a-12+\dfrac{32}{a+4}$ **15.** $y-5$
17. $y^2-2y-1+\dfrac{-8}{y-2}$ **19.** $a^2+4a+15+\dfrac{72}{a-4}$
21. $2y^2+2y-1+\dfrac{8}{5y-2}$
23. $2x^2-x-9+\dfrac{3x+12}{x^2+2}$ **25.** $16x^2+8x+4,\ x\neq\frac{1}{2}$
27. $2x-5,\ x\neq-\frac{2}{3}$ **29.** $x^2+6,\ x\neq-3,\ x\neq3$
31. $x^2-x+1+\dfrac{-4}{x-1}$ **33.** $x^2-5x-23+\dfrac{-43}{x-2}$
35. $3x^2-2x+2+\dfrac{-3}{x+3}$ **37.** $y^2+2y+1+\dfrac{12}{y-2}$
39. $x^4+2x^3+4x^2+8x+16$
41. $3x^2+6x-3+\dfrac{2}{x+\frac{1}{3}}$ **43.** $(x+5)(x-3)(x+1)$
45. $(2x-1)(x+2)(x-5)$
47. $(x+2)(x-2)(x+1)(x+3)$ **49.** 0, 5
51. 12, 13, 14   **53.** $\{-2,5\}$ **55.** ◆
57. ◆   **59.** x^2+2y
61. $a^6-a^5b+a^4b^2-a^3b^3+a^2b^4-ab^5+b^6$
63. $-\frac{3}{2}$ **65.** $0;\ -\frac{7}{2},\frac{5}{3},4$ **67.** 0

EXERCISE SET 6.7

1. $W_1=\dfrac{d_1W_2}{d_2}$ **3.** $v_1=\dfrac{2s-tv_2}{t}$, or $\dfrac{2s}{t}-v_2$
5. $R=\dfrac{r_1r_2}{r_2+r_1}$ **7.** $g=\dfrac{Rs}{s-R}$ **9.** $R=\dfrac{2V}{I}-2r$, or
$\dfrac{2V-2Ir}{I}$ **11.** $p=\dfrac{qf}{q-f}$ **13.** $n=\dfrac{IR}{E-Ir}$
15. $t_1=\dfrac{H}{Sm}+t_2$, or $\dfrac{H+Smt_2}{Sm}$ **17.** $r=\dfrac{eR}{E-e}$
19. $r=1-\dfrac{a}{S}$, or $\dfrac{S-a}{S}$ **21.** $r=\dfrac{A}{P}-1$, or
$\dfrac{A-P}{P}$ **23.** 3:45 A.M.   **25.** 12   **27.** $5\frac{5}{9}$ ohms
29. 6 cm   **31.** $h=\dfrac{2gR^2}{V^2}-R$, or $\dfrac{2gR^2-RV^2}{V^2}$
33. $Q=\dfrac{2Tt-2AT}{A-q}$, or $\dfrac{2AT-2Tt}{q-A}$
35. $(-15,2),(-30,8)$ **37.** Yes   **39.** No

41.

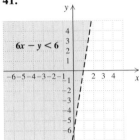

$6x - y < 6$

43. $(x - 105)(x + 28)$ **45.** ◆

47. $M = \dfrac{2ab}{b + a}$

49. $t_1 = t_2 + \dfrac{(d_2 - d_1)(t_4 - t_3)}{a(t_4 - t_2)(t_4 - t_3) + d_3 - d_4}$

REVIEW EXERCISES, CHAPTER 6

1. [6.1] **(a)** $-\dfrac{2}{9}$; **(b)** $-\dfrac{3}{4}$; **(c)** 0 **2.** [6.2] $48x^3$

3. [6.2] $(x + 5)(x - 2)(x - 4)$ **4.** [6.1] $x = -1$

5. [6.2] $x + 3$ **6.** [6.2] $\dfrac{1}{x - 1}$ **7.** [6.1] $\dfrac{b^2c^6d^2}{a^5}$

8. [6.2] $\dfrac{15np + 14m}{18m^2n^4p^2}$ **9.** [6.1] $\dfrac{y - 8}{2}$

10. [6.1] $\dfrac{3a - 1}{a - 3}$ **11.** [6.2] $\dfrac{x - 3}{(x + 1)(x + 3)}$

12. [6.2] $2(x + y)$ **13.** [6.2] $\dfrac{-y}{(y + 4)(y - 1)}$

14. [6.3] $\dfrac{5}{7}$ **15.** [6.3] $\dfrac{ab}{2(b - a)}$

16. [6.3] $\dfrac{(y + 11)(y + 5)}{(y - 5)(y + 2)}$ **17.** [6.3] $\dfrac{(14 - 3x)(x + 3)}{2x^2 + 16x + 6}$

18. [6.4] 2 **19.** [6.4] 6 **20.** [6.4] No solution

21. [6.4] $-1, 4$ **22.** [6.5] $5\frac{1}{7}$ hr   **23.** [6.5] 24 mph

24. [6.5] Motorcycle: 62 mph; car: 70 mph

25. [6.6] $4s^2 + 3s - 2rs^2$

26. [6.6] $y^2 - 5y + 25$

27. [6.6] $4x + 3 + \dfrac{-9x - 5}{x^2 + 1}$

28. [6.6] $(x - 3)(x - 6)(x + 2)$

29. [6.7] $s = \dfrac{Rg}{g - R}$

30. [6.7] $m = \dfrac{H}{S(t_1 - t_2)}$

31. [6.7] $c = \dfrac{b + 3a}{2}$

32. [6.7] $t_1 = t_2 - \dfrac{A}{vT}$, or $\dfrac{-A + vTt_2}{vT}$ **33.** [6.7] Yes

34. [4.4]

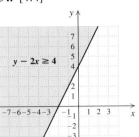

$y - 2x \geq 4$

35. [4.4]

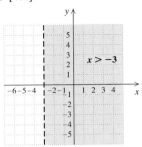

$x > -3$

36. [5.6] $(x^2 - 3)^2$ **37.** [5.5] $(x + 6)(6x - 7)$
38. [5.5] $-6, \frac{7}{6}$ **39.** [2.1] 40
40. [6.2], [6.3], [6.4] ◆ The least common denominator was used to add and subtract rational expressions, to simplify complex rational expressions, and to solve rational equations.   **41.** [6.1], [6.4] ◆ A rational *expression* is a quotient of two polynomials. Expressions can be simplified, multiplied, or added, but they cannot be solved for a variable. A rational *equation* is an equation containing rational expressions. We can often solve for a variable in a rational equation.   **42.** [6.4] All real numbers except 0 and 13   **43.** [6.3], [6.4] 45
44. [6.5] Anna: 56; Franz: 42

TEST, CHAPTER 6

1. [6.1] $\dfrac{3}{4(t + 1)}$ **2.** [6.1] $\dfrac{32x}{(x - 3)(x - 1)}$

3. [6.2] $(x - 3)(x + 11)(x - 9)$ **4.** [6.1] $x = 0$,

$x = 2$ **5.** [6.2] $\dfrac{25x + x^3}{x + 5}$ **6.** [6.2] $3(a - b)$

7. [6.2] $\dfrac{a^3 - a^2b + 4ab + ab^2 - b^3}{(a - b)(a + b)}$

8. [6.2] $\dfrac{3x - 2}{(x - 1)(x + 3)}$ **9.** [6.2] $\dfrac{y - 4}{(y + 3)(y - 2)}$

10. [6.3] $\dfrac{a(2b + 3a)}{5a + b}$ **11.** [6.3] $\dfrac{(x - 9)(x - 6)}{(x + 6)(x - 3)}$

12. [6.4] $-\dfrac{21}{4}$ **13.** [6.4] 15   **14.** [6.1] 5; 0

15. [6.4] $\frac{5}{3}$ **16.** [6.5] $1\frac{31}{32}$ hr

17. [6.6] $\dfrac{4b^2c}{a} - \dfrac{5bc^2}{2a} + 3bc$ **18.** [6.6] $y - 14 + \dfrac{-20}{y - 6}$

19. [6.6] $x^2 + 9x + 40 + \dfrac{153}{x - 4}$

20. [6.6] $(x + 3)(x - 3)(x + 2)(x - 1)$

21. [6.7] $b_1 = \dfrac{2A}{h} - b_2$, or $\dfrac{2A - b_2h}{h}$ **22.** [6.5] 5 and 6;

-6 and -5 **23.** [6.5] $3\frac{3}{11}$ mph   **24.** [5.2] $a^2 + 2a - 2$

25. [5.5] $8(2t + 3)(t - 3)$ **26.** [5.5] $-\frac{3}{2}, 3$

27. [4.4]

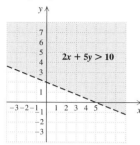

28. [6.4] $-\frac{19}{3}$ **29.** [6.4] $\{x|\ x$ is a real number *and* $x \neq 0$ *and* $x \neq 15\}$ **30.** [6.1], [6.3], [6.4] x-intercept: $(11, 0)$; y-intercept: $\left(0, -\frac{33}{5}\right)$

CUMULATIVE REVIEW, CHAPTERS 1–6

1. [1.2] 10   **2.** [1.4] 5.76×10^9
3. [2.4] Slope: $\frac{7}{4}$; y-intercept: $(0, -3)$
4. [2.6] $y = -\frac{10}{3}x + \frac{11}{3}$ **5.** [3.2] $(-3, 4)$
6. [3.4] $(-2, -3, 1)$ **7.** [3.3] Small: 16; large: 29
8. [3.5] $12, \frac{1}{2}, 7\frac{1}{2}$ **9. (a)** [6.1] $-\frac{1}{2}$; **(b)** [2.7] $\{x|\ x$ is a real number *and* $x \neq 5\}$ **10.** [5.6] $\frac{1}{4}$
11. [5.6] $-\frac{25}{7}, \frac{25}{7}$ **12.** [4.1] $\{x|\ x > -3\}$, or $(-3, \infty)$
13. [4.1] $\{x|\ x \geq -1\}$, or $[-1, \infty)$
14. [4.2] $\{x|\ -10 < x < 13\}$, or $(-10, 13)$
15. [4.2] $\{x|\ x < -\frac{4}{3}$ *or* $x > 6\}$, or $\left(-\infty, -\frac{4}{3}\right) \cup (6, \infty)$
16. [4.3] $\{x|\ x < -6.4$ *or* $x > 6.4\}$, or $(-\infty, -6.4) \cup (6.4, \infty)$ **17.** [4.3] $\left\{x|\ -\frac{13}{4} \leq x \leq \frac{15}{4}\right\}$, or $\left[-\frac{13}{4}, \frac{15}{4}\right]$ **18.** [6.4] $-\frac{5}{3}$ **19.** [6.4] -1
20. [6.4] No solution   **21.** [6.4] $\frac{1}{3}$ **22.** [4.3] $1, \frac{7}{3}$
23. [4.2] $[7, \infty)$ **24.** [2.3] $n = \dfrac{m - 12}{3}$

25. [6.7] $a = \dfrac{Pb}{3 - P}$

26. [4.4]

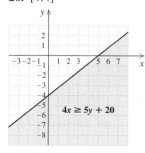

27. [4.4]

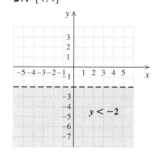

28. [5.1] $-3x^3 + 9x^2 + 3x - 3$ **29.** [5.2] $-15x^4y^4$
30. [5.1] $5a + 5b - 5c$

31. [5.2] $15x^4 - x^3 - 9x^2 + 5x - 2$
32. [5.2] $4x^4 - 4x^2y + y^2$
33. [5.2] $4x^4 - y^2$ **34.** [5.1] $-m^3n^2 - m^2n^2 - 5mn^3$
35. [6.1] $\dfrac{y - 6}{2}$ **36.** [6.1] $x - 1$
37. [6.2] $\dfrac{a^2 + 7ab + b^2}{(a - b)(a + b)}$ **38.** [6.2] $\dfrac{-m^2 + 5m - 6}{(m + 1)(m - 5)}$
39. [6.2] $\dfrac{3y^2 - 2}{3y}$ **40.** [6.3] $\dfrac{y - x}{xy(x + y)}$
41. [6.6] $9x^2 - 13x + 26 + \dfrac{-50}{x + 2}$
42. [5.3] $2x^2(2x + 9)$ **43.** [5.4] $(x - 6)(x + 14)$
44. [5.6] $(4y - 9)(4y + 9)$
45. [5.4] $(x + 13)(x - 2)$ **46.** [5.6] $(t - 8)^2$
47. [5.6] $x^2(x - 1)(x + 1)(x^2 + 1)$
48. [5.4] $(x - 60)(x + 35)$
49. [5.5] $(4x - 1)(5x + 3)$ **50.** [5.5] $(3x + 4)(x - 7)$
51. [5.3] $(x^2 - y)(x^3 + y)$ **52.** [2.7], [5.5] $\{x|\ x$ is a real number *and* $x \neq 2$ *and* $x \neq 5\}$ **53.** [6.5] 1 hr
54. [5.7] 30 ft   **55.** [5.7] All sets of three consecutive even integers satisfy this condition.
56. [2.1] Domain: $\mathbb{R}$; range: $\{3\}$ **57.** [2.1] Domain: $\mathbb{R}$; range: $\mathbb{R}$ **58.** [2.1] Domain: $\mathbb{R}$; range: $[-3, \infty)$
59. [5.1] Domain: $\mathbb{R}$; range: $[-4, \infty)$ **60.** [2.4] (b)
61. [2.1] (a)   **62.** [2.1] (d)   **63.** [6.4] (c)
64. [5.2] $x^3 - 12x^2 + 48x - 64$ **65.** [5.4] $-5, -3, 3, 5$
66. [4.2], [4.3] $\{x|\ -3 \leq x \leq -1$ *or* $7 \leq x \leq 9\}$, or $[-3, -1] \cup [7, 9]$ **67.** [6.4] All real numbers except 9 and -5 **68.** [5.6] $-\frac{1}{4}, 0, \frac{1}{4}$

Chapter 7

INTERACTIVE DISCOVERY, P. 421

1. Domain: $[0, \infty)$; range: $(-\infty, 0]$
2. Domain: $(-\infty, 0]$; range: $[0, \infty)$
3. Domain: $[2, \infty)$; range: $[0, \infty)$
4. Domain: $(-\infty, 2]$; range: $[0, \infty)$
5. Domain: $\mathbb{R}$; range: $[1, \infty)$

INTERACTIVE DISCOVERY, P. 422

(3) and (5) are identities

INTERACTIVE DISCOVERY, P. 424

(1) and (4) are identities.

EXERCISE SET 7.1

1. $4, -4$ **3.** $12, -12$ **5.** $20, -20$ **7.** $7, -7$
9. $-\frac{7}{6}$ **11.** 14   **13.** $-\frac{4}{9}$ **15.** 0.3   **17.** -0.07
19. $p^2 + 4$; 2   **21.** $\dfrac{x}{y + 4}$; 3
23. $\sqrt{20}$; 0; does not exist; does not exist

25. $\sqrt{12}$; does not exist; $\sqrt{30}$; does not exist
27. 1; does not exist; 6   **29.** $5|t|$ **31.** $6|b|$
33. $|5 - b|$ **35.** $|y + 8|$ **37.** $|3x - 5|$ **39.** -4
41. -7 **43.** $-\frac{1}{2}$ **45.** $|y|$ **47.** $7|b|$ **49.** 10
51. $|2a + b|$ **53.** x^6 **55.** $|a^7|$ **57.** $5t$ **59.** $7c$
61. $5 + b$ **63.** $3x + 6$ **65.** -4 **67.** $3x$
69. 10   **71.** $4x$ **73.** a^7 **75.** $(x + 3)^5$
77. 2; 3; -2; -4
79. 2; does not exist; does not exist; 3   **81.** $[5, \infty)$
83. $(-\infty, 5]$ **85.** $(-\infty, \infty)$ **87.** $\left[-\frac{3}{5}, \infty\right)$
89. Domain: $(-\infty, 5]$; range: $[0, \infty)$
91. Domain: $[-1, \infty)$; range: $(-\infty, 1]$
93. Domain: $(-\infty, \infty)$; range: $[5, \infty)$ **95.** (c)   **97.** (d)
99. Yes   **101.** No   **103.** $a^9 b^6 c^{15}$ **105.** $x^2 - 9$
107. $2x^3 - 5x^2 - x + 1$ **109.** ◈   **111.** ◈
113. (a) 13; (b) 15; (c) 18; (d) 20   **115.** $(-4, 5]$

EXERCISE SET 7.2

1. $\sqrt[4]{x}$ **3.** 4   **5.** 3   **7.** $\sqrt[3]{xyz}$ **9.** $\sqrt[5]{a^2 b^2}$
11. $\sqrt[3]{a^2}$ **13.** 8   **15.** 343   **17.** $27\sqrt[4]{x^3}$ **19.** $125x^6$
21. $20^{1/3}$ **23.** $17^{1/2}$ **25.** $x^{3/2}$ **27.** $m^{2/5}$ **29.** $(cd)^{1/4}$
31. $(xy^2 z)^{1/5}$ **33.** $(3mn)^{3/2}$ **35.** $(8x^2 y)^{5/7}$ **37.** $\dfrac{2x}{z^{2/3}}$
39. $\dfrac{1}{x^{1/3}}$ **41.** $\dfrac{1}{(2rs)^{3/4}}$ **43.** $10^{2/3}$ **45.** $a^{5/7}$
47. $\dfrac{2a^{3/4} c^{2/3}}{b^{1/2}}$ **49.** $\left(\dfrac{8yz}{7x}\right)^{3/5}$ **51.** $\dfrac{7x}{z^{1/3}}$ **53.** $\dfrac{5ac^{1/2}}{3}$
55.

$y = (x + 7)^{1/4}$

$Xscl = 5$

57.

$y = (3x - 2)^{1/7}$

59.

$y = x^{3/6}$

$Xscl = 5$

61. 1.552   **63.** 1.778   **65.** -6.240 **67.** $5^{7/8}$
69. $3^{3/4}$ **71.** $10^{6/25}$ **73.** $a^{23/12}$ **75.** $\dfrac{1}{x^{2/7}}$ **77.** $\dfrac{m^{1/3}}{n^{1/8}}$
79. $\sqrt[3]{a}$ **81.** x^5 **83.** $a^5 b^5$ **85.** $\sqrt[4]{3x}$ **87.** $\sqrt{3a}$
89. $\sqrt[8]{x}$ **91.** $a^3 b^3$ **93.** $x^8 y^{20}$ **95.** $\sqrt[12]{xy}$ **97.** $-3, 3$
99. $\frac{1}{3}$ **101.** \$93,500   **103.** ◈   **105.** ◈

107. $\sqrt[10]{x^5 y^3}$ **109.** $\sqrt[4]{2xy^2}$
111. (a) 1.8 m; (b) 3.1 m; (c) 1.5 m; (d) 5.3 m
113. 10 mg

EXERCISE SET 7.3

1. $\sqrt{42}$ **3.** $\sqrt[3]{10}$ **5.** $\sqrt{15ab}$ **7.** $\sqrt[5]{18t^3}$
9. $\sqrt{x^2 - a^2}$ **11.** $\sqrt[3]{0.01x^2}$ **13.** $\sqrt{\dfrac{3x}{5y}}$
15. $\sqrt[7]{\dfrac{5x - 15}{4x + 8}}$ **17.** $\sqrt[6]{5400}$ **19.** $\sqrt[6]{49x^3 y^2}$
21. $\sqrt[6]{x^5 - 4x^4 + 4x^3}$ **23.** $\sqrt[10]{x^9 y^7}$ **25.** $\sqrt[12]{x^{11} y^{10}}$
27. $3\sqrt{3}$ **29.** $2\sqrt{3}$ **31.** $2\sqrt{2}$ **33.** $2\sqrt{11}$
35. $6a^2\sqrt{b}$ **37.** $2x\sqrt[3]{y^2}$ **39.** $-2x^2\sqrt[3]{2}$
41. $f(x) = 5x\sqrt[3]{x^2}$ **43.** $f(x) = |7(x + 5)|$, or $7|x + 5|$
45. $f(x) = |x - 1|\sqrt{5}$ **47.** $ab^2\sqrt{a}$ **49.** $xy^2 z^3\sqrt[3]{x^2 z}$
51. $-2ab^2\sqrt[5]{a^2 b}$ **53.** $ab^2 c\sqrt[5]{ab^2 c^2}$ **55.** $3x^2\sqrt[4]{10x}$
57. $5\sqrt{7}$ **59.** $13\sqrt[3]{y}$ **61.** $7\sqrt{2}$ **63.** $13\sqrt[3]{7} + \sqrt{3}$
65. $21\sqrt{3}$ **67.** $23\sqrt{5}$ **69.** $9\sqrt[3]{2}$ **71.** $(1 + 6a)\sqrt{5a}$
73. $(x + 2)\sqrt[3]{6x}$ **75.** $3\sqrt{a - 1}$ **77.** $(x + 3)\sqrt{x - 1}$
79. $f(x) = -6x\sqrt{5 + x}$ **81.** $f(x) = (x + 3x^2)\sqrt[4]{x - 1}$
83. 8   **85.** $9a^3 b^2 + 5a^2 b^2 - 8a^2 b + 3ab$
87. $(2x + 7)(2x - 7)$ **89.** ◈   **91.** ◈
93. (a) 20 mph; (b) $2\sqrt{450}$ mph; 37.4 mph;
(c) $2\sqrt{350}$ mph; 42.4 mph   **95.** $r^{10} t^3\sqrt{rt}$
97. $ac^2[(3a + 2c)\sqrt{ab} - 2\sqrt[3]{ab}]$ **99.** $(-\infty, \infty)$

EXERCISE SET 7.4

1. $5\sqrt{2}$ **3.** $2\sqrt{21}$ **5.** 2   **7.** $a\sqrt[3]{10}$ **9.** $3x^4\sqrt{2}$
11. $s^2 t^3\sqrt[3]{t}$ **13.** $(x + 5)^2$ **15.** $2ab^3\sqrt[4]{3a}$
17. $x(y + z)^2\sqrt[5]{x}$ **19.** $a\sqrt[4]{a}$ **21.** $b\sqrt[10]{b^9}$
23. $xy\sqrt[6]{xy^5}$ **25.** $3a^2 b\sqrt[4]{ab}$ **27.** $xyz\sqrt[6]{x^5 yz^2}$
29. $9a^2(b + 1)\sqrt[6]{243a^5(b + 1)^5}$ **31.** $\dfrac{5}{6}$
33. $\dfrac{4}{3}$ **35.** $\dfrac{7}{y}$ **37.** $\dfrac{5y\sqrt{y}}{x^2}$ **39.** $\dfrac{3a\sqrt[3]{a}}{2b}$ **41.** $\dfrac{2a}{bc^2}$
43. $\dfrac{ab^2}{c^2}\sqrt[4]{\dfrac{a}{c^2}}$ **45.** $\dfrac{2x}{y^2}\sqrt[5]{\dfrac{x}{y}}$ **47.** $\dfrac{xy}{z^2}\sqrt[6]{\dfrac{y^2}{z^3}}$ **49.** $\sqrt{5}$
51. 3   **53.** $y\sqrt{5y}$ **55.** $2\sqrt[3]{a^2 b}$ **57.** $\sqrt{2ab}$
59. $2x^2 y^3\sqrt[4]{y^3}$ **61.** $\sqrt[12]{a^5}$ **63.** $\sqrt[12]{x^2 y^5}$
65. $\sqrt[10]{ab^9 c^7}$ **67.** $\sqrt[20]{(3x - 1)^3}$ **69.** 8
71. Length: 20; width: 5   **73.** $a_1 = a_2 - \dfrac{m}{A}$, or
$\dfrac{Aa_2 - m}{A}$ **75.** ◈   **77.** ◈
79. (a) 1.62 sec; (b) 1.99 sec; (c) 2.20 sec   **81.** $9\sqrt[3]{9n^2}$
83. $x - y$ **85.** 6

EXERCISE SET 7.5

1. $3\sqrt{7} - 7$ **3.** $\sqrt{6} - \sqrt{10}$ **5.** $2\sqrt{15} - 6\sqrt{3}$
7. -6 **9.** $a + 2a\sqrt[3]{3}$ **11.** 19 **13.** -19
15. $15 - 2\sqrt[3]{10} - \sqrt[3]{100}$ **17.** -6
19. $2\sqrt[3]{9} + 3\sqrt[3]{6} - 2\sqrt[3]{4}$ **21.** $3x + 2\sqrt{3xy} + y$
23. $x\sqrt[6]{xy^5} - \sqrt[15]{x^{13}y^{14}}$
25. $2m^2 + m\sqrt[4]{n} + 2m\sqrt[3]{n^2} + \sqrt[12]{n^{11}}$
27. $(f \cdot g)(x) = \sqrt[4]{2x^2} - x^3$ **29.** $(f \cdot g)(x) = x^2 - 7$
31. $(f \cdot g)(x) = 2 - 3\sqrt{x} + x$ **33.** $27 - 10\sqrt{2}$
35. $8 + 2\sqrt{15}$ **37.** $15 - 10\sqrt{2}$ **39.** $\dfrac{\sqrt{35}}{7}$
41. $\dfrac{\sqrt{14a}}{6}$ **43.** $\dfrac{3\sqrt{5y}}{10xy}$ **45.** $\dfrac{35 + 5\sqrt{2}}{47}$ **47.** $\dfrac{x - \sqrt{xy}}{x - y}$
49. $\dfrac{3 + 6\sqrt{2} + 4\sqrt{15} + 8\sqrt{30}}{-21}$ **51.** $\dfrac{3\sqrt{6} + 4}{2}$
53. $\dfrac{5}{\sqrt{35x}}$ **55.** $\dfrac{2}{\sqrt{6}}$ **57.** $\dfrac{x^2y}{\sqrt{2xy}}$ **59.** $\dfrac{1}{6\sqrt{5} - 12}$
61. $\dfrac{-22}{\sqrt{6} + 5\sqrt{2} + 5\sqrt{3} + 25}$ **63.** $\dfrac{x - y}{x - 2\sqrt{xy} + y}$
65. 6 **67.** $\dfrac{x - 2}{x + 3}$ **69.** ◈
71. $(\sqrt{x} + \sqrt{5})(\sqrt{x} - \sqrt{5})$
73. $(\sqrt{x} + \sqrt{a})(\sqrt{x} - \sqrt{a})$ **75.** $2x - 2\sqrt{x^2 - 4}$
77. $\dfrac{\sqrt[3]{75ac^2}}{5c}$ **79.** $\dfrac{1}{\sqrt{y + 18} + \sqrt{y}}$ **81.** $\dfrac{2a^2}{\sqrt[3]{20ab}}$
83. $\dfrac{(5x + 4y - 3)\sqrt{xy}}{xy}$ **85.** $\dfrac{7\sqrt{3}}{39}$

INTERACTIVE DISCOVERY, P. 456

1. 3; $-3, 3$ **2.** -2; $-2, 2$ **3.** 25; 25
4. No solution; 9

EXERCISE SET 7.6

1. 7 **3.** 12 **5.** 168 **7.** 3 **9.** 19 **11.** $0, 9$
13. 64 **15.** -27 **17.** 125 **19.** No solution
21. 39 **23.** 88 **25.** 5 **27.** $\frac{1}{2}$ **29.** 5 **31.** 7
33. $\frac{80}{9}$ **35.** -1 **37.** 56 **39.** -6 **41.** No solution
43. 1 **45.** 2 **47.** 4 **49.** 2
51.

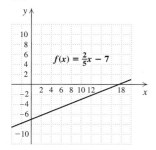

$f(x) = \frac{2}{5}x - 7$

53. ◈   **55.** ◈   **57.** $h = \dfrac{v^2 r}{2gr - v^2}$
59. $22{,}500$ ft   **61.** $-\dfrac{8}{9}$ **63.** $-8, 8$ **65.** $-1, 6$
67. $(2, 0)$ **69.** $(0, 0)$, $\left(\dfrac{125}{4}, 0\right)$

EXERCISE SET 7.7

1. $\sqrt{34}$; 5.831 **3.** $\sqrt{98}$; 9.899 **5.** 5 **7.** $\sqrt{31}$; 5.568
9. $\sqrt{12}$; 3.464 **11.** $\sqrt{n - 1}$ **13.** $\sqrt{325}$ ft; 18.028 ft
15. $\sqrt{14{,}500}$ ft; 120.416 ft   **17.** 20 in.
19. $\sqrt{13{,}000}$ m; 114.018 m
21. $a = 5$; $c = 5\sqrt{2} \approx 7.071$
23. $a = 7$; $b = 7\sqrt{3} \approx 12.124$
25. $a = 5\sqrt{3} \approx 8.660$; $c = 10\sqrt{3} \approx 17.321$
27. $a = \dfrac{13\sqrt{2}}{2} \approx 9.192$; $b = \dfrac{13\sqrt{2}}{2} \approx 9.192$
29. $b = 14\sqrt{3} \approx 24.249$; $c = 28$ **31.** $3\sqrt{3} \approx 5.196$
33. $13\sqrt{2} \approx 18.385$ **35.** $\dfrac{19\sqrt{2}}{2} \approx 13.435$
37. $\sqrt{10{,}561}$ ft; 102.767 ft
39. $2\sqrt{3}$ ft; 3.464 ft   **41.** 8 ft
43. $(0, -4)$ and $(0, 4)$ **45.** 5 **47.** $\sqrt{18}$; 4.243
49. $\sqrt{200}$; 14.142 **51.** 17.8 **53.** $\sqrt{13}$; 3.606
55. $\sqrt{s^2 + t^2}$ **57.** $(2, 5)$; 10 **59.** $(-2, 0)$; 8
61. $(3, -4)$; $\sqrt{7}$ **63.** $x^2 + (y - 3)^2 = 36$
65. $(x + 5)^2 + (y + 7)^2 = 1$
67. $(x - 5)^2 + (y - 7)^2 = 3$ **69.** $3, 8$ **71.** $\left\{4, -\frac{2}{3}\right\}$
73. ◈   **75.** $\sqrt{75}$ cm   **77.** 4 gal. This assumes that the
total area of the doors and windows is at least 164 ft^2.

EXERCISE SET 7.8

1. $5i$ **3.** $i\sqrt{13}$, or $\sqrt{13}i$ **5.** $3i\sqrt{2}$, or $3\sqrt{2}i$
7. $i\sqrt{3}$, or $\sqrt{3}i$ **9.** $9i$ **11.** $10i\sqrt{3}$, or $10\sqrt{3}i$
13. $-7i$ **15.** $4 - 2i\sqrt{15}$, or $4 - 2\sqrt{15}i$
17. $(2 + 2\sqrt{3})i$ **19.** $9 + 5i$ **21.** $3 + 11i$
23. $4 + 5i$ **25.** $-1 - 5i$ **27.** $5 + 5i$ **29.** -30
31. 63 **33.** -35 **35.** $-\sqrt{42}$
37. $-5\sqrt{6}$ **39.** $-6 + 14i$ **41.** $-20 - 24i$
43. $-7 + 26i$ **45.** $40 + 13i$ **47.** $8 + 31i$
49. $7 - 61i$ **51.** $-15 - 30i$ **53.** $39 - 37i$
55. $21 - 20i$ **57.** $12 + 16i$ **59.** $21 + 20i$
61. $\frac{14}{5} + \frac{7}{5}i$ **63.** $\frac{6}{29} + \frac{15}{29}i$ **65.** $-\frac{8}{9}i$ **67.** $-\frac{1}{3} - \frac{7}{6}i$
69. $-\frac{23}{58} + \frac{43}{58}i$ **71.** $\frac{6}{25} - \frac{17}{25}i$ **73.** $-i$ **75.** 1
77. -1 **79.** i **81.** -1 **83.** $-125i$ **85.** 0 **87.** 0
89. $-5, 2$ **91.** $-\frac{4}{3}, 7$ **93.** ◈   **95.** ◈
97. $-4 - 8i$; $-2 + 4i$; $8 - 6i$ **99.** 0
101. $-1 - \sqrt{5}i$ **103.** $-\frac{2}{3}i$

REVIEW EXERCISES, CHAPTER 7

1. [7.1] $\frac{7}{6}$ **2.** [7.1] 0.5 **3.** [7.1] 5

4. [7.1] $\left[\frac{7}{2}, \infty\right)$ **5.** [7.1] $[0, \infty)$ **6.** [7.1] $7|a|$

7. [7.1] $|c + 8|$ **8.** [7.1] $|x - 3|$ **9.** [7.1] $|2x + 1|$

10. [7.1] -2 **11.** [7.4] $-\frac{4x^2}{3}$ **12.** [7.1] $2x^2$

13. [7.2] $(5ab)^{4/3}$, or $5^{4/3}a^{4/3}b^{4/3}$ **14.** [7.2] $8a^6$

15. [7.2] $x^3 y^5$ **16.** [7.2] $a^3 b \sqrt[3]{ab^2}$ **17.** [7.2] $\frac{1}{x^{2/5}}$

18. [7.2] $7^{1/6}$ **19.** [7.1] $f(x) = 5|x - 3|$

20. [7.3] $\sqrt{15xy}$ **21.** [7.4] $3a\sqrt[3]{a^2 b^2}$

22. [7.3] $\sqrt[15]{a^{14}b^{11}}$ **23.** [7.4] $y\sqrt[3]{6}$ **24.** [7.4] $\frac{5\sqrt{x}}{2}$

25. [7.4] $\sqrt[12]{x^5}$ **26.** [7.3] $7\sqrt[3]{x}$ **27.** [7.3] $3\sqrt{3}$

28. [7.3] $(2x + y^2)\sqrt[3]{x}$ **29.** [7.3] $15\sqrt{2}$

30. [7.5] $-43 - 2\sqrt{10}$ **31.** [7.5] $a^2 - 2a\sqrt{2} + 2$

32. [7.5] $\frac{5a - 5\sqrt{ab}}{a - b}$ **33.** [7.5] $\frac{5a}{a + \sqrt{ab}}$

34. [7.6] 4 **35.** [7.6] 14 **36.** [7.7] 9 cm

37. [7.7] $\sqrt{24}$ ft; 4.899 ft

38. [7.7] $a = 10$; $b = 10\sqrt{3} \approx 17.321$

39. [7.7] $\sqrt{13}$; 3.606 **40.** [7.7] $(12, -4)$; 5

41. [7.8] $-2i\sqrt{2}$, or $-2\sqrt{2}i$ **42.** [7.8] $-2 - 9i$

43. [7.8] $1 + i$ **44.** [7.8] 29 **45.** [7.8] i

46. [7.8] $9 - 12i$ **47.** [7.8] $\frac{13}{25} - \frac{34}{25}i$

48. [5.7] $12, 13, 14$ **49.** [6.4] -5 **50.** [5.5] $-\frac{9}{2}, 3$

51. [6.1] $\frac{x(x + 2)}{(x + y)(x - 3)}$

52. [7.1] ◈ An absolute-value sign must be used to simplify $\sqrt[n]{x^n}$ when n is even, since x may be negative. If x is negative while n is even, the radical expression cannot be simplified to x, since $\sqrt[n]{x^n}$ represents the principal, or positive, root. When n is odd, there is only one root, and it will be positive or negative depending on the sign of x. Thus there is no absolute-value sign when n is odd.
53. [7.8] ◈ Every real number is a complex number, but there are complex numbers that are not real. A complex number $a + bi$ is not real if $b \neq 0$.
54. [7.6] 3 **55.** [7.8] $-\frac{2}{5} + \frac{9}{10}i$

TEST, CHAPTER 7

1. [7.3] $5\sqrt{3}$ **2.** [7.4] $-\frac{2}{x^2}$ **3.** [7.1] $10|a|$

4. [7.1] $|x - 4|$ **5.** [7.3] $x^2 y\sqrt[5]{x^2 y^3}$

6. [7.4] $\left|\frac{5x}{6y^2}\right|$, or $\frac{5|x|}{6y^2}$ **7.** [7.3] $\sqrt[3]{10xy^2}$

8. [7.4] $\sqrt[5]{x^2 y^2}$ **9.** [7.4] $xy\sqrt[4]{x}$

10. [7.4] $\sqrt[20]{a^3}$ **11.** [7.3] $5\sqrt{2}$ **12.** [7.3] $(x^2 + 3y)\sqrt{y}$
13. [7.5] $14 - 19\sqrt{x} - 3x$
14. [7.1], Domain: $(-\infty, 2]$; range: $[0, \infty)$
15. [7.5] $27 + 10\sqrt{2}$ **16.** [7.5] $-\frac{3\sqrt{2} + 10}{41}$

17. [7.6] 7
18. [7.7] Longer leg: $10\sqrt{3}$ cm ≈ 17.321 cm; hypotenuse: 20 cm **19.** [7.7] $\sqrt{10,600}$ ft; 102.956 ft
20. [7.7] $\sqrt{35.92}$; 5.993 **21.** [7.7] $(x - 5)^2 + y^2 = 11$
22. [7.8] $5i\sqrt{2}$, or $5\sqrt{2}i$ **23.** [7.8] $10 + 2i$
24. [7.8] -24 **25.** [7.8] $15 - 8i$ **26.** [7.8] $-\frac{13}{53} - \frac{19}{53}i$

27. [7.8] i **28.** [5.5] $-\frac{1}{3}, \frac{5}{2}$ **29.** [6.1] $\frac{x - 3}{x - 4}$

30. [6.4] No solution **31.** [5.7] 16 and 18; -18 and -16
32. [7.6] 3 **33.** [7.8] $-\frac{17}{4}i$

Chapter 8

INTERACTIVE DISCOVERY, P. 488

1. 1 **2.** 1 **3.** 1 **4.** 0 **5.** 0 **6.** 2

EXERCISE SET 8.1

1. 2 **3.** 1 **5.** 0 **7.** $\pm\sqrt{3}$ **9.** $\pm\frac{2}{5}i$

11. $\pm\sqrt{\frac{3}{2}}$ **13.** $-9, 5$ **15.** $-5 \pm 2\sqrt{2}$ **17.** $7 \pm 2i$

19. $\frac{-3 \pm \sqrt{14}}{2}$ **21.** $-7, 13$ **23.** $3, 11$ **25.** $3 \pm \sqrt{13}$

27. $-13, -1$ **29.** $x^2 + 10x + 25, (x + 5)^2$
31. $x^2 - 6x + 9, (x - 3)^2$ **33.** $x^2 + 9x + \frac{81}{4}, \left(x + \frac{9}{2}\right)^2$
35. $x^2 - 3x + \frac{9}{4}, \left(x - \frac{3}{2}\right)^2$ **37.** $x^2 + \frac{2}{3}x + \frac{1}{9}, \left(x + \frac{1}{3}\right)^2$
39. $x^2 - \frac{5}{6}x + \frac{25}{144}, \left(x - \frac{5}{12}\right)^2$ **41.** $-7, 1$ **43.** $-5, -1$
45. $3, 7$ **47.** $-2 \pm \sqrt{3}$ **49.** $-3 \pm 2i$ **51.** $-\frac{1}{2}, 3$

53. $-\frac{3}{2}, -\frac{1}{2}$ **55.** $-\frac{3}{2}, \frac{5}{3}$ **57.** $-1 \pm \sqrt{\frac{1}{2}}$ **59.** 10%

61. 18.75% **63.** 4% **65.** About 10.7 sec
67. About 6.3 sec
69.

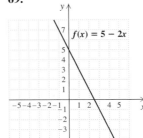

71. $3\sqrt[3]{10}$ **73.** 5   **75.** ◆   **77.** ◆   **79.** ± 18
81. $-\frac{7}{2}, -\sqrt{5}, 0, \sqrt{5}, 8$ **83.** Barge: 8 km/h;
fishing boat: 15 km/h

EXERCISE SET 8.2

1. $\dfrac{-7 \pm \sqrt{33}}{2}$ **3.** $-\dfrac{5}{3}, -1$ **5.** $\dfrac{1 \pm i\sqrt{7}}{2}$ **7.** $3 \pm 2i$

9. $3 \pm \sqrt{5}$ **11.** $\dfrac{-4 \pm \sqrt{19}}{3}$ **13.** $-1, 0$ **15.** $-\dfrac{9}{14}, 0$

17. $\frac{2}{5}$ **19.** $-2, -\frac{3}{4}$ **21.** 5, 10   **23.** $\dfrac{13 \pm \sqrt{509}}{10}$

25. $2 \pm i\sqrt{5}$ **27.** $\frac{2}{3}, \frac{3}{2}$ **29.** $\dfrac{5 \pm \sqrt{37}}{6}$

31. $5 \pm \sqrt{53}$ **33.** $\dfrac{7 \pm \sqrt{85}}{2}$ **35.** Negative   **37.** Zero

39. Two rational   **41.** Two imaginary
43. Two irrational   **45.** One rational
47. Two imaginary   **49.** Two rational
51. Two rational   **53.** Two rational   **55.** Two irrational
57. Two imaginary   **59.** Two irrational
61. Kenyan: 30 lb; Peruvian: 20 lb
63.

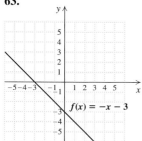

$f(x) = -x - 3$

65. ◆   **67.** ◆
69. $\left(\dfrac{-b + \sqrt{b^2 - 4ac}}{2a}\right)\left(\dfrac{-b - \sqrt{b^2 - 4ac}}{2a}\right) =$

$\dfrac{b^2 - (b^2 - 4ac)}{4a^2} = \dfrac{4ac}{4a^2} = \dfrac{c}{a}$ **71. (a)** 2; **(b)** $1 - i$
73. -1 **75.** ◆   **77.** $(-2, 0), (1, 0)$
79. $-0.42539053, 1.17539053$ **81.** $\dfrac{-5\sqrt{2} \pm \sqrt{34}}{4}$
83. $-i \pm i\sqrt{1 - i}$

EXERCISE SET 8.3

1. First part: 60 mph; second part: 50 mph   **3.** 40 mph
5. Cessna: 150 mph, Beechcraft: 200 mph; or Cessna:
200 mph, Beechcraft: 250 mph
7. To Hillsboro: 10 mph; return trip: 4 mph
9. About 11 mph   **11.** 12 hr   **13.** About 9.34 km/h
15. $r = \dfrac{1}{2}\sqrt{\dfrac{A}{\pi}}$ **17.** $r = \dfrac{-\pi h + \sqrt{\pi^2 h^2 + 2\pi A}}{2\pi}$

19. $s = \sqrt{\dfrac{kQ_1Q_2}{N}}$ **21.** $g = \dfrac{4\pi^2 l}{T^2}$

23. $c = \sqrt{d^2 - a^2 - b^2}$ **25.** $t = \dfrac{-v_0 + \sqrt{v_0^2 + 2gs}}{g}$

27. $n = \dfrac{1 + \sqrt{1 + 8N}}{2}$ **29.** $h = \dfrac{V^2}{12.25}$

31. $r = -1 + \dfrac{-P_2 + \sqrt{P_2^2 + 4AP_1}}{2P_1}$ **33. (a)** 10.1 sec;
(b) 7.49 sec; **(c)** 272.5 m   **35.** 2.9 sec   **37.** 0.87 sec
39. 2.5 m/sec   **41.** 7%   **43.** 1, 5   **45.** $2y\sqrt[3]{9x^2}$

47. ◆   **49.** ◆   **51.** \$2.50   **53.** $l = \dfrac{w + w\sqrt{5}}{2}$

55. $c = \dfrac{vm}{\sqrt{m^2 - m_0^2}}$ **57.** $A(S) = \dfrac{\pi S}{6}$

EXERCISE SET 8.4

1. $\pm 1, \pm 2$ **3.** $\pm\sqrt{3}, \pm 3$ **5.** $\pm\dfrac{\sqrt{3}}{2}, \pm 2$

7. $9 + 4\sqrt{5}$ **9.** $\pm 2\sqrt{2}, \pm 3$ **11.** No solution
13. $-\frac{1}{2}, \frac{1}{3}$ **15.** $-\frac{4}{5}, 1$ **17.** $-27, 8$
19. 729   **21.** 1   **23.** No solution   **25.** $\frac{12}{5}$

27. $\left(\frac{4}{25}, 0\right)$ **29.** $\left(\dfrac{3 + \sqrt{33}}{2}, 0\right), \left(\dfrac{3 - \sqrt{33}}{2}, 0\right),$

$(4, 0), (-1, 0)$ **31.** $(-243, 0), (32, 0)$

33. $\left(\dfrac{9 + \sqrt{89}}{2}, 0\right), \left(\dfrac{9 - \sqrt{89}}{2}, 0\right), (-1 + \sqrt{3}, 0),$

$(-1 - \sqrt{3}, 0)$ **35.** $x^2 + 4x - 21 = 0$
37. $x^2 - 6x + 9 = 0$ **39.** $3x^2 - 14x + 8 = 0$
41. $6x^2 - 5x + 1 = 0$ **43.** $x^2 - 0.08x - 0.84 = 0$
45. $x^2 - 7 = 0$ **47.** $x^2 - 18 = 0$ **49.** $x^2 + 9 = 0$
51. $x^2 - 10x + 29 = 0$ **53.** $x^2 - 4x - 6 = 0$
55. $x^3 - x^2 - 12x = 0$ **57.** $x^3 - 2x^2 - x + 2 = 0$
59. $3x^2\sqrt{x}$
61. (a) $n(t) = 12.5308690523x - 11.61122034342;$

(b) 1957   **63.** ◆   **65.** ◆   **67.** $\pm\sqrt{\dfrac{7 \pm \sqrt{29}}{10}}$

69. $-2, -1, 5, 6$ **71.** $\dfrac{432}{143}$ **73.** 9   **75.** $-2, 1$
77. $a = 8, b = 20, c = -12$
79. $x^4 - 8x^3 + 21x^2 - 2x - 52 = 0$
81. $a = 1, b = 2, c = -3$

EXERCISE SET 8.5

1. $k = 4; y = 4x$ **3.** $k = 1.7; y = 1.7x$ **5.** $k = \frac{15}{4};$
$y = \frac{15}{4}x$ **7.** $k = 1.6; y = 1.6x$ **9.** 6 amperes
11. 241,920,000   **13.** 7,700,000 tons

15. $k = 60$; $y = \dfrac{60}{x}$ **17.** $k = 12$; $y = \dfrac{12}{x}$

19. $k = 36$; $y = \dfrac{36}{x}$ **21.** $k = 9$; $y = \dfrac{9}{x}$ **23.** 27 min

25. 160 cm^3 **27.** Direct   **29.** Neither   **31.** Inverse

33. $y = \dfrac{2}{3}x^2$ **35.** $y = \dfrac{54}{x^2}$ **37.** $y = 0.5xz$

39. $y = 0.3xz^2$ **41.** $y = \dfrac{4wx^2}{z}$ **43.** 36 mph

45. 2.5 m   **47.** 1600 km   **49.** About 57.42 mph

51. $y = -\dfrac{2}{3}x - 5$ **53.** $\dfrac{c - 2a}{3c + 4a}$ **55.** 9   **57.** ◆

59. ◆   **61.** y is multiplied by 8.   **63.** W varies jointly as m_1 and M_1 and inversely as the square of d.

65. (a) $k \approx 0.001$; $N = \dfrac{0.001 P_1 P_2}{d^2}$; (b) 1137 km

67. $d(s) = \dfrac{28}{s}$; 70 yd

INTERACTIVE DISCOVERY, P. 535

1. (a) $(0, 0)$; (b) $x = 0$; (c) up; (d) $[0, \infty)$; (e) minimum;
(f) narrower   **2.** (a) $(0, 0)$; (b) $x = 0$; (c) up; (d) $[0, \infty)$;
(e) minimum; (f) wider   **3.** (a) $(0, 0)$; (b) $x = 0$; (c) up;
(d) $[0, \infty)$; (e) minimum; (f) wider   **4.** (a) $(0, 0)$;
(b) $x = 0$; (c) down; (d) $(-\infty, 0]$; (e) maximum; (f) neither
narrower nor wider   **5.** (a) $(0, 0)$; (b) $x = 0$; (c) down;
(d) $(-\infty, 0]$; (e) maximum; (f) narrower   **6.** (a) $(0, 0)$;
(b) $x = 0$; (c) down; (d) $(-\infty, 0]$; (e) maximum;
(f) wider

INTERACTIVE DISCOVERY, P. 537

1. (a) $(3, 0)$; (b) $x = 3$; (c) same shape; (d) $[0, \infty)$;
(e) minimum of 0 occurs when $x = 3$ **2.** (a) $(-4, 0)$;
(b) $x = -4$; (c) same shape; (d) $[0, \infty)$; (e) minimum of
0 occurs when $x = -4$ **3.** (a) $(-1, 0)$; (b) $x = -1$;
(c) same shape; (d) $[0, \infty)$; (e) minimum of 0 occurs when
$x = -1$ **4.** (a) $(2, 0)$; (b) $x = 2$; (c) same shape;
(d) $(-\infty, 0]$; (e) maximum of 0 occurs when $x = 2$

5. (a) $\left(\frac{3}{2}, 0\right)$; (b) $x = \frac{3}{2}$; (c) same shape; (d) $[0, \infty)$;
(e) minimum of 0 occurs when $x = \frac{3}{2}$ **6.** (a) $(-2, 0)$;
(b) $x = -2$; (c) same shape; (d) $(-\infty, 0]$; (e) maximum of
0 occurs when $x = -2$

INTERACTIVE DISCOVERY, P. 538

1. (a) The value of g is 10 more than the value of f for the
same choice for x. (b) The axis of symmetry is $x = 2$ for
both graphs; the vertex is $(2, 0)$ for f and $(2, 10)$ for g; the
range of f is $[0, \infty)$ and the range of g is $[10, \infty)$.
2. (a) The value of g is 5 less than the value of f for the
same choice for x. (b) The axis of symmetry is $x = 2$ for
both graphs; the vertex is $(2, 0)$ for f and $(2, -5)$ for g;
the range of f is $[0, \infty)$ and the range of g is $[-5, \infty)$.

3. (a) The value of g is 4 more than the value of f for the
same choice for x. (b) The axis of symmetry is $x = -1$ for
both graphs; the vertex is $(-1, 0)$ for f and $(-1, 4)$ for g;
the range of f is $(-\infty, 0]$ and the range of g is $(-\infty, 4]$.

EXERCISE SET 8.6

1. (a) Positive; (b) $(3, 1)$; (c) $x = 3$; (d) $[1, \infty)$
3. (a) Negative; (b) $(-2, -3)$; (c) $x = -2$; (d) $(-\infty, -3]$
5. (a) Positive; (b) $(-3, 0)$; (c) $x = -3$; (d) $[0, \infty)$
7. (f)   **9.** (e)   **11.** (d)

13.

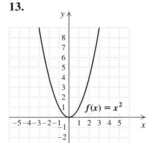

15.

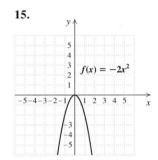

17.

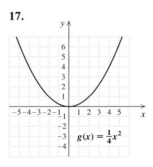

19.

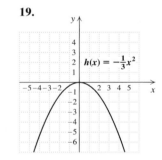

21.

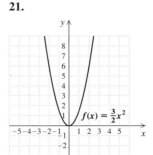

23. Vertex: $(-1, 0)$; axis of symmetry: $x = -1$

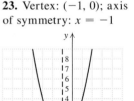

25. Vertex: (2, 0);
axis of symmetry: $x = 2$

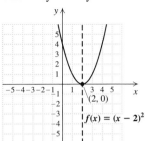

27. Vertex: (−4, 0);
axis of symmetry: $x = -4$

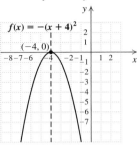

41. Vertex: (−4, 1);
axis of symmetry: $x = -4$;
minimum: 1

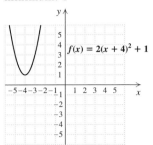

43. Vertex: (1, 2);
axis of symmetry: $x = 1$;
maximum: 2

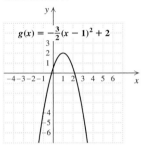

29. Vertex: (−1, 0);
axis of symmetry: $x = -1$

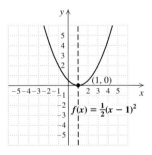

31. Vertex: (3, 0);
axis of symmetry: $x = 3$

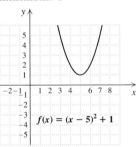

45. Vertex: (9, 5);
axis of symmetry: $x = 9$;
minimum: 5; range: $[5, \infty)$
47. Vertex: (−6, 11);
axis of symmetry: $x = -6$;
maximum: 11; range: $(-\infty, 11]$
49. Vertex: $\left(-\frac{1}{4}, -13\right)$;

axis of symmetry: $x = -\frac{1}{4}$;
minimum: −13; range: $[-13, \infty)$
51. Vertex: (−4.58, 65π);
axis of symmetry: $x = -4.58$;
minimum: 65π; range: $[65\pi, \infty)$
53. (−5, −1)   **55.** $x^2 + 5x + \frac{25}{4}$ **57.** ◈   **59.** ◈
61. $f(x) = 2(x - 5)^2$ **63.** $g(x) = -2(x + 4)^2$
65. $g(x) = -2(x - 3)^2 + 8$ **67.** $F(x) = 3(x - 5)^2 + 1$
69. $F(x) = -\frac{1}{2}(x + 3)^2 - 4$
71.

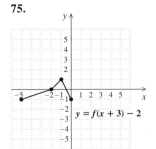

73.

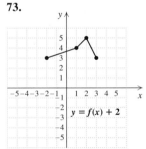

33. Vertex: (1, 0);
axis of symmetry: $x = 1$

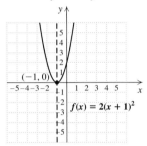

35. Vertex: (5, 1);
axis of symmetry: $x = 5$;
minimum: 1

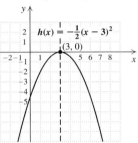

37. Vertex: (−1, −2);
axis of symmetry: $x = -1$;
minimum: −2

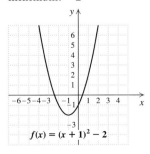

39. Vertex: (1, −3);
axis of symmetry: $x = 1$;
maximum: −3

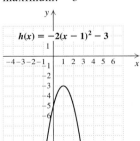

75.

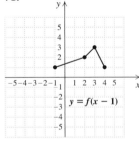

1. (a) $f(x) = (x - 2)^2 + 1$; **(b)** vertex: (2, 1); axis of symmetry: $x = 2$ **3. (a)** $f(x) = -\left(x - \frac{3}{2}\right)^2 - \frac{31}{4}$; **(b)** vertex: $\left(\frac{3}{2}, -\frac{31}{4}\right)$; axis of symmetry: $x = \frac{3}{2}$

5. (a) $f(x) = 2\left(x - \frac{7}{4}\right)^2 - \frac{41}{8}$; **(b)** vertex: $\left(\frac{7}{4}, -\frac{41}{8}\right)$; axis of symmetry: $x = \frac{7}{4}$

7. (a) Vertex: $(-1, -6)$; axis of symmetry: $x = -1$; **(b)**

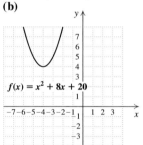

$f(x) = x^2 + 2x - 5$

9. (a) Vertex: $(-4, -4)$; axis of symmetry: $x = -4$; **(b)**

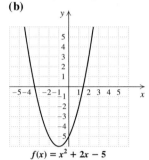

$f(x) = x^2 + 8x + 20$

11. (a) Vertex: $(-4, -7)$; axis of symmetry: $x = -4$; **(b)**

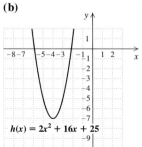

$h(x) = 2x^2 + 16x + 25$

13. (a) Vertex: (1, 6); axis of symmetry: $x = 1$; **(b)**

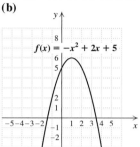

$f(x) = -x^2 + 2x + 5$

15. (a) Vertex: $\left(-\frac{7}{2}, -\frac{53}{4}\right)$; axis of symmetry: $x = -\frac{7}{2}$; **(b)**

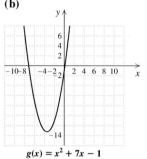

$g(x) = x^2 + 7x - 1$

17. (a) Vertex: $\left(\frac{9}{2}, -\frac{81}{4}\right)$; axis of symmetry: $x = \frac{9}{2}$; **(b)**

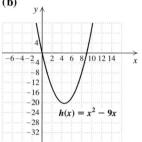

$h(x) = x^2 - 9x$

19. (a) Vertex: $(0, -6)$; axis of symmetry: $x = 0$; **(b)**

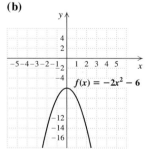

$f(x) = -2x^2 - 6$

21. (a) Vertex: $\left(\frac{5}{6}, \frac{1}{12}\right)$; axis of symmetry: $x = \frac{5}{6}$;
(b)

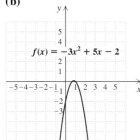

$f(x) = -3x^2 + 5x - 2$

23. (a) Vertex: $\left(-4, -\frac{5}{3}\right)$; axis of symmetry: $x = -4$;
(b)

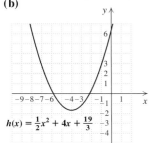

$h(x) = \frac{1}{2}x^2 + 4x + \frac{19}{3}$

25. $(-0.5, -6.25)$ **27.** $(0.1, 0.95)$ **29.** $(3.5, -4.25)$
31. $(3 - \sqrt{6}, 0), (3 + \sqrt{6}, 0); (0, 3)$
33. $(-1, 0), (3, 0); (0, 3)$ **35.** No x-intercepts; $(0, 4)$
37. $(2, 0); (0, -4)$ **39.** $\left(\frac{3 - \sqrt{6}}{2}, 0\right), \left(\frac{3 + \sqrt{6}}{2}, 0\right)$;
$(0, 3)$ **41. (a)** Minimum value: -6.95; **(b)** $(-1.06, 0)$,
$(2.41, 0); (0, -5.89)$ **43. (a)** Maximum value: -0.45;
(b) no x-intercepts; $(0, -2.79)$ **45.** 5   **47.** $n = \dfrac{3Q}{A - T}$
49. ◈   **51.** ◈   **53. (a)** $-3, 1$; **(b)** $-4.5, 2.5$;
(c) $-5.5, 3.5$ **55.** $f(x) = 3\left[x - \left(-\frac{m}{6}\right)\right]^2 + \dfrac{11m^2}{12}$
57. $f(x) = -0.28x^2 - 0.56x + 6.72$
59.

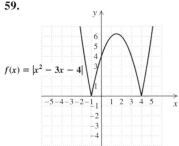

$f(x) = |x^2 - 3x - 4|$

1. 14 ft by 14 ft; 196 ft^2 **3.** -9; -3 and 3
5. 36; -6 and -6 **7.** 450 ft^2; 15 ft by 30 ft (The house
serves as a 30-ft side.)   **9.** 3.5 in.   **11.** 3.5 hundred, or
350   **13.** $P(x) = -x^2 + 980x - 3000$; \$237,100 at
$x = 490$ **15.** Quadratic   **17.** Neither   **19.** Quadratic
21. $f(x) = 2x^2 + 3x - 1$ **23.** $f(x) = -\frac{1}{4}x^2 + 3x - 5$
25. (a) $A(s) = \frac{3}{16}s^2 - \frac{135}{4}s + 1750$; **(b)** about 531
27. $h(d) = -0.0068d^2 + 0.8571d$
29. (a) $f(x) = -\frac{145}{6}x^2 + \frac{1015}{6}x + 35$; **(b)** 277 metric tons;
(c) The regression function gives a closer prediction.
31. (a) $b(x) = 0.1142857143x^2 - 0.4851428571x +$
1.522571429; **(b)** 2.726 per 10,000 passengers
33. $f(x) = -3.820651428x^2 + 53.15961719x +$
499.7568332 **35.** $\dfrac{x - 9}{(x + 9)(x + 7)}$ **37.** ◈
39. The radius of the circular portion of the window and
the height of the rectangular portion should each be
$\dfrac{24}{\pi + 4}$ ft, or approximately 3.36 ft.   **41.** 30
43. 78.4 ft

1. $\left[-4, \frac{3}{2}\right]$ **3.** $(-\infty, -2) \cup (0, 2) \cup (3, \infty)$
5. $\left(-\infty, -\frac{7}{2}\right) \cup (-2, \infty)$ **7.** $(-\infty, -4) \cup (3, \infty)$
9. $[-7, 2]$ **11.** $(-1, 2)$ **13.** $(-\infty, -3] \cup [3, \infty)$
15. $(-\infty, \infty)$ **17.** $(-2, 6)$ **19.** $(-\infty, -2) \cup (0, 2)$
21. $(-3, -1) \cup (2, \infty)$ **23.** $(-\infty, -3) \cup (-2, 1)$
25. $[-0.78, 1.59]$ **27.** $(-\infty, -2) \cup (1, 3)$
29. $(-\infty, -7)$ **31.** $(-\infty, -1] \cup (3, \infty)$ **33.** $\left[-\frac{2}{3}, 3\right]$
35. $\left(\frac{3}{2}, 4\right)$ **37.** $(-\infty, -1] \cup [2, 5)$
39. $(-\infty, -3) \cup [0, \infty)$ **41.** $(0, \infty)$
43. $(-\infty, -4) \cup [1, 3)$ **45.** $\left(0, \frac{1}{4}\right)$ **47.** $\sqrt[15]{a^{11}b^{13}}$
49. ◈   **51.** $(-\infty, -1 - \sqrt{5}) \cup (-1 + \sqrt{5}, \infty)$, or
$(-\infty, -3.24) \cup (1.24, \infty)$ **53.** $\{0\}$ **55. (a)** $(10, 200)$,
or $\{x \mid 10 < x < 200\}$; **(b)** $[0, 10) \cup (200, \infty)$, or
$\{x \mid 0 \le x < 10 \text{ or } x > 200\}$ **57.** $\{n \mid 12 \le n \le 25\}$
59. $f(x)$ has no zeros; $f(x) < 0$ for $(-\infty, 0)$; $f(x) \ge 0$ for
$(0, \infty)$ **61.** $f(x) = 0$ for $x = -2, 1, 2, 3$; $f(x) < 0$ for
$(-2, 1) \cup (2, 3)$; $f(x) \ge 0$ for $(-\infty, -2] \cup [1, 2] \cup [3, \infty)$
63. (a) $p(x) = 34.04540449x^2 - 1242.727373x +$
22659.5843; **(b)** $\{x \mid x \le 1972 \text{ or } x \ge 2004\}$

1. (a) [8.1] 2; **(b)** [8.6] positive; **(c)** [8.6] -3
2. [8.1] $\pm\dfrac{\sqrt{14}}{2}$ **3.** [8.1] $-\dfrac{5}{14}, 0$ **4.** [8.1] 3, 9
5. [8.2] $\dfrac{5 \pm i\sqrt{11}}{5}$ **6.** [8.2] 3, 5

7. [8.2] $\dfrac{5 \pm \sqrt{33}}{2}$; $-0.37, 5.37$ **8.** [8.2] $-\frac{1}{4}, 1$

9. [8.1] $x^2 - 12x + 36$; $(x - 6)^2$

10. [8.1] $x^2 + \frac{3}{5}x + \frac{9}{100}$; $\left(x + \frac{3}{10}\right)^2$ **11.** [8.1] $-4, 2$

12. [8.1] $3 \pm 2\sqrt{2}$ **13.** [8.1] 10%   **14.** [8.1] 6.7 sec

15. [8.2] Two irrational   **16.** [8.2] Two imaginary

17. [8.3] About 2.89 mph   **18.** [8.3] 6 hr

19. [8.4] $(-3, 0), (-2, 0), (2, 0), (3, 0)$ **20.** [8.4] $-5, 3$

21. [8.4] $\pm\sqrt{2}, \pm\sqrt{7}$ **22.** [8.4] $x^2 + 25 = 0$

23. [8.4] $x^2 + 8x + 16 = 0$ **24.** [8.5] 5 amperes

25. [8.5] 64 L   **26.** [8.5] $y = \dfrac{15xz}{2w}$

27. [8.6]

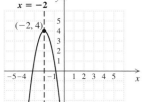

$x = -2$

$(-2, 4)$

$f(x) = -3(x + 2)^2 + 4$
Maximum: 4

28. [8.7] **(a)** Vertex: (3, 5), axis of symmetry: $x = 3$; **(b)**

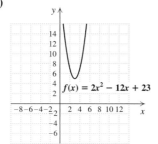

$f(x) = 2x^2 - 12x + 23$

29. [8.7] $(2, 0), (7, 0)$; $(0, 14)$ **30.** [8.3] $p = \dfrac{9\pi^2}{N^2}$

31. [8.3] $T = \dfrac{1 \pm \sqrt{1 + 24A}}{6}$

32. [8.8] -121; 11 and -11

33. [8.8] $f(x) = -x^2 + 6x - 2$ **34.** [8.8] Yes

35. [8.8] No   **36.** [8.8] $f(x) = 0.2727272727x^2 - 3.406060606x + 38.45454545$

37. [8.9] $(-1, 0) \cup (3, \infty)$ **38.** [8.9] $(-3, 5]$

39. [3.3] A: 210 kg; B: 90 kg   **40.** [7.6] 2

41. [6.2] $\dfrac{x + 1}{(x - 3)(x - 1)}$ **42.** [7.4] $3t^5 s\sqrt[3]{s}$

43. [8.7], [8.8] ◈ The x-coordinate of the maximum or minimum point lies halfway between the x-coordinates of the x-intercepts.   **44.** [8.2] ◈ Yes. If the discriminant is a perfect square, then the solutions are rational numbers, p/q and r/s. (Note that if the discriminant is 0, then $p/q = r/s$.) Then the equation can be written in factored form, $(qx - p)(sx - r) = 0$.   **45.** [8.4] ◈ Four; let $u = x^2$. Then $au^2 + bu + c = 0$ has at most two solutions, $u = m$ or $u = n$. Now substitute x^2 for u and obtain $x^2 = m$ or $x^2 = n$. These equations yield the solutions $x = \pm\sqrt{m}$ and $x = \pm\sqrt{n}$. When $m \neq n$, the maximum number of solutions, four, occurs. Also, since the degree of the polynomial is 4, the equation can have at most four roots.   **46.** [8.1], [8.2], [8.7] ◈ Completing the

square was used to solve quadratic equations and to graph quadratic functions by rewriting the function in the form $f(x) = a(x - h)^2 + k$.   **47.** [8.7] $f(x) = \frac{7}{15}x^2 - \frac{14}{15}x - 7$

48. [8.2] $h = 60, k = 60$ **49.** [8.4] 18, 324

TEST, CHAPTER 8

1. **(a)** [8.1] 0; **(b)** [8.6] negative; **(c)** [8.6] -1

2. [8.1] $\pm\dfrac{4}{\sqrt{3}}$ **3.** [8.2] 2, 9   **4.** [8.2] $\dfrac{-1 \pm i\sqrt{3}}{2}$

5. [8.4] $-2, \frac{2}{3}$ **6.** [8.2] $\dfrac{-3 \pm \sqrt{29}}{2}$; $-4.19, 1.19$

7. [8.2] $-\frac{3}{4}, \frac{7}{3}$ **8.** [8.1] $x^2 + 14x + 49$; $(x + 7)^2$

9. [8.1] $x^2 - \frac{2}{7}x + \frac{1}{49}$; $\left(x - \frac{1}{7}\right)^2$ **10.** [8.1] $-6, 3$

11. [8.1] $-5 \pm \sqrt{10}$ **12.** [8.2] Two imaginary

13. [8.3] 16 km/h   **14.** [8.3] 2 hr

15. [8.4] $(-3, 0), (-1, 0), (-2 - \sqrt{5}, 0), (-2 + \sqrt{5}, 0)$

16. [8.4] $x^2 - 13 = 0$ **17.** [8.5] $\frac{833}{125}$, or 6.664 in^2

18. [8.6]

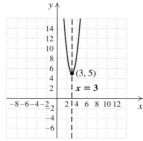

$f(x) = 4(x - 3)^2 + 5$
Minimum: 5

19. [8.7] **(a)** $(-1, -8)$, $x = -1$; **(b)**

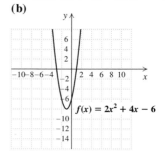

$f(x) = 2x^2 + 4x - 6$

20. [8.7] $(-2, 0), (3, 0)$; $(0, -6)$

21. [8.3] $r = \sqrt{\dfrac{3V}{\pi} - R^2}$ **22.** [8.8] -16

23. [8.8] $f(x) = \frac{1}{5}x^2 - \frac{3}{5}x$ **24.** [8.8] No   **25.** [8.8] Yes

26. [8.8] **(a)** Yes; **(b)** $D(x) = -0.0083843344x^2 + 0.8312075069x + 0.1566556161$ **27.** [8.9] $[-6, 1]$

28. [7.6] 6   **29.** [7.4] $ab\sqrt[4]{2a^2}$
30. [6.2] $\dfrac{x-8}{(x+6)(x+8)}$ **31.** [3.3] Each is 36.
32. [8.2] $\frac{1}{2}$ **33.** [8.4] $x^4 - 14x^3 + 67x^2 - 114x + 26 = 0$;
answers may vary   **34.** [8.4] $x^6 - 10x^5 + 20x^4 +$
$50x^3 - 119x^2 - 60x + 150 = 0$; answers may vary

Chapter 9

INTERACTIVE DISCOVERY, P. 582

1. (a) Increases; **(b)** increases; **(c)** increases; **(d)** decreases;
(e) decreases   **2.** g is steepest.   **3.** r is steeper.

EXERCISE SET 9.1

1. $a > 1$ **3.** $0 < a < 1$
5.
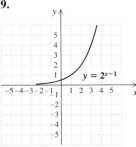

7.
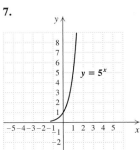

9.
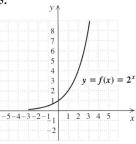

11.
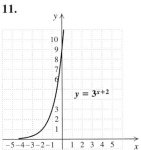

13.
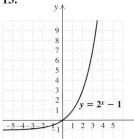

15.
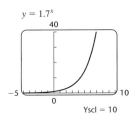

17.
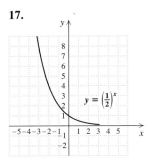

19.
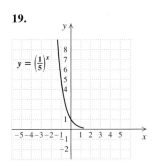

21.
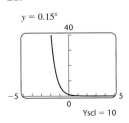

23.
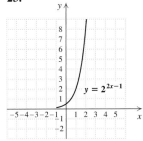

25.
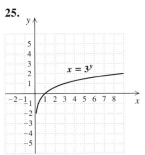

27.
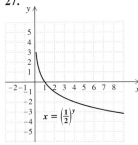

29.
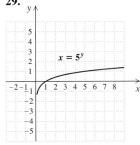

31.
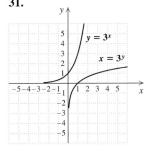

33.

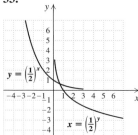

35. (d)   **37.** (f)   **39.** (c)
41. (a) 384,160; (b) 2,066,105;
(c)

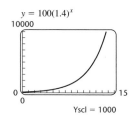

43. (a) 250,000; 166,667; 49,383; 4335;
(b)

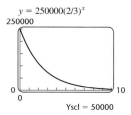

45. (a) 62 billion ft^3; 63.1 billion ft^3;
(b) 65.4 billion ft^3; 78.2 billion ft^3;
(c)

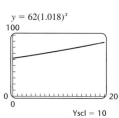

47. x^{-2}, or $\dfrac{1}{x^2}$ **49.** x^{-7}, or $\dfrac{1}{x^7}$ **51.** ◈   **53.** ◈
55. $\pi^{2.4}$

57.

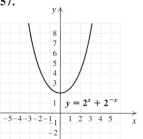

59.

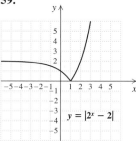

61.

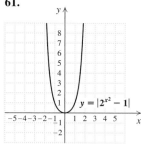

63.

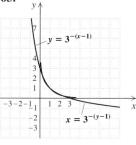

65. $(-5, \infty)$ **67.** $(0, \infty)$

EXERCISE SET 9.2

1. $f \circ g(x) = 12x^2 + 36x + 26$; $g \circ f(x) = 6x^2 + 1$
3. $f \circ g(x) = \dfrac{16}{x^2} - 1$; $g \circ f(x) = \dfrac{2}{4x^2 - 1}$
5. $f \circ g(x) = x^4 + 2x^2 - 2$;
$g \circ f(x) = x^4 - 6x^2 + 10$
7. $f(x) = x^2$; $g(x) = 7 - 5x$
9. $f(x) = x^5$; $g(x) = 3x^2 - 7$ **11.** $f(x) = \dfrac{2}{x}$;
$g(x) = x - 3$ **13.** $f(x) = \dfrac{1}{\sqrt{x}}$; $g(x) = 7x + 2$
15. $f(x) = \dfrac{x + 1}{x - 1}$; $g(x) = x^3$ **17.** Yes   **19.** No
21. Yes   **23.** No   **25.** (a) Yes; (b) $f^{-1}(x) = x - 6$
27. (a) Yes; (b) $f^{-1}(x) = 3 - x$ **29.** (a) Yes;
(b) $g^{-1}(x) = x + 5$ **31.** (a) Yes; (b) $f^{-1}(x) = \dfrac{x}{4}$
33. (a)Yes; (b) $g^{-1}(x) = \dfrac{x - 3}{4}$ **35.** (a)No   **37.** (a)Yes;
(b)$f^{-1}(x) = \dfrac{1}{x}$ **39.** (a) Yes; (b) $f^{-1}(x) = \dfrac{3x - 1}{2}$
41. (a) Yes; (b) $f^{-1}(x) = \sqrt[3]{x + 5}$ **43.** (a) Yes;
(b) $g^{-1}(x) = \sqrt[3]{x} + 2$ **45.** (a) Yes; (b) $f^{-1}(x) = x^2$,
$x \geq 0$

47. (a) Yes; (b) $f^{-1}(x) = \sqrt{\dfrac{x-1}{2}}$

49.

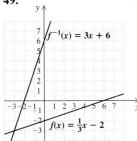

51.

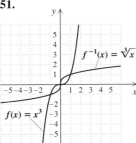

53.

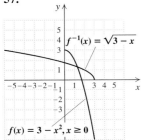

55.

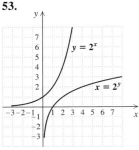

57.

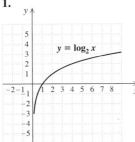

59. No   **61.** Yes   **63.** (1) C; (2) D; (3) B; (4) A

65. (1) $f^{-1} \circ f(x) = f^{-1}(f(x)) = f^{-1}\left(\frac{4}{5}x\right) = \frac{5}{4}\left(\frac{4}{5}\right)x = x$;

(2) $f \circ f^{-1}(x) = f(f^{-1}(x)) = f\left(\frac{5}{4}x\right) = \frac{4}{5}\left(\frac{5}{4}x\right) = x$

67. (1) $f^{-1} \circ f(x) = f^{-1}(f(x))$

$$= f^{-1}\left(\frac{1-x}{x}\right)$$

$$= \frac{1}{\left(\dfrac{1-x}{x}\right)+1}$$

$$= \frac{1}{\dfrac{1-x+x}{x}} = x;$$

(2) $f \circ f^{-1}(x) = f(f^{-1}(x))$

$$= f\left(\frac{1}{x+1}\right)$$

$$= \frac{1-\left(\dfrac{1}{x+1}\right)}{\left(\dfrac{1}{x+1}\right)}$$

$$= \frac{\dfrac{x+1-1}{x+1}}{\dfrac{1}{x+1}} = x$$

69. (a) 40, 42, 46, 50; (b) $f^{-1}(x) = x - 32$;
(c) 8, 10, 14, 18   **71.** $y = 9x$ **73.** $a^{17}b^{17}$

75. ◈   **77.** ◈   **79.** $h(x) = \dfrac{x}{2} + 20$

81. (a) Domain of f: $(-\infty, \infty)$, range of f: $(0, \infty)$, domain of f^{-1}: $(0, \infty)$, range of f^{-1}: $(-\infty, \infty)$; (b) domain of g: $[-1, \infty)$, range of g: $(0, \infty)$, domain of g^{-1}: $(0, \infty)$, range of g^{-1}: $[-1, \infty)$; (c) domain of h: $(-\infty, 2) \cup (2, \infty)$, range of h: $(-\infty, 0) \cup (0, \infty)$, domain of h^{-1}: $(-\infty, 0) \cup (0, \infty)$, range of h^{-1}: $(-\infty, 2) \cup (2, \infty)$. The domain of the function is the range of its inverse, and the range of the function is the domain of its inverse.

EXERCISE SET 9.3

1.

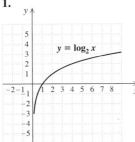

3.

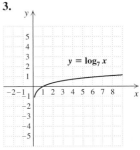

5.

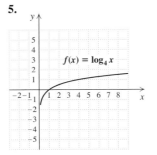

7.

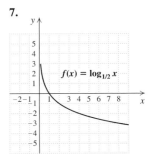

9.

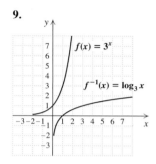

11. 0.6021 **13.** 4.1271 **15.** −0.2782
17. 199.5262 **19.** 0.0011 **21.** 1.0028
23. **25.**

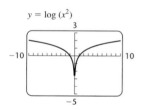 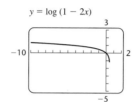

27.

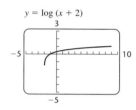

29. $4 = \log_{10} 10{,}000$ **31.** $-3 = \log_5 \frac{1}{125}$
33. $\frac{1}{3} = \log_8 2$ **35.** $0.3010 = \log_{10} 2$ **37.** $n = \log_m r$
39. $2 = \log_e 7.3891$ **41.** $3^t = 8$ **43.** $5^2 = 25$
45. $10^{-1} = 0.1$ **47.** $10^{0.845} = 7$ **49.** $c^{17} = m$
51. $e^{-1.3863} = 0.25$ **53.** $r^{-x} = T$ **55.** 81 **57.** 5
59. 4 **61.** 3 **63.** 8 **65.** 1 **67.** $\frac{1}{2}$ **69.** 4
71. 4 **73.** 0 **75.** 4 **77.** −2 **79.** 1 **81.** 0
83. 15 **85.** $\frac{2}{3}$ **87.** 7 **89.** $\dfrac{x(3y-2)}{2y+x}$ **91.** $\dfrac{1}{4096}$
93. $\dfrac{1}{\sqrt[3]{t}}$ **95.** ◈ **97.** ◈
99. **101.**

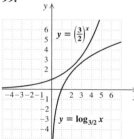 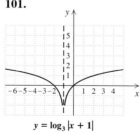

103. 25 **105.** $-\frac{7}{16}$ **107.** 3 **109.** 1 **111.** 1

INTERACTIVE DISCOVERY, P. 609

1. (c) **2.** (b) **3.** (c)

EXERCISE SET 9.4

1. $\log_3 81 + \log_3 27$ **3.** $\log_4 64 + \log_4 16$
5. $\log_c x + \log_c y + \log_c z$ **7.** $\log_a (5 \cdot 14)$, or $\log_a 70$
9. $\log_c (t \cdot y)$ **11.** $7 \log_a t$ **13.** $-3 \log_b C$
15. $\log_2 64 - \log_2 16$ **17.** $\log_b m - \log_b n$
19. $\log_a \frac{15}{7}$ **21.** $2 \log_a x + 3 \log_a y + \log_a z$
23. $\log_b x + 2 \log_b y - 3 \log_b z$
25. $\log_b x + 2 \log_b y - \log_b w - 3 \log_b z$
27. $\frac{1}{2}(6 \log_a x - 5 \log_a y - 8 \log_a z)$
29. $\frac{1}{3}(6 \log_a x + 3 \log_a y - 2 - 7 \log_a z)$
31. $\log_a x^4 y^3$ **33.** $\log_a x$ **35.** $\log_a \dfrac{y^3}{x^{3/2}}$
37. $\log_a (x + 2)$ **39.** 2.708 **41.** −0.51 **43.** −1.609
45. $\frac{3}{2}$ **47.** Cannot be found **49.** 4.317 **51.** 9
53. m **55.** 7 **57.** 3 **59.** $0 + i$, or i **61.** $5 + 10i$
63. $16 - 30i$ **65.** ◈ **67.** ◈ **69.** $\log_a (x^3 + y^3)$
71. $\frac{1}{2} \log_a (c - d) - \frac{1}{2} \log_a (c + d)$ **73.** −2
75. False
77.

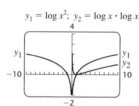

INTERACTIVE DISCOVERY, P. 616

$2.25; $2.370370; $2.441406; $2.613035; $2.704814;
$2.714567; $2.718127

EXERCISE SET 9.5

1. 1.6094 **3.** 4.1271 **5.** 8.3814 **7.** −5.0832
9. 15.0293 **11.** 0.0305 **13.** 1 **15.** 1 **17.** −3.5
19. 2.5702 **21.** 6.6439 **23.** 2.1452 **25.** −2.3219
27. −2.3219 **29.** 3.5471
31. **33.**

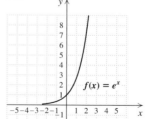 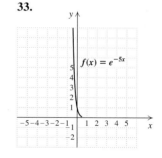

35.
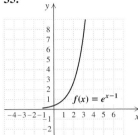

37.
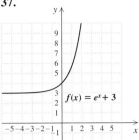

39.
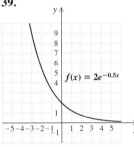

41.
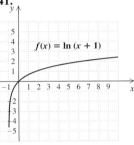

43.
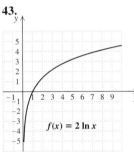

45.
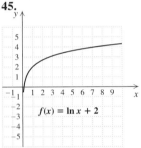

47. $f(x) = \log (x)/\log (5)$, or
$f(x) = \ln (x)/\ln (5)$
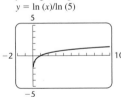

49. $f(x) = \log (x - 5)/\log (2)$, or
$f(x) = \ln (x - 5)/\ln (2)$
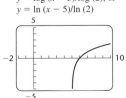

51. $f(x) = \log (x)/\log (3) + x$, or
$f(x) = \ln (x)/\ln (3) + x$
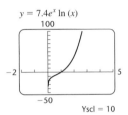

53. $\pm \frac{5}{2}$ **55.** $\frac{15}{17}$ **57.** 16, 256 **59.** ◆ **61.** ◆
63. $\ln M = \dfrac{\log M}{\log e}$ **65.** $\pm 6.0302 \times 10^{17}$ **67.** 1.5893
69. (a) $(0, \infty)$; **(b)** $[-2, 5, -50, 100]$, Yscl = 10; answers may vary;
(c)
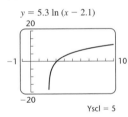

71. (a) $(2.1, \infty)$; **(b)** $[-1, 10, -20, 20]$, Yscl = 5; answers may vary;
(c)

$y = 5.3 \ln (x - 2.1)$

EXERCISE SET 9.6

1. 4   **3.** 4   **5.** 2   **7.** $\frac{4}{3}$ **9.** 2   **11.** $-3, \frac{1}{2}$
13. $\dfrac{\log 13}{\log 2} \approx 3.700$ **15.** $\dfrac{\log 7}{\log 4} \approx 1.404$
17. $\ln 100 \approx 4.605$ **19.** $\dfrac{\ln 0.08}{-0.07} \approx 36.082$
21. $\dfrac{\log 3}{\log 3 - \log 2} \approx 2.710$ **23.** $-6.480, 6.519$
25. $\dfrac{\log 20}{\log 1.7} \approx 5.646$ **27.** 625   **29.** 1000
31. 0.001   **33.** $e \approx 2.718$ **35.** $e^{-3} \approx 0.050$ **37.** -4
39. 1   **41.** No solution   **43.** $\frac{83}{15}$ **45.** 1   **47.** $\sqrt{41}$
49. 4   **51.** 0.0001, 3.445   **53.** 1   **55.** -3

57. 0.00001, 100,000   **59.** $\dfrac{y^{4/3}}{25x^{14/3}z^4}$ **61.** $-16 + 30i$

63. $\pm 2, \pm 10$ **65.** ◈   **67.** ◈   **69.** $\frac{12}{5}$ **70.** $\sqrt[3]{3}$

73. -1 **75.** $-3, -1$ **77.** $\frac{1}{2}$, 5000   **79.** $-\frac{1}{3}$

EXERCISE SET 9.7

1. (a) \$231.16 billion; **(b)** in 2009; **(c)** 4.67 yr
3. (a) 4.2 yr; **(b)** 9.0 yr   **5. (a)** 3.5 yr; **(b)** 13.6 yr
7. 4.9   **9.** 10^{-7} moles per liter   **11.** 65 dB
13. $10^{-1.5}$, or about 3.2×10^{-2} W/m^2
15. (a) $P(t) = P_0 e^{0.06t}$; **(b)** \$5309.18, \$5637.48; **(c)** 11.6 yr
17. (a) $P(t) = 29.1e^{0.008t}$, where $P(t)$ is in millions and t is the number of years after 1997; **(b)** about 32.3 million; **(c)** 2013
19. (a) 2000; **(b)** 2452;
(c)

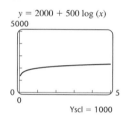

$y = 2000 + 500 \log (x)$

Yscl = 1000

where y is the number of units sold when x thousand dollars is spent on advertising
(d) \$1,000,000 thousand, or \$1,000,000,000
21. (a) $N(t) = 17e^{0.534t}$, where t is the number of years since 1995; **(b)** about 419   **23.** In 2038
25. (a) $k \approx 0.514$, $S(t) = 205e^{-0.514t}$, where $S(t)$ is in millions and t is the number of years since 1983;
(b) 19,664; **(c)** 2021   **27.** About 2103 yr
29. About 7.2 days   **31.** 69.3% per year
33. (a) $V(t) = 451,000e^{0.07t}$, where t is the number of years after 1991; **(b)** \$974,055; **(c)** 9.9 yr; **(d)** 2003
35. No   **37.** Yes
39. (a) $f(x) = 1.809071941(2.316985337)^x$, where $f(x)$ is in millions; **(b)** 84%; **(c)** 18,690 million
41. (a) $P(t) = 0.4728252114e^{0.4334086173t}$;
(b) 2749 million, or 2,749,000,000
43. (a) $f(x) = 647.62971245(0.5602992676)^x$; **(b)** 57.9%;
(c) 152   **45.** $\dfrac{(x-5)(x-3)}{x^2 + 5x - 6}$ **47.** $\dfrac{3a^8b^3}{4}$ **49.** ◈
51. ◈   **53. (a)** Yes;
(b) $f(x) = 34.4497254(1.01987786)^x$; **(c)** 203%; no;
(d) $f(x) = \dfrac{102.870654}{1 + 2.261365843e^{-0.0479171592x}}$; **(e)** 99.8%; yes
55. (6, \$403)

REVIEW EXERCISES, CHAPTER 9

1. [9.1]

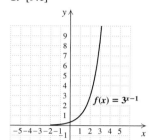

$f(x) = 3^{x-1}$

2. [9.1]

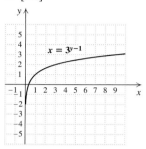

$x = 3^{y-1}$

3. [9.3]

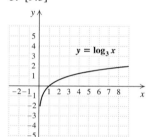

$y = \log_3 x$

4. [9.5]

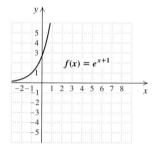

$f(x) = e^{x+1}$

5. [9.2] $f \circ g(x) = 9x^2 - 30x + 25$, $g \circ f(x) = 3x^2 - 5$
6. [9.2] $f(x) = \sqrt{x}$, $g(x) = 4 - 7x$ **7.** [9.2] No
8. [9.2] $f^{-1}(x) = x - 2$ **9.** [9.2] $g^{-1}(x) = \dfrac{7x + 3}{2}$
10. [9.2] $f^{-1}(x) = \dfrac{\sqrt[3]{x}}{2}$ **11.** [9.3] $4^x = 16$
12. [9.3] $\left(\frac{1}{2}\right)^{-3} = 8$ **13.** [9.3] $\log_{10} 10,000 = 4$
14. [9.3] $\log_{25} 5 = \frac{1}{2}$ **15.** [9.3] 2   **16.** [9.3] -1
17. [9.4] $4 \log_a x + 2 \log_a y + 3 \log_a z$
18. [9.4] $\log_a x + \log_a y - 2 \log_a z$
19. [9.4] $\frac{1}{4}(2 \log z - 3 \log x - \log y)$
20. [9.4] $2 \log_q x + \frac{1}{3} \log_q y - 4 \log_q z$
21. [9.4] $\log_a (8 \cdot 15)$, or $\log_a 120$
22. [9.4] $\log_a \frac{72}{12}$, or $\log_a 6$ **23.** [9.4] $\log \dfrac{a^{1/2}}{bc^2}$
24. [9.4] $\log_a \sqrt[3]{\dfrac{x}{y^2}}$ **25.** [9.3] 1   **26.** [9.3] 0
27. [9.4] 17   **28.** [9.4] -7 **29.** [9.4] 6.93
30. [9.4] -3.2698 **31.** [9.4] 8.7601   **32.** [9.4] 3.2698
33. [9.4] 2.54995   **34.** [9.4] -3.6602
35. [9.3] -2.2027 **36.** [9.3] 7.8621   **37.** [9.3] 61.5177
38. [9.3] 2.2387×10^{-4} **39.** [9.5] 10.0821
40. [9.5] -2.6921 **41.** [9.5] 0.3753
42. [9.5] 18.3568   **43.** [9.5] 0.4307

44. [9.5] 1.7097   **45.** [9.6] $\frac{1}{9}$ **46.** [9.6] 2

47. [9.6] $\frac{1}{10,000}$ **48.** [9.6] $e^{-2} \approx 0.1353$ **49.** [9.6] $\frac{7}{2}$

50. [9.6] $-5, 1$ **51.** [9.6] $\frac{\log 8.3}{\log 4} \approx 1.5266$

52. [9.6] $\frac{\ln 0.03}{-0.1} \approx 35.0656$ **53.** [9.6] $-1.8414, 1.1462$

54. [9.6] 8   **55.** [9.6] 20   **56.** [9.6] $\sqrt{43}$

57. [9.7] **(a)** 62; **(b)** 46.8; **(c)** 35 months

58. [9.7] **(a)** $k = \frac{\ln 2}{2} \approx 0.347$, $N(t) = 6e^{0.347t}$, where t is
the number of years after 1994; **(b)** 136; **(c)** 2004

59. [9.7] 4.3% per year   **60.** [9.7] 8.25 yr

61. [9.7] 3463 yr   **62.** [9.7] 90 dB

63. [9.7] **(a)** $f(x) = 42.32738912(1.318506436)^x$;
(b) 27.6%; **(c)** about 24,500

64. [8.3] $T = \frac{-b \pm \sqrt{b^2 + 4aQ}}{2a}$

65. [8.4] $\pm 4, \pm \sqrt{5}$ **66.** [7.8] $\frac{-11}{10} - \frac{17}{10}i$

67. [6.3] $\frac{c - 2a}{2b + 3c}$

68. [9.3] ◈ Negative numbers do not have logarithms
because logarithmic bases are positive, and there is no
power to which a positive number can be raised to yield a
negative number.   **69.** [9.6] ◈ Taking the logarithm on
each side of an equation produces an equivalent equation
because the logarithmic function is one-to-one. If two
quantities are equal, their logarithms must be equal, and if
the logarithms of two quantities are equal, the quantities
must be the same.

70. [9.6] e^{e^3} **71.** [9.6] $-3, -1$ **72.** [9.6] $\left(\frac{8}{3}, -\frac{2}{3}\right)$

TEST, CHAPTER 9

1. [9.1]

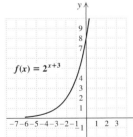

2. [9.3]

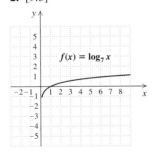

3. [9.2] $f \circ g(x) = 25x^2 - 15x + 2$;
$g \circ f(x) = 5x^2 + 5x - 2$

4. [9.2] No   **5.** [9.2] $f^{-1}(x) = \frac{x + 3}{4}$

6. [9.2] $f^{-1}(x) = \sqrt[3]{x} - 1$ **7.** [9.3] $\log_4 x = -3$

8. [9.3] $\log_{256} 16 = \frac{1}{2}$ **9.** [9.3] $4^2 = 16$

10. [9.3] $7^m = 49$ **11.** [9.4] $3 \log a + \frac{1}{2} \log b - 2 \log c$

12. [9.4] $\log_a x^{1/3}z^2$ **13.** [9.3] 3   **14.** [9.3] 23

15. [9.3] 1   **16.** [9.3] 0   **17.** [9.4] -0.544

18. [9.4] 1.079   **19.** [9.4] 1.322   **20.** [9.3] -1.9101

21. [9.3] 6309.5734   **22.** [9.5] -4.3949

23. [9.5] 107.7701   **24.** [9.5] 1.1881   **25.** [9.6] 5

26. [9.6] 2   **27.** [9.6] 10,000   **28.** [9.6] $\frac{1}{3}$

29. [9.6] $\frac{\log 1.2}{\log 7} \approx 0.0937$ **30.** [9.6] $e^{1/4} \approx 1.2840$

31. [9.6] 4   **32.** [9.7] **(a)** 2.05 ft/sec; **(b)** 984,262

33. [9.7] **(a)** $P(t) = 132e^{0.017t}$, where $P(t)$ is in millions
and t is the number of years after 1997; **(b)** about 165
million; **(c)** in 2021; **(d)** 40.8 yr   **34.** [9.7] 4.6%

35. [9.7] 4684 yr   **36.** [9.7] 7.0

37. [9.7] **(a)** $f(x) = 24.86318448(0.9506288511)^x$; **(b)** 5.1%;

(c) 0.8   **38.** [8.4] 1, 64   **39.** [8.3] $t = \frac{b \pm \sqrt{b^2 + 4aS}}{2a}$

40. [7.8] 29   **41.** [6.3] $\frac{1}{2x}$ **42.** [9.6] $-309, 316$

43. [9.4] 2

CUMULATIVE REVIEW, CHAPTERS 1–9

1. [1.1], [1.4] 2   **2.** [1.1], [1.2] 6   **3.** [1.4] $\frac{y^{12}}{16x^8}$

4. [1.4] $\frac{20x^6z^2}{y}$ **5.** [1.4] $\frac{-y^4}{3z^5}$ **6.** [1.3] $-4x - 1$

7. [1.1] 25   **8.** [2.2] $\frac{11}{2}$ **9.** [3.2] $(3, -1)$

10. [3.4] $(1, -2, 0)$ **11.** [5.4] $-2, 5$ **12.** [6.4] $\frac{9}{2}$

13. [7.6] 5   **14.** [7.6] $\frac{1}{2}$ **15.** [8.1] $\pm 5i$

16. [8.4] $\pm 3, \pm 2$ **17.** [9.3] 8   **18.** [9.3] 7

19. [9.6] $\frac{3}{2}$ **20.** [9.6] $\frac{\log 7}{5 \log 3} \approx 0.3542$

21. [8.9] $(-\infty, -5) \cup (1, \infty)$ **22.** [9.6] $\frac{80}{9}$

23. [8.2] $-3 \pm 2\sqrt{5}$ **24.** [4.3] $(-\infty, -3] \cup [6, \infty)$

25. [6.7] $a = \frac{Db}{b - D}$ **26.** [2.3] $B = \frac{3M - 2A}{2}$, or

$B = \frac{3}{2}M - A$ **27.** [2.4] **(c)**   **28.** [8.7] **(b)**

29. [6.4] **(a)**   **30.** [9.5] **(d)**   **31.** [5.5] $\left\{x\middle| x \text{ is a real}\right.$
$\left. \text{number } and \ x \neq -\frac{1}{3} \ and \ x \neq 2\right\}$

32. [2.6] **(a)** $A(t) = -225t + 1400$, where t is the number
of years after 1994; **(b)** 50 acres   **33.** [2.6] **(a)** Linear;
(b) $f(x) = 0.3395x + 2.606$, where $f(x)$ is in trillions of
dollars and x is the number of years after 1988

34. [8.8] **(a)** Quadratic; **(b)** $f(x) = 0.2814529618x^2 - 4.723504582x + 65.86338269$, where x is the number of years after 1977   **35.** [2.3] Length: 36 m; width: 20 m

36. [2.3] A: 15°; B: 45°; C: 120°   **37.** [6.5] $5\frac{5}{11}$ hr

38. [3.3] Swim clean: 60 L; Pure Swim: 40 L

39. [6.5] $2\frac{7}{9}$ km/h   **40.** [8.8] -49; -7 and 7

41. [9.7] 78   **42.** [9.7] 67.5

43. [9.7] $P(t) = 51e^{-0.001t}$, where P is in millions and t is the number of years after 1997

44. [9.7] 50.6 million   **45.** [8.5] 18

46. [5.1] $7p^2q^3 - 2p^3q + pq - 6$

47. [5.1] $8x^2 - 11x - 1$ **48.** [5.2] $9x^4 - 12x^2y + 4y^2$

49. [5.2] $10a^2 - 9ab - 9b^2$ **50.** [6.1] $\dfrac{(x+4)(x-3)}{2(x-1)}$

51. [6.3] $\dfrac{1}{x-4}$ **52.** [6.2] $\dfrac{7x+4}{(x+6)(x-6)}$

53. [5.3] $x(y - 2z + w)$

54. [5.6] $(1 - x)^2$, or $(x - 1)^2$

55. [5.5] $2(3x - 2y)(x + 2y)$ **56.** [5.3] $(x - 4)(x^3 + 7)$

57. [5.6] $(a - 5 + 9b)(a - 5 - 9b)$

58. [5.6] $2(m + 3n)^2$

59. [5.4] $3(x - 3)(x + 2)$ **60.** [2.1] -12

61. [6.6] $x^3 - 2x^2 - 4x - 12 + \dfrac{-42}{x-3}$

62. [1.4] 1.82×10^{-1} **63.** [1.4] 5.0×10^{13}

64. [7.4] $2y^2 \sqrt[3]{y}$ **65.** [7.4] $14xy^2 \sqrt{x}$

66. [7.2] $81a^8b \sqrt[3]{b}$ **67.** [7.5] $\dfrac{6 + \sqrt{y} - y}{4 - y}$

68. [7.7] $(5, -1)$; 7   **69.** [7.8] $12 + 4\sqrt{3}\,i$

70. [7.8] $\dfrac{18}{25} + \dfrac{1}{25}i$ **71.** [9.2] $f^{-1}(x) = \dfrac{x-7}{-2}$, or $\dfrac{7-x}{2}$

72. [2.6] $f(x) = -5x - 3$ **73.** [2.5] $y = \frac{1}{2}x - 1$

74. [2.5]

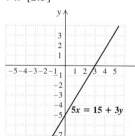

75. [8.7]

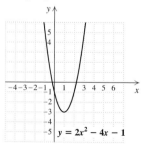

76. [9.3]

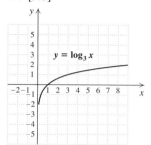

77. [9.1]

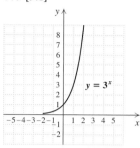

78. [4.4]

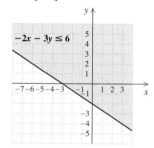

79. [8.6]

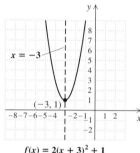

$f(x) = 2(x + 3)^2 + 1$
Minimum: 1

80. [9.4] $2 \log a + 3 \log c - \log b$

81. [9.4] $\log \left(\dfrac{x^3}{y^{1/2}z^2} \right)$ **82.** [9.3] $a^x = 5$

83. [9.3] $\log_x t = 3$ **84.** [9.3] -1.2545

85. [9.3] 776.2471   **86.** [9.5] 2.5479   **87.** [9.5] 0.2466

88. [6.4] All real numbers except 1 and -2

89. [9.6] $\frac{1}{3}$, $\frac{10{,}000}{3}$ **90.** [8.3] 35 mph

Chapter 10

EXERCISE SET 10.1

1. 2, 7, 12, 17; 47; 72

$u(n) = 5n - 3$

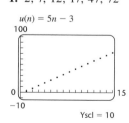

Yscl = 10

3. $\frac{1}{3}$, $\frac{1}{2}$, $\frac{3}{5}$, $\frac{2}{3}$; $\frac{5}{6}$; $\frac{15}{17}$

$u(n) = n/(n + 2)$

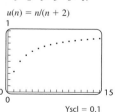

Yscl = 0.1

5. −1, 0, 3, 8; 80; 195

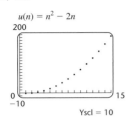

$u(n) = n^2 - 2n$
Yscl = 10

7. $2, 2\frac{1}{2}, 3\frac{1}{3}, 4\frac{1}{4}; 10\frac{1}{10}; 15\frac{1}{15}$

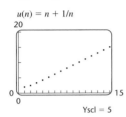

$u(n) = n + 1/n$
Yscl = 5

9. −1, 4, −9, 16; 100; −225

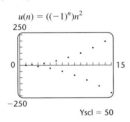

$u(n) = ((-1)^n)n^2$
Yscl = 50

11. −2, −1, 4, −7; −25; 40

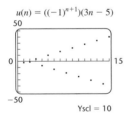

$u(n) = ((-1)^{n+1})(3n - 5)$
Yscl = 10

13. 403   **15.** −23.5   **17.** 43   **19.** $2n - 1$
21. $(-1)^n 2(3)^{n-1}$ **23.** $\dfrac{n}{n+1}$ **25.** $(\sqrt{3})^n$, or $3^{n/2}$
27. $-3n + 2$, or $-(3n - 2)$ **29.** 4   **31.** 30
33. $\frac{1}{2} + \frac{1}{4} + \frac{1}{6} + \frac{1}{8} + \frac{1}{10} = \frac{137}{120}$
35. $1 + 3 + 9 + 27 + 81 = 121$
37. $\frac{1}{2} + \frac{2}{3} + \frac{3}{4} + \frac{4}{5} + \frac{5}{6} + \frac{6}{7} + \frac{7}{8} + \frac{8}{9} = \frac{15{,}551}{2520}$
39. $-1 + 1 - 1 + 1 - 1 = -1$
41. $2 - 4 + 8 - 16 + 32 - 64 + 128 - 256 = -170$
43. $3 + 2 + 3 + 6 + 11 + 18 = 43$ **45.** $\sum\limits_{k=1}^{5} \dfrac{k+1}{k+2}$
47. $\sum\limits_{k=1}^{6} k^2$ **49.** $\sum\limits_{k=2}^{n} (-1)^k k^2$ **51.** $\sum\limits_{k=1}^{\infty} 5k$

53. $\sum\limits_{k=1}^{\infty} \dfrac{1}{k(k+1)}$ **55.** 29   **57.** 1   **59.** 7   **61.** ◈
63. ◈   **65.** 0, 3, 12, 147, 21,612, 467,078,547
67. $5200, $3900, $2925, $2193.75, $1645.31, $1233.98, $925.49, $694.12, $520.59, $390.44
69. $\frac{3}{2}, \frac{3}{2}, \frac{9}{8}, \frac{3}{4}, \frac{15}{32}; \frac{171}{32}$
71. $\{x \mid x = 4n - 1,$ where n is a natural number$\}$
73. 1225

EXERCISE SET 10.2

1. $a_1 = 3, d = 5$ **3.** $a_1 = 6, d = -4$
5. $a_1 = \frac{3}{2}, d = \frac{3}{4}$ **7.** $a_1 = \$2.12, d = \$0.12$
9. 47   **11.** −41   **13.** −$1628.16   **15.** 27th
17. 102nd   **19.** 82   **21.** 5   **23.** 28
25. $a_1 = 8; d = -3; 8, 5, 2, -1, -4$ **27.** 780
29. 45,150   **31.** 2550   **33.** 918   **35.** 990
37. 42; 420   **39.** 1260   **41.** $31,000   **43.** 722
45. $a^k = P$ **47.** $\ln 0.1579 = t$ **49.** ◈   **51.** $S_n = n^2$
53. $8760, $7961.77, $7163.54, $6365.31, $5567.08, $4768.85, $3970.62, $3172.39, $2374.16, $1575.93
55. Let $d =$ the common difference. Since $p, m,$ and q form an arithmetic sequence, $m = p + d$ and $q = p + 2d$. Then

$$\frac{p+q}{2} = \frac{p + (p + 2d)}{2} = p + d = m.$$

57. Yes; $a_n = 23 + (n - 1)$ where $n = 1$ corresponds to 1993.   **59.** No

INTERACTIVE DISCOVERY, P. 677

1. **(a)** $\frac{1}{3}$; **(b)** 1, 1.33333, 1.44444, 1.48148, 1.49383, 1.49794, 1.49931, 1.49977, 1.49992, 1.49997 (sums are rounded to 5 decimal places); **(c)** 1.5
2. **(a)** $-\frac{7}{2}$; **(b)** 1, −2.5, 9.75, −33.125, 116.9375, −408.28125, 1429.984375, −5003.945313, 17514.86859, −61300.83008; **(c)** does not exist
3. **(a)** 2; **(b)** 4, 12, 28, 60, 124, 252, 508, 1020, 2044, 4092; **(c)** does not exist
4. **(a)** −0.4; **(b)** 1.3, 0.78, 0.988, 0.9048, 0.93808, 0.92477, 0.93009, 0.92796, 0.92881, 0.92847 (sums are rounded to 5 decimal places); **(c)** 0.928

EXERCISE SET 10.3

1. 2   **3.** −1   **5.** $-\frac{1}{2}$ **7.** $\frac{1}{5}$ **9.** $\dfrac{3}{m}$ **11.** 320
13. 48   **15.** 648   **17.** $2331.64   **19.** $a_n = 3^{n-1}$
21. $a_n = (-1)^{n-1}$ **23.** $a_n = \dfrac{1}{x^n}$ **25.** 889   **27.** $\dfrac{547}{18}$
29. $\dfrac{1 - x^8}{1 - x}$, or $(1 + x)(1 + x^2)(1 + x^4)$ **31.** $5134.51

33. $\frac{27}{2}$ **35.** $\frac{49}{4}$ **37.** No **39.** No **41.** $\frac{43}{99}$
43. $25,000 **45.** $\frac{7}{9}$ **47.** $\frac{830}{99}$ **49.** $\frac{5}{33}$ **51.** $\frac{5}{1024}$ ft
53. 155,797 **55.** $43,318.94 **57.** 25 min
59. 3100.35 ft **61.** 20.48 in. **63.** Arithmetic
65. Geometric **67.** Geometric **69.** $\left(-\frac{63}{29}, -\frac{114}{29}\right)$

71. ◆ **73.** ◆ **75.** $\frac{1 - x^n}{1 - x}$

EXERCISE SET 10.4

1. 40,320 **3.** 3,628,800 **5.** 210 **7.** 720 **9.** 28
11. 210 **13.** 190 **15.** 595
17. $m^5 + 5m^4n + 10m^3n^2 + 10m^2n^3 + 5mn^4 + n^5$
19. $x^6 - 6x^5y + 15x^4y^2 - 20x^3y^3 + 15x^2y^4 - 6xy^5 + y^6$
21. $x^{10} - 15x^8y + 90x^6y^2 - 270x^4y^3 + 405x^2y^4 - 243y^5$
23. $729c^6 - 1458c^5d + 1215c^4d^2 - 540c^3d^3 +$
$135c^2d^4 - 18cd^5 + d^6$ **25.** $x^3 - 3x^2y + 3xy^2 - y^3$
27. $x^9 + \dfrac{18x^8}{y} + \dfrac{144x^7}{y^2} + \dfrac{672x^6}{y^3} + \dfrac{2016x^5}{y^4} + \dfrac{4032x^4}{y^5} +$

$\dfrac{5376x^3}{y^6} + \dfrac{4608x^2}{y^7} + \dfrac{2304x}{y^8} + \dfrac{512}{y^9}$
29. $a^{10} - 5a^8b^3 + 10a^6b^6 - 10a^4b^9 + 5a^2b^{12} - b^{15}$
31. $9 - 12\sqrt{3}t + 18t^2 - 4\sqrt{3}t^3 + t^4$
33. $x^{-8} + 4x^{-4} + 6 + 4x^4 + x^8$ **35.** $15a^4b^2$
37. $-64,481,508a^3$ **39.** $1120x^{12}y^2$
41. $-1,959,552u^5v^{10}$ **43.** 4 **45.** $\ln 280 \approx 5.6348$
47. ◆ **49.** $\binom{5}{2}(0.313)^3(0.687)^2 \approx 0.145$

51. $\binom{5}{2}(0.313)^3(0.687)^2 + \binom{5}{3}(0.313)^2(0.687)^3 +$

$\binom{5}{4}(0.313)(0.687)^4 + \binom{5}{5}(0.687)^5 \approx 0.964$

53. $\binom{n}{n - r} = \dfrac{n!}{[n - (n - r)]!(n - r)!}$

$= \dfrac{n!}{r!(n - r)!} = \binom{n}{r}$

55. $-4320x^6y^{9/2}$ **57.** $-\dfrac{35}{x^{1/6}}$

REVIEW EXERCISES, CHAPTER 10

1. [10.1] 1, 5, 9, 13; 29; 45 **2.** [10.1] $0, \frac{1}{5}, \frac{1}{5}, \frac{3}{17}; \frac{7}{65}; \frac{11}{145}$
3. [10.1] $a_n = -2n$ **4.** [10.1] $a_n = n^2$
5. [10.1] $-2 + 4 + (-8) + 16 + (-32) = -22$

6. [10.1] $-3 + (-5) + (-7) + (-9) + (-11) +$

$(-13) = -48$ **7.** [10.1] $\displaystyle\sum_{k=1}^{5} 4k$ **8.** [10.1] $\displaystyle\sum_{k=1}^{5} \dfrac{1}{(-2)^k}$

9. [10.2] 85 **10.** [10.2] $\frac{8}{3}$
11. [10.2] $d = 1.25$, $a_1 = 11.25$ **12.** [10.2] -544
13. [10.2] 8580 **14.** [10.3] $1024\sqrt{2}$ **15.** [10.3] $\frac{2}{3}$

16. [10.3] $a_n = 2(-1)^n$ **17.** [10.3] $a_n = 3\left(\dfrac{x}{4}\right)^{n-1}$

18. [10.3] 4095 **19.** [10.3] $-4095x$ **20.** [10.3] 12
21. [10.3] $\frac{49}{11}$ **22.** [10.3] No **23.** [10.3] No
24. [10.3] $40,000 **25.** [10.3] $\frac{5}{9}$ **26.** [10.3] $\frac{46}{33}$
27. [10.2] $17.80 **28.** [10.2] 903
29. [10.3] $22,521.92 **30.** [10.3] 6 m
31. [10.4] 40,320 **32.** [10.4] 56
33. [10.4] $190a^{18}b^2$
34. [10.4] $x^4 - 8x^3y + 24x^2y^2 - 32xy^3 + 16y^4$
35. [3.2] $\left(\frac{26}{11}, \frac{1}{11}\right)$ **36.** [9.6] $\frac{5}{9}$ **37.** [9.3] 13
38. [9.3] -1
39. [10.3] ◆ For a geometric sequence with $|r| < 1$, as n gets larger, the absolute value of the terms gets smaller, since $|r^n|$ gets smaller. **40.** [10.4] ◆ The first form of the binomial theorem draws the coefficients from Pascal's triangle; the second form uses factorial notation. The second form avoids the need to compute all preceding rows of Pascal's triangle, and is generally easier to use when only one term of an expansion is needed. When several terms of an expansion are needed and n is not large (say, $n \leq 8$), it is often easier to use Pascal's triangle.
41. [10.3] $\dfrac{1 - (-x)^n}{x + 1}$
42. [10.4] $x^{-15} + 5x^{-9} + 10x^{-3} + 10x^3 + 5x^9 + x^{15}$

TEST, CHAPTER 10

1. [10.1] 1, 7, 13, 19, 25; 91 **2.** [10.1] $a_n = 4\left(\frac{1}{3}\right)^n$
3. [10.1] $1 - 1 - 5 - 13 - 29 = -47$
4. [10.1] $\displaystyle\sum_{k=1}^{5} k^3$ **5.** [10.2] -46 **6.** [10.2] $\frac{3}{8}$
7. [10.2] $a_1 = 31.2$, $d = -3.8$ **8.** [10.2] 2508
9. [10.3] $\frac{9}{128}$ **10.** [10.3] $\frac{2}{3}$ **11.** [10.3] $(-1)^{n+1}3^n$
12. [10.3] $511 + 511x$ **13.** [10.3] 1 **14.** [10.3] No
15. [10.3] $\dfrac{\$25,000}{23} \approx \$1086.96$ **16.** [10.3] $\frac{85}{99}$
17. [10.2] 63 **18.** [10.2] 1378 **19.** [10.3] $8981.05
20. [10.3] 36 m **21.** [10.4] 78
22. [10.4] $x^{10} - 15x^8y + 90x^6y^2 - 270x^4y^3 +$
$405x^2y^4 - 243y^5$ **23.** [10.4] $220a^9x^3$ **24.** [9.6] 3

25. [9.3] 3   **26.** [3.2] $(-1, 2)$ **27.** [9.3] $10^{1.5} = x$
28. [10.2] $n(n + 1)$

29. [10.3] $\dfrac{1 - \left(\dfrac{1}{x}\right)^n}{1 - \dfrac{1}{x}}$, or $\dfrac{x^n - 1}{x^{n-1}(x - 1)}$

48. [4.4]

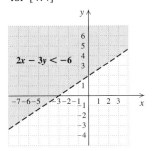

$2x - 3y < -6$

49. [7.1]

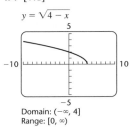

$y = \sqrt{4 - x}$

Domain: $(-\infty, 4]$
Range: $[0, \infty)$

CUMULATIVE REVIEW, CHAPTERS 1–10

1. [1.4] $-45x^6y^{-4}$, or $\dfrac{-45x^6}{y^4}$ **2.** [1.2] 6.3

3. [1.3] $-3y + 17$ **4.** [1.1] 280   **5.** [1.1], [1.2] $\frac{7}{6}$
6. [5.1] $3a^2 - 8ab - 15b^2$
7. [5.1] $13x^3 - 7x^2 - 6x + 6$
8. [5.2] $6a^2 + 7a - 5$ **9.** [5.2] $9a^4 - 30a^2y + 25y^2$
10. [6.2] $\dfrac{4}{x + 2}$ **11.** [6.1] $\dfrac{x - 4}{3(x + 2)}$

12. [6.1] $\dfrac{3}{(x + 3)(x + 1)}$ **13.** [6.3] $x - a$
14. [5.6] $(2x - 3)^2$ **15.** [5.4] $-2(t - 3)(t + 2)$
16. [5.3] $(a + 3)(a^2 - b)$ **17.** [5.5] $3(y^2 + 3)(5y^2 - 4)$
18. [2.1] 20   **19.** [6.6] $7x^3 + 9x^2 + 19x + 38 + \dfrac{72}{x - 2}$
20. [2.2] $\frac{2}{3}$ **21.** [6.4] $-\frac{6}{5}, 4$ **22.** [4.2] $\mathbb{R}$
23. [3.2] $(1, -1)$ **24.** [3.4] $(2, -1, 1)$ **25.** [7.6] 2
26. [8.4] $\pm 2, \pm 5$
27. [9.6] $-\ln 1.703 \approx -0.5324$ **28.** [5.5] $-2, \frac{1}{3}$
29. [9.6] $\dfrac{\ln 8}{\ln 5} \approx 1.2920$ **30.** [9.6] 1005
31. [9.3] $\frac{1}{16}$ **32.** [9.6] $-\frac{1}{2}$ **33.** [4.3] $[-2, 3]$
34. [8.1] $\pm i\sqrt{2}$ **35.** [8.2] $-2 \pm \sqrt{7}$
36. [8.9] $(-\infty, -5) \cup (2, \infty)$ **37.** [5.4] $-6, 8$
38. [4.1] More than 4   **39.** [2.3] 65, 66, 67
40. [3.3] $11\frac{3}{7}$ **41.** [3.3] 24 L of A; 56 L of B
42. [6.5] 350 mph   **43.** [6.5] $8\frac{2}{5}$ min or 8 min, 24 sec
44. [8.5] 20   **45.** [8.8] 1250 ft²
46. [2.5]

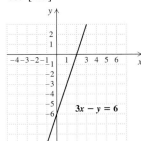

$3x - y = 6$

47. [9.3]

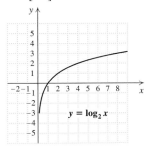

$y = \log_2 x$

50. [8.6]

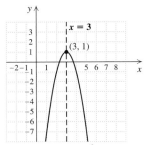

$x = 3$

$(3, 1)$

$f(x) = -2(x - 3)^2 + 1$
Maximum: 1
Domain: $(-\infty, \infty)$
Range: $(-\infty, 1]$

51. [2.3] $r = \dfrac{V - P}{-Pt}$ **52.** [6.7] $R = \dfrac{Ir}{1 - I}$
53. [2.6] $y = -\frac{1}{3}x + \frac{11}{3}$ **54.** [4.2] $\left(-\infty, \frac{5}{3}\right]$
55. [5.4] $\{x \mid x$ is a real number $and \ x \neq 1\}$, or
$(-\infty, 1) \cup (1, \infty)$ **56.** [1.4] 6.764×10^{-12}
57. [7.3] $8x^2\sqrt{y}$ **58.** [7.2] $125x^2y^{3/4}$ **59.** [7.4] $\dfrac{\sqrt[3]{5xy}}{y}$
60. [7.5] $\dfrac{1 - 2\sqrt{x} + x}{1 - x}$ **61.** [7.4] $\sqrt[6]{x + 1}$
62. [7.8] $26 - 13i$ **63.** [8.4] $x^2 - 50 = 0$
64. [9.4] $\log_a \dfrac{\sqrt[3]{x^2} \cdot z^5}{\sqrt{y}}$ **65.** [9.3] $a^5 = c$
66. [9.7] $S(t) = 75e^{1.1836t}$, where S is in millions of dollars
and t is the number of years since 1993
67. [9.7] \$297,357 million, or about \$297 billion
68. [7.7] 5   **69.** [10.2] -121 **70.** [10.2] 875
71. [10.3] $16\left(\frac{1}{4}\right)^{n-1}$ **72.** [10.4] $13,440a^4b^6$
73. [10.3] $74.88671875x$ **74.** [10.3] \$168.95
75. [9.7] **(a)** Exponential; **(b)** $f(x) = 1879.955876$
$(1.295904158)^x$, where x is the number of years since
1987

76. [2.6] **(a)** Linear; **(b)** $f(x) = 28.265x + 181.0511111$, where x is the number of years since 1990
77. [8.8] **(a)** Quadratic; **(b)** $f(x) = -2.554464286x^2 + 20.34321429x + 92.96428571$, where x is the number of years since 1990

78. [6.4] All real numbers except 0 and -12
79. [9.6] 81   **80.** [8.5] y gets divided by 8
81. [7.8] $-\dfrac{7}{13} + \dfrac{2\sqrt{30}}{13}i$ **82.** [3.5] 84 yr

Index

Index of Applications

Geometry

Plane Geometry

Rectangle
Area: $A = lw$
Perimeter: $P = 2l + 2w$

Square
Area: $A = s^2$
Perimeter: $P = 4s$

Triangle
Area: $A = \frac{1}{2}bh$

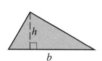

Sum of Angle Measures
$A + B + C = 180°$

Right Triangle
Pythagorean theorem
(equation):
$a^2 + b^2 = c^2$

Parallelogram
Area: $A = bh$

Trapezoid
Area: $A = \frac{1}{2}h(a + b)$

Circle
Area: $A = \pi r^2$
Circumference:
$\quad C = \pi d = 2\pi r$
$\left(\frac{22}{7} \text{ and } 3.14 \text{ are different}\right.$
approximations for π)

Solid Geometry

Rectangular Solid
Volume: $V = lwh$

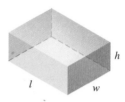

Cube
Volume: $V = s^3$

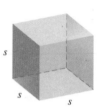

Right Circular Cylinder
Volume: $V = \pi r^2 h$
Lateral surface area:
$\quad L = 2\pi rh$
Total surface area:
$\quad S = 2\pi rh + 2\pi r^2$

Right Circular Cone
Volume: $V = \frac{1}{3}\pi r^2 h$
Lateral surface area:
$\quad L = \pi rs$
Total surface area:
$\quad S = \pi r^2 + \pi rs$
Slant height:
$\quad s = \sqrt{r^2 + h^2}$

Sphere
Volume: $V = \frac{4}{3}\pi r^3$
Surface area: $S = 4\pi r^2$